Neutrosophic and Plithogenic Inventory Models for Applied Mathematics

Florentin Smarandache
University of New Mexico, USA

Maissam Jdid
Damascus University, Syria

Maikel Leyva-Vazquez
Regional Autonomous University of Los Andes, Ecuador

IGI Global
Publishing Tomorrow's Research Today

Published in the United States of America by
IGI Global Scientific Publishing
701 East Chocolate Avenue
Hershey, PA, 17033, USA
Tel: 717-533-8845
Fax: 717-533-8661
Website: https://www.igi-global.com E-mail: cust@igi-global.com

Library of Congress Cataloging-in-Publication Data

Names: Smarandache, Florentin, editor. | Jdid, Maissm Ahmad, editor. | Leyva Vázquez, Maikel, 1979- editor.
Title: Neutrosophic and plithogenic inventory models for applied mathematics / edited by Florentin Smarandache, Maissm Jdid, Maikel Leyva-Vazquez.
Description: Hershey : IGI Global Scientific Publishing, [2025] | Includes bibliographical references and index. | Summary: "The scope of this book encompasses a new development based on the concepts of neutrosophic logic to reformulate some operations research methods and their applications in practical life, through what was presented in the first six chapters and a detailed presentation of some static inventory models through what was presented in the seventh and eighth chapters and important studies"-- Provided by publisher.
Identifiers: LCCN 2024053002 (print) | LCCN 2024053003 (ebook) | ISBN 9798369332047 (hardcover) | ISBN 9798369350188 (paperback) | ISBN 9798369332054 (ebook)
Subjects: LCSH: Inventory control--Mathematical models. | Neutrosophic logic--Mathematical models. | Fuzzy sets. | Mathematical optimization.
Classification: LCC TS160 .N47 2025 (print) | LCC TS160 (ebook) | DDC 658.7/87015118--dc23/eng/20250210
LC record available at https://lccn.loc.gov/2024053002
LC ebook record available at https://lccn.loc.gov/2024053003

British Cataloguing in Publication Data
A Cataloguing in Publication record for this book is available from the British Library.

Table of Contents

Preface xx

Chapter 1
Advancements in Plithogenics Exploring New Dimensions of Soft Sets 1
Sima Das, Bengal College of Engineering and Technology, India
Monojit Manna, RCC Institute of Information Technology, India
Subrata Modak, Global College of Science and Technology, India

Chapter 2
Pythagorean Anti-Neutrosophic Linear Space 19
Sambandan Sivaramakrishnan, Manakula Vinayagar Institute of Technology, India
Parthasarathy Balaji, Measi Academy of Architecture, India
M. Sivasakthi, Krishnasamy College of Science, Arts, and Management for Women, India

Chapter 3
Pentapartitioned Neutrosophic Vague Number Assignment Problem: A Novel Approach Using Interval-Valued Trapezoidal Pentapartitioned Neutrosophic Vague 35
V. R. Radhika, Nirmala college for Women, Coimbatore, India
K. Mohana, Nirmala College for Women, Coimbatore, India

Chapter 4
Significance and Applications of Neutrosophic Generalized Feebly Connected Topology in Diverse Realms 61
Santhi P., The Standard Fireworks Rajaratnam College for Women, India
Yuvarani A., The American College, India
Vijaya S., Thiagarajar College, India

Chapter 5
On Fuzzy Hypersoft Relations 99
Adem Yolcu, Kafkas University, Turkey
Taha Yasin Ozturk, Kafkas University, Turkey

Chapter 6
Applications of an Interval-Valued Anti-Neutrosophic Semigroup 131
Sambandan Sivaramakrishnan, Manakula Vinayagar Institute of Technology, Puducherry, India
Parthasarathy Balaji, Measi Academy of Architecture, Royapettah, India

Chapter 7
Neutrosophic Values and Sensitivity Analysis to Study Static Models Without Deficit 147
Maissam Ahmad Jdid, Damascus University, Syria
Florentin Smarandache, New Mexico University, USA

Chapter 8
Neutrosophical Study of the Two Models: Storage Static With Deficit and Storage Static With Safety Reserve and Important Economic Indicators.......... 167
Maissam Ahmad Jdid, Damascus University, Syria

Chapter 9
Mathematical Programming for Neutrosophic Supply Chain Management 185
Hadi Basirzadeh, Faculty of Mathematical Sciences and Computer, Shahid Chamran University of Ahvaz, Iran
Madineh Farnam, Shohadaye Hoveizeh Campus of Technology, Shahid Chamran University of Ahvaz, Iran
Roohollah Abbasi Shureshjani, Department of Management, Humanities College, Hazrat-e Masoumeh University, Qom, Iran
Sara Ahmadi, Department of Mathematic, Faculty of Basic Sciences, University of Qom, Iran

Chapter 10
An Approach to Solve Non-Linear Neutrosophic Transportation Problem With Volume Discount 219
Aakanksha Singh, Indira Gandhi Delhi Technical University for Women, Delhi, India & Aryabhatta College, University of Delhi, India
Ritu Arora, Keshav Mahavidyalaya, University of Delhi, India
Shalini Arora, Indira Gandhi Delhi Technical University for Women, Delhi, India

Chapter 11
Neutrosophic Inventory Model With Coordinated Rework Stations and Distribution Centers With Ant Colony Optimization 247
M. Renee Miriam, Madurai Kamaraj University, India
Nivetha Martin, Arul Anandar College (Autonomous), Karumathur, India
Akbar Rezaei, Payame Noor University, Iran
Seyyed Ahmad Edalatpanah, Ayandegan Institute of Higher Education, Iran

Chapter 12
A Probabilistic Approach for Renewable Energy Alternative Selection Through Correlation-Based Neutrosophic TOPSIS Approach 287
Biplab Sinha Mahapatra, Haldia Institute of Technology, India
Mihir Baran Bera, Haldia Institute of Technology, India
Manoj Kumar Mondal, Haldia Institute of Technology, India
Pinaki Pratim Acharjya, Haldia Institute of Technology, India

Chapter 13
A Proposed Neutrosophic Probability Model for Normalized DifferenceVegetation Index Using Remote Sensing: Model Building on Climate 315
Shan E. Fatima, Government College University, Lahore, Pakistan
Hina Khan, Government College University, Lahore, Pakistan
Kanwal Javaid, Governemnt College University, Lahore, Pakistan

Chapter 14
Development of Some New Hybrid Structures of Hypersoft Set With Possibility-Degree Settings 337
Atiqe Ur Rahman, University of Management and Technology, Lahore, Pakistan
Florentin Smarandache, University of New Mexico, USA
Muhammad Saeed, University of Management and Technology, Lahore, Pakistan
Khuram Ali Khan, University of Sargodha, Pakistan

Chapter 15
MCDM Using Normalized Weighted Bonferroni Mean Operator in Fermatean Neutrosophic Environment 387
A. Revathy, Sri Eshwar College of Engineering, India
V. Inthumathi, NGM College, Pollachi, India
S. Krishnaprakash, Sri Krishna College of Engineering and Technology, India
S. Gomathi, Sri GVG Vishalakshi College for Women, India
N. Akiladevi, Sri Eshwar College of Engineering, India

Chapter 16
Neutrosophic Optimization and Its Uncertainty Quantification 411
Srinivasan Vijayabalaji, University College of Engineering, Anna University, Panruti, India
Parthasarathy Balaji, Measi Academy of Architecture, India
Gunalan Venkadesh, Krishnasamy College of Engineering and Technology, Cuddalore, India

Chapter 17
Interval Valued Neutrosophic Information System and Its Applications to Decision Making 433
V. Lakshmana Gomathi Nayagam, National Institute of Technology, Tiruchirappalli, India
Daniel P., St. Xavier's College, India
Bharanidharan R., National Institute of Technology, Tiruchirappalli, India

Chapter 18
On Multi-Criteria Job Sequencing Decision-Making Problem via Fermatean Pentapartitioned Neutrosophic Set 459
R. Subha, Nirmala College for Women, Coimbatore, India
K. Mokana, Nirmala College for Women, Coimbatore, India

Chapter 19
An Innovative Approach to Group Decision-Making Based on Weighted Hypersoft Expert System 483
Ajoy Kanti Das, Tripura University, India
Rakhal Das, ICFAI University, India
Rupak Datta, Tripura University, India
Carlos Granados, Universidad de Antioquia, Colombia

Chapter 20
Generalized Plithogenic Sets in Multi-Attribute Decision Making 519
Nivetha Martin, Arul Anandar College (Autonomous), Karumathur, India
R. Priya, Sethu Institute of Technology, India
Florentin Smarandache, University of New Mexico, USA

Chapter 21
Solving Neutrosophic Minimum Spanning Tree Problem by Least Edge Weight Algorithm 547
Shayathri Linganathan, Vellore Institute of Techhnology, India
Purusotham Singamsetty, Vellore Institute of Technology, India

Chapter 22
Algorithms of Designing Decision Trees From Indeterm Soft Sets 561
Erick González Caballero, Asociación Latinoamericana de Ciencias Neutrosóficas, Cuba
Ketty Marilú Moscoso-Paucarchuco, Universidad Nacional Autónoma Altoandina de Tarma, Peru
Noel Batista Hernandez, Universidad Regional Autónoma de los Andes, Ecuador
Lorenzo Jovanny Cevallos Torres, Universidad Bolivariana del Ecuador, Ecuador
Maikel Leyva, Universidad de Guayaquil, Ecuador
Victor Gustavo Gómez Rodríguez, Universidad Bolivariana del Ecuador, Ecuador

Compilation of References 587

About the Contributors 641

Index 649

Detailed Table of Contents

Preface xx

Chapter 1
Advancements in Plithogenics Exploring New Dimensions of Soft Sets 1
Sima Das, Bengal College of Engineering and Technology, India
Monojit Manna, RCC Institute of Information Technology, India
Subrata Modak, Global College of Science and Technology, India

In this chapter, the authors delve into the concept of Plithogenics and its contributions to exploring new dimensions of soft sets. They begin by providing a brief introduction to soft sets and highlighting their applications in real-world scenarios. They then delve into the motivations behind exploring Plithogenics, focusing on the need for a more expressive and versatile framework to handle complex data structures. The core of this chapter lies in presenting the advancements in Plithogenics, where the authors introduce the concept of multilayer soft sets.

Chapter 2
Pythagorean Anti-Neutrosophic Linear Space 19
Sambandan Sivaramakrishnan, Manakula Vinayagar Institute of Technology, India
Parthasarathy Balaji, Measi Academy of Architecture, India
M. Sivasakthi, Krishnasamy College of Science, Arts, and Management for Women, India

This chapter introduces the concept of a Pythagorean anti-neutrosophic linear space, expanding on the principles of Pythagorean, anti-neutrosophic, and fuzzy linear space. These advancements offer new approaches for addressing uncertainty, indeterminacy, and contradiction within traditional linear space. The authors prove that the union of two Pythagorean anti-neutrosophic linear spaces forms another Pythagorean anti-neutrosophic linear space. However, they provide a counterexample to show that the intersection of two such spaces are not necessarily result in a Pythagorean anti-neutrosophic linear space. Furthermore, the authors demonstrate that the Cartesian product of two Pythagorean anti-neutrosophic linear spaces creates a new Pythagorean anti-neutrosophic linear space. Additionally, we define and explore the concepts of homomorphic image and inverse image of Pythagorean anti-neutrosophic linear spaces. This chapter aims to enhance the theoretical understanding of these advanced linear spaces in handling complex mathematical uncertainties.

Chapter 3
Pentapartitioned Neutrosophic Vague Number Assignment Problem: A Novel Approach Using Interval-Valued Trapezoidal Pentapartitioned Neutrosophic Vague 35
V. R. Radhika, Nirmala college for Women, Coimbatore, India
K. Mohana, Nirmala College for Women, Coimbatore, India

An extensively researched and significant field in optimization is assignment problems (APs), this research manuscript introduces the pentapartitioned neutrosophic vague assignment problem (PNVAP), an assignment problem in a pentapartitioned neutrosophic vague environment. Utilizing interval-valued trapezoidal pentapartitioned neutrosophic vague numbers in the cost matrix's elements, the problem is put forth. This interval-valued trapezoidal pentapartitioned neutrosophic vague assignment problem (IVTPNVAP) is converted to an interval-valued AP in accordance with the idea of a score function. The authors apply the order relations to optimize the objective function in interval form. The choices made by the decision maker are represented by these relations. The decision makers' preference introduces order relations, which transform the maximization (or minimization) model with an objective function in interval form into a multi-objective model in the case of interval profits (or costs). Finally, the authors resolve a numerical example to validate the suggested methodology for solving the problem.

Chapter 4
Significance and Applications of Neutrosophic Generalized Feebly Connected Topology in Diverse Realms 61
Santhi P., The Standard Fireworks Rajaratnam College for Women, India
Yuvarani A., The American College, India
Vijaya S., Thiagarajar College, India

The study of connectedness in topology plays a vital role in understanding and demonstrating the overall structure of several geometric matters in real-life scenarios. This induces us to introduce feebly connectedness and semi connectedness concept in neutrosophic generalized topological spaces (Ng-TS), as it is the rudimentary attributes that correlates the innate ability to decipher the structure of several geometric matters. Thereafter, Ng-Feebly Ci-disconnectedness ($i = 1, 2, 3, 4$), Ng-Feebly irresolute function, Ng-Feebly strongly connectedness, Ng-Feebly separated sets, Ng-Semi connectedness, Ng-Semi irresolute function, and Ng-Semi separated sets notions are developed and analyzed with befitting illustrations. Furthermore, its influence across diverse realms such as knot theory, chaos theory, epidemiology, cosmology is reviewed.

Fuzzy hypersoft set theory is a current topic that has been intensively studied by researchers in recent years. In this chapter, the concept of fuzzy hypersoft Cartesian product is defined differently from other studies. In the definition presented, both parameters and alternatives are multiplied. Based on the concept of Cartesian product, the concept of relation is constructed. In addition to studying a few related features, the authors have proposed various theories on fuzzy hypersoft cartesian products, relations, and functions together with supporting data and examples. Finally, injective, surjective, bijective, and constant fuzzy hypersoft function ideas have also investigated.

In this chapter, the authors explore the notion of an interval-valued anti-neutrosophic semigroup, which combines the theories of semigroups and interval-valued anti-neutrosophic sets. They begin by introducing the basic definitions, proving some results and their generalizations. Furthermore, they investigate the practical applications of an interval-valued anti-neutrosophic semigroup in various domains. One such application involves utilizing superior-subordinate roles, where the semigroup operations model the interactions and relationships between different roles within an organization or system. Another practical application the authors explore is the use of DNA sequences in the context of an interval-valued anti-neutrosophic semigroup and they provide some results on it.

Inventory management is one of the most important management functions that plays an important role in production and marketing processes, especially in production facilities and commercial institutions that have warehouses in which they keep their equipment and goods. They are concerned with determining the appropriate size of inventory for each material, to secure demand for it during the duration of

under neutrosophic data is considered. However, how to deal with the uncertainty caused by neutrosophic numbers in an optimization problem can be a fundamental challenge in finding appropriate answers in the problem-solving process. Hence, a weighted ranking method is introduced to find a range of responses. Then, the accuracy and reasonableness of the proposed ranking method for neutrosophic numbers is tested by proving several theorems. In addition, the parametric nature of the method gives the decision maker access to a desirable level of answers in real-world problems. In the following, to demonstrate the efficiency of the method, a practical problem of a multi-product three-level supply chain has been modeled under neutrosophic data.

Chapter 10
An Approach to Solve Non-Linear Neutrosophic Transportation Problem With Volume Discount 219
Aakanksha Singh, Indira Gandhi Delhi Technical University for Women, Delhi, India & Aryabhatta College, University of Delhi, India
Ritu Arora, Keshav Mahavidyalaya, University of Delhi, India
Shalini Arora, Indira Gandhi Delhi Technical University for Women, Delhi, India

In real-life transportation problems (TPs), the supply, demand, and costs parameters are uncertain in nature. During transportation due to unavoidable conditions like accidents, road conditions, poor handling, etc., the commodity gets damaged resulting in damage cost. Henceforth, the decision maker (DM) aims at minimizing both the transportation cost and the damage cost. Also, there are shipping policies where discounts associated with each shipment are applied and are directly proportional to the commodity transported. Neutrosophic numbers of the type b+b'I are capable of representing uncertainties. In the chapter, the authors formulate a non-linear TP with neutrosophic parameters. The considered objective function will simultaneously minimize the transportation cost; damage cost and volume discounts are also incorporated. A solution methodology based on interval programming is propounded. A solved numerical establishes the efficiency of the approach. Sensitivity analysis is performed on different values of I and their solutions are compared.

In the realm of supply chain management, achieving optimal inventory control and efficient rework processes within distribution centers are critical for maintaining competitiveness and customer satisfaction. This chapter introduces an approach that integrates a neutrosophic inventory model (NIM) with coordinated rework stations and distribution centers, bolstered by ant colony optimization (ACO). Coordinated rework stations within distribution centers aim to minimize cycle times and improve product quality. ACO, inspired by the foraging behavior of real ants, is employed to adaptively optimize the dynamic and uncertain nature of supply chain and rework processes. Simulation studies validate the effectiveness of the proposed model, showcasing its ability to improve overall supply chain performance, reduce costs, and enhance resilience to uncertainties. This research contributes to advancing supply chain methodologies by offering a holistic approach to inventory and rework optimization through neutrosophic logic and swarm intelligence.

The selection of renewable energy alternatives becomes critical due to several quantitative and qualitative factors describing the criteria. Often these criteria cannot be appropriately quantified due to linguistic assessment. Hence, selecting renewable energy alternatives becomes a complex, uncertain multi-criteria decision-making (MCDM) problem. Several soft set-based TOPSIS approaches are used to solve uncertain MCDM problems. In this chapter, the authors introduce a neutrosophic soft set-based TOPSIS approach to solve MCDM problems in an uncertain situation. The TOPSIS methods calculate the relative measure of distance between its positive ideal solution (PIS) and negative ideal solution (NIS). This distance is often measured by Euclidean or Hamming techniques which create ambiguity and computational complexities. The authors introduce a correlation-based TOPSIS approach to solve an MCDM problem to avoid this ambiguity and computational hazards. A numerical discussion of renewable energy alternative selection is given to establish the proposed approach.

Shan E. Fatima, Government College University, Lahore, Pakistan
Hina Khan, Government College University, Lahore, Pakistan
Kanwal Javaid, Governemnt College University, Lahore, Pakistan

In this research work, a new neutrosophic probability model named a neutrosophic Topp-Leone exponentiated generalized exponential distribution is proposed to model the normalized difference vegetation index (NDVI) of Pakistan in the year 2022. The MODIS-Merra-2 satellite was used in this research, which gave spatial and spectral information on NDVI in the Earth. NDVI maps were produced from the acquired data, and NDVI sensors computed the normalized difference vegetation index. Mathematical characteristics of the proposed model of NDVI are derived, including central tendency measures, dispersion estimators, inferential statistics, moment generating function, order statistics, reliability measures, quantile-based statistical measures, Rényi entropy, and graphical representations of the density function are also presented. The maximum likelihood technique is used to estimate the parameters for the proposed probability model.

Atiqe Ur Rahman, University of Management and Technology, Lahore, Pakistan
Florentin Smarandache, University of New Mexico, USA
Muhammad Saeed, University of Management and Technology, Lahore, Pakistan
Khuram Ali Khan, University of Sargodha, Pakistan

The concept of a hypersoft membership function is introduced in the extension of a soft set known as a hypersoft set, permitting it to handle complicated and uncertain information in a more powerful and flexible manner. Many academics have already become fascinated with this new area of study, leading to the development of a number of hybrid structures. This chapter develops some new hybrid hypersoft set structures by taking into account multiple fuzzy set-like settings and possibility degree-based settings collectively. Additionally, numerical examples are included to clarify the concept of these structures. Researchers can utilize this work to better understand and apply a variety of mathematical ideas.

A. Revathy, Sri Eshwar College of Engineering, India
V. Inthumathi, NGM College, Pollachi, India
S. Krishnaprakash, Sri Krishna College of Engineering and Technology, India
S. Gomathi, Sri GVG Vishalakshi College for Women, India
N. Akiladevi, Sri Eshwar College of Engineering, India

Fermatean fuzzy set (FFS) is a comprehensive form of intuitionistic fuzzy set (IFS) which has wide range for truth values (TV) and false values (FV). The neutrosophic set (NS) can quantify the indeterminacy of fuzzy characteristics of the dataset beyond the TV and FV independently. Fermatean neutrosophic set (FNS) set is an effective tool to handle the uncertainty in multi criteria decision making (MCDM) since it incorporates the important aspects of NS as well as FFS. For MCDM, the Bonferroni mean operator, which addresses the interdependencies between attributes, is a beneficial tool in certain circumstances because of its easy accessibility and stability. In this chapter Fermatean neutrosophic normalized weighted Bonferroni mean (FNNWBM) operator is presented and their features are examined. The Fermatean neutrosophic numbers (FNN) are aggregated using FNNWBM operator and the alternatives are ranked by the FNN extended score function in MCDM. The effectiveness of the proposed operator is checked through the obtained results' simulation and comparison analysis with other existing methods.

Srinivasan Vijayabalaji, University College of Engineering, Anna University, Panruti, India
Parthasarathy Balaji, Measi Academy of Architecture, India
Gunalan Venkadesh, Krishnasamy College of Engineering and Technology, Cuddalore, India

The purpose of this chapter is to introduce the critical path problem in generalized Fermatean neutrosophic set. An algorithm to find the critical path using generalized Fermatean neutrosophic set is also provided with an example. The chapter is conducted by comparing this work with the existing algorithm to validate the new work. Furthermore, the authors intend to expand the Fermatean neutrosophic set to include previously defined score, accuracy, and certainty functions. In operations research, they want to present an algorithm for detecting the critical path using the

Fermatean neutrosophic set and compare the results. They intend to compare the currently available traditional critical path method to the Fermatean neutrosophic set, along with suitable circumstances.

Chapter 17
Interval Valued Neutrosophic Information System and Its Applications to Decision Making 433
V. Lakshmana Gomathi Nayagam, National Institute of Technology, Tiruchirappalli, India
Daniel P., St. Xavier's College, India
Bharanidharan R., National Institute of Technology, Tiruchirappalli, India

Smarandache introduced neutrosophic sets, interval-valued / n-valued refined neutrosophic sets due to their utility in various research fields. Multi criteria decision making problem in which evaluations of the alternatives based on criteria are given in the form of single-valued / interval-valued neutrosophic triplets (SVNT / IVNT) is an important area in neutrosophic research. Nayagam, et al have developed total ordering method to rank SVNT / IVNT. Few dominance-based ranking techniques have also been developed in the recent years to identify the best alternative in an information system. In this article, single valued and interval valued neutrosophic information systems (SVNIS and IVNIS) are introduced by defining generalized weighted dominance relation on objects and generalized weighted dominance degree of objects. Further, IVNIS is compared with Intuitionistic fuzzy interval information system (IFIIS) by an example. Further, algorithms to identify optimal objects from SVNIS / IVNIS are developed and are illustrated through numerical examples.

Chapter 18
On Multi-Criteria Job Sequencing Decision-Making Problem via Fermatean Pentapartitioned Neutrosophic Set 459
R. Subha, Nirmala College for Women, Coimbatore, India
K. Mokana, Nirmala College for Women, Coimbatore, India

Every strategy for making decisions involves some degree of uncertainty. Numerous theories exist to tackle this uncertainty in real-world models. The word "scheduling" appears so often in our daily lives. We can organize and arrange tasks in the right order with the aid of scheduling. It is highly appropriate for handling production management since it helps to boost earnings and shorten the total amount of time needed to do the assigned tasks. In the natural sequencing problem, which is a specific kind of scheduling problem, a timetable is entirely determined by the order in which the jobs are completed. One significant use of operations research is job sequencing, which identifies the best order for tasks in order to reduce the overall amount of time that has passed. This chapter solves a machine sequencing problem with suitable numerical applications using a Feramatean pentapartitioned neutrosophic set technique. The findings give room for more investigation into the problems this study identified and future research directions.

Chapter 19
An Innovative Approach to Group Decision-Making Based on Weighted Hypersoft Expert System 483
Ajoy Kanti Das, Tripura University, India
Rakhal Das, ICFAI University, India
Rupak Datta, Tripura University, India
Carlos Granados, Universidad de Antioquia, Colombia

The aim of this chapter is to explore the concept of weighted hypersoft expert sets (WHSES) and analyze their core properties, extending from hypersoft expert sets (HSES), also to introduce innovative notions like agree-WHSES, disagree-WHSES, and weighted scores, enhancing understanding of decision consensus in WHSES frameworks. This proposed decision-making method (DMM) is adaptable and suited for real-life decision-making problems (DMPs), demonstrated with practical examples. Furthermore, authors conducted a comparative analysis to validate the effectiveness of our approach in advancing decision science methodologies.

Chapter 20
Generalized Plithogenic Sets in Multi-Attribute Decision Making 519
Nivetha Martin, Arul Anandar College (Autonomous), Karumathur, India
R. Priya, Sethu Institute of Technology, India
Florentin Smarandache, University of New Mexico, USA

The developments in the field of Plithogeny find extensive applications in multi-attribute decision making. This chapter proposes the generalized version of plithogenic sets as an extension of extended plithogenic sets. The generalized plithogenic sets are of 9-tuple form comprising realistic representations of different attribute types. The attributes are categorized into dominant, recessive, and satisfactory in a generalized plithogenic set. A multi-attribute decision making model on material selection with representations of generalized Plithogenic sets is formulated in this chapter. The proposed model is compared with the representations of extended plithogenic sets and basic plithogenic sets. It is observed that the proposed model is more comprehensive in nature. This chapter suggests decision making models based on generalized plithogenic models to the researchers for determining optimal decisions to the real-life problems.

Chapter 21
Solving Neutrosophic Minimum Spanning Tree Problem by Least Edge Weight Algorithm 547
Shayathri Linganathan, Vellore Institute of Techhnology, India
Purusotham Singamsetty, Vellore Institute of Technology, India

A minimum spanning tree problem (MST) is a tree that identifies a subset of edges which connects all the vertices of a connected, undirected, and edge-weighted graph with the least total weight. In general, the edge weight can be distance, cost, time, etc. However, in case of any ambiguity of information, the edge weight may not be a deterministic value. A neutrosophic set is a powerful tool for complexity, dealing with imprecise, ambiguous, and inconsistent information in the actual wide world. In neutrosophic MST, the edge weight is represented by a neutrosophic number set. When compared to the fuzzy MST, neutrosophic MST graphs give more accuracy and compatibility in a neutrosophic environment. To solve the neutrosophic MST, least edge weight algorithm is proposed and executed in MATLAB.

Chapter 22
Algorithms of Designing Decision Trees From Indeterm Soft Sets 561
Erick González Caballero, Asociación Latinoamericana de Ciencias Neutrosóficas, Cuba
Ketty Marilú Moscoso-Paucarchuco, Universidad Nacional Autónoma Altoandina de Tarma, Peru
Noel Batista Hernandez, Universidad Regional Autónoma de los Andes, Ecuador
Lorenzo Jovanny Cevallos Torres, Universidad Bolivariana del Ecuador, Ecuador
Maikel Leyva, Universidad de Guayaquil, Ecuador
Victor Gustavo Gómez Rodríguez, Universidad Bolivariana del Ecuador, Ecuador

Since their creation, soft sets have served as a technique to model different situations of uncertainty. Furthermore, they have been extended to or hybridized with other theories of uncertainty. One of them is the introduction of IndetermSoft sets, where indeterminacy is incorporated into at least one of the components of the soft sets. Decision trees are well-known tools that allow an object or entity to be classified by the data measured according to some attributes. The simplest algorithms for creating decision trees are ID3 and C4.5. The purpose of this chapter is to extend these two algorithms in the area of uncertainty and indeterminacy for data represented in the

Preface

In an era where artificial intelligence is profoundly changing our daily lives and scientific endeavors, we find it necessary to provide innovative research and studies that bridge the gap between theoretical foundations and practical applications. A concerted effort to explore and develop cutting-edge methodologies in optimization, framed by leading theories of Neutrosophy and Plithogeny.

Neutrosophic and plithogenic theories provide powerful frameworks for dealing with uncertainty, imprecision, and indeterminacy, which are intrinsic to real-world problems. These concepts extend traditional set, logic, probability, and statistical theories to more effectively understand and manage complex and dynamic systems. Our goal is to provide a comprehensive overview of how these advanced theories can be applied across fields, Medical, industrial, technical, commercial....

The scope of this book encompasses a new development based on the concepts of neutrosophic logic to reformulate some operations research methods and their applications in practical life, through what was presented in the first six chapters and a detailed presentation of some static inventory models through what was presented in the seventh and eighth chapters and important studies. It helps decision makers make ideal decisions based on scientific foundations through what was presented in the remaining chapters.

This collective research effort is designed to serve a broad audience, including students, faculty, researchers, engineers, and technicians. It is also a valuable resource for participants in the educational process - advanced school students, college students, and graduate students - who seek to integrate these advanced concepts into their studies and research. Moreover, professionals and decision-makers in consumer and production organizations will find this book indispensable for its practical applications and innovative solutions.

We express our gratitude to contributors from various fields whose work has greatly enriched this volume. Their experiences and insights have been useful in illustrating the diversity and applicability of neutrosophic and plithogeny theories.

We also commend the support provided by academic institutions and research centers that provided the necessary environment and resources for this collaborative project.

We invite you, the reader, to engage with the material presented in this book, to explore the potential of neutrosophic and plithogenic optimization, and to contribute to the ongoing dialogue and development in this dynamic field.

As editors of this compendium, we are pleased to present *Neutrosophic and Plithogenic Inventory Models for Applied Mathematics*. This volume summarizes a pivotal moment in the development of optimization technologies, against the backdrop of neutrosophic and plithogenic theories. The chapters in this book display a profound combination of theoretical rigor and practical innovation, and address the ubiquitous challenges of uncertainty, imprecision, and indeterminacy in contemporary problem solving. Each chapter offers a unique perspective, enhanced by robust mathematical frameworks and detailed case studies that underscore practical utility. These advanced concepts.

In the chapters of this book, the contributors presented a neutrosophic study of some operations research methods and their applications in practical life. It included a presentation of important concepts in neutrosophic logic through what was presented in the first six chapters and a presentation of an extremely important field for the smooth operation of various sectors, which is neutrosophic static inventory models, in the two chapters. Seventh and eighth, and important neutrosophical studies that help decision makers make ideal decisions based on scientific foundations through what was presented in the remaining chapters.

The audience for this book is broad, and includes students, teachers, researchers, engineers, and industry professionals. By bridging the gap between academic research and practical application, we aim to provide our readers with the tools and knowledge needed to effectively navigate and solve real-world challenges.

Florentin Smarandache
University of New Mexico, USA

Maissam Jdid
Damascus University, Syria

Maikel Leyva-Vazquez
Regional Autonomous University of Los Andes, Ecuador

Chapter 1
Advancements in Plithogenics Exploring New Dimensions of Soft Sets

Sima Das
https://orcid.org/0000-0001-8048-6597
Bengal College of Engineering and Technology, India

Monojit Manna
RCC Institute of Information Technology, India

Subrata Modak
Global College of Science and Technology, India

ABSTRACT

In this chapter, the authors delve into the concept of Plithogenics and its contributions to exploring new dimensions of soft sets. They begin by providing a brief introduction to soft sets and highlighting their applications in real-world scenarios. They then delve into the motivations behind exploring Plithogenics, focusing on the need for a more expressive and versatile framework to handle complex data structures. The core of this chapter lies in presenting the advancements in Plithogenics, where the authors introduce the concept of multilayer soft sets.

DOI: 10.4018/979-8-3693-3204-7.ch001

I. INTRODUCTION

In recent years, the field of soft sets has gained significant attention and applications in various domains due to its flexibility in handling uncertain and imprecise data. Soft sets were introduced as a mathematical framework by Molodtsov in 1999 as a generalization of classical sets to accommodate the concept of vagueness and uncertainty. Soft sets have proven to be effective in various real-world scenarios, such as decision-making, pattern recognition, data mining, and information fusion, among others. However, traditional soft sets are limited in their ability to capture complex relationships and interdependencies within datasets, especially when dealing with multidimensional data. To address these limitations, researchers have been exploring the concept of Plithogenics – a novel approach that extends soft sets into higher dimensions. Plithogenics introduces a more comprehensive and versatile framework for handling complex data structures and exploring new dimensions in soft set theory. The concept of Plithogenics revolves around the idea of utilizing multiple layers or dimensions in soft sets to represent a more intricate understanding of uncertain data. Traditional soft sets are based on a single-layer approach, where each element is associated with a degree of membership to a particular set. However, Plithogenics goes beyond this by incorporating multiple layers of soft sets, each layer contributing to a more nuanced representation of the data. Plithogenics introduces the notion of multilayer soft sets, wherein each layer represents a different aspect or characteristic of the data. For example, in a decision-making scenario, one layer might represent the preferences of individual decision-makers, while another layer could capture the collective preferences of a group. These multilayer soft sets enable a more comprehensive analysis of the data from various perspectives. Plithogenics explores the relationships between different layers of soft sets. By identifying connections and dependencies across the layers, researchers can gain insights into the interplay between various aspects of the uncertain data. This is particularly useful in applications where complex interactions between factors influence the overall decision-making process. Developing efficient algorithms for processing and manipulating Plithogenic soft sets is a crucial aspect of the advancements. Researchers have been working on devising algorithms to perform operations such as intersection, union, and complement on multilayer soft sets, enabling better data analysis and decision support.

Applications in Diverse Fields: The advancements in Plithogenics have opened up new possibilities for applications in various fields, such as healthcare, finance, engineering, and social sciences. For instance, in healthcare, Plithogenic soft sets can be used to model patient preferences across different treatment options, leading to more personalized and effective medical decisions.

Uncertainty Management: Plithogenics provides a powerful tool for managing uncertainty in data. By capturing uncertainty across multiple dimensions, decision-makers can have a more comprehensive view of the potential outcomes and risks associated with their choices.

The objectives of exploring advancements in Plithogenics and new dimensions of soft sets can be summarized as follows:

1. **Enhance Soft Set Representations:** The primary objective is to extend the traditional soft set framework to higher dimensions, enabling a more comprehensive and expressive representation of uncertain data. By incorporating multilayer soft sets, researchers aim to capture complex relationships and interdependencies within datasets, leading to a richer understanding of the underlying data structure.
2. **Develop Efficient Algorithms:** Another key objective is to develop efficient algorithms and computational methods for processing and manipulating Plithogenic soft sets. These algorithms should facilitate essential operations such as set operations, similarity measures, and decision-making support, ensuring practical applicability of the Plithogenics framework.
3. **Explore Interlayer Relations:** Investigating the interlayer relationships within Plithogenic soft sets is crucial to understanding how different layers of data interact and influence each other. Identifying and analyzing these relationships can provide valuable insights into the complexities of uncertain data, leading to better decision-making and problem-solving capabilities.
4. **Applications in Diverse Fields:** The objective is to explore and demonstrate the applicability of Plithogenics in various domains, including healthcare, finance, engineering, social sciences, and more. By showcasing successful applications, researchers aim to promote the adoption of Plithogenics as a powerful tool for managing uncertainty and making informed decisions.
5. **Comparison with Existing Models:** Another objective is to compare Plithogenics with other existing models for handling uncertainty, such as fuzzy sets, rough sets, and intuitionistic fuzzy sets. Understanding the advantages and limitations of Plithogenics relative to these models can provide valuable insights into its unique strengths and potential use cases.
6. **Uncertainty Management and Risk Analysis:** Plithogenics can offer effective uncertainty management and risk analysis capabilities. The objective is to develop methodologies to assess and manage uncertainty in decision-making processes and real-world applications, ensuring robustness and reliability in the face of uncertain data.

7. **Theoretical Foundations:** Advancing the theoretical foundations of Plithogenics is an essential objective. Researchers aim to establish rigorous mathematical frameworks for Plithogenic soft sets, including axioms, properties, and theorems, to ensure the validity and coherence of the proposed extensions.
8. **Educational and Community Awareness:** Researchers also aim to promote awareness and education about Plithogenics and its potential in the scientific community. Workshops, seminars, and research publications are means to disseminate knowledge and encourage collaboration in this emerging area.

These objectives, researchers can unlock the full potential of Plithogenics, leading to innovative solutions in handling uncertain and complex data, improving decision-making processes, and advancing our understanding of uncertainty in various fields of application.

II. MOTIVATION AND CONTRIBUTIONS

Motivation and contributions are as follows:

Motivation

The motivation behind exploring advancements in Plithogenics and new dimensions of soft sets stems from several key factors:

1. **Handling Uncertainty in Complex Data:** In many real-world scenarios, data is often uncertain, imprecise, or incomplete. Traditional mathematical frameworks may struggle to handle such complexities effectively. The motivation for Plithogenics arises from the need to develop a more robust and versatile approach to manage uncertainty in complex data structures.
2. **Flexibility and Expressiveness:** Soft sets have already demonstrated their flexibility in handling uncertain data. The motivation for advancing Plithogenics lies in extending this flexibility to higher dimensions, allowing for a more expressive representation of data and capturing intricate relationships between different aspects of uncertainty.
3. **Decision Support and Risk Analysis:** Decision-making processes often involve considering multiple factors and evaluating various alternatives. Plithogenics can provide enhanced decision support by enabling a comprehensive analysis of uncertain data, leading to better risk assessment and informed decision-making.

4. **Applications in Diverse Fields:** The motivation arises from the potential applications of Plithogenics in various fields such as healthcare, finance, engineering, social sciences, and more. These fields deal with complex and uncertain data, and the adoption of Plithogenics can lead to improved outcomes and solutions.

Contributions

The exploration of advancements in Plithogenics and new dimensions of soft sets can make several significant contributions to the field of uncertainty modeling and decision-making:

1. **Novel Mathematical Framework:** Plithogenics introduces a novel mathematical framework that extends soft sets to multilayer representations. This advancement contributes to a more comprehensive way of representing and managing uncertainty in data, allowing for a deeper understanding of complex relationships.
2. **Improved Data Analysis:** Plithogenics enhances the capability to analyze uncertain data by incorporating interlayer relations. This contribution enables researchers to explore dependencies and correlations between different layers, leading to more insightful data analysis and knowledge discovery.
3. **Practical Algorithms:** The development of efficient algorithms for performing operations on Plithogenic soft sets is a significant contribution. These algorithms facilitate practical applications of Plithogenics, making it feasible to process and manipulate complex uncertain data in real-world scenarios.
4. **Applications in Various Domains:** Demonstrating successful applications of Plithogenics in diverse fields contributes to the wider adoption of the framework. The practical use of Plithogenics in healthcare, finance, engineering, and other domains showcases its potential to revolutionize decision-making and uncertainty management.
5. **Theoretical Advancements:** The exploration of Plithogenics also contributes to the advancement of theoretical foundations. This includes defining axioms, properties, and theorems for Plithogenic soft sets, ensuring rigor and coherence in the framework's mathematical underpinnings.
6. **Community Awareness and Collaboration:** Research in Plithogenics fosters community awareness and collaboration among researchers interested in uncertainty modeling and decision support. The exchange of ideas, workshops, and research publications encourage interdisciplinary efforts and knowledge dissemination.

The exploration of advancements in Plithogenics and new dimensions of soft sets is motivated by the need to address uncertainty in complex data and improve decision-making processes. The contributions of this research lie in the development of a novel mathematical framework, practical algorithms, successful applications, theoretical advancements, and community engagement, all leading to a more robust and effective approach to uncertainty management.

III. LITERATURE SURVEY

Literature survey shown in **Table 1.**

Table 1. Literature survey

Citation	Technology	Advantages	Limitations	Application
(Martin et al., 2022)	Plithogenic number	Introduction to plithogenic sociogram	N/A	Sociogram representation using plithogenic numbers
(Quek et al., 2020)	N/A	Entropy measures for plithogenic sets	Application limited to entropy measures	Multi-attribute decision-making using plithogenic sets
(Smarandache, 2018, pp. 153-166)	N/A	Extension of crisp, fuzzy, intuitionistic fuzzy, and neutrosophic sets	N/A	N/A
(Gomathy et al., 2002)	N/A	Application of plithogenic sets in decision making	Application specific to decision making	Decision-making using plithogenic sets
(Abdel-Basset et al., 2020)	Best-Worst Method	Solving the supply chain problem using a plithogenic model	Limited to supply chain problem	Supply chain problem solving using the best-worst method
(Sultana et al.,2022)	N/A	Study of plithogenic graphs and applications	Application specific to studying the spread of COVID-19	Study of the spread of COVID-19 using plithogenic graphs
(Gayen et al., 2020)	N/A	Introduction to Plithogenic Subgroup	N/A	Subgroup analysis using plithogenic sets

IV. METHODOLOGY

Exploring Plithogenics and new dimensions in soft sets requires the development of specialized algorithms to handle multilayer soft sets and leverage their potential in uncertainty modeling and decision support. Below is an outline of the main steps involved in developing such algorithms as shown in Figure 1:

Figure 1. Methodology of proposed system

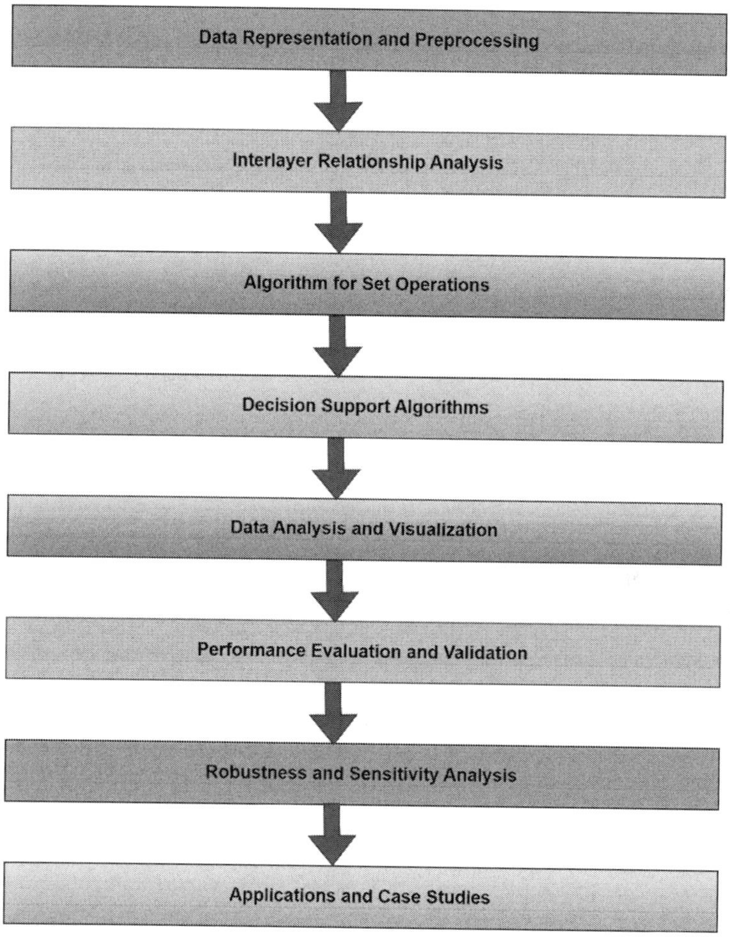

Step 1: Data Representation and Preprocessing

Convert the available data into a multilayer soft set representation, where each layer represents a different aspect or dimension of uncertainty. Handle missing data and outliers in each layer, employing appropriate imputation and data cleaning techniques. Consider data fusion methods to integrate information from various sources into a coherent multilayer soft set.

Step 2: Interlayer Relationship Analysis

Investigate the relationships between different layers to understand how uncertainties in one layer affect other layers. Develop methods to quantify the dependencies and interlayer interactions, allowing for a deeper analysis of complex data structures.

Step 3: Algorithm for Set Operations

Develop algorithms for basic set operations, such as union, intersection, complement, and Cartesian product, in the context of multilayer soft sets. Ensure that these operations capture the interlayer relationships and provide meaningful results.

Step 4: Decision Support Algorithms

Design decision support algorithms to aid in decision-making based on multilayer soft sets.

Incorporate uncertainty propagation methods to assess the impact of uncertainty on decision outcomes.

Step 5: Data Analysis and Visualization

Develop techniques for analyzing and visualizing multilayer soft sets to gain insights into the complex relationships and patterns within the data. Create visual representations that allow for intuitive interpretation and understanding.

Step 6: Performance Evaluation and Validation

Evaluate the performance of the developed algorithms using real-world datasets and synthetic data with known characteristics. Validate the results against ground truth or expert opinions to assess the accuracy and reliability of the algorithm.

Step 7: Robustness and Sensitivity Analysis

Perform sensitivity analysis to study the robustness of the algorithms under different parameter settings and data scenarios. Assess the algorithm's performance when handling different levels of uncertainty and variability.

Step 8: Applications and Case Studies

Apply the developed algorithms to diverse real-world applications, such as healthcare, finance, engineering, and social sciences. Demonstrate the practical utility of Plithogenics and new dimensions in soft sets through case studies and comparisons with other methods.

Step 9: Documentation and Dissemination

Document the algorithm design, implementation, and results thoroughly to facilitate reproducibility and further research. Disseminate the findings through research papers, conference presentations, and open-source software repositories.

By following this algorithm development process, researchers can unlock the potential of Plithogenics and new dimensions in soft sets, leading to innovative solutions in uncertainty modeling and decision support across various domains.

V. ADVANTAGES

Advantages of Plithogenics and the exploration of new dimensions in soft sets which is shown in Figure 2 include:

1. **Enhanced Uncertainty Modeling:** Plithogenics allows for the representation of uncertainties across multiple layers, providing a more comprehensive and nuanced understanding of complex data. This enables better uncertainty modeling and management in various real-world applications.
2. **Improved Decision Support:** By considering various layers of data and their interdependencies, Plithogenics offers richer insights for decision support. Decision-makers can make more informed and reliable decisions, leading to improved outcomes and reduced risks.
3. **Flexible Framework:** Plithogenics is a flexible framework that can be tailored to different domains and applications. It can handle diverse types of data, making it applicable in various fields, such as healthcare, finance, engineering, social sciences, and more.
4. **Holistic Analysis:** The multilayer approach of Plithogenics allows for a holistic analysis of complex systems and relationships. It takes into account the interactions between different factors, leading to a more complete understanding of the underlying dynamics.
5. **Data Fusion and Integration:** Plithogenics enables the integration of data from various sources and modalities. It can fuse heterogeneous data and provide a unified representation, facilitating better data integration and knowledge discovery.
6. **Robustness to Incomplete Information:** Plithogenics can handle incomplete and uncertain information effectively. It allows for the inclusion of partial or missing data, making it robust in situations where complete data may not be available.
7. **Interpretability and Visualization:** The multilayer structure of Plithogenics facilitates interpretability and visualization. Decision-makers can understand how different layers contribute to overall outcomes, aiding in explaining results to stakeholders.
8. **Scalability and Efficiency:** Plithogenics algorithms have been developed to efficiently process and manipulate multilayer soft sets. This enables scalability to handle large datasets, making it practical for real-world applications.
9. **Cross-Domain Applications:** Plithogenics can be applied across diverse domains, including engineering, economics, environmental sciences, and more. Its versatility allows researchers to explore novel applications and solve complex problems.

10. **Advancement of Soft Set Theory:** The exploration of new dimensions in soft sets, such as Plithogenics, contributes to the advancement of soft set theory itself. It enriches the theoretical foundations and expands the capabilities of soft sets as a powerful tool for uncertainty management.

Figure 2. Advantages of proposed work

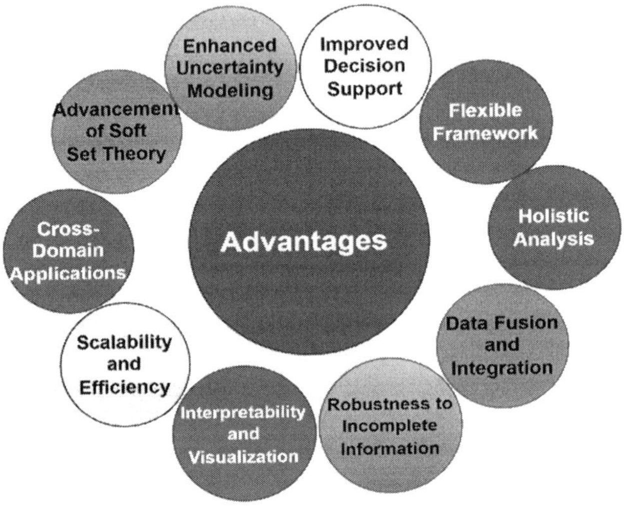

The advantages of Plithogenics lie in its ability to tackle uncertainty in a more holistic and sophisticated manner, leading to better decision-making, improved analysis, and enhanced applications across various disciplines. As research continues, Plithogenics is expected to unlock new dimensions of insights and open up new opportunities for uncertainty modeling and data analysis.

VI. LIMITATIONS

While Plithogenics and the exploration of new dimensions in soft sets offer several advantages, there are also some potential limitations to consider, which is shown in Figure 3:

1. **Complexity:** Plithogenics introduces additional complexity due to its multilayer representation. Handling and processing multilayer soft sets may require more computational resources and time compared to traditional soft sets.

2. **Data Requirements:** Plithogenics may require more extensive data collection efforts to build and populate the multiple layers adequately. In cases where data availability is limited or costly, this can be a significant disadvantage.
3. **Interlayer Relationships:** The interlayer relationships in Plithogenics can be challenging to define and model accurately. Determining the dependencies and interactions between different layers may require domain expertise and careful consideration.
4. **Interpretability:** While Plithogenics offers rich insights, interpreting the relationships between layers may be challenging, especially in complex systems. Understanding how specific factors in one layer impact other layers may not always be straightforward.
5. **Algorithmic Complexity:** Developing efficient algorithms for processing multilayer soft sets can be more complex than traditional soft sets. Researchers need to address the computational challenges arising from the increased dimensionality.
6. **Data Fusion Challenges:** Integrating and fusing data from various sources into a coherent multilayer soft set can be difficult. Data fusion techniques must be carefully designed to ensure consistency and reliability.
7. **Risk of Overfitting:** In cases where the number of layers is high relative to the available data, there is a risk of overfitting the model to the training data. Overfitting may lead to less generalizable results and biased conclusions.
8. **Applicability to Small Datasets:** Plithogenics may not be as effective when applied to small datasets with limited variability across layers. The benefits of the multilayer approach may not be fully realized in such scenarios.
9. **Lack of Standardization:** As Plithogenics is a relatively new concept, there may be a lack of standardization in terms of terminology, methodology, and evaluation metrics. This can hinder comparisons and replication of studies.
10. **Learning Curve:** The adoption of Plithogenics in practice may require a learning curve for researchers and practitioners. Familiarizing oneself with the concept and implementation may take time and effort.

Figure 3. Limitations of proposed work

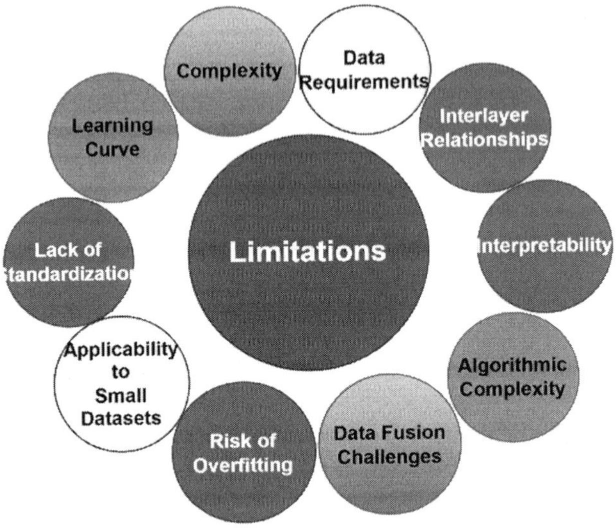

Despite these potential limitations, Plithogenics remains a promising framework for uncertainty modeling and decision support. Addressing these challenges through further research and development can lead to more effective and practical applications of Plithogenics in various domains.

VII. APPLICATIONS

The advancements in Plithogenics and the exploration of new dimensions in soft sets have led to diverse applications across various fields. Some of the key applications of Plithogenics are shown in Figure 4:

1. **Decision Support Systems:** Plithogenic soft sets offer a powerful framework for decision support systems. They allow decision-makers to model uncertainty and complex relationships between different factors, leading to more informed and reliable decision-making.
2. **Healthcare:** In the healthcare domain, Plithogenics can be used for medical diagnosis and treatment planning. By considering multiple layers of patient data, such as medical history, symptoms, and test results, healthcare professionals can make more personalized and accurate decisions.

3. **Financial Analysis:** Plithogenics can be applied in financial analysis and investment decision-making. By considering economic conditions, sector performance, and individual stock data, investors can better assess risks and opportunities in their portfolios.
4. **Supply Chain Management:** Plithogenics can aid in optimizing supply chain operations. By considering various factors like demand, supplier performance, and transportation constraints, supply chain managers can improve efficiency and reduce uncertainty.
5. **Environmental Modeling:** Plithogenics can be applied in environmental modeling to assess the impact of different factors on the environment. For instance, it can help in understanding the influence of climate conditions, human activities, and natural events on environmental changes.
6. **Social Sciences:** Plithogenics can be used in social sciences to analyze complex social relationships and decision-making processes. It can aid in understanding the interplay of various factors that influence social behaviors and outcomes.
7. **Marketing and Customer Analysis:** Plithogenics can be applied in marketing to analyze customer preferences and behaviors. By considering multiple layers of data, marketers can tailor their strategies to specific customer segments.
8. **Smart Cities:** Plithogenics can contribute to smart city planning and management. It can help in optimizing resource allocation, transportation systems, and infrastructure development in urban environments.
9. **Risk Management:** Plithogenics can assist in risk management by considering multiple layers of data to assess and mitigate risks in various scenarios, such as insurance underwriting and project planning.
10. **Education:** Plithogenics can be applied in educational contexts to personalize learning experiences based on individual student preferences, abilities, and learning styles.
11. **Smart Grid Management:** Plithogenics can be utilized in managing smart grid systems, where factors like electricity demand, renewable energy generation, and grid stability can be considered across different layers to optimize power distribution and ensure grid reliability.
12. **Transportation Planning:** In transportation planning, Plithogenics can be used to model various aspects like traffic flow, infrastructure capacity, and environmental impacts, leading to more efficient transportation systems and reduced congestion.
13. **Robotics and Autonomous Systems:** Plithogenics can play a role in robotics and autonomous systems by considering multiple layers of sensor data, environmental conditions, and task requirements, enabling more sophisticated decision-making and behavior adaptation.

14. **Climate Change Adaptation:** Plithogenics can aid in climate change adaptation strategies by incorporating multiple layers of climate data, socioeconomic factors, and vulnerability assessments to devise more resilient and adaptive measures.
15. **Natural Disaster Management:** Plithogenics can be applied in natural disaster management to analyze various layers of data related to hazards, exposure, and vulnerability, helping in disaster preparedness and response planning.
16. **Quality Control and Manufacturing:** Plithogenics can be used in manufacturing to consider multiple layers of data related to process variables, quality parameters, and equipment performance, leading to improved quality control and productivity.
17. **Bioinformatics:** In bioinformatics, Plithogenics can be utilized to analyze multi-layered biological data, such as genomics, proteomics, and transcriptomics, aiding in the understanding of complex biological processes and diseases.
18. **Internet of Things (IoT) Applications:** Plithogenics can be integrated into IoT applications to analyze multi-dimensional sensor data from various devices and optimize IoT systems for improved performance and resource utilization.
19. **Natural Language Processing (NLP):** In NLP, Plithogenics can be used to represent multiple layers of linguistic features and context, leading to more accurate and context-aware language processing.
20. **Social Media Analysis:** Plithogenics can be applied in social media analysis to model user behavior, sentiment, and content preferences across different layers, facilitating targeted marketing and personalized recommendations.

Figure 4. Applications

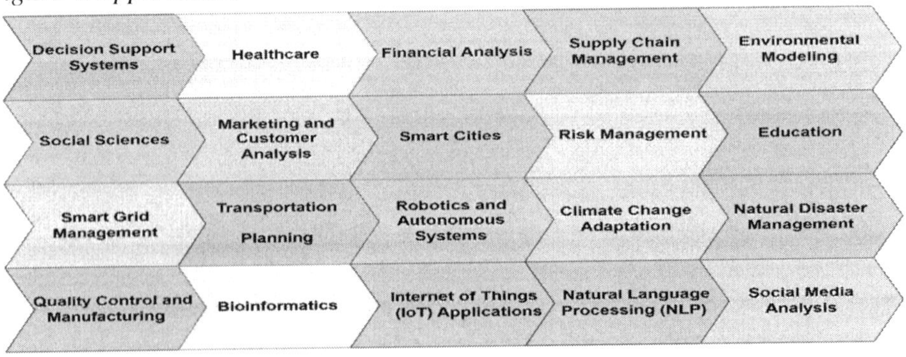

These applications demonstrate the versatility and potential impact of Plithogenics in addressing uncertainty and making better decisions across a wide range of domains. As research continues, the scope of applications for Plithogenics is likely to expand further, leading to more sophisticated and effective solutions in various fields.

VIII. CONCLUSION AND FUTURE WORK

In conclusion, the exploration of advancements in Plithogenics and the extension of soft sets into new dimensions have presented a promising direction in handling uncertainty and complex data. By introducing multilayer soft sets, researchers have achieved a more comprehensive representation of uncertain data, capturing intricate relationships and dependencies across different aspects. The interlayer relations offer valuable insights into the interplay between various dimensions of uncertainty, leading to more nuanced data analysis and decision-making support.

The development of efficient algorithms for processing Plithogenic soft sets has further enhanced the practical applicability of this framework in real-world scenarios. Applications in diverse fields such as healthcare, finance, engineering, and social sciences have demonstrated the potential of Plithogenics to revolutionize decision-making processes and improve uncertainty management.

The theoretical foundations of Plithogenics have been established, defining essential axioms, properties, and theorems, ensuring the validity and coherence of the extended soft set framework.

Future Work

As the field of Plithogenics continues to evolve, there are several avenues for future research and development: Further research can focus on enhancing the scalability and efficiency of algorithms for processing large-scale multilayer soft sets. This will enable the application of Plithogenics to big data and real-time decision-making scenarios. Investigating the integration of Plithogenics with machine learning techniques can lead to enhanced predictive modeling and pattern recognition capabilities, providing more accurate and robust solutions in uncertain data analysis. Studying the propagation of uncertainty across layers in Plithogenic soft sets can provide insights into how uncertainty propagates through interconnected factors, contributing to risk analysis and resilience assessment. Exploring decision fusion techniques with multilayer soft sets can lead to more informed and reliable decision-making processes, especially in group decision-making scenarios. Extending the range of real-world applications for Plithogenics, such as in environmental modeling, supply chain management, and smart cities, can demonstrate its versatility and utility in diverse domains. Investigating hybrid approaches that combine Plithogenics with other uncertainty modeling techniques like fuzzy sets, rough sets, and interval-valued fuzzy sets can lead to more comprehensive uncertainty management strategies. Developing visualization techniques for multilayer soft sets can aid in interpreting and presenting complex data relationships, facilitating better understanding and communication of results. The ongoing research in Plithogenics and new dimensions of soft sets holds tremendous promise in advancing uncertainty modeling, decision support, and data analysis. The continued exploration of these advancements will pave the way for more sophisticated and effective tools to tackle uncertainty in various fields, revolutionizing how we handle complex data and make informed decisions.

REFERENCES

Abdel-Basset, M., Mohamed, R., Zaied, Abd El-Nasser H., Gamal, A., & Smarandache, F. (2020). Solving the supply chain problem using the best-worst method based on a novel Plithogenic model. DOI: 10.1016/B978-0-12-819670-0.00001-9

Gayen, S., Smarandache, F., Jha, S., Singh, M. K., Broumi, S., & Kumar, R. (2020). Introduction to Plithogenic Subgroup. In Smarandache, F., & Broumi, S. (Eds.), *Neutrosophic Graph Theory and Algorithms* (pp. 213–259). IGI Global. DOI: 10.4018/978-1-7998-1313-2.ch008

Gomathy, S., Deivanayagampillai, N., & Said, B. (2002). Plithogenic sets and their application in decision making. *Neutrosophic Sets and Systems., 38*. Advance online publication. DOI: 10.5281/zenodo.4300565

Martin, N., Smarandache, F., & Priya, R. (2022). Introduction to plithogenic sociogram with preference representations by plithogenic number. *Journal of Fuzzy Extension and Applications*, *3*(1), 96–108. DOI: 10.22105/jfea.2021.288057.1151

Quek, S. G., Selvachandran, G., Smarandache, F., Vimala, J., Le, S. H., Bui, Q.-T., & Gerogiannis, V. C. (2020). Entropy Measures for Plithogenic Sets and Applications in Multi-Attribute Decision Making. *Mathematics*, *8*(6), 965. DOI: 10.3390/math8060965

Smarandache, F. (2018). Plithogenic Set, an Extension of Crisp, Fuzzy, Intuitionistic Fuzzy, and Neutrosophic Sets - Revisited. *Neutrosophic Sets and Systems.*, *21*, 153–166.

Sultana, F., Gulistan, M., Ali, M., Yaqoob, N., Khan, M., Rashid, T., & Ahmed, T. (2022, April 4). A study of plithogenic graphs: Applications in spreading coronavirus disease (COVID-19) globally. *Journal of Ambient Intelligence and Humanized Computing*, 1–21. Advance online publication. DOI: 10.1007/s12652-022-03772-6 PMID: 35401852

Chapter 2
Pythagorean Anti-Neutrosophic Linear Space

Sambandan Sivaramakrishnan
https://orcid.org/0000-0002-8647-1174
Manakula Vinayagar Institute of Technology, India

Parthasarathy Balaji
Measi Academy of Architecture, India

M. Sivasakthi
Krishnasamy College of Science, Arts, and Management for Women, India

ABSTRACT

This chapter introduces the concept of a Pythagorean anti-neutrosophic linear space, expanding on the principles of Pythagorean, anti-neutrosophic, and fuzzy linear space. These advancements offer new approaches for addressing uncertainty, indeterminacy, and contradiction within traditional linear space. The authors prove that the union of two Pythagorean anti-neutrosophic linear spaces forms another Pythagorean anti-neutrosophic linear space. However, they provide a counterexample to show that the intersection of two such spaces are not necessarily result in a Pythagorean anti-neutrosophic linear space. Furthermore, the authors demonstrate that the Cartesian product of two Pythagorean anti-neutrosophic linear spaces creates a new Pythagorean anti-neutrosophic linear space. Additionally, we define and explore the concepts of homomorphic image and inverse image of Pythagorean anti-neutrosophic linear spaces. This chapter aims to enhance the theoretical understanding of these advanced linear spaces in handling complex mathematical uncertainties.

DOI: 10.4018/979-8-3693-3204-7.ch002

INTRODUCTION

Zadeh's (Zadeh,1965) development of fuzzy sets has revolutionized mathematical modelling and problem-solving by providing a more flexible and nuanced approach. The field of fuzzy logic continues to evolve and find applications in various domains, enabling the effective handling of uncertainty and vagueness in real-world problems. Interval-valued fuzzy sets provide a more flexible representation of uncertainty and vagueness in real-world applications compared to traditional fuzzy sets. Instead of assigning a single membership value to an element, a range of possible membership values is assigned, making it a generalization of traditional fuzzy sets. By introducing interval-valued fuzzy sets, Zadeh (Zadeh,1975) provided a more robust framework for dealing with imprecise and uncertain information, leading to advancements in fields such as decision-making, pattern recognition, and control systems. This extension of fuzzy set theory has opened up new possibilities for modelling complex systems and handling ambiguity in a more nuanced way.

Intuitionistic fuzzy sets (IFS) were first proposed by in the article (Atanassov,1986) as an extension of classical fuzzy sets. IFS originated to overcome the constraints of traditional fuzzy sets when handling uncertain, imprecise and incomplete data. Atanassov's idea behind IFS was to provide a more sophisticated and adaptable approach to modelling uncertainty by incorporating membership and non-membership degrees and a hesitation degree.

The researcher (Yager, 2013) introduced a groundbreaking concept in fuzzy set theory known as Pythagorean fuzzy sets. These sets are considered to be more comprehensive than intuitionistic fuzzy sets due to their unique characteristics. Unlike traditional fuzzy sets that only have a degree of membership, Pythagorean fuzzy sets also incorporate a degree of non-membership. This allows for a more nuanced representation of uncertainty and ambiguity in decision-making processes.

The defining feature of Pythagorean fuzzy sets is that the sum of the squares of their membership and non-membership degrees does not exceed 1. This constraint ensures that the total uncertainty associated with an element's membership and non-membership is appropriately balanced. By incorporating both membership and non-membership degrees, Pythagorean fuzzy sets provide a more accurate and flexible representation of complex and uncertain information.

Building upon Yager's work, The researchers (Peng and Yang, 2016) extended the concept of Pythagorean fuzzy sets by introducing the notion of interval-valued Pythagorean fuzzy sets (IVPFS). These sets allow for the representation of uncertain information in the form of intervals, rather than precise values. This extension further enhances the ability of Pythagorean fuzzy sets to handle complex and imprecise data, making them even more applicable in real-world decision-making scenarios.

In a recent development, The researchers (Asif et al, 2020) proposed the concept of Pythagorean fuzzy linear space. This concept combines the principles of Pythagorean fuzzy sets with linear algebra, providing a framework for handling fuzzy linear equations and systems. By incorporating the notion of Pythagorean fuzzy sets into linear algebra, this concept offers a powerful tool for solving problems involving uncertainty and imprecision in linear systems.

This idea of Neutrosophic set (NS) introduced by the researcher (Smarandache , 2005) is revolutionized the field of mathematics by incorporating an intermediate membership concept. This concept allows for a more nuanced understanding of uncertainty and ambiguity in mathematical modelling, providing a new framework for dealing with complex and contradictory information. The Neutrosophic set theory has been utilized in diverse domains including decision-making, artificial intelligence and image processing, thereby presenting novel opportunities to tackle previously deemed unsolvable issues. Smarandache's innovative idea continues to inspire further research and development in the field of mathematics and beyond.

The field of fuzzy linear space is enriched with a fresh and inventive perspective through the introduction of the Neutrosophic interval-valued anti-fuzzy linear space concept by the researchers (Sivaramakrishnan et al, , 2022) proposed the concept of an Neutrosophic interval-valued anti-fuzzy linear space. This concept combines the ideas of Neutrosophic set, interval-valued fuzzy sets and anti-fuzzy sets to provide a more comprehensive and flexible framework for dealing with uncertainty and imprecision in linear spaces. The concept of the Pythagorean Neutrosophic set was first introduced by the researcher Jansi et all in their article (Jansi et al, 2019) to further elaborate on the Neutrosophic set.

In this chapter, we provide a proof that the union of two Pythagorean anti-neutrosophic linear spaces results in another Pythagorean anti-neutrosophic linear space. Additionally, we demonstrate through a counter example that the intersection of two Pythagorean anti-neutrosohoic linear spaces need not always yield a Pythagorean anti-neutrosophic linear space. Moreover, the cartesian product of two Pythagorean anti-neutrosophic linear spaces gives rise to a new Pythagorean anti-neutrosohic linear space. Furthermore, we establish the definitions of homomorphism and isomorphism between two Pythagorean anti-neutrosophic linear spaces.

PRELIMINARIES

Definition 2.1 (Yager, 2013). Let $\mathscr{P}$ be the set. The Pythagorean fuzzy set (PFS) is defined as $\boldsymbol{\psi} = \{ < \boldsymbol{p}, \boldsymbol{\mu}_{\psi}(\boldsymbol{p}), \ \boldsymbol{v}_{\psi}(\boldsymbol{p}) > |\boldsymbol{p} \in \mathscr{P}\}$, such that $\mu_{\psi} \in [0,1]$ and $v_{\psi} \in [0,1]$ satisfy the following condition $0 \leq (\mu_{\psi})^2 + (v_{\psi})^2 \leq 1$, for every $\boldsymbol{p} \in \mathscr{P}$. The set of all Pythagorean fuzzy sets over $\mathscr{P}$ is denoted by $\boldsymbol{PF}^{\mathscr{P}}$.

Definition 2.2 (Yager, 2013). X should be regarded as the universe of discourse. A Neutrosophic set (NS), denoted as $\Omega = \{s, \xi_{\Omega}(s), \Psi_{\Omega}(s), \zeta_{\Omega}(s) | s \in X\}$ is defined as follows: $\Omega = (\xi_{\Omega}(s), \Psi_{\Omega}(s), \zeta_{\Omega}(s))$. Here, ξ: $\mathbf{X} \rightarrow [0,1]$, Ψ: $\mathbf{X} \rightarrow [0,1]$ and ζ: $\mathbf{X} \rightarrow [0,1]$ show the truth value, membership degree, indeterminacy level, membership of falsehood for the object s in X, in that order. It should be noted that $0 \leq \xi_{\Omega}(v) + \Psi_{\Omega}(v) + \zeta_{\Omega}(v) \leq 3$.

Definition 2.3 (Prema & Radha, 2022). Let **X** denote a universal set, characterized by a Pythagorean neutrosophic set (PNS) $P_N = \{w, \mu_p(w), \Psi_p(w), \zeta_p(w) | w \in \mathbf{X}\}$. The functions μ: $\mathbf{X} \rightarrow [0,1]$, Ψ: $\mathbf{X} \rightarrow [0,1]$ and ζ: $\mathbf{X} \rightarrow [0,1]$ indicate the levels of membership, non-membership, and indeterminacy of the element $w \in \mathbf{X}$ with respect to the subset P_N, subject to the constraint $0 \leq (\mu_p(w))^2 + (\Psi_p(w))^2 + (\zeta_p(w))^2 \leq 2$ for all $w \in \mathbf{X}$. To simplify, a PNS is represented as $P_N = (\mu_p(w), \Psi_p(w), \zeta_p(w))$.

Remark 2.4. Neutrosophic systems are a generalization of intuitionistic fuzzy systems designed to handle imprecise, indeterminate and inconsistent information. The key difference between single valued and interval-valued neutrosophic systems lies in how they represent the degrees of truth, indeterminacy and falsity.

Single Valued Neutrosophic Systems (SVNS)

In SVNS, each element is associated with a single-valued triple, representing the degrees of truth, indeterminacy, and falsity. Specifically, for any element x, its Neutrosophic value is given by (T(x), I(x), F(x)), where:

T(x): Degree of truth of the element x, which is a value in the range [0,1].
I(x): Degree of indeterminacy of the element x, which is a value in the range [0,1].
F(x): Degree of falsity of the element x, which is a value in the range [0,1].

The constraints are that $T(x), I(x), F(x) \in [0,1]$.

Interval-Valued Neutrosophic Systems (IVNS)

In IVNS, each element is associated with an interval for each of the three components: truth, indeterminacy, and falsity. Specifically, for any element x, its Neutrosophic value is given by (T(x), I(x), F(x)), where:

T(x): Degree of truth of the element x, which is represented as an interval $[T_1(x), T_2(x)]$ with $T_1(x) \leq T_2(x)$ and $T_1(x), T_2(x) \in [0,1]$.
I(x): Degree of indeterminacy of the element x, which is represented as an interval $[I_1(x), I_2(x)]$ with $I_1(x) \leq I_2(x)$ and $I_1(x), I_2(x) \in [0,1]$.
F(x): Degree of falsity of the element x, which is represented as an interval $[F_1(x), F_2(x)]$ with $F_1(x) \leq F_2(x)$ and $F_1(x), F_2(x) \in [0,1]$.

Example:

Single Valued Neutrosophic Set (SVNS)

Let x be an element with a Neutrosophic value of (0.81, 0.35, 0.12).
Degree of truth: 0.81
Degree of indeterminacy: 0.35
Degree of falsity: 0.12

Interval-Valued Neutrosophic Set (IVNS)

Let x be an element with a Neutrosophic value of ([0.6, 0.8], [0.1, 0.3], [0.0, 0.2]).
Degree of truth: between 0.6 and 0.8
Degree of indeterminacy: between 0.1 and 0.3
Degree of falsity: between 0.0 and 0.2.

PYTHAGOREAN ANTI-NEUTROSOPHIC LINEAR SPACE

Definition 3.1. Let $\mathbf{V_p}$ is a crisp linear space over a field $\mathbf{F_p}$. A non-empty Pythagorean anti-neutrosophic set $P_N = (\mu_p(w), \Psi_p(w), \zeta_p(w))$ of $\mathbf{V_p}$ is called Pythagorean anti-neutrosophic linear space (**PANLS**) of $\mathbf{V_p}$ over $\mathbf{F_p}$ denoted as $\mathbf{V_p(F_p)}$ if it satisfies:

(i) $\mu\mathbf{p}(fw_1 * gw_2) \leq \max\{\mu\mathbf{p}(w_1), \mu\mathbf{p}(w_2)\}$,
(ii) $\Psi_p(fw_1 * gw_2) \leq \max\{\Psi_p(w_1), \Psi_p(w_2)\}$,
(iii) $\zeta_p(fw_1 * gw_2) \geq \min\{\zeta_p(w_1), \zeta_p(w_2)\}$, for all w_1, w_2 in $\mathbf{V_p}$ and f,g in $\mathbf{F_p}$.

Example 3.2. $\mathbf{V_p}$ is defined as a crisp linear space over $\mathbf{F_p}$ with elements {v_w_1, v_w_2, v_w_3, v_w_4 } and operated by "∗".

Table 1. Example for Pythagorean anti-neutrosophic linear space using Cayley table.

∗	v_w_1	v_w_2	v_w_3	v_w_4
v_w_1	v_w_1	v_w_2	v_w_3	v_w_4
v_w_2	v_w_2	v_w_1	v_w_4	v_w_3
v_w_3	v_w_3	v_w_4	v_w_1	v_w_2
v_w_4	v_w_4	v_w_3	v_w_2	v_w_1

Assume $\mathbf{F_p}$ to be a $G_F(2)$. Suppose that (0)w = e, (1)w = w for all w in $\mathbf{V_p}$.
Define the mappings $\mu\mathbf{p}$: $\mathbf{V_p}\to[0,1]$, Ψ_p: $\mathbf{V_p}\to[0,1]$ and ζ_p: $\mathbf{V_p}\to[0,1]$ by

$$\mu\mathbf{p}(w) = \begin{cases} [0.51 \; if\, w = v\_w1, \\ \quad 1, otherwise. \end{cases}$$

$$\Psi_p(\Psi_p)\begin{cases} 0.32 \; if\, w = \mathbf{v}_{w1}, \\ 0.91, \; otherwise. \end{cases} \text{ and}$$

$$\zeta_p = \begin{cases} 0.92 \; if\, w = v\_w1, \\ \quad 0.4, \; otherwise. \end{cases}$$

Clearly $P_N = (\mu_p(w), \Psi_p(w), \zeta_p(w))$ is a **PANLS** of $\mathbf{V_p(F_p)}$.

Theorem 3.3. If $(P_N)_1$ and $(P_N)_2$ are **PANLSs** of $\mathbf{V_p(F_p)}$, then the union $(P_N)_1 \cup (PN)_2$ so is.

Proof. Let w_1 and w_2 in $\mathbf{V_p}$ and f, g in $\mathbf{F_p}$.

Define

$(P_N)_1 \cup (PN)_2$

$= \{<w, \mu((P_N)_1 \cup (P_N)_2)(w), \Psi((P_N)_1 \cup (P_N)_2)(w), \zeta((P_N)_1 \cup (P_N)_2)(w)>: w\in\mathbf{V_p}\}$

Now,

$\mu((P_N)_1 \cup (PN)_2)(fw_1 * gw_2)$

$= \max\{\mu(P_N)_1(fw_1 * gw_2), \mu(P_N)_2(fw_1 * gw_2)\}$

$\le \max\{\max[\mu(P_N)_1(w_1), \mu(P_N)_1(w_2)], \max[\mu(P_N)_2(w_1), \mu(P_N)_2(w_2)]\}$

$= \max\{ \max[\mu(P_N)_1(w_1), \mu(P_N)_2(w_1)], \max[\mu(P_N)_1(w_2), \mu(P_N)_2(w_2)]\}$

$\leq \max\{\mu((P_N)_1 \cup (P_N)_2)(w_1), \mu((P_N)_1 \cup (P_N)_2)(w_2)\}$

Similarly, we can prove the other inequalities.

Hence, $(P_N)_1 \cup (PN)_2$ is a **PANLS** of $\mathbf{V_p(F_p)}$.

Remark 3.4. The intersection of two **PANLSs** of $\mathbf{V_p(F_p)}$ need not be a **PANLS.**

Proof. An illustration will be utilized to showcase the aforementioned assertion.

Consider **Vp** = {v_w$_1$, v_ w$_2$, v_ w$_3$, v_w$_4$} be the Klein 4-group as in Example 3.2.

Assume **Fp** to be a $G_F(2)$. Suppose that (0)w = e, (1)w = w for all w in **Vp**. Then, $\mathbf{V_p(F_p)}$.

Define $\boldsymbol{\mu}(P_N)_1$ and $\mu(P_N)_2$ as follows:

$\mu(P_N)_1(v\_w_1) = 0.23$, $\mu(P_N)_1(v\_w_2) = 0.72 = \mu(P_N)_1(v\_w_3)$, $\mu(P_N)_1(v\_w_4) = 0.53$

&

$\mu(P_N)_2(v\_w_1) = 0.33$, $\mu(P_N)_2(v\_w_2) = 0.42$, $\mu(P_N)_2(v\_w_3) = \mu(P_N)_2(v\_w_4) = 0.61$.

Define

$\mu((P_N)_1 \cap (PN)_2)$ by $\mu((P_N)_1 \cap (PN)_2)(w)$

$= \min\{\mu(P_N)_1(w), \mu(P_N)_2(w)\}$ for all w in **Vp.**

Thus,

$\mu((P_N)_1 \cap (PN)_2)\ (v\_w_1) = 0.23$, $\mu((P_N)_1 \cap (PN)_2)\ (v\_w_2) = 0.42$,

$\mu((P_N)_1 \cap (PN)_2)\ (v\_w_3) = 0.61$, $\mu((P_N)_1 \cap (PN)_2)\ (v\_w_4) = 0.53$.

When f = g = 1, then the Definition 3.1 in (i) becomes

$\mu((P_N)_1 \cap (PN)_2)\ (v\_w_2 * v\_w_4) \leq \max\{\mu((P_N)_1 \cap (PN)_2)\ (v\_w_2), \mu((P_N)_1 \cap (PN)_2)\ (v\_w_4)\}$

$\mu((P_N)_1 \cap (PN)_2)\ (v\_w_3) \leq \max\ \{0.42, 0.53\}$

But $\mu((P_N)_1 \cap (PN)_2)\ (v\_w_3) = 0.61] \leq 0.53$.

This is absurd.

The other inequalities are similarly proved.

Hence, the intersection of two **PANLSs** of $\mathbf{V_p(F_p)}$ need not be a **PANLS.**

Definition 3.5. Let $(P_N)_1$ and $(P_N)_2$ be Pythagorean anti-neutrosophic of $\mathbf{V_{p1}}$ and $\mathbf{V_{p2}}$ respectively. Then the cartesian product of $(P_N)_1$ and $(P_N)_2$ denoted by $(P_N)_1\times(P_N)_2$ is defined by

$$(P_N)_1 \times (P_N)_2 = \{ < (w_1, w_2), \mu_{(PN)1 \times (PN)2}(w_1, w_2), \Psi_{(PN)1 \times (PN)2}(w_1, w_2),$$

$$\zeta_{(PN)1 \times (PN)2}(w_1, w_2)) > w_1 \in \mathbf{V}_{p1}, w_2 \in \mathbf{V}_{p2}\},$$

where

$$\mu_{(PN)1 \times (PN)2}(w_1, w_2) = \max\{\mu(\mathbf{P_N})_1(w_1), \mu(P_N)_2(w_2)\},$$

$$\Psi_{(PN)1 \times (PN)2}(w_1, w_2) = \max\{\Psi(P_N)_1(w_1), \Psi(P_N)_2(w_2)\}$$

and $\zeta_{(PN)1 \times (PN)2}(w_1, w_2) = \min\{\zeta(P_N)_1(w_1), \zeta(P_N)_2(w_2)\}$

Theorem 3.6. If $(P_N)_1$ and $(P_N)_2$ are **PANLSs** of $\mathbf{V_p}$, then $(P_N)_1\times(P_N)_2$ is a **PANLS** of $\mathbf{V_{p1}}\times\mathbf{V_{p2}}$.

Proof. Let $w = (w_1, w_2)$, $z = (z_1, z_2) \in \mathbf{V_{p1}}\times\mathbf{V_{p2}}$. Then

$$\mu_{(PN)1 \times (PN)2}(fw * gz)$$

$$= \mu_{(PN)1 \times (PN)2}(f(w_1, w_2) * g(z_1, z_2))$$

$$= \mu_{(PN)1 \times (PN)2}((f w_1 * g z_1), (f w_2 * g z_2),$$

$$= \max\{\mu(\mathbf{PN})_1(f w_1 * g z_1), \mu(\mathbf{PN})_2(f w_2 * g z_2)\}$$

$$\le \max\{\max[\mu(\mathbf{PN})_1(w_1), \mu(\mathbf{PN})_1(z_1)], \max[\mu(\mathbf{PN})_2(w_2), \mu(\mathbf{PN})_2(z_2)\}$$

$$= \max\{\max[\mu(\mathbf{PN})_1(w_1), \mu(\mathbf{PN})_2(w_2)], max[\mu(\mathbf{PN})_1(z_1), \mu(\mathbf{PN})_2(z_2)\}$$

$$= \max\{\mu_{(PN)1 \times (PN)2}(w_1, w_2), \mu_{(PN)1 \times (PN)2}(z_1, z_2)\}$$

$$= \max\{\mu_{(PN)1 \times (PN)2}(w), \mu_{(PN)1 \times (PN)2}(z)\}$$

$$\Psi_{(PN)1 \times (PN)2}(fw * gz)$$

$$= \Psi_{(PN)1 \times (PN)2}(f(w_{1,} w_2) *g (z_{1,} z_2))$$

$$= \Psi_{(PN)1 \times (PN)2}((f w_1 * g z_1), (f w_2 * g z_2),$$

$$= \max \{\Psi(PN)_1 (f w_1 * g z_1), \Psi(PN)_2 (f w_2 * g z_2)\}$$

$$\leq \max\{\max[\Psi(PN)_1(w_1), \Psi(PN)_1(z_1)], \max[\Psi(PN)_2(w_2), \Psi(PN)_2(z_2)\}$$

$$= \max\{\max[\Psi(PN)_1(w_1), \Psi(PN)_2(w_2)], \max[\Psi(PN)_1(z_1), \Psi(PN)_2(z_2)\}$$

$$= \max\{ \Psi_{(PN)1 \times (PN)2}(w_{1,} w_2), \Psi_{(PN)1 \times (PN)2}(z_{1,} z_2)\}$$

$$= \max\{ \Psi_{(PN)1 \times (PN)2}(w), \Psi_{(PN)1 \times (PN)2}(z)\}$$

$$\zeta_{(PN)1 \times (PN)2}(fw * gz)$$

$$= \zeta_{(PN)1 \times (PN)2}(f(w_{1,} w_2) *g (z_{1,} z_2))$$

$$= \zeta_{(PN)1 \times (PN)2}((f w_1 * g z_1), (f w_2 * g z_2),$$

$$= \min \{\zeta(PN)_1 (f w_1 * g z_1), \zeta(PN)_2 (f w_2 * g z_2)\}$$

$$\geq \min\{\min[\zeta(PN)_1(w_1), \zeta(PN)_1(z_1)], \min[\zeta(PN)_2(w_2), \zeta(PN)_2(z_2)\}$$

$$= \min\{\min[\zeta(PN)_1(w_1), \zeta(PN)_2(w_2)], \min[\zeta(PN)_1(z_1), \zeta(PN)_2(z_2)\}$$

$$= \min\{ \zeta\Psi_{(PN)1 \times (PN)2}(w_{1,} w_2), \zeta_{(PN)1 \times (PN)2}(z_{1,} z_2)\}$$

$$= \min\{ \zeta_{(PN)1 \times (PN)2}(w), \zeta_{(PN)1 \times (PN)2}(z)\}$$

So, $(P_N)_1 \times (P_N)_2$ is a **PANLS** of $\mathbf{V_{p1}} \times \mathbf{V_{p2}}$.

The definition and study of the image and inverse image of a PANLS are established, with a focus on different results related to them.

Definition 3.7. If there exists a mapping **f**: $\mathbf{V_{p1}} \rightarrow \mathbf{V_{p2}}$ between linear spaces $\mathbf{V_{p1}}$ and $\mathbf{V_{p2}}$ over field $\mathbf{F_p}$, and $P_N = (\mu_p, \Psi_p, \zeta_p)$ is a **PANLS** of $\mathbf{V_{p2}}$ over $\mathbf{F_p}$, then the inverse image of P_N under **f,** denoted by $\mathbf{f^{-1}}(P_N) = (f^{-1}(\mu_p), \mathbf{f^{-1}}(\Psi_p), f^{-1}(\zeta_p))$, is a **PANLS** of $\mathbf{V_{p1}}$. This inverse image is defined by $f^{-1}(P_N)(w) = P_N(\mathbf{f}(w)) = (\mu_p(\mathbf{f}(w)), \Psi_p(\mathbf{f}(w)), \zeta_p(\mathbf{f}(w))$ for all $w \in \mathbf{V_{p1}}$.

Theorem 3.8. If **f**: $\mathbf{V_{p1}} \rightarrow \mathbf{V_{p2}}$ is a linear space homomorphism of $\mathbf{V_p(F_p)}$, and $P_N = (\mu_p, \Psi_p, \zeta_p)$ is a **PANLS** of $\mathbf{V_{p2}}$, then the inverse image of P_N under f, denoted as $\mathbf{f}^{-1}(P_N)(w)$ is equal to P_N $(\mathbf{f}(w)) = (\mu_p(\mathbf{f}(w)), \Psi_p(\mathbf{f}(w)), \zeta_p(\mathbf{f}(w))$ for all $w \in \mathbf{V_{p1}}$.

Proof. Assume that $P_N = (\mu_p, \Psi_p, \zeta_p)$ is a **PANLS** of $\mathbf{V_{p2}}$ and w, z in $\mathbf{V_{p1}}$ and f, g in $\mathbf{F_p}$.

Then we have

$$f^{-1}(\mu_p)(fw * gz) = \mu_p(\mathbf{f}(fw * gz))$$

$$= \mu_p(f(w)f(z)) \text{ (since } \mathbf{f} \text{ is homomorphism)}$$

$$\leq \max\{\mu_p(f(w)), \mu_p(f(z))\}$$

$$= \max\{\mathbf{f}^{-1}(\mu_p(w)), \mathbf{f}^{-1}(\mu_p(z))\}$$

$$\mathbf{f}^{-1}(\mu_p)(fw * gz) \leq \max\{\mathbf{f}^{-1}(\mu_p(w)), \mathbf{f}^{-1}(\mu_p(z))\}$$

Therefore $\mathbf{f}^{-1}(\mu_p)$ is a **PANLS** of $\mathbf{V_{p1}}$.

$$f^{-1}(\Psi_p)(fw * gz) = \Psi_p(\mathbf{f}(fw * gz))$$

$$= \Psi_p(f(w)f(z)) \text{ (since } \mathbf{f} \text{ is homomorphism)}$$

$$\leq \max\{\Psi_p(f(w)), \Psi_p(f(z))\}$$

$$= \max\{\mathbf{f}^{-1}(\Psi_p(w)), \mathbf{f}^{-1}(\Psi_p(z))\}$$

$$\mathbf{f}^{-1}(\Psi_p)(fw * gz) \leq \max\{\mathbf{f}^{-1}(\Psi_p(w)), \mathbf{f}^{-1}(\Psi_p(z))\}$$

Hence, $\mathbf{f}^{-1}\Psi(_p)$ is a **PANLS** of $\mathbf{V_{p1}}$.

$$f^{-1}(\zeta_p)(fw * gz) = \zeta_p(\mathbf{f}(fw * gz))$$

$$= \zeta_p(f(w)f(z)) \text{ (since } \mathbf{f} \text{ is homomorphism)}$$

$$\geq \min\{\zeta_p(f(w)), \zeta_p(f(z))\}$$

$$= \min\{\mathbf{f}^{-1}(\zeta_p(w)), \mathbf{f}^{-1}(\zeta_p(z))\}$$

$$\mathbf{f}^{-1}(\zeta_p)(fw * gz) \leq \min\{\mathbf{f}^{-1}(\zeta_p(w)), \mathbf{f}^{-1}(\zeta_p(z))\}$$

So, $\mathbf{f}^{-1}(\zeta_p)$ is a **PANLS** of $\mathbf{V}_{p1}$.

Theorem 3.9. Consider $P_N = (\mu_p, \Psi_p, \zeta_p)$ as a **PANLS** of $\mathbf{V}_p$, and let $\mathbf{f}$: $\mathbf{V}_p \rightarrow \mathbf{V}_p$ be a surjective homomorphism. In this case, the mapping $\mathbf{PN}^f$: $\mathbf{V}_p \rightarrow [0, 1]$, which is defined as $P_N^f(w) = P_N(f(w))$ for all w in $\mathbf{V}_p$, can be identified as another **PANLS** of $\mathbf{V}_p$.

Proof. For any w, z in $\mathbf{V}_p$ and f, g in $\mathbf{F}_p$.

$$\mu_p^f(fw * gz) = \mu_p(\mathbf{f}(fw * gz))$$

$$= \mu_p(\mathbf{f}(w)\,\mathbf{f}(z)) \text{ (since } \mathbf{f} \text{ is homomorphism)}$$

$$\leq \max\{\mu_p(\mathbf{f}(w)),\ \mu_p(\mathbf{f}(z))\}$$

$$= \max\{\mu_p^f(f(w)), \mu_p^f(f(z))\}$$

$$\mu_p^f(fw * gz) \leq \max\{\mu_p^f(f(w)), \mu_p^f(f(z))\}$$

$$\Psi_p^f(fw * gz) = \Psi_p(\mathbf{f}(fw * gz))$$

$$= \Psi_p(\mathbf{f}(w)\,\mathbf{f}(z)) \text{ (since } \mathbf{f} \text{ is homomorphism)}$$

$$\leq \max\{\Psi_p(\mathbf{f}(w)),\ \Psi_p(\mathbf{f}(z))\}$$

$$= \max\{\Psi_p^f(f(w)), \Psi_p^f(f(z))\}$$

$$\Psi_p^f(fw * gz) \leq \max\{\Psi_p^f(f(w)), \Psi_p^f(f(z))\}$$

$$\zeta_p^f(fw * gz) = \zeta_p(\mathbf{f}(fw * gz))$$

$$= \zeta_p(\mathbf{f}(w)\,\mathbf{f}(z)) \text{ (since } \mathbf{f} \text{ is homomorphism)}$$

$$\geq \min\{\zeta_p(\mathbf{f}(w)),\ \zeta_p(\mathbf{f}(z))\}$$

$$= \min\{\zeta_p^f(f(w)), \zeta_p^f(f(z))\}$$

$$\zeta_p^f(fw * gz) \geq \min\{\zeta_p^f(f(w)), \zeta_p^f(f(z))\}$$

Theorem 3.10. If $\mathbf{f}$: $\mathbf{V}_{p1} \rightarrow \mathbf{V}_{p2}$ is an epimorphism of linear spaces $\mathbf{V}_p(\mathbf{F}_p)$, and $P_N = (\mu_p, \Psi_p, \zeta_p)$ is an **f**-invariant **PANLS** of $\mathbf{V}_{p1}$, then $\mathbf{f}(P_N)$ is a **PANLS** of $\mathbf{V}_{p2}$.

Proof. Let w_1, z_1 in Vp_2 and f, g in $\mathbf{F}_p$. Then there exists w, z in $\mathbf{V}_{p1}$ such that $\mathbf{f}(w) = w_1$ and $\mathbf{f}(z) = z_1$.

Also $fw_1 * gz_1 = \mathbf{f}(fw * gz)$. Since P_N is **f**-invariant,

$$\mathbf{f}(\mu_p)(fw * gz) = \mu_p(fw_1 * gz_1) \leq \max\{\mu_p(w_1), \mu_p(z_1)\}$$

$$= \max\{\mathbf{f}(\mu_p)(w), \mathbf{f}(\mu_p)(z)\}$$

$$\mathbf{f}(\mu_p)(fw * gz) \leq \max\{\mathbf{f}(\mu_p)(w), \mathbf{f}(\mu_p)(z)\}$$

$$\mathbf{f}(\Psi_p)(fw * gz) = \Psi_p(fw_1 * gz_1) \leq \max\{\Psi_p(w_1), \Psi_p(z_1)\}$$

$$= \max\{\mathbf{f}(\Psi_p)(w), \mathbf{f}(\Psi_p)(z)\}$$

$$\mathbf{f}(\Psi_p)(fw * gz) \leq \max\{\mathbf{f}(\Psi_p)(w), \mathbf{f}(\Psi_p)(z)\}$$

$$\mathbf{f}(\zeta_p)(fw * gz) = \zeta_p(fw_1 * gz_1) \leq \max\{\zeta_p(w_1), \zeta_p(z_1)\}$$

$$= \max\{\mathbf{f}(\zeta_p)(w), \mathbf{f}(\zeta_p)(z)\}$$

$$\mathbf{f}(\zeta_p)(fw * gz) \leq \max\{\mathbf{f}(\zeta_p)(w), \mathbf{f}(\zeta_p)(z)\}$$

CONCLUSION

In this paper, we present the notion of a Pythagorean anti-neutrosophic linear space. The advancements achieved in comprehending Pythagorean, anti-neutrosophic, and fuzzy linear spaces has opened up new opportunities for addressing uncertainty, indeterminacy, and contradiction in classical linear spaces. We offer a proof that combining two Pythagorean anti-neutrosophic linear spaces results in another Pythagorean anti-neutrosophic linear space. Additionally, we illustrate with a counter example that the intersection of two Pythagorean anti-neutrosophic linear spaces may not always produce a Pythagorean anti-neutrosophic linear space. Furthermore, the cartesian product of two Pythagorean anti-neutrosophic linear spaces leads to the emergence of a novel Pythagorean anti-neutrosophic linear space. Moreover, we establish the definitions of homomorphic image and inverse image between two Pythagorean anti-neutrosophic linear spaces.

This idea will be extended to various algebraic structures in the further research work.

- M-semigroup
- R-semigroup
- Ring theory

- Vector space
- Metric space and
- Decision making problems of rough and soft sets.

REFERENCES

Asif, M., Akram, M., & Ali, G. (2020). Pythagorean fuzzy matroids with application. *Symmetry, 12*(3), 423. DOI: 10.3390/sym12030423

Atanassov, K. T. (1986). Intuitionistic fuzzy sets. *Fuzzy Sets and Systems, 20*(1), 87–96. DOI: 10.1016/S0165-0114(86)80034-3

Biswas, R. (1990). Fuzzy subgroups and anti fuzzy subgroups. *Fuzzy Sets and Systems, 35*(1), 121–124. DOI: 10.1016/0165-0114(90)90025-2

Chinnadurai, V., & Arulselvam, A. (2021). Pythagorean Neutrosophic Ideals in Semigroups. *Neutrosophic Sets and Systems, 41*, 258–269.

Jansi, R., Mohana, K., & Smarandache, F. (2019). Correlation Measure for Pythagorean Neutrosophic Sets with T and F as Dependent Neutrosophic Components. *Neutrosophic Sets and Systems, 30*, 202–212.

Peng, X., & Yang, Y. (2016). Fundamental Properties of Interval-Valued Pythagorean Fuzzy Aggregation Operators. *International Journal of Intelligent Systems, 31*(5), 444–487. DOI: 10.1002/int.21790

Prema, R., & Radha, R. (2022). Generalized Neutrosophic Pythagorean Set. *International Research Journal of Modernization in Engineering Technology and Science, 4*(11), 1571–1575. https://www.doi.org/10.56726/IRJMETS31596

Sivaramakrishnan, S., & Suresh, K. (2019). Interval-valued anti fuzzy subring. *Journal of Applied Science and Computations, 6*(1), 521–526.

Sivaramakrishnan, S., & Vijayabalaji, S. (2020). Interval-valued anti fuzzy linear space. *IEEE Xplore, 839-841*, 1–3. DOI: 10.1109/ICSCAN49426.2020.9262371

Sivaramakrishnan, S., Vijayabalaji, S., & Balaji, P. (2024). Neutrosophic interval-valued anti-fuzzy linear space. *Neutrosophic Sets and Systems, 63*, 271–284. DOI: 10.5281/zenodo.10531827

Smarandache, F. (2006). Neutrosophic set - A generalization of the intuitionistic fuzzy set. In *2006 IEEE International Conference on Granular Computing* (pp. 38–42). IEEE. DOI: 10.1109/GRC.2006.1635754

Songsaeng, M., & Iampan, A. (2020). Neutrosophic cubic set theory applied to up- algebras. *Thai Journal of Mathematics, 18*(3), 1447–1474.

Vijayabalaji, S. (2012). Cartesian Product and Homomorphism of Interval-Valued Fuzzy Linear Space. *International Journal of Open Problems in Computer Science and Mathematics*, *5*(4), 93–103. DOI: 10.12816/0006141

Vijayabalaji, S., & Sivaramakrishnan, S. (2012, September 26). Sivaramakrishnan; Anti fuzzy M-semigroup. *AIP Conference Proceedings*, *1482*(1), 446–448. DOI: 10.1063/1.4757511

Yager, R. R. (2013). Pythagorean fuzzy subsets. In *Proceedings of the 2013 Joint IFSA World Congress and NAFIPS Annual Meeting, IFSA/NAFIPS 2013* (pp. 57–61). IEEE. DOI: 10.1109/IFSA-NAFIPS.2013.6608375

Zadeh, L. A. (1965). Fuzzy sets. *Information and Control*, *8*(3), 338–353. DOI: 10.1016/S0019-9958(65)90241-X

Zadeh, L. A. (1975). The concept of a linguistic variable and its application to approximate reasoning-I. *Information Sciences*, *8*(3), 199–249. DOI: 10.1016/0020-0255(75)90036-5

Chapter 3 Pentapartitioned Neutrosophic Vague Number Assignment Problem: A Novel Approach Using Interval-Valued Trapezoidal Pentapartitioned Neutrosophic Vague

V. R. Radhika
Nirmala college for Women, Coimbatore, India

K. Mohana
Nirmala College for Women, Coimbatore, India

ABSTRACT

An extensively researched and significant field in optimization is assignment problems (APs), this research manuscript introduces the pentapartitioned neutrosophic vague assignment problem (PNVAP), an assignment problem in a pentapartitioned neutrosophic vague environment. Utilizing interval-valued trapezoidal pentapartitioned neutrosophic vague numbers in the cost matrix's elements, the problem is put forth. This interval-valued trapezoidal pentapartitioned neutrosophic vague assignment problem (IVTPNVAP) is converted to an interval-valued AP in accordance with the idea of a score function. The authors apply the order relations to optimize the objective function in interval form. The choices made by the decision maker are represented by these relations. The decision makers' preference introduces order

DOI: 10.4018/979-8-3693-3204-7.ch003

relations, which transform the maximization (or minimization) model with an objective function in interval form into a multi-objective model in the case of interval profits (or costs). Finally, the authors resolve a numerical example to validate the suggested methodology for solving the problem.

INTRODUCTION

An AP can be found in numerous significant real-world applications, including production planning, resource scheduling, transportation planning and distribution, economics, plant location, and flexible manufacturing systems. These applications have drawn increasing interest from researchers (Dahiya & Verma, 2007; Dubois 1980; Rajarajeswari et al., 2013), as they address the problem of allocating n personnel or machines to m tasks in a way that maximizes profit while minimizing costs.

In order to achieve these research goals, the DM must try to optimize models from linear AP to nonlinear AP. Given this, the linear AP is a particular type of linear programming problem (LPP) in which individuals or machines are assigned to different tasks according to a one-to-one rule in order to maximize the assignment profit (or cost). When the number of rows and the number of columns is equal, as stated in (Ehrgott et al., 2016), the AP is well described as having an optimal assignee. (Bao et al., 2007) developed a new method to study the assignment problem with multiple objectives. (Geetha & Nair, 1993) then applied this method to determine the cost-time AP problem as a multiple criteria decision-making problem.

A few decades back, the fundamental concept of fuzzy sets was extensively studied by numerous writers and decision-makers worldwide. The theory of fuzzy sets was first introduced by Zadeh (1965), and it has since been widely used to investigate a number of real-world issues, such as financial risk management. The fuzzy concept is then further represented by fuzzy quantities and/or fuzzy constraints. With the support of the fuzzification method, (Dubois, 1980) proposed implementing algebraic operations on crisp numbers to fuzzy numbers. On the other hand, the set of parameters that make up the AP represents a real-life scenario. Decision-makers determine these parameters' values. When using the conventional approach, DMs needed to fix exact values to the parameters. The model parameters are then typically defined in an uncertain way because DMs are unable to precisely estimate the value of parameters. The first LP model with multiple objectives that was solved using appropriate membership functions was (Zimmermann, 1978). (Bellman & Zadeh, 1970) applied the fuzzy set notion to a decision-making problem involving uncertainty and imprecision.

Fuzzy multiobjective linear programming (MOLP) problems are theoretically introduced by (Sakawa & Yano, 1989). Using the objective function's coefficients as triangular fuzzy numbers, (Hamadameen, 2018) developed a method for obtaining the fuzzy MOLP model's optimal solution. According to (Wang et al., 2010) explanation, the fuzzy MOLP problem was reduced to crisp MOLP with the aid of ranking functions. The fuzzy programming technique was then used to solve the issue. (Leberling, 1981) used a specific type of nonlinear membership functions to solve the vector maximum LP problem. Fuzzy methodology was used by (Bit et al., 1992) for a multiple objective transportation model. A fuzzy AP with multiple criteria was studied by Belacela & Boulasselb (2001). An method for handling the fuzzy AP problem was created by Lin & Wen, (2004). We talked about the interval numbers cost coefficients MOAP problem with (Kagade & Bajaj 2010). Fuzzy simulation was used by Yang & Liu (2005) to create a Tabu search technique that finds the best answer to the fuzzy AP. Furthermore, using the fuzzy goal programming technique, (De & Yadav, 2011) presented a solution approach to MOAP. (Yager's, 1981) ranking method was used by Mukherjee & Basu (2010) to solve the fuzzy cost AP problem. Using imprecise costs, time, and ineffectiveness, (Pramanik & Biswas 2012) investigated multi-objective AP. A few generalized AP models in an imprecise environment were examined by Haddad et al., (2002). For the fuzzy AP, an alternative development was proposed with fuzzy profits or fuzzy costs for all possible assignments, as explained by Emrouznejad et al., (2012). A methodology to solve fuzzy AP and fuzzy travelling salesman problem under different membership functions and ranking index introduced by (Yager, 1981) was studied by Kumar & Gupta (2011). To find the best answer to a MOAP, (Medvedeva & Medvedev, 2018) used the primal and dual concepts. The branch and bound approach was used by Hamou & Mohamed (2018) to produce the set of all effective solutions for MOAP. A novel technique known as the optimal flowing method that provides the ideal and set of all efficient solutions was studied by Jayalakshmi & Sujatha (2018). In order to find the best solution to the problem, (Pandian & Anuradha, 2010) looked into a novel methodology that used the zero-point method that (Pandian & Natarajan, 2010) had introduced.

The multi-objective assignment problem with trapezoidal fuzzy numbers was examined by Khalifa et al., (2018). After introducing an interactive method for solving it, they identified the first-kind stability set that went along with the solution. The Weighting Tchebycheff program is the foundation of (Khalifa, 2020) method, which addresses the multi-objective assignment problem in a neutrosophic environment.

The neutrosophic set is an extension of the intuitionistic fuzzy set. There are three defining functions in the neutrosophic set. These three functions are the indeterminacy membership,and non-membership functions. These tasks are completely unrelated to one another. (Abdel et al., 2020) presented a novel solution approach for the FN-LPP with practical applications. In their investigation of a novel method of solving the fuzzy Pythagorean transportation problem, (Kumar et al., 2019) extended the interval basic feasible solution and then used the current optimality method to determine the transportation cost.

Using a neutrosophic concept, Khalifa et al., 2020) investigated a complex programming problem. To find the best solution for neutrosophic complex programming, they used the lexicographic order. The neutrosophic MOLP problem was studied by Vidhya et al., (2017).

A goal programming methodology for the MOLP problem under neutrosophic numbers was proposed by Pramanik & Banerjee (2018). Some new operations for interval neutrosophic sets in terms of geometrical, harmonic, and arithmetic means were introduced by Broumi & Smarandache (2015). A novel compromise approach for many objective transportation problems was proposed by Rizk-Allah et al., (2018). Zimmermann's fuzzy programming approach and the terminology of neutrosophic sets were used to further investigate this approach. A plithogenic multi-criteria decision-making model based on a neutrosophic analytical hierarchy process was presented by Abdel-Basset et al., (2020) and ranked performance according to resemblance to the optimal financial performance solution. In the face of uncertainty, (Abdel-Basset et al., 2020) assessed a set of metrics for supplying sustainable supply chain financing to the gas sector. An integrated approach based on the neutrosophic set was presented by (Abdel-Basset et al., 2020) to assess the innovation value of smart product-service systems.

In actuality, there are situations when it is not possible to precisely define the degree of truth, falsity, or indeterminacy of a given statement; instead, it can be indicated by a range of possible interval values. Thus, just like with IVIFS, the interval neutrosophic set (INS) was needed. The set-theoretic operators of INS were provided by Wang et al. along with the concept of INS. Although the INS's operations were covered in the comparison techniques were not mentioned there. Ye also proposed similarity measures between INSs based on the relationship between similarity measures and distances, and he defined the Hamming and Euclidean distances between INSs. On the other hand, the INS may occasionally operate irrationally.

The term "Neutrosophic vague set," which combines the terms "Neutrosophic set" and "vague set," was first introduced by Alkhazaleh, (2015). Neutrosophic vague theory is a useful method for analysing ambiguous, conflicting, and incomplete data. Indeterminacy was divided into three categories by Smarandache (2010): unknown, contradiction, and ignorance. He also developed Five Symbol Valued Neutrosophic Logic (FSVNL).

A development of fuzzy set theory, vague sets were first proposed by Gau & Buehrar (1993) .Given that they offer more information than fuzzy sets, vague sets are thought to be an effective tool for dealing with uncertainty (Zhang et al., 2009).

Many researchers have blended nebulous sets with other theories, according to a number of studies. A study by (Xu et al., 2020) looked at the properties of vague soft sets. Later, Hassan combined a soft expert set with a vague set, and its operations were revealed. To further address the issue of uncertainty in decision-making, additional hybrid theories including the possibility vague soft set (Alhazaymeh et al., 2013), generalized interval valued vague set (Alhazaymeh et al., 2013), complex vague soft set (Selvachandran et al., 2016), and interval valued vague soft set (Alhazaymeh & Hassan, 2012). have been proposed.

This paper presents an assignment problem where all parameters have interval-valued trapezoidal pentapatitioned neutrosophic numbers. This problem is transformed into an assignment problem with two objectives, and it is then solved by applying the Weighting Tchebycheff program with the ideal targets.

The suggested research article's layout is set up as follows: We provide some preliminary information in the next section, which is crucial to the current investigation. The interval-valued trapezoidal Pentapartitioned Neutrosophic assignment problem is formulated in Section 3. The solution approach for determining the preferred solution is proposed in Section 4.

The effectiveness of the solution approach is demonstrated by solving a numerical example in Section 5. The final thoughts are summed up in Section 6, along with some recommendations for future research.

PRELIMINARIES

The fundamental ideas and findings pertaining to fuzzy numbers, neutrosophic sets, pentapartitioned neutrosophic vague sets, and their arithmetic operations are presented in this section.

Definition 2.1 (Zadeh, (1965): A fuzzy set $\widetilde{P}$ defined on the set of real numbers $\mathbb{R}$ is called fuzzy number when the membership function $\mu_{\widetilde{P}}(x):\mathbb{R} \rightarrow [0,1]$, have the following properties:

1. $\mu_{\widetilde{P}}(x)$ is an upper semi-continuous membership function;
2. $\widetilde{P}$ is convex fuzzy set, i.e., $\mu_{\widetilde{P}}(\delta x + (1-\delta)y) \geq \min\{\mu_{\widetilde{P}}(x), \mu_{\widetilde{P}}(y)\}$ for all

$x, y \in \mathbb{R}; 0 \leq \delta \leq 1$;

3. $\widetilde{P}$ is normal, i.e., $\exists x_0 \in \mathbb{R}$ for which $\mu_{\widetilde{P}}(x_0) = 1$;
4. $Supp(\widetilde{P}) = \{x \in \mathbb{R}: \mu_{\widetilde{P}}(x) > 0\}$ is the support of $\widetilde{P}$, and the closure $cl(Supp(\widetilde{P}))$ sure is compact set.

Definition 2.2 (Atanassov, (2012): An Intuitionistic Fuzzy Set (IFS), A in X is given by $A = \{\langle x, \mu_A(x), \nu_A(x)\rangle / x \in X\}$ where $\mu_A: X \rightarrow [0,1]$ and $\nu_A: X \rightarrow [0,1]$ with the condition $0 \leq \mu_A(x) + \nu_A(x) \leq 1, \forall x \in X$.Here $\mu_A(x)$ and $\nu_A(x) \in [0,1]$ denote the membership and non-membership functions of the Fuzzy set A.

Definition 2.3 (Margaret et al., 2019): A vague set is defined by a truth-membership function t_v and a false membership function f_v, where $t_v(x)$ is a lower bound on the grade of membership of x derived from the evidence for x, and $f(x)$ is a lower bound on the negation of x derived from the evidence against x. The values of $t_v(x)$ and $f_v(x)$ are both defined on the closed interval [0, 1] with each point in a basic set X, where $t_v(x) + f_v(x) \leq 1$.

Definition 2.4 (Ishibuchi & Tanaka, 1990): An interval on $\mathbb{R}$ is defined as

$$A = [a^L, a^R] = \{a; a^L \leq a \leq a^R, a \in \mathbb{R}\},$$

where a^L is left limit and a^R is right limit of A.

Definition 2.5 (Ishibuchi & Tanaka, 1990): The interval is also defined by

$$A = a_c, a_w = \{a : a_c - a_w \leq a \leq a_c + a_w,\ a \in \mathbb{R}\},$$

where $a_c = \frac{1}{2}(a^R + a^L)$ is center and $a_w = \frac{1}{2}(a^R - a^L)$ is width of A.

Definition 2.6 (Wang et al., 2010): Let X be a nonempty set. Then a neutrosophic set $\overline{P}_N$ of nonempty set X is defined as

$$\overline{P}_N = \{x;\ T_{\overline{P}_N}, I_{\overline{P}_N}, F_{\overline{P}_N} : x \in X\},$$

where $T_{\overline{P}_N}, I_{\overline{P}_N}, F_{\overline{P}_N} : X \rightarrow [0\_, 1^+]$ define respectively the degree of membership function, the degree of indeterminacy, and the degree of non-membership of element $x \in X$ to the set $\overline{P}_N$ with the condition:

$$0\_ \leq T_{\overline{P}_N} + I_{\overline{P}_N} + F_{\overline{P}_N} \leq 3^+$$

Definition 2.7 (Broumi & Smarandache, 2015): Let X be a nonempty set. Then an interval valued neutrosophic vague (IVNV) set $\overline{P}_{NV}^{IV}$ of X is defined as:

$$\overline{P}_{NV}^{IV} = \{x; [\widetilde{T}_{\overline{P}_{NV}}^L, \widetilde{T}_{\overline{P}_{NV}}^U], [\widetilde{I}_{\overline{P}_{NV}}^L, \widetilde{I}_{\overline{P}_{NV}}^U], [\widetilde{F}_{\overline{P}_{NV}}^L, \widetilde{F}_{\overline{P}_{NV}}^U]; x \in X\},$$

where

$$\widetilde{T}_{\overline{P}_{NV}}^L = [T^{L+}, T^{L-}];\ \widetilde{T}_{\overline{P}_{NV}}^U = [T^{U+}, T^{U-}]$$

$$\tilde{I}^{L}_{\overline{P}_{NV}} = [I^{L+}, I^{L-}];\ \tilde{I}^{U}_{\overline{P}_{NV}} = [I^{U+}, I^{U-}]$$

$$\tilde{F}^{L}_{\overline{P}_{NV}} = [F^{L+}, F^{L-}];\ \tilde{F}^{U}_{\overline{P}_{NV}} = [F^{U+}, F^{U-}]$$

Also

$$\left[\tilde{T}^{L}_{\overline{P}_{NV}}, \tilde{T}^{U}_{\overline{P}_{NV}}\right], \left[\tilde{I}^{L}_{\overline{P}_{NV}}, \tilde{I}^{U}_{\overline{P}_{NV}}\right] \text{ and } \left[\tilde{F}^{L}_{\overline{P}_{NV}}, \tilde{F}^{U}_{\overline{P}_{NV}}\right] \subset [0,\ 1]$$

for each $x \in X$.

Definition 2.8 (Broumi & Smarandache, 2015): Let

$$\overline{P}^{IV}_{NV} = \left\{x; \left[\tilde{T}^{L}_{\overline{P}_{NV}}, \tilde{T}^{U}_{\overline{P}_{NV}}\right], \left[\tilde{I}^{L}_{\overline{P}_{NV}}, \tilde{I}^{U}_{\overline{P}_{NV}}\right], \left[\tilde{F}^{L}_{\overline{P}_{NV}}, \tilde{F}^{U}_{\overline{P}_{NV}}\right]; x \in X\right\}$$

be IVNVS, then $\overline{P}^{IV}_{NV}$ is empty if $\tilde{T}^{L}_{\overline{P}_{NV}} = \tilde{T}^{U}_{\overline{P}_{NV}} = 0$, $\tilde{I}^{L}_{\overline{P}_{NV}} = \tilde{I}^{U}_{\overline{P}_{NV}} = 1$, $\tilde{F}^{L}_{\overline{P}_{NV}} = \tilde{F}^{U}_{\overline{P}_{NV}} = 1$, for all $x \in \overline{P}_{NV}$,

Let $0\_ = x;\ 0,\ 1,\ 1$, and $1\_ = x;\ 1,\ 0,\ 0$.

Definition 2.9 (Sakawa & Yano, 1989): Let X be a universe. A Pentapartitioned Neutrosophic set A on X is defined as

$$A = \{\langle x, T_A(x), C_A(x), G_A(x), U_A(x), F_A(x)\rangle : x \in X\}$$

Where $T, C, G, U, F: X \rightarrow [0,1]$ and $0 \leq T_A(x)+C_A(x)+G_A(x)+U_A(x)+F_A(x) \leq 5$. Here $T_A(x)$ is the truth membership, $C_A(x)$ is contradiction membership, $G_A(x)$ is ignorance membership, $U(x)$ is unknown and $F_A(x)$ is the false membership.

Definition 2.10 (Radhika & Mohana, n.d): Pentapartitioned Neutrosophic Vague Set $A_{\overline{PNV}}$ ($\overline{PNVS}$ in short form) on the universe of discourse written as

$$A_{\overline{PNV}} = \left\{x;\ \hat{T}_{A_{\overline{PNV}}}(x); \hat{C}_{A_{\overline{PNV}}}(x);\ \hat{G}_{A_{\overline{PNV}}}(x); \hat{U}_{A_{PNV}}(x); \hat{F}_{A_{\overline{PNV}}}(x); x \in X\right\}$$

whose truth membership, contradiction membership, ignorance membership, unknown membership, and false membership function is defined as:

$$\hat{T}_{A_{\overline{PNV}}}(x) = [T^{-}, T^{+}],\ \hat{C}_{A_{\overline{PNV}}}(x) = [C^{-}, C^{+}],$$

$$\hat{G}_{A_{\overline{PNV}}}(x) = [G^{-}, G^{+}],\ \hat{U}_{A_{\overline{PNV}}}(x) = [U^{-}, U^{+}],\ \hat{F}_{A_{\overline{PNV}}}(x) = [F^{-}, F^{+}],$$

Where (1) $T^{+} = 1-F^{-}$ (2) $F^{+} = 1-T^{-}$ (3) $C^{+} = 1-U^{-}$ (4) $U^{+} = 1-C^{-}$ (5) $0 \leq T^{-}+C^{-}+G^{-}+U^{-}+F^{-} \leq 4^{+}$

Definition 2.11 (Pramanik, 2023): Let D be a universe discourse U. Then an interval pentapertitioned neutrosophic vague set denoted as D_{IPNV} is written as:

$$D_{IPNV} = \left\{ \begin{array}{c} \left(w, \begin{bmatrix} \widetilde{T}_D^L(w), \\ \widetilde{T}_D^U(w) \end{bmatrix}, \begin{bmatrix} \widetilde{C}_D^L(w), \\ \widetilde{C}_D^U(w) \end{bmatrix}, \begin{bmatrix} \widetilde{G}_D^L(w), \\ \widetilde{G}_D^U(w) \end{bmatrix}, \begin{bmatrix} \widetilde{U}_D^L(w), \\ \widetilde{U}_D^U(w) \end{bmatrix}, \begin{bmatrix} \widetilde{F}_D^L(w), \\ \widetilde{F}_D^U(w) \end{bmatrix} \right) ; \\ w \in W \end{array} \right\}$$

Whose truth, contradiction, ignorance, unknown, and falsity membership functions are defined as:

$$\begin{array}{l} \widetilde{T}_D^L(w) = [T^{L-}, T^{L+}],\ \widetilde{T}_D^U(w) = [T^{U-}, T^{U+}],\ \widetilde{C}_D^L(w) = [C^{L-}, C^{L+}], \\ \widetilde{C}_D^U(w) = [C^{U-}, C^{U+}],\ \widetilde{G}_D^L(w) = [G^{L-}, G^{L+}],\ \widetilde{G}_D^U(w) = [G^{U-}, G^{U+}], \\ \widetilde{U}_D^L(w) = [U^{L-}, U^{L+}],\ \widetilde{U}_D^U(w) = [U^{U-}, U^{U+}], \\ \widetilde{F}_D^L(w) = [F^{L-}, F^{L+}],\ \widetilde{F}_D^U(w) = [F^{U-}, F^{U+}] \end{array},$$

Where $T^{L+} = 1 - F^{L-}$; $F^{L+} = 1 - T^{L-}$; $C^{L+} = 1 - U^{L-}$; $U^{L+} = 1 - C^{L-}$; $T^{U+} = 1 - F^{U-}$; $F^{U+} = 1 - T^{U-}$; $C^{U+} = 1 - U^{U-}$; $U^{L+} = 1 - C^{U-}$

$$0 \le T^{+} + C^{+} + G^{+} + U^{+} + F^{+} \le 5^{+}; 0 \le T^{-} + C^{-} + G^{-} + U^{-} + F^{-} \le 5^{+}$$

Definition 2.12 (Pramanik, 2023): Let A_{IPNV} be a *IPNVS* of the universe U where $\forall w_i \in U$,

$$\widetilde{T}^L_{A_{IPNV}}(w) = [1,\ 1],\ \widetilde{T}^U_{A_{IPNV}}(w) = [1,\ 1],$$

$$\widetilde{C}^L_{A_{IPNV}}(w) = [1,\ 1],\ \widetilde{C}^U_{A_{IPNV}}(w) = [1,\ 1],$$

$$\widetilde{G}^L_{A_{IPNV}}(w) = [0,\ 0],\ \widetilde{T}^U_{A_{IPNV}}(w) = [0,\ 0],$$

$$\widetilde{U}^L_{A_{IPNV}}(w) = [0,\ 0],\ \widetilde{U}^U_{A_{IPNV}}(w) = [0,\ 0],$$

$$\widetilde{F}^L_{A_{IPNV}}(w) = [0,\ 0],\ \widetilde{F}^U_{A_{IPNV}}(w) = [0,\ 0]$$

Then, *a* unit IPNVS is denoted as A_{IPNV} where $1 \le i \le n$.

Definition 2.13 (Pramanik, 2023): Let B_{IPNV} be a *IPNVS* of the universe U where $\forall w_i \in U$,

$$\widetilde{T}^L_{A_{IPNV}}(w) = [0,\ 0],\ \widetilde{T}^U_{A_{IPNV}}(w) = [0,\ 0],$$

$$\widetilde{C}^{L}_{A_{IPNV}}(w) = [1, 1],\ \widetilde{C}^{U}_{A_{IPNV}}(w) = [0, 0],$$

$$\widetilde{G}^{L}_{A_{IPNV}}(w) = [0, 0],\ \widetilde{T}^{U}_{A_{IPNV}}(w) = [1, 1],$$

$$\widetilde{U}^{L}_{A_{IPNV}}(w) = [0, 0],\ \widetilde{U}^{U}_{A_{IPNV}}(w) = [1, 1],$$

$$\widetilde{F}^{L}_{A_{IPNV}}(w) = [0, 0],\ \widetilde{F}^{U}_{A_{IPNV}}(w) = [1, 1]$$

Then, *a* zero IPNVS is denoted as B_{IPNV} where $1 \leq i \leq n$.

Definition 2.14 (Broumi & Smarandache, 2015): (Interval-valued trapezoidal neutrosophic number).

Let $u_{\tilde{a}}, v_{\tilde{a}}, w_{\tilde{a}} \subset [0,1]$, and $a_1, a_2, a_3, a_4 \in \mathbb{R}$ such that $a_1 \leq a_2 \leq a_3 \leq a_4$. Then an interval-valued trapezoidal fuzzy neutrosophic vague number,

$$\tilde{a} = (a_1, a_2, a_3, a_4); [\tilde{u}^{L}_{\tilde{a}}, \tilde{u}^{U}_{\tilde{a}}], [\tilde{v}^{L}_{\tilde{a}}, \tilde{v}^{U}_{\tilde{a}}], [\tilde{w}^{L}_{\tilde{a}}, \tilde{w}^{U}_{\tilde{a}}],$$

whose degrees of membership function, the degrees of indeterminacy, and the degrees of non-membership are:

$$\vartheta_{\tilde{a}}(x) = \begin{cases} u_{\tilde{a}}\left(\dfrac{x - a_1}{a_2 - a_1}\right), & \textit{for } a_1 \leq x \leq a_2, \\ u_{\tilde{a}}, & \textit{for } a_2 \leq x \leq a_3, \\ u_{\tilde{a}}\left(\dfrac{a_4 - x}{a_4 - a_3}\right), & \textit{for } a_3 \leq x \leq a_4, \\ 0, & \textit{otherwise}, \end{cases}$$

$$\mu_{\tilde{a}}(x) = \begin{cases} \left(\dfrac{a_2 - x + v_{\tilde{a}}(x - a_1)}{a_2 - a_1}\right) & \textit{for } a_1 \leq x \leq a_2, \\ v_{\tilde{a}} & \textit{for } a_2 \leq x \leq a_3, \\ \left(\dfrac{x - a_3 + v_{\tilde{a}}(a_4 - x)}{a_4 - a_3}\right) & \textit{for } a_3 \leq x \leq a_4, \\ 1 & \textit{otherwise} \end{cases}$$

$$\varphi_{\tilde{a}}(x) = \begin{cases} \left(\frac{a_2 - x + w_{\tilde{a}}(x - a_1)}{a_2 - a_1}\right), & for\ a_1 \le x \le a_2, \\ w_{\tilde{a}}, & for\ a_2 \le x \le a_3, \\ \left(\frac{x - a_3 + w_{\tilde{a}}(a_4 - x)}{a_4 - a_3}\right), & for\ a_3 \le x \le a_4, \\ 1, & otherwise, \end{cases}$$

Where $\vartheta_{\tilde{a}}(x)$, $\mu_{\tilde{a}}(x)$ and $\varphi_{\tilde{a}}(x)$ are the upper bound of membership degree, lower bound of indeterminacy degree, and lower bound of non – membership degree respectively.

Definition 2.15 (Radhika & Mohana, n.d): The union of two $\widetilde{PNVSs}\, A_{\widetilde{PNV}}$ and B_{PNV} is a $\widetilde{PNVS}$ $K_{\widetilde{PNV}}$ written as $K_{\widetilde{PNV}} = A_{\widetilde{PNV}} \bigcup B_{\widetilde{PNV}}$ whose truth-membership, contradiction-membership, ignorance membership and false membership functions are related to those of $A_{\widetilde{PNV}}$ and $B_{\widetilde{PNV}}$ by

$$\widehat{T}_{K_{\widetilde{PNV}}}(x) = \left[max\left(\widehat{T}^{-}_{A_{\widetilde{PNV}}}, \widehat{T}^{-}_{B_{\widetilde{PNV}}}\right), max\left(\widehat{T}^{+}_{A_{\widetilde{PNV}}}, \widehat{T}^{+}_{B_{\widetilde{PNV}}}\right)\right]$$

$$\widehat{C}_{K_{\widetilde{PNV}}}(x) = \left[max\left(\widehat{C}^{-}_{A_{\widetilde{PNV}}}, \widehat{C}^{-}_{B_{\widetilde{PNV}}}\right), max\left(\widehat{C}^{+}_{A_{\widetilde{PNV}}}, \widehat{C}^{+}_{B_{\widetilde{PNV}}}\right)\right]$$

$$\widehat{G}_{K_{\widetilde{PNV}}}(x) = \left[min\left(\widehat{G}^{-}_{A_{\widetilde{PNV}}}, \widehat{G}^{-}_{B_{\widetilde{PNV}}}\right), min\left(\widehat{G}^{+}_{A_{\widetilde{PNV}}}, \widehat{G}^{+}_{B_{\widetilde{PNV}}}\right)\right]$$

$$\widehat{U}_{K_{\widetilde{PNV}}}(x) = \left[min\left(\widehat{U}^{-}_{A_{\widetilde{PNV}}}, \widehat{U}^{-}_{B_{\widetilde{PNV}}}\right), min\left(\widehat{U}^{+}_{A_{\widetilde{PNV}}}, U^{+}_{B_{\widetilde{PNV}}}\right)\right]$$

$$\widehat{F}_{K_{\widetilde{PNV}}}(x) = \left[min\left(\widehat{F}^{-}_{A_{\widetilde{PNV}}}, \widehat{F}^{-}_{B_{\widetilde{PNV}}}\right), min\left(\widehat{F}^{+}_{A_{\widetilde{PNV}}}, F^{+}_{B_{\widetilde{PNV}}}\right)\right]$$

Definition 2.16 (Radhika & Mohana, n.d): The intersection of two $\widetilde{PNVSs}\, A_{\widetilde{PNV}}$ and $B_{\widetilde{PNV}}$ is a $\widetilde{PNVS}$ $H_{\widetilde{PNV}}$, written as $H_{\widetilde{PNV}} = A_{\widetilde{PNV}} \bigcap B_{\widetilde{PNV}}$ whose truth-membership, contradiction-membership, ignorance membership, unknown membership and false-membership functions are related to those of $A_{\widetilde{PNV}}$ and $B_{\widetilde{PNV}}$ by

$$\widehat{T}_{H_{\widetilde{PNV}}}(x) = \left[min\left(\widehat{T}^{-}_{A_{\widetilde{PNV}}}, \widehat{T}^{-}_{B_{\widetilde{PNV}}}\right), min\left(\widehat{T}^{+}_{A_{\widetilde{PNV}}}, \widehat{T}^{+}_{B_{\widetilde{PNV}}}\right)\right]$$

$$\widehat{C}_{H_{\widetilde{PNV}}}(x) = \left[min\left(\widehat{C}^{-}_{A_{\widetilde{PNV}}}, \widehat{C}^{-}_{B_{\widetilde{PNV}}}\right), min\left(\widehat{C}^{+}_{A_{\widetilde{PNV}}}, \widehat{C}^{+}_{B_{\widetilde{PNV}}}\right)\right]$$

$$\widehat{G}_{H_{\widetilde{PNV}}}(x) = \left[max\left(\widehat{G}^{-}_{A_{\widetilde{PNV}}}, \widehat{G}^{-}_{B_{\widetilde{PNV}}}\right), max\left(\widehat{G}^{+}_{A_{\widetilde{PNV}}}, \widehat{G}^{+}_{B_{\widetilde{PNV}}}\right)\right]$$

$$\widehat{U}_{H_{\overline{PNV}}}(x) = \left[max\left(\widehat{U}^{-}_{A_{\overline{PNV}}},\ \widehat{U}^{-}_{B_{\overline{PNV}}}\right),\ max\left(\widehat{U}^{+}_{A_{\overline{PNV}}},\ U^{+}_{B_{\overline{PNV}}}\right)\right]$$

$$\widehat{F}_{H_{\overline{PNV}}}(x) = \left[max\left(\widehat{F}^{-}_{A_{\overline{PNV}}},\ \widehat{F}^{-}_{B_{\overline{PNV}}}\right),\ max\left(\widehat{F}^{+}_{A_{\overline{PNV}}},\ F^{+}_{B_{\overline{PNV}}}\right)\right]$$

Definition 2.17 (Broumi & Smarandache, 2015). **(Arithmetic operations)**. Let

$$\tilde{a} = (a_1,\ a_2,\ a_3,\ a_4);\ [\tilde{u}^L_{\tilde{a}}, \tilde{u}^U_{\tilde{a}}],\ [\tilde{v}^L_{\tilde{a}}, \tilde{v}^U_{\tilde{a}}],\ [\tilde{w}^L_{\tilde{a}}, \tilde{w}^U_{\tilde{a}}],$$

and

$$\tilde{b} = (b_1,\ b_2,\ b_3,\ b_4);\ [\tilde{u}^L_{\tilde{b}}, \tilde{u}^U_{\tilde{b}}],\ [\tilde{v}^L_{\tilde{b}}, \tilde{v}^U_{\tilde{b}}],\ [\tilde{w}^L_{\tilde{b}}, \tilde{w}^U_{\tilde{b}}],$$

be two IVNV numbers. Then,

$$\tilde{a} \oplus \tilde{b} = (a_1 + b_1, a_2 + b_2, a_3 + b_3, a_4 + b_4) : A,\ B,\ C,$$

$$\tilde{a} \ominus \tilde{b} = (a_1 - 4, a_2 - b_3, a_3 - b_2, a_4 - b_1) : A,\ B,\ C,$$

$$\tilde{a} \odot \tilde{b} = \begin{cases} \langle (a_1b_1, a_2b_2,\ a_3b_3,\ a_4b_4) : A,\ B,\ C \rangle & if\ a_4 > 0,\ b_4 > 0 \\ \langle (a_1b_4, a_2b_3,\ a_3b_2,\ a_4b_1) : A,\ B,\ C \rangle & if\ a_4 \langle 0,\ b_4 \rangle 0 \\ \langle (a_4b_4, a_3b_3,\ a_2b_2,\ a_1b_1) : A,\ B,\ C \rangle & if\ a_4 < 0,\ b_4 < 0 \end{cases}$$

$$\tilde{a} \oslash \tilde{b} =$$
$$\begin{cases} \langle (a_1 /\ b_4, a_2 /\ b_3,\ a_3 / b_2,\ a_4 / b_1) : A,\ B,\ C \rangle & if\ a_4 > 0,\ b_4 > 0, \\ \langle (a_4 / b_4, a_3 / b_3,\ a_2 / b_2,\ a_1 / b_1) : A,\ B,\ C \rangle & if\ a_4 \langle 0,\ b_4 \rangle 0, \\ \langle (a_4 / b_1, a_3 / b_2,\ a_2 / b_3,\ a_1 / b_4) : A,\ B,\ C \rangle & if\ a_4 < 0,\ b_4 < 0. \end{cases}$$

$$k\,\tilde{a} = \begin{cases} (ka_1, ka_2, ka_3, ka_4);\ [\tilde{u}^L_{\tilde{a}}, \tilde{u}^U_{\tilde{a}}],\ [\tilde{v}^L_{\tilde{a}}, \tilde{v}^U_{\tilde{a}}],\ [\tilde{w}^L_{\tilde{a}}, \tilde{w}^U_{\tilde{a}}], & if\ k > 0 \\ (ka_1, ka_2, ka_3, ka_4);\ [\tilde{u}^L_{\tilde{a}}, \tilde{u}^U_{\tilde{a}}],\ [\tilde{v}^L_{\tilde{a}}, \tilde{v}^U_{\tilde{a}}],\ [\tilde{w}^L_{\tilde{a}}, \tilde{w}^U_{\tilde{a}}], & if\ k < 0 \end{cases}$$

$$\tilde{a}^{-1} = (1/a_4,\ 1/a_3, 1/a_2, 1/a_1);\ [\tilde{u}^L_{\tilde{a}}, \tilde{u}^U_{\tilde{a}}],\ [\tilde{v}^L_{\tilde{a}}, \tilde{v}^U_{\tilde{a}}],\ [\tilde{w}^L_{\tilde{a}}, \tilde{w}^U_{\tilde{a}}], \tilde{a} \neq 0.$$

Where

$$A = \left[min(\tilde{u}^L_{\tilde{a}}, \tilde{u}^L_{\tilde{b}}),\ min(\tilde{u}^U_{\tilde{a}}, \tilde{u}^U_{\tilde{b}})\right],$$

$$B = \left[max\left(\tilde{v}^{L}_{\tilde{a}}, \tilde{v}^{L}_{\tilde{b}}\right), max\left(\tilde{v}^{U}_{\tilde{a}}, \tilde{v}^{U}_{\tilde{b}}\right)\right],$$

and

$$C = \left[max\left(\tilde{w}^{L}_{\tilde{a}}, \tilde{w}^{L}_{\tilde{b}}\right), max\left(\tilde{w}^{U}_{\tilde{a}}, \tilde{w}^{U}_{\tilde{b}}\right)\right]$$

Definition 2.18. (Mukherjee & Basu, 2010). The score function for the IVNV number

$$\tilde{a} = (a_1, a_2, a_3, a_4); \left[\tilde{u}^{L}_{\tilde{a}}, \tilde{u}^{U}_{\tilde{a}}\right], \left[\tilde{v}^{L}_{\tilde{a}}, \tilde{v}^{U}_{\tilde{a}}\right], \left[\tilde{w}^{L}_{\tilde{a}}, \tilde{w}^{U}_{\tilde{a}}\right]$$

is defined as

$$S(\tilde{a}) = \frac{1}{16}(a_1 + a_2 + a_3 + a_4) \times \left[\vartheta_{\tilde{a}} + (1 - \mu_{\tilde{a}}) + (1 - \varphi_{\tilde{a}})\right].$$

PROBLEM DESCRIPTION AND SOLUTION CONCEPTS

Assumption, Index, and Notation

Assumption

We make the assumption that there are m jobs that need to be completed by m people, and that the costs vary depending on the particular tasks. Each person must be given a specific task to complete, and each task must be assigned to exactly one person.

Index

i: Persons
j: Jobs

Notation

$\left(\tilde{b}_{ij}\right)^{IV}_{PNV}$: Interval – valued trapezoidal Pentapartitioned neutrosophic vague cost of ith person assigned to j^{th} job.

x_{ij}: Number of jth jobs assigned to i^{th} person.

Consider the following interval-valued trapezoidal Pentapartitioned neutrosophic vague assignment problem (IVTPNVAP)

(IVTPNVAP) $\min \widetilde{Z}^{IV}_{PNV} = \sum_{i=1}^{m} \sum_{j=1}^{m} \left(\widetilde{b}_{ij}\right)^{IV}_{PNV} x_{ij}$

Subject to

$\sum_{i=1}^{m} \dot{x}_{ij} = 1, j=1,2,\ldots,m$ (only one person would be assigned the jth job)

$\sum_{i=1}^{m} x_{ij} = 1, i=1,2,\ldots,m$ (only one job selected by i^{th} person)

$x_{ij} = 0$ or 1.

It obvious that $\left(\widetilde{b}_{ij}\right)^{IV}_{PNV}$ $(i=j= 1,2,\ldots,m; 1,2,\ldots,k)$ are interval-valued trapezoidal Pentapartitioned neutrosophic vague numbers.

Based on score function, the IVTPNVAP in converted into the following interval-valued assignment problem (IVAP)

(IVAP) $\min Z^{IV} = \sum_{i=1}^{m} \sum_{j=1}^{m} \left[b_{ij}^{L-}, b_{ij}^{L+}\right] \left[b_{ij}^{U-}, b_{ij}^{U+}\right] x_{ij}, x \in X$

$\sum_{i=1}^{m} x_{ij} = 1, \ j = 1, 2, \ldots, m$

$\sum_{i=1}^{m} x_{ij} = 1, \ i = 1, 2, \ldots, m$

$x_{ij} = 0$ or 1

Definition 3.2. $x \in X$ is solution of problem IVAP if and only if there is no $\hat{x} \in X$ satisfies $Z(\hat{x}) \leq_{LR} Z(x)$ or $Z(\hat{x}) <_{CW} Z(x)$.

Definition 3.3. $x \in X$ is solution of problem IVAP if and only if there is no $\hat{x} \in X$ satisfies that $Z(\hat{x}) \leq_{LR} Z(x)$

The solution set of problem IVAP can be obtained as the efficient solution of the following MOAP:

$\min(Z^R, Z^C)$

Subject to $x \in X$.

Using the Weighting Tchebycheff problem,the Problem (3) is described in the following form $\min \Psi$ Subject to

$w_1\left[Z^R - \widehat{Z}^R\right] \leq w_2\left[Z^C - \widehat{Z}^C\right] \leq \psi$

$x \in X$

Where w_1, $w_2 \geq 0$; $\hat{Z}^R$ and $\hat{Z}^C$ are defined as the ideal targets.

SOLUTION PROCEDURE

The steps involved in solving the IVTPNVAP solution procedure can be summed up as follows:

Step 1: Formulate the IVTPNVAP

Step 2: Convert the IVTPNVAP using the score function into the IVAP using following score function:

The score function for the IVNV number

$$\tilde{a} = (a_1, a_2, a_3, a_4); [\tilde{u}^L_{\tilde{a}}, \tilde{u}^U_{\tilde{a}}], [\tilde{v}^L_{\tilde{a}}, \tilde{v}^U_{\tilde{a}}], [\tilde{w}^L_{\tilde{a}}, \tilde{w}^U_{\tilde{a}}], [\tilde{s}^L_{\tilde{a}}, \tilde{s}^U_{\tilde{a}}], [\tilde{t}^L_{\tilde{a}}, \tilde{t}^U_{\tilde{a}}]$$

is defined as

$$S(\tilde{a}) = \frac{1}{20}(a_1 + a_2 + a_3 + a_4) \times [\vartheta_{\tilde{a}} + \mu_{\tilde{a}} + (1 - \varphi_{\tilde{a}}) + (1 - \phi_{\tilde{a}}) + (1 - \omega_{\tilde{a}})].$$

$$\vartheta_{\tilde{a}}(x) = \begin{cases} u_{\tilde{a}}\left(\frac{x - a_1}{a_2 - a_1}\right), & for\ a_1 \leq x \leq a_2, \\ u_{\tilde{a}}, & for\ a_2 \leq x \leq a_3, \\ u_{\tilde{a}}\left(\frac{a_4 - x}{a_4 - a_3}\right), & for\ a_3 \leq x \leq a_4, \\ 0, & otherwise, \end{cases}$$

$$\mu_{\tilde{a}}(x) = \begin{cases} v_{\tilde{a}}\left(\frac{x - a_1}{a_2 - a_1}\right), & for\ a_1 \leq x \leq a_2, \\ v_{\tilde{a}}, & for\ a_2 \leq x \leq a_3, \\ v_{\tilde{a}}\left(\frac{a_4 - x}{a_4 - a_3}\right), & for\ a_3 \leq x \leq a_4, \\ 0, & otherwise, \end{cases}$$

$$\varphi_{\tilde{a}}(x) = \begin{cases} \left(\frac{a_2 - x + w_{\tilde{a}}(x - a_1)}{a_2 - a_1}\right) & for\ a_1 \leq x \leq a_2 \\ w_{\tilde{a}} & for\ a_2 \leq x \leq a_3 \\ \left(\frac{x - a_3 + w_{\tilde{a}}(a_4 - x)}{a_4 - a_3}\right) & for\ a_3 \leq x \leq a_4 \\ 1 & otherwise \end{cases}$$

$$\phi_{\tilde{a}}(x) = \begin{cases} \left(\frac{a_2 - x + s_{\tilde{a}}(x - a_1)}{a_2 - a_1}\right) & for\ a_1 \le x \le a_2 \\ s_{\tilde{a}} & for\ a_2 \le x \le a_3 \\ \left(\frac{x - a_3 + s_{\tilde{a}}(a_4 - x)}{a_4 - a_3}\right) & for\ a_3 \le x \le a_4 \\ 1 & otherwise \end{cases}$$

$$\omega_{\tilde{a}}(x) = \begin{cases} \left(\frac{a_2 - x + t_{\tilde{a}}(x - a_1)}{a_2 - a_1}\right) & for\ a_1 \le x \le a_2 \\ t_{\tilde{a}} & for\ a_2 \le x \le a_3 \\ \left(\frac{x - a_3 + t_{\tilde{a}}(a_4 - x)}{a_4 - a_3}\right) & for\ a_3 \le x \le a_4 \\ 1 & otherwise \end{cases}$$

Step 3: Estimate the ideal points $\hat{Z}^R$ and $\hat{Z}^C$ for the *IVAP* from the following relation

$$\hat{Z}^R = \min Z^R$$

Subject to $x \in X$, and

$$\hat{Z}^C = \min Z^C$$

Subject to $x \in X$.

Step 4: For each objective function, ascertain the value of each individual maximum and minimum within the given restrictions.

Step 5: Compute the weights from the relation

$$w_1 = \frac{\bar{z}^R - \underline{z}^R}{(\bar{z}^R - \underline{z}^R) + (\bar{z}^C - \underline{z}^C)}$$

$$w_2 = \frac{\bar{z}^C - \underline{z}^C}{(\bar{z}^R - \underline{z}^R) + (\bar{z}^C - \underline{z}^C)}$$

Here $\bar{z}^R$, $\bar{z}^C$ and $\underline{z}^R$, $\underline{z}^C$ are the value of individual maximum and minimum of the Z^R, and Z^C, respectively.

Step 6: Stop.

NUMERICAL EXAMPLE

Consider the following IVTPNVAP

$$\min Z(x)_{PNV}^{IV} =$$

$$\begin{cases} \langle(15, 18, 22, 29); [0.3, 0.5], [0.2, 0.4], [0.1, 0.6], [0.5, 0.5], [0.4, 0.7]\rangle x_{11} \\ \oplus \langle(14, 19, 21, 25); [0.4, 0.6], [0.2, 0.3], [0.3, 0.5], [0.2, 0.8], [0.5, 0.6]\rangle x_{12} \\ \oplus \langle(21, 26, 30, 35); [0.9, 1.0], [0.3, 0.5], [0.2, 0.5], [0.3, 0.9], [0.6, 0.7]\rangle x_{13} \\ \oplus \langle(16, 19, 24, 31); [0.8, 1.0], [0.3, 0.4], [0.3, 0.6], [0.4, 0.7], [0.2, 0.5]\rangle x_{21} \\ \oplus \langle(7, 10, 13, 15); [0.7, 0.9], [0.2, 0.5], [0.3, 0.7], [0.1, 0.4], [0.4, 0.8]\rangle x_{22} \\ \oplus \langle(16, 19, 24, 31); [0.7, 0.9], [0.1, 0.4], [0.4, 0.6], [0.3, 0.5], [0.2, 0.4]\rangle x_{23} \\ \oplus \langle(14, 19, 21, 25); [0.4, 0.8], [0.2, 0.5], [0.4, 0.8], [0.1, 0.4], [0.3, 0.7]\rangle x_{31} \\ \oplus \langle(15, 20, 22, 26); [0.2, 0.7], [0.3, 0.6], [0.4, 0.7], [0.1, 0.5], [0.5, 0.8]\rangle x_{32} \\ \oplus \langle(15, 17, 22, 24); [0.7, 0.9], [0.4, 0.7], [0.3, 0.5], [0.6, 0.8], [0.3, 0.6]\rangle x_{33} \end{cases}$$

Subject to

$$\sum_{i=1}^{3} x_{ij} = 1, \quad j = 1, 2, 3; \quad \sum_{j=1}^{3} x_{ij} = 1, \quad i = 1, 2, 3;$$

$x_{ij} = 0$ or 1.

$$\min Z(x)^{IV} = \begin{pmatrix} [8.82, 10.5]x_{11} + [7.9, 10.27]x_{12} + [13.44, 17.36]x_{13} + \\ [11.7, 14.4]x_{21} + [5.625, 6.975]x_{22} + [12.6, 13.05]x_{23} + \\ [9.48, 11.06]x_{31} + [9.545, 10.375]x_{32} + [10.53, 11.31]x_{33} \end{pmatrix}$$

Subject to

$$\sum_{i=1}^{3} x_{ij} = 1, \quad j = 1, 2, 3; \quad \sum_{j=1}^{3} x_{ij} = 1, \quad i = 1, 2, 3;$$

$x_{ij} = 0$ or 1.

We determine the optimal solution for the following problems individually with respect to the given constraints:

$$\hat{Z}^{R} = \min Z^{R} = \begin{pmatrix} 10.5x_{11} + 10.27x_{12} + 17.36x_{13} + 14.4x_{21} + 6.975x_{22} \\ 13.05x_{23} + 11.06x_{31} + 10.375x_{32} + 11.31x_{33} \end{pmatrix}$$

$$\hat{Z}^C = \min Z^C = \begin{pmatrix} 9.66x_{11} + 9.085x_{12} + 15.4x_{13} + 13.05x_{21} + 6.3x_{22} + \\ 12.825x_{23} + 10.27x_{31} + 9.96x_{32} + 10.92x_{33} \end{pmatrix}$$

$$\text{Max } Z^R = \begin{pmatrix} 10.5x_{11} + 10.27x_{12} + 17.36x_{13} + 14.4x_{21} + 6.975x_{22} \\ 13.05x_{23} + 11.06x_{31} + 10.375x_{32} + 11.31x_{33} \end{pmatrix}$$

$$\text{Max } Z^C = \begin{pmatrix} 9.66x_{11} + 9.085x_{12} + 15.4x_{13} + 13.05x_{21} + 6.3x_{22} + \\ 12.825x_{23} + 10.27x_{31} + 9.96x_{32} + 10.92x_{33} \end{pmatrix}$$

Subject to

$$\sum_{i=1}^{3} x_{ij} = 1, \quad j = 1, 2, 3; \quad \sum_{j=1}^{3} x_{ij} = 1, \quad i = 1, 2, 3;$$

$x_{ij} = 0$ or 1.

The optimal compromise solution is $x_{11}=1$, $x_{22}=1$, $x_{33}=1$, $x_{12}=0$, $x_{13}=0$, $x_{21}=0$, $x_{22}=0$, $x_{31}=0$, and $x_{32}=0$.

So, the interval valued trapezoidal Pentapartitioned neutrosophic vague optimum value is

$$Z(X)^{IV}_{PNV} = (35, 44, 56, 67); [0.7, 0.9], [0.4, 0.7], [0.3, 0.5], [0.6, 0.8], [0.3, 0.6]$$

It is apparent that the minimum assigned cost will exceed 35 but fall below 67 overall.

The entire minimum assigned cost falls within the range of 43 to 55.

Then, for the remaining of total minimum assigned cost, the truthfulness degree is

$$\begin{cases} [0.7,0.9]\left(\frac{x-35}{44-35}\right) & for\ 35 \le x \le 44 \\ [0.7,0.9] & for\ 44 \le x \le 56 \\ [0.7,0.9]\left(\frac{67-x}{67-56}\right) & for\ 56 \le x \le 67 \\ 0 & otherwise \end{cases}$$

The contradiction degree is

$$\mu_{\tilde{a}}(x) = \begin{cases} [0.4,\ 0.7]\left(\frac{x-35}{44-35}\right) & for\ 35 \le x \le 44 \\ [0.4,\ 0.7] & for\ 44 \le x \le 56 \\ [0.4,\ 0.7]\left(\frac{67-x}{67-56}\right) & for\ 56 \le x \le 67 \\ 0 & otherwise \end{cases}$$

The ignorance degree is

$$\varphi_{\tilde{a}}(x) = \begin{cases} \left(\frac{44-x+[0.3,0.5](x-35)}{44-35}\right) & for\ 35 \le x \le 44 \\ [0.3,0.5] & for\ 44 \le x \le 56 \\ \left(\frac{x-56+[0.3,0.5](67-x)}{67-56}\right) & for\ 56 \le x \le 67 \\ 1 & otherwise \end{cases}$$

The unknown degree is

$$\phi_{\tilde{a}}(x) = \begin{cases} \left(\frac{44-x+[0.6,0.8](x-35)}{44-35}\right) & for\ 35 \le x \le 44 \\ [0.6,0.8] & for\ 44 \le x \le 56 \\ \left(\frac{x-56+[0.6,0.8](67-x)}{67-56}\right) & for\ 56 \le x \le 67 \\ 1 & otherwise \end{cases}$$

The falsity degree is

$$\omega_{\tilde{a}}(x) = \begin{cases} \left(\frac{44 - x + [0.3, 0.6](x - 35)}{44 - 35}\right) & for 35 \leq x \leq 44 \\ [0.3, 0.6] & for 44 \leq x \leq 56 \\ \left(\frac{x - 56 + [0.3, 0.6](67 - x)}{67 - 56}\right) & for 56 \leq x \leq 67 \\ 1 & otherwise \end{cases}$$

Consequently, the DM deduces that the total interval valued trapezoidal Pentapartitioned neutrosophic vague assigned cost, with degrees of truth, contradiction, ignorance, unknown, and falsity is contained between 35 and 67.

CONCLUSION

In the current study, a novel approach to solving the assignment problem with objective function coefficients that are interval-valued trapezoidal pentapartitioned neutrosophic vague numbers was explored. The issue is converted into the multi-objective optimization problem (MOOP) by way of the corresponding interval-valued problem. Subsequently, the obtained MOOP is used to solve the Weighting Tchebycheff problem. This approach has the advantage of being more flexible than the standard assignment problem in that the DM can select the targets he is willing to work with.

This idea could be included in a transportation model for future study. It is also possible to take into account the assignment problem's stochastic nature and use the same approach to solve it. Furthermore, a potential extension could be investigated by taking into account fuzzy-random,fuzzy-stochastic, etc. Furthermore, the suggested methodology for solving problems can be implemented in various fields (such as financial management, management science, and decision science) where the assignment problems arise in a pentapartitioned neutrosophic environment.

GLOSSARY

AP: Assignment problem.
DM: Decision makers.
FN-LPP: Fuzzy neutrosophic LPP.
GAMS: General Algebraic Modeling System.
IVN: Interval-valued neutrosophic.
IVTNAP: Interval- valued trapezoidal neutrosophic assignment problem.

IVTPNVAP: Interval- valued trapezoidal Pentapartitioned neutrosophic vague assignment problem.

LP: Linear programming.

MOAP: Multi-objective assignment problem

MOLP: Multi-objective linear programming

MOOP: Multi-objective optimization problem.

NAP: Neutrosophic assignment problem.

REFERENCES

Abdel-Basset, M., Ding, W., Mohamed, R., & Metawa, N. (2020). An integrated plithogenic MCDM approach for financial performance evaluation of manufacturing industries. *Risk Management*, *22*, 192–218.

Abdel-Basset, M., Mohamed, R., Sallam, K., & Elhoseny, M. (2020). A novel decision-making model for sustainable supply chain finance under uncertainty environment. *Journal of Cleaner Production*, *269*, 122324.

Abdel-Basst, M., Mohamed, R., & Elhoseny, M. (2020). A novel framework to evaluate innovation value proposition for smart product–service systems. *Environmental Technology & Innovation*, *20*, 101036.

Al-Quran, A., & Hassan, N. (2016). Neutrosophic vague soft expert set theory. *Journal of Intelligent & Fuzzy Systems*, *30*(6), 3691–3702.

Alhazaymeh, K., & Hassan, N. (2012). Interval-valued vague soft sets and its application. *Advances in Fuzzy Systems*, *2012*, 15–15.

Alhazaymeh, K., Hassan, N., & Alhazaymeh, K. (2013). Generalized interval-valued vague soft set. *Applied Mathematical Sciences*, *7*(140), 6983–6988.

Alkhazaleh, S. (2015). Neutrosophic vague set theory. *Critical Review*, *10*, 29–39.

Atanassov, K. T. (2012). *On intuitionistic fuzzy sets theory* (Vol. 283). Springer.

Bao, C. P., Tsai, M. C., & Tsai, M. I. (2007). A new approach to study the multi-objective assignment problem. *WHAMPOA-An Interdisciplinary Journal, 53*, 123-132.

Belacela, N., & Boulasselb, M. R. (2001). Multi-criteria fuzzy assignment problem: A useful tool to assist medical diagnosis. *Artificial Intelligence in Medicine*, *21*, 201–207.

Bellman, R. E., & Zadeh, L. A. (1970). Decision-making in a fuzzy environment. *Management Science*, *17*(4), B-141.

Bit, A. K., Biswal, M. P., & Alam, S. (1992). Fuzzy programming approach to multicriteria decision making transportation problem. *Fuzzy Sets and Systems*, *50*(2), 135–141.

Broumi, S., & Smarandache, F. (2015). New operations on interval neutrosophic sets. *Journal of New Theory*, (1), 24-37.

Chu, P. C., & Beasley, J. E. (1997). A general algorithm for the generalized assignment problem. *Computers & Operations Research, 24*, 17–23.

Dahiya, K., & Verma, V. (2007). Capacitated transportation problem with bounds on RIM conditions. *European Journal of Operational Research, 178*(3), 718–737.

De, P. K., & Yadav, B. (2011). An algorithm to solve multi-objective assignment problem using interactive fuzzy goal programming approach. *International Journal of Contemporary Mathematical Sciences, 6*(34), 1651–1662.

Dıaz, J. A., & Fernández, E. (2001). A tabu search heuristic for the generalized assignment problem. *European Journal of Operational Research, 132*(1), 22–38.

Dubois, D. J. (1980). *Fuzzy sets and systems: theory and applications* (Vol. 144). Academic press.

Ehrgott, M., Gandibleux, X., & Przybylski, A. (2016). Exact methods for multi-objective combinatorial optimisation. *Multiple criteria decision analysis: State of the art surveys*, 817-850.

Emrouznejad, A., & Zerafat Angiz,, L. M., & Ho, W. (2012). An alternative formulation for the fuzzy assignment problem. *The Journal of the Operational Research Society, 63*(1), 59–63.

Gau, W. L., & Buehrer, D. J. (1993). Vague sets. *IEEE Transactions on Systems, Man, and Cybernetics, 23*(2), 610–614.

Geetha, S., & Nair, K. P. K. (1993). A variation of the assignment problem. *European Journal of Operational Research, 68*(3), 422–426.

Haddad, H., Mohammadi, H., & Pooladkhan, H. (2002). Two models for the generalized assignment problem in uncertain environment. *Management Science Letters, 2*(2), 623-630.

Hamadameen, O. A. (2018). A noval technique for solving multi- objective linear programming problems. *Aro - The Scientific Journal of Koya University, 5*(2), 1-8

Hamou, A., & El-Amine, C. (2018) An exact method for the multi- objective assignment problem. *Les Annales RECITS, 5*, 31-36.

Ishibuchi, H., & Tanaka, H. (1990). Multiobjective programming in optimization of the interval objective function. *European Journal of Operational Research, 48*(2), 219–225.

Jayalakshmi, M., & Sujatha, V. (2018). A new algorithm to solve multi-objective assignment problem. *International Journal of Pure and Applied Mathematics, 119*(16), 719–724.

Kagade, K. L., & Bajaj, V. H. (2010). Fuzzy method for solving multi-objective assignment problem with interval cost. *Journal of Statistics and Mathematics, 1*(1), 1-9.

Khalifa, H. A. (2020). An approach to the optimization of multi- objective assignment problem with neutrosophic numbers. *International Journal of Industrial Engineering& Production Research, 31*(2), 287–294.

Khalifa, H. A., Al-Shabi, M., & Mukherjee, S. (2018). An interactive approach for solving fuzzy multi-objective assignment problems. *Journal of advances in mathematics and computer science, 28*(6), 1-12.

Khalifa, H. A. E. W., Kumar, P., & Smarandache, F. (2020). *On optimizing neutrosophic complex programming using lexicographic order*. DOI: 10.5281/zenodo.3723173

Kiruthiga, M., & Loganathan, C. (2015). Fuzzy multi-objective linear programming problem using membership function. *International Journal of Science, Engineering, and Technology, Applied Sciences, 5*(8), 1171-1178.

Kumar, A., & Gupta, A. (2011). Methods for solving fuzzy assignment problems and fuzzy travelling salesman problems with different membership functions. *Fuzzy Information and Engineering, 3*, 3–21.

Kumar, R., Edalatpanah, S. A., Jha, S., & Singh, R. (2019). A Pythagorean fuzzy approach to the transportation problem. *Complex & Intelligent Systems, 5*(2), 255–263.

Leberling, H. (1981). On finding compromise solutions in multicriteria problems using the fuzzy min-operator. *Fuzzy Sets and Systems, 6*(2), 105–118.

Lin, C. J., & Wen, U. P. (2004). A labeling algorithm for the fuzzy assignment problem. *Fuzzy Sets and Systems, 142*(3), 373–391.

Lohgaonkar, M. H., & Bajaj, V. H. (2010). Fuzzy approach to solve multi-objective capacitated transportation problem. *International Journal of Bioinformatics Research, 2*(1), 10–14.

Lupiáñez, F. G. (2009). Interval neutrosophic sets and topology. *Kybernetes, 38*(3/4), 621–624.

Margaret, M. A., Pricilla, T. M., & Alkhazaleh, S. (2019). *Neutrosophic Vague Topological Spaces, Neutrosophic Sets and Systems, 28*. University of New Mexico.

Medvedeva, O. A., & Medvedev, S. N. (2018, March). A dual approach to solving a multi-objective assignment problem. *Journal of Physics: Conference Series, 973*(1), 012039.

Mukherjee, S., & Basu, K. (2010). Application of fuzzy ranking method for solving assignment problems with fuzzy costs. *International Journal of Computational and Applied Mathematics, 5*(3), 359-369.

Pandian, P., & Anuradha, D. (2010). A new approach for solving solid transportation problems. *Applied Mathematical Sciences, 4*(72), 3603–3610.

Pandian, P., & Natarajan, G. (2010). A new method for finding an optimal solution for transportation problems. *International Journal of Mathematical Sciences and engineering applications, 4*(2), 59-65.

Pramanik, S. (2023). Interval pentapartitioned neutrosophic sets. *Neutrosophic Sets and Systems, 55*, 232–246.

Pramanik, S., & Banerjee, D. (2018). *Neutrosophic number goal programming for multi-objective linear programming problem in neutrosophic number environment.*

Pramanik, S., & Biswas, P. (2012). Multi-objective assignment problem with generalized trapezoidal fuzzy numbers. *International Journal of Applied Information Systems, 2*(6), 13–20.

Rajarajeswari, P., Sudha, A. S., & Karthika, R. (2013). A new operation on hexagonal fuzzy number. *International Journal of Fuzzy Logic Systems, 3*(3), 15-26.

Rizk-Allah, R. M., Hassanien, A. E., & Elhoseny, M. (2018). A multi-objective transportation model under neutrosophic environment. *Computers & Electrical Engineering, 69*, 705–719.

Rodríguez, A., Ortega, F., & Concepción, R. (2017). An intuitionistic method for the selection of a risk management approach to information technology projects. *Information Sciences, 375*, 202–218.

Rodríguez, R. M., Labella, Á., & Martínez, L. (2016). An overview on fuzzy modelling of complex linguistic preferences in decision making. *International Journal of Computational Intelligence Systems, 9*(Suppl 1), 81–94.

Sakawa, M., & Yano, H. (1989). Interactive decision making for multiobjective nonlinear programming problems with fuzzy parameters. *Fuzzy Sets and Systems, 29*(3), 315–326.

Selvachandran, G., Maji, P. K., Abed, I. E., & Salleh, A. R. (2016). Complex vague soft sets and its distance measures. *Journal of Intelligent & Fuzzy Systems, 31*(1), 55–68.

Smarandache, F. (1998). *A unifying field of logics. Neutrosophy: neutrosophic probability, set and logic*. American Research Press.

Smarandache, F. (2006, May). *Neutrosophic set-a generalization of the intuitionistic fuzzy set. In 2006 IEEE international conference on granular computing*. IEEE.

Smarandache, F. (2010, July). α-discounting method for multi-criteria decision making (α-d MCDM). In *2010 13th International Conference on Information Fusion* (pp. 1-7). IEEE.

Thamaraiselvi, A., & Santhi, R. (2015). Optimal solution of fuzzy transportation problem using hexagonal fuzzy numbers. *International Journal of Scientific and Engineering Research, 6*(3), 40–45.

Vidhya, R., Hepzibah, I., & Gani, N. (2017). Neutrosophic multi-objective linear programming problems. *Global Journal of Pure and Applied Mathematics, 13*(2), 265–280.

Wang, J. J., Jing, Y. Y., Zhang, C. F., & Zhao, J. H. (2009). Review on multi-criteria decision analysis aid in sustainable energy decision-making. *Renewable & Sustainable Energy Reviews, 13*(9), 2263–2278.

Xu, W., Ma, J., Wang, S., & Hao, G. (2010). Vague soft sets and their properties. *Computers & Mathematics with Applications (Oxford, England), 59*(2), 787–794.

Yager, R. R. (1981). A procedure for ordering fuzzy subsets of the unit interval. *Information Sciences, 24*(2), 143–161.

Yang, L., & Liu, B. (2005, May). A multi-objective fuzzy assignment problem: New model and algorithm. In *The 14th IEEE International Conference on Fuzzy Systems, 2005. FUZZ'05.* (pp. 551-556). IEEE.

Ye, J. (2014). Similarity measures between interval neutrosophic sets and their applications in multicriteria decision-making. *Journal of Intelligent & Fuzzy Systems, 26*(1), 165–172.

Zadeh, L. A. (1965). Fuzzy sets. *Information and Control, 8*(3), 338–353.

Zhang, D., Zhang, J., Lai, K. K., & Lu, Y. (2009). An novel approach to supplier selection based on vague sets group decision. *Expert Systems with Applications, 36*(5), 9557–9563.

Chapter 4

Significance and Applications of Neutrosophic Generalized Feebly Connected Topology in Diverse Realms

Santhi P.
The Standard Fireworks Rajaratnam College for Women, India

Yuvarani A.
http://orcid.org/0000-0002-5755-3429
The American College, India

Vijaya S.
http://orcid.org/0000-0002-4173-0451
Thiagarajar College, India

ABSTRACT

The study of connectedness in topology plays a vital role in understanding and demonstrating the overall structure of several geometric matters in real-life scenarios. This induces us to introduce feebly connectedness and semi connectedness concept in neutrosophic generalized topological spaces (Ng-TS), as it is the rudimentary attributes that correlates the innate ability to decipher the structure of several geometric matters. Thereafter, Ng-Feebly Ci-disconnectedness (i = 1, 2, 3, 4), Ng-Feebly irresolute function, Ng-Feebly strongly connectedness, Ng-Feebly separated

DOI: 10.4018/979-8-3693-3204-7.ch004

sets, Ng-Semi connectedness, Ng-Semi irresolute function, and Ng-Semi separated sets notions are developed and analyzed with befitting illustrations. Furthermore, its influence across diverse realms such as knot theory, chaos theory, epidemiology, cosmology is reviewed.

1. INTRODUCTION

The existence of indeterminacy in various aspects of the real world, the Neutrosophic sense, has come into view as a valuable means of research. After incorporating neutrosophic ideas, challenges posed by uncertainties can be addressed proficiently. Initially, (Zadeh. L.A, 1965) proposed the fuzzy set theory. Then, (Atanassov K.T, 1986) considered non-membership along with membership in intuitionistic fuzzy sets during the year 1983.

In (Florentin Smarandache, 1999; Florentin Smarandache, 2002; Florentin Smarandache, 2010; Wadel Faris et al., 2016) focused degree of indeterminacy that eventually led in the evolution of Neutrosophic Sets (NS). Subsequently, (Salama A.A & Albowi S.A, 2012) launched Neutrosophic Topological Spaces (NTS) in 2012. Previously (Csaszar A, 2002; Csaszar A, 2004) introduced Generalized topological spaces and look into its properties.

The initiative of Ng-TS is presented by (Murad Arar & Saeid Jafari, 2020; Raksha Ben N & Hari Siva Annam G, 2021). Neutrosophic feebly open and closed sets are defined by (Jeya Puvaneswari P & Bageerathi K, 2017; Jeya Puvaneswari P & Bageerathi K, 2019). Further, (Bageerathi K & Jeya Puvaneswari P, 2019) introduced the concept of feebly connectedness via neutrosophic sets.

In this article, our intention is to introduce Feebly Connectedness and Semi Connectedness in Ng-TS, as it is the rudimentary attributes that correlates the innate ability to decipher the structure of several geometric matters. Thereafter, Ng-Feebly C_i-disconnectedness (i = 1, 2, 3, 4), Ng-Feebly irresolute function, Ng-Feebly strongly connectedness, Ng-Feebly separated sets, Ng-Semi connectedness, Ng-Semi irresolute function and Ng-Semi separated sets notions developed and analysed their characteristics with befitting illustrations. Furthermore, its influence across diverse realms such as Knot theory, Chaos theory, Epidemiology, Cosmology is reviewed.

2. PRELIMINARIES

Initially, we will make known to the fundamental definitions. The next one is obviously influenced by (Florentin Smarandache, 1999; Florentin Smarandache, 2002; Florentin Smarandache, 2010; Wadel Faris et al., 2016) and (Salama A.A & Albowi S.A, 2012).

2.1 Definition (Salama A.A & Albowi S.A, 2012)

Consider Λ as the universe. A Ng-Set H in Λ is described by Y_H- a truth membership function, U_H- an indeterminacy membership function and N_H- a falsity membership function. Also, Y_H, U_H and N_H are real standard elements of [0, 1], which is written as

$$H = \{ < \lambda, (Y_H(\lambda), U_H(\lambda), N_H(\lambda)) > : \lambda \in \Lambda, Y_H, U_H, N_H \in]^-0, 1^+[\}$$

.

There is no restriction on the sum of $Y_H(\lambda)$, $U_H(\lambda)$, and $N_H(\lambda)$. So, $0 \leq Y_H(\lambda)+U_H(\lambda)+N_H(\lambda) \leq 3$.

2.2 Remark (Salama A.A & Albowi S.A, 2012)

$$0_N = \{ < \lambda,0,0,1 > : \lambda \in \Lambda\} \; or \{ < \lambda,0,1,1 > : \lambda \in \Lambda\} \; or \{ < \lambda,0,1,0 > : \lambda \in \Lambda\} \; or \{ < \lambda,0,0,0 > : \lambda \in \Lambda\}$$

$$1_N = \{ < \lambda,1,0,0 > : \lambda \in \Lambda\} \; or \{ < \lambda,1,0,1 > : \lambda \in \Lambda\} \; or \{ < \lambda,1,1,0 > : \lambda \in \Lambda\} \; or \{ < \lambda,1,1,1 > : \lambda \in \Lambda\}$$

Quite a few relations and operations among neutrosophic sets can be demarcated as follows:

2.3 Definition (Salama A.A & Albowi S.A, 2012)

Let

$\Lambda \neq \phi$ *and let*

$$J = \{ < \lambda, (Y_J(\lambda), U_J(\lambda), N_J(\lambda)) > : \lambda \in \Lambda\}$$

and

$$H = \{ < \lambda, (Y_H(\lambda), U_H(\lambda), N_H(\lambda)) > : \lambda \in \Lambda\}$$

.

Then- Complement of H is

$$H^c = \{ < \lambda, N_H(\lambda), 1 - U_H(\lambda), Y_H(\lambda) > : \lambda \in \Lambda\}$$

$$J \leq H \Rightarrow Y_J(\lambda) \leq Y_H(\lambda), U_J(\lambda) \leq U_H(\lambda), N_J(\lambda) \geq N_H(\lambda), \forall \lambda \in \Lambda$$

$$J \leq H \Rightarrow Y_J(\lambda) \leq Y_H(\lambda), U_J(\lambda) \geq U_H(\lambda), N_J(\lambda) \geq N_H(\lambda), \forall \lambda \in \Lambda$$

$$J \wedge H = < \lambda, Y_J(\lambda) \wedge Y_H(\lambda), U_J(\lambda) \vee U_H(\lambda), N_J(\lambda) \vee N_H(\lambda) >$$

$$J \wedge H = < \lambda, Y_J(\lambda) \wedge Y_H(\lambda), U_J(\lambda) \wedge U_H(\lambda), N_J(\lambda) \vee N_H(\lambda) >$$

$$J \vee H = < \lambda,\ Y_J(\lambda) \vee Y_H(\lambda),\ U_J(\lambda) \wedge U_H(\lambda),\ N_J(\lambda) \wedge N_H(\lambda) >$$

$$J \vee H = < \lambda,\ Y_J(\lambda) \vee Y_H(\lambda),\ U_J(\lambda) \vee U_H(\lambda),\ N_J(\lambda) \wedge N_H(\lambda) >$$

In the following definitions, the initiative of Ng-topology by (Murad Arar & Saeid Jafari, 2020; Raksha Ben N & Hari Siva Annam G, 2021) is delineated.

2.4 Definition (Raksha Ben N & Hari Siva Annam G, 2021)

Let (Λ,λ) be a Ng-TS and $\Lambda \neq \Phi$. A family of Ng-subsets of Λ is Ng-topology if it satisfies $0_N \in \lambda$

$H_1 \vee H_2 \in \lambda$ for any $H_1,\ H_2 \in \lambda$.

Each Ng-set in λ is called a Ng-open set (Ng-OS) and A^c, the complement of a Ng-OS A is called a Ng-closed set (Ng-CS). Set which is both Ng-OS and Ng-CS is a Ng-clopen set in (Λ,λ).

2.5 Definition (Raksha Ben N & Hari Siva Annam G, 2021)

Let (Λ,λ) be a Ng-TS and $J = \{ < \lambda,\ (Y_J(\lambda),\ U_J(\lambda),\ N_J(\lambda)) > : \lambda \in \Lambda \}$ be a Ng-subset. Then

- Ng-Closure of J (Ng-*cr*(J)) $= \wedge\ \{X : J \leq X$, X is Ng-CS$\}$
- Ng-Interior of J (Ng-*ir*(J)) $= \vee\ \{Y : Y \leq J$, Y is Ng-OS$\}$

2.6 Definition (Raksha Ben N & Hari Siva Annam G, 2021)

A Ng-subset, J in a Ng-TS $(\Lambda,\lambda)$is said to be

- Ng-σ-open set (Ng-σ-OS) if J ≤ Ng-*cr*(Ng-*ir*(J)),
- Ng-π-open set (Ng-π-OS) if J ≤ Ng-*ir*(Ng-*cr*(J)),
- Ng-α-open set (Ng-α-OS) if J ≤ Ng-*ir*(Ng-*cr*(Ng-*ir*(J))),
- Ng-β-open set (Ng-β-OS) if J ≤ Ng-*cr*(Ng-*ir*(Ng-*cr*(J))),
- Ng-regular open set (Ng-r-OS) if J = Ng-*ir*(Ng-*cr*(J)).

3. NG-FEEBLY CONNECTED SPACES

3.1 Definition

A subset A of a Ng-TS (Λ,λ) is called as

(i) Ng-Feebly open set (Ng-F-OS) if $\exists$ a Ng-OS U in $\Lambda \ni U \leq A \leq Ng-\sigma-cr(U)$.
(ii) Ng-Feebly closed set (Ng-F-CS) if $\exists$ a Ng-CS U in $\Lambda \ni Ng-\sigma-ir(U) \leq A \leq U$.
(iii) Ng-Feebly regular open set (Ng-F-ROS) if $\exists$ a Ng-subset U in $\Lambda \ni U = Ng-ir(Ng-cr(U))$

In general, a subset A of a Ng-TS $(\Lambda,\lambda)$is Ng-F-CS iff A^c is Ng-F-OS.

3.2 Lemma

(i) Every Ng-OS is a Ng-F-OS.
(ii) A subset A of a Ng-TS $(\Lambda,\lambda)$is Ng-F-CS iff Ng-*cr*(Ng-*ir*(Ng-*cr*(A))) $\leq$ A.
(iii) A subset A of a Ng-TS $(\Lambda,\lambda)$is Ng-F-CS iff Ng-σ-*ir*(Ng-*cr*(A)) $\leq$ A.
(iv) Every Ng-CS is a Ng-F-CS.

3.3 Definition

Let (Λ,λ) be a Ng-TS and $J = \{ < \lambda, (Y_J(\lambda), U_J(\lambda), N_J(\lambda)) > : \lambda \in \Lambda\}$ be a Ng-subset in Λ. Then Ng-Feebly interior of J is defined by Ng-F-*ir*(J) = $\vee \{W : W \leq J$, W is Ng-F-OS$\}$ and Ng-Feebly closure of J is defined by Ng-F-*cr*(J) $= \wedge \{T : J \leq T$, T is Ng-F-CS$\}$.

3.4 Lemma

Let (Λ,λ) be a Ng-TS. Then for any Ng-Feebly subsets A & B of a Ng-TS Λ,

(i) Ng-F-ir(A) $\leq$ A
(ii) A is Ng-F-OS in Λ iff Ng-F-ir(A) = A
(iii) $(\text{Ng-F-ir}(A))^c = \text{Ng-F-cr}(A^c)$
(iv) $(\text{Ng-F-cr}(A))^c = \text{Ng-F-ir}(A^c)$
(v) A $\leq$ Ng-F-cr(A^c)
(vi) A is Ng-F-CS in Λ iff Ng-F-cr(A) = A

(vii) If $A \leq B$, then Ng-F-cr(A) $\leq$ Ng-F-cr(B)

3.5 Definition

A Ng-TS (Λ,λ) is Ng-Feebly disconnected if $\exists$ Ng-F-OS A, B in Λ,

$$A \neq 0_N, B \neq 0_N \ni A \vee B = 1_N \,\&\, A \wedge B = 0_N$$

.

That is

(i) $< \Lambda, Y_A \vee Y_B, U_A \wedge U_B, N_A \wedge N_B > = 1_N$
(ii) $< \Lambda, Y_A \vee Y_B, U_A \vee U_B, N_A \wedge N_B > = 1_N$
(iii) $< \Lambda, Y_A \wedge Y_B, U_A \wedge U_B, N_A \vee N_B > = 0_N$
(iv) $< \Lambda, Y_A \wedge Y_B, U_A \vee U_B, N_A \vee N_B > = 0_N$

If Λ is not Ng-Feebly disconnected, then it is Ng-Feebly connected.

3.6 Illustration

Let $\Lambda = \{p, q, r\}$, $\lambda = \{0_N, K, L, M, N\}$ where $K = \{< (0,0,1), (0,0,0), (1,0,1) >\}$, $L = \{< (1,1,0), (1,0,0), (0,0,0) >\}$, $M = \{< (1,1,0), (1,0,0), (1,0,0) >\}$, $N = \{< (0,0,1), (0,0,0), (0,0,1) >\}$. K & L are Ng-F-OS in Λ, $K \neq 0_N$, $L \neq 0_N$ & $K \vee L = M = 1_N$, $K \wedge L = N = 0_N$. Hence Λ is Ng-Feebly disconnected.

3.7 Illustration

Let $\Lambda = \{u, v\}$, $\lambda = \{0_N, K\}$ where $K = \{< (0.2, 0.2, 0.7), (0.2, 0.4, 0.5) >\}$. Then (Λ,λ) is a Ng-TS. Let $L = \{< (0.4, 0.7, 0.6), (0.3, 0.5, 0.4) >\}$. Then K & L are Ng-F-OS in Λ, $K \neq 0_N$, $L \neq 0_N$ & $K \vee L \neq 1_N$, $K \wedge L \neq 0_N$. Hence Λ is Ng-Feebly connected.

3.8 Definition

Let A be a Ng-subset in Ng-TS (Λ,λ). If $\exists$ Ng-F-OS U & W in Λ sustaining the ensuing properties, then A is called Ng-Feebly c_i-disconnected ($i = 1, 2, 3, 4$):

C_1: $A \leq U \vee W$, $U \wedge W \leq A^c$, $A \wedge U \neq 0_N$, $A \wedge W \neq 0_N$
C_2: $A \leq U \vee W$, $A \wedge U \wedge W = 0_N$, $A \wedge U \neq 0_N$, $A \wedge W \neq 0_N$

$C_3: A \leq U \vee W, U \wedge W \leq A^c, U \geq A^c, W \geq A^c$
$C_4: A \leq U \vee W, A \wedge U \wedge W = 0_N, U \geq A^c, W \geq A^c$

A is said to be Ng-Feebly c_i-connected (i = 1, 2, 3, 4) if A is not Ng-Feebly c_i-disconnected (i = 1, 2, 3, 4).

Clearly, the ensuing implications are obtained among the types of Ng-Feebly c_i-connected (i = 1, 2, 3, 4).

1. Ng-Feebly C_1-connected ⇒ Ng-Feebly C_2-connected
2. Ng-Feebly C_1-connected ⇒ Ng-Feebly C_3-connected
3. Ng-Feebly C_3-connected ⇒ Ng-Feebly C_4-connected
4. Ng-Feebly C_2-connected ⇒ Ng-Feebly C_4-connected *Figure 1.*

3.9 Illustration

Let Λ = {a, b}, λ = {0_N, K, L} where K = {< (0.5, 0.4, 0.7), (0.1, 0.4, 0.9) >}, L = {< (0.6, 0.5, 0.7), (0.4, 0.4, 0.8) >} Consider the Ng-subset H = {< (0.4, 0.3, 0.8), (0.1, 0.4, 0.9) >} and Ng-F-OS U = {< (0.6, 0.5, 0.7), (0.5, 0.5, 0.7) >}, W = {< (0.7, 0.5, 0.6), (0.6, 0.6, 0.5) >}. Then H is Ng-Feebly c_i-connected (i = 2, 3, 4) but Ng-Feebly C_1-disconnected.

3.10 Illustration

Let Λ = {a, b}, λ = {0_N, K, L} where K = {< (0.2, 0.1, 0.8), (0.1, 0.1,0.9) >}, L = {< (0.3, 0.2, 0.7), (0.2, 0.1, 0.8) >} Consider the Ng-subset H = {< (0.4, 0.4, 0.5), (0.4, 0.2, 0.6) >} and Ng-F-OS U = {< (0.6, 0.7, 0.3), (0.7, 0.3, 0.3) >}, V = {< (0.5, 0.6, 0.4), (0.6, 0.2, 0.4) >}. Then H is Ng-Feebly C_4-connected but Ng-Feebly C_3-disconnected.

3.11 Definition

A Ng-TS (Λ, λ) is Ng-Feebly C_5-disconnected if ∃ Ng-subset J in Λ which is both Ng-F-OS & Ng-F-CS in Λ ∍ J≠0_N, J≠1_N. If J is not Ng-Feebly C_5-disconnected, then it is said to be Ng-Feebly C_5-connected.

3.12 Illustration

Let Λ = {a, b}, λ = {0_N, M} where M = {< (0.3, 0.5, 0.3), (0.7, 0.5, 0.7) >}. Then M is Ng-F-OS & Ng-F-CS in Λ, M≠0_N, M≠1_N. Thus, Λ is Ng-Feebly C_5-disconnected.

3.13 Definition

A Ng-TS (Λ, λ) is called as Ng-connected if Λ has no proper Ng-clopen subsets.

3.14 Theorem

Ng-Feebly C_5-connectedness $\Rightarrow$ Ng-Feebly connectedness.

Proof

On the contrary, suppose that $\exists$ Ng-F-OS

$$A \neq 0_N \,\&\, B \neq 0_N \ni A \vee B = 1_N \,\&\, A \wedge B = 0_N$$

Then

$$Y_A \vee Y_B = 1_N,\ U_A \wedge U_B = 0_N,\ N_A \wedge N_B = 0_N \,\&\, Y_A \wedge Y_B = 0_N,\ U_A \vee U_B = 1_N,\ N_A \vee N_B = 1_N$$

In other words, $B^c = A$. By theorem 5.14, Λ is Ng-Feebly C_5-disconnected. This contradiction proves the result.

The ensuing example shows that the converse of Theorem 3.14 need not be true.

3.15 Illustration

Let $\Lambda = \{e, f\}$, $\lambda = \{0_N, E, F\}$ where $E = \{< (0.4, 0.5, 0.4), (0.6, 0.5, 0.6) >\}$ and $F = \{< (0.7, 0.3, 0.3), (0.7, 0.7, 0.4) >\}$. Then E & F are Ng-F-OS in Λ, $E \vee F \neq 1_N$. Then $E \wedge F \neq 0_N$. Hence Λ is Ng-Feebly connected. Since Ng-subset E in Λ is both Ng-F-OS and Ng-F-CS, Λ is Ng-Feebly c_5-disconnected.

3.16 Definition

Let (Λ_1,λ_1) and (Λ_2,λ_2) be Ng-TSs. Then $\phi:\Lambda_1 \rightarrow \Lambda_2$ is defined to be Ng-Feebly irresolute if the inverse image of Ng-F-OS in $(\Lambda_2,\lambda_2)$is a Ng-F-OS in (Λ_1,λ_1).

3.17 Theorem

Let $\phi:(\Lambda_1,\lambda_1)\rightarrow (\Lambda_2,\lambda_2)$ be a Ng-Feebly irresolute surjection and let Λ_1 be Ng-Feebly connected. Then Λ_2 is Ng-Feebly connected.

Proof

Presume that Λ_2 is not Ng-Feebly connected, then $\exists$ Ng-F-OS

$$A \neq 0_N \;\&\; B \neq 0_N \ni A\vee B = 1_N \;\&\; A\wedge B = 0_N$$

Since Φ is Ng-Feebly irresolute mapping, $U=\phi^{-1}(A) \neq 0_N$, $V=\phi^{-1}(B) \neq 0_N$, which are Ng-F-OS in Λ_1 and

$$\phi^{-1}(A)\vee\phi^{-1}(B) = \phi^{-1}(1_N) = 1_N, \Rightarrow U\vee V = 1_N$$

Also, $\phi^{-1}(A)\wedge\phi^{-1}(B) = \phi^{-1}(0_N) = 0_N$, which implies $U\wedge V = 0_N$. Thus, Λ_1 is Ng-Feebly disconnected, which contradicts the hypothesis. Hence Λ_2 is Ng-Feebly connected.

3.18 Theorem

A Ng-TS (Λ, λ) is Ng-Feebly c_5-connected iff $\exists$ no nonempty Ng-F-OS K & L in $\Lambda \ni K = L^c$.

Proof

Suppose that K & L are Ng-F-OS in $\Lambda \ni K \neq 0_N$, $L \neq 0_N$ and $K = L^c$. Since $K = L^c$, L^c is a Ng-F-OS and follows that L is Ng-F-CS. Also, K, $L \neq 0_N$ & $K = L^c$ $\Rightarrow K \neq 1_N$, which contradicts the fact that Λ is Ng-Feebly c_5-connected.

Conversely, let K be both Ng-F-OS & Ng-F-CS and L is Ng-F-CS in $\Lambda \ni K \neq 0_N$, $L \neq 1_N$. Now take $K^c = L$ which is Ng-F-OS and $K \neq 1_N \Rightarrow K^c = L \neq 0_N$ which is a contradiction. Hence Λ is Ng-Feebly c_5-connected.

3.19 Theorem

A Ng-TS (Λ, λ) is Ng-Feebly connected iff $\exists$ no non-empty Ng-F-OS K & L in $\Lambda \ni K = L^c$.

Proof

Necessity: Let K and L are Ng-F-OS in $\Lambda \ni K \neq 0_N$, $L \neq 0_N$ and $K = L^c$. Consequently, L^c is a Ng-F-CS. Since $K \neq 0_N$, $K \neq 1_N \Rightarrow K$ is a proper Ng-clopen subset in Λ. Hence Λ is not Ng-Feebly connected space, which contradicts the hypothesis. Thus, $\exists$ no non-zero Ng-F-OS K & L in $\Lambda \ni K = L^c$.

Sufficiency: Let K be both Ng-F-OS and Ng-F-CS in $\Lambda \ni K \neq 0_N$, $L \neq 1_N$. Now let $L = K^c$. Then L is a Ng-F-OS and $L \neq 1_N \Rightarrow L = K^c \neq 0_N$. This contradiction to the hypothesis leads that (Λ, λ) is a Ng-Feebly connected space.

3.20 Theorem

A Ng-TS (Λ, λ) is Ng-Feebly connected space iff $\exists$ no non-zero Ng-F-OS K and L in $\Lambda \ni K = L^c$, $L = (Ng - F - cr(K))^c$ and $K = (Ng - F - cr(L))^c$.

Proof

Necessity: Let K and L are Ng-subsets in $\Lambda \ni K \neq 0_N$, $L \neq 0_N$ and $K = L^c$, $L = (Ng - F - cr(K))^c$ and $K = (Ng - F - cr(L))^c$. Since $(Ng - F - cr(K))^c$ and $(Ng - F - cr(L))^c$ are Ng-F-OS in Λ, K and L are Ng-F-OS in Λ. By Theorem 3.19, Λ is not a Ng-Feebly connected space. This contradiction implies that, $\exists$ no non-zero Ng-F-OS K & L in $\Lambda \ni K = L^c$, $L = (Ng - F - cr(K))^c$ and $K = (Ng - F - cr(L))^c$.

Sufficiency: Let K be both Ng-F-OS and Ng-F-CS in $\Lambda \ni K \neq 0_N$ and $K \neq 1_N$. Now by taking $L = K^c$, a contradiction to the hypothesis is obtained. Hence Λ is Ng-Feebly connected space.

3.21 Definition

A Ng-TS (Λ, λ) is Ng-Feebly strongly connected space, if $\exists$ no nonempty Ng-F-CS A & B in

$$\Lambda \ni Y_A + Y_B \geq 1_N,\ U_A + U_B \geq 1_N,\ N_A + N_B \leq 1_N$$

or

$$Y_A + Y_B \geq 1_N, U_A + U_B \leq 1_N, N_A + N_B \leq 1_N$$

In other words, a Ng-TS (Λ, λ) is Ng-Feebly strongly connected, if $\exists$ no nonempty Ng-F-CS A & B in $\Lambda \ni A \wedge B = 0_N$.

3.22 Theorem

A Ng-TS (Λ, λ) is Ng-Feebly strongly connected, if $\exists$ no nonempty Ng-F-OS A & B in Λ, $A \neq 1_N \neq B \ni$

$$Y_A + Y_B \geq 1_N, U_A + U_B \geq 1_N, N_A + N_B \leq 1_N$$

Proof

Let A and B be Ng-F-OS in $\Lambda \ni A \neq 1_N \neq B$ and $Y_A + Y_B \geq 1_N, U_A + U_B \geq 1_N, N_A + N_B \leq 1_N$. If we take $C = A^c$ and $D = B^c$, then C and D become Ng-F-CS in Λ and $C \neq 0_N \neq D$,

$$N_C + N_D = Y_A + Y_B \geq 1_N$$

$$Y_C + Y_D = N_A + N_B \leq 1_N$$

a contradiction. Converse part can be proved using a similar technique as above.

3.23 Theorem

Let $\phi:(\Lambda_1, \lambda_1) \rightarrow (\Lambda_2, \lambda_2)$ be a Ng-Feebly irresolute surjection and let Λ_1 be Ng-Feebly strongly connected. Then Λ_2 is also Ng-Feebly strongly connected.

Proof

Presume that Λ_2 is not Ng-Feebly strongly connected, then ∃ Ng-FCS A and B in $\Lambda_2 \ni A \neq 0_N$, $B \neq 0_N$ and $A \wedge B = 0_N$. Since Φ is Ng-Feebly irresolute mapping, $U = \phi^{-1}(A) \neq 0_N$, $V = \phi^{-1}(B) \neq 0_N$, which are Ng-F-CS in Λ_1. Since Φ is a surjection, $\phi^{-1}(A) \wedge \phi^{-1}(B) = \phi^{-1}(0_N) = 0_N$, which implies $U \wedge V = 0_N$. Thus, Λ_1 is not Ng-Feebly strongly connected. This contradiction to the hypothesis proved that Λ_2 is Ng-Feebly strongly connected.

3.24 Remark

The ensuing examples show that Ng-Feebly strongly connected and Ng-Feebly c_5-connected are independent.

3.25 Illustration

Let $\Lambda = \{a, b\}$, $\lambda = \{0_N, U\}$ where $U = \{<(0.2, 0.2, 0.7), (0.2, 0.4, 0.8)>\}$. Consider Ng-F-OS $V = \{<(0.4, 0.5, 0.4), (0.6, 0.5, 0.6)>\}$ and $W = \{<(0.7, 0.7, 0.3), (0.6, 0.6, 0.3)>\}$. There exists no nonempty Ng-F-CS in Λ ∋ their intersection is empty. Hence Λ is Ng-Feebly strongly connected. But here V is both Ng-F-OS and Ng-F-CS in Λ, $V \neq 0_N$, $V \neq 1_N$. Thus, Λ is Ng-Feebly c_5-disconnected.

3.26 Illustration

Let $\Lambda = \{a, b\}$, $\lambda = \{0_N, A, B\}$ where $A = \{<(1, 0, 0.7), (0.4, 0, 0)>\}$ and $B = \{<(0.6, 0, 0), (1, 0, 0.3)>\}$. Here Λ is Ng-Feebly c_5-connected, but not Ng-Feebly strongly connected, since A and B are Ng-F-CS in Λ such that their intersection is empty.

4. NG-FEEBLY SEPARATED SETS

4.1 Definition

If $K \neq 0_N$ & $L \neq 0_N$ are Ng-subsets in Ng-TS (Λ, λ), then K & L are said to be

(i) Ng-Feebly weakly separated if Ng-F-*cr*(K) $\leq L^c$ and Ng-F-*cr*(L) $\leq K^c$
(ii) Ng-Feebly separated if Ng-F-*cr*(K) $\wedge$ L = K $\wedge$ Ng-F-*cr*(L) $= 0_N$.

4.2 Remark

Any two disjoint non-empty Ng-F-CS are Ng-Feebly separated.

Proof

Suppose K $\neq 0_N$ and L $\neq 0_N$ be two Ng-F-CS $\ni$ K $\wedge$ L $= 0_N$. Then Ng-F-*cr*(K) $\wedge$ L = K $\wedge$ Ng-F-*cr*(L) = K $\wedge$ L $= 0_N$. This confirms that K & L are Ng-Feebly separated.

4.3 Theorem

(i) If K & L are Ng-Feebly separated and $M \leq K, N \leq L, M \leq K, N \leq L$ then M & N are also Ng-Feebly separated.
(ii) If K & L are both Ng-F-OS and if $H = K \wedge L^c$ and $G = L \wedge K^c$, then H & G are Ng- Feebly separated sets.

Proof

(i) Let K & L are Ng-Feebly separated sets in Ng-TS(Λ, λ). Then

$$Ng - F - cr(K) \wedge L = 0_N = K \wedge Ng - F - cr(L)$$

Since

$$M \leq K \ \& \ N \leq L, M \leq K \& N \leq L, Ng - F - cr(M) \leq Ng - F - cr(K)$$

&

$$Ng - F - cr(N) \leq Ng - F - cr(L)$$

This implies,

$Ng - F - cr(M) \wedge N \leq Ng - F - cr(K) \wedge L = 0_N$

and hence

$Ng - F - cr(M) \wedge N = 0_N$

Likewise,

$Ng - F - cr(N) \wedge M \leq Ng - F - cr(L) \wedge K = 0_N$

and hence $Ng - F - cr(N) \wedge M = 0_N$. Therefore M & N are Ng-Feebly separated.

(ii) Let K & L are both Ng-F-OS in Ng-TS(Λ, λ). Then K^c & L^c are Ng-F-CS. Since $H \leq L^c$,

$Ng - F - cr(H) \leq Ng - F - cr(L^c) = L^c$

and consequently $Ng - F - cr(H) \wedge L = 0_N$. Since G≤L,

$Ng - F - cr(H) \wedge G \leq Ng - F - cr(H) \wedge L = 0_N$

and so $Ng - F - cr(H) \wedge G = 0_N$. Correspondingly, $Ng - F - cr(G) \wedge H = 0_N$. Hence H & G are Ng-Feebly separated sets.

4.4 Theorem

The Ng-subsets K & L in Ng-TS (Λ, λ) are Ng-Feebly separated ⇔∃ Ng-F-OS M &

$N \ni K \leq M, L \leq N \,\&\, K \wedge N = 0_N, L \wedge M = 0_N$

Proof

Let K & L be Ng-Feebly separated. Then

$$K \wedge Ng - F - cr(L) = 0_N = Ng - F - cr(K) \wedge L$$

Consider $N = (Ng - F - cr(K))^c$ and $M = (Ng - F - cr(L))^c$. Then M & N are Ng-F-OS ∋ K≤M,

$$L \leq N \;\&\; K \wedge N = 0_N, L \wedge M = 0_N$$

Conversely let M & N are Ng-F-OS ∋ $K \leq M, L \leq N \;\&\; K \wedge N = 0_N, L \wedge M = 0_N$. Then

$$K \leq N^c, L \leq M^c$$

and N^c,M^c are Ng-F-CS. This implies

$$Ng - F - cr(K) \leq Ng - F - cr(N^c) = N^c \leq L^c$$

and

$$Ng - F - cr(L) \leq Ng - F - cr(M^c) = M^c \leq K^c$$

i.e., $Ng - F - cr(K) \leq L^c$ and $Ng - F - cr(L) \leq K^c$.

Consequently $K \wedge Ng - F - cr(L) = 0_N = Ng - F - cr(K) \wedge L$. Hence K & L are Ng-Feebly separated sets.

4.5 Proposition

Each two Ng-Feebly separated sets are always disjoint.

Proof

Let K & L be Ng-Feebly separated. Then,

$K \wedge Ng - F - cr(L) = 0_N = Ng - F - cr(K) \wedge L \Rightarrow K \wedge L = K \wedge Ng - F - cr(L) = 0_N$

and hence K & L are disjoint.

4.6 Theorem

A Ng-TS (Λ,λ) is Ng-Feebly connected $\Leftrightarrow$ $1_N \neq K \vee L$, where K & L are Ng-Feebly separated sets.

Proof

Take on, (Λ,λ) is Ng-Feebly connected space. Presume $1_N = K \vee L$, where K & L are Ng-Feebly separated sets. Then

$K \wedge Ng - F - cr(L) = 0_N = Ng - F - cr(K) \wedge L$

Since

$K \leq Ng - F - cr(K), K \wedge L \leq Ng - F - cr(K) \wedge L = 0_N$

Consequently $Ng - F - cr(K) \leq L^c = K$ &

$Ng - F - cr(L) \leq K^c = L \Rightarrow K = Ng - F - cr(K)$ & $L = Ng - F - cr(L)$

Therefore K & L are Ng-F-CS and hence $K=L^c$ & $L=K^c$ are disjoint Ng-F-OS. Thus

$$K \neq 0_N, L \neq 0_N, K \vee L = 1_N$$

and $K \wedge L = 0_N$, K & L are Ng-F-OS, i.e., (Λ,λ) is not Ng-Feebly connected, which ⇒⇐ the hypothesis. Thus, $1_N \neq K \vee L$.

Conversely, presume that $1_N \neq K \vee L$where K & L are Ng-Feebly separated sets. Suppose (Λ,λ) is not Ng- Feebly connected space, then $1_N = K \vee L$, where $K \neq 0_N, L \neq 0_N, K \wedge L = 0_N$, K & L are Ng-F-OS in Λ. Since

$$K \leq L^c \& L \leq K^c, Ng - F - cr(K) \wedge L \leq L^c \wedge L = 0_N$$

&

$$K \wedge Ng - F - cr(L) \leq K^c \wedge K = 0_N$$

i.e., K & L are Ng-Feebly separated sets, which ⇒⇐ the assumption. Therefore (Λ,λ) is Ng-Feebly connected space.

4.7 Definition

A Ng-TS (Λ,λ) is Ng-Feebly super disconnected if ∃ a Ng-F-ROS J in Λ, $J \neq 0_N$ & $J \neq 1_N$. A Ng-TS Λ is Ng-Feebly super connected if Λ is not Ng-Feebly super disconnected.

4.8 Illustration

Let $\Lambda=\{p,q\}$, $\lambda=\{0_N,K,L\}$ where K= {<(0.3, 0.8, 0.7), (0.2, 0.6, 0.9)>} and L= {<(1,0,0), (0,0,1)>}. Here L is Ng-F-ROS in Λ. Hence Λ is Ng-Feebly super disconnected.

4.9 Theorem

For a Ng-TS (Λ,λ), the following are equivalent:

(a) Λ is Ng-Feebly super connected.

(b) For each Ng-F-OS $K \neq 0_N$ in $\Lambda, Ng - F - cr(K) = 1_N$.
(c) For each Ng-F-CS $L \neq 1_N$ in $\Lambda, Ng - F - ir(L) = 0_N$.
(d) There exists no Ng-F-OS K & L in (Λ,λ), $\ni K \neq 0_N, L \neq 0_N$ & $K \leq L^c$.
(e) There exists no Ng-F-OS K & L in (Λ,λ), $\ni K \neq 0_N, L \neq 0_N, L = (Ng - F - cr(K))^c$ & $K = (Ng - F - cr(L))^c$.
(f) There exists no Ng-F-CS K & L in (Λ,λ), $\ni K \neq 1_N, L \neq 1_N, L = (Ng - F - cr(K))^c$ & $K = (Ng - F - cr(L))^c$.

Proof

(a)⇒(b) Presume ∃ a Ng-F-OS

$$K \neq 0_N, Ng - F - cr(K) \neq 1_N$$

Now take

$$L = Ng - F - ir(Ng - F - cr(K))$$

Then L is proper NG-F-ROS in Λ which ⇒⇐ that Λ is Ng-Feebly super connected.

(b)⇒(c) Let $K \neq 1_N$ be a Ng-F-CS in Λ. If $L = K^c$, then L is Ng-F-OS in Λ and $L \neq 0_N$. Hence

$$Ng - F - cr(K) = 1_N, (Ng - F - cr(L))^c = 0_N \Rightarrow Ng - F - ir(L^c) = 0_N \Rightarrow Ng - F - ir(K) = 0_N$$

(c)⇒(d) Let K & L is Ng-F-OS in Λ $\ni K \neq 0_N \neq L$ & $K \leq L^c$. Since L^c is Ng-F-CS in Λ &

$$L \neq 0_N \Rightarrow L^c \neq 1_N, Ng - F - ir(L^c) = 0_N$$

is obtained. But, from $K \leq L^c$,

$$0_N \neq K = Ng - F - ir(K) \leq Ng - F - ir(L^c) = 0_N$$

which is a ⇒⇐ to the hypothesis.

(d)⇒(a) Let $0_N \neq K \neq 1_N$ be Ng-F-ROS in Λ. If $L = (Ng - F - cr(K))^c$, then $L \neq 0_N$. Otherwise,

$$L \neq 0_N \Rightarrow (Ng - F - cr(K))^c = 0_N$$

That ⇒ $Ng - F - cr(K) = 1_N$. That confirms

$$K = Ng - F - ir(Ng - F - cr(K)) = Ng - F - ir(1_N) = 1_N$$

But this ⇒⇐ to $K \neq 1_N$. Further, $K \leq L^c$, is also a contradiction.

(a)⇒(e) Let K & L be Ng-F-OS in $\Lambda \ni K \neq 0_N \neq L$ &

$$L = (Ng - F - cr(K))^c, K = (Ng - F - ir(L))^c$$

Now

$$Ng - F - ir(Ng - F - cr(K)) = Ng - F - ir(L^c) = (Ng - F - cr(L))^c = K$$

& $0_N \neq K \neq 1_N$. Suppose not, if $K = 1_N$, then

$$1_N = (Ng - F - cr(L))^c \Rightarrow 0_N = Ng - F - cr(L) \Rightarrow L = 0_N$$

which ⇒⇐ the hypothesis.

(e)⇒(a)Let K be Ng-F-OS in

$$\Lambda \ni K = Ng - F - ir(Ng - F - cr(K)), 0_N \neq K \neq 1_N$$

Now $L = (Ng - F - cr(K))^c$ and

$(Ng - F - cr(L))^c = (Ng - F - cr(Ng - F - cr(K))^c)^c = Ng - F - ir(Ng - F - cr(K)) = K$

which ⇒⇐ the hypothesis.

(e)⇒(f)Let K & L be Ng-F-CS in $\Lambda \ni K \neq 1_N \neq L$

$L = (Ng - F - ir(K))^c, K = (Ng - F - ir(L))^c$

Taking M=K^c & N=L^c, M & N become Ng-F-OS in Λ & $M \neq 0_N \neq N$

$(Ng - F - cr(M))^c = (Ng - F - cr(K^c))^c = (Ng - F - ir(K))^c)^c = Ng - F - ir(K) = L^c = N$

& similarly$(Ng - F - cr(N))^c = M$. But this ⇒⇐ the hypothesis.

(f)⇒(e) Proof is similar as above.

5. Ng-SEMI CONNECTED SPACES

5.1 Definition

Let (Λ,λ) be a Ng-TS and

$J = \{ < \lambda, (Y_J(\lambda), U_J(\lambda), N_J(\lambda)) > : \lambda \in \Lambda\}$

be a Ng-subset in Λ. Then Ng-semi Interior of J is defined by

$$Ng - \sigma - ir(J) = \vee \{Z : Z \, is \, Ng - \sigma - OS \, and \, Z \leq J\}$$

and Ng-semi Closure of J is defined by

$$Ng - \sigma - cr(J) = \wedge \{D : D \, is \, Ng - \sigma - CS \, and \, J \leq D\}$$

5.2 Remark

For any Ng-subsets K and L in a Ng-TS(Λ,λ),

(i) $Ng - \sigma - ir(K)$ is the largest Ng-σ-OS in Λ which contained in K.
(ii) $Ng - \sigma - cr(L)$ is the smallest Ng-σ-CS in Λ which contains L.
(iii) K is Ng-σ-OS in Λ $\Leftrightarrow$ $Ng - \sigma - ir(K) = K.$
(iv) If K≤L, then $Ng - \sigma - ir(K) \leq Ng - \sigma - ir(L).$
(v) L is Ng-σ-CS in Λ $\Leftrightarrow$ $Ng - \sigma - cr(L) = L.$
(vi) $Ng - \sigma - cr(Ng - \sigma - cr(L)) = Ng - \sigma - cr(L).$
(vii) If K≤L, then $Ng - \sigma - cr(K) \leq Ng - \sigma - cr(L).$

5.3 Definition

A Ng-set J in Ng-TS(Λ,λ) is called as a Ng-σ-disconnected set if ∃ Ng-σ-OS K, L inΛ $\ni K \vee L = J \; \& \; K \wedge L = 0_{N}.$
If J is not Ng-σ-disconnected, then it is Ng-σ-connected.

5.4 Illustration

Let Λ={p,q}, λ= {0_N, K,L,M} where

$$K = \{< (1, 0.4, 0.4), (0.6, 0.8, 0.5) >\}, \; L = \{< (0, 0.4, 1), (0.6, 0.8, 0.5) >\}$$

and $M = \{< (1, 0, 0.4), (0, 0, 1) >\}$. Here, K, L & M are Ng-σ-OS in Λ, $L \vee M = K$ and $L \wedge M = 0_{N}$. Hence K is a Ng-σ-disconnected set.

5.5 Illustration

Let $\Lambda=\{u,v\}$, $\lambda= \{0_N, K,L,M\}$ where K= {<(0.3, 0.6, 0.5), (0.7, 0.1, 0.2)>}, L= {<(1,0,0.4), (0,0.4,1)>}, M= {<(0,0.3,1), (1,0,0.7)>}. Here, $L \wedge M = 0_N$ and $L \vee M \neq K$. Hence K is a Ng-σ-connected set.

5.6 Definition

Let (Λ_1,λ_1) and (Λ_2,λ_2) be Ng-TSs. Then $\Phi: \Lambda_1 \to \Lambda_2$ is defined to be Ng-σ-irresolute if the inverse image of Ng-σ-OS in (Λ_2,λ_2) is a Ng-σ-OS in (Λ_1,λ_1).

5.7 Theorem

Let Φ: $(\Lambda_1,\lambda_1)\to(\Lambda_2,\lambda_2)$ be Ng-σ-irresolute and K be the Ng-set in Λ_1. If K is Ng-σ-connected, then Φ(K) is Ng-σ-connected.

Proof

Presume that Φ(K) is not Ng-σ-connected, then $\exists$ Ng-σ-OS L, $M \ni L \vee M = \phi(K)$ & $L \wedge M = 0_N$. Thus,

$$K \leq \phi^{-1}(\phi(K)) = \phi^{-1}(L \vee M) = \phi^{-1}(L) \vee \phi^{-1}(M)$$

and

$$\phi^{-1}(L) \wedge \phi^{-1}(M) = \phi^{-1}(L \wedge M) = \phi^{-1}(0_N) = 0_N$$

i.e., $K = \phi^{-1}(L) \vee \phi^{-1}(M)$ & $\phi^{-1}(L) \wedge \phi^{-1}(M) = 0_N$ which $\Rightarrow\Leftarrow$ the hypothesis. Hence Φ(K) is Ng-σ-connected.

5.8 Definition

A Ng-TS (Λ,λ) is Ng-σ-disconnected space if $\exists$ Ng-σ-OS A, B in Λ, $A\neq 0_N$, $B \neq 0_N \ni A \vee B = 1_N$ & $A \wedge B = 0_N$. If Λ is not Ng-σ-disconnected space, then it is Ng-σ-connected space.

5.9 Illustration

Let Λ= {p,q,r}, λ= {0_N, K,L} where K= {<(1,0,0), (0,1,0), (0,1,1)>}, L= {<(0,1,1), (1,0,1), (1,0,0)>}. Here K & L are Ng-σ-OS in Λ, K≠0_N, L≠0_N & $K \vee L = 1_N$, $K \wedge L = 0_N$. Hence Λ is Ng-σ-disconnected space.

5.10 Illustration

Let Λ={u}, λ= {0_N, K,L} where K= {<(0.5, 0.7, 0.3)>} and L= {<0.1, 0.6, 0.8)>}. Then K & L are Ng-σ-OS in Λ, K≠0_N, L≠0_N & $K \vee L \neq 1_N$,$K \wedge L \neq 0_N$. Hence Λ is Ng-σ-connected space.

5.11 Theorem

Let (Λ,λ_1) be a Ng-σ-connected space and $\lambda_2 \leq \lambda_1$. Then (Λ,λ_2) is a Ng-σ-connected space.

Proof

Presume that (Λ,λ_2) is a Ng-σ-disconnected space, then ∃ Ng-σ-OS L, M in λ_2, L≠0_N, $M \neq 0_N \ni L \vee M = 1_N$ & $L \wedge M = 0_N$. As $\lambda_2 \leq \lambda_1$ is given, L & M are in λ_2, L≠0_N, $M \neq 0_N \ni L \vee M = 1_N$ & $L \wedge M = 0_N$, i.e., (Λ,λ_1) is a Ng-σ-disconnected space, which ⇒⇐ the hypothesis. Hence (Λ,λ_2) is Ng-σ-connected space.

5.12 Theorem

Let Φ: $(\Lambda_1,\lambda_1) \rightarrow (\Lambda_2,\lambda_2)$ be Ng-σ-irresolute surjective mapping. If Λ_1 is Ng-σ-connected, then Λ_2 is Ng-σ-connected.

Proof

Presume that Λ_2 is not Ng-σ-connected, then ∃ $Ng - \sigma - OS\ K \neq 0_N$ and

$$L \neq 0_N \ni K \vee L = 1_N \& K \wedge L = 0_N.$$

Thus,

$$\phi^{-1}(K) \vee \phi^{-1}(L) = \phi^{-1}(K \vee L) = \phi^{-1}(1_N) = 1_N$$

and

$$\phi^{-1}(K) \wedge \phi^{-1}(L) = \phi^{-1}(K \wedge L) = \phi^{-1}(0_N) = 0_N.$$

Accordingly, Λ_1 is Ng-σ-disconnected, which ⇒⇐ the hypothesis. Hence Λ_2 is Ng-σ-connected.

6. Ng-SEMI SEPARATED SETS

6.1 Definition

Two Ng-sets $K \neq 0_N$ and $L \neq 0_N$ in a Ng-TS(Λ,λ) are called Ng-Semi Separated (Ng-σ-separated) Sets if

$$Ng - \sigma - cr(K) \wedge L = K \wedge Ng - \sigma - cr(L) = 0_N$$

6.2 Theorem

Any two disjoint non-empty Ng-σ-CS in a Ng-TS(Λ,λ) are Ng-σ-separated sets.

Proof

Let $K \neq 0_N$ and $L \neq 0_N$ be two $Ng - \sigma - CS \ni K \wedge L = 0_N$. Then

$$Ng - \sigma - cr(K) \wedge L = K \wedge Ng - \sigma - cr(L) = K \wedge L = 0_N.$$

Thus K & L are Ng-σ-separated sets.
However, Ng-σ-separated sets are not Ng-σ-CS as shown below.

6.3 Illustration

Let Λ={u,v}, λ= {0_N, K, L, M, N} where K= {<(0.5,1,0.3), (0.6,1,0.3)>}, L= {<(0,0.2,1), (0,0,1)>}, M= {<(1,0.7,0), (1,0.5,0)>}, N= {<(0.5,0.7,0.3), (0.6,0.5,0.3)>}. Here,

$Ng-\sigma-cr(K^c) \;=\; K^c$ and $Ng-\sigma-cr(L) \;=\; M^c$.

Thus

$Ng-\sigma-cr(K^c)\wedge L \;=\; K^c\wedge L \;=\; 0_N$ and $K^c\wedge Ng-\sigma-cr(L) \;=\; K^c\wedge M^c \;=\; 0_N$.

Hence K^c & L are Ng-σ-separated sets but L is not a Ng-σ-CS.

6.4 Theorem

(i) If K & L are Ng-σ-separated sets and M≤K, N≤L, then M & N are also Ng-σ-separated sets.
(ii) If K & L are both Ng-σ-OS and if $H \;=\; K\wedge L^c$ and $G \;=\; L\wedge K^c$, then H & G are Ng-σ-separated sets.

Proof

(i) Let K & L are Ng-σ-separated sets in Ng-TS(Λ,λ). Then

$Ng-\sigma-cr(K)\wedge L \;=\; 0_N$

$=\; K\wedge Ng-\sigma-cr(L).$

Since M≤K and N≤L, by Remark 5.2 (viii),

$Ng - \sigma - cr(M) \le Ng - \sigma - cr(K)$ and $Ng - \sigma - cr(N) \le Ng - \sigma - cr(L)$

This

$\Rightarrow Ng - \sigma - cr(M) \wedge N \le Ng - \sigma - cr(K) \wedge L = 0_N$

and hence $Ng - \sigma - cr(M) \wedge N = 0_N$. Likewise,

$Ng - \sigma - cr(N) \wedge M \le Ng - \sigma - cr(L) \wedge K = 0_N \Rightarrow Ng - \sigma - cr(N) \wedge M = 0_N$.

Therefore M & N are Ng-σ-separated.

(ii) Let K & L are both Ng-σ-OS in Ng-TS(Λ,λ). Then K^c & L^c are Ng-σ-CS. Since $H \le L^c$,

$Ng - \sigma - cr(H) \le Ng - \sigma - cr(L^c) = L^c$

and so$Ng - \sigma - cr(H) \wedge L = 0_N$. Since

$G \le L, cr(H) \wedge G \le Ng - \sigma - cr(H) \wedge L = 0_N \Rightarrow Ng - \sigma - cr(H) \wedge G = 0_N$.

$Ng - \sigma -$

Correspondingly, $Ng - \sigma - cr(G) \wedge H = 0_N$. Hence H & G are Ng-σ-separated sets.

6.5 Theorem

The Ng-subsets K & L in Ng-TS(Λ,λ) are Ng-σ-separated

$\Leftrightarrow \exists Ng - \sigma - OS\, M \ \& \ N \ni K \leq M, L \leq N \ \& \ K \wedge N = 0_N, L \wedge M = 0_N.$

Proof

Let K & L be Ng-σ-separated. Then

$K \wedge Ng - \sigma - cr(L) = 0_N = Ng - \sigma - cr(K) \wedge L.$

Consider $N = (Ng - \sigma - cr(K))^c$ and $M = (Ng - \sigma - cr(L))^c$.Then M & N are Ng-σ-OS, K≤M,

$L \leq N \ \& \ K \wedge N = 0_N, L \wedge M = 0_N.$

Conversely let M & N are Ng-σ-OS∋$K \leq M, L \leq N \ \& \ K \wedge N = 0_N$, $L \wedge M = 0_N$. Then $K \leq N^c$, $L \leq M^c$ and N^c, M^c are Ng-σ-CS. This $\Rightarrow Ng - \sigma - cr(K) \leq Ng - \sigma - cr(N^c) = N^c \leq L^c$ and

$Ng - \sigma - cr(L) \leq Ng - \sigma - cr(M^c) = M^c \leq K^c,$

i.e.$Ng - \sigma - cr(K) \leq L^c$ and $Ng - \sigma - cr(L) \leq K^c$.
Consequently $K \wedge Ng - \sigma - cr(L) = 0_N = Ng - \sigma - cr(K) \wedge L$. Hence K & L are Ng-σ-separated sets.

6.6 Proposition

Each two Ng-σ-separated sets are always disjoint.

Proof

Let K & L be Ng-σ-separated. Then,

$K \wedge Ng - \sigma - cr(L) = 0_N = Ng - \sigma - cr(K) \wedge L \Rightarrow K \wedge L \leq K \wedge Ng - \sigma - cr(L) = 0_N$

& hence $K \wedge L = 0_N$ i.e., K & L are disjoint.

The reverse of the above proposition need not be true as shown by the ensuing example.

6.7 Illustration

Let Λ={u,v}, λ= {0_N, K, L, M, N} where K= {<(0.3,0,0.5), (0,0.3,1)>}, L= {<(0,0.4,1),(0.6,0.3,0.8)>},M={<0.3,0.4,0.5),(0.6,0.3,0.8)>},N={<(0.5,0.7,0.3), (0.6,0.5,0.3)>}. Here, $K \wedge L = 0_N$ i.e., K & L are disjoint.

Now $Ng - \sigma{-}cr(K) = M^c$ and $Ng - \sigma - cr(L) = M^c$. Thus

$Ng - \sigma - cr(K) \wedge L = M^c \wedge L = L \neq 0_N$ and $K \wedge Ng - \sigma - cr(L) = K \wedge M^c = K \neq 0_N$.

Hence K & L are not Ng-σ-separated sets.

6.8 Theorem

A Ng-subset M of Ng-TS(Λ,λ) is Ng-σ-connected ⇔ $M \neq K \vee L$, where K & L are Ng-σ-separated sets.

Proof

Take on, (Λ,λ) is Ng-σ-connected space. Presume $M = K \vee L$,where K & L are Ng-σ-separated sets. Then

$$K \wedge Ng - \sigma - cr(L) = 0_N = Ng - \sigma - cr(K) \wedge L.$$

Since $K \leq Ng - \sigma - cr(K), K \wedge L \leq Ng - \sigma - cr(K) \wedge L = 0_N$.

Consequently $K \wedge L = 0_N, Ng - \sigma - cr(K) \leq L^c = K$,

$$Ng - \sigma - cr(L) \leq K^c = L \Rightarrow K = Ng - \sigma - cr(K) \;\&\; L = Ng - \sigma - cr(L).$$

So, K & L are Ng-σ-CS and hence $K=L^c$ & $L=K^c$ are disjoint Ng-σ-OS which ⇒⇐ the hypothesis M is not Ng-σ-connected. Thus, $M \neq K \vee L$.

Conversely, presume that $M \neq K \vee L$,where K & L are Ng-Feebly separated sets. Suppose (Λ,λ) is not Ng-σ-connected space, then $M = K \vee L$,where K & L are $Ng - \sigma - OS \ni K \neq 0_N$, $L \neq 0_N$ & $K \wedge L = 0_N$. Since $K \leq L^c$ & $L \leq K^c$,

$$Ng - \sigma - cr(K) \wedge L \leq L^c \wedge L = 0_N \ \& \ K \wedge Ng - \sigma - cr(L) \leq K^c \wedge K = 0_N$$

i.e., K & L are Ng-σ-separated sets, which ⇒⇐ the assumption. Therefore (Λ,λ) is Ng-σ-connected space.

6.9 Theorem

A Ng-TS(Λ,λ) is Ng-σ-connected ⇔ $1_N \neq K \vee L$, where K & L are Ng-σ-separated sets.

Proof

Take on, (Λ,λ) is Ng-σ-connected space. Presume $1_N = K \vee L$, where K & L are Ng-σ-separated sets. Then

$$K \wedge Ng - \sigma - cr(L) = 0_N = Ng - \sigma - cr(K) \wedge L.$$

Since $K \leq Ng - \sigma - cr(K), K \wedge L \leq Ng - \sigma - cr(K) \wedge L = 0_N$.
Consequently $K \wedge L = 0_N, Ng - \sigma - cr(K) \leq L^c = K$,

$$Ng - \sigma - cr(L) \leq K^c = L \Rightarrow K = Ng - \sigma - cr(K)$$

& L=Ng-σ-*cr*(L). So, K & L are Ng-σ-CS and hence $K=L^c$ & $L=K^c$ are disjoint Ng-σ-OS, i.e., 1_N is not Ng-σ-connected, which is a ⇒⇐ to (Λ,λ) is Ng-σ-connected space. Thus, $1_N \neq K \vee L$.

Conversely, presume that $1_N \neq K \vee L$, where K & L are Ng-σ-separated sets. Suppose (Λ,λ) is not Ng-σ-connected space, then $1_N = K \vee L$, where K & L are Ng-σ-OS ∋ $K \neq 0_N$, $L \neq 0_N$ & $K \wedge L = 0_N$. Since $K \leq L^c$ & $L \leq K^c$,

$$Ng - \sigma - cr(K) \wedge L \leq L^c \wedge L = 0_N \;\&\; K \wedge Ng - \sigma - cr(L) \leq K^c \wedge K = 0_N$$

i.e., K & L are Ng-σ-separated sets, which ⇒⇐ the assumption. Therefore (Λ,λ) is Ng-σ-connected space.

6.10 Theorem

If $M \leq K \vee L$, where M is a Ng-σ-connected set and K, L are Ng-σ-separated sets, then either M≤K or M≤L.

Proof

Take on M≰K & M≰K. Let $M_1 = K \wedge M$ & $M_2 = L \wedge M$. Then M_1 & M_2 are non-empty Ng-sets and

$$M_1 \vee M_2 = (K \wedge M) \vee (L \wedge M) = (K \vee L) \wedge M = M,$$

because $M \leq K \vee L$. Since $M_1 \leq K$, $M_2 \leq K$, K & L are Ng-σ-separated sets,

$$Ng - \sigma - cr(M_1) \wedge M_2 \leq Ng - \sigma - cr(K) \wedge L = 0_N$$

&

$$M_1 \wedge Ng - \sigma - cr(M_2) \subseteq K \wedge Ng - \sigma - cr(L) = 0_N.$$

Therefore M_1 & M_2 are Ng-σ-separated sets $\exists M = M_1 \vee M_2$. Hence by Theorem 6.8, M is Ng-σ-disconnected. This is a ⇒⇐ to M is Ng-σ-connected. Thus, either $M \leq K$ or $M \leq L$.

6.11 Theorem

If M is a Ng-σ-connected set, then Ng-σ-*cr*(M) is also a Ng-σ-connected set.

Proof

Let M be a Ng-σ-connected set and Ng-σ-*cr*(M) be Ng-σ-disconnected. Then by Theorem 6.8, ∃Ng-σ-separated sets H & K ∋ $Ng - \sigma - cr(M) = H \vee K$. Since M is Ng-σ-connected & $M \leq Ng - \sigma - cr(M) = H \vee K$, by Theorem 6.10, either M≤H or M≤K. Since M≤H,

$$Ng - \sigma - cr(M) \leq Ng - \sigma - cr(H).$$

Again, since H & K are Ng-σ-separated sets, $H \neq 0_N \neq K$ &

$$Ng - \sigma - cr(M) \wedge K \leq Ng - \sigma - cr(H) \wedge K = 0_N.$$

Thus $K \leq (Ng - \sigma - cr(M))^c$. Also, $K \leq H \vee K = Ng - \sigma - cr(M)$. Consequently

$$K \subseteq (Ng - \sigma - cr(M))^c \wedge Ng - \sigma - cr(M) = 0_N$$

which is a ⇒⇐ to $K \neq 0_N$. Likewise if M≤K, which is a ⇒⇐ to $H \neq 0_N$ is obtained. Therefore Ng-σ-*cr*(M) is Ng-σ-connected.

6.12 Theorem

Let K be a Ng-σ-connected in Ng-TS(Λ,λ). If K≤L≤Ng-σ-*cr*(K), then L is Ng-σ-connected.

Proof

Presume L is Ng-σ-disconnected. Then ∃Ng-σ-separated sets M & N ∋M≠0_N, N≠0_N & $L = M \vee N$. Since $K \leq L = M \vee N$ and by Theorem 6.10, K≤M or K≤N. Let K≤M. Then, Ng-σ-*cr*(K)≤Ng-σ-*cr*(M). Now,

$$Ng - \sigma - cr(K) \wedge N \leq Ng - \sigma - cr(M) \wedge N = 0_N.$$

Thus $Ng - \sigma - cr(K) \wedge N = 0_N$. Also, $M \vee N = L \leq Ng - \sigma - cr(K)$, $N \leq L \subseteq Ng - \sigma - cr(K)$.Therefore, $Ng - \sigma - cr(K) \wedge N = N$. Hence N=$0_N$, which is a ⇒⇐ to N≠$0_N$. Hence L is Ng-σ-connected.

6.13 Theorem

If K & L are Ng-σ-connected in Λ ∋ $K \wedge L \neq 0_N$, then $K \vee L$ is Ng-σ-connected in Λ.

Proof

Let $K \vee L$ be Ng-σ-disconnected. Using Theorem 6.8, ∃Ng-σ-separated sets M, N∋ $K \vee L = M \vee N$. Since M & N are Ng-σ-separated sets, M & N are non-empty Ng-sets and $M \wedge N \leq Ng - \sigma - cr(M) \wedge N = 0_N$. Since

$$K \leq K \vee L = M \vee N, L \leq K \vee L = M \vee N$$

and K, L are Ng-σ-connected, by Theorem 6.10, $K \leq M \; or \; K \leq N \; \& \; L \leq M \; or \; L \leq N$.

(1) If K≤M & L≤M, then $K \vee L \leq M$. Thus $K \vee L = M$. Since $M \wedge N = 0_N$, N=0_N, which ⇒⇐ N≠0_N. Likewise, if K≤N & L≤N, ⇒⇐ will be occurred.
(2) If K≤M & L≤N, then $K \wedge L \leq M \wedge N = 0_N$. Therefore $K \wedge L = 0_N$. This is a ⇒⇐ to $K \wedge L \neq 0_N$. Correspondingly, a ⇒⇐ will be occurred if K≤N & L≤M. So, $K \vee L$ is Ng-σ-connected in Λ.

6.14 Theorem

Let $\phi:(\Lambda_1, \lambda_1) \rightarrow (\Lambda_2, \lambda_2)$ be a Ng-σ-irresolute surjection and let Λ_1 be Ng-σ-connected. Then Λ_2 is Ng-σ-connected.

Proof

Presume that Λ_2 is not Ng-Feebly connected, then $\Lambda_2 = K \vee L$, where K & L are Ng-σ-OS in $\Lambda_2 \ni K \neq 0_N \neq L$ & $K \wedge L = 0_N$. Since Φ is Ng-σ-irresolute mapping, $\phi^{-1}(K) \vee \phi^{-1}(L) = 1_N$, where $\phi^{-1}(K)$ & $\phi^{-1}(L)$ are Ng-σ-OS in $\Lambda_1 \ni \phi^{-1}(K) \neq 0_N \neq \phi^{-1}(L)$ & $\phi^{-1}(K) \wedge \phi^{-1}(L) = 0_N$. Thus, Λ_1 is Ng-σ-disconnected, which $\Rightarrow\Leftarrow$ the hypothesis. Hence Λ_2 is Ng-σ-connected.

7. APPLICATIONS IN DIVERSE REALMS

The significance and applications of connectedness and feebly connectedness in Generalized Neutrosophic Topological spaces over various fields, like Knot theory, Chaos theory, Epidemiology, Cosmology are highlighted.

Knot Theory

In mathematics, Knot theory is the study of closed curves in 3D, and their probable deformations without one part cutting through another. The Ambient space refers to a larger space in which a particular mathematical object or system is embedded. Topological methods are crucial for classification, identification of invariants, and understanding the fundamental characteristics of knots and links.

Significance of Feebly Connectedness in Knot Theory

The spaces are connected if there are no disjoint non-empty open sets whose union is the space. In knot theory, feebly connectedness plays a crucial role in determining the properties and behaviors of knots and links through continuous deformations or isotopies within their ambient spaces. If two knots can be continuously deformed into each other within the same connected topological space, they are considered equivalent or isotopic.

The space in which knots and links are embedded is often considered to be three-dimensional Euclidean space. When studying knots, mathematicians often focus on the complement of the knot or link within its ambient space. The knot complement is obtained by removing the interior of the knot or link from the ambient space.

The resulting space is still connected unless the knot/link is unknotted (trivial), in which case the complement consists of two disconnected components (the removed knot and the rest of the space).

Applications for Knot Theory in Various Fields

Specifically, because of Knot theory's theoretical nature, it is frequently used to frame mathematical models of more tangible concepts in various fields such as DNA modeling, effects of enzymes, and statistical mechanics. Knot theory helps in making these concrete perceptions more abstract and easier to handle.

DNA Modeling: The strands of DNA contain numerous knots and links owing to its physical properties and connections. Certain DNA knots can obstruct some biological processes and lead to genomic variability or ailments. Feebly connectedness in Knot theory gives a way to categorize, examine these diverse configurations and to understand genomic stability.

Effects of Enzymes: Topoisomerases enzymes employ the DNA topology by creating or eradicating knots as well as links. Understanding the existence and prevalence of knots in DNA helps to identify how these enzymes affect DNA topology during replication, transcription, and re-amalgamation.

Statistical Mechanics: Statistical mechanism frequently utilizes Monte Carlo simulations to frame the behavior of polymer. Connectedness idea in Knot theory supports in developing algorithms in order to simulate polymer chains and forecast the possibility of knotting under various circumstances.

Chaos Theory

In chaos theory, connected topology mentions the physical and geometric properties of dynamical systems and reconnoiters the performance of nonlinear systems that expose unpredictable habit over time.

Significance of Feebly Connectedness in Chaos Theory

Chaos theory deliberates the topology of the phase space that signifies all probable states a system can occupy, and it determines the system's performance and nature of its attractors. Understanding the feebly connectivity in topology assists in describing chaotic comfort.

In chaos theory, the "butterfly effect", which means the minor changes on initial conditions lead to extremely distinct trajectories in phase space, is essential. The connected topology of the phase space impacts the sensitivity of the system related to the initial perturbations.

Applications of Chaos Theory in Various Fields

Ergodic Theory: Based on time, how a system reconnoiters its phase space, and the affinity of its orbits is explained by this theory. Chaos theory looks into topological characteristics like ergodicity and mixing rates in defining how trajectories travel in the phase space, provide a route to the chaotic behavior of the system.

Poincare Sections: An effective method of examining chaotic motion is to refer what is called the Poincare section. Analyzing the topological connectivity of these sections supports learning the trajectories structure of the system and recognizing patterns linked with chaos.

Epidemiology

Topological Connectedness in epidemiology aids in analyzing the super-spreader disease and transmission within clusters, considering the interconnectedness, communication patterns and internal connections among people. Mathematical models used in epidemiology foresee the evolution of epidemics to measure intervention tactics such as social distancing or immunization.

Cosmology

In cosmology, space exhibits a multi-connected topology, where distant regions of the universe is correlated by spatial frames. Experimental and theoretical surveys proceed to discover the attainability of connected topology, targeting to decrypt the factual structure of cosmos beyond its noticeable boundaries.

8. CONCLUSION

In this article, initially, Feebly Connectedness in Neutrosophic Generalized Topological Spaces (Ng-TS) is discussed. Then, Ng-Feebly C_i-disconnectedness ($i = 1, 2, 3, 4$), Ng-Feebly irresolute function, Ng-Feebly strongly connectedness is established and their characteristics are analyzed with apt paradigms. Thereafter, its impact across various realms such as Knot theory, Chaos theory, Epidemiology, Cosmology is deliberated. The researchers can make use of these notions for further analysis in different fields.

REFERENCES

Arar, M., & Jafari, S. (2020). Neutrosophic μ-Topological spaces. *Neutrosophic Sets and Systems*, *38*, 51–66.

Atanassov, K. T. (1986). Intuitionistic fuzzy sets. *Fuzzy Sets and Systems*, *20*, 87–96.

Bageerathi, K., & Jeya Puvaneswari, P. (2019). Neutrosophic Feebly Connectedness and Compactness. *IOSR Journal of Polymer and Textile Engineering*, *6*(3), 7–13.

Csaszar, A. (2002). Generalized topology, generalized continuity. *Acta Mathematica Hungarica*, *96*, 351–357.

Csaszar, A. (2004). Extremally disconnected generalized topologies. *Annales Univ. Sci. Budapest.*, *47*, 91–96.

Faris, & Smarandache. (2016). New Neutrosophic Sets via Neutrosophic Topological Spaces. *New Trends in Neutrosophic Theory and Applications*, (2), 1–10.

Jeya Puvaneswari, P., & Bageerathi, K. (2017). On Neutrosophic Feebly open sets in Neutrosophic topological spaces. *International Journal of Mathematics Trends and Technology*, *41*(3), 230–237.

Jeya Puvaneswari, P., & Bageerathi, K. (2019). Some Functions Concerning Neutrosophic Feebly Open & Closed Sets. *International Journal of Scientific Research and Reviews*, *8*(2), 1546–1559.

Raksha Ben, N., & Hari Siva Annam, G. (2021). Some new open sets in μ_N topological space. *Malaya Journal of Matematik*, *9*(1), 89–94.

Raksha Ben, N., & Hari Siva Annam, G. (2021). Generalized Topological Spaces via Neutrosophic Sets. *J. Math. Comput. Sci.*, *11*, 716–734.

Salama, A. A., & Albowi, S. A. (2012). Neutrosophic set and Neutrosophic Topological Space. *ISOR J. Mathematics*, *3*(4), 31–35.

Smarandache, F. (1999). *A Unifying Field in Logic: Neutrosophic Logic, Neutrosophy, Neutrosophic set, Neutrosophic Probability*. American Research Press.

Smarandache, F. (2002). Neutrosophy and Neutrosophic Logic. *First International Conference on Neutrosophy*, Neutrosophic Logic, Set, Probability, and Statistics University of New Mexico.

Smarandache, F. (2010). Neutrosophic Set: A Generalization of Intuitionistic Fuzzy set. *Journal of Defense Resources Management*, *1*, 107–116.

Zadeh, L. A. (1965). Fuzzy set. *Information and Control*, *8*, 338–353.

Chapter 5
On Fuzzy Hypersoft Relations

Adem Yolcu
https://orcid.org/0000-0002-4317-652X
Kafkas University, Turkey

Taha Yasin Ozturk
https://orcid.org/0000-0003-2402-6507
Kafkas University, Turkey

ABSTRACT

Fuzzy hypersoft set theory is a current topic that has been intensively studied by researchers in recent years. In this chapter, the concept of fuzzy hypersoft Cartesian product is defined differently from other studies. In the definition presented, both parameters and alternatives are multiplied. Based on the concept of Cartesian product, the concept of relation is constructed. In addition to studying a few related features, the authors have proposed various theories on fuzzy hypersoft cartesian products, relations, and functions together with supporting data and examples. Finally, injective, surjective, bijective, and constant fuzzy hypersoft function ideas have also investigated.

1. INTRODUCTION

Today, most formal modelling, computing, and reasoning tools in mathematics are of a clear, accurate, and predictable character. However, in real-world situations, issues in a variety of disciplines including economics, engineering, the environment, social science, and medicine do not always involve crystal-clear facts. Because of the several types of uncertainties that are present in traditional classical procedures, we

DOI: 10.4018/979-8-3693-3204-7.ch005

are unable to resolve these problems using them. These worries have been dispelled by a number of ideas, including the theory of fuzzy sets (Zadeh, 1965), intuitionistic fuzzy sets (Atanassov, 1999), and rough sets (Pawlak, 1982). These theories could be utilized as mathematical instruments to assist us in navigating uncertainties. However, as Molodtsov pointed out in (Molodtsov, 1999), each of these concepts has inherent problems of its own. These difficulties can be brought on by the inadequate parametrization tool for the theories. Molodtsov created the concept of soft set theory (Molodtsov, 1999) as a novel mathematical strategy for dealing with ambiguity and uncertainty that is free from the aforementioned problems as a result.

A set-valued map is what a soft set is, which is a description of the domain of discourse based on a few criteria. Since it is more general than the formers, the theory of soft sets has attracted a great deal of attention from academics and has undergone significant development. Around 2000, Maji et. al. (Maji et. al., 2002; Maji et.al., 2003) started researching soft set theory after being introduced to it by Molodtsov. The authors employed this strategy in (Maji et. al., 2002) to successfully address a problem involving decision-making after defining and studying certain basic concepts in maji, such as complement, union, and intersection of soft sets. Ali et al. (Ali et.al., 2009) and Aktas and Cagman (Aktas&Cagman, 2007), Sezgin and Atagun (Sezgin&Atagun, 2011) and Maji et al. (Maji et.al., 2003) improved the work of Maji et al. (Maji et.al., 2001), who also proposed several new operations and features of soft set theory. Although soft set theory may offer solutions to numerous problems, its inherent complexity still exists. This theory cannot be applied to resolve real-world problems since they entail unclear information and an uncertain environment, which fuzzy set theory models. They combined fuzzy set theory and soft set theory to produce a theory of soft sets, and they came up with the phrase "fuzzy soft set" to describe a fuzzy version of soft sets. Then, in (Roy&Maji, 2007) they discussed actual applications of fuzzy soft sets in problem-solving contexts. Ahmad and Kharal (Ahmad&Kharal, 2009; Kharal&Ahmad, 2009) made additional contributions to the features of fuzzy soft sets and fuzzy soft mappings. Numerous scholars, including (Ali, 2011; Cagman et.al., 2011; Roy&Maji, 2007; Yolcu & Ozturk, 2021), have previously illustrated some of the numerous uses for soft set and fuzzy soft set theories. Numerous academics, including Tanay and Kandemir (Tanay&Kandemir, 2011), Kharal and Ahmad (Kharal&Ahmad, 2009), Celik et. al. (Celik et.al 2013), examined the idea of fuzzy soft function.

Samarandache (Smarandache, 2018) introduced the hypersoft set (HSS) technique in 2018 as an improvement to a soft set. The primary reason for utilizing Hypersoft Set (HSS) is because the SS environment cannot be used to handle scenarios when the characteristics are more than one and further bisected. Saeed et al. (Saeed et.al, 2021), Saeed et al. (Saeed et.al, 2022), Rahman et al. (Rahman et.al, 2020) and Yolcu and Ozturk (Yolcu & Ozturk, 2021) explored the fundamental principles of the HSS and some of their entire setups in an HSS. There have been many studies on hypersoft, fuzzy hypersoft and other hybrid structures (Ahsan et.al, 2021; Saeed et. al., 2021b; Saeed et. al., 2022; Saeed et. al., 2023; Yolcu et.al., 2021; Yolcu and Ozturk, 2022; Yolcu A., 2023)

These facts serve as our inspiration as we create a unique method for fuzzy hypersoft relations in this research. With the use of numerical examples, authors describe it. Authors also illustrate some of its key characteristics. To sum up, the following are the driving forces behind this chapter:

1. The Cartesian product serves as the foundation for the idea of function, as is well known. The shift from Cartesian product to relation and from relation to function is well defined in this paper.
2. The main drawbacks of the current fuzzy hypersoft and other hybrid fuzzy models are: There is no description of the idea of function that successfully uses both alternatives and parameters, despite the fact that functions are built on Cartesian products. The proposed structure is made more functional in terms of relations and functions with the binary multiplication of both parameters and alternatives, and it will be a useful tool for applications.

The rest of this chapter is structured as follows. Section 2 reviews some basic notions, including fuzzy hypersoft sets. Section 3 proposes a new approach to fuzzy hypersoft cartesian product, fuzzy hypersoft relations and fuzzy hypersoft functions. Also, based on this new approach, some important theorems are given and several examples are presented.

2. PRELIMINARIES

Definition 2.1 (Smarandache, 2018) (Hypersoft Set) Let Z be the universal set and $P(Z)$ be the power set of Z. Consider $e_1,e_2,e_3,\ldots,e_n$ for $n\geq1$, be n well-defined attributes, whose corresponding attribute values are respectively the sets $E_1,E_2,\ldots,E_n$ with $E_i \cap E_j = \varnothing$, for $i\neq j$ and $i,j\in\{1,2,\ldots,n\}$, then the pair $(L, E_1 \times E_2 \times \ldots \times E_n)$ is said to be Hypersoft set over Z where $L{:}E_1 \times E_2 \times \ldots \times E_n \rightarrow P(Z)$.

Definition 2.2 (Yolcu & Ozturk, 2021) (Fuzzy Hypersoft Set) Let Z be the universal set and $FP(Z)$ be a family of all fuzzy set over Z and $E_1,E_2,\ldots,E_n$ the pairwise disjoint sets of parameters. Let A_i be the nonempty subset of E_i for each $i=1,2,\ldots,n$. A fuzzy hypersoft set defined as the pair $(L,A_1 \times A_2 \times \ldots \times A_n)$ where; $L{:}A_1 \times A_2 \times \ldots \times A_n \rightarrow FP(Z)$ and

$$L(A_1 \times A_2 \times \ldots \times A_n)$$
$$= \{\langle u, L(\alpha)(u)\rangle : u \in Z, \alpha \in A_1 \times A_2 \times \ldots \times A_n \subseteq E_1 \times E_2 \times \ldots \times E_n\}$$

For sake of simplicity, we write the symbols Σ for $E_1 \times E_2 \times \ldots \times E_n$, Λ for $A_1 \times A_2 \times \ldots \times A_n$ and α for an element of the set Λ. The set of all fuzzy hypersoft sets over Z will be denoted by $FHS(Z,\Sigma)$. Here after, FHS will be used for short instead of fuzzy hypersoft sets.

Definition 2.3 (Yolcu & Ozturk, 2021)

i) A fuzzy hypersoft set (L,Σ) over the universe Z is said to be null fuzzy hypersoft set and denoted by $0_{(Z_{FH},\Sigma)}$ if for all $u \in Z$ and all $\alpha \in \Sigma$, $L(\alpha)(u)=0$.
ii) A fuzzy hypersoft set (L,Σ) over the universe Z is said to be absolute fuzzy hypersoft set and denoted by $1_{(Z_{FH},\Sigma)}$ if for all $u \in Z$ and $\alpha \in \Sigma$, $L(\alpha)(u)=1$.

Definition 2.4 (Yolcu & Ozturk, 2021) Let Z be an initial universe set (L_1,Λ_1), (L_2,Λ_2) be two fuzzy hypersoft sets over the universe Z. We say that (L_1,Λ_1) is a fuzzy hypersoft subset of (L_2,Λ_2) and denote $(L_1,\Lambda_1)\tilde{\subseteq}(L_2,\Lambda_2)$ if

i) $\Lambda_1 \subseteq \Lambda_2$
ii) For any $\alpha \in \Lambda_1$, $L_1(\alpha) \subseteq L_2(\alpha)$.

Definition 2.5 (Yolcu & Ozturk, 2021) Let Z be an initial universe set (L_1,Λ_1), (L_2,Λ_2) be two fuzzy hypersoft sets over the universe Z. The union of (L_1,Λ_1) and (L_2,Λ_2) is denoted by $(L_1,\Lambda_1)\tilde{\cup}(L_2,\Lambda_2) = (L_3,\Lambda_3)$ where $\Lambda_3 = \Lambda_1 \cup \Lambda_2$ and

$$L_3(\alpha) = \begin{cases} L_1(\alpha) & if\ \alpha \in \Lambda_1 - \Lambda_2 \\ L_2(\alpha) & if\ \alpha \in \Lambda_2 - \Lambda_1 \\ \max\{ L_1(\alpha), L_2(\alpha)\} & if\ \alpha \in \Lambda_1 \cap \Lambda_2 \end{cases}$$

Definition 2.6 (Yolcu & Ozturk, 2021) Let Z be an initial universe set and (L_1,Λ_1), (L_2,Λ_2) be fuzzy hypersoft sets over the universe Z. The intersection of (L_1,Λ_1) and (L_2,Λ_2) is denoted by $(L_1,\Lambda_1)\tilde{\cap}(L_2,\Lambda_2) = (L_3,\Lambda_3)$ where $\Lambda_3 = \Lambda_1 \cap \Lambda_2$ and each

$$\alpha \in \Lambda_3, L_3(\alpha)(u) = \min\{L_1(\alpha)(u), L_2(\alpha)(u)\}.$$

Definition 2.7 (Yolcu & Ozturk, 2021) The complement of fuzzy hypersoft set (L,Λ) over the universe Z is denoted by $(L,\Lambda)^c$ and defined as $(L,\Lambda)^c=(L^c,\Lambda)$, where

$$L^c(\Lambda) = \{\langle u, 1 - L(\alpha)(u)\rangle : u \in Z, \alpha \in \Lambda\}.$$

Definition 2.8 (Abbas et.al., 2020) Let $\Lambda \subset \Sigma$, $\alpha \in \Lambda$ and $x \in Z$. A FHS (L,Λ) is said to be a fuzzy hypersoft point (briefly, FHP) if $L(\alpha')$ is a null fuzzy set for every $\alpha' \in \Lambda \backslash \{\alpha\}$ and $L(y)=0$ for all $y \neq x$. We will denote (L,Λ) simply by $P_{FH}^{(\alpha,x)}$ and denote all the fuzzy hypersoft points over Z simply by $FHP(Z,\Sigma)$.

Definition 2.9 (Abbas et.al., 2020) A FHP $P_{FH}^{(\alpha,x)}$ is said to belong to a FHS (L,Λ) if $P_{FH}^{(\alpha,x)} \tilde{\subseteq} (L,\Lambda)$. We write it as $P_{FH}^{(\alpha,x)} \tilde{\in} (L,\Lambda)$.

It is clear that the fuzzy hypersoft union of FHPs of a FHS (L,Λ) returns the FHS (L,Λ), that is,

$$(L,\Lambda) = \tilde{\cup}\left\{P_{FH}^{(\alpha,x)} : P_{FH}^{(\alpha,x)} \tilde{\in} (L,\Lambda)\right\}.$$

Example 2.1 Let Z be the set of mobile phones as $Z = \{u_1, u_2, u_3\}$ also consider the set of attributes given as;

$$E_1 = Ram = \{8GB(\alpha_1), 4GB(\alpha_2)\}$$

$$E_2 = ScreenSize = \{3.9in(\beta_1), 4.5in(\beta_2), 6.2in(\beta_3)\}$$

$$E_3 = \text{Camera Resolution}$$
$$= \{\text{Up to } 6.9MP(\gamma_1), 7.0 \text{ to } 9.9MP(\gamma_2), 13.0MP \text{ and above}(\gamma_3)\}$$

Suppose that

$$A_1 = \{\alpha_2\}, A_2 = \{\beta_2, \beta_3\}, A_3 = \{\gamma_1, \gamma_2\}$$

$$B_1 = \{\alpha_1,\alpha_2\}, B_2 = \{\beta_1,\beta_2\}, B_3 = \{\gamma_1\}$$

are subset of E_i for each i=1,2,3. Then the fuzzy hypersoft sets (L_1,Λ_1) and (L_2,Λ_2) defined as follows;

$$(L_1,\Lambda_1) = \left\{\begin{array}{c} \left\langle(\alpha_2,\beta_2,\gamma_1),\left\{\frac{u_1}{0,2},\frac{u_2}{0,5}\right\}\right\rangle, \\ \left\langle(\alpha_2,\beta_2,\gamma_2),\left\{\frac{u_1}{0,3},\frac{u_2}{0,4},\frac{u_3}{0,3}\right\}\right\rangle, \\ \left\langle(\alpha_2,\beta_3,\gamma_1),\left\{\frac{u_1}{0,4},\frac{u_3}{0,3}\right\}\right\rangle \end{array}\right\},$$

$$(L_2,\Lambda_2) = \left\{\begin{array}{c} \left\langle(\alpha_1,\beta_1,\gamma_1),\left\{\frac{u_1}{0,3},\frac{u_3}{0,8}\right\}\right\rangle, \\ \left\langle(\alpha_1,\beta_2,\gamma_1),\left\{\frac{u_2}{0,1},\frac{u_3}{0,4}\right\}\right\rangle, \end{array}\right\}$$

3. FUZZY HYPERSOFT RELATIONS

Definition 3.1 (Fuzzy Hypersoft Cartesian Product) Let Z be a universe (L_1,Λ_1) and (L_2,Λ_2) be two fuzzy hypersoft set over the common universe Z. Then their cartesian product is $(L_1,\Lambda_1)\times(L_2,\Lambda_2)=(N,C)$ where $C=\Lambda_1\times\Lambda_2$, N: $\Lambda_1\times\Lambda_2\rightarrow FHS(Z\times Z)$. Then $(N.C)$ is defined as follows;

$$(N,C) = \left\{\left(\begin{array}{c}(a,b),\langle(u_i,u_j),\mu_{N(a,b)}(u_i,u_j)\rangle : \\ u_i \in L_1(\Lambda_1), u_j \in L_2(\Lambda_2)\end{array}\right): a \in \Lambda_1, b \in \Lambda_2\right\}$$

where

$$\mu_{N(a,b)}(u_i,u_j) = \min\left\{\mu_{L_1(a)}(u_i),\mu_{L_2(b)}(u_j)\right\}.$$

Example 3.1 We consider that Example-2.1. Let's assume $(\alpha_2,\beta_2,\gamma_1) = \pi_1$, $(\alpha_2,\beta_2,\gamma_2) = \pi_2$, $(\alpha_2,\beta_3,\gamma_1) = \pi_3$ in (L_1,Λ_1) and, $(\alpha_1,\beta_1,\gamma_1) = \tau_1$, in (L_2,Λ_2) for easier operation. Then $(L_1,\Lambda_1) \times (L_2,\Lambda_2) = (N,C)$ calculated as follows:

$$(N,C)=\left\{\begin{matrix}\left[\begin{matrix}(\pi_1,\tau_1),\\ \left(\begin{matrix}\langle(u_1,u_1),0.2\rangle,\langle(u_1,u_3),0.2\rangle,\\ \langle(u_2,u_1),0.3\rangle,\langle(u_2,u_3),0.5\rangle\end{matrix}\right)\end{matrix}\right],\left[\begin{matrix}(\pi_1,\tau_2),\\ \left(\begin{matrix}\langle(u_1,u_2),0.1\rangle,\langle(u_1,u_3),0.2\rangle,\\ \langle(u_2,u_2),0.1\rangle,\langle(u_2,u_3),0.4\rangle\end{matrix}\right)\end{matrix}\right],\\ \left[\begin{matrix}(\pi_2,\tau_1),\\ \left(\begin{matrix}\langle(u_1,u_1),0.3\rangle,\langle(u_1,u_3),0.3\rangle,\\ \langle(u_2,u_1),0.3\rangle,\langle(u_2,u_3),0.4\rangle,\\ \langle(u_3,u_1),0.3\rangle,\langle(u_3,u_3),0.3\rangle\end{matrix}\right)\end{matrix}\right],\left[\begin{matrix}(\pi_2,\tau_2),\\ \left(\begin{matrix}\langle(u_1,u_2),0.1\rangle,\langle(u_1,u_3),0.3\rangle,\\ \langle(u_2,u_2),0.1\rangle,\langle(u_2,u_3),0.4\rangle,\\ \langle(u_3,u_2),0.1\rangle,\langle(u_3,u_3),0.3\rangle\end{matrix}\right)\end{matrix}\right],\\ \left[\begin{matrix}(\pi_3,\tau_1),\\ \left(\begin{matrix}\langle(u_1,u_1),0.3\rangle,\langle(u_1,u_3),0.4\rangle,\\ \langle(u_3,u_1),0.3\rangle,\langle(u_3,u_3),0.3\rangle\end{matrix}\right)\end{matrix}\right],\left[\begin{matrix}(\pi_3,\tau_2),\\ \left(\begin{matrix}\langle(u_1,u_2),0.1\rangle,\langle(u_1,u_3),0.4\rangle,\\ \langle(u_3,u_2),0.1\rangle,\langle(u_3,u_3),0.3\rangle\end{matrix}\right)\end{matrix}\right]\end{matrix}\right\}$$

Remark 3.1 Definition 10 is also provided in case of taking different universal sets and different parameter set. Assuming that in Example 2, there are two different universal sets, Z and Y, and two different parameters, Σ and Σ', we get $N:\Sigma\times\Sigma'\rightarrow Z\times Y$. As a result, it is clear that definition 10 has been provided.

Definition 3.2 (Fuzzy Hypersoft Relation) (L_1,Λ_1) and (L_2,Λ_2) be two fuzzy hypersoft set over the common universe Z. If (R,C) (briefly R) is a fuzzy hypersoft subset of $(L_1,\Lambda_1)\times(L_2,\Lambda_2)$, where $C\subseteq\Lambda_1\times\Lambda_2$ and $(R,C)\subseteq(L_1,\Lambda_1)\times(L_2,\Lambda_2)$, then (R,C) is called a fuzzy hypersoft relation on Z.

Example 3.2 We consider the Example 3.1. Then R is fuzzy hypersoft relation and defined as follows.

$$R=\left\{\begin{matrix}\left[\begin{matrix}(\pi_1,\tau_1),\\ \left(\begin{matrix}\langle(u_1,u_1),0.2\rangle,\langle(u_1,u_3),0.2\rangle,\\ \langle(u_2,u_1),0.3\rangle,\langle(u_2,u_3),0.5\rangle\end{matrix}\right)\end{matrix}\right], & \left[\begin{matrix}(\pi_1,\tau_2),\\ \left(\begin{matrix}\langle(u_1,u_2),0.1\rangle,\langle(u_1,u_3),0.2\rangle,\\ \langle(u_2,u_2),0.1\rangle,\langle(u_2,u_3),0.4\rangle\end{matrix}\right)\end{matrix}\right],\\ \left[\begin{matrix}(\pi_2,\tau_2),\\ \left(\begin{matrix}\langle(u_1,u_2),0.1\rangle,\langle(u_1,u_3),0.3\rangle,\\ \langle(u_2,u_2),0.1\rangle,\langle(u_2,u_3),0.4\rangle,\\ \langle(u_3,u_2),0.1\rangle,\langle(u_3,u_3),0.3\rangle\end{matrix}\right)\end{matrix}\right], & \left[\begin{matrix}(\pi_3,\tau_1),\\ \left(\begin{matrix}\langle(u_1,u_1),0.3\rangle,\langle(u_1,u_3),0.4\rangle,\\ \langle(u_3,u_1),0.3\rangle,\langle(u_3,u_3),0.3\rangle\end{matrix}\right)\end{matrix}\right]\end{matrix}\right\}$$

Definition 3.3 Let (L_1,Λ_1) and (L_2,Λ_2) be two fuzzy hypersoft sets over the universe Z and let R be fuzzy hypersoft relation from (L_1,Λ_1) to (L_2,Λ_2). Then R^{-1} is a inverse fuzzy hypersoft relation from (L_2,Λ_2) to (L_1,Λ_1) and defined as follows.

$$R^{-1} = \left\{ \begin{pmatrix} (b,a), \langle (u_j, u_i), \mu_{R^{-1}(b,a)}(u_j,u_i) \rangle : \\ u_i \in L_1(\Lambda_1), u_j \in L_2(\Lambda_2) \end{pmatrix} : a \in \Lambda_1, b \in \Lambda_2 \right\}$$

where

$$\mu_{R^{-1}(b,a)}\left(u_j,u_i\right) = \min\left\{ \mu_{L_2(b)}(u_j), \mu_{L_1(a)}(u_i) \right\}$$

Example 3.3 We consider the Example 3.2 Then we get R^{-1} as follows.

$$R^{-1} = \left\{ \begin{matrix} \begin{bmatrix} (\tau_1,\pi_1), \\ \begin{pmatrix} \langle (u_1,u_1),0.2\rangle, \langle (u_3,u_1),0.2\rangle, \\ \langle (u_1,u_2),0.3\rangle, \langle (u_3,u_2),0.5\rangle \end{pmatrix} \end{bmatrix}, & \begin{bmatrix} (\tau_2,\pi_1), \\ \begin{pmatrix} \langle (u_2,u_1),0.1\rangle, \langle (u_3,u_1),0.2\rangle, \\ \langle (u_2,u_2),0.1\rangle, \langle (u_3,u_2),0.4\rangle \end{pmatrix} \end{bmatrix}, \\ \begin{bmatrix} (\tau_2,\pi_2), \\ \begin{pmatrix} \langle (u_2,u_1),0.1\rangle, \langle (u_3,u_1),0.3\rangle, \\ \langle (u_2,u_2),0.1\rangle, \langle (u_3,u_2),0.4\rangle, \\ \langle (u_2,u_3),0.1\rangle, \langle (u_3,u_3),0.3\rangle \end{pmatrix} \end{bmatrix}, & \begin{bmatrix} (\tau_1,\pi_3), \\ \begin{pmatrix} \langle (u_1,u_1),0.3\rangle, \langle (u_3,u_1),0.4\rangle, \\ \langle (u_1,u_3),0.3\rangle, \langle (u_3,u_3),0.3\rangle \end{pmatrix} \end{bmatrix}, \end{matrix} \right\}$$

Proposition 3.1 Let (L_1,Λ_1) and (L_2,Λ_2) be two fuzzy hypersoft sets over the common universe Z and R be fuzzy hypersoft relation from (L_1,Λ_1) to (L_2,Λ_2). Then R^{-1} is a fuzzy hypersoft relation from (L_2,Λ_2) to (L_1,Λ_1).

Proof. For $u_i \in L_1(\Lambda_1)$, $u_j \in L_2(\Lambda_2)$, $a \in \Lambda_1$ and $b \in \Lambda_2$, we have

$$\mu_{R^{-1}(b,a)}\left(u_j,u_i\right) = \min\left\{ \mu_{L_2(b)}(u_i), \mu_{L_1(a)}(u_j) \right\}$$
$$= \min\left\{ \mu_{L_1(a)}(u_i), \mu_{L_1(a)}(u_j) \right\} = \mu_{R(a,b)}\left(u_i,u_j\right)$$

Then, R^{-1} is a fuzzy hypersoft relation from (L_2,Λ_2) to (L_1,Λ_1).

Proposition 3.2 Let (L_1,Λ_1) and (L_2,Λ_2) be two fuzzy hypersoft sets over the common universe Z and R_1,R_2 be fuzzy hypersoft relation from (L_1,Λ_1) to (L_2,Λ_2). Then

1. $(R_1^{-1})^{-1} = R_1$
2. $R_1 \subseteq R_2 \Rightarrow R_1^{-1} \subseteq R_2^{-1}$

Proof.

1. Let $u_i \in L_1(\Lambda_1)$, $u_j \in L_2(\Lambda_2)$, $a \in \Lambda_1$ and $b \in \Lambda_2$. From the Proposition 1, $\mu_{R_1(a,b)}(u_i, u_j) = \mu_{R_1^{-1}(b,a)}(u_j, u_i)$. Then we have,

$$\mu_{R_1(a,b)}(u_i, u_j) = \mu_{R_1^{-1}(b,a)}(u_j, u_i) = \mu_{(R_1^{-1})^{-1}(b,a)}(u_j, u_i)$$

It follows that $(R_1^{-1})^{-1} = R_1$.

2. If $R_1 \subseteq R_2$, then $\mu_{R_1(a,b)}(u_i, u_j) \le \mu_{R_2(a,b)}(u_i, u_j)$. Thus

$$\mu_{R_1(a,b)}(u_i, u_j) = \mu_{R_1^{-1}(b,a)}(u_j, u_i) \le \mu_{R_2(a,b)}(u_i, u_j) = \mu_{R_2^{-1}(b,a)}(u_j, u_i).$$

So, $\mu_{R_1^{-1}(b,a)}(u_j, u_i) \le \mu_{R_2^{-1}(b,a)}(u_j, u_i)$. Hence, we have $R_1^{-1} \subseteq R_2^{-1}$. This completes of the condition.

Definition 3.4 Let Z be a universe, (L_1,Λ_1), (L_2,Λ_2) and (L_3,Λ_3) be fuzzy hypersoft sets over the common universe $Z.R$ and S be two fuzzy hypersoft relations from (L_1,Λ_1) to (L_2,Λ_2) and (L_2,Λ_2) to (L_3,Λ_3), respectively. Then their fuzzy hypersoft composition (briefly composition) is denoted by $R \circ S$ and defined by

$$R \circ S = \left\{ \begin{pmatrix} (a,c), \langle (u_i, u_j), \mu_{R\circ S(a,c)}(u_i, u_j) \rangle \\ : u_i \in L_1(a), u_j \in L_3(c) \end{pmatrix} : a \in \Lambda_1, c \in \Lambda_3 \right\}$$

where

$$\mu_{R\circ S(a,c)}(u_i, u_j) = \max\left\{ \min\{\mu_{R(a,b)}(u_i, u_j), \mu_{S(b,c)}(u_i, u_j)\} \right\}$$

$\forall a \in \Lambda_1$, $\forall b \in \Lambda_2$ and $\forall c \in \Lambda_3$.

Remark 3.2 When examining the composition of two relations, the following situations should be considered.

1. If a parameter pair such as (τ_p, π_j) in the composition consists of a single match, the common alternative pairs in the relations forming the (τ_p, π_j) parameter are taken into account. Other alternative pairs take the value of membership 0.
2. If a parameter pair such as (τ_p, π_j) in the composition consists of more than one match, all the alternative pairs in the relations forming the (τ_p, π_j) parameter are taken into account.

The following example illustrates this more clearly.

Example 3.4 We consider the Example 2.1 and 3.1. Let (L_3, Λ_3) be given as follows.

$$(L_2, \Lambda_2) = \left\{ \left\langle (\alpha_1, \beta_1, \gamma_1), \left\{ \frac{u_1}{0,3}, \frac{u_3}{0,8} \right\} \right\rangle, \left\langle (\alpha_1, \beta_2, \gamma_1), \left\{ \frac{u_2}{0,1}, \frac{u_3}{0,4} \right\} \right\rangle, \right\}$$

$$(L_3, \Lambda_3) = \left\{ \begin{array}{c} <(\alpha_1, \beta_1, \gamma_1), \{\frac{u_1}{0,3}, \frac{u_3}{0,8}\}>, <(\alpha_1, \beta_2, \gamma_1), \{\frac{u_2}{0,1}, \frac{u_3}{0,4}\}>, \\ <(\alpha_2, \beta_1, \gamma_1), \{\frac{u_1}{0,2}, \frac{u_3}{0,3}\}> \end{array} \right\}$$

We assume that $(\alpha_1, \beta_1, \gamma_1) = \tau_1$, $(\alpha_1, \beta_2, \gamma_1) = \tau_2$, in (L_2, Λ_2) and $(\alpha_2, \beta_1, \gamma_1) = \rho_1$, $(\alpha_1, \beta_2, \gamma_1) = \rho_2$, $(\alpha_2, \beta_1, \gamma_1) = \rho_3$ for easier operation. Let $R \subseteq (L_1, \Lambda_1) \times (L_2, \Lambda_2)$, $S \subseteq (L_2, \Lambda_2) \times (L_3, \Lambda_3)$ relations be defined as below.

$$R = \left\{ \begin{array}{cc} \left[\begin{array}{c} (\pi_1, \tau_1), \\ \left(\begin{array}{c} \langle(u_1, u_1), 0.2\rangle, \langle(u_1, u_3), 0.2\rangle, \\ \langle(u_2, u_1), 0.3\rangle, \langle(u_2, u_3), 0.5\rangle \end{array} \right) \end{array} \right], & \left[\begin{array}{c} (\pi_1, \tau_2), \\ \left(\begin{array}{c} \langle(u_1, u_2), 0.1\rangle, \langle(u_1, u_3), 0.2\rangle, \\ \langle(u_2, u_2), 0.1\rangle, \langle(u_2, u_3), 0.4\rangle \end{array} \right) \end{array} \right], \\ \left[\begin{array}{c} (\pi_2, \tau_2), \\ \left(\begin{array}{c} \langle(u_1, u_2), 0.1\rangle, \langle(u_1, u_3), 0.3\rangle, \\ \langle(u_2, u_2), 0.1\rangle, \langle(u_2, u_3), 0.4\rangle, \\ \langle(u_3, u_2), 0.1\rangle, \langle(u_3, u_3), 0.3\rangle \end{array} \right) \end{array} \right], & \left[\begin{array}{c} (\pi_3, \tau_1), \\ \left(\begin{array}{c} \langle(u_1, u_1), 0.3\rangle, \langle(u_1, u_3), 0.4\rangle, \\ \langle(u_3, u_1), 0.3\rangle, \langle(u_3, u_3), 0.3\rangle \end{array} \right) \end{array} \right], \end{array} \right\}$$

$$S=\left\{\begin{bmatrix}(\tau_1,\rho_1),\\ \left(\begin{matrix}\langle(u_1,u_1),0.3\rangle,\langle(u_1,u_3),0.8\rangle,\\ \langle(u_3,u_1),0.3\rangle,\langle(u_3,u_3),0.8\rangle\end{matrix}\right)\end{bmatrix},\begin{bmatrix}(\tau_1,\rho_2),\\ \left(\begin{matrix}\langle(u_1,u_2),0.1\rangle,\langle(u_1,u_3),0.3\rangle,\\ \langle(u_3,u_2),0.1\rangle,\langle(u_3,u_3),0.4\rangle\end{matrix}\right)\end{bmatrix},\begin{bmatrix}(\tau_2,\rho_2),\\ \left(\begin{matrix}\langle(u_1,u_1),0.1\rangle,\langle(u_1,u_3),0.1\rangle,\\ \langle(u_3,u_1),0.4\rangle,\langle(u_3,u_3),0.4\rangle\end{matrix}\right)\end{bmatrix}\right\}$$

The parameters of $R \circ S$ system is provided as;

$$\{(\pi_1,\tau_1)\circ(\tau_1,\rho_1)\} \Rightarrow (\pi_1,\rho_1),$$

$$\{[(\pi_1,\tau_1)\circ(\tau_1,\rho_2)],[(\pi_1,\tau_2)\circ(\tau_2,\rho_2)]\} \Rightarrow (\pi_1,\rho_2),$$

$$\{(\pi_3,\tau_1)\circ(\tau_1,\rho_1)\} \Rightarrow (\pi_3,\rho_1)$$

$$\{(\pi_3,\tau_1)\circ(\tau_1,\rho_2)\} \Rightarrow (\pi_3,\rho_2)$$

$$\{(\pi_2,\tau_2)\circ(\tau_2,\rho_2)\} \Rightarrow (\pi_2,\rho_2)$$

Here, according to Remark 2 (1. condition), we examine $R \circ S$ to (π_1,ρ_1). Parameter pair (π_1,ρ_1) in the composition consists of a single match $\{(\pi_1,\tau_1)\circ(\tau_1,\rho_1)\}$. For the (π_1,ρ_1) parameter, we should investigate the (π_1,τ_1) parameter in R relation and (τ_1,ρ_1) parameter in S relation. Common alternative pairs for the (π_1,τ_1) parameter in R relation and (τ_1,ρ_1) parameter in S relation are (u_1,u_1) and (u_1,u_3). The membership values of other alternatives according to (π_1,ρ_1) parameter becomes 0 by the definition of composition.

Some calculations are provided as follow,

$$\mu_{R\circ S(\pi_1,\rho_1)}(u_1,u_1)$$
$$=\max\left\{\min\{\mu_{R(\pi_1,\tau_1)}(u_1,u_1),\mu_{S(\pi_1,\tau_1)}(u_1,u_1)\}\right\}$$
$$=\max\{\min(0.2,0.3)\} = 0.2$$

$$\mu_{R\circ S(\pi_1,\rho_1)}(u_1,u_3)$$
$$= \max\left\{ \min\{\mu_{R(\pi_1,\tau_1)}(u_1,u_3),\mu_{S(\pi_1,\tau_1)}(u_1,u_3)\} \right\}$$
$$= \max\{\min(0.2,0.8)\} \;=\; 0.2$$

Now, we examine $R \circ S$ according to (π_1,ρ_2). Parameter pair (π_1,ρ_2) in the composition consists of more than one match. We see that it consist of product $\{(\pi_1,\tau_1)\circ(\tau_1,\rho_2)\}$ and $\{(\pi_1,\tau_2)\circ(\tau_2,\rho_2)\}$. For parameters consisting of such multi-parameter product, any alternatives cannot be ignored. Some calculations for the (π_1,ρ_2) parameter are as follows.

$$\mu_{R\circ S(\pi_1,\rho_2)}(u_1,u_3)$$

$$= \max\left\{ \begin{matrix} \min\{\mu_{R(\pi_1,\tau_1)}(u_1,u_3),\mu_{S(\tau_1,\rho_2)}(u_1,u_3)\}, \\ \min\{\mu_{R(\pi_1,\tau_2)}(u_1,u_3),\mu_{S(\tau_2,\rho_2)}(u_1,u_3)\} \end{matrix} \right\}$$

$$= \max\{\min\{0.2,0.3\},\min\{0.2,0.1\}\}$$

$$= \max\{0.2,0.1\} \;=\; 0,2$$

Other calculations can easily be done similarly. The composition $R \circ S$ is shown in Table 1 below.

Table 1. The composition

$R \circ S$	$(_1,_1)$	$(_1,_2)$	$(_3,_1)$	$(_3,_2)$	$(_2,_2)$
(u_1,u_1)	0.2	0	0.3	0	0
(u_1,u_2)	0	0	0	0	0
(u_1,u_3)	0.2	0.2	0.4	0.3	0.1

continued on following page

Table 1. Continued

$R \circ S$	$(_{1}, _{1})$	$(_{1}, _{2})$	$(_{3}, _{1})$	$(_{3}, _{2})$	$(_{2}, _{2})$
(u_2,u_1)	0	0	0	0	0
(u_2,u_2)	0	0	0	0	0
(u_2,u_3)	0	0	0	0	0
(u_3,u_1)	0	0	0.3	0	0
(u_3,u_2)	0	0	0	0	0
(u_3,u_3)	0	0	0.3	0.3	0.3

Proposition 3.3 Let Z be a universe. (L_1,Λ_1), (L_2,Λ_2) and (L_3,Λ_3) be two fuzzy hypersoft set over the common universe Z. R and S be two fuzzy hypersoft relations from (L_1,Λ_1) to (L_2,Λ_2) and (L_2,Λ_2) to (L_3,Λ_3) , respectively. Then $R \circ S$ is fuzzy hypersoft relation from (L_1,Λ_1) to (L_3,Λ_3) .

Proof. By Definition 3.4, for all $a \in \Lambda_1$, $b \in \Lambda_2$ and $c \in C$,

$$\mu_{R \circ S(a,c)}\left(u_i, u_j\right)$$

$$= \max\left\{ \min\left\{\mu_{R(a,b)}(u_i, u_j), \mu_{S(b,c)}(u_i, u_j)\right\} \right\}$$

$$= \max\left\{ \begin{array}{l} \min\left\{\mu_{L_1(a)}(u_i, u_j), \mu_{L_2(b)}(u_i, u_j)\right\}, \\ \min\left\{\mu_{L_2(b)}(u_i, u_j), \mu_{L_3(c)}(u_i, u_j)\right\} \end{array} \right\}$$

$$= \max\left\{ \min\left\{\mu_{L_1(a)}(u_i, u_j), \mu_{L_2(b)}(u_i, u_j), \mu_{L_3(c)}(u_i, u_j)\right\} \right\}$$

$$\leq \max\left\{ \min\left\{\mu_{L_1(a)}(u_i, u_j), 1, \mu_{L_3(c)}(u_i, u_j)\right\} \right\}$$

$$= \max\left\{ \min\left\{\mu_{L_1(a)}(u_i, u_j), \mu_{L_3(c)}(u_i, u_j)\right\} \right\}$$

$$= \min\left\{ \mu_{L_1(a)}(u_i, u_j), \mu_{L_3(c)}(u_i, u_j) \right\}$$

Then, $\mu_{R \circ S(a,c)}(u_i, u_j) \leq \min\left\{ \mu_{L_1(a)}(u_i, u_j), \mu_{L_3(c)}(u_i, u_j) \right\}$. Hence $R \circ S \subseteq (L_1, \Lambda_1) \times (L_3, \Lambda_3)$.

Proposition 3.4 The fuzzy hypersoft composition is associate. If R, S, T are three fuzzy hypersoft relation from (L_1, Λ_1) to (L_2, Λ_2), (L_2, Λ_2) to (L_3, Λ_3) and (L_3, Λ_3) to (L, D). Then $(R \circ S) \circ T = R \circ (S \circ T)$.

Proof. For all $a \in \Lambda_1$, $b \in \Lambda_2$, $c \in C$ and $d \in D$.

$$\mu_{(R \circ S) \circ T(a,d)}(u_i, u_j)$$

$$= \max\left\{ \min\{ \mu_{(R \circ S)(a,c)}(u_i, u_j), \mu_{T(c,d)}(u_i, u_j) \} \right\}$$

$$= \max\left\{ \min\{ \mu_{R(a,b)}(u_i, u_j), \mu_{S(b,c)}(u_i, u_j), \mu_{T(c,d)}(u_i, u_j) \} \right\}$$

$$= \max\left\{ \min\{ \mu_{R(a,b)}(u_i, u_j), \mu_{(S \circ T)(b,d)}(u_i, u_j) \} \right\}$$

$$= \mu_{R \circ (S \circ T)(a,d)}(u_i, u_j)$$

Hence, $(R \circ S) \circ T = R \circ (S \circ T)$.

Proposition 3.5 Let $R \subseteq (L_1, \Lambda_1) \times (L_2, \Lambda_2)$ and $S, T \subseteq (L_2, \Lambda_2) \times (L_3, \Lambda_3)$ be three fuzzy hypersoft relations defined over Z. Then $R \circ (S \cup T) = (R \circ S) \cup (R \circ T)$.

Proof. Let $S \cup T \neq \emptyset$. For all $a \in \Lambda_1$, $b \in \Lambda_2$ and $c \in C$, we have

$$\mu_{(R \circ (S \cup T))(a,c)}(u_i, u_j)$$

$$= \max\left\{ \min\left\{ \mu_{R(a,b)}(u_i, u_j), \mu_{(S\cup T)(b,c)}(u_i, u_j) \right\} \right\} \quad (3.1)$$

$$= \max\left\{ \min\left\{ \max\left\{ \mu_{R(a,b)}(u_i, u_j), \mu_{S(b,c)}(u_i, u_j), \mu_{T(b,c)}(u_i, u_j) \right\} \right\} \right\}$$

$$= \max\left\{ \min\left\{ \begin{array}{l} \min\left\{ \mu_{L_1(a)}(u_i, u_j), \mu_{L_2(b)}(u_i, u_j) \right\}, \\ \min\left\{ \mu_{L_2(b)}(u_i, u_j), \mu_{L_3(c)}(u_i, u_j) \right\} \end{array} \right\} \right\}$$

$$= \max\left\{ \min\left\{ \mu_{L_1(a)}(u_i, u_j), \mu_{L_2(b)}(u_i, u_j), \mu_{L_3(c)}(u_i, u_j) \right\} \right\}. \quad (3.2)$$

Then, $\mu_{(R\circ(S\cup T))(a,c)}(u_i, u_j) = \max\left\{ \min\left\{ \mu_{L_1(a)}(u_i, u_j), \mu_{L_2(b)}(u_i, u_j), \mu_{L_3(c)}(u_i, u_j) \right\} \right\}$.
On the other hand, Let $(R \circ S) \cap (R \circ T) \neq \varnothing$. Then

$$\mu_{[(R\circ S)\cup(R\circ T)](a,c)}(u_i, u_j)$$

$$= \max\left\{ \mu_{(R\circ S)(a,c)}(u_i, u_j), \mu_{(R\circ T)(a,c)}(u_i, u_j) \right\} \quad (3.3)$$

$$= \max\left\{ \begin{array}{l} \max\left\{ \min\left\{ \mu_{R(a,b)}(u_i, u_j), \mu_{S(b,c)}(u_i, u_j) \right\} \right\}, \\ \max\left\{ \min\left\{ \mu_{R(a,b)}(u_i, u_j), \mu_{T(b,c)}(u_i, u_j) \right\} \right\} \end{array} \right\}$$

$$= \max\left\{ \min\left\{ \begin{array}{l} \min\left\{ \mu_{L_1(a)}(u_i, u_j), \mu_{L_2(b)}(u_i, u_j) \right\}, \\ \min\left\{ \mu_{L_2(b)}(u_i, u_j), \mu_{L_3(c)}(u_i, u_j) \right\} \end{array} \right\} \right\}$$

$$= \max\left\{ \min\left\{ \mu_{L_1(a)}(u_i, u_j), \mu_{L_2(b)}(u_i, u_j), \mu_{L_3(c)}(u_i, u_j) \right\} \right\} \quad (3.4)$$

By (3.2) and (3.4), we have $R \circ (S \cup T) = (R \circ S) \cup (R \circ T)$.

Definition 3.5 Let Z be a universe and Σ be a parameter sets. The relation $\Delta \in FSR(Z \times Z)_{\Sigma\times\Sigma}$ is called identity relation if

$$\mu_{\Delta}\left(u_{\beta},\psi_{\beta'}\right) = \begin{cases} 1 & if \quad u_{\beta} = \psi_{\beta'} \\ 0 & if \quad u_{\beta} \neq \psi_{\beta'} \end{cases}$$

where $u_{\beta},\psi_{\beta'} \in (Z,\Sigma) \times (Z,\Sigma)$.

Definition 3.6 Let Z be a universe, Σ be a parameter sets and $FSS(Z,\Sigma)$ be all fuzzy hypersoft sets over the (Z,Σ).

1. The relation $R \in FSR\,(Z \times Z)_{\Sigma\times\Sigma}$ is called reflexive if for every $u_{\beta} \in FSS(Z,\Sigma), \mu_R\left(u_{\beta},u_{\beta}\right) = 1.$
2. The relation $R \in FSR\,(Z \times Z)_{\Sigma\times\Sigma}$ is called antireflexive if for every $u_{\beta} \in FSS(Z,\Sigma), \mu_R\left(u_{\beta},u_{\beta}\right) = 0.$

Definition 3.7 Let Z be a universe, Σ be a parameter sets and $FSS(Z,\Sigma)$ be all fuzzy hypersoft sets over the (Z,Σ). A relation $R \in FSR(Z \times Z)_{\Sigma\times\Sigma}$ is called symmetric if $R=R^{-1}$, that is, for every $u_{\beta},\psi_{\beta'} \in FSS(Z,\Sigma), \mu_R\left(u_{\beta},\psi_{\beta'}\right) = \mu_R\left(\psi_{\beta'},u_{\beta}\right)$.

Definition 3.8 A fuzzy hypersoft relation R on (L_1,Λ_1) is said to be fuzzy hypersoft transitive relation if $R \circ R \subseteq R$.

Definition 3.9 A fuzzy hypersoft relation R is called to be fuzzy hypersoft equivalance relation if it is symmetric, transitive and reflective.

Example 3.5 We consider that attributes in Example-2.1. Let $(\pi_1,\tau_1)=k_1$, $(\pi_2,\tau_2)=k_2$ and a fuzzy hypersoft set on Z following as;

$$(L,\Lambda) = \begin{Bmatrix} \left(k_1,\langle u_1,0.2\rangle,\langle u_2,0.5\rangle,\langle u_3,0.4\rangle\right) \\ \left(k_2,\langle u_1,0.5\rangle,\langle u_2,0.4\rangle,\langle u_3,0.7\rangle\right) \end{Bmatrix}$$

Then we get a fuzzy hypersoft relation R on (L,Λ) as follows;

$$R = \left\{ \begin{array}{l} \left((k_1,k_1), \left(\begin{array}{l} <(x_1,x_1),0.2>,<(x_1,x_2),0.2>,<(x_1,x_3),0.2>, \\ <(x_2,x_1),0.2>,<(x_2,x_2),0.5>,<(x_2,x_3),0.4>, \\ <(x_3,x_1),0.2>,<(x_3,x_2),0.4>,<(x_3,x_3),0.4>, \end{array} \right) \right) \\ \left((k_1,k_2), \left(\begin{array}{l} <(x_1,x_1),0.2>,<(x_1,x_2),0.2>,<(x_1,x_3),0.2>, \\ <(x_2,x_1),0.5>,<(x_2,x_2),0.4>,<(x_2,x_3),0.5>, \\ <(x_3,x_1),0.4>,<(x_3,x_2),0.4>,<(x_3,x_3),0.4>, \end{array} \right) \right) \\ \left((k_2,k_1), \left(\begin{array}{l} <(x_1,x_1),0.2>,<(x_1,x_2),0.5>,<(x_1,x_3),0.4>, \\ <(x_2,x_1),0.2>,<(x_2,x_2),0.4>,<(x_2,x_3),0.4>, \\ <(x_3,x_1),0.2>,<(x_3,x_2),0.5>,<(x_3,x_3),0.4>, \end{array} \right) \right) \\ \left((k_2,k_2), \left(\begin{array}{l} <(x_1,x_1),0.5>,<(x_1,x_2),0.4>,<(x_1,x_3),0.5>, \\ <(x_2,x_1),0.4>,<(x_2,x_2),0.4>,<(x_2,x_3),0.4>, \\ <(x_3,x_1),0.5>,<(x_3,x_2),0.4>,<(x_3,x_3),0.7>, \end{array} \right) \right) \end{array} \right\}$$

R on (L,Λ) is a fuzzy hypersoft equivalence relation because it is neutrosophic soft symmetric, fuzzy hypersoft soft transitive and fuzzy hypersoft reflexive.

Definition 3.10 Let $FSS(Z,\Sigma)$ and $FSS(Y,\Sigma')$ be two classes of fuzzy hypersoft sets over the Z and Y respectively. Let $f: Z \to Y$ and $g:\Sigma \to \Sigma'$ be mappings and (f,g) be a fuzzy hypersoft relation on $FSS(Z,\Sigma) \times FSS(Y,\Sigma')$. Then (f,g) is called a fuzzy hypersoft function if (f,g) associates each elements of $FSS(Z,\Sigma)$ with the unique element of $FSS(Y,\Sigma')$. Then a mapping $(f,g):FSS(Z,\Sigma) \to FSS(Y,\Sigma')$ is defined as follows:

For all $u^{\mu}_{\beta} \in FSS(Z,\Sigma)$, The images of u^{μ}_{β} is $(f,g)\left(u^{\mu}_{\beta}\right) = f(u)^{\mu_{u_\beta}}_{g(\beta)}$.

Also, the set

$$(f,g)\left(L_1,\Lambda_1\right) = \left\{f(u)^{\max\{\mu_{u_\beta}\}}_{g(\beta)} : \forall u_\beta \in \left(L_1,\Lambda_1\right)\right\} \subseteq FSS(Y,\Sigma')$$

for $\left(L_1,\Lambda_1\right) \subseteq FSS(Z,\Sigma)$ is called the image of the set (L_1,Λ_1) under (f,g).

Definition 3.11 Let $FSS(Z,\Sigma)$ and $FSS(Y,\Sigma')$ be two fuzzy hypersoft classes and $f: Z \to Y$ and $g:\Sigma \to \Sigma'$ be mappings. $(f,g)^{-1}:FSS(Y,\Sigma') \to FSS(Z,\Sigma)$ is defined as follows:

For all $\psi^{\mu}_{\beta'} \in FSS(Y,\Sigma')$, The inverse image of $\psi^{\mu}_{\beta'}$ is $(f,g)^{-1}\left(\psi_{\beta'}\right) = f^{-1}(\psi)^{\delta_{\psi_{\Sigma 1}}}_{g^{-1}(\beta')}$.

Also, the set

$$(f,g)^{-1}(L_2,\Lambda_2) = \left\{f^{-1}(\psi)^{\delta_{\psi_{\varepsilon}}}_{g^{-1}(\beta')} : \forall\psi_{\beta'} \in (L_2,\Lambda_2)\right\} \subseteq FSS(Z,\Sigma)$$

for $(L_2,\Lambda_2) \subseteq FSS(Y,\Sigma')$ is called the inverse image of the set (L_2,Λ_2).

Example 3.6 Let $Z=\{u_1,u_2,u_3\}$, $Y=\{\psi_1,\psi_2,\psi_3\}$, $\Sigma=E_1\times E_2$ and $\Sigma' = E'_1 \times E'_2$.

$$E_1 = \{\alpha_1,\alpha_2\}$$

$$E_2 = \{\alpha_3,\alpha_4\}$$

$$E'_1 = \{\beta_1,\beta_2\}$$

$$E'_2 = \{\beta_3,\beta_4\}$$

We have define (L_1,Λ_1) and (L_2,Λ_2) in (Z,Σ) and (Y,Σ'), respectively, as follows,

$$(L_1,\Lambda_1) = \begin{Bmatrix} ((\alpha_1,\alpha_3),(\langle u_1,0.5\rangle,\langle u_2,0.1\rangle,\langle u_3,0.8\rangle)), \\ ((\alpha_2,\alpha_3),(\langle u_1,0.1\rangle,\langle u_2,0.9\rangle,\langle u_3,0.5\rangle)), \\ ((\alpha_1,\alpha_4),(\langle u_1,0.4\rangle,\langle u_2,0.3\rangle,\langle u_3,0.6\rangle)), \end{Bmatrix},$$

$$(L_2,\Lambda_2) = \begin{Bmatrix} ((\beta_1,\beta_3),(\langle \psi_1,0.3\rangle,\langle \psi_2,0.5\rangle,\langle \psi_3,0.1\rangle)), \\ ((\beta_1,\beta_4),(\langle \psi_1,0.9\rangle,\langle \psi_2,0.1\rangle,\langle \psi_3,0.5\rangle)), \\ ((\beta_2,\beta_3),(\langle \psi_1,0.7\rangle,\langle \psi_2,0.5\rangle,\langle \psi_3,0.6\rangle)), \end{Bmatrix}$$

We assume that $(\alpha_1,\alpha_3)=\pi_1$, $(\alpha_1,\alpha_4)=\pi_2$, $(\alpha_2,\alpha_3)=\pi_3$, $(\alpha_2,\alpha_4)=\pi_4$ and $(\beta_1,\beta_3)=\tau_1$, $(\beta_1,\beta_4)=\tau_2$, $(\beta_2,\beta_3)=\tau_3$, $(\beta_2,\beta_3)=\tau_4$ for easier operation. Let f: $Z\rightarrow Y$ and $g:\Sigma \rightarrow \Sigma'$ be mappings defined as

$$f(u_1) = \psi_3$$

$f(u_2) = \psi_2$

$f(u_3) = \psi_2$

$g(\pi_1) = \tau_1$

$g(\pi_2) = \tau_3$

$g(\pi_3) = \tau_2$

$g(\pi_4) = \tau_2$

We can also write the fuzzy hypersoft sets (L_1,Λ_1) and (L_2,Λ_2) pointwise as follows.

$$(L_1,\Lambda_1) = \left\{ \begin{matrix} P^{(\pi_1,u_1)}_{FH(0.5)}, P^{(\pi_1,u_2)}_{FH(0.1)}, P^{(\pi_1,u_3)}_{FH(0.8)}, P^{(\pi_2,u_1)}_{FH(0.4)}, P^{(\pi_2,u_2)}_{FH(0.3)}, P^{(\pi_2,u_3)}_{FH(0.6)} \\ P^{(\pi_3,u_1)}_{FH(0.1)}, P^{(\pi_3,u_2)}_{FH(0.9)}, P^{(\pi_3,u_3)}_{FH(0.5)} \end{matrix} \right\},$$

$$(L_2,\Lambda_2) = \left\{ \begin{matrix} P^{(\tau_1,\psi_1)}_{FH(0.3)}, P^{(\tau_1,\psi_2)}_{FH(0.5)}, P^{(\tau_1,\psi_3)}_{FH(0.1)}, P^{(\tau_3,\psi_1)}_{FH(0.7)}, P^{(\tau_3,\psi_2)}_{FH(0.5)}, P^{(\tau_3,\psi_3)}_{FH(0.6)} \\ P^{(\tau_2,\psi_1)}_{FH(0.9)}, P^{(\tau_2,\psi_2)}_{FH(0.1)}, P^{(\tau_2,\psi_3)}_{FH(0.5)} \end{matrix} \right\}$$

Then the fuzzy hypersoft image of (L_1,Λ_1) under $(f, g): FSS(Z, \Sigma) \rightarrow FSS(Y, \Sigma')$ is obtained as

$$(f,g)((L_1,\Lambda_1))=\left\{\begin{array}{c}P_{FH_{(0.5)}}^{(\tau_1,\psi_3)},P_{FH_{(0.1)}}^{(\tau_1,\psi_2)},P_{FH_{(0.8)}}^{(\tau_1,\psi_2)},P_{FH_{(0.4)}}^{(\tau_3,\psi_3)},P_{FH_{(0.3)}}^{(\tau_3,\psi_2)},P_{FH_{(0.6)}}^{(\tau_3,\psi_2)},\\ P_{FH_{(0.1)}}^{(\tau_2,\psi_3)},P_{FH_{(0.9)}}^{(\tau_2,\psi_2)},P_{FH_{(0.5)}}^{(\tau_2,\psi_2)}\end{array}\right\},$$

$$=\left\{P_{FH_{(0.5)}}^{(\tau_1,\psi_3)},P_{FH_{(0.8)}}^{(\tau_1,\psi_2)},P_{FH_{(0.4)}}^{(\tau_3,\psi_3)},P_{FH_{(0.6)}}^{(\tau_3,\psi_2)},P_{FH_{(0.1)}}^{(\tau_2,\psi_3)},P_{FH_{(0.9)}}^{(\tau_2,\psi_2)}\right\}$$

Then, we have

$$(f,g)((L_1,\Lambda_1))=\left\{\begin{array}{l}((\beta_1,\beta_3),(\langle\psi_1,0\rangle,\langle\psi_2,0.8\rangle,\langle\psi_3,0.5\rangle)),\\ ((\beta_1,\beta_4),(\langle\psi_1,0\rangle,\langle\psi_2,0.9\rangle,\langle\psi_3,0.1\rangle)),\\ ((\beta_2,\beta_3),(\langle\psi_1,0\rangle,\langle\psi_2,0.6\rangle,\langle\psi_3,0.4\rangle))\end{array}\right\}$$

Now, the fuzzy hypersoft inverse image of (L_2,Λ_2) under $(f,g)^{-1}:FSS(Y,\Sigma')\rightarrow FSS(Z,\Sigma)$ is obtained as

$$(f,g)^{-1}\left(P_{FH_{(0.3)}}^{(\tau_1,\psi_1)}\right)=\varnothing,$$

$$(f,g)^{-1}\left(P_{FH_{(0.5)}}^{(\tau_1,\psi_2)}\right)=\left\{P_{FH_{(0.5)}}^{(\pi_1,u_2)},P_{FH_{(0.5)}}^{(\pi_1,u_3)}\right\},$$

$$(f,g)^{-1}\left(P_{FH_{(0.1)}}^{(\tau_1,\psi_3)}\right)=\left\{P_{FH_{(0.1)}}^{(\pi_1,u_1)}\right\},$$

$$(f,g)^{-1}\left(P_{FH_{(0.7)}}^{(\tau_3,\psi_1)}\right)=\varnothing,$$

$$(f,g)^{-1}\left(P_{FH_{(0.5)}}^{(\tau_3,\psi_2)}\right)=\left\{P_{FH_{(0.5)}}^{(\pi_2,u_2)},P_{FH_{(0.5)}}^{(\pi_2,u_3)}\right\},$$

$$(f,g)^{-1}\left(P_{FH_{(0.6)}}^{(\tau_3,\psi_3)}\right)=\left\{P_{FH_{(0.6)}}^{(\pi_2,u_1)}\right\},$$

$$(f,g)^{-1}\left(P^{(\tau_2,\psi_1)}_{FH_{(0.9)}}\right) = \varnothing,$$

$$(f,g)^{-1}\left(P^{(\tau_2,\psi_2)}_{FH_{(0.1)}}\right) = \left\{P^{(\pi_3,u_2)}_{FH_{(0.1)}}, P^{(\pi_3,u_3)}_{FH_{(0.1)}}, P^{(\pi_4,u_2)}_{FH_{(0.1)}}\right\},$$

$$(f,g)^{-1}\left(P^{(\tau_2,\psi_3)}_{FH_{(0.5)}}\right) = \left\{P^{(\pi_1,u_3)}_{FH_{(0.5)}}, P^{(\pi_1,u_4)}_{FH_{(0.5)}}, P^{(\pi_4,u_3)}_{FH_{(0.5)}}\right\}$$

Then;

$$(f,g)^{-1}((L_2,\Lambda_2)) = \left\{\begin{matrix} P^{(\pi_1,u_2)}_{FH_{(0.5)}}, P^{(\pi_1,u_3)}_{FH_{(0.5)}}, P^{(\pi_1,u_1)}_{FH_{(0.1)}}, P^{(\pi_2,u_2)}_{FH_{(0.5)}}, P^{(\pi_2,u_3)}_{FH_{(0.5)}}, P^{(\pi_2,u_1)}_{FH_{(0.6)}}, \\ P^{(\pi_3,u_2)}_{FH_{(0.1)}}, P^{(\pi_3,u_3)}_{FH_{(0.1)}}, P^{(\pi_4,u_2)}_{FH_{(0.1)}}, P^{(\pi_1,u_3)}_{FH_{(0.5)}}, P^{(\pi_1,u_4)}_{FH_{(0.5)}}, P^{(\pi_4,u_3)}_{FH_{(0.5)}} \end{matrix}\right\}.$$

Theorem 3.1 Let $(f,g):FSS(Z,\Sigma) \rightarrow FSS(Y,\Sigma')$ be a fuzzy hypersoft function. Then for fuzzy hypersoft sets (L_1,Λ_1) and (L_2,Λ_2) in the fuzzy class $FSS(Z,\Sigma)$, we have,

1. $(f,g)\left(0_{(Z,\Sigma)}\right) = 0_{(Y,\Sigma)}$
2. $(f,g)\left(1_{(Z,\Sigma)}\right) \subseteq 1_{(Y,\Sigma)}$
3. $(f,g)\left((L_1,\Lambda_1) \cup (L_2,\Lambda_2)\right) = (f,g)\left((L_1,\Lambda_1)\right) \cup (f,g)\left((L_2,\Lambda_2)\right)$
4. $(f,g)\left((L_1,\Lambda_1) \cap (L_2,\Lambda_2)\right) \subseteq (f,g)\left((L_1,\Lambda_1)\right) \cap (f,g)\left((L_2,\Lambda_2)\right)$
5. If $(L_1,\Lambda_1) \subseteq (L_2,\Lambda_2)$, then $(f,g)\left((L_1,\Lambda_1)\right) \subseteq (f,g)\left((L_2,\Lambda_2)\right)$

Proof. We only prove (3)-(5)

(3) Suppose that

$$(L_1,\Lambda_1) \cup (L_2,\Lambda_2) = (N,\Lambda_1 \cup \Lambda_2)$$

and

$$(f,g)\big((L_1,\Lambda_1)\big) \cup (f,g)\big((L_2,\Lambda_2)\big) = \big(f(L),g(\Lambda_1)\big) \cup \big(f(S),g(\Lambda_2)\big) = \big(S,g(\Lambda_1) \cup g(\Lambda_2)\big).$$

Then

$$(f,g)\big((L_1,\Lambda_1) \cup (L_2,\Lambda_2)\big) = \big(f(N),g(\Lambda_1 \cup \Lambda_2)\big) = \big(f(N),g(\Lambda_1) \cup g(\Lambda_2)\big).$$

For any $\psi \in Y$ and $\beta' \in g(\Lambda_1) \cup g(\Lambda_2)$, if $f^{-1}(\psi) = \varnothing$, then

$$\mu_{S(\beta')}(\psi) = \mu_{f(N)(\beta')}(\psi) = 0.$$

Otherwise, we consider the following cases.

Case 1: $\beta' \in g(\Lambda_1) - g(\Lambda_2)$. Then $S(\beta') = f(L)(\beta')$. On the other hand, $\beta' \in g(\Lambda_1) - g(\Lambda_2)$ implies that there does not exist $\beta \in \Lambda_2$ such that $g(\beta) = \beta'$, that is, for any $\beta = g^{-1}(\beta') \cap (\Lambda_1 \cup \Lambda_2)$, we have $\beta \in g^{-1}(\beta') \cap (\Lambda_1 - \Lambda_2)$. Hence by Definition-19, we have

$$\mu_{f(N)(\beta')}(\psi) = \mu_{S(\beta')}(\psi)$$

Case 2: $\beta' \in g(\Lambda_2) - g(\Lambda_1)$. Analogous to case 1, we have $\mu_{f(N)(\beta')}(\psi) = \mu_{S(\beta')}(\psi)$.

Case 3: $\beta' \in g(\Lambda_1) \cap g(\Lambda_2)$. Then

$$\mu_{f(N)(\beta')}(\psi) = \max\left\{\mu_{f(L)(\beta')}(\psi), \mu_{f(S)(\beta')}(\psi)\right\} = \mu_{S(\beta')}(\psi)$$

Thus, in any case, $\mu_{f(N)(\beta')}(\psi) = \mu_{S(\beta')}(\psi)$. Therefore,

$$(f,g)\big((L_1,\Lambda_1) \cup (L_2,\Lambda_2)\big) = (f,g)\big((L_1,\Lambda_1)\big) \cup (f,g)\big((L_2,\Lambda_2)\big).$$

(4) Suppose that

$$(L_1,\Lambda_1) \cap (L_2,\Lambda_2) = (N,\Lambda_1 \cup \Lambda_2)$$

and

$$(f,g)((L_1,\Lambda_1)) \cap (f,g)((L_2,\Lambda_2)) = (f(L),g(\Lambda_1)) \cap (f(S),g(\Lambda_2)) = (S,g(\Lambda_1) \cup g(\Lambda_2)).$$

Then

$$(f,g)((L_1,\Lambda_1) \cap (L_2,\Lambda_2)) = (f(N),g(\Lambda_1 \cup \Lambda_2)) = (f(N),g(\Lambda_1) \cup g(\Lambda_2)).$$

For any $\psi \in Y$ and $\beta' \in g(\Lambda_1 \cup \Lambda_2)$, if $f^{-1}(\psi) = \varnothing$, then

$\mu_{S(\beta')}(\psi) = \mu_{f(N)(\beta')}(\psi) = 0.$

Otherwise, we consider the following cases.

Case 1: $\beta' \in g(\Lambda_1) - g(\Lambda_2)$. Then $S(\beta') = f(L)(\beta')$. On the other hand, $\beta' \in g(\Lambda_1) - g(\Lambda_2)$ implies that there does not exist $\beta \in \Lambda_2$ such that $g(\beta) = \beta'$, that is, for any $\beta = g^{-1}(\beta') \cap (\Lambda_1 \cup \Lambda_2)$, we have $\beta \in g^{-1}(\beta') \cap (\Lambda_1 - \Lambda_2)$. We have

$$\mu_{f(N)(\beta')}(\psi) \leq \min\{\mu_{f(L)(\beta')}(\psi), \mu_{f(S)(\beta')}(\psi)\} = \mu_{S(\beta')}(\psi)$$

Case 2: $\beta' \in g(\Lambda_2) - g(\Lambda_1)$. Analogous to case 1, we have $\mu_{f(N)(\beta')}(\psi) \leq \mu_{S(\beta')}(\psi)$.

Case 3: $\beta' \in g(\Lambda_1) \cap g(\Lambda_2)$. Then

$$\mu_{f(N)(\beta')}(\psi) \leq \max\{\mu_{f(L)(\beta')}(\psi), \mu_{f(S)(\beta')}(\psi)\} = \mu_{S(\beta')}(\psi)$$

Therefore

$$(f,g)\big((L_1,\Lambda_1)\cap(L_2,\Lambda_2)\big)\subseteq(f,g)\big((L_1,\Lambda_1)\big)\cap(f,g)\big((L_2,\Lambda_2)\big).$$

(5) Let $(L_1,\Lambda_1)\subseteq(L_2,\Lambda_2)$. Then $\Lambda_1\subseteq\Lambda_2$ and for any $\beta\in\Lambda_1$ and $u\in Z$, we have $f(\Lambda_1)\subseteq f(\Lambda_2)$, we have,

$$\mu_{f(L)(\beta')}(\psi)=\begin{cases}\sup\mu_{L(\beta)}(u), & if\ \ f^{-1}(\psi)\neq\emptyset\\ 0, & otherwise\end{cases}$$

$$\leq\begin{cases}\sup\mu_{S(\beta)}(u), & if\ \ f^{-1}(\psi)\neq\emptyset\\ 0, & otherwise\end{cases}$$

$$=\mu_{f(S)(\beta')}(\psi)$$

Similarly $I_{f(L)(\beta')}(\psi)\leq I_{f(S)(\beta')}(\psi)$ and $L_{f(L)(\beta')}(\psi)\geq L_{f(S)(\beta')}(\psi)$. Therefore, $f\big((L_1,\Lambda_1)\big)\subseteq f\big((L_2,\Lambda_2)\big)$.

Theorem 3.2 Let $(f,g):FSS(Z,\Sigma)\rightarrow FSS(\tilde{Y},\Sigma)$ be a fuzzy hypersoft function. Then for fuzzy hypersoft sets (L_1,Λ_1) and (L_2,Λ_2) in the fuzzy hypersoft class $FSS(Y,\Sigma')$, we have,

1. $(f,g)^{-1}\big(0_{(Y,\Sigma)}\big)=0_{(Z,\Sigma)}$
2. $(f,g)^{-1}\big(1_{(Y,\Sigma)}\big)\subseteq 1_{(Z,\Sigma)}$
3. $(f,g)^{-1}\big((L_1,\Lambda_1)\tilde{\cup}(L_2,\Lambda_2)\big)=(f,g)^{-1}\big((L_1,\Lambda_1)\big)\tilde{\cup}(f,g)^{-1}\big((L_2,\Lambda_2)\big)$
4. $(f,g)^{-1}\big((L_1,\Lambda_1)\tilde{\cap}(L_2,\Lambda_2)\big)=(f,g)^{-1}\big((L_1,\Lambda_1)\big)\tilde{\cap}(f,g)^{-1}\big((L_2,\Lambda_2)\big)$
5. If $(L_1,\Lambda_1)\subseteq(L_2,\Lambda_2)$, then $(f,g)^{-1}\big((L_1,\Lambda_1)\big)\subseteq(f,g)^{-1}\big((L_2,\Lambda_2)\big)$
6. $(L_1,\Lambda_1)\subseteq(f,g)^{-1}\big((f,g)((L_1,\Lambda_1))\big),(f,g)\big((f,g)^{-1}((L_2,\Lambda_2))\big)=(L_2,\Lambda_2)\cap(f,g)\big(1_{(Z,\Sigma)}\big)$

Proof. Straightforward.

Definition 3.12 Let $FSS(Z,\Sigma)$, $FSS(Y,\Sigma)$ be two fuzzy hypersoft classes, $(L_1,\Lambda_1)\in FSS(Z,\Sigma),(L_2,\Lambda_2)\in FSS(Y,\Sigma')$. Then $(f,g)=(f,g):FSS(Z,\Sigma)\rightarrow FSS(Y,\Sigma')$ be a fuzzy hypersoft mapping such that $f:Z\rightarrow Y,g:\Sigma\rightarrow\Sigma'$.

a) The fuzzy hypersoft mapping (f,g) is called a fuzzy hypersoft injective function if for every $P^{(\pi,u_1)}_{FH_{(\mu)}},P^{(\pi,u_2)}_{FH_{(\mu)}}\in(L_1,\Lambda_1),P^{(\pi,u_1)}_{FH_{(\mu)}}\neq P^{(\pi,u_2)}_{FH_{(\mu)}}$ implies

$$(f,g)\left(P_{FH_{(\mu)}}^{(\pi,u_1)}\right) = (f(u_1),g(\beta)) \neq (f,g)\left(P_{FH_{(\mu)}}^{(\pi,u_2)}\right) = (f(u_2),g(\beta)).$$

b) The fuzzy hypersoft mapping (f,g) is called a fuzzy hypersoft surjective function if there exists a fuzzy hypersoft point $u_\beta^\mu \in (L_1,\Lambda_1)$, such that $(f,g)(u_\beta^\mu) = \psi_{\beta'}^{\mu'}$ for every $\psi_{\beta'}^{\mu'} \in (L_2,\Lambda_2)$.
c) The fuzzy hypersoft mapping (f,g) is called a fuzzy hypersoft bijective function if (f,g) is both injective and surjective.
d) The fuzzy hypersoft mapping (f,g) is called a fuzzy hypersoft constant function if $(f,g)(u_\beta^\mu) = \psi_{\beta'}^{\mu'}$ is provided for $\forall u_\beta^\mu \in (L_1,\Lambda_1), \exists \psi_{\beta'}^{\mu'} \in (L_2,\Lambda_2)$.

Example 3.7 Let $Z=\{u_1,u_2\}$, $Y = \{\psi_1,\psi_2\}$, $\Sigma = E_1 \times E_2$ and $\Sigma' = E'_1 \times E'_2$. Let f: $Z \rightarrow Y$ and g: $\Sigma \rightarrow \Sigma'$.

$$E_1 = \{\alpha_1,\alpha_2\}$$

$$E_2 = \{\alpha_3\}$$

$$E'_1 = \{\beta_1,\beta_2\}$$

$$E'_2 = \{\beta_3\}$$

We have define (L_1,Λ_1) and (L_2,Λ_2) in (Z,Σ) and (Y,Σ'), respectively, as follows,

$$(L_1,\Lambda_1) = \left\{\begin{matrix} ((\alpha_1,\alpha_3),(\langle u_1,0.3\rangle,\langle u_2,0.4\rangle)), \\ ((\alpha_2,\alpha_3),(\langle u_1,0.2\rangle,\langle u_2,0.8\rangle)), \end{matrix}\right\},$$

$$(L_2,\Lambda_2) = \left\{\begin{matrix} ((\beta_1,\beta_3),(\langle \psi_1,0.4\rangle,\langle \psi_2,0.6\rangle)), \\ ((\beta_2,\beta_3),(\langle \psi_1,0.2\rangle,\langle \psi_2,0.5\rangle)), \end{matrix}\right\}$$

We assume that $(\alpha_1,\alpha_3)=\pi_1$, $(\alpha_2,\alpha_3)=\pi_2$, and $(\beta_1,\beta_3)=\tau_1$, $(\beta_2,\beta_3)=\tau_2$ for easier operation. We can also write the fuzzy hypersoft sets (L_1,Λ_1) and (L_2,Λ_2) pointwise as follows.

$$(L_1,\Lambda_1) = \left\{ P^{(\pi_1,u_1)}_{FH_{(0.3)}}, P^{(\pi_1,u_2)}_{FH_{(0.4)}}, P^{(\pi_2,u_1)}_{FH_{(0.2)}}, P^{(\pi_2,u_2)}_{FH_{(0.8)}} \right\},$$

$$(L_2,\Lambda_2) = \left\{ P^{(\tau_1,\psi_1)}_{FH_{(0.4)}}, P^{(\tau_1,\psi_2)}_{FH_{(0.6)}}, P^{(\tau_2,\psi_1)}_{FH_{(0.2)}}, P^{(\tau_2,\psi_2)}_{FH_{(0.5)}} \right\}$$

Let f: $Z \rightarrow Y$ and g:$\Sigma \rightarrow \Sigma'$ be mappings defined as

$$f(u_1) = \psi_1,\ f(u_2) = \psi_2$$

$$g(\pi_1) = \tau_2,\ g(\pi_2) = \tau_1$$

Then the fuzzy hypersoft image of (L_1,Λ_1) under (f,g):$(Z,\Sigma) \rightarrow (Y,\Sigma')$ is obtained as

$$(f,g)((L_1,\Lambda_1)) = \left\{ P^{(\tau_1,\psi_1)}_{FH_{(0.2)}}, P^{(\tau_1,\psi_2)}_{FH_{(0.8)}}, P^{(\tau_2,\psi_1)}_{FH_{(0.3)}}, P^{(\tau_2,\psi_2)}_{FH_{(0.4)}} \right\}$$

Then, we have

$$(f,g)((L_1,\Lambda_1)) = \left\{ \begin{array}{l} ((\beta_1,\beta_3),(\langle\psi_1,0.2\rangle,\langle\psi_2,0.8\rangle), \\ ((\beta_2,\beta_3)),(\langle\psi_1,0.3\rangle,\langle\psi_2,0.4\rangle), \end{array} \right\}$$

It is clear that, for any two different points selected from the fuzzy hypersoft set (L_1,Λ_1), the images of these points under (f,g) mapping are different from each other. Therefore this mapping is fuzzy hypersoft injective mapping.

For any hypersoft fuzzy hypersoft point from the selected the fuzzy hypersoft set (L_2,Λ_2), there exist a fuzzy hypersoft point $P^{(\pi,u)}_{FH_{(\mu)}} \in \left(L_1,\Lambda_1\right)$ such that $(f,g)\left(P^{(\pi,u)}_{FH_{(\mu)}}\right) = P^{(\tau,\psi)}_{FH_{(\mu)}}$. Therefore this mapping is also fuzzy hypersoft surjective mapping.

Since the fuzzy hypersoft mapping (f,g) is both injective and surjective, it is bijective.

Definition 3.13 Let $(f,g):FSS(Z,\Sigma) \rightarrow FSS(Y,\Sigma'),(h,z):FSS(Z,\Sigma) \rightarrow FSS(Y,\Sigma')$ be fuzzy hypersoft mappings. Fuzzy hypersoft mappings f and g is called two fuzzy hypersoft equal mappings if for every $P^{(\pi,u)}_{FH_{(\mu)}} \in (Z,\Sigma)$ implies $(f,g)\left(P^{(\pi,u)}_{FH_{(\mu)}}\right) = (h,z)\left(P^{(\pi,u)}_{FH_{(\mu)}}\right)$. It is denoted by $(f,g) = (h,z)$.

Definition 3.14 Let $(f,g):FSS(Z,\Sigma) \rightarrow FSS(Y,\Sigma'),(h,z):FSS(Y,\Sigma') \rightarrow FSS(Z,\Sigma'')$ be fuzzy hypersoft mappings. Then the composition of (f,g) and (h,z), denoted by

$$(h,z)o(f,g):FSS(Z,\Sigma) \rightarrow FSS(Z,\Sigma''),$$

is a fuzzy hypersoft mapping and defined by $((h,z)o(f,g))\left(u^{\mu}_{\beta}\right) = (h,z)\left((f,g)\left(u^{\mu}_{\beta}\right)\right)$ for $u^{\mu}_{\beta} \in FSS(Z,\Sigma)$.

Proposition 3.6 Let $(f,g):FSS(Z,\Sigma) \rightarrow FSS(Y,\Sigma'),(h,z):FSS(Y,\Sigma') \rightarrow FSS(Z,\Sigma'')$ be two fuzzy hypersoft mappings. We have;

1. If the fuzzy hypersoft mappings (f,g) and (h,z) are two fuzzy hypersoft injective mappings then the composition of (f,g) and (h,z), $(h,z)o(f,g):FSS(Z,\Sigma) \rightarrow FSS(Z,\Sigma'')$ is also fuzzy hypersoft injective mappings.
2. If the fuzzy hypersoft mappings (f,g) and (h,z) are two fuzzy hypersoft suurjective mappings then the composition of (f,g) and (h,z), $(h,z)o(f,g):FSS(Z,\Sigma) \rightarrow FSS(Z,\Sigma'')$ is also fuzzy hypersoft surjective mappings.
3. If the fuzzy hypersoft mappings (f,g) and (h,z) are two fuzzy hypersoft bijective mappings then the composition of (f,g) and (h,z), $(h,z)o(f,g):FSS(Z,\Sigma) \rightarrow FSS(Z,\Sigma'')$ is also fuzzy hypersoft bijective mappings.

Proof. Straightforward.

CONCLUSION

In this chapter, the authors define the fuzzy hypersoft cartesian product. Using this definition, they then investigated the concepts of fuzzy hypersoft relation and fuzzy hypersoft function. Unlike previous works in the literature, all the defined concepts had many dimensions and alternatives. The authors talked about fuzzy hypersoft images, inverse images of the given set, and examined their many attributes. The ideas of injective, surjective, bijective and constant fuzzy hypersoft functions are also discussed. The paper is enriched with numerous examples provided. In the future, based on the idea presented in this chapter, the concepts of cartesian product, relation and function on other structures can be redefined and various applications can be presented.

REFERENCES

Abbas, M., Murtaza, G., & Smarandache, F. (2020). Basic operation on hypersoft sets and hypersoft point. *Neutrosophic Sets and System, 35*, 407–421.

Ahmad, B., & Kharal, A. (2009). On Fuzzy Soft Sets. *Advances in Fuzzy Systems, 2009*, 586507.

Ahsan, M., Saeed, M., & Rahman, A. U. (2021). A theoretical and analytical approach for fundamental framework of composite mappings on fuzzy hypersoft classes. *Neutrosophic Sets and Systems, 45*(1), 18.

Aktaş, H., & Çağman, N. (2007). Soft sets and soft groups. *Information Sciences, 177*(13), 2726–2735.

Ali, M. I. (2011). A note on soft sets, rough soft sets and fuzzy soft sets. *Applied Soft Computing, 11*(4), 3329–3332.

Ali, M. I., Feng, F., Liu, X., Min, W. K., & Shabir, M. (2009). On some new operations in soft set theory. *Computers & Mathematics with Applications (Oxford, England), 57*(9), 1547–1553.

Atanassov, K. T. (1999). Intuitionistic fuzzy sets. *Fuzzy Sets and Systems, 20*, 87–96.

Cagman, N., Enginoglu, S., & Citak, F. (2011). Fuzzy soft set theory and its applications. *Iranian Journal of Fuzzy Systems, 8*(3), 137-147.

Celik, Y., Ekiz, C., & Yamak, S. (2013). Applications of fuzzy soft sets in ring theory. *Annals Fuzzy Mathematics and Informatics, 5*(3), 451–462.

Haiyan, Z., & Jingjing, J. (2015, December). Fuzzy soft relation and its application in decision making. In 2015 7th International Conference on Modelling, Identification and Control (ICMIC) (pp. 1-4). IEEE.

Khameneh, A. Z., Kiliçman, A., & Salleh, A. R. (2014). Fuzzy soft product topology. *Ann. Fuzzy Math.Inform (Champaign, Ill.), 7*(6), 935–947.

Kharal, A., & Ahmad, B. (2009). Mappings on fuzzy soft classes. *Advances in Fuzzy Systems, 2009*, 1–6.

Maji, P. K., Biswas, R., & Roy, A. R. (2001). Fuzzy soft sets. *Journal of Fuzzy Mathematics., 9*(3), 589–602.

Maji, P. K., Roy, A. R., & Biswas, R. (2002). An application of soft sets in a decision making problem. *Computers & Mathematics with Applications (Oxford, England), 44*(8-9), 1077–1083.

Maji, P. K., Biswas, R., & Roy, A. R. (2003). Soft set theory. *Computers & Mathematics with Applications (Oxford, England)*, *45*(4-5), 555–562.

Manikantan, T., & Ramkumar, S. (2021). Cartesian product of the extensions of fuzzy soft ideals over near-rings. *International Journal of Dynamical Systems and Differential Equations*, *11*(5-6), 426–447.

Molodtsov, D. (1999). Soft set theory-first results. *Computers & Mathematics with Applications (Oxford, England)*, *37*(4-5), 19–31.

Pawlak, Z. (1982). Rough sets. *International Journal of Computer & Information Sciences*, *11*(5), 341-356.

Rahman, A. U., Saeed, M., Smarandache, F., & Ahmad, M. R. (2020). Development of Hybrids of Hypersoft Set with Complex Fuzzy Set, Complex Intuitionistic Fuzzy set and Complex Neutrosophic Set. *Neutrosophic Sets and Systems*, *1*, 334.

Roy, A. R., & Maji, P. K. (2007). A fuzzy soft set theoretic approach to decision making problems. *Journal of Computational and Applied Mathematics*, *203*(2), 412–418.

Roy, S., & Samanta, T. K. (2013). An introduction to open and closed sets on fuzzy soft topological spaces. *Annals of Fuzzy Mathematics and Informatics*, *6*(2), 425–431.

Saeed, M., Rahman, A. U., Ahsan, M., & Smarandache, F. (2021a). An inclusive study on fundamentals of hypersoft set. *Theory and Application of Hypersoft Set*, *1*, 1–23.

Saeed, M., Ahsan, M., Ur Rahman, A., Saeed, M. H., & Mehmood, A. (2021b). An application of neutrosophic hypersoft mapping to diagnose brain tumor and propose appropriate treatment. *Journal of Intelligent & Fuzzy Systems*, *41*(1), 1677–1699.

Saeed, M., Rahman, A. U., Ahsan, M., & Smarandache, F. (2022). Theory of hypersoft sets: Axiomatic properties, aggregation operations, relations, functions and matrices. *Neutrosophic Sets and Systems*, *51*(1), 46.

Saeed, M., Ahsan, M., Saeed, M. H., Rahman, A. U., Mehmood, A., Mohammed, M. A., & Damaševičius, R. (2022). An optimized decision support model for COVID-19 diagnostics based on complex fuzzy hypersoft mapping. *Mathematics*, *10*(14), 2472.

Saeed, M., Ahsan, M., Saeed, M. H., Rahman, A. U., Mohammed, M. A., Nedoma, J., & Martinek, R. (2023). An algebraic modeling for tuberculosis disease prognosis and proposed potential treatment methods using fuzzy hypersoft mappings. *Biomedical Signal Processing and Control*, *80*, 104267.

Sezgin, A., & Atagün, A. O. (2011). On operations of soft sets. *Computers & Mathematics with Applications (Oxford, England)*, *61*(5), 1457–1467.

Smarandache, F. (2018). Extension of soft set to hypersoft set, and then to plithogenic hypersoft set. *Neutrosophic Sets and Systems, 22*(1), 168-170.

Tanay, B., & Kandemir, M. B. (2011). Topological structure of fuzzy soft sets. *Computers & Mathematics with Applications (Oxford, England)*, *61*(10), 2952–2957.

Yolcu, A., & Ozturk, T. Y. (2021). Fuzzy Hypersoft Sets and It's Application to Decision-Making. *Theory and Application of Hypersoft Set* (pp. 50-64), Pons Publishing House, Brussel.

Yolcu, A., Smarandache, F., & Öztürk, T. Y. (2021). Intuitionistic fuzzy hypersoft sets. *Communications Faculty of Sciences University of Ankara Series A1 Mathematics and Statistics, 70*(1), 443-455.

Yolcu, A., & Ozturk, T. Y. (2022). On fuzzy hypersoft topological spaces. *Caucasian Journal of Science*, *9*(1), 1–19.

Yolcu, A. (2023). Intuitionistic fuzzy hypersoft topology and ıts applications to multi-criteria decision-making. *Sigma*, *41*(1), 106–118.

Zadeh, L. A. (1965). Fuzzy sets. *Information and Control*, *8*, 338–353.

Chapter 6
Applications of an Interval-Valued Anti-Neutrosophic Semigroup

Sambandan Sivaramakrishnan
https://orcid.org/0000-0002-8647-1174
Manakula Vinayagar Institute of Technology, Puducherry, India

Parthasarathy Balaji
Measi Academy of Architecture, Royapettah, India

ABSTRACT

In this chapter, the authors explore the notion of an interval-valued anti-neutrosophic semigroup, which combines the theories of semigroups and interval-valued anti-neutrosophic sets. They begin by introducing the basic definitions, proving some results and their generalizations. Furthermore, they investigate the practical applications of an interval-valued anti-neutrosophic semigroup in various domains. One such application involves utilizing superior-subordinate roles, where the semigroup operations model the interactions and relationships between different roles within an organization or system. Another practical application the authors explore is the use of DNA sequences in the context of an interval-valued anti-neutrosophic semigroup and they provide some results on it.

1. INTRODUCTION

In 1965, Zadeh (Zadeh,1965) initially proposed the concept of a fuzzy set. Fuzzy sets offer a highly dependable mathematical tool for addressing ambiguity and unpredictability, enabling more nuanced and flexible representations of intricate

DOI: 10.4018/979-8-3693-3204-7.ch006

systems and phenomena. The applications of fuzzy set theory span across various fields, such as artificial intelligence, decision-making, control systems, pattern recognition and data mining. It presents a versatile framework for managing imprecise, incomplete or uncertain information, making it well-suited for modelling real-world problems where precise boundaries are challenging to define. Zadeh (Zadeh,1975), a renowned mathematician and computer scientist, introduced the groundbreaking idea of interval-valued fuzzy sets in the realm of fuzzy logic in 1975. The introduction of interval-valued fuzzy sets has opened up new avenues for research and development in the field of fuzzy logic. This, in turn, has led to further advancements in algorithms and techniques for handling uncertainty and imprecision, contributing to the progress of fuzzy logic as an efficient tool for dealing with complex and unpredictable data. Subsequently, Rosenfeld's work has played a significant role in the advancement of fuzzy group concepts (Rosenfeld, A.,1971). In addition, Kuroki provided (Kuroki, 1981,1982 & 1991) detailed descriptions, characterizations of fuzzy semigroups and various types of fuzzy ideals. The researchers (Narayanan & Meenakshi, 2003) presented the concept of an M-semigroup and its relevance in the context of kinship relationships. Further, the researchers (Vijayabalaji & Sivaramakrishnan, 2012) put forth the idea to create an anti-fuzzy M-semigroup.

The author (Biswas.1990) introduced the theoretical concept of an anti-fuzzy subgroup, The researchers (Yassein & Mohammed, 2011) proposed the idea of anti-fuzzy semigroups. The eminent researcher Atanassov developed intuitionistic fuzzy sets (IFS) (Atanassov, 1986) as a generalization of classical fuzzy sets. The purpose of IFS was to address the limitations of traditional fuzzy sets in handling uncertain, imprecise and incomplete information. Atanassov's proposal aimed to model uncertainty in a more nuanced and flexible manner by considering both membership and non-membership degrees, along with a hesitation degree. The Pythagorean fuzzy set, introduced by Yager (Yager, 2013 & 2014), serves as an interpretation of the fuzzy set. Initially proposed by Yager and Abbasov (Yager & Abbasov,2013), this notion can be viewed as an effective conjecture of intuitionistic fuzzy sets. In 2005, The researcher (Smarandache, F.,2006) introduced the innovative concept of Neutrosophic set (NS), which incorporates an intermediate membership. The Pythagorean Neutrosophic set was first introduced by Jansi et al (Jansi et al, 2019) as an extension of the Neutrosophic set. Sivaramakrishnan et al (Sivaramakrishnan et al, 2024) presented the concept of Neutrosophic interval-valued anti-fuzzy linear space.

A semigroup is acting as an intermediate structure between a monoid and a group. Understanding semigroups is vital in various mathematical disciplines, including algebra, combinatorics, and theoretical computer science. Employing semigroup theory to simulate basic functions in superior-subordinate relationships can provide organizations with a systematic understanding of their internal dynamics, potentially leading to improvements in operational efficiency and effectiveness.

As a result, the Subordinate Set (S) and Manager Set (M) share authority and responsibility. The manager is responsible for allocating duties, providing guidance, and overseeing the work of subordinates. Subordinates, in turn, are expected to follow orders, complete tasks in a timely manner and report to their managers.

The manager-subordinate relationship is critical in every business, ensuring that tasks are executed efficiently and successfully. To effectively manage their subordinates, managers must possess strong leadership skills, excellent communication abilities and a thorough understanding of their team's strengths and weaknesses. Subordinates, in turn, must be able to follow instructions, collaborate with team members and report any issues or concerns to their managers. Overall, the subordinate set and manager set must embody mutual respect, trust and collaboration. When both parties collaborate effectively, the organization can successfully achieve its aims and objectives.

The researcher (Zhang et al, 2006) effectively defined the function f(x) through the process of mapping elements from the set {A, G, T, C} to the set {1, -1, i, -i}. This implies that a value f(x) exists in the set {1, -1, i, -i}, corresponding to each element x in the set {A, G, T, C}. Zhang and his colleagues introduced a methodical and transparent approach to assigning values to the elements in the set. In genetics, for instance, the set {A, G, T, C} denotes the four nucleotides comprising DNA. Utilizing the function f(x) to assign values to these nucleotides enables scientists to conduct calculations and analyse genetic data more efficiently. This can potentially facilitate comprehension of genetic variations, patterns of gene expression, and the prognostication of specific mutations' consequences. Moreover, the function f(x) can be utilized to examine the interactions between proteins and DNA in molecular biology. The assignment of values to the nucleotides enables scientists to ascertain the intensity and characteristics of these interactions, thereby providing valuable information regarding various biological processes, including DNA replication, transcription and translation.

The function f(x) can be employed in the domain of bioinformatics to analyse and compare DNA sequences across various organisms. Researchers have the ability to generate algorithms and computational tools that discern similarities and distinctions in DNA sequences through the assignment of values to the nucleotides. This may be utilized to design new drugs or therapies, identify functional elements in the genome, or even investigate evolutionary relationships. In general, the development of the function f(x) by Zhang et al has introduced novel opportunities to the fields of molecular biology, genetics, and bioinformatics. Through the process of designating values to the elements comprising the set.

In the case of DNA sequences, the alphabet set typically consists of adenine (A), cytosine (C), thymine (T) and guanine (G), which together form a semigroup under specific operations. Neutrosophic sets enhance the functionalities of fuzzy sets and intuitionistic fuzzy sets by explicitly incorporating degrees of indeterminacy and contradiction. Fuzzy sets are designed to handle uncertainty through membership degrees, while intuitionistic fuzzy sets are specifically designed to address uncertainty and hesitation. On the other hand, Neutrosophic sets simultaneously account for contradiction, ambiguity, and indeterminacy, providing a more comprehensive methodology. Neutrosophic set theory offers a versatile framework for representing and reasoning about uncertain and contradictory information, making it highly valuable for addressing real-world problems where ambiguity and complexity are prevalent.

A fuzzy semigroup allows elements to combine in a flexible or fuzzy manner, departing from the precise binary operation of classical semigroups. This concept originates from fuzzy set theory, where membership degrees replace traditional set membership. In contrast, an anti-fuzzy semigroup emphasizes crisp, precise operations rather than dealing with degrees of membership and uncertainty, diverging from the principles of fuzzy semigroups. Combining Neutrosophic and single-valued sets yields Neutrosophic single-valued sets, where each element's truth value is determined within Neutrosophic logic. This allows elements to be true, false, or indeterminate, reflecting inherent uncertainties and contradictions. Neutrosophic single-valued sets are applied in fields with prevalent uncertainty and ambiguity, such as pattern recognition, artificial intelligence and decision-making, offering formal methods for systematic reasoning about uncertain and contradictory information. Several researchers offer significant contributions to the enhancement of hybrid structures of the above-mentioned models ((Chinnadurai, V., & Arulselvam, A, 2021), (Prema, R., & Radha, R. 2022), (Sivaramakrishnan, S., & Vijayabalaji, S. ,2020) (Sivaramakrishnan, S., & Suresh, K. ,2019)). Based upon the inspiration from the above-mentioned literature survey we put forth this paper.

Furthermore, our investigation focuses on the practical use of an interval-valued anti-neutrosophic semigroup in different fields. One example of this use involves integrating superior-subordinate fundamental roles, where the semigroup functions are employed to model the relationships and interactions among various roles within an entity or structure. Furthermore, we are examining the practical application of DNA sequences within the framework of interval-valued anti-neutrosophic semigroup and we are sharing some outcomes related to it.

2. PRELIMINARIES

Definition 2.1 (Yassein, H.R., & Mohammed, A. H. 2011)

Let the semigroup be **S**. The fuzzy set δ: **S** → [0,1] in which is an anti-fuzzy semigroup if $\delta(xy) \leq \max\{\delta(x), \delta(y)\}$, ∀ x, y ∈ **S**.

Definition 2.2 (Smarandache, F. 2005)

Let X represent the domain under consideration. A Neutrosophic set (NS), denoted as $\Omega = \{s, \xi_\Omega(s), \psi_\Omega(s), \zeta_\Omega(s) \mid s \in X\}$ is defined as follows: $\Omega = \{\xi_\Omega(s), \psi_\Omega(s), \zeta_\Omega(s)\}$. Here, $\xi_\Omega: X \to [0,1]$, $\psi_\Omega : X \to [0,1]$ and $\zeta_\Omega : X \to [0,1]$ represent the degree of truth - membership and indeterminacy - membership and false - membership of the object s∈ X respectively. It should be noted that $0 \leq \xi_\Omega(s) + \psi_\Omega(s) + \zeta_\Omega(s) \leq 3$.

Definition 2.3 (Rudolf Lidl & Günter Pilz. 1984)

A Kinship system is represented as a semigroup **S** = [X, R], where

(i) X denotes a collection of "elementary kinship relationships",
(ii) R represents a relation on X*, signifying the equality of kinship relationships.

The product in S is consistently interpreted as a relational product.

Remark 2.4. S is defined as Subordinate Set and M is defined as Manager Set. Subordinate means" Subject to or under the authority of superior" (Dictionary.com). Hence it is understood that subordinate works under the authority of the Superior. Here the superior has been referred as Manager. A person who is assigned with the work can be subordinate Here the person who assigns the work has been referred as manager. Manager is someone who controls and organizes someone. Here someone has been referred as Subordinate.

Remark 2.5. Kinship relationships offer crucial social and emotional backing, but the superior-subordinate framework proves to be a more efficient approach for tackling sociological issues pertaining to organizational efficiency, equality and conflict resolution. With its structured, meritocratic and flexible characteristics, it aligns better with the demands of modern society and organizations, promoting a more equitable, effective and vibrant atmosphere.

3. APPLICATIONS OF AN INTERVAL-VALUED ANTI- NEUTROSOPHIC SEMIGROUP

Definition 3.1. Let **S** be a crisp semigroup and let $\overline{\Omega} = (\xi_{\overline{\Omega}}, \psi_{\overline{\Omega}}, \zeta_{\overline{\Omega}})$ be an interval-valued anti-neutrosophic semigroup (IVANS) where,

$\xi_{\overline{\Omega}}: S \to D[0,1]$, $\psi_{\overline{\Omega}}: S \to D[0,1]$, and $\zeta_{\overline{\Omega}}: S \to D[0,1]$

represent the degree of membership, indeterminacy and non-membership, subject to the condition $0 \leq \xi_{\overline{\Omega}}(s) + \psi_{\overline{\Omega}}(s) + \zeta_{\overline{\Omega}}(s) \leq 3$ if it holds.

(i) $\xi_{\overline{\Omega}}(s_1 s_2) \leq \max\{\xi_{\overline{\Omega}}(s_1), \xi_{\overline{\Omega}}(s_2)\}$,
(ii) $\psi_{\overline{\Omega}}(s_1 s_2) \leq \max\{\psi_{\overline{\Omega}}(s_1), \psi_{\overline{\Omega}}(s_2)\}$,
(iii) $\zeta_{\overline{\Omega}}(s_1 s_2) \geq \min\{\psi_{\overline{\Omega}}(s_1), \psi_{\overline{\Omega}}(s_2)\}$ *for every* $s_1, s_2 \in S$

Example 3.2. S is defined as a crisp semigroup with elements $\{e_s, a_s, b_s, c_s\}$ and operated by "•".

Table 1. Example for an interval-valued anti-neutrosophic semigroup - Cayley table under operation

•	e_s	a_s	b_s	c_s
e_s	e_s	a_s	b_s	c_s
a_s	e_s	a_s	b_s	c_s
b_s	b_s	c_s	e_s	a_s
c_s	b_s	c_s	e_s	a_s

Define the mappings $\xi_{\overline{\Omega}}: S \rightarrow D[0,1], \psi_{\overline{\Omega}}: S \rightarrow D[0,1]$, and $\zeta_{\overline{\Omega}}: S \rightarrow D[0,1]$ by

$$\xi_{\overline{\Omega}}(s) = \begin{cases} [0.43, 0.5], & \text{if } s = e_s, a_s, \\ [0.91, 1], & \text{otherwise.} \end{cases}$$

$$\psi_{\overline{\Omega}}(s) = \begin{cases} [0.24, 0.33], & \text{if } s = e_s, a_s, \\ [0.72, 0.9], & \text{otherwise.} \end{cases}$$

$$\zeta_{\overline{\Omega}}(s) = \begin{cases} [0.82, 0.93], & \text{if } s = e_s, a_s, \\ [0.5, 0.42], & \text{otherwise.} \end{cases}$$

Clearly $\overline{\Omega} = (\xi_{\overline{\Omega}}, \psi_{\overline{\Omega}}, \zeta_{\overline{\Omega}})$ is an IVANS of **S.**

Example 3.3. A Kinship system can be defined as a semigroup $\mathfrak{R} = [X, L]$, where X represents a set of kinship relationships and L is a relation on X*L that signifies equality among kinship relationships. Within X*, a free semigroup is formed where each combination of relationships from X is distinct.

We take manager and subordinate in its term meaning defined in the Dictionary. In a Department S and M Jointly works for attaining the defined objectives. M and M works together the outcome or product would result relationship set with the subordinate set as they arrive at set of instruction to be given for subordinates. MS is Equal to SM, as they join as a team to work for the common Goal. In the organization both Top down as well as Bottom-up approach is existing. The product decision arises in either way and remains neutral. S and S work together for the attainment of team (group of subordinates) targets as subordination of individual interests with group interest in applicable (As per Mc Kinsey 7S Model). In general, the defined relationships between the Manager and the subordinate would be formal and directed toward the attainment of common objective. Hence forth, Manager Set and Subordinate can be identified as the Kinship set with defined relationships.

Let M = *is manager of*, S = *is subordinate of*, MS = *is manager of the subordinate*, SM = *is subordinate of the manager*.

Let X = {M, S, MS, SM}. The kinship relationship equality set is denoted as L and it consists of the following pairs: {(MM, M), (SS, S), (MS, SM)}. Let the operation of the relation product is denoted by the symbol ◇. In the result (MM, M), which is the first pair of L, depicts that the" manager of the manager" results in the relationship set" manager".

That is, M ◇ M = M.

Table 2. Example for an interval-valued anti-neutrosophic semigroup - Cayley table with kinship relationship under the operation ◇

◇	M	S	MS	SM
M	M	MS	MS	SM
S	SM	S	MS	SM
MS	SM	MS	MS	SM
SM	SM	SM	MS	SM

Note that $\mathfrak{R}$ = [X, L] is a semigroup.

Define the mappings $\xi_{\overline{\Omega}}:S \rightarrow D[0,1]$, $\psi_{\overline{\Omega}}:S \rightarrow D[0,1]$, and $\zeta_{\overline{\Omega}}:S \rightarrow D[0,1]$ by

$$\xi_{\overline{\Omega}}(S) = [0,0], \xi_{\overline{\Omega}}(M) = \xi_{\overline{\Omega}}(MS) = \xi_{\overline{\Omega}}(SM) = [0.9, 1],$$

$$\psi_{\overline{\Omega}}(S) = [0,0], \psi_{\overline{\Omega}}(M) = \psi_{\overline{\Omega}}(MS) = \psi_{\overline{\Omega}}(SM) = [0.7,0.9],$$

$\zeta_{\overline{\Omega}}(S) = [0.9,1], \zeta_{\overline{\Omega}}(M) = \zeta_{\overline{\Omega}}(MS) = \zeta_{\overline{\Omega}}(SM) = [0.2,0.3].$

Clearly $\overline{\Omega} = (\xi_{\overline{\Omega}}, \psi_{\overline{\Omega}}, \zeta_{\overline{\Omega}})$ is an IVANS of $\mathfrak{R}$.

The second example relates DNA sequence with IVANS.

Example 3.4. Zhang et al have defined f(x) as follows: f(x) maps the set $\{A_S, G_S, T_S, C_S\}$ to the set $\{1, -1, i, -i\}$.

$$f(x) = \begin{cases} 1, \text{ if } x = G_S, \\ -1, \text{ if } x = T_S, \\ i \text{ if } x = A_S, \\ -i \text{ if } x = C_S. \end{cases}$$

A denotes Adenine, G denotes Guanine, C denotes Cytosine, and T denotes Thymine, while x denotes any of the four nucleotides.

The matrices associated with each nucleotide are as follows:

$$A_S = \begin{pmatrix} 0 & 1 \\ -1 & 0 \end{pmatrix}, T_S = \begin{pmatrix} -1 & 0 \\ 0 & -1 \end{pmatrix}, G_S = \begin{pmatrix} 1 & 0 \\ 0 & 1 \end{pmatrix} \text{ and } C_S = \begin{pmatrix} 0 & -1 \\ 1 & 0 \end{pmatrix}$$

Consider the semigroup **S**, defined as **S** = $\{A_S, G_S, T_S, C_S\}$, with the operation denoted by "•".

Table 3. Example for an interval-valued anti-neutrosohic semigroup - Cayley table with DNA sequence under the operation·

•	A_S	G_S	T_S	C_S
A_S	T_S	A_S	C_S	G_S
G_S	A_S	G_S	T_S	C_S
T_S	C_S	T_S	G_S	A_S
C_S	G_S	C_S	A_S	T_S

Define the mappings $\xi_{\overline{\Omega}}: S \to D[0,1], \psi_{\overline{\Omega}}: S \to D[0,1]$, and $\zeta_{\overline{\Omega}}: S \to D[0,1]$ by

$$\xi_{\overline{\Omega}}(s) = \begin{cases} [0,0], & \text{if } s = G, \\ [0.8,0.9], & \text{otherwise.} \end{cases}$$

$$\psi_{\overline{\Omega}}(s) = \begin{cases} [0,0], & \text{if } s = G, \\ [0.7,0.89], & \text{otherwise.} \end{cases}$$

$$\zeta_{\overline{\Omega}}(s) = \begin{cases} [1,1], & if \quad s = G, \\ [0.5, 0.42], & otherwise. \end{cases}$$

We see that $\overline{\Omega} = (\xi_{\overline{\Omega}}, \psi_{\overline{\Omega}}, \zeta_{\overline{\Omega}})$ is an IVANS of **S.**

Theorem 3.5. If **S** is a semigroup and $\overline{\Omega} = (\xi_{\overline{\Omega}}, \psi_{\overline{\Omega}}, \zeta_{\overline{\Omega}})$ represents an interval-valued anti-neutrosophic set, then $\overline{\Omega}$ can be considered as an IVANS of **S** if and only if $\overline{\Omega}^c$ is an interval-valued neutrosophic semigroup within **S.**

Proof.

First, we assume that $\overline{\Omega}$ is an IVANS of **S.**

Let $s_1, s_2 \in$ **S.**

$$\begin{aligned} \xi_{\overline{\Omega}^c}(s_1 s_2) &= 1 - \xi_{\overline{\Omega}}(s_1 s_2) \\ &\geq 1 - \max\{\xi_{\overline{\Omega}}(s_1), \xi_{\overline{\Omega}}(s_2)\} \\ \text{(i)} \quad &= \min\{1 - \xi_{\overline{\Omega}}(s_1), 1 - \xi_{\overline{\Omega}}(s_2)\} \\ &= \min\{\xi_{\overline{\Omega}^c}(s_1), \xi_{\overline{\Omega}^c}(s_2)\} \\ \xi_{\overline{\Omega}^c}(s_1 s_2) &\geq \min\{\xi_{\overline{\Omega}^c}(s_1), \xi_{\overline{\Omega}^c}(s_2)\} \end{aligned}$$

$$\begin{aligned} \psi_{\overline{\Omega}^c}(s_1 s_2) &= 1 - \psi_{\overline{\Omega}}(s_1 s_2) \\ &\geq 1 - \max\{\psi_{\overline{\Omega}}(s_1), \psi_{\overline{\Omega}}(s_2)\} \\ \text{(ii)} \quad &= \min\{1 - \psi_{\overline{\Omega}}(s_1), 1 - \psi_{\overline{\Omega}}(s_2)\} \\ &= \min\{\psi_{\overline{\Omega}^c}(s_1), \psi_{\overline{\Omega}^c}(s_2)\} \\ \psi_{\overline{\Omega}^c}(s_1 s_2) &\geq \min\{\psi_{\overline{\Omega}^c}(s_1), \psi_{\overline{\Omega}^c}(s_2)\} \end{aligned}$$

$$\begin{aligned} \zeta_{\overline{\Omega}^c}(s_1 s_2) &= 1 - \zeta_{\overline{\Omega}}(s_1 s_2) \\ &\leq 1 - \min\{\zeta_{\overline{\Omega}}(s_1), \zeta_{\overline{\Omega}}(s_2)\} \\ \text{(iii)} \quad &= \max\{1 - \zeta_{\overline{\Omega}}(s_1), 1 - \zeta_{\overline{\Omega}}(s_2)\} \\ &= \max\{\zeta_{\overline{\Omega}^c}(s_1), \zeta_{\overline{\Omega}^c}(s_2)\} \\ \zeta_{\overline{\Omega}^c}(s_1 s_2) &\leq \max\{\zeta_{\overline{\Omega}^c}(s_1), \zeta_{\overline{\Omega}^c}(s_2)\} \end{aligned}$$

Therefore, $\overline{\Omega}^c$ is an interval-valued neutrosophic semigroup in **S**.

Conversely, we consider $\overline{\Omega}^c$ be an interval-valued neutrosophic semigroup in **S**.

$$\xi_{\overline{\Omega}}(s_1 s_2) = 1 - \xi_{\overline{\Omega}^c}(s_1 s_2)$$
$$\leq 1 - \min\{\xi_{\overline{\Omega}^c}(s_1), \xi_{\overline{\Omega}^c}(s_2)\}$$
$$= \max\{1 - \xi_{\overline{\Omega}^c}(s_1), 1 - \xi_{\overline{\Omega}^c}(s_2)\}$$
$$= \max\{\xi_{\overline{\Omega}}(s_1), \xi_{\overline{\Omega}}(s_2)\}$$
$$\xi_{\overline{\Omega}}(s_1 s_2) \leq \max\{\xi_{\overline{\Omega}}(s_1), \xi_{\overline{\Omega}}(s_2)\}$$

$$\psi_{\overline{\Omega}}(s_1 s_2) = 1 - \psi_{\overline{\Omega}^c}(s_1 s_2)$$
$$\leq 1 - \min\{\psi_{\overline{\Omega}^c}(s_1), \psi_{\overline{\Omega}^c}(s_2)\}$$
$$= \max\{1 - \psi_{\overline{\Omega}^c}(s_1), 1 - \psi_{\overline{\Omega}^c}(s_2)\}$$
$$= \max\{\psi_{\overline{\Omega}}(s_1), \psi_{\overline{\Omega}}(s_2)\}$$
$$\psi_{\overline{\Omega}}(s_1 s_2) \leq \max\{\psi_{\overline{\Omega}}(s_1), \psi_{\overline{\Omega}}(s_2)\}$$

$$\zeta_{\overline{\Omega}}(s_1 s_2) = 1 - \zeta_{\overline{\Omega}^c}(s_1 s_2)$$
$$\geq 1 - \max\{\zeta_{\overline{\Omega}^c}(s_1), \zeta_{\overline{\Omega}^c}(s_2)\}$$
$$= \min\{1 - \zeta_{\overline{\Omega}^c}(s_1), 1 - \zeta_{\overline{\Omega}^c}(s_2)\}$$
$$= \min\{\zeta_{\overline{\Omega}}(s_1), \zeta_{\overline{\Omega}}(s_2)\}$$
$$\zeta_{\overline{\Omega}}(s_1 s_2) \geq \min\{\zeta_{\overline{\Omega}}(s_1), \zeta_{\overline{\Omega}}(s_2)\}$$

Hence, $\overline{\Omega}$ is an IVANS of **S.**

Definition 3.6. Let $\overline{\Omega}_1$ and $\overline{\Omega}_2$ be Neutrosophic anti fuzzy subsets of S_1 and $\mathbf{S}_2$ respectively. Then the cartesian product of $\overline{\Omega}_1$ and $\overline{\Omega}_2$ denoted by $\overline{\Omega}_1 \times \overline{\Omega}_2$ is defined by

$$\overline{\Omega}_1 \times \overline{\Omega}_2 = \left\{\left\langle (s_1 s_2), \xi_{\overline{\Omega}_1 \times \overline{\Omega}_2}(s_1 s_2), \psi_{\overline{\Omega}_1 \times \overline{\Omega}_2}(s_1 s_2), \zeta_{\overline{\Omega}_1 \times \overline{\Omega}_2}(s_1 s_2)\right\rangle s_1 \in s_1, s_2 \in s_2\right\}$$

where,

$$\xi_{\overline{\Omega}_1 \times \overline{\Omega}_2}(s_1 s_2) = \max\{\xi_{\overline{\Omega}_1}(s_1), \xi_{\overline{\Omega}_2}(s_2)\},$$

$$\psi_{\overline{\Omega}_1 \times \overline{\Omega}_2}(s_1 s_2) = \max\left\{ \psi_{\overline{\Omega}_1}(s_1), \psi_{\overline{\Omega}_2}(s_2) \right\}$$

.

And

$$\zeta_{\overline{\Omega}_1 \times \overline{\Omega}_2}(s_1 s_2) = \min\left\{ \zeta_{\overline{\Omega}_1}(s_1), \zeta_{\overline{\Omega}_2}(s_2) \right\}$$

.

Theorem 3.9. If $\overline{\Omega}_1$ and $\overline{\Omega}_2$ are IVANSs of S_1 and $\mathbf{S}_2$, then $\overline{\Omega}_1 \times \overline{\Omega}_2$ is an IVANS of $\mathbf{S}_1 \times \mathbf{S}_2$.

Proof.

Let $s = (s_1, s_2), t = (t_1, t_2) \in \mathbf{S}_1 \times \mathbf{S}_2$. *Then*

$$\xi_{\overline{\Omega}_1 \times \overline{\Omega}_2}(st) = \xi_{\overline{\Omega}_1 \times \overline{\Omega}_2}\left((s_1, s_2)(t_1, t_2)\right)$$

$$= \xi_{\overline{\Omega}_1 \times \overline{\Omega}_2}\left((s_1 t_1), (s_2 t_2)\right)$$

$$= \max\left\{ \xi_{\overline{\Omega}_1}(s_1 t_1), \xi_{\overline{\Omega}_2}(s_2 t_2) \right\}$$

$$\leq \max\left\{ \max\left[\xi_{\overline{\Omega}_1}(s_1), \xi_{\overline{\Omega}_1}(t_1)\right], \max\left[\xi_{\overline{\Omega}_2}(s_2), \xi_{\overline{\Omega}_2}(t_2)\right] \right\}$$

$$= \max\left\{ \max\left[\xi_{\overline{\Omega}_1}(s_1), \xi_{\overline{\Omega}_2}(s_2)\right], \max\left[\xi_{\overline{\Omega}_1}(t_1), \xi_{\overline{\Omega}_2}(t_2)\right] \right\}$$

$$= \max\left\{ \xi_{\overline{\Omega}_1 \times \overline{\Omega}_2}(s_1, s_2), \xi_{\overline{\Omega}_1 \times \overline{\Omega}_2}(t_1, t_2) \right\}$$

$$= \max\left\{ \xi_{\overline{\Omega}_1 \times \overline{\Omega}_2}(s), \xi_{\overline{\Omega}_1 \times \overline{\Omega}_2}(t) \right\}$$

$$\xi_{\overline{\Omega}_1 \times \overline{\Omega}_2}(st) \leq \max\left\{ \xi_{\overline{\Omega}_1 \times \overline{\Omega}_2}(s), \xi_{\overline{\Omega}_1 \times \overline{\Omega}_2}(t) \right\}$$

$$\zeta_{\overline{\Omega}_1 \times \overline{\Omega}_2}(st) = \zeta_{\overline{\Omega}_1 \times \overline{\Omega}_2}\left((s_1, s_2)(t_1, t_2)\right)$$

$$= \zeta_{\overline{\Omega}_1 \times \overline{\Omega}_2}\left((s_1 t_1), (s_2 t_2)\right)$$

$$= \min\left\{ \zeta_{\overline{\Omega}_1}(s_1 t_1), \zeta_{\overline{\Omega}_2}(s_2 t_2) \right\}$$

$$\geq \min\left\{ \min\left[\zeta_{\overline{\Omega}_1}(s_1), \zeta_{\overline{\Omega}_2}(t_1)\right], \min\left[\zeta_{\overline{\Omega}_2}(s_2), \zeta_{\overline{\Omega}_2}(t_2)\right] \right\}$$

$$= \min\left\{ \min\left[\zeta_{\overline{\Omega}_1}(s_1), \zeta_{\overline{\Omega}_2}(s_2)\right], \min\left[\zeta_{\overline{\Omega}_1}(t_1), \zeta_{\overline{\Omega}_2}(t_2)\right] \right\}$$

$$= \min\left\{ \zeta_{\overline{\Omega}_1 \times \overline{\Omega}_2}(s_1, s_2), \zeta_{\overline{\Omega}_1 \times \overline{\Omega}_2}(t_1, t_2) \right\}$$

$$= \min\left\{ \zeta_{\overline{\Omega}_1 \times \overline{\Omega}_2}(s), \zeta_{\overline{\Omega}_1 \times \overline{\Omega}_2}(t) \right\}$$

$$\zeta_{\overline{\Omega}_1 \times \overline{\Omega}_2}(st) \geq \min\left\{ \zeta_{\overline{\Omega}_1 \times \overline{\Omega}_2}(s), \zeta_{\overline{\Omega}_1 \times \overline{\Omega}_2}(t) \right\}$$

So, $\overline{\Omega}_1 \times \overline{\Omega}_2$ is an IVANS of $\mathbf{S}_1 \times \mathbf{S}_2$.

4. CONCLUSION

The present study extensively analyses the concept of an interval-valued anti-neutrosophic semigroup, which integrates the principles of semigroups and interval-valued anti-neutrosophic sets. We investigate the practical applications of an interval-valued anti-neutrosophic semigroup across various domains. One such application involves utilizing superior-subordinate roles, where semigroup operations model the interactions and relationships between different roles within an organization or a system. We also investigate the practical application of DNA sequences in the context of an interval-valued anti-neutrosophic semigroup.

This idea will be implemented in various algebraic structures in the further research work:

- M-semigroup,
- Ring theory
- Vector space
- Metric space and
- Decision making problems on rough and soft sets.

REFERENCES

Atanassov, K. T. (1986). Intuitionistic fuzzy sets. *Fuzzy Sets and Systems, 20*(1), 87–96. DOI: 10.1016/S0165-0114(86)80034-3

Biswas, R. (1990). Fuzzy subgroups and anti fuzzy subgroups. *Fuzzy Sets and Systems, 35*(1), 121–124. DOI: 10.1016/0165-0114(90)90025-2

Chinnadurai, V., & Arulselvam, A. (2021). Pythagorean Neutrosophic Ideals in Semigroups. *Neutrosophic Sets and Systems, 41*, 258–269.

Fletcher, C. R. (1986). Applied abstract algebra, by R. Lidl and G. Pilz. Pp 545. DM 136. 1984. ISBN 3-540-96035-X (Springer). *Mathematical Gazette, 70*(453), 246–247. DOI: 10.2307/3615715

Jansi, R., Mohana, K., & Smarandache, F. (2019). Correlation Measure for Pythagorean Neutrosophic Sets with T and F as Dependent Neutrosophic Components. *Neutrosophic Sets and Systems, 30*, 202–212.

Kuroki, N. (1981). On fuzzy ideals and fuzzy bi-ideals in semigroups. *Fuzzy Sets and Systems, 5*(2), 203–215. DOI: 10.1016/0165-0114(81)90018-X

Kuroki, N. (1982). Fuzzy semiprime ideals in semigroups. *Fuzzy Sets and Systems, 8*(1), 71–79. DOI: 10.1016/0165-0114(82)90031-8

Kuroki, N. (1991). On fuzzy semigroups. *Information Sciences, 53*(3), 203–236. DOI: 10.1016/0020-0255(91)90037-U

Munn, W. D. (1964). The Algebraic Theory of Semigroups, Vol. I. By A. H. Clifford and G. B. Preston. Pp. xv + 224. $10.60. 1961. (American Mathematical Society, Providence.). *Mathematical Gazette, 48*(363), 122–122. DOI: 10.2307/3614367

Narayanan, A. L., & Meenakshi, A. R. (2003). Fuzzy M-semigroup. *The Journal of Fuzzy Mathematics, 11*(1), 41–52.

Prema, R., & Radha, R. (2022). Generalized Neutrosophic Pythagorean Set. *International Research Journal of Modernization in Engineering Technology and Science, 4*(11), 1571–1575. https://www.doi.org/10.56726/IRJMETS31596

Rosenfeld, A. (1971). Fuzzy groups. *Journal of Mathematical Analysis and Applications, 35*(3), 512–517. DOI: 10.1016/0022-247X(71)90199-5

Sivaramakrishnan, S., & Suresh, K. (2019). Interval-valued anti fuzzy subring. *Journal of Applied Science and Computations, 6*(1), 521–526.

Sivaramakrishnan, S., & Vijayabalaji, S. (2020). Interval-valued anti fuzzy linear space. *IEEE Xplore, 839-841*, 1–3. DOI: 10.1109/ICSCAN49426.2020.9262371

Sivaramakrishnan, S., Vijayabalaji, S., & Balaji, P. (2024). Neutrosophic interval-valued anti-fuzzy linear space. *Neutrosophic Sets and Systems*, *63*, 271–284. DOI: 10.5281/zenodo.10531827

Smarandache, F. (2006). Neutrosophic set - A generalization of the intuitionistic fuzzy set. In *2006 IEEE International Conference on Granular Computing* (pp. 38–42). DOI: 10.1109/GRC.2006.1635754

Songsaeng, M., & Iampan, A. (2020). Neutrosophic cubic set theory applied to up- algebras. *Thai Journal of Mathematics*, *18*(3), 1447–1474.

Vijayabalaji, S. (2012). Cartesian Product and Homomorphism of Interval-Valued Fuzzy Linear Space. *International Journal of Open Problems in Computer Science and Mathematics*, *5*(4), 93–103. DOI: 10.12816/0006141

Vijayabalaji, S., & Sivaramakrishnan, S. (2012, September 26). Sivaramakrishnan; Anti fuzzy M-semigroup. *AIP Conference Proceedings*, *1482*(1), 446–448. DOI: 10.1063/1.4757511

Weihrich, H., Koontz, H., Cannice, M., & SDR Printers. (2013). *Management : a global, innovative, and entrepreneurial perspective.* McGraw-Hill Education Publishing Company.

Yager, R. R. (2013). Pythagorean fuzzy subsets. In *Proceedings of the 2013 Joint IFSA World Congress and NAFIPS Annual Meeting, IFSA/NAFIPS 2013* (pp. 57–61). IEEE. DOI: 10.1109/IFSA-NAFIPS.2013.6608375

Yager, R. R. (2014). Pythagorean membership grades in multicriteria decision making. *IEEE Transactions on Fuzzy Systems*, *22*(4), 958–965. DOI: 10.1109/TFUZZ.2013.2278989

Yager, R. R., & Abbasov, A. M. (2013). Pythagorean membership grades, complex numbers, and decision making. *International Journal of Intelligent Systems*, *28*(5), 436–452. DOI: 10.1002/int.21584

Yassein, H. R., & Mohammed, A. H. (2011). *Antifuzzy bi-Γ-ideals of Γ- semigroups.* Research Gate. https://www.researchgate.net/publication/321706124

Zadeh, L. A. (1965). Fuzzy sets. *Information and Control*, *8*(3), 338–353. DOI: 10.1016/S0019-9958(65)90241-X

Zadeh, L. A. (1975). The concept of a linguistic variable and its application to approximate reasoning-I. *Information Sciences*, *8*(3), 199–249. DOI: 10.1016/0020-0255(75)90036-5

Zhang, M., Cheng, M. X., & Tarn, T. J. (2006). A mathematical formulation of DNA computation. *IEEE Transactions on Nanobioscience*, *5*(1), 32–40. DOI: 10.1109/TNB.2005.864017 PMID: 16570871

Chapter 7
Neutrosophic Values and Sensitivity Analysis to Study Static Models Without Deficit

Maissam Ahmad Jdid
https://orcid.org/0000-0003-4413-4783
Damascus University, Syria

Florentin Smarandache
https://orcid.org/0000-0002-5560-5926
New Mexico University, USA

ABSTRACT

Inventory management is one of the most important management functions that plays an important role in production and marketing processes, especially in production facilities and commercial institutions that have warehouses in which they keep their equipment and goods. They are concerned with determining the appropriate size of inventory for each material, to secure demand for it during the duration of the storage cycle and achieve the greatest profit. possible or the least possible loss. Therefore, it was necessary to provide a scientific study of these models that would help decision-makers in establishments make an ideal decision regarding the size of inventory needed for each storage cycle. In this chapter, the authors present an extensive study of static inventory models without deficits and the most important economic indicators of these models using neutrosophic values and the sensitivity analysis method with the aim of obtaining mathematical relationships, through which we can determine the ideal size of the stock and provide a comprehensive economic study of it.

DOI: 10.4018/979-8-3693-3204-7.ch007

1. INTRODUCTION

There are many reasons to keep an inventory, including saving time and avoiding the cost and inconvenience resulting from constant compensation. Inventory theory has gone through various stages since its inception in 1920. In the beginning, these models were simple and used a limited number of variables to capture the main factors. With the addition of more variables, it became more complex and contained many details, but it still neglected the impact of changes that could occur in the work environment, which in turn had a significant impact on the primary goal of the storage process, which is to secure the establishments' need for manufactured materials or materials. Manufacturable at the lowest possible cost. We know that if the size of the inventory is very large, this guarantees the availability of the material on the one hand, but in return it may cause the organization to suffer losses because the value of the inventory is frozen capital and the large quantity requires a marketing period that depends on the rate of demand. However, if the quantity of inventory is small This may lead to a bottleneck in securing materials and to various disturbances such as high prices and others. It is necessary to reconcile the two matters and determine the ideal size of the inventory that meets the facility's need at the lowest possible cost, and since the size of the inventory is affected by the size of the demand for it and the frequency of orders at the same time and when these two If the two factors are constant, we get what are called static inventory models, meaning that the size of the remaining inventory in the warehouse will decrease over time until it reaches a certain level at which it will be restocked again. Researchers in this field have presented the following classifications for static inventory models: Static model without deficit - The static model with a shortage - the static model with a safety reserve - the variable price model - the static model for several materials without a shortage. These models were studied according to classical logic by adopting a set of hypotheses called the basic hypotheses of static inventory models (Alali., 2004), and (Bukajh J.S et al., 1998) from which we determine data. The issue under study is that this data is subject to increase or decrease, and this depends on the conditions in the work environment. To avoid losses, the researchers accompanied the solution with a sensitivity analysis, the purpose of which is to clarify the extent of the impact of the change that could occur on the data of the issue in determining the optimal solution and obtaining good results. And it is acceptable for this issue. Also, in previous research

(Jdid et al., 2021) The static model of inventory management without a deficit with Neutrosophic logic,(Jdid, 2022) Important Neutrosophic Economic Indicators of the Static Model of Inventory Management without Deficit, (Jdid et al., 2022) Neutrosophic Treatment of the Static Model of Inventory Management with Deficit, The Neutrosophic Treatment of the Static Model for the Inventory Management with

Safety Reserve, and, The Neutrosophic Treatment for Multiple Storage Problem of Finite Materials and Volumes,(Jdid et al., 2023) ,we presented Neutrosophic Static Model without Deficit and Variable Prices, we presented a neutrosophic study of these models in which we took the rate of demand for inventory as a neutrosophic value that takes into account the worst and best conditions that the work environment can experience during the duration of the storage cycle, and since the value that indicates The rate of demand for inventory is not the only value in the data that can affect the total cost, and it is not the only value that is subject to increase or decrease. In this chapter, we will reformulate the static inventory model without a deficit of its basic indicators using the concepts of neutrosophic logic, as we will take all the data that affects the cost. The total neutrosophic values take into account all conditions and achieve the facility's goal of the storage process. Returning to the subject of the sensitivity analysis attached to the classic study, the aim of which was to determine the percentage of change that occurs in the total cost against any change in one of the model's data, we find that we can determine the percentage of change through the neutrosophic value. For the storage cost we obtain after finding the optimal solution for the model that we will build based on neutrosophic data.

2. MATHEMATICAL BACKGROUND

2.1. Classical Operations Research Methods Have Provided a Study of Static Inventory Models: (Alali., 2004), and (Bukajh J.S et al., 1998), Based on the Following Hypotheses:

1- Order size Q
2- The rate of demand for inventory is continuous and at a constant rate of λ over one period of time, that is, it is subject to a uniform probability distribution.
3- The fixed cost of preparing the order is $C_1=K$.
4- The cost of purchasing, delivering and receiving is equal to C per unit, the total cost of this cost for each order is $C_2=C.Q$. It is called the variable cost of the order.
5- The storage cost is equal to a "fixed" amount h for each item present in the warehouse during one period of time, and includes the cost of frozen liquidity, the cost of occupied space, the cost of protection, security, insurance, taxes, and various fees.
6- The cost of the shortage resulting from a shortage in inventory is equal to a "fixed" amount for each item that is not present in the warehouse during one period of time, such as late fines, interest on the amount paid, the loss of a customer as a missed opportunity, or loss of confidence in the warehouse...

The previous hypotheses were used to determine the ideal size of inventory. In the following cases:

- Static model of inventory management without deficit.
- Static model without deficit and variable prices.
- Static model of inventory management with deficit.
- Static model for the inventory management with safety reserve.
- Multiple storage of finite materials and volumes.

To clarify what was presented in the classic study of static inventory models and sensitivity analysis, we present the following study:

2.2. Classic Study of the Static Model Without a Deficit and for One Substance: (Alali., 2004), and (Bukajh J.S et al., 1998).

The basic assumptions of this model:

1- Order size Q.
2- The rate of demand for inventory is continuous and at a constant rate of λ during one period of time.
3- The fixed cost of preparing the demand is $C_1 = K$.
4- The cost of purchasing, delivering and receiving $C_2 = C.Q$.
5- Storage cost for the quantity remaining in the warehouse during one time C_3
6- The duration of running out of the stored quantity is $\frac{Q}{\lambda}$ (or storage cycle duration).

Using the previous hypotheses, we obtained the following nonlinear model:
Find:

$$C(Q) = \frac{K\lambda}{Q} + C\lambda + \frac{hQ}{2} \rightarrow Min$$

$$Q > 0$$

The optimal solution to the previous model gives us the ideal size of the order and is calculated through the following relation:

$$Q^* = \sqrt{\frac{2K\lambda}{h}} \qquad (1)$$

Practical Example 1

In a book printing press whose paper warehouse must be replaced periodically, we will assume that the paper comes in wide rolls, that printers consume this paper at a rate of λ=32 rolls per week, and that the compensation cost (which includes the cost of maintaining the rolls, loading and handling) is estimated at \$25, in addition As for the cost of paper, the cost of storage, including the rent for the occupied space, insurance, and interest on the frozen capital, is estimated at \$1 per roll within a week.

To calculate the ideal order size that makes the storage cost as small as possible, we substitute the relation (1):

$$Q = \sqrt{\frac{2K\lambda}{h}} = \sqrt{\frac{2.25.32}{1}} = 40 \ \ roll$$

2.3. Sensitivity Analysis: (Bukajh J.S et al., 1998)

The static model is characterized by robustness if it seeks to give good and acceptable results, even if the values of the variables are wrong. To understand the reason for robustness, we assume that one of the values in the previous example was 100% wrong. For example, the fixed cost of preparing students C_1=K was K=50\$ instead of K=25\$ Then the value of Q becomes as follows:

$$Q = \sqrt{\frac{2K\lambda}{h}} = \sqrt{\frac{2.50.32}{1}} = \frac{\sqrt{2.2.25.32}}{1}\sqrt{2}.40 \ \ roll$$

That is, the value of Q is 1.41 times greater than the previous value. In other words, an error in the input of 100% led to an error in the result of 41%. Likewise, we can determine the percentage of error if the rest of the values that interfere in the relation through which we calculate the ideal size are wrong values: The cost function is sensitive to errors, and the reason for this is the sign of the square root, which gives this model the characteristic of robustness, which indicates that this model is applied fairly reliably in the case where the values of the variables are not known accurately, which is the most common case. It is reassuring to know that although these values are inaccurate, the final results are acceptable or close to ideal. This is provided that the hypotheses are appropriate, and it must be emphasized that marketing requests must be truly specific.

3. FUTURE RESEARCH DIRECTIONS

In previous research, we presented a study of static inventory models using the concepts of neutrosophic logic. We took the rate of demand for inventory as a neutrosophic value λ_N, which is one of the values of a range whose lowest limit expresses the rate of demand for inventory in the worst conditions, and whose upper limit expresses the highest value of the rate of demand for inventory in the best conditions, that is $\lambda_N \in [\lambda_1, \lambda_2]$, according to the definition of neutrosophic values, see(Smarandache et al., 2023) On Overview of Neutrosophic and Plithogenic Theories and Applications, Prospects for Applied Mathematics and Data Analysis. In each of the static inventory models, we reached a relation through which we can determine the size of the inventory, and the resulting values are neutrosophic values, which are ranges whose lowest limits express the size of the inventory in the worst conditions, and the highest represents the size of the inventory in the best conditions. In light of this neutrosophic study, we find that the probability that the demand rate for inventory is a wrong value, i.e., a value that does not belong to the range $[\lambda_1, \lambda_2]$. The probability is very small, if not non-existent. In addition, if we take into account the robustness of inventory models, we find that we can dispense with studying the sensitivity analysis regarding the demand rate. On the stock, and this matter prompted us to update the neutrosophic study presented in (Jdid et al., 2021) The static model of inventory management without a deficit with Neutrosophic logic,(Jdid, 2022) Important Neutrosophic Economic Indicators of the Static Model of Inventory Management without Deficit, (Jdid et al., 2022) Neutrosophic Treatment of the Static Model of Inventory Management with Deficit .

We will take all the values that affect the selection of the ideal size as neutrosophic values, thus dispensing with the sensitivity analysis.

3.1. Neutrosophic Static Model Without Deficit and for One Substance

The basic assumptions of this model:

1. Order size Q_N.
2. The rate of demand for inventory at one time is λ_N, where $\lambda_N \in [\lambda_1, \lambda_2]$.
3. The fixed cost of preparing the order is $K_N \in [\mu_1, \mu_2]$.
4. The cost of purchasing, delivering and receiving per unit of inventory is $C_N \in [\delta_1, \delta_2]$.
5. The storage cost is equal to a "fixed" amount $h_N \in [\beta_1, \beta_2]$ for each unit present in the warehouse during one time.

6. The duration of running out of the stored quantity is $\frac{Q_N}{\lambda_N}$ (or storage cycle duration).

Based on the study presented in (Jdid et al., 2021) The static model of inventory management without a deficit with Neutrosophic logic , and using the previous hypotheses, we obtain the following neutrosophical nonlinear model:

Find:

$$C(Q_N) = \frac{K_N \lambda_N}{Q_N} + C_N \lambda_N + \frac{h_N Q_N}{2} \rightarrow Min$$

$$Q_N > 0$$

In the references, (Jdid, 2023), Neutrosophic Nonlinear Models, (Jdid et al., 2023), Lagrange Multipliers and Neutrosophic Nonlinear Programming Problems Constrained by Equality Constraints, and Graphical Method for Solving Neutrosophical Nonlinear Programming Models, we presented a study of nonlinear neutrosophic models and some methods for finding the optimal solution for them. The optimal solution to the previous model gives us the ideal size of the order and is calculated through the following relation:

$$Q_N^* = \sqrt{\frac{2 K_N \lambda_N}{h_N}} \tag{2}$$

The cost of storage per time is calculated from the following relation:

$$C(Q_N) = \frac{K_N \lambda_N}{Q_N} + C_N \lambda_N + \frac{h_N Q_N}{2} \tag{3}$$

3.2. Important Neutrosophic Economic Indicators for the Static Stock Model Without Deficit

Based on the study presented in (Jdid, 2023) Important Neutrosophic Economic Indicators of the Static Model of Inventory Management without Deficit, and using the relation (2) we find:

1. The ideal time period between one order and another order is replaced by the relation:

$$T_N^* = \frac{Q_N^*}{\lambda_N^*}$$

Substituting Q_N^* by what it equals, we get the relation:

$$T_N^* = \frac{Q_N^*}{\lambda_N^*} = \sqrt{\frac{2K_N}{\lambda_N h_N}}$$

2. The number of orders needed during one time:

$$n_N^* = \frac{\lambda_N}{Q_N}$$

3. Reorder quantity:

In this form, the deficit is not allowed, so the quantity that is available in the warehouse must be known, and then the order must be repeated, and we will symbolize it with the code Q_N^1 so that the deficit does not occur, we can calculate it after knowing the time required to receive the order, which we will denote by the code d, then we get:

$$\frac{d}{T_N^*} = \frac{Q_N^1}{Q_N^*} \Rightarrow Q_N^1 = \frac{d}{T_N^*}Q_N^* = \frac{d}{\frac{Q_N^*}{\lambda_N}}Q_N^* \Rightarrow Q_N^1 = d\lambda_N$$

4. Total stock accumulation in warehouse during storage cycle duration:

Calculated from the relation:

$$Y_N^* = \int_0^{T_N^*} (Q_N^* - \lambda_N t)\,dt$$

$$Y_N^* = \int_0^{T_N^*} (Q_N^* - \lambda_N t)\,dt = \left[Q_N^* - \lambda_N \frac{t^2}{2}\right]_0^{T_N^*} = \left[Q_N^* T_N^* - \lambda_N \frac{(Q_N^*)^2}{2}\right]$$

$$= \left[\frac{(Q_N^*)^2}{\lambda_N} - \lambda_N \frac{(Q_N^*)^2}{2\lambda_N^2}\right] = \frac{(Q_N^*)^2}{2\lambda_N} \Rightarrow Y_N^* = \frac{(Q_N^*)^2}{2\lambda_N}$$

5. Average volume of stock in the warehouse during unit of time in the period $[0, T_N^*]$:

Calculated from the relation:

$$\overline{Y_N} = \frac{Y_N^*}{T_N^*} = \frac{Q_N^*}{2}$$

6. Smallest storage cost:

Calculated from the relation:

$$C(Q_N^*) = \frac{K_N \lambda_N}{Q_N^*} + C_N \lambda_N + \frac{h_N Q_N^*}{2}$$

7. Total storage cost during storage cycle duration:

Calculated from the relation:

$$TC(Q_N^*) = C(Q_N^*) \,.\, T_N^*$$

Substituting $.T_N^*$ by its equivalent, we get the relation:

$$TC(Q_N^*) = K_N + C_N Q_N^* + \frac{h_N (Q_N^*)^2}{2\lambda_N}$$

8. Total sales during the unit of time:

If the selling price of one unit of the stored material equals P and we symbolize the total sales during one time by V_N, then it is calculated from the relation: $V_N = P\lambda_N$

9. Total sales during the storage cycle period:

We symbolize the total sales during the storage cycle period with W_N, then W_N is calculated from the relation:

$$W_N = P * \lambda_N * T_N^* \Rightarrow P * \lambda_N * \frac{Q_N^*}{\lambda_N} \Rightarrow W_N = P * Q_N^*$$

10. Average profit during the unit of time:

We denote the average profit during one with the symbol B_N and calculate it from the relation:

$$B_N = V_N - C(Q_N^*)$$

11. Average profit during the storage cycle period:

We get the average profit over the storage cycle period B'_N from the relation:

$$B'_N = W_N - TC(Q^*_N)$$

12. We get the average profit per year from the relation:

$$B'_N = 12 * B_N$$

We declare the above with the following example:
Practical example according to neutrosophic logic:
A merchant wishes to invest his existing capital in storing and selling iron used in construction. Before starting the work, he conducted a study of the market movement in the area in which he wanted to invest, and obtained the following information:

1- The monthly demand rate for iron ranges between 250 and 300 tons, meaning there is no exact value for the demand rate. Therefore, the demand rate is a neutrosophic value and equals $\lambda_N \in [250, 300]$ ton.
2- The cost of preparing Q_N ton order is $K_N \in [150, 250]$\$.
3- The cost of storing one ton of iron during the unit of time (month) is $h_N \in [5, 15]$\$.
4- The cost of purchasing, delivering, receiving and arranging an order of size Q equals $C_N \in [5000, 7000]$\$
5- The period of receiving the order is $d_N \in [4,6]$ *days.*
6- The selling price of one ton is $P_N \in [6500, 7500]$\$

From the data we find:
The ideal size of the order we substitute in the following relation:

$$Q^*_N = \sqrt{\frac{2K_N\lambda_N}{h_N}}$$

$$Q^*_N = \sqrt{\frac{2K_N\lambda_N}{h_N}}$$

$$= \sqrt{\frac{2 * [150, 250] * [250, 300]}{[5, 15]}} = \sqrt{[10000, 15000]} \in [100, 122.5] \ \textit{Tons}$$

Let's calculate the basic indicators that give this investor a future vision of the business situation over time:

1. The ideal time interval between one order and another, we substitute in the relation:

$$T_N^* = \frac{Q_N^*}{\lambda_N^*} = \frac{[100, 122.5]}{[250, 300]} \in [0.4, 0.41] \; month$$

And it equals:

$$T_N^* = \frac{Q_N^*}{\lambda_N^*} = [0.4, 0.41] \; * 30 \in [12, 12.3] \; day$$

2. The number of orders needed during a month; we substitute in the relation:

$$n_N^* = \frac{1}{T_N^*} = \frac{\lambda_N}{Q_N}$$

$$n_N^* = \frac{1}{T_N^*} = \frac{[250, 300]}{[100, 122.5]} \in [2.45, 2.5] \; order$$

$$n_N^* = \frac{\lambda_N}{Q_N} = \frac{[250, 300]}{[100, 122.5]} \in [2.45, 2.5] \; order$$

3. The quantity that is available in the warehouse, and then the order must be repeated, we substitute in the relation:

$$Q_N^1 = d_N * \lambda_N = [4, 6] * [250, 300] \in [1000, 1800] \; Tons$$

4. The total accumulation of inventory in the warehouse during the period of the storage cycle, we substitute in the relation:

$$Y_N^* = \frac{(Q_N^*)^2}{2\lambda_N} \Rightarrow Y_N^* = \frac{([100, 122.5])^2}{2 * [250, 300]} \in [20, 25] \; Tons$$

5. The average volume of stock held in the warehouse during one time in the period we substitute in the relation:

$$\overline{Y_N} = \frac{Y_N^*}{T_N^*} = \frac{Q_N^*}{2}$$

$$\overline{Y_N} = \frac{Q_N^*}{2} = \frac{[100, 122.5]}{2} \in [50, 61.25] \; Tons$$

6. The smallest storage cost we substitute in the relation:

$$C(Q_N^*) = \frac{K_N \lambda_N}{Q_N^*} + C_N \lambda_N + \frac{h_N Q_N^*}{2}$$

$$C(Q_N^*) = \frac{[150,250]*[250,300]}{[100,122.5]} + [5000,7000]*[250,300] + [5,15]*\frac{[100,122.5]}{2}$$

$$C(Q_N^*) \in [1269000, 2101530.75] \ \$$$

7. The total cost of storage during the storage cycle period, we substitute in the relation:

$$TC(Q_N^*) = C(Q_N^*) * T_N^*$$

$$TC(Q_N^*) = [1269000, 2101530.75]*[0.4, 0.41] \in [507600, 861627.6] \ \$$$

8. The total value of sales during a month, we substitute in the relation:

$$V_N = P*\lambda_N$$

$$V_N = [6500,7500]*[250,300] \in [1625000, 2250000] \ \$$$

9. Total sales during the storage cycle period, we compensate with the relation:

$$W_N = P*\lambda_N*T_N^* \Rightarrow P*\lambda_N*\frac{Q_N^*}{\lambda_N} \Rightarrow W_N = P*Q_N^*$$

$$W_N = [6500,7500]*[100,122.5] \in [650000, 918750] \ \$$$

10. Average profit during the unit of time (month) we substitute in the relation:

$$B_N = V_N - C(Q_N^*)$$

$$B_N = [1625000, 2250000] - [1269000, 2101530.75] \in [148469.25, 356000] \ \$$$

11. The average profit during the storage cycle period, we substitute in the relation:

$$B_N^{'} = W_N - TC(Q_N^*)$$

$$B'_N = [650000, 918750] - [507600, 861627.6] \in [57122.4, 142400]\ \$$$

12. Average profit per year we substitute in the relation:

$$B'_N = 12 * B_N$$

$$B'_N = 12 * [148469.25, 356000] \in [1781631, 4272000]\ \$$$

3.3. Neutrosophic Static Model of Variable Price Discounts Without Deficit

Based on the study presented in, (Jdid et al., 2023), we presented Neutrosophic Static Model without Deficit and Variable Prices. In this paragraph, we will study the static model without deficit the previous, with the addition of a new hypothesis, which is to benefit from the offers offered by the producing companies to market their products, and these offers are linked to the size of the order. Here, those responsible for the storage process are required to determine the appropriate and ideal size of stock and make the most of the offers. We explain what previously through the following study:

Basic assumptions of the model:

1- Order size Q_N.
2- The rate of demand for inventory at one time is λ_N, where $\lambda_N \in [\lambda_1, \lambda_2]$.
3- The fixed cost of preparing the order is $K_N \in [\mu_1, \mu_2]$.
4- The cost of purchasing, delivering and receiving per unit of inventory is $C_N \in [\delta_1, \delta_2]$.
5- The storage cost is equal to a "fixed" amount $h_N \in [\beta_1, \beta_2]$ for each unit present in the warehouse during one time.
6- The duration of running out of the stored quantity is $\frac{Q_N}{\lambda_N}$ (or storage cycle duration).
7- One of the companies producing materials to be stored provided four price levels that are inversely proportional to the size of the order:

$$\text{Price for one item} \begin{cases} C_1 & \text{If } 0 \le Q \le Q_1 \\ C_2 & \text{If } Q_1 \le Q \le Q_2 \\ C_3 & \text{If } Q_2 \le Q \le Q_3 \\ C_4 & \text{If } Q_3 \le Q \le Q_4 \end{cases}$$

Where $C_1 > C_2 > C_3 > C_4$

From the study presented in the research (Jdid et al., 2021) The static model of inventory management without a deficit with Neutrosophic logic, we arrived at the following:

In this model, we build the mathematical model for this issue within hypotheses 1-6, which in themselves are the basic hypotheses of the static model without neutrosophic deficit that was studied in the previous paragraph, where we arrived at the following nonlinear model:

Find:

$$C(Q_N) = \frac{K_N \lambda_N}{Q_N} + C_N \lambda_N + \frac{h_N Q_N}{2} \rightarrow Min$$

$$Q_N > 0$$

The ideal size of students is given by the following relation:

$$Q_N^* = \sqrt{\frac{2K_N \lambda_N}{h_N}}$$

The minimum storage cost is calculated from the following relation:

$$C(Q_N) = \frac{K_N \lambda_N}{Q_N} + C_N \lambda_N + \frac{h_N Q_N}{2}$$

To address the issue at hand and choose the optimal size of the order, taking into account hypothesis No. (7) the offers presented,

We determine price levels based on discounts by calculating the following storage costs:

$$C_i(Q_N) = \frac{K_N \lambda_N}{Q_N} + C_N \lambda_N + \frac{h_N Q_N}{2} \tag{4}$$

Where i=1,2,3,4,… for all price cases.

Since $C_1 > C_2 > C_3 > C_4$, the function $C_i(Q)$ satisfies the following inequality

$$C_1(Q) > C_2(Q) > C_3(Q) > C_4(Q)$$

For each of the previous functions, a minimum limit is the optimal solution to the nonlinear model, i.e., the ideal size of the inventory. We symbolize it as Q_{0N}, calculated from the following relation:

$$Q_{N0} = \sqrt{\frac{2K_N\lambda_N}{h_N}}$$

After calculating Q_{N0}, we compare it with the offers provided, assuming that $Q_1 < Q_{N0} < Q_2$. We calculate the cost corresponding to this size from the relationship (4). We obtain $C_2(Q_{0N})$, and to determine the ideal size of the order Q_N^*, we calculate the cost functions for the lower limits of the quantity ranges specified in the discounts table. Then we choose the smallest of these costs. The corresponding ideal size is the size that secures the inventory for the system during the duration of the storage cycle at the lowest possible cost and while taking advantage of the offers provided.

Practical Example 3

A production institution wants to secure its need for a certain material. If it knows that the rate of demand for this material is [250, 330] units per year, the cost of purchasing one unit in the market is [350, 450] monetary units, the cost of storing one unit per year is 10% of its price, and the cost of preparing the order is equal to [100, 200] monetary units. The company producing this material offers the following offers:

2% discount if quantity $50 \leq Q \leq 100$.
3% discount if quantity $100 \leq Q \leq 200$.
5% discount if quantity is $200 \leq Q$.

Who here needs to do the math?

Required: Find the optimal quantityQ_N^* that makes the total costs of storage as small as possible.

Solution:

From our data:

$$K_N \in [100, 200]\ ,\ C_N \in [350, 450]\ ,\ \lambda_N \in [250, 350]$$

h_N is the storage cost and amounts to 10% of the price of one unit of stock in the market. Therefore:

$$h_N = \frac{10}{100} * [250, 300] \in [35, 45]$$

We determine price levels by discounts:

a. When $Q \leq 50$ then $C_{N1} = C_N \in [350, 450]$ there is no discount.

b. When $50 \leq Q \leq 100$ the discount is 2% and the purchase price is equal to:

$$C_{N2} \in [350,450]\left(1-\tfrac{2}{100}\right) = [343,441]$$

c. When $100 \leq Q \leq 200$ the discount is 3% and the purchase price equal to:

$$C_{N3} \in [350,450]\left(1-\tfrac{3}{100}\right) = [339,436.5]$$

d. When $Q \geq 200$ the discount is 5% and the purchase price is equal to:

$$C_{N4} \in [350,450]\left(1-\tfrac{5}{100}\right) = [332.5,427.5]$$

We calculate the initial quantity of inventory:

We study the issue based on hypotheses 1-6, and here we are faced with a storage model without a neutrosophic deficit. We calculate the ideal size of the order through the following relation:

$$Q_{N0} = \sqrt{\frac{2K_N\lambda_N}{h_N}} = \sqrt{\frac{2*[100,200]*[250,350]}{[35,45]}} \in [223.6,374]$$

We note that $Q_{N0} \in [50, 100]$, meaning that this quantity deserves a 2% discount, and the purchase price for one unit is $C_{N2} \in [343, 441]$.

The total storage cost is calculated from the relation:

$$C_2(Q_{N0}) = \frac{K_N\lambda_N}{Q_{N0}} + \frac{h_N Q_{N0}}{2} + C_{N2}\lambda_N$$

$$C_2([223.6,374]) = \frac{[100,200]*[250,350]}{[223.6,374]}$$
$$+\frac{[35,45]*[223.6,374]}{2} + [343,441]*[250,350] \in [89774.8,162952.2]$$

To benefit more from the offers presented, we calculate costs for the minimum limits of the offer areas:

For the range $100 \leq Q \leq 200$ we find:

$$C_{N3}(100) = \frac{[100,200]*[250,350]}{100}$$
$$+\frac{[35,45]*100}{2} + [339,436.5]*[250,350] \in [86875,155725]$$

For the range $Q \geq 200$ we find:

$$C_{N4}(200) = \frac{[100,200]*[250,350]}{200} + \frac{[35,45]*200}{2} + [332.5,427.5]*[250,350] \in [86750,154475]$$

Studying the results we obtained, we note the following:

As for the cost $C_{N1}=C_N \in [350, 450]$ it is offset by the order size $Q \leq 50$. In return for this size, there is no discount. In addition, this size is less than the minimum required, meaning that it does not suit the company because the company works on the basis of no shortage, and this quantity is for the order. It will cause a deficit for which the company will pay fines, which will be reflected in the total cost. Therefore, we rule out this solution, and for the rest of the costs, we choose the lowest cost.

That is, we take

$$Min\{C_2(Q), C_3(Q), C_4(Q)\}$$

We find

$$Min\{[89774.8,162952.2],[86875,155725],[86750,154475]\} = [86750,154475] \quad (5)$$

$$Min\{C_2(Q), C_3(Q), C_4(Q)\} \in [86750,154475]$$

which corresponds to an order size $Q=200$. This size is suitable for the company's workflow. In order to achieve the maximum benefit from the offers presented, we calculate the costs corresponding to the largest size that the company can adopt if the storage cost is appropriate, which corresponds to an order size equal to the rate of demand for inventory, i.e., $Q_N \in [250, 350]$. We find that the price of one unit will be $C_4 \in [332.5, 427.5]$. The total storage costs equal:

$$C([250,350]) = \frac{[100,200]*[250,350]}{[250,350]} + \frac{[35,45]*[250,350]}{2} + [250,350]*[332.5,427.5] \in [87600,157700]$$

We compare this cost with the cost we obtained in (5). We find that the cost is greater, meaning that there is no interest for the company in requesting this size of the order because it can ensure a safe workflow and benefit from the offers provided at a lower cost when the order size is $Q=200$.

From the above, we note that the minimum value of storage costs is [99187.5, 137262.5] and corresponds to the ideal size of the order, which is equal to $Q^*=200$.

4. CONCLUSION

The solution that we obtain using one of the methods for solving nonlinear models is an ideal solution that fulfills the conditions given in the problem. If any change occurs in the data of the problem, this leads to changing the optimal decision. Therefore, we resort to analyzing the sensitivity of this solution. Sensitivity analysis techniques enable us to measure the effects of these changes. And take the necessary measures without resolving the issue. Through the neutrosophical study that was presented in this chapter of the static inventory model without a deficit and the variable price model without a deficit, we find that the possibility that the values used during the search for the optimal size are wrong values, that is, values that do not belong to the areas that were determined Determining it after taking into account the worst and best conditions is a very small possibility, if not non-existent. In addition to that, the robustness of inventory models makes the error rate in the results small. We find that we can dispense with studying sensitivity analysis when studying inventory models using concepts neutrosophic logic, meaning that the neutrosophic study of inventory models is a good and acceptable study and provides us with much better results than the results provided by the classical study. It should be noted that the same method presented in this chapter can be followed to prepare a neutrosophic study similar to other static inventory models.

REFERENCES

Bukajh J.S. (1998). *Operations Research Book translated into Arabic*. The Arab Center for Arabization, Translation, Authoring and Publishing -Damascus.

Jdid, M. (2023). Important Neutrosophic Economic Indicators of the Static Model of Inventory Management without Deficit. *Journal of Neutrosophic and Fuzzy Systems, 5*(1), 08-14. DOI: 10.54216/JNFS.050101

Jdid, M., Alhabib, R., & Salama, A. (2021). The static model of inventory management without a deficit with Neutrosophic logic. *International Journal of Neutrosophic Science, 16*(1). DOI: 10.54216/IJNS.160104

Jdid, M., Alhabib, R., & Bahbouh, O. (2022). The Neutrosophic Treatment for Multiple Storage Problem of Finite Materials and Volumes. *International Journal of Neutrosophic Science, 18*(1), 42-56. DOI: 10.54216/IJNS.180105

Jdid, M., Alhabib, R., Khalid, H. & Salama, A. (2022). The Neutrosophic Treatment of the Static Model for the Inventory Management with Safety Reserve. *InternationalJournal of Neutrosophic Science, 18*(2), 262-271. DOI: 10.54216/IJNS.180209

Jdid, M., & Models, N. N. (2023). *Prospects for Applied Mathematics and Data Analysis, 2*(1), 42-46. DOI: 10.54216/PAMDA.020104

Jdid, M., & Smarandache, F. (2023). Neutrosophic Static Model without Deficit and Variable Prices. *Neutrosophic Sets and Systems, 60,* 124-132. https://fs.unm.edu/nss8/index.php/111/article/view/3744

Jdid, M., & Smarandache, F. (2023). Lagrange Multipliers and Neutrosophic Nonlinear Programming Problems Constrained by Equality Constraints. *Neutrosophic Systems with Applications, 6*, 25–31. DOI: 10.61356/j.nswa.2023.35

Jdid, M., & Smarandache, F. (2023). Graphical Method for Solving Neutrosophical Nonlinear Programming Models, / Int.J.Data. *Sci. & Big Data Anal., 3*(2), 66–72. DOI: 10.51483/IJDSBDA.3.2.2023.66-72

Jdid, M. (2022). Neutrosophic Treatment of the Static Model of Inventory Management with Deficit. *International Journal of Neutrosophic Science, 18*(1), 20-29. DOI: 10.54216/IJNS.180103

Smarandache, F. (2014). *Introduction to Neutrosophic statistics*. Sitech & Education Publishing.

Smarandache, F., & Jdid, M. (2023). On Overview of Neutrosophic and Plithogenic Theories and Applications. *Prospects for Applied Mathematics and Data Analysis, 2*(1), 19-26. DOI: 10.54216/PAMDA.020102

Chapter 8
Neutrosophical Study of the Two Models: Storage Static With Deficit and Storage Static With Safety Reserve and Important Economic Indicators

Maissam Ahmad Jdid
https://orcid.org/0000-0003-4413-4783
Damascus University, Syria

ABSTRACT

Inventory management is considered one of the most important functions of management in terms of determining the mechanism used in the storage process. The nature of the material to be stored determines the ideal size of the inventory. In perishable materials, the authors use the static model with a deficiency. Here the authors allow the acceptance of a shortage and the accumulation of orders until storage is restocked again. Decision makers in these facilities must make an example decision that is proportional to the rate of demand for inventory during the storage cycle and takes into account the shelf life. As for materials that have a long expiration date, they can determine the ideal size of the inventory using the static inventory model with a safety reserve. In this chapter, they present a neutrosophic study of two types of static inventory models: the static model with a deficit and the static model with a safety reserve. The authors will also calculate some economic indicators.

DOI: 10.4018/979-8-3693-3204-7.ch008

1. DISCUSSION

In this chapter, we present an extensive neutrosophical study of two static inventory models and some economic indicators for each, the purpose of which is to determine the ideal size of the inventory during the cycle period. Storage for perishable materials that do not have a long expiration date, using the static model without shortages that was classically studied in references (Alali., 2004), and (Bukajh J.S et al., 1998) A neutrosophic vision was presented for it that took into account the rate of demand for inventory, a neutrosophic value in. Since the demand rate Inventory is not the only value that is subject to change depending on the surrounding conditions. In this chapter, we will take all the data, neutrosophic values, to determine the ideal size of the inventory appropriate for all conditions and the nature of the materials to be stored. Then we will calculate some important economic indicators that provide decision makers with a future vision of the workflow. Likewise, we will study the static model with a safety reserve used to secure the necessary materials for emergency and first aid situations: such as medicines, food, fuel, etc.

1.1. Neutrosophic Static Stock Model With Deficit for One Matter

The basic assumptions of this model:

- Order size Q_N.
- The rate of demand for inventory at one time is λ_N, where $\lambda_N \in [\lambda_1, \lambda_2]$.
- The fixed cost of preparing the order is $K_N \in [\mu_1, \mu_2]$.
- The cost of purchasing, delivering and receiving one unit of inventory is: $C_N \in [\delta_1, \delta_2]$
- The storage cost is equal to a "fixed" amount $h_N \in [\beta_1, \beta_2]$ for each unit present in the warehouse during one time.
- The amount of the deficit allowed in each storage cycle is equal to $S_N \in [\alpha_1, \alpha_2]$ and the cost of the deficit is equal to $b_N \in [\varphi_1, \varphi_2]$ for each unfulfilled demand during one time, including delay fines, loss of customers, etc.
- The duration of running out of the stored quantity is $\frac{Q_N}{\lambda_N}$ (or storage cycle duration).

Based on the study mentioned in Reference (Jdid, 2023) Neutrosophic Nonlinear Models, the mathematical model:

We symbolize the size of the stock available at the beginning of the storage cycle with the symbol R_N, and the volume of the order that comes into the warehouse at the end of each cycle with the symbol Q_N, where $R_N < Q_N$, then the amount of the deficit is $S_N = Q_N - R_N$. Find:

$$C(Q_N, R_N) = \frac{K_N \lambda_N}{Q_N} + C_N \lambda_N + \frac{h_N (R_N)^2}{2Q_N} + \frac{P_N (Q_N - R_N)^2}{2Q_N} \rightarrow Min$$

Within the conditions:

$$Q_N > R_N$$

$$Q_N \geq 0$$

$$R_N \geq 0$$

This model is a non-linear model. To find the optimal solution, we find the partial derivatives of the objective function, then eliminate these derivatives. We obtain a set of two equations with two unknowns, R_N, Q_N. By solving the set of two equations we get:

$$Q_N^* = \sqrt{\left(\frac{2K_N \lambda_N}{h_N}\right) * \left(\frac{P_N + h_N}{P_N}\right)}$$

$$R_N^* = \sqrt{\left(\frac{2K_N \lambda_N}{h_N}\right) * \left(\frac{P_N}{P_N + h_N}\right)}$$

We calculate the value of the total costs and obtain the following relation:

$$C(Q_N^*, R_N^*) = \frac{K_N \lambda_N}{Q_N^*} + C_N \lambda_N + \frac{h_N (R_N^*)^2}{2Q_N^*} + \frac{P_N (Q_N^* - R_N^*)^2}{2Q_N^*}$$

The value of the total costs during the storage cycle is $T_N^* = \frac{Q_N^*}{\lambda_N}$.

$$TC(Q_N^*, R_N^*) = K_N + C_N Q_N^* + \frac{h_N (R_N^*)^2}{2\lambda_N} + \frac{P_N (Q_N^* - R_N^*)^2}{2\lambda_N}$$

Practical Example 1

A merchant wanted to invest his existing capital in storing and selling iron used in construction. Before starting work, he conducted a study of the market movement in the area in which he wanted to invest, and obtained the following information:

- The monthly demand rate for iron ranges between 250 and 300 tons, meaning there is no exact value for the demand rate and therefore.
- The demand rate is a neutrosophic value and equals $\lambda_N \in [250, 350]$ tons.
- The cost of preparing an order consisting of Q_N tons is equal to $K_N \in [150, 250]$\$.
- The cost of storing one ton of iron during one time (month) is equal to $h_N \in [8, 12]$\$.
- The cost of purchasing, delivering, receiving and arranging an order of size Q equals $C_N \in [4000, 6000]$\$.
- The duration of receiving the order is $d_N \in [3, 7]$ *days*.
- The selling price of one ton is $b_N \in [5000, 6000]$\$.
- The cost of the shortfall is $P_N \in [40, 60]$ for each unfulfilled demand over time and includes late fines, loss of customers, etc.

Required:

- Calculate the ideal size for the first order R_N^* and for the recurring orderQ_N^*.
- Calculating the amount of the allowable deficit.
- Calculate the smallest cost of storage during a single period of time (month), and the smallest cost of storage during the storage cycle.

From the text of the problem, we have the following data:

$\lambda_N \in [250, 350]$, $d_N \in [3, 7]$, $K_N \in [150, 250]$, $C_N \in [4000, 6000]$, $h_N \in [8, 12]$, $P_N \in [40, 60]$, $b_N \in [5000, 6000]$

Mathematical Model

Find:

$$C(Q_N^*, R_N^*) = \frac{K_N \lambda_N}{Q_N^*} + C_N \lambda_N + \frac{h_N (R_N^*)^2}{2Q_N^*} + \frac{P_N \left(Q_N^* - R_N^*\right)^2}{2Q_N^*} \rightarrow Min$$

$$C(Q_N^*, R_N^*) = \frac{[150,250]*[250,350]}{Q_N} + [250,350]$$

$$*[4000,6000] + \frac{[8,12](R_N^*)^2}{2Q_N} + \frac{[40,60](Q_N^* - R_N^*)^2}{2Q_N} \rightarrow Min$$

Within the conditions:

$$Q_N > R_N$$

$$Q_N \geq 0$$

$$R_N \geq 0$$

Calculate the ideal size for the first order R_N^* and for the recurring order Q_N^*.

$$Q_N^* = \sqrt{\left(\frac{2K_N\lambda_N}{h_N}\right) * \left(\frac{P_N + h_N}{P_N}\right)}$$

$$R_N^* = \sqrt{\left(\frac{2K_N\lambda_N}{h_N}\right) * \left(\frac{P_N}{P_N + h_N}\right)}$$

We find:

$$Q_N^* = \sqrt{\left(\frac{2[150,250]*[250,350]}{[8,12]}\right) * \left(\frac{[40,60]+[8,12]}{[40,60]}\right)} \in [106,132]$$

$$R_N^* = \sqrt{\left(\frac{2[150,250]*[250,300]}{[5,15]}\right) * \left(\frac{[40,60]}{[40,60]+[5,15]}\right)} \in [88,110]$$

Therefore, the allowable deficit is:

$$S_N^* = Q_N^* - R_N^* = [106,132] - [88,110] \in [18,22]$$

1.2. The Basic Indicators That Give the Investor a Future Vision of the Business Situation Over Time

1- Duration of the storage cycle:

$$T_N^* = \frac{Q_N^*}{\lambda_N} = \frac{[106,132]}{[250,350]} \in [0.4,0.4] \ \ month$$

$$T_N^* = [0.4,0.4]*30 = 12 \ \ day$$

2- Duration of stock out:

$$t_N^* = \frac{R_N^*}{\lambda_N} = \frac{[88,110]}{[250,350]} \in [0.3,0.37] \ \ month$$
$$t_N^* = [0.3,0.37]*30 \in [9,11] \ \ day$$

3- The number of orders needed to fulfill the order during the storage cycle period (month):

$$n_N^* = \frac{\lambda_N}{Q_N^*} = \frac{[250,350]}{[106,132]} \in [2.4,2.6] \ \ order$$

4- Reorder Quantity:

$$Q_{N1} = \lambda_N * d_N - S_N^* = \frac{[3,7]}{30} * [250,350] \ - [18,22] \ \in [7,60]$$

5- The smallest cost of storage during one time that reaches its smallest value at the two values Q_N^*, R_N^* we substitute into the relation:

$$C(Q_N^*, R_N^*) = \frac{K_N \lambda_N}{Q_N^*} + C_N \lambda_N + \frac{h_N (R_N^*)^2}{2Q_N^*} + \frac{P_N (Q_N^* - R_N^*)^2}{2Q_N^*}$$

We find:

$$C(Q_N^*, R_N^*) = \frac{[150,250]*[250,350]}{[106,132]} + [250,350]$$
$$*[4000,6000] + \frac{[8,12,60]([88,110])^2}{2[106,132]} + \frac{[40,60]([18,22])^2}{2[106,132]} \in [1001876, 2103523]$$

6- The smallest cost of storage during a storage cycles whose duration is $T_N^* = \frac{Q_N^*}{\lambda_N}$, we substitute with the relation:

$$TC(Q_N^*, R_N^*) = K_N + C_N Q_N^* + \frac{h_N (R_N^*)^2}{2\lambda_N} + \frac{P_N (Q_N^* - R_N^*)^2}{2\lambda_N}$$

We find:

$$TC(Q_N^*, R_N^*) = [150,250] + [4000,6000]*[106,132]$$
$$+ \frac{[8,12]*([88,110])^2}{2[250,300]} + \frac{[40,60]([18,22])^2}{2[250,300]} \in [424300, 792540]$$

7- Total sales value during one storage cycle:

$$V_N^* = b_N * Q_N^* = [5000, 7000] * [106, 132] \in [530000, 924000]$$

8- Profit during the storage cycle:

$$B_N^* = V_N^* - TC\left(Q_{N'}^*, R_N^*\right) \in [53000 - 924000] - [424300, 792540] \in [105700, 131460]$$

9- His monthly profit:

$$B_1^* = n_N^* * B_N^* \in [2.4, 2.6] * [105700, 131460] \in [253680, 341796]$$

1.3. Neutrosophic Static Stock Model With Safety Reserve

We study this model in the following cases:

First case: The rate of demand for inventory is not specified, and the amount of the safety reserve is also not specified

Second case: The rate of demand for inventory is not specified and the amount of safety reserve is specified with a fixed value S

Third case: The rate of demand for inventory is set at a fixed value and the amount of safety reserve is not specified

1.3.1. First Case (Where the Rate of Demand for Inventory Is Not Specified and the Amount of the Safety Reserve Is Also Not Specified)

The basic assumptions of this model:

- Order size Q_N.
- The rate of demand for inventory at one time is $\lambda_{N'}$ where $\lambda_N \in [\lambda_1, \lambda_2]$.
- The fixed cost of preparing the order is $K_N \in [\mu_1, \mu_2]$.
- The cost of purchasing, delivering and receiving one unit of inventory is: $C_N \in [\delta_1, \delta_2]$
- The storage cost is equal to a "fixed" amount $h_N \in [\beta_1, \beta_2]$ for each unit present in the warehouse during one time.
- The amount of the safety reserve in each storage cycle and over time is equal to S_N (indefinite) $S_N \in [S_1, S_2]$ such that S_1 is the minimum limit of the safety reserve, S_2 is the upper limit of the safety reserve, and the cost of storing one unit of it during one time is $h_{N'}$ then the cost is That reserve equals $B_n = h_n * S_N$.

- The duration of running out of the stored quantity is $\frac{Q_N}{\lambda_N}$ (or storage cycle duration).

Using the previous hypotheses and benefiting from the results we reached in the study of the static model without a deficit according to the neutrosophic logic (Jdid et al., 2021) The static model of inventory management without a deficit with Neutrosophic logic, we find that the ideal size for the first order is given by the following relation:

$$Q_N^{**} = Q_N^* + S_N$$

Where Q_N^* is the ideal neutrosophic size of the order that we arrived at by studying the static model without deficit according to neutrosophic logic, and it is given by the following relation:

$$Q_N^* = \sqrt{\frac{2K_N\lambda_N}{h_N}}$$

From the above we conclude that:

1- We conclude that the ideal neutrosophic volume of the first order is equal to:

$$Q_N^{**} = \sqrt{\frac{2K_N\lambda_N}{h_N}} + S_N$$

After determining the quantity of the first order, the inventory movement proceeds until it reaches the level $S_N \in [S_1, S_2]$ and then is renewed with a new order Q_N^*, The purchase price of the safety reserve amount is $C_N * S_N$ and is calculated once.

2- The total cost of inventory during each successive cycle is equal to:

$$TC(Q_N^*) = K_N + C_N Q_N^* + \frac{h_N Q_N^*}{2\lambda_N} + h_N S_N T^*$$

3- As for the total cost of inventory during one time of successive cycles:

$$C(Q_N^*) = \frac{K_N\lambda_N}{Q_N^*} + C_N\lambda_N + \frac{h_N Q_N^*}{2} + h_N S_N$$

4- The reorder quantity is calculated from the relation:

$Q_{N1} = d_N\lambda_N + S_N$

Practical Example 2

A merchant wanted to invest his existing capital in storing and selling iron used in construction. Before starting work, he conducted a study of the market movement in the area in which he wanted to invest, and obtained the following information:

- The demand rate is a neutrosophic value and equals $\lambda_N \in [250, 350]$ tons.
- The cost of preparing an order consisting of Q_N tons is equal to $K_N \in [150, 250]$\$.
- The cost of storing one ton of iron during one time (month) is equal to $h_N \in [8,12]$\$.
- The cost of purchasing, delivering, receiving and arranging an order of size Q equals $C_N \in [4000, 6000]$\$.
- The duration of receiving the order is $d_N \in [3,7]$ *days.*
- The selling price of one ton is $b_N \in [5000, 6000]$\$.
- The safety reserve to ensure the continuity of work is $S_N \in [0,100]$ tons per month

Required:

- Calculate the ideal size for the first orderR_N^* and for the recurring order Q_N^*.
- Calculating the amount of the allowable deficit.
- Calculate the smallest cost of storage during a single period of time (month), and the smallest cost of storage during the storage cycle.

From the text of the problem, we have the following data:

$\lambda_N \in [250, 350]$, $d_N \in [3,7]$, $K_N \in [150, 250]$, $C_N \in [4000, 6000]$, $h_N \in [8,12]$, $P_N \in [40,60]$, $b_N \in [5000, 6000]$, $S_N \in [0,100]$

We calculate Q_N^*:

$$Q_N^* = \sqrt{\frac{2K_N\lambda_N}{h_N}} = \sqrt{\frac{2 * [150, 250] * [250, 350]}{[8, 12]}} \in [97, 121]$$

From this we find that the optimal size for successive orders is:

$$Q_N^* = [97, 121]$$

1- Ideal size for the first order:

$$Q_N^{**} = Q_N^* + S_N = [97, 121] + [0, 100] \in [97, 221]$$

2- The reorder quantity is given by the following relation:

$$Q_{N1} = d_N \lambda_N + S_N = [3, 7] * [250, 350] + [0, 100] \in [750, 2550]$$

3- Calculate the total storage cost during one month:

$$C(Q_N^*) = \frac{K_N \lambda_N}{Q_N^*} + C_N \lambda_N + \frac{h_N Q_N^*}{2} + h_N S_N$$

$$C(Q_N^*) = \frac{[150, 250] * [250, 350]}{[97, 121]} + [4000, 6000] * [250, 350]$$

$$+\frac{[8, 12]}{2} * [97, 121] + [8, 12] * [0, 100] \in [1000775, 2102649]$$

4- Calculate the investment cost of storage within a year:

$$CI_N = C(Q_N^*) * 12 + C_N * S_N$$

$$CI_N = [1000775, 2102649] * 12 + [4000, 6000] * [0, 100] \in [1209300, 25831788]$$

Second case: (where the rate of demand for inventory is not specified and the amount of safety reserve is specified with a fixed value S)
Basic assumptions:

- Order size Q_N.
- The rate of demand for inventory at one time is λ_N, where $\lambda_N \in [\lambda_1, \lambda_2]$.
- The fixed cost of preparing the order is $K_N \in [\mu_1, \mu_2]$.
- The cost of purchasing, delivering and receiving one unit of inventory is: $C_N \in [\delta_1, \delta_2]$
- The storage cost is equal to a "fixed" amount $h_N \in [\beta_1, \beta_2]$ for each unit present in the warehouse during one time.
- The amount of the safety reserve in each storage cycle and over time is equal to S, and the cost of storing one unit of it during one time is h_N, then the cost is That reserve equals $B_n = h_n * S_N$.
- The duration of running out of the stored quantity is $\frac{Q_N}{\lambda_N}$ (or storage cycle duration).

Using the previous hypotheses and benefiting from the results we reached in the study of the static model without a deficit according to the neutrosophic logic (Jdid et al., 2021) The static model of inventory management without a deficit with Neutrosophic logic, we find that the ideal size for the first order is given by the following relation:

$$Q_N^{**} = Q_N^* + S_N$$

Where Q_N^* is the ideal neutrosophic size of the order that we arrived at by studying the static model without deficit according to neutrosophic logic, and it is given by the following relation:

$$Q_N^* = \sqrt{\frac{2K_N\lambda_N}{h_N}}$$

From the above we conclude that:

1- We conclude that the ideal neutrosophic volume of the first order is equal to:

$$Q_N^{**} = \sqrt{\frac{2K_N\lambda_N}{h_N}} + S$$

2- Total cost of inventory during each successive cycle is equal to:

$$TC(Q_N^*) = K_N + C_N . Q_N^* + \frac{h_N Q_N^*}{2\lambda_N} + h_N . S . T^*$$

3- As for the total cost of inventory during one time of successive cycles:

$$C(Q_N^*) = \frac{K_N\lambda_N}{Q_N^*} + C_N\lambda_N + \frac{h_N Q_N^*}{2} + h_N S$$

4- The reorder quantity is calculated from the relation:

$$Q_{N1} = d_N\lambda_N + S$$

Practical Example 3

A merchant wanted to invest his existing capital in storing and selling iron used in construction. Before starting work, he conducted a study of the market movement in the area in which he wanted to invest, and obtained the following information:

- The demand rate is a neutrosophic value and equals $\lambda_N \in [250, 350]$ tons.
- The cost of preparing an order consisting of Q_N tons is equal to $K_N \in [150, 250]\$$.
- The cost of storing one ton of iron during one time (month) is equal to $h_N \in [8,12]\$$.
- The cost of purchasing, delivering, receiving and arranging an order of size Q equals $C_N \in [4000, 6000]\$$.
- The duration of receiving the order is $d_N \in [3,7]$ *days*.
- The selling price of one ton is $b_N \in [5000, 6000]\$$.
- The safety reserve to ensure the continuity of work is S=100 tons per month

Required:

- Calculate the ideal size for the first order R_N^* and for the recurring order Q_N^*.
- Calculating the amount of the allowable deficit.
- Calculate the smallest cost of storage during a single period of time (month), and the smallest cost of storage during the storage cycle.

From the text of the problem, we have the following data:

$\lambda_N \in [250, 350]$, $d_N \in [3,7]$, $K_N \in [150, 250]$, $C_N \in [4000, 6000]$, $h_N \in [8,12]$, $P_N \in [40,60]$, $b_N \in [5000, 6000]$, S=

100

We calculate Q_N^*:

$$Q_N^* = \sqrt{\frac{2K_N\lambda_N}{h_N}} = \sqrt{\frac{2*[150,250]*[250,350]}{[8,12]}} \in [97,121]$$

From this we find that the optimal size for successive orders is:

$$Q_N^* = [97,121]$$

1- Ideal size for the first order:

$$Q_N^{**} = Q_N^* + S = [97,121] + 100 \in [197,221]$$

2- The reorder quantity is given by the following relation:

$$Q_{N1} = d_N \lambda_N + S_N = [3,7]*[250,350] + 100 \in [850,2550]$$

3- Calculate the total storage cost during one month:

$$C(Q_N^*) = \frac{K_N \lambda_N}{Q_N^*} + C_N \lambda_N + \frac{h_N Q_N^*}{2} + h_N S$$

$$C(Q_N^*) = \frac{[150,250]*[250,350]}{[97,121]} + [4000,6000]*[250,350]$$

$$+\frac{[8,12]}{2}*[97,121] + [8,12]*100 \in [1001575,2102649]$$

4- Calculate the investment cost of storage within a year:

$$CI_N = C(Q_N^*)*12 + C_N * S$$

$$CI_N = [1000775,2102649]*12 + [4000,6000]*100 \in [1249300,25831788]$$

The third case: (where the rate of demand for inventory is specified at a fixed value and the amount of the safety reserve is not specified)

The basic assumptions of this model:

- Order size Q_N.
- The rate of demand for inventory at one time is λ.
- The fixed cost of preparing the order is $K_N \in [\mu_1, \mu_2]$.
- The cost of purchasing, delivering and receiving one unit of inventory is: $C_N \in [\delta_1, \delta_2]$
- The storage cost is equal to a "fixed" amount $h_N \in [\beta_1, \beta_2]$ for each unit present in the warehouse during one time.
- The amount of the safety reserve in each storage cycle and over time is equal to S_N (indefinite) $S_N \in [S_1, S_2]$ such that S_1 is the minimum limit of the safety reserve, S_2 is the upper limit of the safety reserve, and the cost of storing one unit of it during one time is h_N, then the cost is That reserve equals $B_n = h_n * S_N$.
- The duration of running out of the stored quantity is $\frac{Q_N}{\lambda}$ (or storage cycle duration).

Using the previous hypotheses and benefiting from the results we reached in the study of the static model without a deficit according to the neutrosophic logic (Jdid et al., 2021) The static model of inventory management without a deficit with Neutrosophic logic, we find that the ideal size for the first order is given by the following relation:

$$Q_N^{**} = Q_N^* + S_N$$

Where Q_N^* is the ideal neutrosophic size of the order that we arrived at by studying the static model without deficit according to neutrosophic logic, and it is given by the following relation:

$$Q_N^* = \sqrt{\frac{2K_N\lambda}{h_N}}$$

From the above we conclude that:

1- We conclude that the ideal neutrosophic volume of the first order is equal to:

$$Q_N^{**} = \sqrt{\frac{2K_N\lambda}{h_N}} + S$$

2- Total cost of inventory during each successive cycle is equal to:

$$TC(Q_N^*) = K_N + C_N.Q_N^* + \frac{h_N Q_N^*}{2\lambda} + h_N.S.T^*$$

3- As for the total cost of inventory during one time of successive cycles:

$$C(Q_N^*) = \frac{K_N\lambda}{Q_N^*} + C_N\lambda + \frac{h_N Q_N^*}{2} + h_N S$$

4- The reorder quantity is calculated from the relation:

$$Q_{N1} = d_N\lambda + S$$

Practical Example 4

A merchant wanted to invest his existing capital in storing and selling iron used in construction. Before starting work, he conducted a study of the market movement in the area in which he wanted to invest, and obtained the following information:

- The demand rate is a neutrosophic value and equals λ=300 tons.
- The cost of preparing an order consisting of Q_N tons is equal to $K_N \in [150, 250]$\$.
- The cost of storing one ton of iron during one time (month) is equal to $h_N \in [8,12]$\$.
- The cost of purchasing, delivering, receiving and arranging an order of size Q equals $C_N \in [4000, 6000]$\$.
- The duration of receiving the order is $d_N \in [3,7]$ *days.*
- The selling price of one ton is $b_N \in [5000, 6000]$\$.
- The safety reserve to ensure the continuity of work is $S_N \in [0,100]$ tons per month

Required:

- Calculate the ideal size for the first orderR_N^* and for the recurring order Q_N^*.
- Calculating the amount of the allowable deficit.
- Calculate the smallest cost of storage during a single period of time (month), and the smallest cost of storage during the storage cycle.

From the text of the problem, we have the following data:

$\lambda=300$, $d_N \in [3,7]$, $K_N \in [150, 250]$, $C_N \in [4000, 6000]$, $h_N \in [8,12]$, $P_N \in [40,60]$, $b_N \in [5000, 6000]$, $S_N \in [0,100]$

We calculate Q_N^*:

$$Q_N^* = \sqrt{\frac{2K_N\lambda}{h_N}} = \sqrt{\frac{2*[150,250]*300}{[8,12]}} \in [106,112]$$

From this we find that the optimal size for successive orders is:

$$Q_N^* = [106,112]$$

1- Ideal size for the first order:

$$Q_N^{**} = Q_N^* + S_N = [97,121] + [0,100] \in [97,221]$$

2- The reorder quantity is given by the following relation:

$$Q_{N1} = d_N \lambda + S_N = [3,7]*300 + [0,100] \in [900,2200]$$

3- Calculate the total storage cost during one month:

$$C(Q_N^*) = \frac{K_N \lambda}{Q_N^*} + C_N \lambda + \frac{h_N Q_N^*}{2} + h_N S_N$$

$$C(Q_N^*) = \frac{[150,250]*300}{[97,121]} + [4000,6000]*300$$

$$+\frac{[8,12]}{2}*[97,121] + [8,12]*[0,100] C(Q_N^*) \in [1206049,1808122]$$

4- Calculate the investment cost of storage within a year:

$$CI_N = C(Q_N^*)*12 + C_N * S_N$$

$$CI_N = [1206049,1808122]*12 + [4000,6000]*[0,100] \in [14472588,22297464]$$

CONCLUSION AND RESULTS

In this chapter, we presented an extensive neutrosophic study of two static inventory models and calculated some important economic indicators for each of them. Through the neutrosophic values that we obtained, which take into account all the conditions that the work environment may experience, decision makers can determine a future vision for the workflow in the facility. And determine the ideal size of the stock in the case of perishable materials by using the study of the static model with deficit and the economic indicators specific to this model. They can also determine the ideal size of the stock of materials needed in emergency situations by using the study provided for the static model with safety reserve and the economic indicators specific to this model.

REFERENCES

Bukajh J.S. (1998). *Operations Research Book translated into Arabic.* The Arab Center for Arabization, Translation, Authoring and Publishing -Damascus.

Jdid, M. (2023). Important Neutrosophic Economic Indicators of the Static Model of Inventory Management without Deficit. *Journal of Neutrosophic and Fuzzy Systems, 5*(1), 08-14. doi:10.54216/JNFS.050101DOI: 10.54216/JNFS.050101

Jdid, M., Alhabib, R., & Salama, A. (2021). The static model of inventory management without a deficit with Neutrosophic logic. *International Journal of Neutrosophic Science, 16*(1). doi:10.54216/IJNS.160104DOI: 10.54216/IJNS.160104

Jdid, M., Alhabib, R., & Bahbouh, O. (2022). The Neutrosophic Treatment for Multiple Storage Problem of Finite Materials and Volumes. *International Journal of Neutrosophic Science, 18*(1), 42-56. doi:10.54216/IJNS.180105DOI: 10.54216/IJNS.180105

Jdid, M., Alhabib, R., Khalid, H. & Salama, A. (2022). The Neutrosophic Treatment of the Static Model for the Inventory Management with Safety Reserve. *InternationalJournal of Neutrosophic Science, 18*(2), 262-271. doi:10.54216/IJNS.180209DOI: 10.54216/IJNS.180209

Jdid, M., & Models, N. N. (2023). *Prospects for Applied Mathematics and Data Analysis, 2*(1), 42-46. doi:10.54216/PAMDA.020104DOI: 10.54216/PAMDA.020104

Jdid, M., & Smarandache, F. (2023). Neutrosophic Static Model without Deficit and Variable Prices. *Neutrosophic Sets and Systems, 60,* 124-132. https://fs.unm.edu/nss8/index.php/111/article/view/3744

Jdid, M., & Smarandache, F. (2023). Lagrange Multipliers and Neutrosophic Nonlinear Programming Problems Constrained by Equality Constraints. *Neutrosophic Systems with Applications, 6,* 25–31. doi:10.61356/j.nswa.2023.35DOI: 10.61356/j.nswa.2023.35

Jdid, M., & Smarandache, F. (2023). Graphical Method for Solving Neutrosophical Nonlinear Programming Models, / Int.J.Data. *Sci. & Big Data Anal., 3*(2), 66–72. doi:10.51483/IJDSBDA.3.2.2023.66-72DOI: 10.51483/IJDSBDA.3.2.2023.66-72

Jdid, M. (2022). Neutrosophic Treatment of the Static Model of Inventory Management with Deficit. *International Journal of Neutrosophic Science, 18*(1), 20-29. doi:10.54216/IJNS.180103DOI: 10.54216/IJNS.180103

Smarandache, F. (2014). *Introduction to Neutrosophic statistics*. Sitech & Education Publishing.

Chapter 9
Mathematical Programming for Neutrosophic Supply Chain Management

Hadi Basirzadeh
Faculty of Mathematical Sciences and Computer, Shahid Chamran University of Ahvaz, Iran

Madineh Farnam
Shohadaye Hoveizeh Campus of Technology, Shahid Chamran University of Ahvaz, Iran

Roohollah Abbasi Shureshjani
Department of Management, Humanities College, Hazrat-e Masoumeh University, Qom, Iran

Sara Ahmadi
Department of Mathematic, Faculty of Basic Sciences, University of Qom, Iran

ABSTRACT

Modeling the supply chain problem under neutrosophic data can show a more flexible and realistic notion of the problem's findings than the deterministic state. Therefore, in this chapter, the modeling of a three-level multi-product supply chain problem under neutrosophic data is considered. However, how to deal with the uncertainty caused by neutrosophic numbers in an optimization problem can be a fundamental challenge in finding appropriate answers in the problem-solving process. Hence, a weighted ranking method is introduced to find a range of responses. Then, the accuracy and reasonableness of the proposed ranking method for neutrosophic

DOI: 10.4018/979-8-3693-3204-7.ch009

numbers is tested by proving several theorems. In addition, the parametric nature of the method gives the decision maker access to a desirable level of answers in real-world problems. In the following, to demonstrate the efficiency of the method, a practical problem of a multi-product three-level supply chain has been modeled under neutrosophic data.

1. INTRODUCTION

Today, the fast and growing changes in industrial and commercial environments, as well as customer needs, have created significant competitive challenges for industrial managers. Therefore, the ability of organizations to respond promptly and adequately to customer demands is a crucial factor in gaining a competitive advantage in today's markets. Supply chain management has attracted the attention of many researchers as a predictive strategy for business success. Organizations aim to define an integrated performance to achieve their goals by identifying the production process of a product from its starting point to its delivery to the customer, taking into account the middle rings. The supply chain structure includes different links of production, distribution, customers, and complementary structures. Figure 1 shows an example of a three-tier supply chain with two manufacturers, three distributors, and four customers. Any insignificant performance at any level can negatively impact the overall performance of the supply chain. A review of multi-level supply chains is provided by Gümüs and Güneri (2007).

Figure 1. Three-level (manufactures, distributors, and customers) supply chain structure

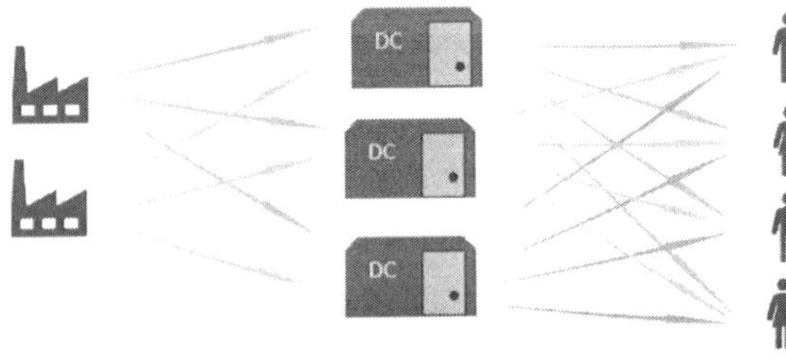

Tayur et al. (2012) have developed some quantitative models for supply chain management. Later, Ivanov et al. (2021) focused on aspects of the supply chain management solution process, including simulation-based optimization techniques, results from large deviation theory, and practical algorithms applicable to industrial-scale problems. Their research aimed to bridge theory and practice in problems with discriminative data. In this regard, the authors modelled examples of supply chains in various areas, including manufacturing, services, and e-operations, and used different methods and strategies to solve them. However, it is not possible to model all the problems that arise in the real world with their approach.

As Chopra and Meindl (2007) suggest, the supply chain encompasses all the parts that are directly or indirectly involved in fulfilling a customer request. Therefore, the presence and needs of customers are essential to the function of the supply chain. Customer satisfaction should be the top priority in all supply chain activities. Improving performance by meeting or exceeding customer expectations is crucial to staying competitive. Other priorities, such as reducing overall costs and increasing income, should not be ignored. Weaknesses in supply chain management can lead to decreased product quality, shipping delays, increased prices, and damage to customer trust. In essence, integrated operation within a supply chain can either attract and retain customers or drive them away from manufactured products. Integrated performance aims to achieve optimality and coordination between various and sometimes contradictory goals, resulting in increased profitability, reduced costs, and improved customer satisfaction. In addition to the goals, it is important to consider the limitations that govern the problem. For instance, to achieve favorable results, it is crucial to take into account factors such as matching supply with customer demand, the production capacity of each factory, and the overall demand of distribution centers. Businesses can gain a significant advantage over their competitors by identifying customer needs and planning accordingly.

In today's and modern economy, accurately pricing products, shipping costs, storage, and the duration of sending goods can be challenging due to turbulence. Therefore, managers need to be flexible in supply chain management to deal with uncertainty. The concept of fuzzy sets has been used to model problems in various decision-making areas, including the supply chain, due to the emergence of fuzzy theory and the expansion of operators on them, which allows for greater flexibility in dealing with real-world problems (Bellman and Zadeh, 1970). Furthermore, in recent decades, there have been more novel extensions of fuzzy numbers that include different structural features of membership functions, non-membership functions, and more. These extensions have been invented to match real-world problems with human thought and knowledge.

Petrovic et al. (1999), examine a production supply chain (SC) in which all facilities are connected in a serial manner. The SC operates in an uncertain environment, where uncertainty is associated with customer demand, supply deliveries along the SC, and external or market supply. To describe these uncertainties, imprecise phrases are used and represented by fuzzy sets. Tang et al. (2003) proposed a new approach for a medium-term production planning model of multi-product with fuzzy demand.

The objective of the problem was to minimize total costs, which included quadratic production costs and linear inventory holding costs. For this aim, the authors modeled the medium-term production planning problem of multiple products with fuzzy demand into a fuzzy quadratic planning model with fuzzy goals and constraints.

Vendor selection problem (VSP) has been treated by Kumar et al. (2006) as a "Fuzzy Multi-objective Integer Programming Vendor Selection Problem" (F-MIP_VSP) formulation that incorporates the three important goals: cost-minimization, quality-maximization and maximization of on-time-delivery-with the realistic constraints such as meeting the buyers' demand, vendors' capacity, vendors' quota flexibility, etc. The proposed model treats various input parameters as vague using a linear membership function of fuzzy type. Chen and Chang (2006) conducted a study on a three-level supply chain that involved multiple suppliers, manufacturers, distribution centers, and sales centers across various time periods. Their mathematical model takes into account the costs of purchasing, production, and distribution.

Liang (2008) proposed a fuzzy multi-objective model for the production-distribution problem. The model aims to minimize the costs of purchasing, distribution, total product returns, and total shipping times which was solved using the max-min method. in another research, Peidro et al. (2009) propose a fuzzy mathematical programming model for supply chain planning that considers uncertainties in supply, demand, and processes. The model is formulated as a fuzzy mixed-integer linear programming model, where data are ill-known and modeled by triangular fuzzy numbers. Bilgen (2010) addresses the production and distribution planning problem in a supply chain system that involves allocating production volumes among different production lines in manufacturing plants and delivering products to distribution centers. Due to the highly dynamic and uncertain nature of real supply chains, the model is transformed into fuzzy models by them that consider the fuzziness in capacity constraints and the aspiration level of costs using different aggregation operators. Gholamian et al. (2015) presented a mathematical model for production planning in a supply chain under demand uncertainty. To account for the sensitivity of the system to different uncertainty values, fuzziness is also considered in the constraints that include uncertain demand. Nasseri et al. (2017) used a new mixed integer multi-objective linear programming model to solve the fully fuzzy multi-objective supplier selection problem, which is a crucial aspect of supply chain management.

In another work, Arasteh (2020) discussed the combination of fuzzy multi-objective planning and real options approaches in supply chain management under uncertainty. The aim of his paper is to evaluate and analyze supply chain planning decisions in the presence of market and/or technical uncertainty. Shafi Salimi and Edalatpanah (2020) evaluated and compared suppliers for a manufacturing company using two methods of fuzzy hierarchical analysis with D-numbers in their research.

To express a more complete adaptation of the human mental decision-making process, Smarandache (1998) introduced the Neutrosophic set to deal with uncertain and inconsistent situations. Wang et al. (2010) proposed the concept of single-valued Neutrosophic set (SVN). Over time neutrosophic sets have been widely extended in both theoretical and practical applications since their introduction. In the context of mathematical modelling for real-world problems, further research can be conducted based on this type of uncertainty, particularly in optimization problems related to the supply chain. Some effective research has already been conducted on topics related to this field. For instance, Ahmad et al. (2018) provide a construction of the manufacturing system of SCM using single-valued neutrosophic hesitant fuzzy data. In a separate study, Han et al. (2020) proposed an entropy-based equivalent model to achieve a reasonable degree of constraint satisfaction when dealing with uncertainty caused by single-valued neutrosophic data in the system. They applied this concept to transform the zinc electrowinning process application problem into a multi-objective optimization problem. Abdel-Basset et al. (2020) developed a linear model to handle scheduling problems with neutrosophic information duration times. The authors analyzed the key aspects of the trade-off between time and cost in their approach. Ahmad et al. (2020) used a neutrosophic fuzzy programming method to design a neutrosophic-CLSC network. They also developed a new hybrid method, modified NFPA, with intuitionistic fuzzy importance relations for uncertain networks.

Edalatpanah (2020) developed a process for evaluating decision-making units in existing Data Envelopment Analysis models using a ranking system based on triangular neutrosophic numbers. In 2020, Pratihar et al. (2020) developed principles of linear programming to solve transportation problems under neutrosophic data for supply, demand, and cell cost. The proposed methods, such as Vogel's approximation method and the northwest principle, were also considered. Later, Kamal et al. (2021) formulated the multi-objective selective maintenance allocation problem with fuzzy parameters under neutrosophic data. The model was formulated with fuzzy parameters and a new defuzzification technique based on beta distribution was considered to convert uncertain parameters into crisp values.

In another study, Fallah and Nozari (2021) designed a multi-objective sustainable biomass supply chain network to optimize total costs, greenhouse gas emissions, number of potential hires, and product delivery time under neutrosophic information. They evaluated the performance of each goal and made strategic and tactical decisions in different fields. Ahmad and John (2022) introduced the definition of the neutrosophic hesitant fuzzy Pareto optimal solution to overcome uncertainty and hesitation in parameters. They applied this definition in two different optimization methods to demonstrate its superiority. In a relevant study, Giri and Roy (2022) proposed two compromise methods for solving a multi-objective green 4-dimensional fixed-charge transportation problem under single valued trapezoidal neutrosophic number (SVTNN). They used the cut of SVTNN to transform the parameters in interval form of the suggested model to minimize transportation time, carbon emissions, and transportation costs.

Nozari et al. (2022) employed the neutrosophic fuzzy programming approach to locate warehouses, production centers, and route vehicles for distributing medical materials to hospitals during the COVID-19 pandemic. Later Kousar et al. (2023) applied a neutrosophic multi-objective fuzzy linear programming (MONFLP) model to optimize the production of Kharif and Rabbi crops in Pakistan. Their model includes the maximization of net profit and product output to optimize production planning. Ghosh and Roy (2023) established a multi-objective waste management (MOWM) problem under a neutrosophic hesitant fuzzy (NHF) environment. They applied a novel ranking technique to defuzzify NHF numbers and optimize objectives such as maximum profit, minimum carbon emissions under policies, and minimum workload deviation to maintain sustainability of the MOWM problem. Recently, Jdid and Smarandache (2023) presented a new structure for the optimal use of agricultural land under the concepts of neutrosophic science, where they obtained allocable levels in each of the areas for each crop.

Although ranking methods have been used in various fields of mathematical optimization, including supply chain management under neutrosophic data. In this research, we propose a parametric ranking method based on a new perspective to solve a three-level multi-objective supply chain problem with neutrosophic objectives.

For this purpose, other parts of this chapter are arranged as follows:

Section 2 presents some preliminaries and operators required for neutrosophic numbers.

Section 3 explains the structure of the proposed ranking method for numbers.

Section 4 demonstrates that the method applies to many logical features beyond flexibility, making it effective for multi-objective optimization problems.

Section 5 includes the mathematical modeling of the three-level multi-objective supply chain problem.

A practical example is used to validate the mathematical model and ascertain its effectiveness in section 6.

Finally, in section 7, some conclusions and future works are highlighted based on the present study.

2. PRELIMINARIES

In this section, we consider a brief required concept of neutrosophic logic, along with some essential operations which are related to the subsequent sections of this chapter.

Definition 1. (Wang et al., 2010) let $\mathfrak{U}$is a universe of discourse, then a neutrosophic set in $\mathfrak{U}$ is defined by the following representation

$$\widetilde{N} = \left\{\langle\mu_{\widetilde{N}}(u),\eta_{\widetilde{N}}(u),\tau_{\widetilde{N}}(u)\rangle \middle| 0 \leq \mu_{\widetilde{N}}(u),\eta_{\widetilde{N}}(u),\tau_{\widetilde{N}}(u) \leq 1, u \in \mathfrak{U}\right\}, \quad (1)$$

Where $\mu_{\widetilde{N}}:\mathfrak{U} \rightarrow [0,1]$ is truth-membership function, $\eta_{\widetilde{N}}:\mathfrak{U} \rightarrow [0,1]$is an indeterminacy-membership function, and $\mu_{\widetilde{N}}:U \rightarrow [0,1]$ is falsity-membership function. Furthermore.

$$0 \leq \mu_{\widetilde{N}}(u)+\eta_{\widetilde{N}}(u)+\tau_{\widetilde{N}}(u) \leq 3$$

Definition 2. (Şahin et al., 2018) Assume

$$\widetilde{n}_1 = \langle(a_{\widetilde{n}_1},b_{\widetilde{n}_1},c_{\widetilde{n}_1})\rangle;\omega_{\widetilde{n}_1},\phi_{\widetilde{n}_1},\psi_{\widetilde{n}_1}$$

is a triangular single valued neutrosophic number in the set of real numbers where

$$\omega_{\widetilde{n}_1},\phi_{\widetilde{n}_1},\psi_{\widetilde{n}_1} \in [0,1]$$

Then, its truth membership function is

$$\mu_{\widetilde{n}_1}(u)=\begin{cases}\dfrac{(u-a_{\widetilde{n}_1})\omega_{\widetilde{n}_1}}{b_{\widetilde{n}_1}-a_{\widetilde{n}_1}}, a_{\widetilde{n}_1} \leq u \leq b_{\widetilde{n}_1}\\ \omega_{\widetilde{n}_1},\quad u = b_{\widetilde{n}_1}\\ \dfrac{(c_{\widetilde{n}_1}-u)}{c_{\widetilde{n}_1}-b_{\widetilde{n}_1}} b_{\widetilde{n}_1} \leq u \leq c_{\widetilde{n}_1}\\ 0,\quad 0.w\end{cases} \quad (2)$$

Its indeterminacy membership function is

$$\eta_{\tilde{n}_1}(u)=\begin{cases}\dfrac{(b_{\tilde{n}_1}-u+(u-a_{\tilde{n}_1})\phi_{\tilde{n}_1})}{b_{\tilde{n}_1}-a_{\tilde{n}_1}}, & a_{\tilde{n}_1}\le u\le b_{\tilde{n}_1}\\ \phi_{\tilde{n}_1}, & u=b_{\tilde{n}_1}\\ \dfrac{(u-b_{\tilde{n}_1}+(c_{\tilde{n}_1}-u)\phi_{\tilde{n}_1})}{c_{\tilde{n}_1}-b_{\tilde{n}_1}}, & b_{\tilde{n}_1}\le u\le c_{\tilde{n}_1}\\ 0, & 0.w\end{cases} \tag{3}$$

and its falsity membership function is

$$\tau_{\tilde{n}_1}(u)=\begin{cases}\dfrac{(b_{\tilde{n}_1}-u+(u-a_{\tilde{n}_1})\psi_{\tilde{n}_1})}{b_{\tilde{n}_1}-a_{\tilde{n}_1}}, & a_{\tilde{n}_1}\le u\le b_{\tilde{n}_1}\\ \psi_{\tilde{n}_1}, & u=b_{\tilde{n}_1}\\ \dfrac{(u-b_{\tilde{n}_1}+(c_{\tilde{n}_1}-u)\psi_{\tilde{n}_1})}{c_{\tilde{n}_1}-b_{\tilde{n}_1}}, & b_{\tilde{n}_1}\le u\le c_{\tilde{n}_1}\\ 0, & 0.w\end{cases} \tag{4}$$

Where $0 \le \omega_{\tilde{n}_1}+\phi_{\tilde{n}_1}+\psi_{\tilde{n}_1}\le 3$.

Figure 2 shows the representation of a triangular single valued neutrosophic (Tri-SVN) number.

Figure 2. Tri-SVN number

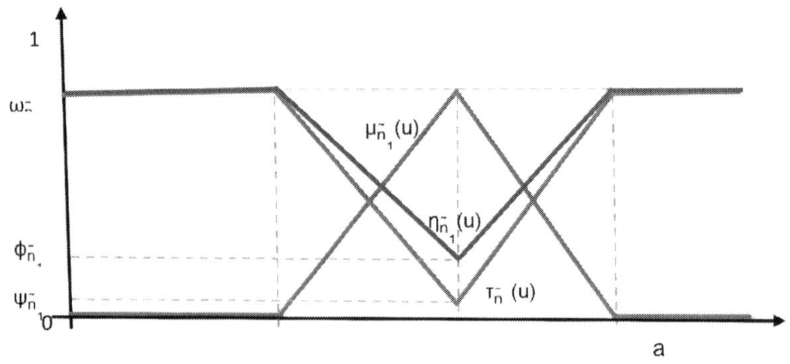

Remark 1. a) If $a_{\tilde{n}_1} \geq 0$ and $c_{\tilde{n}_1} > 0$, then

$$\tilde{n}_1 = \langle (a_{\tilde{n}_1}, b_{\tilde{n}_1}, c_{\tilde{n}_1}); \omega_{\tilde{n}_1}, \phi_{\tilde{n}_1}, \psi_{\tilde{n}_1} \rangle$$

is a positive Tri-SVN number.

If $a_{\tilde{n}_1} < 0$ and $c_{\tilde{n}_1} \leq 0$, then $\tilde{n}_1 = \langle (a_{\tilde{n}_1}, b_{\tilde{n}_1}, c_{\tilde{n}_1}); \omega_{\tilde{n}_1}, \phi_{\tilde{n}_1}, \psi_{\tilde{n}_1} \rangle$ is a negative Tri-SVN number.

Definition 3. (Şahin et al., 2018) assume

$$\tilde{n}_1 = \langle (a_{\tilde{n}_1}, b_{\tilde{n}_1}, c_{\tilde{n}_1}); \omega_{\tilde{n}_1}, \phi_{\tilde{n}_1}, \psi_{\tilde{n}_1} \rangle \text{ and } \tilde{n}_2 = \langle (a_{\tilde{n}_2}, b_{\tilde{n}_2}, c_{\tilde{n}_2}); \omega_{\tilde{n}_2}, \phi_{\tilde{n}_2}, \psi_{\tilde{n}_2} \rangle$$

are two Tri-SVN numbers and ς is an actual number. Then,

1) $$\tilde{n}_1 \oplus \tilde{n}_2 = \langle (a_{\tilde{n}_1} + a_{\tilde{n}_2}, b_{\tilde{n}_1} + b_{\tilde{n}_2}, c_{\tilde{n}_1} + c_{\tilde{n}_2}); \omega_{\tilde{n}_1} \wedge \omega_{\tilde{n}_2}, \phi_{\tilde{n}_1} \vee \phi_{\tilde{n}_2}, \psi_{\tilde{n}_1} \vee \psi_{\tilde{n}_2} \rangle,$$

2) $$\varsigma\tilde{n}_1 = \begin{cases} \langle (\varsigma a_{\tilde{n}_1}, \varsigma b_{\tilde{n}_1}, \varsigma c_{\tilde{n}_1}); \omega_{\tilde{n}_1}, \phi_{\tilde{n}_1}, \psi_{\tilde{n}_1} \rangle, \varsigma \geq 0 \\ \langle (\varsigma c_{\tilde{n}_1}, \varsigma b_{\tilde{n}_1}, \varsigma a_{\tilde{n}_1}); \omega_{\tilde{n}_1}, \phi_{\tilde{n}_1}, \psi_{\tilde{n}_1} \rangle, \varsigma \leq 0 \end{cases}$$

3. THE NOVEL WEIGHTED PARAMETRIC RANKING APPROACH FOR TRI-SVN NUMBER

In this section, we want to express a new weighted parametric ranking function for continuous neutrosophic numbers, specifically for triangular neutrosophic numbers. For this purpose, we first state the following definition based on (Farnam et al., 2024).

Definition 4. (Farnam et al., 2024) Assume

$$\tilde{n}_1 = \langle (a_{\tilde{n}_1}, b_{\tilde{n}_1}, c_{\tilde{n}_1}); \omega_{\tilde{n}_1}, \phi_{\tilde{n}_1}, \psi_{\tilde{n}_1} \rangle$$

is a Tri-SVN number. α-cut, β-cut, γ-cut and (α,β,γ)-cut sets of $\tilde{n}_1$, are defined by the following intervals, respectively,

$$1)\ \tilde{n}_{1\alpha} = \{u | \mu_{\tilde{n}_1}(u) \geq \alpha\}$$
$$= \left[a_{\tilde{n}_1} + \frac{\alpha}{\omega_{\tilde{n}_1}}(b_{\tilde{n}_1} - a_{\tilde{n}_1}), c_{\tilde{n}_1} + \frac{\alpha}{\omega_{\tilde{n}_1}}(b_{\tilde{n}_1} - c_{\tilde{n}_1})\right], 0 \leq \alpha \leq \omega_{\tilde{n}_1}, \tag{5}$$

$$2)\ \tilde{n}_{1\beta} = \{u | \eta_{\tilde{n}_1}(u) \leq \beta\}$$
$$= \left[a_{\tilde{n}_1} + \frac{1-\beta}{1-\phi_{\tilde{n}_1}}(b_{\tilde{n}_1} - a_{\tilde{n}_1}), c_{\tilde{n}_1} + \frac{1-\beta}{1-\phi_{\tilde{n}_1}}(b_{\tilde{n}_1} - c_{\tilde{n}_1})\right], \phi_{\tilde{n}_1} \leq \beta \leq 1, \tag{6}$$

$$3)\ \tilde{n}_{1\gamma} = \left\{u \mid \tau_{\tilde{n}_1}(u) \le \gamma\right\}$$

$$= \left[a_{\tilde{n}_1} + \frac{1-\gamma}{1-\psi_{\tilde{n}_1}}\left(b_{\tilde{n}_1} - a_{\tilde{n}_1}\right), c_{\tilde{n}_1} + \frac{1-\gamma}{1-\psi_{\tilde{n}_1}}\left(b_{\tilde{n}_1} - c_{\tilde{n}_1}\right)\right], \psi_{\tilde{n}_1} \le \gamma \le 1, \tag{7}$$

$$4)\ \tilde{n}_{1\alpha,\beta,\gamma} = \left\{x \mid \mu_{\tilde{n}_1}(u) \ge \alpha, \eta_{\tilde{n}_1}(u) \le \beta, \tau_{\tilde{n}_1}(u) \le \gamma\right\}$$

$$= \left[a_{\tilde{n}_1} + \frac{\alpha}{\omega_{\tilde{n}_1}}\left(b_{\tilde{n}_1} - a_{\tilde{n}_1}\right), c_{\tilde{n}_1} + \frac{\alpha}{\omega_{\tilde{n}_1}}\left(b_{\tilde{n}_1} - c_{\tilde{n}_1}\right)\right]$$

$$\left[a_{\tilde{n}_1} + \frac{1-\beta}{1-\phi_{\tilde{n}_1}}\left(b_{\tilde{n}_1} - a_{\tilde{n}_1}\right), c_{\tilde{n}_1} + \frac{1-\beta}{1-\phi_{\tilde{n}_1}}\left(b_{\tilde{n}_1} - c_{\tilde{n}_1}\right)\right]$$

$$\cap \left[a_{\tilde{n}_1} + \frac{1-\gamma}{1-\psi_{\tilde{n}_1}}\left(b_{\tilde{n}_1} - a_{\tilde{n}_1}\right), c_{\tilde{n}_1} + \frac{1-\gamma}{1-\psi_{\tilde{n}_1}}\left(b_{\tilde{n}_1} - c_{\tilde{n}_1}\right)\right],$$

Where $0 \le \alpha \le \omega_{\tilde{n}_1}, \phi_{\tilde{n}_1} \le \beta \le 1, \psi_{\tilde{n}_1} \le \gamma \le 1$ and

$$0 \le \alpha + \beta + \gamma \le 2 + \omega_{\tilde{n}_1} - \phi_{\tilde{n}_1} - \psi_{\tilde{n}_1}. \tag{8}$$

Definition 5. (Farnam et al., 2024) Assume

$$\tilde{n}_1 = \langle \left(a_{\tilde{n}_1}, b_{\tilde{n}_1}, c_{\tilde{n}_1}\right); \omega_{\tilde{n}_1}, \phi_{\tilde{n}_1}, \psi_{\tilde{n}_1} \rangle$$

is a Tri-SVN number. β'–cut, γ'–cut and $(\alpha, \beta', \gamma')$–cut sets of $\tilde{n}_1$, are defined by the following intervals, respectively,

$$1)\ \tilde{n}_{1\beta'} = \left\{u \mid \eta_{\tilde{n}_1}(u) \ge \beta'\right\}$$

$$= \left[a_{\tilde{n}_1} + \frac{\beta'}{1-\phi_{\tilde{n}_1}}\left(b_{\tilde{n}_1} - a_{\tilde{n}_1}\right), c_{\tilde{n}_1} + \frac{\beta'}{1-\phi_{\tilde{n}_1}}\left(b_{\tilde{n}_1} - c_{\tilde{n}_1}\right)\right], 0 \le \beta' \le 1 - \phi_{\tilde{n}_1}, \tag{9}$$

$$2)\ \tilde{n}_{1\gamma'} = \left\{u \mid \tau_{\tilde{n}_1}(u) \ge \gamma'\right\}$$

$$= \left[a_{\tilde{n}_1} + \frac{\gamma'}{1-\psi_{\tilde{n}_1}}\left(b_{\tilde{n}_1} - a_{\tilde{n}_1}\right), c_{\tilde{n}_1} + \frac{\gamma'}{1-\psi_{\tilde{n}_1}}\left(b_{\tilde{n}_1} - c_{\tilde{n}_1}\right)\right], 0 \le \gamma' \le 1 - \psi_{\tilde{n}_1} \tag{10}$$

$$3)\ \tilde{n}_{\alpha,\beta',\gamma'} = \left\{u \mid \mu_{\tilde{n}_1}(u) \ge \alpha, \eta_{\tilde{n}_1}(u) \ge \beta', \tau_{\tilde{n}_1}(u) \ge \gamma'\right\}$$

$$= \left[a_{\tilde{n}_1} + \frac{\alpha}{\omega_{\tilde{n}_1}}\left(b_{\tilde{n}_1} - a_{\tilde{n}_1}\right), c_{\tilde{n}_1} + \frac{\alpha}{\omega_{\tilde{n}_1}}\left(b_{\tilde{n}_1} - c_{\tilde{n}_1}\right)\right]$$

$$\cap \left[a_{\tilde{n}_1} + \frac{\beta'}{1-\phi_{\tilde{n}_1}}\left(b_{\tilde{n}_1} - a_{\tilde{n}_1}\right), c_{\tilde{n}_1} + \frac{\beta'}{1-\phi_{\tilde{n}_1}}\left(b_{\tilde{n}_1} - c_{\tilde{n}_1}\right)\right]$$

$$\cap \left[a_{\tilde{n}_1} + \frac{\gamma'}{1-\psi_{\tilde{n}_1}}\left(b_{\tilde{n}_1} - a_{\tilde{n}_1}\right), c_{\tilde{n}_1} + \frac{\gamma'}{1-\psi_{\tilde{n}_1}}\left(b_{\tilde{n}_1} - c_{\tilde{n}_1}\right)\right]$$

where $0 \le \alpha \le \omega_{\tilde{n}_1}, 0 \le \beta' \le 1 - \phi_{\tilde{n}_1}, 0 \le \gamma' \le 1 - \psi_{\tilde{n}_1}$ and

$$0 \le \alpha + \beta' + \gamma' \le 2 + \omega_{\tilde{n}_1} - \phi_{\tilde{n}_1} - \psi_{\tilde{n}_1}. \tag{11}$$

Pouresmaeil et al. (2022) for a discrete neutrosophic number such as

$$\widetilde{N} = \langle \mu_{\widetilde{N}}(u), \eta_{\widetilde{N}}(u), \tau_{\widetilde{N}}(u) \rangle$$

defined the following score function.

$$\mathfrak{R}^{w}(\widetilde{N}) = \frac{2}{3} + \frac{\mu_{\widetilde{N}}(u)}{3} - w_1 \frac{\eta_{\widetilde{N}}(u)}{3} - w_2 \frac{\tau_{\widetilde{N}}(u)}{3}, \tag{12}$$

Where $0<w_1<w_2\le1$ and $R^w(u)\in[0,1]$.

Inspired by equation (12) and definition 6, we define the following weighted parametric ranking function.

Definition 6. Assume $\tilde{n}_1 = \langle (a_{\tilde{n}_1}, b_{\tilde{n}_1}, c_{\tilde{n}_1}); \omega_{\tilde{n}_1}, \phi_{\tilde{n}_1}, \psi_{\tilde{n}_1} \rangle$ is a Tri-SVN number. Then

$$\mathfrak{R}^{w}_{\alpha,\beta',\gamma'}(\tilde{n}_1) = \frac{2}{3} + \frac{S^{\omega}_{\alpha}(\tilde{n}_1)}{3} + w_1 \frac{S^{\phi}_{\beta'}(\tilde{n}_1)}{3} + w_2 \frac{S^{\psi}_{\gamma'}(\tilde{n}_1)}{3} \tag{13}$$

Where $0<w_2<w_1\le1$ and $\mathfrak{R}^w \in [0,1]$. Also,

$$S^{\omega}_{\alpha}(\tilde{n}_1) = \frac{1}{2}\left(S^{\underline{\omega}}_{\alpha}(\tilde{n}_1) + S^{\overline{\omega}}_{\alpha}(\tilde{n}_1)\right)$$

$$= \frac{1}{2}\left(\int_{\alpha}^{\omega_{\tilde{n}_1}} (a_{\tilde{n}_1} + \frac{r}{\omega_{\tilde{n}_1}}(b_{\tilde{n}_1} - a_{\tilde{n}_1}))dr + \int_{\alpha}^{\omega_{\tilde{n}_1}} (c_{\tilde{n}_1} + \frac{r}{\omega_{\tilde{n}_1}}(b_{\tilde{n}_1} - c_{\tilde{n}_1}))dr\right)$$

$$= \frac{1}{2}\left\{ (a_{\tilde{n}_1} + c_{\tilde{n}_1})\left(\omega_{\tilde{n}_1} - \alpha\right) + \frac{1}{2\omega_{\tilde{n}_1}}\left(\left(b_{\tilde{n}_1} - a_{\tilde{n}_1}\right) + \left(b_{\tilde{n}_1} - c_{\tilde{n}_1}\right)\right)\left(\omega_{\tilde{n}_1}{}^{2} - \alpha^2\right)\right\} \tag{14}$$

$$S^{\phi}_{\beta'}(\tilde{n}_1) = \frac{1}{2}\left(S^{\underline{\phi}}_{\beta'}(\tilde{n}_1) + S^{\overline{\phi}}_{\beta'}(\tilde{n}_1)\right)$$

$$= \frac{1}{2}\left(\int_{\beta'}^{1-\phi_{\tilde{n}_1}} (a_{\tilde{n}_1} + \frac{r}{(1-\phi_{\tilde{n}_1})}(b_{\tilde{n}_1} - a_{\tilde{n}_1}))dr + \int_{\beta'}^{1-\phi_{\tilde{n}_1}} (c_{\tilde{n}_1} + \frac{r}{(1-\phi_{\tilde{n}_1})}(b_{\tilde{n}_1} - c_{\tilde{n}_1}))dr\right)$$

$$= \frac{1}{2}\Big\{ (a_{\tilde{n}_1} + c_{\tilde{n}_1})\big((1 - \phi_{\tilde{n}_1}) - \beta'\big) + \frac{1}{2(1 - \phi_{\tilde{n}_1})}\Big(\big(b_{\tilde{n}_1} - a_{\tilde{n}_1}\big) + \big(b_{\tilde{n}_1} - c_{\tilde{n}_1}\big)\Big)\Big((1 - \phi_{\tilde{n}_1})^2 - \beta'^2\Big) \Big\} \tag{15}$$

$$S^{\psi}_{\gamma'}(\tilde{n}_1) = \frac{1}{2}\big(S^{\underline{\psi}}_{\gamma'}(\tilde{n}_1) + S^{\overline{\psi}}_{\gamma'}(\tilde{n}_1)\big)$$

$$= \frac{1}{2}\Big(\int_{\gamma'}^{\psi_{\tilde{n}_1}} (a_{\tilde{n}_1} + \frac{r}{(1 - \psi_{\tilde{n}_1})}(b_{\tilde{n}_1} - a_{\tilde{n}_1}))dr + \int_{\gamma'}^{\psi_{\tilde{n}_1}} (c_{\tilde{n}_1} + \frac{r}{(1 - \psi_{\tilde{n}_1})}(b_{\tilde{n}_1} - c_{\tilde{n}_1}))dr \Big)$$

$$= \frac{1}{2}\Big\{ (a_{\tilde{n}_1} + c_{\tilde{n}_1})\big((1 - \psi_{\tilde{n}_1}) - \gamma'\big) + \frac{1}{2(1 - \psi_{\tilde{n}_1})}\Big(\big(b_{\tilde{n}_1} - a_{\tilde{n}_1}\big) + \big(b_{\tilde{n}_1} - c_{\tilde{n}_1}\big)\Big)\Big((1 - \psi_{\tilde{n}_1})^2 - \gamma'^2\Big) \Big\} \tag{16}$$

It is important to mention that instead of differential operators in (12), for the continuous state according to equation (13), the effect of functions (9) and (10) has been used.

4. THEOREMS

Theorem 1. If $\tilde{n}_1 = \langle (a_{\tilde{n}_1}, b_{\tilde{n}_1}, c_{\tilde{n}_1}); \omega_{\tilde{n}_1}, \phi_{\tilde{n}_1}, \psi_{\tilde{n}_1} \rangle$ and

$\tilde{n}_2 = \langle (a_{\tilde{n}_2}, b_{\tilde{n}_2}, c_{\tilde{n}_2}); \omega_{\tilde{n}_2}, \phi_{\tilde{n}_2}, \psi_{\tilde{n}_2} \rangle$

be two Tri-SVN numbers where

$$a_{\tilde{n}_2} = a_{\tilde{n}_1} + \varepsilon_1, b_{\tilde{n}_2} = b_{\tilde{n}_1}, c_{\tilde{n}_2} = c_{\tilde{n}_1}, \omega_{\tilde{n}_2} = \omega_{\tilde{n}_1}, \phi_{\tilde{n}_2} = \phi_{\tilde{n}_1}, \psi_{\tilde{n}_2} = \psi_{\tilde{n}_1}$$

Also, ε_1 is an actual number and $0 < \varepsilon_1 \leq b_{\tilde{n}_1} - a_{\tilde{n}_1}$. Then, we have

$$\mathfrak{R}^{w}_{\alpha,\beta',\gamma'}(\tilde{n}_2) \geq \mathfrak{R}^{w}_{\alpha,\beta',\gamma'}(\tilde{n}_1).$$

Proof. It is sufficient to show

$$\mathfrak{R}^{w}_{\alpha,\beta',\gamma'}(\tilde{n}_2) - \mathfrak{R}^{w}_{\alpha,\beta',\gamma'}(\tilde{n}_1) \geq 0$$

From (13) we have

$$\Re^{w}_{\alpha,\beta',\gamma'}(\tilde{n}_2) - \Re^{w}_{\alpha,\beta',\gamma'}(\tilde{n}_1) = \frac{2}{3} + \frac{S^{\omega}_{\alpha}(\tilde{n}_2)}{3}$$

$$+w_1\frac{S^{\phi}_{\beta'}(\tilde{n}_2)}{3} + w_2\frac{S^{\psi}_{\gamma'}(\tilde{n}_2)}{3} - \frac{2}{3} - \frac{S^{\omega}_{\alpha}(\tilde{n}_2)}{3} - w_1\frac{S^{\phi}_{\beta'}(\tilde{n}_2)}{3} - w_2\frac{S^{\psi}_{\gamma'}(\tilde{n}_2)}{3}$$

Using (14-16) we can write

$$\Re^{w}_{\alpha,\beta',\gamma'}(\tilde{n}_2) - \Re^{w}_{\alpha,\beta',\gamma'}(\tilde{n}_1)$$

$$=\frac{1}{6}\Big[(a_{\tilde{n}_2} + c_{\tilde{n}_2})(\omega_{\tilde{n}_2} - \alpha)$$

$$+\frac{1}{2\omega_{\tilde{n}_2}}\Big((b_{\tilde{n}_2} - a_{\tilde{n}_2}) + (b_{\tilde{n}_2} - c_{\tilde{n}_2})\Big)(\omega_{\tilde{n}_2}{}^2 - \alpha^2) - (a_{\tilde{n}_1} + c_{\tilde{n}_1})(\omega_{\tilde{n}_1} - \alpha)$$

$$-\frac{1}{2\omega_{\tilde{n}_1}}\Big((b_{\tilde{n}_1} - a_{\tilde{n}_1}) + (b_{\tilde{n}_1} - c_{\tilde{n}_1})\Big)(\omega_{\tilde{n}_1}{}^2 - \alpha^2)\Big]+\frac{w_1}{6}\Big[(a_{\tilde{n}_2} + c_{\tilde{n}_2})(\phi_{\tilde{n}_2} - \beta')$$

$$+\frac{1}{2\phi_{\tilde{n}_2}}\Big((b_{\tilde{n}_2} - a_{\tilde{n}_2}) + (b_{\tilde{n}_2} - c_{\tilde{n}_2})\Big)(\phi_{\tilde{n}_2}{}^2 - \beta'^2) - (a_{\tilde{n}_1} + c_{\tilde{n}_1})(\phi_{\tilde{n}_1} - \beta') - \frac{1}{2\phi_{\tilde{n}_1}}((b_{\tilde{n}_1} - a_{\tilde{n}_1})$$

$$-(b_{\tilde{n}_1} - c_{\tilde{n}_1}))(\phi_{\tilde{n}_1}{}^2 - \beta'^2)\Big]+\frac{w_2}{6}\Big[(a_{\tilde{n}_2} + c_{\tilde{n}_2})(\psi_{\tilde{n}_2} - \gamma') + \frac{1}{2\psi_{\tilde{n}_2}}((b_{\tilde{n}_2} - a_{\tilde{n}_2})$$

$$+(b_{\tilde{n}_2} - c_{\tilde{n}_2}))(\psi_{\tilde{n}_2}{}^2 - \gamma'^2)$$

$$-(a_{\tilde{n}_1} + c_{\tilde{n}_1})(\psi_{\tilde{n}_1} - \gamma') - \frac{1}{2\psi_{\tilde{n}_1}}\Big((b_{\tilde{n}_1} - a_{\tilde{n}_1}) + (b_{\tilde{n}_1} - c_{\tilde{n}_1})\Big)(\psi_{\tilde{n}_1}{}^2 - \gamma'^2)].$$

By substituting

$$a_{\tilde{n}_2} = a_{\tilde{n}_1} + \varepsilon_1, b_{\tilde{n}_2} = b_{\tilde{n}_1}, c_{\tilde{n}_2} = c_{\tilde{n}_1}, \omega_{\tilde{n}_2} = \omega_{\tilde{n}_1}, \phi_{\tilde{n}_2} = \phi_{\tilde{n}_1} and \psi_{\tilde{n}_2} = \psi_{\tilde{n}_1},$$

into a above expressions, we get

$$\Re^{w}_{\alpha,\beta',\gamma'}(\tilde{n}_2) - \Re^{w}_{\alpha,\beta',\gamma'}(\tilde{n}_1)$$

$$=\frac{1}{6}\Big[(a_{\tilde{n}_1} + \varepsilon_1 + c_{\tilde{n}_1})(\omega_{\tilde{n}_1} - \alpha)$$

$$+\frac{1}{2\omega_{\tilde{n}_1}}\Big((b_{\tilde{n}_1} - a_{\tilde{n}_1} - \varepsilon_1) + (b_{\tilde{n}_1} - c_{\tilde{n}_1})\Big)(\omega_{\tilde{n}_1}{}^2 - \alpha^2) - (a_{\tilde{n}_1} + c_{\tilde{n}_1})(\omega_{\tilde{n}_1} - \alpha)$$

$$-\frac{1}{2\omega_{\tilde{n}_1}}\Big((b_{\tilde{n}_1} - a_{\tilde{n}_1}) + (b_{\tilde{n}_1} - c_{\tilde{n}_1})\Big)(\omega_{\tilde{n}_1}{}^2 - \alpha^2)\Big]+\frac{w_1}{6}\Big[(a_{\tilde{n}_1} + \varepsilon_1 + c_{\tilde{n}_2})(\phi_{\tilde{n}_1} - \beta')$$

$$+\frac{1}{2\phi_{\tilde{n}_1}}\Big((b_{\tilde{n}_1} - a_{\tilde{n}_1} - \varepsilon_1) + (b_{\tilde{n}_1} - c_{\tilde{n}_1})\Big)(\phi_{\tilde{n}_1}{}^2 - \beta'^2) - (a_{\tilde{n}_1} + c_{\tilde{n}_1})(\phi_{\tilde{n}_1} - \beta')$$

$$-\frac{1}{2\phi_{\tilde{n}_1}}\Big(\big(b_{\tilde{n}_1}-a_{\tilde{n}_1}\big)-\big(b_{\tilde{n}_1}-c_{\tilde{n}_1}\big)\Big)\Big(\phi_{\tilde{n}_1}{}^2-\beta'^2\Big)\Big]+\frac{w_2}{6}\Big[\big(a_{\tilde{n}_1}+\varepsilon_1+c_{\tilde{n}_1}\big)\big(\psi_{\tilde{n}_1}-\gamma'\big)$$

$$+\frac{1}{2\psi_{\tilde{n}_1}}\Big(\big(b_{\tilde{n}_2}-a_{\tilde{n}_1}-\varepsilon_1\big)+\big(b_{\tilde{n}_1}-c_{\tilde{n}_1}\big)\Big)\Big(\psi_{\tilde{n}_1}{}^2-\gamma'^2\Big)$$

$$-\big(a_{\tilde{n}_1}+c_{\tilde{n}_1}\big)\big(\psi_{\tilde{n}_1}-\gamma'\big)-\frac{1}{2\psi_{\tilde{n}_1}}\Big(\big(b_{\tilde{n}_1}-a_{\tilde{n}_1}\big)+\big(b_{\tilde{n}_1}-c_{\tilde{n}_1}\big)\Big)\Big(\psi_{\tilde{n}_1}{}^2-\gamma'^2\Big)\Big]$$

After simplifying, we have

$$\mathfrak{R}^{w}_{\alpha,\beta',\gamma'}(\tilde{n}_2)-\mathfrak{R}^{w}_{\alpha,\beta',\gamma'}(\tilde{n}_1)$$

$$=\frac{1}{6}\Big[\varepsilon_1(\omega_{\tilde{n}_1}-\alpha)-\frac{\varepsilon_1}{2\omega_{\tilde{n}_1}}\big(\omega_{\tilde{n}_1}{}^2-\alpha^2\big)\Big]+\frac{w_1}{6}\Big[\varepsilon_1\big(\phi_{\tilde{n}_1}-\beta'\big)-\frac{\varepsilon_1}{2\phi_{\tilde{n}_1}}\big(\phi_{\tilde{n}_1}{}^2-\beta'^2\big)\Big]$$

$$+\frac{w_2}{6}\Big[\varepsilon_1\big(\psi_{\tilde{n}_1}-\gamma'\big)-\frac{\varepsilon_1}{2\psi_{\tilde{n}_1}}\big(\psi_{\tilde{n}_1}{}^2-\gamma'^2\big)\Big]$$

$$=\frac{1}{6}\varepsilon_1\big(\omega_{\tilde{n}_1}-\alpha\big)\Big[1-\frac{1}{2\omega_{\tilde{n}_1}}\big(\omega_{\tilde{n}_1}+\alpha\big)\Big]$$

$$+\frac{w_1}{6}\varepsilon_1\big(\phi_{\tilde{n}_1}-\beta'\big)\Big[1-\frac{1}{2\phi_{\tilde{n}_1}}\big(\phi_{\tilde{n}_1}+\beta'\big)\Big]+\frac{w_2}{6}\varepsilon_1\big(\psi_{\tilde{n}_1}-\gamma'\big)\Big[1-\frac{1}{2\psi_{\tilde{n}_1}}\big(\psi_{\tilde{n}_1}+\gamma'\big)\Big]$$

From (5) it can be inferred that $\alpha \le \omega_{\tilde{n}_1}$. Therefore, $\alpha+\omega_{\tilde{n}_1} \le 2\omega_{\tilde{n}_1}$ dividing both side by $2\omega_{\tilde{n}_1}$, we have $\frac{1}{2\omega_{\tilde{n}_1}}\big(\omega_{\tilde{n}_1}+\alpha\big) \le 1$ thus $1-\frac{1}{2\omega_{\tilde{n}_1}}\big(\omega_{\tilde{n}_1}+\alpha\big) \ge 0$

Similarly based on relations (6) and (7) it can be demonstrated

$$1-\frac{1}{2\phi_{\tilde{n}_1}}\big(\phi_{\tilde{n}_1}+\beta'\big) \le 0$$

And

$$1-\frac{1}{2\psi_{\tilde{n}_1}}\big(\psi_{\tilde{n}_1}+\gamma'\big) \le 0$$

So, we can conclude

$$\frac{1}{6}\varepsilon_1\big(\omega_{\tilde{n}_1}-\alpha\big)\Big[1-\frac{1}{2\omega_{\tilde{n}_1}}\big(\omega_{\tilde{n}_1}+\alpha\big)\Big]+\frac{w_1}{6}\varepsilon_1\big(\phi_{\tilde{n}_1}-\beta'\big)\Big[1-\frac{1}{2\phi_{\tilde{n}_1}}\big(\phi_{\tilde{n}_1}+\beta'\big)\Big]$$

$$+\frac{w_2}{6}\varepsilon_1\big(\psi_{\tilde{n}_1}-\gamma'\big)\Big[1-\frac{1}{2\psi_{\tilde{n}_1}}\big(\psi_{\tilde{n}_1}+\gamma'\big)\Big] \ge 0$$

That is mean

$$\mathfrak{R}^{w}_{\alpha,\beta',\gamma'}(\tilde{n}_2) - \mathfrak{R}^{w}_{\alpha,\beta',\gamma'}(\tilde{n}_1) \geq 0$$

The proof is complete.

Theorem 2. If $\tilde{n}_1 = \langle (a_{\tilde{n}_1}, b_{\tilde{n}_1}, c_{\tilde{n}_1}); \omega_{\tilde{n}_1}, \phi_{\tilde{n}_1}, \psi_{\tilde{n}_1} \rangle$ and

$$\tilde{n}_2 = \langle (a_{\tilde{n}_2}, b_{\tilde{n}_2}, c_{\tilde{n}_2}); \omega_{\tilde{n}_2}, \phi_{\tilde{n}_2}, \psi_{\tilde{n}_2} \rangle$$

be two Tri-SVN numbers where

$$a_{\tilde{n}_2} = a_{\tilde{n}_1}, b_{\tilde{n}_2} = b_{\tilde{n}_1} + \varepsilon_2, c_{\tilde{n}_2} = c_{\tilde{n}_1}, \omega_{\tilde{n}_2} = \omega_{\tilde{n}_1}, \phi_{\tilde{n}_2} = \phi_{\tilde{n}_1}, \psi_{\tilde{n}_2} = \psi_{\tilde{n}_1}.$$

Also, ε_2 is an actual number and $0 < \varepsilon_2 \leq c_{\tilde{n}_1} - b_{\tilde{n}_1}$. Then, we have

$$\mathfrak{R}^{w}_{\alpha,\beta',\gamma'}(\tilde{n}_2) \geq \mathfrak{R}^{w}_{\alpha,\beta',\gamma'}(\tilde{n}_1).$$

Proof. The proof of this theorem is similar to the proof of the previous one and is left to the reader.

Theorem 3. If $\tilde{n}_1 = \langle (a_{\tilde{n}_1}, b_{\tilde{n}_1}, c_{\tilde{n}_1}); \omega_{\tilde{n}_1}, \phi_{\tilde{n}_1}, \psi_{\tilde{n}_1} \rangle$ and

$$\tilde{n}_2 = \langle (a_{\tilde{n}_2}, b_{\tilde{n}_2}, c_{\tilde{n}_2}); \omega_{\tilde{n}_2}, \phi_{\tilde{n}_2}, \psi_{\tilde{n}_2} \rangle$$

be two Tri-SVN numbers where

$$a_{\tilde{n}_2} = a_{\tilde{n}_1}, b_{\tilde{n}_2} = b_{\tilde{n}_1}, c_{\tilde{n}_2} = c_{\tilde{n}_1} + \varepsilon_3, \omega_{\tilde{n}_2} = \omega_{\tilde{n}_1}, \phi_{\tilde{n}_2} = \phi_{\tilde{n}_1}, \psi_{\tilde{n}_2} = \psi_{\tilde{n}_1}.$$

Also, ε_3 is an actual number and $\varepsilon_3>0$. Then, we have

$$\mathfrak{R}^{w}_{\alpha,\beta',\gamma'}(\tilde{n}_2) \geq \mathfrak{R}^{w}_{\alpha,\beta',\gamma'}(\tilde{n}_1).$$

Proof. The proof of this theorem is similar to the proof of Theorem (1) and is left to the reader.

Theorem 4. If $\tilde{n}_1 = \langle (a_{\tilde{n}_1}, b_{\tilde{n}_1}, c_{\tilde{n}_1}); \omega_{\tilde{n}_1}, \phi_{\tilde{n}_1}, \psi_{\tilde{n}_1} \rangle$ and

$$\tilde{n}_2 = \langle (a_{\tilde{n}_2}, b_{\tilde{n}_2}, c_{\tilde{n}_2}); \omega_{\tilde{n}_2}, \phi_{\tilde{n}_2}, \psi_{\tilde{n}_2} \rangle$$

be two Tri-SVN numbers where

$a_{\tilde{n}_2} = a_{\tilde{n}_1}, b_{\tilde{n}_2} = b_{\tilde{n}_1}, c_{\tilde{n}_2} = c_{\tilde{n}_1}, \omega_{\tilde{n}_2} = \omega_{\tilde{n}_1} + \varepsilon_4, \phi_{\tilde{n}_2} = \phi_{\tilde{n}_1}, \psi_{\tilde{n}_2} = \psi_{\tilde{n}_1}.$

Also, ε_4 is an actual number and $0 < \varepsilon_4 \leq 1 - \omega_{\tilde{n}_1}$. Then, we have

$$\mathfrak{R}^{w}_{\alpha,\beta',\gamma'}(\tilde{n}_2) \geq \mathfrak{R}^{w}_{\alpha,\beta',\gamma'}(\tilde{n}_1).$$

Proof. We should show

$$\mathfrak{R}^{w}_{\alpha,\beta',\gamma'}(\tilde{n}_2) - \mathfrak{R}^{w}_{\alpha,\beta',\gamma'}(\tilde{n}_1) \geq 0$$

From (13) we have

$$\mathfrak{R}^{w}_{\alpha,\beta',\gamma'}(\tilde{n}_2) - \mathfrak{R}^{w}_{\alpha,\beta',\gamma'}(\tilde{n}_1)$$

$$= \frac{2}{3} + \frac{S^{\omega}_{\alpha}(\tilde{n}_2)}{3} + w_1\frac{S^{\phi}_{\beta'}(\tilde{n}_2)}{3} + w_2\frac{S^{\psi}_{\gamma'}(\tilde{n}_2)}{3} - \frac{2}{3} - \frac{S^{\omega}_{\alpha}(\tilde{n}_2)}{3} - w_1\frac{S^{\phi}_{\beta'}(\tilde{n}_2)}{3} - w_2\frac{S^{\psi}_{\gamma'}(\tilde{n}_2)}{3}$$

Since $a_{\tilde{n}_2} = a_{\tilde{n}_1}, b_{\tilde{n}_2} = b_{\tilde{n}_1}, c_{\tilde{n}_2} = c_{\tilde{n}_1}, \phi_{\tilde{n}_2} = \phi_{\tilde{n}_1}, \psi_{\tilde{n}_2} = \psi_{\tilde{n}_1}$ and only

$\omega_{\tilde{n}_2}$ and $\omega_{\tilde{n}_1}$

are not equal, then we can write

$$\mathfrak{R}^{w}_{\alpha,\beta',\gamma'}(\tilde{n}_2) - \mathfrak{R}^{w}_{\alpha,\beta',\gamma'}(\tilde{n}_1) = \frac{S^{\omega}_{\alpha}(\tilde{n}_2)}{3} - \frac{S^{\omega}_{\alpha}(\tilde{n}_1)}{3}$$

So, it is sufficient to show

$$S^{\omega}_{\alpha}(\tilde{n}_2) - S^{\omega}_{\alpha}(\tilde{n}_1) \geq 0$$

using (14) we can write

$$S^{\omega}_{\alpha}(\tilde{n}_2) - S^{\omega}_{\alpha}(\tilde{n}_1)$$

$$= \frac{1}{2}\Big[(a_{\tilde{n}_2} + c_{\tilde{n}_2})\big(\omega_{\tilde{n}_2} - \alpha\big)$$
$$+ \frac{1}{2\omega_{\tilde{n}_2}}\Big((b_{\tilde{n}_2} - a_{\tilde{n}_2}) + (b_{\tilde{n}_2} - c_{\tilde{n}_2})\Big)\big(\omega_{\tilde{n}_2}{}^2 - \alpha^2\big) - (a_{\tilde{n}_1} + c_{\tilde{n}_1})\big(\omega_{\tilde{n}_1} - \alpha\big)$$
$$- \frac{1}{2\omega_{\tilde{n}_1}}\Big((b_{\tilde{n}_1} - a_{\tilde{n}_1}) + (b_{\tilde{n}_1} - c_{\tilde{n}_1})\Big)\big(\omega_{\tilde{n}_1}{}^2 - \alpha^2\big)\Big]$$

By substituting

$$a_{\tilde{n}_2} = a_{\tilde{n}_1}, b_{\tilde{n}_2} = b_{\tilde{n}_1}, c_{\tilde{n}_2} = c_{\tilde{n}_1}, \omega_{\tilde{n}_2} = \omega_{\tilde{n}_1} + \varepsilon_4$$

into a above expressions, we get

$$S_\alpha^\omega(\tilde{n}_2) - S_\alpha^\omega(\tilde{n}_1)$$

$$= \frac{1}{2}\Big[(a_{\tilde{n}_1} + c_{\tilde{n}_1})\left(\omega_{\tilde{n}_1} + \varepsilon_4 - \alpha\right)$$

$$+ \frac{1}{2\left(\omega_{\tilde{n}_1} + \varepsilon_4\right)}\left(\left(b_{\tilde{n}_1} - a_{\tilde{n}_1}\right) + \left(b_{\tilde{n}_1} - c_{\tilde{n}_1}\right)\right)\left((\omega_{\tilde{n}_1} + \varepsilon_4)^2 - \alpha^2\right) - (a_{\tilde{n}_1} + c_{\tilde{n}_1})\left(\omega_{\tilde{n}_1} - \alpha\right)$$

$$- \frac{1}{2\omega_{\tilde{n}_1}}\left(\left(b_{\tilde{n}_1} - a_{\tilde{n}_1}\right) + \left(b_{\tilde{n}_1} - c_{\tilde{n}_1}\right)\right)\left(\omega_{\tilde{n}_1}{}^2 - \alpha^2\right)]$$

After simplifying we have

$$S_\alpha^\omega(\tilde{n}_2) - S_\alpha^\omega(\tilde{n}_1)$$

$$= \frac{1}{2}\Big[(a_{\tilde{n}_1} + c_{\tilde{n}_1})(\varepsilon_4)$$

$$+ \frac{1}{2}((b_{\tilde{n}_1} - a_{\tilde{n}_1}) + (b_{\tilde{n}_1} - c_{\tilde{n}_1}))\left(\frac{1}{(\omega_{\tilde{n}_1} + \varepsilon_4)}((\omega_{\tilde{n}_1} + \varepsilon_4)^2 - \alpha^2) - \frac{1}{\omega_{\tilde{n}_1}}(\omega_{\tilde{n}_1}{}^2 - \alpha^2)\right)\Big]$$

$$= \frac{1}{2}[\left(a_{\tilde{n}_1} + c_{\tilde{n}_1}\right)(\varepsilon_4) + \frac{1}{2}\left(2b_{\tilde{n}_1} - (a_{\tilde{n}_1} + c_{\tilde{n}_1})\right)(\omega_{\tilde{n}_1} + \varepsilon_4 - \frac{\alpha^2}{\left(\omega_{\tilde{n}_1} + \varepsilon_4\right)} - \omega_{\tilde{n}_1} + \frac{\alpha^2}{\omega_{\tilde{n}_1}}]$$

$$= \frac{1}{2}\left[(a_{\tilde{n}_1} + c_{\tilde{n}_1})(\varepsilon_4) + \frac{1}{2}(\varepsilon_4)(2b_{\tilde{n}_1} - (a_{\tilde{n}_1} + c_{\tilde{n}_1}))\left(1 + \frac{\alpha^2}{(\omega_{\tilde{n}_1} + \varepsilon_4)(\omega_{\tilde{n}_1})}\right)\right)$$

$$= \frac{1}{2}\left[(a_{\tilde{n}_1} + c_{\tilde{n}_1})(\varepsilon_4)\left(1 - \frac{1}{2} - \frac{\alpha^2}{2(\omega_{\tilde{n}_1} + \varepsilon_4)(\omega_{\tilde{n}_1})}\right) + (b_{\tilde{n}_1})(\varepsilon_4)\left(1 + \frac{\alpha^2}{(\omega_{\tilde{n}_1} + \varepsilon_4)(\omega_{\tilde{n}_1})}\right)\right]$$

Using the definition, we know that all the values inside the parentheses in the above relation are non-negative so

$$S_\alpha^\omega(\tilde{n}_2) - S_\alpha^\omega(\tilde{n}_1) \geq 0$$

Note that to prove the non-negativity of $1 - \frac{\alpha^2}{\left(\omega_{\tilde{n}_1} + \varepsilon_4\right)\left(\omega_{\tilde{n}_1}\right)}$, from (5) we have $\alpha \leq \omega_{\tilde{n}_1}$ also $\alpha \leq \omega_{\tilde{n}_1} + \varepsilon_4$, so $\alpha^2 \leq \left(\omega_{\tilde{n}_1} + \varepsilon_4\right)\left(\omega_{\tilde{n}_1}\right)$

dividing both side by $\left(\omega_{\tilde{n}_1} + \varepsilon_4\right)\left(\omega_{\tilde{n}_1}\right)$, we have $\frac{\alpha^2}{\left(\omega_{\tilde{n}_1} + \varepsilon_4\right)\left(\omega_{\tilde{n}_1}\right)} \leq 1$, thus

$$1 - \frac{\alpha^2}{\left(\omega_{\tilde{n}_1} + \varepsilon_4\right)\left(\omega_{\tilde{n}_1}\right)} \geq 0$$

Theorem 5. If $\tilde{n}_1 = \langle\left(a_{\tilde{n}_1}, b_{\tilde{n}_1}, c_{\tilde{n}_1}\right); \omega_{\tilde{n}_1}, \phi_{\tilde{n}_1}, \psi_{\tilde{n}_1}\rangle$and

$$\tilde{n}_2 = \langle\left(a_{\tilde{n}_2}, b_{\tilde{n}_2}, c_{\tilde{n}_2}\right); \omega_{\tilde{n}_2}, \phi_{\tilde{n}_2}, \psi_{\tilde{n}_2}\rangle$$

be two Tri-SVN numbers where

$$a_{\tilde{n}_2} = a_{\tilde{n}_1}, b_{\tilde{n}_2} = b_{\tilde{n}_1}, c_{\tilde{n}_2} = c_{\tilde{n}_1}, \omega_{\tilde{n}_2} = \omega_{\tilde{n}_1}, \phi_{\tilde{n}_2} = \phi_{\tilde{n}_1} - \varepsilon_5, \psi_{\tilde{n}_2} = \psi_{\tilde{n}_1}$$

Also, ε_5 is an actual number and $0 < \varepsilon_5 \leq 1 - \phi_{\tilde{n}_1}$. Then, we have

$$\mathfrak{R}^{w}_{\alpha,\beta',\gamma'}\left(\tilde{n}_2\right) \geq \mathfrak{R}^{w}_{\alpha,\beta',\gamma'}\left(\tilde{n}_1\right).$$

Proof. The proof of this theorem is similar to the proof of the previous one and is left to the reader.

Theorem 6. If $\tilde{n}_1 = \langle\left(a_{\tilde{n}_1}, b_{\tilde{n}_1}, c_{\tilde{n}_1}\right); \omega_{\tilde{n}_1}, \phi_{\tilde{n}_1}, \psi_{\tilde{n}_1}\rangle$and

$$\tilde{n}_2 = \langle\left(a_{\tilde{n}_2}, b_{\tilde{n}_2}, c_{\tilde{n}_2}\right); \omega_{\tilde{n}_2}, \phi_{\tilde{n}_2}, \psi_{\tilde{n}_2}\rangle$$

be two Tri-SVN numbers where

$$a_{\tilde{n}_2} = a_{\tilde{n}_1}, b_{\tilde{n}_2} = b_{\tilde{n}_1}, c_{\tilde{n}_2} = c_{\tilde{n}_1}, \omega_{\tilde{n}_2} = \omega_{\tilde{n}_1}, \phi_{\tilde{n}_2} = \phi_{\tilde{n}_1}, \psi_{\tilde{n}_2} = \psi_{\tilde{n}_1} - \varepsilon_6$$

Also, ε_6 is an actual number and $0 < \varepsilon_6 \leq 1 - \psi_{\tilde{n}_1}$. Then, we have

$$\mathfrak{R}^{w}_{\alpha,\beta',\gamma'}\left(\tilde{n}_2\right) \geq \mathfrak{R}^{w}_{\alpha,\beta',\gamma'}\left(\tilde{n}_1\right).$$

Proof. The proof of this theorem is similar to the proof Theorem (4) one and is left to the reader.

Corollary. If $\tilde{n}_1 = \langle\left(a_{\tilde{n}_1}, b_{\tilde{n}_1}, c_{\tilde{n}_1}\right); \omega_{\tilde{n}_1}, \phi_{\tilde{n}_1}, \psi_{\tilde{n}_1}\rangle$and

$$\tilde{n}_2 = \langle\left(a_{\tilde{n}_2}, b_{\tilde{n}_2}, c_{\tilde{n}_2}\right); \omega_{\tilde{n}_2}, \phi_{\tilde{n}_2}, \psi_{\tilde{n}_2}\rangle$$

be two Tri-SVN numbers where

$a_{\tilde{n}_2} = a_{\tilde{n}_1} + \varepsilon_1, b_{\tilde{n}_2} = b_{\tilde{n}_1} + \varepsilon_2, c_{\tilde{n}_2} = c_{\tilde{n}_1} + \varepsilon_3, \omega_{\tilde{n}_2} = \omega_{\tilde{n}_1} + \varepsilon_4, \phi_{\tilde{n}_2} = \phi_{\tilde{n}_1} - \varepsilon_5, \psi_{\tilde{n}_2} = \psi_{\tilde{n}_1} - \varepsilon_6.$

Also, $\varepsilon_1, \varepsilon_2, \varepsilon_3, \varepsilon_4, \varepsilon_5$, and ε_6 are actual number where
$0 < \varepsilon_1 \leq b_{\tilde{n}_1} - a_{\tilde{n}_1}, 0\langle \varepsilon_2 \leq c_{\tilde{n}_1} - b_{\tilde{n}_1}, \varepsilon_3 \rangle 0, 0 < \varepsilon_4 \leq 1 - \omega_{\tilde{n}_1}, 0 < \varepsilon_5 \leq \phi_{\tilde{n}_1}, 0 < \varepsilon_6 \leq$

Then, we have

$$\mathfrak{R}^{w}_{\alpha,\beta',\gamma'}(\tilde{n}_2) \geq \mathfrak{R}^{w}_{\alpha,\beta',\gamma'}(\tilde{n}_1)$$

5. GENERAL STRUCTURE OF MULTI-OBJECTIVE MULTI-PRODUCT THREE-LEVEL SUPPLY CHAIN PROBLEM UNDER NEUTROSOPHIC DATA

This section examines a three-tier supply chain consisting of factories, distribution centers, and end customers, where the shipping process is viewed as a direct chain. This implies that products cannot be transferred from one factory to another or from one distributor to another. The manufacturing plants make the decision to produce a certain number of products within a given time period. Each factory has a specific capacity to produce goods due to raw material limitations. Products are supplied to distribution centers based on demand.

Conflicting goals often arise among the different levels of the supply chain management structure, and the key to success for decision makers is to integrate planning to achieve all goals while ensuring customer satisfaction. Therefore, this section considers coordination in different stages of the supply chain and between different goals to create an integrated and effective system.

The following goals are contradictory:

- The contradiction between reducing production costs and increasing product quality
- Contradiction to send products faster and reduce shipping cost

When modelling the problem, it is important to consider not only the goals but also the limitations such as matching supply with customer demand, production capacity of each factory, and overall demand of distribution centers.

Deterministic Model

This section presents the sets, parameters, decision variables, functions, and limitations of the deterministic model in detail.

Sets

I: Set of factories producing products – $i=\{1,2,\ldots,I\}$

J: Set of product distribution centers – $j=\{1,2,\ldots,J\}$

K: Set of final customers (consumers) – $k=\{1,2,\ldots,K\}$

PL Set of manufactured goods – $p=\{1,2,\ldots,P\}$

Parameters

QU_i^p: The quality of product p produced in factory i

TR_{ij}^p: The cost of transporting product p from factory i to distribution center j

VA_i^p: The transport vehicles' capacity to deliver product p from factory i

HO_j^p: The cost of storing each unit of product p at distribution center j

PE_{jk}^p: A fine will be paid for each unit of product p that is returned by distributor j to customer k

RP_{jk}^p: Provide the percentage of product unit p sent by distributor j to customer k

ST_{jk}^p: Time to transport each unit of product p from distributor j to customer k

VA_j^p: The transport vehicles' capacity to deliver product p from distribution center j

SE_{jk}^p: The price charged for each unit of product p sold by distributor j to customer k.

X_i^p: The maximum amount of product p that can be produced in factory i

Y_j^p: The minimum amount of demand for product p from distribution center j

Decision Variables

x_{ij}^p the quantity of product p transported from factory I to distribution center j

y_{jk}^p the quantity of product p transported from distribution center j to customer k

Objective Functions and Constraints: Model 1

$$max\left\{ F_q = \sum_{i=1}^{I} \sum_{j=1}^{J} \sum_{p=1}^{P} QU_i^p x_{ij}^p \right\} \quad (1\text{-}1)$$

$$min\left\{ F_p = \sum_{i=1}^{I} \sum_{j=1}^{J} \sum_{p=1}^{P} TR_{ij}^p (x_{ij}^p / VA_{ij}^p) + \sum_{i=1}^{I} \sum_{j=1}^{J} \sum_{p=1}^{P} HO_j^p (x_{ij}^p / 2) + \sum_{j=1}^{J} \sum_{k=1}^{K} \sum_{p=1}^{P} RP_{jk}^p (PE_{jk}^p y_{jk}^p) \right\} \quad (1\text{-}2)$$

$$min\left\{ F_t = \sum_{j=1}^{J} \sum_{k=1}^{K} \sum_{p=1}^{P} ST_{jk}^p (y_{jk}^p / VA_{jk}^p) \right\} \quad (1\text{-}3)$$

$$max\left\{ F_s = \sum_{j=1}^{J} \sum_{i=1}^{I} \sum_{p=1}^{P} SE_{jk}^p y_{jk}^p \right\} \quad (1\text{-}4)$$

$$\sum_{i=1}^{I} x_{ij}^{p} = \sum_{k=1}^{K} y_{jk}^{p}, (j = 1,2,\ldots,J), (p = 1,2,\ldots,P) \quad (1\text{-}5)$$

$$\sum_{j=1}^{J} x_{ij}^{p} \leq X_{i}^{p}, \quad (i = 1,2,\ldots,I), (p = 1,2,\ldots,P) \quad (1\text{-}6)$$

$$\sum_{k=1}^{K} y_{jk}^{p} \geq Y_{j}^{p}, \quad (j = 1,2,\ldots,J), (p = 1,2,\ldots,P) \quad (1\text{-}7)$$

$$x_{ij}^{p}, y_{jk}^{p} \geq 0 \ (i = 1,2,\ldots,I), (j = 1,2,\ldots,J), (k = 1,2,\ldots,K), (p = 1,2,\ldots,P), \quad (1\text{-}8)$$

Equation 1-1 is the objective function related to the quality of products produced in factories that are sent to existing distribution centers. This goal that occurs in the circle of producers should be maximized so that, in this way, the producers provide quality goods to the distribution centers and, as a result, to the final customers. Equation 1-2 is related to the objective function of the total cost in the supply chain, which includes several terms. This relation has happened in different links of the supply chain and consists of the following sentences:

The cost of sending products from manufacturers to distributors (first sentence)

The cost of keeping the products sent from the production centers to the warehouses of the distribution centers (second sentence)

The cost related to the penalty for returned goods that is paid to customers by distribution centers. (third sentence)

According to the explanations mentioned each of the sentences in equation 1-2, the aim of supply chain management is to minimize equation 1-2, so that the transportation costs related to sending goods, the cost of keeping goods in warehouses, and the price of fines related to Reduce returned goods.

Equation 1-3 is the function related to the time of sending the goods from the distribution centers to the final customers. As much as the time to respond to customers' requests to receive products is reduced, the level of customer service for that product has increased, and hence, the level of customer satisfaction in receiving that product will be higher. It is clear that this goal is in conflict with goal 1-2, because there the goal is to reduce costs, and meeting this goal can affect the increase in delivery time and the final response to customer demand. Equation 1-4 shows the objective function of the revenue from the sale of products from distribution centers to final customers. We have to maximize this goal that happens between the circles of distributors and customers to make more money as one of the main goals in supply chain management. Relations 1-5 guarantees the non-shortage constraints of products. This relationship has been proposed so that the manufacturers can supply the number of orders for each of the products issued by the distributors at the right time. It is clear that the timely fulfillment of customer demand leads to increased growth and overall credibility of the system. Relationships 1-6 show the limits of the maximum production capacity of products by factories. Relationships 1-7 show the constraints related to the minimum number of demands related to each of the products to be sent to the distribution centers.

Neutrosophic Model

In the real world, quality, cost, delivery time, and revenue are all things that cannot be determined with certainty. The level of precision in the production of products, the quality of the raw materials used in the production of the main products, the conditions governing the transportation systems, individual factors, weather and road conditions, the economic situation, inflation and policy changes can be among the factors influencing the lack of Certainty counts in each of the supply chain objectives. Therefore, in this section we consider the supply chain problem presented in the previous section with neutrosophic objectives (triangular data). In this case, the mathematical model of the supply chain management problem is expressed as follows.

Model 2

$$max\left\{ F_q = \sum_{i=1}^{I} \sum_{j=1}^{J} \sum_{p=1}^{P} \widetilde{QU}_i^p x_{ij}^p \right\} \tag{2-1}$$

$$min\left\{ F_p = \sum_{i=1}^{I} \sum_{j=1}^{J} \sum_{p=1}^{P} \widetilde{TR}_{ij}^p (x_{ij}^p / VA_i^p) + \sum_{i=1}^{I} \sum_{j=1}^{J} \sum_{p=1}^{P} \widetilde{HO}_j^p (x_{ij}^p / 2) + \sum_{j=1}^{J} \sum_{k=1}^{K} \sum_{p=1}^{P} RP_{jk}^p (\widetilde{PE}_{jk}^p . y_{jk}^p) \right\} \tag{2-2}$$

$$min\left\{ F_t = \sum_{j=1}^{J} \sum_{k=1}^{K} \sum_{p=1}^{P} \widetilde{ST}_{jk}^p (y_{jk}^p / VA_j^p) \right\} \tag{2-3}$$

$$max\left\{ F_s = \sum_{j=1}^{J} \sum_{i=1}^{I} \sum_{p=1}^{P} \widetilde{SE}_{jk}^p y_{jk}^p \right\} \tag{2-4}$$

s.t. (1–5: 1:8)

Solving Process

In general, the use of multi-objective planning is an undeniable necessity for modeling existing processes in supply chain problems. So far, effective tools and techniques for solving multi-objective problems have been investigated by researchers. A set of general methods related to solving processes for multi-objective problems can be found in (Ehrgott, 2005). Here, we want to apply a hybrid solution process based on the proposed ranking method for Tri-SVNs and the weighted sum method of the multi-objective problem to solve Model 2. For this purpose, two general steps are described as follows:

Step 1: Adjust model 2 based on problem information. In model 2, the values of

$$\widetilde{QU}^p_i,\ \widetilde{TR}^p_{ij},\ \widetilde{HO}^p_j,\ \widetilde{PE}^p_{jk},\ \widetilde{ST}^p_{jk},\ \widetilde{ST}^p_{jk}$$

are all Tri-SVNs and other variables and coefficients are definite. Using proposed ranking method (equation 13), convert the parameters represented by Tri-SVNs into definite numbers. Based on the expectation level of the decision maker, the values α,β,γ and the coefficients w_1 and w_2 are determined.

Step 2: Use the weighted sum method to transform the multi-objective problem into a single-objective one. The weights for each objective are considered the same here.

In the next part, a practical example for interpreting the proposed model presented in this part is given under neutrosophic data for all goals.

6. PRACTICAL EXAMPLE

Two types of products are produced in two factories with different qualities. In order to send goods to four customers, the products are first sent to three distribution centers. Items such as product quality, transportation cost, product storage cost, fine paid for each returned product, product delivery time, selling price of each product are considered as a triangular neutrosophic number, and items such as vehicle carrying capacity, return percentage for each unit of goods, the amount of supply and demand are given as definite numbers. All the values are given in Tables 1-17.

Table 1. The quality of the product produced by the manufacturer

	i	**Manufacturer**	
p		**1**	**2**
Product	1	$\langle(0.87, 0.90, 0.93);\ 0.9, 0.2, 0.2\rangle$	$\langle(0.89, 0.92, 0.95);\ 0.9, 0.2, 0.2\rangle$
	2	$\langle(0.91, 0.94, 0.97);\ 0.9, 0.2, 0.2\rangle$	$\langle(0.85, 0.88, 0.91);\ 0.9, 0.2, 0.2\rangle$

Table 2. Capacity of carriers used by production center (in units of goods)

	i	**Manufacturer**	
p		**1**	**2**
Product	1	40	45
	2	30	35

Table 3. Shipping cost from manufacturer to distributor warehouse, p=1 (in currency)

	i	**Distributor**		
p		**1**	**2**	**3**
Manufacturer	1	$\langle (0.03, 0.06, 0.09); 0.9, 0.2, 0.2 \rangle$	$\langle (0.01, 0.04, 0.07); 0.9, 0.2, 0.2 \rangle$	$\langle (0.03, 0.06, 0.09); 0.9, 0.2, 0.2 \rangle$
	2	$\langle (0.01, 0.04, 0.07); 0.9, 0.2, 0.2 \rangle$	$\langle (0.03, 0.06, 0.09); 0.9, 0.2, 0.2 \rangle$	$\langle (0.01, 0.04, 0.07); 0.9, 0.2, 0.2 \rangle$

Table 4. Shipping cost from manufacturer to distributor warehouse, p=2 (in currency)

	i	**Distributor**		
p		**1**	**2**	**3**
Manufacturer	1	$\langle (0.04, 0.07, 0.1); 0.9, 0.2, 0.2 \rangle$	$\langle (0.03, 0.06, 0.09); 0.9, 0.2, 0.2 \rangle$	$\langle (0.04, 0.07, 0.1); 0.9, 0.2, 0.2 \rangle$
	2	$\langle (0.03, 0.06, 0.09); 0.9, 0.2, 0.2 \rangle$	$\langle (0.04, 0.07, 0.1); 0.9, 0.2, 0.2 \rangle$	$\langle (0.03, 0.06, 0.09); 0.9, 0.2, 0.2 \rangle$

Table 5. Cost of keeping the manufacturer's goods in the distributor's warehouse, p=1 (in currency)

	j	**Distributor**		
p		**1**	**2**	**3**
Manufacturer	1	$\langle (0.03, 0.06, 0.09); 0.9, 0.2, 0.2 \rangle$	$\langle (0.06, 0.09, 0.12); 0.9, 0.2, 0.2 \rangle$	$\langle (0.06, 0.09, 0.12); 0.9, 0.2, 0.2 \rangle$
	2	$\langle (0.03, 0.06, 0.09); 0.9, 0.2, 0.2 \rangle$	$\langle (0.06, 0.09, 0.12); 0.9, 0.2, 0.2 \rangle$	$\langle (0.06, 0.09, 0.12); 0.9, 0.2, 0.2 \rangle$

Table 6. Cost of keeping the manufacturer's goods in the distributor's warehouse, p=2 (in currency)

	j	**Distributor**		
p		**1**	**2**	**3**
Manufacturer	1	$\langle (0.07, 0.1, 0.13);\ 0.9, 0.2, 0.2 \rangle$	$\langle (0.08, 0.11, 0.14);\ 0.9, 0.2, 0.2 \rangle$	$\langle (0.08, 0.11, 0.14);\ 0.9, 0.2, 0.2 \rangle$
	2	$\langle (0.07, 0.1, 0.13);\ 0.9, 0.2, 0.2 \rangle$	$\langle (0.08, 0.11, 0.14);\ 0.9, 0.2, 0.2 \rangle$	$\langle (0.08, 0.11, 0.14);\ 0.9, 0.2, 0.2 \rangle$

Table 7. Capacity of carriers used by distribution center (in commodity units)

	j	**Distributor**		
p		**1**	**2**	**3**
Product	1	50	45	50
	2	40	35	40

Table 8. Distributor return percentage rate, p=1.

	k	**Customer**			
j		**1**	**2**	**3**	**4**
Distributor	1	0.03	0.03	0.04	0.03
	2	0.04	0.05	0.03	0.03
	3	0.04	0.03	0.04	0.04

Table 9. Distributor return percentage rate, p=2.

	k	**Customer**			
j		**1**	**2**	**3**	**4**
Distributor	1	0.01	0.01	0.02	0.01
	2	0.02	0.03	0.01	0.01
	3	0.02	0.01	0.02	0.02

Table 10. Amount of the fine paid by the distributor to the customer, p=1.

	k	Customer			
j		1	2	3	4
Distributor	1	$\langle (0.40, 0.43, 0.46); 0.9, 0.2, 0.2 \rangle$	$\langle (0.46, 0.49, 0.52); 0.9, 0.2, 0.2 \rangle$	$\langle (0.51, 0.54, 0.57); 0.9, 0.2, 0.2 \rangle$	$\langle (0.40, 0.43, 0.46); 0.9, 0.2, 0.2 \rangle$
	2	$\langle (0.51, 0.54, 0.57); 0.9, 0.2, 0.2 \rangle$	$\langle (0.40, 0.43, 0.46); 0.9, 0.2, 0.2 \rangle$	$\langle (0.56, 0.59, 0.62); 0.9, 0.2, 0.2 \rangle$	$\langle (0.46, 0.49, 0.52); 0.9, 0.2, 0.2 \rangle$
	3	$\langle (0.40, 0.43, 0.46); 0.9, 0.2, 0.2 \rangle$	$\langle (0.40, 0.43, 0.46); 0.9, 0.2, 0.2 \rangle$	$\langle (0.40, 0.43, 0.46); 0.9, 0.2, 0.2 \rangle$	$\langle (0.51, 0.54, 0.57); 0.9, 0.2, 0.2 \rangle$

Table 11. Amount of the fine paid by the distributor to the customer, p=2.

	k	Customer			
j		1	2	3	4
Distributor	1	$\langle (0.54, 0.57, 0.60); 0.9, 0.2, 0.2 \rangle$	$\langle (0.60, 0.63, 0.66); 0.9, 0.2, 0.2 \rangle$	$\langle (0.66, 0.69, 0.72); 0.9, 0.2, 0.2 \rangle$	$\langle (0.54, 0.57, 0.60); 0.9, 0.2, 0.2 \rangle$
	2	$\langle (0.66, 0.69, 0.72); 0.9, 0.2, 0.2 \rangle$	$\langle (0.54, 0.57, 0.60); 0.9, 0.2, 0.2 \rangle$	$\langle (0.70, 0.73, 0.76); 0.9, 0.2, 0.2 \rangle$	$\langle (0.60, 0.63, 0.66); 0.9, 0.2, 0.2 \rangle$
	3	$\langle (0.54, 0.57, 0.60); 0.9, 0.2, 0.2 \rangle$	$\langle (0.54, 0.57, 0.60); 0.9, 0.2, 0.2 \rangle$	$\langle (0.66, 0.69, 0.72); 0.9, 0.2, 0.2 \rangle$	$\langle (0.66, 0.69, 0.72); 0.9, 0.2, 0.2 \rangle$

Table 12. Delivery time from the distributor's warehouse to the customer, p=1 (in units of time)

	k	Customer			
j		1	2	3	4
Distributor	1	$\langle (0.22, 0.25, 0.28); 0.9, 0.2, 0.2 \rangle$	$\langle (0.35, 0.38, 0.41); 0.9, 0.2, 0.2 \rangle$	$\langle (0.72, 0.75, 0.78); 0.9, 0.2, 0.2 \rangle$	$\langle (0.47, 0.50, 0.53); 0.9, 0.2, 0.2 \rangle$
	2	$\langle (0.51, 0.54, 0.57); 0.9, 0.2, 0.2 \rangle$	$\langle (0.47, 0.50, 0.53); 0.9, 0.2, 0.2 \rangle$	$\langle (0.47, 0.50, 0.53); 0.9, 0.2, 0.2 \rangle$	$\langle (0.72, 0.75, 0.78); 0.9, 0.2, 0.2 \rangle$
	3	$\langle (0.47, 0.50, 0.53); 0.9, 0.2, 0.2 \rangle$	$\langle (0.47, 0.50, 0.53); 0.9, 0.2, 0.2 \rangle$	$\langle (0.22, 0.25, 0.28); 0.9, 0.2, 0.2 \rangle$	$\langle (0.35, 0.38, 0.41); 0.9, 0.2, 0.2 \rangle$

Table 13. Delivery time from the distributor's warehouse to the customer, p=2 (in units of time)

	k	Customer			
j		1	2	3	4
Distributor	1	$\langle (0.72, 0.75, 0.78);\ 0.9, 0.2, 0.2 \rangle$	$\langle (0.47, 0.50, 0.53);\ 0.9, 0.2, 0.2 \rangle$	$\langle (0.97, 1.0, 1.3);\ 0.9, 0.2, 0.2 \rangle$	$\langle (0.72, 0.75, 0.78);\ 0.9, 0.2, 0.2 \rangle$
	2	$\langle (0.47, 0.50, 0.53);\ 0.9, 0.2, 0.2 \rangle$	$\langle (0.72, 0.75, 0.78);\ 0.9, 0.2, 0.2 \rangle$	$\langle (0.72, 0.75, 0.78);\ 0.9, 0.2, 0.2 \rangle$	$\langle (0.97, 1.0, 1.3);\ 0.9, 0.2, 0.2 \rangle$
	3	$\langle (0.72, 0.75, 0.78);\ 0.9, 0.2, 0.2 \rangle$	$\langle (0.72, 0.75, 0.78);\ 0.9, 0.2, 0.2 \rangle$	$\langle (0.47, 0.50, 0.53);\ 0.9, 0.2, 0.2 \rangle$	$\langle (0.47, 0.50, 0.53);\ 0.9, 0.2, 0.2 \rangle$

Table 14. Sales price per unit of distribution to customer, p=1 (in units of time)

	k	Customer			
j		1	2	3	4
Distributor	1	$\langle (0.54, 0.57, 0.60);\ 0.9, 0.2, 0.2 \rangle$	$\langle (0.61, 0.64, 0.67);\ 0.9, 0.2, 0.2 \rangle$	$\langle (0.68, 0.71, 0.74);\ 0.9, 0.2, 0.2 \rangle$	$\langle (0.54, 0.57, 0.60);\ 0.9, 0.2, 0.2 \rangle$
	2	$\langle (0.68, 0.71, 0.74);\ 0.9, 0.2, 0.2 \rangle$	$\langle (0.54, 0.57, 0.60);\ 0.9, 0.2, 0.2 \rangle$	$\langle (0.75, 0.78, 0.81);\ 0.9, 0.2, 0.2 \rangle$	$\langle (0.61, 0.64, 0.67);\ 0.9, 0.2, 0.2 \rangle$
	3	$\langle (0.54, 0.57, 0.60);\ 0.9, 0.2, 0.2 \rangle$	$\langle (0.54, 0.57, 0.60);\ 0.9, 0.2, 0.2 \rangle$	$\langle (0.68, 0.71, 0.74);\ 0.9, 0.2, 0.2 \rangle$	$\langle (0.68, 0.71, 0.74);\ 0.9, 0.2, 0.2 \rangle$

Table 15. Sales price per unit of distribution to customer, p=2 (in units of time) Customer k

	k	Customer			
j		1	2	3	4
Distributor	1	$\langle (0.75, 0.78, 0.81);\ 0.9, 0.2, 0.2 \rangle$	$\langle (0.83, 0.86, 0.89);\ 0.9, 0.2, 0.2 \rangle$	$\langle (0.90, 0.93, 0.96);\ 0.9, 0.2, 0.2 \rangle$	$\langle (0.75, 0.78, 0.81);\ 0.9, 0.2, 0.2 \rangle$
	2	$\langle (0.90, 0.93, 0.96);\ 0.9, 0.2, 0.2 \rangle$	$\langle (0.75, 0.78, 0.81);\ 0.9, 0.2, 0.2 \rangle$	$\langle (0.97, 1.0, 1.3);\ 0.9, 0.2, 0.2 \rangle$	$\langle (0.83, 0.86, 0.89);\ 0.9, 0.2, 0.2 \rangle$
	3	$\langle (0.75, 0.78, 0.81);\ 0.9, 0.2, 0.2 \rangle$	$\langle (0.75, 0.78, 0.81);\ 0.9, 0.2, 0.2 \rangle$	$\langle (0.90, 0.93, 0.96);\ 0.9, 0.2, 0.2 \rangle$	$\langle (0.90, 0.93, 0.96);\ 0.9, 0.2, 0.2 \rangle$

Table 16. Maximum production capacity (in units of commodity)Manufacturer i

	i	**Manufacturer**	
p		**1**	**2**
Product	1	1500	2000
	2	1000	1200

Table 17. Minimum customer demand from distribution centers (in units)

	j	**Distributor**		
p		**1**	**2**	**3**
Product	1	700	650	700
	2	500	500	600

Modeling the problem based on form 2 presented in the previous section is interpreted as follows:

$$max\Big\{ F_q = \widetilde{QU}_1^1 x_{11}^1 + \widetilde{QU}_1^1 x_{12}^1 + \widetilde{QU}_1^1 x_{13}^1 + \widetilde{QU}_2^1 x_{21}^1 + \widetilde{QU}_2^1 x_{22}^1 + \widetilde{QU}_2^1 x_{23}^1 + \widetilde{QU}_1^2 x_{11}^2$$

$$+\widetilde{QU}_1^2 x_{12}^2 + \widetilde{QU}_1^1 x_{13}^2 + \widetilde{QU}_2^2 x_{21}^2 + \widetilde{QU}_2^2 x_{22}^2 + \widetilde{QU}_2^2 x_{23}^2 \Big\}$$

$$min\Big\{ F_p = \widetilde{TR}_{11}^1\left(\frac{x_{11}^1}{VA_1^1}\right) + \widetilde{TR}_{12}^1\left(\frac{x_{12}^1}{VA_1^1}\right)$$

$$+\widetilde{TR}_{13}^1\left(\frac{x_{13}^1}{VA_1^1}\right) + \widetilde{TR}_{21}^1\left(\frac{x_{21}^1}{VA_2^1}\right) + \widetilde{TR}_{22}^1\left(\frac{x_{22}^1}{VA_2^1}\right)$$

$$+\widetilde{TR}_{23}^1\left(\frac{x_{23}^1}{VA_2^1}\right) + \widetilde{TR}_{11}^2\left(\frac{x_{11}^2}{VA_1^2}\right) + \widetilde{TR}_{12}^2\left(\frac{x_{12}^2}{VA_1^2}\right)$$

$$+\widetilde{TR}_{13}^2\left(\frac{x_{13}^2}{VA_1^2}\right) + \widetilde{TR}_{21}^2\left(\frac{x_{21}^2}{VA_2^2}\right) + \widetilde{TR}_{22}^2\left(\frac{x_{22}^2}{VA_2^2}\right)$$

$$+\widetilde{TR}_{23}^2\left(\frac{x_{23}^2}{VA_2^2}\right) + \widetilde{HO}_1^1\left(\frac{x_{11}^1}{2}\right) + \widetilde{HO}_2^1\left(\frac{x_{12}^1}{2}\right)$$

$$+\widetilde{HO}_3^1\left(\frac{x_{13}^1}{2}\right) + \widetilde{HO}_1^1\left(\frac{x_{21}^1}{2}\right) + \widetilde{HO}_2^1\left(\frac{x_{22}^1}{2}\right) + \widetilde{HO}_3^1\left(\frac{x_{23}^1}{2}\right)$$

$$+\widetilde{HO}_1^2\left(\frac{x_{11}^2}{2}\right) + \widetilde{HO}_2^2\left(\frac{x_{12}^2}{2}\right) + \widetilde{HO}_3^2\left(\frac{x_{13}^2}{2}\right) + \widetilde{HO}_1^2\left(\frac{x_{21}^2}{2}\right) + \widetilde{HO}_2^2\left(\frac{x_{22}^2}{2}\right) + \widetilde{HO}_3^2\left(\frac{x_{23}^2}{2}\right)$$

$$+RP_{11}^1\left(\widetilde{PE}_{11}^1 y_{11}^1\right) + RP_{12}^1\left(\widetilde{PE}_{12}^1 y_{12}^1\right)$$

$$+RP_{13}^1\left(\widetilde{PE}_{13}^1 y_{13}^1\right) + RP_{14}^1\left(\widetilde{PE}_{14}^1 y_{14}^1\right) + RP_{21}^1\left(\widetilde{PE}_{21}^1 y_{21}^1\right)$$

$$+RP_{22}^{1}\left(\widetilde{PE}_{22}^{1}y_{22}^{1}\right)+RP_{23}^{1}\left(\widetilde{PE}_{23}^{1}y_{23}^{1}\right)$$
$$+RP_{24}^{1}\left(\widetilde{PE}_{24}^{1}y_{24}^{1}\right)+RP_{31}^{1}\left(\widetilde{PE}_{31}^{1}y_{31}^{1}\right)+RP_{32}^{1}\left(\widetilde{PE}_{32}^{1}y_{32}^{1}\right)$$
$$+RP_{33}^{1}\left(\widetilde{PE}_{33}^{1}y_{33}^{1}\right)+RP_{34}^{1}\left(\widetilde{PE}_{34}^{1}y_{34}^{1}\right)$$
$$+RP_{11}^{2}\left(\widetilde{PE}_{11}^{2}y_{11}^{2}\right)+RP_{12}^{2}\left(\widetilde{PE}_{12}^{2}y_{12}^{2}\right)+RP_{13}^{2}\left(\widetilde{PE}_{13}^{2}y_{13}^{2}\right)$$
$$+RP_{14}^{2}\left(\widetilde{PE}_{14}^{2}y_{14}^{2}\right)+RP_{21}^{2}\left(\widetilde{PE}_{21}^{2}y_{21}^{2}\right)$$
$$+RP_{22}^{2}\left(\widetilde{PE}_{22}^{2}y_{22}^{2}\right)+RP_{23}^{2}\left(\widetilde{PE}_{23}^{2}y_{23}^{2}\right)+RP_{24}^{2}\left(\widetilde{PE}_{24}^{2}y_{24}^{2}\right)$$
$$+RP_{31}^{2}\left(\widetilde{PE}_{31}^{2}y_{31}^{2}\right)+RP_{32}^{2}\left(\widetilde{PE}_{32}^{2}y_{32}^{2}\right)+RP_{33}^{2}\left(\widetilde{PE}_{33}^{2}y_{33}^{2}\right)+RP_{34}^{2}\left(\widetilde{PE}_{34}^{2}y_{34}^{2}\right)$$

$$min\left\{F_{t}=\widetilde{ST}_{11}^{1}\left(\frac{y_{11}^{1}}{VA_{1}^{1}}\right)+\widetilde{ST}_{12}^{1}\left(\frac{y_{12}^{1}}{VA_{1}^{1}}\right)\right.$$
$$+\widetilde{ST}_{13}^{1}\left(\frac{y_{13}^{1}}{VA_{1}^{1}}\right)+\widetilde{ST}_{14}^{1}\left(\frac{y_{14}^{1}}{VA_{1}^{1}}\right)+\widetilde{ST}_{21}^{1}\left(\frac{y_{21}^{1}}{VA_{2}^{1}}\right)+\widetilde{ST}_{22}^{1}\left(\frac{y_{22}^{1}}{VA_{2}^{1}}\right)$$
$$+\widetilde{ST}_{23}^{1}\left(\frac{y_{23}^{1}}{VA_{2}^{1}}\right)+\widetilde{ST}_{14}^{1}\left(\frac{y_{24}^{1}}{VA_{2}^{1}}\right)$$
$$\left.+\widetilde{ST}_{31}^{1}\left(\frac{y_{31}^{1}}{VA_{2}^{1}}\right)+\widetilde{ST}_{32}^{1}\left(\frac{y_{32}^{1}}{VA_{2}^{1}}\right)+\widetilde{ST}_{33}^{1}\left(\frac{y_{33}^{1}}{VA_{2}^{1}}\right)+\widetilde{ST}_{34}^{1}\left(\frac{y_{34}^{1}}{VA_{2}^{1}}\right)\right\}$$

$$max\left\{F_{s}=\widetilde{SE}_{11}^{1}y_{11}^{1}+\widetilde{SE}_{12}^{1}y_{12}^{1}+\widetilde{SE}_{13}^{1}y_{13}^{1}+\widetilde{SE}_{14}^{1}y_{14}^{1}+\widetilde{SE}_{21}^{1}y_{21}^{1}+\widetilde{SE}_{22}^{1}y_{22}^{1}\right.$$
$$\left.+\widetilde{SE}_{23}^{1}y_{23}^{1}+\widetilde{SE}_{14}^{1}y_{24}^{1}+\widetilde{SE}_{31}^{1}y_{31}^{1}+\widetilde{SE}_{32}^{1}y_{32}^{1}+\widetilde{SE}_{33}^{1}y_{33}^{1}+\widetilde{SE}_{34}^{1}y_{34}^{1}\right\}$$

$$\text{s.t.}x_{11}^{1}+x_{21}^{1}=y_{11}^{1}+y_{12}^{1}+y_{13}^{1}+y_{14}^{1}$$

$$x_{12}^{1}+x_{22}^{1}=y_{21}^{1}+y_{22}^{1}+y_{23}^{1}+y_{24}^{1}$$

$$x_{13}^{1}+x_{23}^{1}=y_{31}^{1}+y_{32}^{1}+y_{33}^{1}+y_{34}^{1}$$

$$x_{11}^{2}+x_{21}^{2}=y_{11}^{2}+y_{12}^{2}+y_{13}^{2}+y_{14}^{2}$$

$$x_{12}^{2}+x_{22}^{2}=y_{21}^{2}+y_{22}^{2}+y_{23}^{2}+y_{24}^{2}$$

$$x_{13}^{2}+x_{23}^{2}=y_{31}^{2}+y_{32}^{2}+y_{33}^{2}+y_{34}^{2}$$

$$x_{11}^{1}+x_{12}^{1}+x_{13}^{1}\leq 1500$$

$$x_{21}^{1}+x_{22}^{1}+x_{23}^{1}\leq 2000$$

$$x_{11}^{2}+x_{12}^{2}+x_{13}^{2}\leq 1500$$

$$x_{21}^2 + x_{22}^2 + x_{23}^2 \leq 2000$$

$$y_{11}^1 + y_{12}^1 + y_{13}^1 + y_{14}^1 \geq 700;\ y_{21}^1 + y_{22}^1 + y_{23}^1 + y_{24}^1 \geq 650;\ y_{31}^1 + y_{32}^1 + y_{33}^1 + y_{34}^1 \geq 700$$

$$y_{11}^2 + y_{12}^2 + y_{13}^2 + y_{14}^2 \geq 500;\ y_{21}^2 + y_{22}^2 + y_{23}^2 + y_{24}^2 \geq 500;\ y_{31}^2 + y_{32}^2 + y_{33}^2 + y_{34}^2 \geq 600$$

$$x_{11}^1, x_{12}^1, x_{13}^1, x_{21}^1, x_{22}^1, x_{23}^1, y_{11}^1, y_{12}^1, y_{13}^1, y_{14}^1, y_{21}^1, y_{22}^1, y_{23}^1, y_{24}^1, y_{31}^1, y_{32}^1, y_{33}^1, y_{34}^1 \geq 0$$

$$x_{11}^2, x_{12}^2, x_{13}^2, x_{21}^2, x_{22}^2, x_{23}^2, y_{11}^2, y_{12}^2, y_{13}^2, y_{14}^2, y_{21}^2, y_{22}^2, y_{23}^2, y_{24}^2, y_{31}^2, y_{32}^2, y_{33}^2, y_{34}^2 \geq 0$$

Table 18. Results of the model with respect to the objectives

Objective function	Optimal value	Optimal variables			
OBJF1	-4.9755*e*+03	p=1	p=2	p=1	p=2
OBJF2	2.1796*e*+03	$x_{11}^1 = 0$	$x_{11}^2 = 0$	$y_{11}^1 = 0, y_{12}^1 = 0$	$y_{11}^2 = 0, y_{12}^2 = 0$
OBJF3	107.4927	$x_{12}^1 = 1500$	$x_{12}^2 = 1000$	$y_{13}^1 = 700, y_{14}^1 = 0$	$y_{13}^2 = 500, y_{14}^2 = 0$
OBJF4	-4.8778*e*+03	$x_{13}^1 = 0$	$x_{13}^2 = 0$	$y_{21}^1 = 0, y_{22}^1 = 0$	$y_{21}^2 = 0, y_{22}^2 = 0$
weighted sum method	-1.8915*e*+03	$x_{21}^1 = 700$	$x_{21}^2 = 500$	$y_{23}^1 = 2100, y_{24}^1 = 0$	$y_{23}^2 = 1100, y_{24}^2 = 0$
		$x_{22}^1 = 600$	$x_{22}^2 = 100$	$y_{31}^1 = 0, y_{32}^1 = 0$	$y_{31}^2 = 0, y_{32}^2 = 0$
		$x_{23}^1 = 700$	$x_{23}^2 = 600$	$y_{33}^1 = 700, y_{34}^1 = 0$	$y_{33}^2 = 0, y_{34}^2 = 600$

The problem includes 4 objectives, 16 adverbs and 36 variables. Using equation 13, we convert each of the triangular neutrosophic coefficients into a definite number for this purpose α,β,γ=0.4 and w_1=0.4, w_2=0.5 have been assumed. After converting the neutrosophic problem into a deterministic problem, based on step 2, using the weighted sum method, we convert the multi-objective problem into a single-objective one.

The calculation results for each of the variables, the total objective function and each of the objective functions are given in Table 18.

REFERENCES

Abdel-Basset, M., Ali, M., & Atef, A. (2020). Uncertainty assessments of linear time-cost tradeoffs using neutrosophic set. *Computers & Industrial Engineering, 141*, 106286. DOI: 10.1016/j.cie.2020.106286

Ahmad, F., Adhami, A. Y., & Smarandache, F. (2018). Single valued neutrosophic hesitant fuzzy computational algorithm for multiobjective nonlinear optimization problem. *Neutrosophic sets and systems, 22*, 76-86.

Ahmad, F., Adhami, A. Y., & Smarandache, F. (2020). Modified neutrosophic fuzzy optimization model for optimal closed-loop supply chain management under uncertainty. In *Optimization theory based on neutrosophic and plithogenic sets* (pp. 343–403). Elsevier. DOI: 10.1016/B978-0-12-819670-0.00015-9

Ahmad, F., & John, B. (2022). Modeling and optimization of multiobjective programming problems in neutrosophic hesitant fuzzy environment. *Soft Computing, 26*(12), 5719–5739. DOI: 10.1007/s00500-022-06953-9

Arasteh, A. (2020). Supply chain management under uncertainty with the combination of fuzzy multi-objective planning and real options approaches. *Soft Computing, 24*(7), 5177–5198. DOI: 10.1007/s00500-019-04271-1

Bellman, R. E., & Zadeh, L. A. (1970). Decision-making in a fuzzy environment. *Management Science, 17*(4), B-141–B-164. DOI: 10.1287/mnsc.17.4.B141

Bilgen, B. (2010). Application of fuzzy mathematical programming approach to the production allocation and distribution supply chain network problem. *Expert Systems with Applications, 37*(6), 4488–4495. DOI: 10.1016/j.eswa.2009.12.062

Chen, S.-P., & Chang, P.-C. (2006). A mathematical programming approach to supply chain models with fuzzy parameters. *Engineering Optimization, 38*(6), 647–669. DOI: 10.1080/03052150600716116

Chopra, S., & Meindl, P. (2007). *Supply chain management. Strategy, planning & operation*. Springer.

Edalatpanah, S. (2020). Data envelopment analysis based on triangular neutrosophic numbers. *CAAI Transactions on Intelligence Technology, 5*(2), 94–98. DOI: 10.1049/trit.2020.0016

Ehrgott, M. (2005). *Multicriteria optimization* (Vol. 491). Springer Science & Business Media.

Fallah, M., & Nozari, H. (2021). Neutrosophic mathematical programming for optimization of multi-objective sustainable biomass supply chain network design. *Computer Modeling in Engineering & Sciences*, *129*(2), 927–951. DOI: 10.32604/cmes.2021.017511

Farnam, M., Darehmiraki, M., & Behdani, Z. (2024). Neutrosophic data envelopment analysis based on parametric ranking method. *Applied Soft Computing*, *153*, 111297. DOI: 10.1016/j.asoc.2024.111297

Gholamian, N., Mahdavi, I., Tavakkoli-Moghaddam, R., & Mahdavi-Amiri, N. (2015). Comprehensive fuzzy multi-objective multi-product multi-site aggregate production planning decisions in a supply chain under uncertainty. *Applied Soft Computing*, *37*, 585–607. DOI: 10.1016/j.asoc.2015.08.041

Ghosh, S., & Roy, S. K. (2023). Closed-loop multi-objective waste management through vehicle routing problem in neutrosophic hesitant fuzzy environment. *Applied Soft Computing*, *148*, 110854. DOI: 10.1016/j.asoc.2023.110854

Giri, B. K., & Roy, S. K. (2022). Neutrosophic multi-objective green four-dimensional fixed-charge transportation problem. *International Journal of Machine Learning and Cybernetics*, *13*(10), 3089–3112. DOI: 10.1007/s13042-022-01582-y

Gümüs, A. T., & Güneri, A. F. (2007). Multi-echelon inventory management in supply chains with uncertain demand and lead times: Literature review from an operational research perspective. *Proceedings of the Institution of Mechanical Engineers. Part B, Journal of Engineering Manufacture*, *221*(10), 1553–1570. DOI: 10.1243/09544054JEM889

Han, J., Yang, C., Lim, C.-C., Zhou, X., Shi, P., & Gui, W. (2020). Power scheduling optimization under single-valued neutrosophic uncertainty. *Neurocomputing*, *382*, 12–20. DOI: 10.1016/j.neucom.2019.11.089

Ivanov, D., Tsipoulanidis, A., & Schönberger, J. (2021). *Global supply chain and operations management*. Springer.

Jdid, M., & Smarandache, F. (2023). *Optimal Agricultural Land Use: An Efficient Neutrosophic Linear Programming Method*. Neutrosophic Systems With Applications.

Kamal, M., Modibbo, U. M., AlArjani, A., & Ali, I. (2021). Neutrosophic fuzzy goal programming approach in selective maintenance allocation of system reliability. *Complex & Intelligent Systems*, *7*(2), 1045–1059. DOI: 10.1007/s40747-021-00269-1

Kousar, S., Sangi, M. N., Kausar, N., Pamučar, D., Ozbilge, E., & Cagin, T. (2023). Multi-objective optimization model for uncertain crop production under neutrosophic fuzzy environment: A case study. *AIMS Mathematics, 8*(3), 7584–7605. DOI: 10.3934/math.2023380

Kumar, M., Vrat, P., & Shankar, R. (2006). A fuzzy programming approach for vendor selection problem in a supply chain. *International Journal of Production Economics, 101*(2), 273–285. DOI: 10.1016/j.ijpe.2005.01.005

Liang, T.-F. (2008). Integrating production-transportation planning decision with fuzzy multiple goals in supply chains. *International Journal of Production Research, 46*(6), 1477–1494. DOI: 10.1080/00207540600597211

Nasseri, H., Morteznia, M., & Mirmohseni, M. (2017). A new method for solving fully fuzzy multi objective supplier selection problem. *International journal of research in industrial engineering, 6*(3), 214-227.

Nozari, H., Tavakkoli-Moghaddam, R., & Gharemani-Nahr, J. (2022). A neutrosophic fuzzy programming method to solve a multi-depot vehicle routing model under uncertainty during the covid-19 pandemic. *International Journal of Engineering, 35*(2), 360–371.

Peidro, D., Mula, J., Poler, R., & Verdegay, J.-L. (2009). Fuzzy optimization for supply chain planning under supply, demand and process uncertainties. *Fuzzy Sets and Systems, 160*(18), 2640–2657. DOI: 10.1016/j.fss.2009.02.021

Petrovic, D., Roy, R., & Petrovic, R. (1999). Supply chain modelling using fuzzy sets. *International Journal of Production Economics, 59*(1-3), 443–453. DOI: 10.1016/S0925-5273(98)00109-1

Pouresmaeil, H., Khorram, E., & Shivanian, E. (2022). A parametric scoring function and the associated method for interval neutrosophic multi-criteria decision-making. *Evolving Systems, 13*(2), 347–359. DOI: 10.1007/s12530-021-09394-1

Pratihar, J., Kumar, R., Dey, A., & Broumi, S. (2020). Transportation problem in neutrosophic environment. In *Neutrosophic graph theory and algorithms* (pp. 180–212). IGI Global. DOI: 10.4018/978-1-7998-1313-2.ch007

Şahin, M., Kargın, A., & Smarandache, F. (2018). *Generalized single valued triangular neutrosophic numbers and aggregation operators for application to multi-attribute group decision making*. Infinite Study.

Shafi Salimi, P., & Edalatpanah, S. A. (2020). Supplier selection using fuzzy AHP method and D-Numbers. *Journal of fuzzy extension and applications, 1*(1), 1-14.

Smarandache, F. (1998). *Neutrosophy: neutrosophic probability, set, and logic: analytic synthesis & synthetic analysis.*

Tang, J., Fung, R. Y., & Yung, K.-L. (2003). Fuzzy modelling and simulation for aggregate production planning. *International Journal of Systems Science, 34*(12-13), 661–673. DOI: 10.1080/00207720310001624113

Tayur, S., Ganeshan, R., & Magazine, M. (2012). *Quantitative models for supply chain management, 17.* Springer Science & Business Media.

Wang, H., Smarandache, F., Zhang, Y., & Sunderraman, R. (2010). Single valued neutrosophic sets. *Infinite study, 12*, 20110.

Chapter 10
An Approach to Solve Non-Linear Neutrosophic Transportation Problem With Volume Discount

Aakanksha Singh
Indira Gandhi Delhi Technical University for Women, Delhi, India & Aryabhatta College, University of Delhi, India

Ritu Arora
https://orcid.org/0000-0002-6078-739X
Keshav Mahavidyalaya, University of Delhi, India

Shalini Arora
Indira Gandhi Delhi Technical University for Women, Delhi, India

ABSTRACT

In real-life transportation problems (TPs), the supply, demand, and costs parameters are uncertain in nature. During transportation due to unavoidable conditions like accidents, road conditions, poor handling, etc., the commodity gets damaged resulting in damage cost. Henceforth, the decision maker (DM) aims at minimizing both the transportation cost and the damage cost. Also, there are shipping policies where discounts associated with each shipment are applied and are directly proportional to the commodity transported. Neutrosophic numbers of the type b+b'I are capable of representing uncertainties. In the chapter, the authors formulate a non-linear TP with neutrosophic parameters. The considered objective function will simultaneously minimize the transportation cost; damage cost and volume discounts are also incorporated. A solution methodology based on interval programming is propounded. A

DOI: 10.4018/979-8-3693-3204-7.ch010

solved numerical establishes the efficiency of the approach. Sensitivity analysis is performed on different values of I and their solutions are compared.

AN APPROACH TO SOLVE NON-LINEAR NEUTROSOPHIC TRANSPORTATION PROBLEM WITH VOLUME DISCOUNT

Transportation problem (TP) is an optimization problem which deals in finding out how much of the commodity should be transported from a set of supply points to another set of destination points such that the total expenditure on transportation is minimum. In some cases, due to unavoidable reasons like road and weather conditions, spillage, infestation, mishandling it may happen that while transporting some fraction of goods gets damaged. Then, cost of depreciation or damage cost comes into picture and it becomes imperative for the DM to minimise both the costs simultaneously. Also, in the physical world, the transportation parameters such as supply-demand and costs are also uncertain and cannot be denoted by crisp numbers. Factors such as fuel price, repair and maintenance of tyres and engines etc. are reasons for uncertain costs while factors such as shifted consumer preference, weather conditions, availability of substitutes, natural disasters etc. lead to an uncertain supply and demand situation.

It is a general assumption that the transportation cost (TC) per unit commodity transported from a given supply point to a given destination is constant or fixed, irrespective of the amount transported. However, in certain shipping policies, this cost is not fixed as volume discounts (directly related to the unit commodity or same rate for some amount) are available. This can be termed as a TP with volume discounts. Transportation is one of the major factors which affects the pricing system and supply-demand of any commodity. Reasons why DMs offer these discounts as more goods are carried with fewer trips, fewer vehicles and employees. Thus, the cost per delivery decreases. These discounts incentivize consistent business as the buyer receives the commodity at a more favourable price.

The above scenario motivates us to develop a transportation model which

1. Bridges the gap between supply and demand by discounting.
2. Simultaneously minimizes the TC and the damage cost.
3. Handles the uncertainty in the transportation parameters viz. cost, supply and demand with an appropriate representation.

Research gap: Upon delving into the literature, it was found that no work has been done on TPs with NNs with all of the above three considerations together. The propounded work attempts to fill this research gap by

1. Developing a new non-linear neutrosophic TP (NLNTP) with neutrosophic numbers (NNs) of the type $b + b'I$ for the first time (Here, I denotes indeterminacy, $b'$is the determinate part and $b'I$ is the indeterminate part).
2. Minimising the above said costs using an indefinite quadratic (IQ) objective function. This function is a product of two linear factors and its minimisation means simultaneously minimising each factor. IQ comprises of one factor as a cost function with neutrosophic TC coefficients and the second factor as a cost function with neutrosophic damage cost coefficients. Also, crisp discount costs are incorporated. In an entirety, a minimisation type non-linear objective function is formulated and solved. (Discounts are in % and directly related to the unit commodity)
3. Representing the uncertain supplies at the supply points and demands at destinations in the constraints as NNs.
4. Proposing a less complicated solution strategy based on interval programming and fuzzy programming.
5. Performing sensitivity analysis on the range of indeterminacy.
6. Comparing the obtained solutions and drawing inferences.

The solution strategy begins by taking $I \in [I_L, I_U]$ in the NNs $b + b'I$ and converting them into intervals. The resultant problem is a TP with interval costs, supply and demand parameters. Using the concept of half-width and centre of an interval, a bi-objective TP (Ishibuchi & Tanaka, 1990) is formulated and the interval supply-demand in the constraints are modified using a proposition (Shaocheng, 1994). To arrive at a compromise solution to the NLNTP, a fuzzy programming model is proposed. The compromise solution so obtained is then substituted into two objective functions and the minimised TC is calculated. To establish the efficiency of the model, a solved numerical example followed by sensitivity analysis on different values of I is also performed and results are compared.

LITERATURE REVIEW

The initial development of TP is attributed to Hitchcock (1941) and further, independent development to Koopmans (1949). In a standard TP, the objective function is a linear function with crisp supply, demand and cost parameters and solution methodologies include methods like row minima, column minima, matrix minima, north-west corner rule (for initial basic feasible solution) and Vogel's approximation method (VAM) (for the optimal solution). However, this linear function is at times not capable of depicting the strategies, aims and policies of a DM in a real-life transportation scenario. Hence, different variations/formulation in constraints, objective functions, parameters and solution strategies are desired. For instance, in case of multiple conflicting objectives, a multi-objective TP (MOTP) (Charnes and Cooper, 1957; Lee and Moore, 1973) is considered whereas to represent ratios of profit/cost or profit/time, a fractional objective function is considered (Stancu-Minasian, 1997). In a time-minimization TP, the total time necessary for transporting the goods is minimized (Hammer, 1969, 1971) and an IQ objective function (Sharma & Swarup, 1977; Bhatia, 1981) is used to minimise two types of costs simultaneously. Some other variations include that of fixed-charge TP (Balinski, 1961; Gray, 1971), constrained TP (Klingman & Russell, 1975), bicriteria TP (Aneja & Nair, 1979), bottleneck TP (Garfinkel & Rao, 1971) and stochastic TP (Williams, 1963). A branch and bound methodology-based solution to a transportation type problem was presented in Balachandran & Perry (1976). A MOTP with interval transportation parameters was worked upon by Das et al. (1999). By varying the supply and demand between some specified intervals, Ahmad and Adhami (2019) developed optimization models for TPs. Recently, a modified approach without adding a dummy source or destination for solving an unbalanced TP was proposed by Khandelwal & Kumar (2024).

The transportation parameters are not certain in the physical world and a representation other than a crisp number is solicited. Zadeh (1965) conceptualised fuzzy sets which were characterised by the membership degree (or degree of truth). A TP formulated in a fuzzy environment is called a fuzzy TP (FTP) (ÓhÉigeartaigh,1982; Chanas & Kuchta,1996). Verma et al. (1997) used non-linear membership functions to solve the MOTP whereas an interactive fuzzy goal programming approach (FGPA) for a MOTP was proposed by Abd El-Wahed & Lee, (2006). Yang & Liu (2007) designed a hybrid intelligent algorithm to solve a fuzzy fixed-charge solid TP (STP). Kaur & Kumar (2011) solved a FTP using ranking function approach and Liu (2016) applied the concept of α-cut on triangular fuzzy supply, demand and cost parameters in a fractional TP. A modified VAM was used to solve a TP using interval type 2 fuzzy sets in Pratihar et al. (2021) and FGPA to solve a multi-choice fractional stochastic MOTP was recently proposed by (El Sayed & Baky, 2023).

To describe the ambiguity of the situation when degree of membership was not sufficient, Atanassov (1986) developed intuitionistic fuzzy sets (IFS) which apart from degree of membership were also described by degree of non-membership (or degree of falsity). A TP formulated in an intuitionistic fuzzy (IF) environment is called an intuitionistic fuzzy TP (IFTP). Since conception, various researchers have formulated and solved IFTP. A new ranking technique based on signed distance was proposed to solve an intuitionistic fuzzy STP by Aggarwal & Gupta (2016). An attempt to provide an IF optimal solution instead of a crisp solution to a fully IFTP was made by Ebrahimnejad & Verdegay (2018). Interval-valued IFS were used to solve a TP pertaining to COVID-19 scenario (Bharati, 2021). Non-linear membership functions were employed to solve a fixed-charge STP under IF environment in Chhibber et al. (2021). Some other recent works in an IFTP include that of Choudhary & Yadav (2022); Radhika & Arun Prakash (2022); Jansi Rani et al. (2023); Singh et al. (2023); Rani & Manivannan (2024).

To overcome some situations where degree of truth and falsity are incapable or insufficient to describe a situation aptly, Smarandache (1998) extended the IFS theory to neutrosophic set theory. In this theory, a grade independent of the membership and the non-membership is incorporated viz. the indeterminacy membership grade. In the field of TPs with neutrosophic set theory, initial literature was based on solving TP in a neutrosophic environment. Kour & Basu (2015) proposed a solution methodology for multi-objective multi-index TP in neutrosophic environment. A TP with varying demand and supplies was investigated by Xie et al. (2017). Singh et al. (2017) formulated a TP in neutrosophic environment and proposed its solution approach based on score functions. Rizk-Allah et al. (2018) proposed a compromise algorithm to solve MOTP inspired by Zimmermann's fuzzy programming. Kumar et al. (2021) applied the (α,β,γ)-cut concept on single-valued NNs (See Definition 4) to obtain an optimal solution of a TP. Veeramani et al. (2021) solved a multi-objective fractional TP and worked upon a problem which minimises total TCs, minimises deterioration and total delivery time. The concept of centre and half width of an interval was applied to solve a neutrosophic TP with pentagonal NNs by Habiba & Quddoos (2022). Ahmad et al. (2022) proposed an extended neutrosophical programming approach to solve a multi-objective four-index TP which considers environmental impact of transportation. A fixed-charge STP with budget constraints based on carbon emission in neutrosophic environment was proposed by Ghosh et al. (2022). Bera & Mahapatra (2022) employed single-valued trapezoidal NNs to solve a TP using ranking function. A neutrosophic logic-based methodology is used to solve a STP with triangular NNs in Qiuping et al. (2023). Samanta et al. (2024) formalised a multi-period two stage 4D TP for breakable items and employed Generalised Reduced Gradient method to solve it. Khalifa et al. (2023) modelled a neutrosophic fractional TP with non-linear discounting cost by taking single- valued trapezoidal NNs as supply, demand and cost parameters. Kar et al. (2023) solved a fully neutrosophic incompatible multi-item fixed charge 4D TP with vehicle volume constraints. Neutrosophic goal programming approach was employed to solve a multi-objective fixed-charge TP with single-valued trapezoidal NNs parameters by Gupta et al. (2024). The contribution of neutrosophic sets is not only restricted in the field of TPs but other optimization fields as well. Some of them being in the field of bimatrix games (Gaber et al., 2021), data envelopment analysis (Tapia, 2021), multi-objective linear programming problem (MOLPP) (Wang et al., 2021), route selection problem (Simić et al., 2023).

Neutrosophic set theory was further extended to NNs (Smarandache 2013, 2014) which are of the form $B = b + b'I$ (where I denotes literal or non-numerical indeterminacy, b is the determinate part and $b'I$ is the indeterminate part (Maiti, 2020)). In the best scenario i.e. case of no indeterminacy $B=b$, and in the worst case scenario i.e. full indeterminacy, $B = b'I$. Since conception, NNs established its footing in various domains some of them being multiple-attribute group decision-making (Ye, 2016), multi-level MOLPP (Maiti et al., 2020), production planning problem (Tu et al., 2021) and water data (Albassam et al., 2021). Singh et al. (2022) formulated and solved a bilevel linear fractional TP and a bilevel IQ TP in neutrosophic environment with an application to vaccine transportation.

The above citations have showcased the extensive usage of single-valued NNs in field of TPs but the field of TPs with NNs of the type $b + b'I$ is still at a very nascent stage and is yet to be explored. Thus, in the proposed work, we amalgamate TPs and NNs in a real-life transportation scenario and propose a mathematical formulation to depict a specific circumstance and policy. From here onwards, the paper is organised in the following manner. Firstly, the *Mathematical Model of NLNTP* is defined. This section describes the notations and steps of formulation of the TP to a crisp one. This is followed by a *Solution Methodology* section in which the proposed fuzzy programming model is described. In the next section, *Numerical Example,* a NLNTP pertaining to a real-life scenario is solved as per the steps of the fuzzy model described in the previous section. Once the solution is obtained, in the *Sensitivity Analysis* section the solution obtained by taking different values of I is tabulated. *Results and Discussion* section discusses the results and their interpretation. *Conclusion and Future work* section concludes the work in an entirety and future work is suggested.

MATHEMATICAL MODEL OF NLNTP

Real-life TPs have uncertain cost, supply and demand parameters. During transportation, due to unavoidable conditions such as accidents, road-conditions, poor-handling etc., the commodity gets damaged and adds up to the depreciation cost to the already existing TC. Thus, the DM has to minimize them both simultaneously while handling the uncertainties in the transportation parameters. Mathematically, this can be achieved by formulating an IQ objective function which is a product of two linear factors and its minimisation means simultaneously minimising each factor. We take one factor as a cost function with TC coefficients and the second factor as a cost function with damage cost coefficients. Also, there are shipping policies where discounts associated with each shipment are applied and are directly proportional to the commodity transported.

In the proposed work, we incorporate these along with representing the uncertain transportation parameters (supply, demand and cost) using NNs (of the type $b + b'I$ and formulate a non-linear transportation problem. In an entirety, we aim at simultaneous minimization of TC and depreciation cost while considering volume discounts subject to uncertain supply and demand constraints.

Definition

We first state the notations which will be used in defining the NLNTP mathematically.

t: Number of sources.
w: Number of destinations.
x_{rs}: Quantity to be supplied from r^{th} source to s^{th} destination.
$e_{rs} + e'_{rs}I$: Neutrosophic TC during transportation of one unit from r^{th} source to s^{th} destination.
$f_{rs} + f'_{rs}I$: Neutrosophic depreciation cost during transportation of one unit from r^{th} source to
s^{th} destination.
v_{rs}: Volume discount offered (in % in transporting one unit from r^{th} source to s^{th} destination.
$l_r + l'_rI$: Neutrosophic supply available at the rth source.
$m_s + m'_sI$: Neutrosophic demand required at the sth destination.

The mathematical representation of NLNTP is given by,

$$Min Z = \left[\sum_{r=1}^{t}\sum_{s=1}^{w}(e_{rs} + e'_{rs}I)x_{rs}\right]\left[\sum_{r=1}^{t}\sum_{s=1}^{w}(f_{rs} + f'_{rs}I)x_{rs}\right] - \left(\sum_{r=1}^{t}\sum_{s=1}^{w}v_{rs}x_{rs}^2\right)$$

subject to

$$\sum_{s=1}^{w}x_{rs} \leq l_r + l'_r l \quad \forall r = 1,...,t \ ;$$

(Supply constraints)

$$\sum_{r=1}^{t}x_{rs} \geq m_s + m'_s l \quad \forall s = 1,...,w \ ;$$

(Demand constraints)

$x_{rs} \geq 0$ (Non-negativity constraints)
Feasibility conditions: Total supply = Total demand i.e.

$$\sum_r l_r + l'_r l = \sum_s m_s + m'_s l$$

Equivalent Crisp Model of NLNTP

We begin the solution process to the given NLNTP by converting all the neutrosophic parameters into an interval form for a given $I \in [I_L, I_U]$ which will be further converted into crisp numbers.

Construction of Interval Objective Function and Interval Constraints

Consider any NN $b + b'I$. For any $I \in [I_L, I_U]$,

$$b + b'I = b + b'[I_L, I_U] = [b + b' I_L, b + b' I_U] = [b_L, b_U].$$

Thus, each neutrosophic parameter can be converted into an interval form. Then, an equivalent interval NLNTP is given by

$$Min\, Z = \left(\sum_{r=1}^{t} \sum_{s=1}^{w} [e_{rsL} + e_{rsU}] x_{rs} \right) \left(\sum_{r=1}^{t} \sum_{s=1}^{w} [f_{rsL} + f_{rsU}] x_{rs} \right) - \left(\sum_{r=1}^{t} \sum_{s=1}^{w} v_{rs} x_{rs}^2 \right) \quad (1)$$

subject to

$$\sum_{s=1}^{w} x_{rs} \leq [I_{rL}, I_{rU}], \quad (2)$$

$$\sum_{r=1}^{t} x_{rs} \geq [m_{sL}, m_{sU}], \quad (3)$$

$$x_{rs} \geq 0 \forall r = 1, ..., t;\ s = 1, ..., w.$$

Construction of Crisp Objective Function and Crisp Constraints

We now use the concept of centre and half-width of an interval (Ishibuchi & Tanaka, 1990) which converts the objective function with interval cost coefficients into two distinct functions with crisp cost coefficients. The resulting problem is a multi-objective TP of minimization type.

Mathematically, it can be represented as (See Remark 1):

$$Min\ Z_c = \left(\sum_{r=1}^{t}\sum_{s=1}^{w} E_{rsc} x_{rs}\right)\left(\sum_{r=1}^{t}\sum_{s=1}^{w} F_{rsc} x_{rs}\right) - \left(\sum_{r=1}^{t}\sum_{s=1}^{w} v_{rs} x_{rs}^2\right) \tag{4}$$

$$Min\ Z_h = \left(\sum_{r=1}^{t}\sum_{s=1}^{w} E_{rsh} x_{rs}\right)\left(\sum_{r=1}^{t}\sum_{s=1}^{w} F_{rsh} x_{rs}\right) - \left(\sum_{r=1}^{t}\sum_{s=1}^{w} v_{rs} x_{rs}^2\right) \tag{5}$$

For the interval constraints in (2) and (3), the best optimal solution will be obtained over the maximum feasible region. To obtain this feasible region the following proposition is employed:

Proposition 1 Shaocheng (1994) Suppose $\sum_{s=1}^{w}[p_1^s,\ p_2^s]x_s \geq [g_1, g_2]$ then $\sum_{s=1}^{w}[\ p_2^s]x_s \geq [g_1]$ and $\sum_{s=1}^{w}[\ p_1^s]x_s \geq [g_2]$ are the maximum and minimum values range inequalities for the constraint condition respectively.

Implementing *Proposition 1* on the interval type constraints set (2) and (3) by taking $[p_1^s,\ p_2^s] = 1$ the maximum feasible region (Shaocheng, 1994; Chinneck and Ramadan, 2000) is obtained resulting in the following crisp constraints:

$$\sum_{s=1}^{w} x_{rs} \leq l_{rU} \forall r = 1, \ldots t\ ; \tag{6}$$

$$\sum_{r=1}^{t} x_{rs} \geq m_{sL} \forall s = 1, \ldots w. \tag{7}$$

SOLUTION METHODOLOGY

We have now converted the interval NLNTP into a crisp bi-objective NLNTP (CNLNTP) given by (4)-(7).

(CNLNTP):

$$Min\ Z_c = \left(\sum_{r=1}^{t}\sum_{s=1}^{w} E_{rsc} x_{rs}\right)\left(\sum_{r=1}^{t}\sum_{s=1}^{w} F_{rsc} x_{rs}\right) - \left(\sum_{r=1}^{t}\sum_{s=1}^{w} v_{rs} x_{rs}^2\right)$$

$$Min\ Z_h = \left(\sum_{r=1}^{t}\sum_{s=1}^{w} E_{rsh}x_{rs}\right)\left(\sum_{r=1}^{t}\sum_{s=1}^{w} F_{rsh}x_{rs}\right) - \left(\sum_{r=1}^{t}\sum_{s=1}^{w} v_{rs}x_{rs}^2\right)$$

subject to

$$\sum_{s=1}^{w} x_{rs} \le l_{rU} \forall r$$

$$\sum_{r=1}^{t} x_{rs} \ge m_{sL} \forall s$$

$$x_{rs} \ge 0 \forall r = 1, ..., t;\ s = 1, ..., w.$$

We solve this bi-objective CNLNTP by fuzzy programming approach using the following steps.

Step 1: Solve CNLNTP by taking Z_c as the sole objective (ignoring Z_h and obtain the respective optimal solution. If needed, dummy supply or dummy destination should be added to balance the CNLNTP. Repeat the same process with Z_h as the objective and ignoring Z_c.

Step 2: Find the payoff matrix and obtain the lower bound L_{Z_k} (the aspired achievement level) and the upper bound U_{Z_k} (the highest acceptable achievement level) for each objective function Z_k, $k= c,h$ respectively.

Step 3: Define a fuzzy membership function $\mu_k(Z_k)$, $k = c, h$ defined as

$$\mu_k(Z_k) = \begin{cases} 1 & if\ Z_k \le L_{Z_k} \\ 1 - \dfrac{Z_k - L_{Z_k}}{U_{Z_k} - L_{Z_k}} & if\ L_{Z_k} \le Z_k \le U_{Z_k} \\ 0 & if\ Z_Z \ge U_{Z_k} \end{cases}$$

Step 4: The fuzzy optimization model based on maximizing the membership degree is given by Max λ

subject to

$$\lambda \le \frac{U_{Z_k} - Z_k}{U_{Z_k} - L_{Z_k}};\ k = c, h$$

$$\sum_s x_{rs} \le l_{rU}; \sum_t x_{rs} \ge m_{sL}\ ;$$

$$x_{rs} \ge 0 \forall r = 1, ..., t;\ s = 1, ..., w;\ \lambda \in [0, 1]$$

On further simplification, the fuzzy model to solve the NLNTP (FM-NLNTP) is given by

(FM-NLNTP): Max λ

subject to

$$Z_k + \lambda(U_{Z_k} - L_{Z_k}) \le U_{Z_k}; k = c, h$$

$$\sum_s x_{rs} \le l_{rU}; \sum_t x_{rs} \ge m_{sL}\ ;$$

$$x_{rs} \ge 0 \forall r = 1, ..., t;\ s = 1, ..., w;\ \lambda \in [0, 1]$$

Step 5: Solve the FM-NLNTP model to find the pareto optimal solution x_{rs}. The solution is such that it provides the highest degree of membership λ and all the constraints are also satisfied.

Step 6: Substitute x_{rs} in both the objective functions to obtain their corresponding values Z_c and Z_h. Then, using Remark 1, $< Z_c, Z_h > \ = \ [Z_c - Z_h, Z_c + Z_h]$ will give the cost incurred in interval form from which neutrosophic cost can be obtained.

NUMERICAL EXAMPLE FROM FMCG SECTOR

Problem Description

Consider a transportation scenario in FMCG domain in a city with four supply points R_1, R_2, R_3, R_4. From these points, a commodity is shipped in packed cartons (each carton with 50 units of the commodity) to four destinations S_1, S_2, S_3, S_4 located in different parts of the city. Products like medicines, dairy, electronics, fruits and vegetables are highly prone to damage during transit due to unavoidable reasons like road conditions, temperature fluctuations and mishandling and adds up to the damage cost apart from the already existing TC. It is now imperative for the DM to minimize these costs simultaneously.

To represent the uncertainties in supply, demand and cost parameters we represent them using NNs. For the objective function, the neutrosophic TC coefficients (e_{rs} + $e'_{rs}I$) and neutrosophic depreciation cost coefficients ($f_{rs} + f'_{rs}I$) where r,s= {1,2, 3, 4} are tabulated in Table 1 and Table 2.

Table 1. Neutrosophic transportation cost coefficients

Supply point	*Destination*			
	R_1	R_2	R_3	R_4
S_1	$6 + 4I$	$14 + 2I$	$6 + 4I$	$20 + 4I$
S_2	$8 + 3I$	$16 + 4I$	$13 + 5I$	$5 + 2I$
S_3	$6 + 5I$	$9 + 6I$	$6 + 7I$	$15 + 8I$
S_4	$10 + 5I$	$12 + 7I$	$14 + 9I$	$9 + I$

Table 2. Neutrosophic depreciation cost coefficients

Supply point	*Destination*			
	R_1	R_2	R_3	R_4
S_1	$5 + 4I$	$8 + 6I$	$11 + 6I$	$14 + I$
S_2	$2 + I$	$5 + 3I$	$9 + 3I$	$18 + 4I$
S_3	$16 + 5I$	$7 + 2I$	$10 + 3I$	$14 + 4I$
S_4	$15 + 3I$	$6 + 3I$	$7 + I$	$9 + 8I$

With the supposition that the distributor or the DM gives discount on each box transported from r^{th} source to s^{th} destination then, the discount values (in %) are given by Table 3 as mentioned below.

Table 3. Volume discount cost coefficients

Supply point	Destination			
	R_1	R_2	R_3	R_4
S_1	0.02	0.01	0.03	0.07
S_2	0.05	0.08	0.01	0.002
S_3	0.08	0.001	0.06	0.02
S_4	0.005	0.07	0.08	0.01

The uncertain availabilities at the four supply points and uncertain demands at the four destinations are given by the following supply and demand constraints (r,s = 1, 2, 3, 4):

$$\sum_s x_{1s} \le 4 + 3I, \sum_s x_{2s} \le 6 + 3I, \sum_s x_{3s} \le 7 + 4I, \sum_s x_{4s} \le 10 + I,$$

$$\sum_r x_{r1} \ge 4 + 2I, \sum_r x_{r2} \ge 5 + 2I, \sum_r x_{r3} \ge 10 + 4I, \sum_r x_{r4} \ge 8 + 3I.$$

We have considered a balanced TP with total supply = total demand (= 27 + 11I).

Solution Methodology

Formation of Interval Objective Function and Constraints

Let the DM gives the value $I \in [I_L, I_U] = [0, 0.15]$. Then, all the neutrosophic parameters get converted into an interval form. The interval transportation and depreciation cost coefficients are tabulated in Table 4 and Table 5.

Table 4. Interval transportation cost coefficients

Supply point	Destination			
	R_1	R_2	R_3	R_4
S_1	[6, 6.6]	[14, 14.3]	[9, 9.6]	[20, 20.6]
S_2	[8, 8.45]	[16, 16.6]	[13, 13.75]	[5, 5.3]
S_3	[6, 6.75]	[9, 9.9]	[6, 7.05]	[15, 16.2]
S_4	[10, 10.75]	[12, 13.05]	[14, 15.35]	[9, 9.15

Table 5. Interval depreciation cost coefficients

Supply point	Destination			
	R_1	R_2	R_3	R_4
S_1	[5, 5.6]	[8, 8.3]	[11, 11.9]	[14, 14.15]
S_2	[2, 2.15]	[5, 5.45]	[9, 9.45]	[18, 18.6]
S_3	[16, 16.75]	[7, 7.3]	[10, 10.45]	[14, 14.6]
S_4	[15, 15.45]	[6, 6.45]	[7, 7.15]	[9, 10.2]

The interval supply and the demand constraints are given by

$$\sum_s x_{1s} \le [4,\ 4.45], \sum_s x_{2s} \le [6,\ 6.45], \sum_s x_{3s} \le [7,\ 7.6], \sum_s x_{4s} \le [10,\ 10.15],$$

$$\sum_r x_{r1} \ge [4,\ 4.3], \sum_r x_{r2} \ge [5,\ 5.3], \sum_r x_{r3} \ge [10,\ 10.6], \sum_r x_{r4} \ge [8,\ 8.45]$$

Formation of Crisp Objective Function and Constraints (CNLNTP)

Using (4) and (5) and allowing the volume discounts on each product shipped from source r to destination s as given in Table 3, the non-linear crisp cost functions are given by

$$\begin{aligned}
\text{Min } Z_{c'} = &((6.3x_{11} + 14.15x_{12} + 9.3x_{13} + 20.3x_{14} + 8.225x_{21} + 16.3x_{22} \\
&+13.375x_{23} + 5.15x_{24} + 6.375x_{31} + 9.45x_{32} + 6.525x_{33} + 15.6x_{34} \\
&+10.375x_{41} + 12.525x_{42} + 14.67x_{43} + 9.075x_{44})(5.3x_{11} + 8.15x_{12} \\
&+11.45x_{13} + 14.075x_{14} + 2.075x_{21} + 5.225x_{22} + 9.225x_{23} + 18.3x_{24} \\
&+16.375x_{31} + 7.15x_{32} + 10.225x_{33} + 14.3x_{34} + 15.225x_{41} + 6.225x_{42} \\
&+7.075x_{43} + 9.6x_{44})) - 0.02x_{11}^2 + 0.01x_{12}^2 + 0.03x_{13}^2 + 0.07x_{14}^2 + 0.05x_{21}^2 \\
&+0.08x_{22}^2 + 0.01x_{23}^2 + 0.002x_{24}^2 + 0.08x_{31}^2 + 0.001x_{32}^2 + 0.06x_{33}^2 + 0.02x_{34}^2 \\
&+0.005x_{41}^2 + 0.07x_{42}^2 + 0.08x_{43}^2 + 0.01x_{44}^2)
\end{aligned}$$

$$
\begin{aligned}
\text{Min } Z_h = {} & ((0.3x_{11} + 0.15x_{12} + 0.3x_{13} + 0.3x_{14} + 0.225x_{21} + 0.3x_{22} + 0.375x_{23} \\
& +0.15x_{24} + 0.375x_{31} + 0.45x_{32} + 0.525x_{33} + 0.6x_{34} + 0.375x_{41} \\
& +0.525x_{42} + 0.675x_{43} + 0.075x_{44})(0.3x_{11} + 0.15x_{12} + 0.45x_{13} \\
& +0.075x_{14} + 0.075x_{21} + 0.225x_{22} + 0.225x_{23} + 0.3x_{24} + 0.375x_{31} \\
& +0.15x_{32} + 0.225x_{33} + 0.3x_{34} + 0.225x_{41} + 0.225x_{42} + 0.075x_{43} \\
& +0.6x_{44})) - 0.02x_{11}^2 + 0.01x_{12}^2 + 0.03x_{13}^2 + 0.07x_{14}^2 + 0.05x_{21}^2 \\
& +0.08x_{22}^2 + 0.01x_{23}^2 + 0.002x_{24}^2 + 0.08x_{31}^2 + 0.001x_{32}^2 + 0.06x_{33}^2 \\
& +0.02x_{34}^2 + 0.005x_{41}^2 + 0.005x_{41}^2 + 0.07x_{42}^2 + 0.08x_{43}^2 + 0.01x_{44}^2)
\end{aligned}
$$

Using (6) and (7), the crisp supply and the demand constraints are given by

$$\sum_s x_{1s} \le 4.45, \sum_s x_{2s} \le 6.45, \sum_s x_{3s} \le 7.6, \sum_s x_{4s} \le 10.15,$$

$$\sum_r x_{r1} \ge 4, \sum_r x_{r2} \ge 5, \sum_r x_{r3} \ge 10, \sum_r x_{r4} \ge 8$$

Fuzzy Programming Approach

After obtaining the crisp objective functions and constraints we now solve the CNLNTP by fuzzy programming approach.

Step1: Solve CNLNTP by taking Z_c as the sole objective (ignoring Z_h) and obtain the respective optimal solution. Repeat the same process with Z_h as the objective and ignoring Z_c. As the CNLNTP is unbalanced, we added a dummy destination.

Step2: Find the payoff matrix to obtain $U_{Z_c} = 82578.29$, $L_{Z_c} = 54550.19$, $U_{Z_h} = 64.70$ and $L_{Z_h} = 28.97$.

Step3: Using Z_c and Z_h as mentioned above, define fuzzy membership functions $\mu_c(Z_c)$ and $\mu_h(Z_h)$ as follows:

$$
\mu_c(Z_c) = \begin{cases} 1 & \text{if } Z_c \le 54550.19 \\ 1 - \dfrac{Z_c - 54550.19}{28028.10} & \text{if } 54550.19 \le Z_c \le 82578.29 \\ 0 & \text{if } Z_c \ge 82578.29 \end{cases}
$$

$$\mu_h(Z_h) = \begin{cases} 1 & if\, Z_h \le 28.97 \\ 1 - \dfrac{Z_h - 28.97}{35.72} & if\, 28.97 \le Z_h \le 64.70 \\ 0 & if\, Z_h \ge 64.70 \end{cases}$$

Step4: Solve the FM-NLNTP model

Max λ

subject to

$$\begin{aligned}
&((6.3x_{11} + 14.15x_{12} + 9.3x_{13} + 20.3x_{14} + 8.225x_{21} + 16.3x_{22} + 13.375x_{23} \\
&+5.15x_{24} + 6.375x_{31} + 9.45x_{32} + 6.525x_{33} + 15.6x_{34} + 10.375x_{41} + 12.525x_{42} \\
&+14.67x_{43} + 9.075x_{44})(5.3x_{11} + 8.15x_{12} + 11.45x_{13} + 14.075x_{14} + 2.075x_{21} \\
&+5.225x_{22} + 9.225x_{23} + 18.3x_{24} + 16.375x_{31} + 7.15x_{32} + 10.225x_{33} + 14.3x_{34} \\
&+15.225x_{41} + 6.225x_{42} + 7.075x_{43} + 9.6x_{44})) - 0.02x_{11}^2 + 0.01x_{12}^2 + 0.03x_{13}^2 \\
&+0.07x_{14}^2 + 0.05x_{21}^2 + 0.08x_{22}^2 + 0.01x_{23}^2 + 0.002x_{24}^2 + 0.08x_{31}^2 + 0.001x_{32}^2 + 0.06x_{33}^2 \\
&+0.02x_{34}^2 + 0.005x_{41}^2 + 0.07x_{42}^2 + 0.08x_{43}^2 + 0.01x_{44}^2) + \lambda(28028.10) \le 82578.29,
\end{aligned}$$

$$\begin{aligned}
&((0.3x_{11} + 0.15x_{12} + 0.3x_{13} + 0.3x_{14} + 0.225x_{21} + 0.3x_{22} + 0.375x_{23} \\
&+0.15x_{24} + 0.375x_{31} + 0.45x_{32} + 0.525x_{33} + 0.6x_{34} + 0.375x_{41} + 0.525x_{42} \\
&+0.675x_{43} + 0.075x_{44})(0.3x_{11} + 0.15x_{12} + 0.45x_{13} + 0.075x_{14} + 0.075x_{21} \\
&+0.225x_{22} + 0.225x_{23} + 0.3x_{24} + 0.375x_{31} + 0.15x_{32} + 0.225x_{33} + 0.3x_{34} \\
&+0.225x_{41} + 0.225x_{42} + 0.075x_{43} + 0.6x_{44})) - 0.02x_{11}^2 + 0.01x_{12}^2 + 0.03x_{13}^2 \\
&+0.07x_{14}^2 + 0.05x_{21}^2 + 0.08x_{22}^2 + 0.01x_{23}^2 + 0.002x_{24}^2 + 0.08x_{31}^2 + 0.001x_{32}^2 \\
&+0.06x_{33}^2 + 0.02x_{34}^2 + 0.005x_{41}^2 + 0.005x_{41}^2 + 0.07x_{42}^2 + 0.08x_{43}^2 + 0.01x_{44}^2) \\
&+\lambda(35.72) \le 64.70,
\end{aligned}$$

$$\sum_s x_{1s} \le 4.45, \sum_s x_{2s} \le 6.45, \sum_s x_{3s} \le 7.6, \sum_s x_{4s} \le 10.15,$$

$$\sum_r x_{r1} \ge 4, \sum_r x_{r2} \ge 5, \sum_r x_{r3} \ge 10, \sum_r x_{r4} \ge 8$$

$$x_{rs} \ge 0,\ 0 \le \lambda \le 1$$

Step5: The pareto optimal solution obtained after solving through the computing software MATLAB is x_{12}=2.0793, x_{14}=2.3707, x_{21}=4, x_{24}=2.45, x_{32}=2.9207, x_{33}=4.6793, x_{43}=5.3207, x_{44}=3.1793, λ=0.4757.

Step6: Substitute x_{rs} in both the objective functions to obtain Z_c=69245.51, Z_h=47.70. The value of the objective function in interval form is given by $<Z_c, Z_h>$=[69197.81, 69293.21]. For $I \in [0, 0.15]$, the total neutrosophic shipping cost is given by 69197.81 + 636I where 69197.81 is the determined part and 636I is the indetermined part and that total cost incurred can be any value within this interval.

SENSITIVITY ANALYSIS

A sensitivity analysis was performed on different ranges of indeterminacy. These ranges of $I \in [I_L, I_U]$ are provided by the DM as per his acceptance range. The corresponding solutions of Z_c, Z_h, $<Z_c, Z_h>$ along with their neutrosophic representation are tabulated in Table 6.

Table 6. Objective function values for different ranges of I $[I_L, I_U]$

Value of I	Z_c	Z_h	$<Z_c, Z_h>$	Minimized shipping cost in neutrosophic form
$I \in [0, 0.15]$	69245.51	47.70	[69197.81, 69293.21]	69197.81 + 636I
$I \in [0, 0.25]$	70017.60	143.39	[69874.20, 70161.00]	69874.20 + 1147.19I
$I \in [0, 0.35]$	71950.25	283.12	[71667.13, 72233.37]	71667.13 + 1617.83I
$I \in [0, 0.45]$	74145.05	464.71	[73680.33, 74609.77]	73680.33 + 2065.41I
$I \in [0, 0.55]$	76437.89	685.58	[75752.31, 77123.47]	75752.31 + 2493.02I
$I \in [0, 0.65]$	78781.37	942.92	[77838.44, 79724.29]	77838.44 + 2901.31I
$I \in [0, 0.75]$	81160.13	1233.86	[79926.26, 82393.99]	79926.26 + 3290.31I
$I \in [0, 0.85]$	83766.69	1564.00	[82202.69, 85330.70]	82202.69+ 3680.01I
$I \in [0, 0.95]$	86606.70	1937.16	[84669.54, 88543.86]	84669.54 + 4078.24I

RESULTS AND DISCUSSION

The solution methodology of the propounded model is based on fuzzy programming approach which maximises the degree of membership λ and all the constraints are also satisfied. All the solutions are obtained using the computing software MATLAB. Using this approach provides a pareto optimal solution to the NLNTP for any $I \in [I_L, I_U]$. The sensitivity analysis for different ranges of I shows us that the total shipping cost increases as indeterminacy in costs, supplies and demand increases. Suppose that the indeterminacy range of [0, 0.85] is suitable to the DM, then after referring the above table he is sure that he has to pay a minimum amount of 82202.69. Additionally, a variable amount $\leq 85330.70 - 82202.6 = 3128.01$ might also be spent. In this way, for any suitable range DM can ensure his determinate and indeterminate expenditure. Additionally, if the DM wants an overall viewpoint of the expenditure, the minimum cost ranges from a minimum value of 69197.80 to a maximum value of 88543.86.

Advantages and Disadvantages

The advantage of using the propounded methodology is that it is computationally less burdening as we have used linear membership functions and the DM can choose the minimised shipping cost as per the range of indeterminacy acceptable to him. Some disadvantages of the methodology are that finding solution becomes more cumbersome if non-linear membership functions are used (our objective function is already non-linear). Further, in case of a multi-objective NLNTP the number of objectives gets doubled and it becomes difficult to handle.

CONCLUSION AND FUTURE WORK

We have analysed a realistic non-linear transportation problem in a neutrosophic environment which considers volume discounts, neutrosophic transportation and depreciation costs along with neutrosophic supply and demand parameters. As a NN $b + b'I$ can be converted to an interval for a chosen $I \in [I_L, I_U]$, we approach the solution to our formulated problem via the concept of centre and half-width of an interval. The usage of the concept transforms the NLNTP into a crisp bi-objective problem. After appropriate adjustments of the interval constraints, a fuzzy programming approach is applied. The solution obtained (x_{rs}'s) is a pareto optimal solution of the NLNTP. Substituting this solution into the objective function gives the centre and width of the total cost incurred from which we can obtain the final minimised cost in interval and in neutrosophic form as well.

In future, researchers can explore the formulated NLNTP using other interval programming techniques to solve the interval NLNTP problem. The problem can be extended to a three-dimensional TP as well.

ACKNOWLEDGMENT

The authors are thankful to the reviewers for their valuable suggestions, incorporating which has helped us in improving the quality of the chapter to a great extent.

REFERENCES

Abd El-Wahed, W. F., & Lee, S. M. (2006). Interactive fuzzy goal programming for multi-objective transportation problems. *Omega*, *34*(2), 158–166. DOI: 10.1016/j.omega.2004.08.006

Aggarwal, S., & Gupta, C. (2016). Solving intuitionistic fuzzy solid transportation problem via new ranking method based on signed distance. *International Journal of Uncertainty, Fuzziness and Knowledge-based Systems*, *24*(04), 483–501. DOI: 10.1142/S0218488516500240

Ahmad, F., & Adhami, A. Y. (2019). Total cost measures with probabilistic cost function under varying supply and demand in transportation problem. *Opsearch*, *56*(2), 583–602. DOI: 10.1007/s12597-019-00364-5

Ahmad, F., Smarandache, F., & Das, A. K. (2022). Neutrosophical fuzzy modeling and optimization approach for multiobjective four-index transportation problem.

Albassam, M., Khan, N., & Aslam, M. (2021). Neutrosophic D'Agostino Test of Normality: An Application to Water Data. *Journal of Mathematics*, *2021*, 1–5. DOI: 10.1155/2021/5582102

Alefeld, G., & Herzberger, J. (2012). *Introduction to interval computation*. Academic press.

Aneja, Y. P., & Nair, K. P. (1979). Bicriteria transportation problem. *Management Science*, *25*(1), 73–78. DOI: 10.1287/mnsc.25.1.73

Atanassov, K. T. (1986). Intuitionistic fuzzy sets. *Fuzzy Sets and Systems*, *20*(1), 87–96. DOI: 10.1016/S0165-0114(86)80034-3

Balachandran, V., & Perry, A. (1976). Transportation type problems with quantity discounts. *Naval Research Logistics Quarterly*, *23*(2), 195–209. DOI: 10.1002/nav.3800230203

Balinski, M. L. (1961). Fixed-cost transportation problems. *Naval Research Logistics Quarterly*, *8*(1), 41–54. DOI: 10.1002/nav.3800080104

Bera, T., & Mahapatra, N. K. (2022). Neutrosophy-based transportation problem and its solution approach. *International Journal of Mathematics in Operational Research, 22*(2), 252–281. DOI: 10.1504/IJMOR.2022.124041

Bharati, S. K. (2021). Transportation problem with interval-valued intuitionistic fuzzy sets: Impact of a new ranking. *Progress in Artificial Intelligence, 10*(2), 129–145. DOI: 10.1007/s13748-020-00228-w

Bhatia, H. L. (1981). Indefinite quadratic solid transportation problem. *Journal of Information and Optimization Sciences, 2*(3), 297–303. DOI: 10.1080/02522667.1981.10698711

Chanas, S., & Kuchta, D. (1996). A concept of the optimal solution of the transportation problem with fuzzy cost coefficients. *Fuzzy Sets and Systems, 82*(3), 299–305. DOI: 10.1016/0165-0114(95)00278-2

Charnes, A., & Cooper, W. W. (1957). Management models and industrial applications of linear programming. *Management Science, 4*(1), 38–91. DOI: 10.1287/mnsc.4.1.38

Chhibber, D., Bisht, D. C., & Srivastava, P. K. (2021). Pareto-optimal solution for fixed-charge solid transportation problem under intuitionistic fuzzy environment. *Applied Soft Computing, 107*, 107368. DOI: 10.1016/j.asoc.2021.107368

Chinneck, J. W., & Ramadan, K. (2000). Linear programming with interval coefficients. *The Journal of the Operational Research Society, 51*(2), 209–220. DOI: 10.1057/palgrave.jors.2600891

Choudhary, A., & Yadav, S. P. (2022). An approach to solve interval valued intuitionistic fuzzy transportation problem of Type-2. *International Journal of System Assurance Engineering and Management, 13*(6), 2992–3001. DOI: 10.1007/s13198-022-01771-6

Das, S. K., Goswami, A., & Alam, S. S. (1999). Multiobjective transportation problem with interval cost, source and destination parameters. *European Journal of Operational Research, 117*(1), 100–112. DOI: 10.1016/S0377-2217(98)00044-7

Ebrahimnejad, A., & Verdegay, J. L. (2018). A new approach for solving fully intuitionistic fuzzy transportation problems. *Fuzzy Optimization and Decision Making, 17*(4), 447–474. DOI: 10.1007/s10700-017-9280-1

El Sayed, M. A., & Baky, I. A. (2023). Multi-choice fractional stochastic multi-objective transportation problem. *Soft Computing, 27*(16), 11551–11567. DOI: 10.1007/s00500-023-08101-3

Gaber, M., Alharbi, M. G., Dagestani, A. A., & Ammar, E. S. (2021). Optimal Solutions for Constrained Bimatrix Games with Payoffs Represented by Single-Valued Trapezoidal Neutrosophic Numbers. *Journal of Mathematics, 2021*, 1–13. DOI: 10.1155/2021/5594623

Garfinkel, R. S., & Rao, M. R. (1971). The bottleneck transportation problem. *Naval Research Logistics Quarterly, 18*(4), 465–472. DOI: 10.1002/nav.3800180404

Ghosh, S., Roy, S. K., & Verdegay, J. L. (2022). Fixed-charge solid transportation problem with budget constraints based on carbon emission in neutrosophic environment. *Soft Computing, 26*(21), 11611–11625. DOI: 10.1007/s00500-022-07442-9

Gray, P. (1971). Exact solution of the fixed-charge transportation problem. *Operations Research, 19*(6), 1529–1538. DOI: 10.1287/opre.19.6.1529

Gupta, G., Shivani, & Rani, D. (2024). Neutrosophic goal programming approach for multi-objective fixed-charge transportation problem with neutrosophic parameters. *OPSEARCH*, 1-27. DOI: 10.1007/s12597-024-00747-3

Habiba, U., & Quddoos, A. (2022). Pentagonal Neutrosophic Transportation Problems with Interval Cost. *Neutrosophic Sets and Systems, 51*(i), 896–90. DOI: 10.5281/zenodo.7135436

Hammer, P. L. (1969). Time-minimizing transportation problems. *Naval Research Logistics Quarterly, 16*(3), 345–357. DOI: 10.1002/nav.3800160307

Hammer, P. L. (1971). Communication on "the bottleneck transportation problem" and "some remarks on the time transportation problem". *Naval Research Logistics Quarterly, 18*(4), 487–490. DOI: 10.1002/nav.3800180406

Hitchcock, F. L. (1941). The distribution of a product from several sources to numerous localities. *Journal of Mathematics and Physics, 20*(1-4), 224–230. DOI: 10.1002/sapm1941201224

Ishibuchi, H., & Tanaka, H. (1990). Multiobjective programming in optimization of the interval objective function. *European Journal of Operational Research, 48*(2), 219–225. DOI: 10.1016/0377-2217(90)90375-L

Jansi Rani, J., Dhanasekar, S., Micheal, D. R., & Manivannan, A. (2023). On solving fully intuitionistic fuzzy transportation problem via branch and bound technique. *Journal of Intelligent & Fuzzy Systems, 44*(4), 6219–6229. DOI: 10.3233/JIFS-221345

Kar, C., Samim Aktar, M., Maiti, M., & Das, P. (2023). Solving Fully Neutrosophic Incompatible Multi-Item Fixed Charge Four-Dimensional Transportation Problem with Volume Constraints. *New Mathematics and Natural Computation*, 1-29. DOI: 10.1142/S1793005724500054

Kaur, A., & Kumar, A. (2011). A new method for solving fuzzy transportation problems using ranking function. *Applied Mathematical Modelling*, *35*(12), 5652–5661. DOI: 10.1016/j.apm.2011.05.012

Khalifa, H. A. E. W., Broumi, S., Edalatpanah, S. A., & Alburaikan, A. (2023). A novel approach for solving neutrosophic fractional transportation problem with non-linear discounting cost. *Neutrosophic Sets and Systems*, *61*(1), 10.

Khandelwal, A., & Kumar, A. (2024). A modified method for solving the unbalanced TP. *International Journal of Operations Research*, *49*(1), 1–18. DOI: 10.1504/IJOR.2024.136005

Klingman, D., & Russell, R. (1975). Solving constrained transportation problems. *Operations Research*, *23*(1), 91–106. DOI: 10.1287/opre.23.1.91

Koopmans, T. C. (1949). Optimum utilization of the transportation system. *Econometrica*, *17*, 136–146. DOI: 10.2307/1907301

Kour, D., & Basu, K. (2015). Application of extended fuzzy programming technique to a real-life transportation problem in neutrosophic environment. *Neutrosophic Sets and Systems*, *10*, 74–86.

Kumar, A., Chopra, R., & Saxena, R. R. (2021). An Efficient Enumeration Technique for a Transportation Problem in Neutrosophic Environment. *Neutrosophic Sets and Systems*, *47*, 354–365.

Lee, S. M., & Moore, L. J. (1973). Optimizing transportation problems with multiple objectives. *AIIE Transactions*, *5*(4), 333–338. DOI: 10.1080/05695557308974920

Liu, S. T. (2016). Fractional transportation problem with fuzzy parameters. *Soft Computing*, *20*(9), 3629–3636. DOI: 10.1007/s00500-015-1722-5

Maiti, I., Mandal, T., & Pramanik, S. (2020). Neutrosophic goal programming strategy for multi-level multi-objective linear programming problem. *Journal of Ambient Intelligence and Humanized Computing*, *11*(8), 3175–3186. DOI: 10.1007/s12652-019-01482-0

Moore, R. E. (1979). *Method and Applications of Interval Analysis*. SLAM. DOI: 10.1137/1.9781611970906

ÓhÉigeartaigh, M. (1982). A fuzzy transportation algorithm. *Fuzzy Sets and Systems*, *8*(3), 235–243. DOI: 10.1016/S0165-0114(82)80002-X

Pratihar, J., Kumar, R., Edalatpanah, S. A., & Dey, A. (2021). Modified Vogel's approximation method for transportation problem under uncertain environment. *Complex & Intelligent Systems*, *7*(1), 29–40. DOI: 10.1007/s40747-020-00153-4

Qiuping, N., Yuanxiang, T., Broumi, S., & Uluçay, V. (2023). A parametric neutrosophic model for the solid transportation problem. *Management Decision*, *61*(2), 421–442. DOI: 10.1108/MD-05-2022-0660

Radhika, K., & Arun Prakash, K. (2022). Multi-objective optimization for multi-type transportation problem in intuitionistic fuzzy environment. *Journal of Intelligent & Fuzzy Systems*, *43*(1), 1439–1452. DOI: 10.3233/JIFS-213517

Rani, J. J., & Manivannan, A. (2024). An application of generalized symmetric type-2 intuitionistic fuzzy variables to a transportation problem with the effect of a new ranking function. *Expert Systems with Applications*, *237*, 121384. DOI: 10.1016/j.eswa.2023.121384

Rizk-Allah, R. M., Hassanien, A. E., & Elhoseny, M. (2018). A multi-objective transportation model under neutrosophic environment. *Computers & Electrical Engineering*, *69*, 705–719. DOI: 10.1016/j.compeleceng.2018.02.024

Samanta, S., Chakraborty, D., & Jana, D. K. (2024). Neutrosophic multi-period two stage four-dimensional transportation problem for breakable items. *Expert Systems with Applications*, *246*, 123266. DOI: 10.1016/j.eswa.2024.123266

Shaocheng, T. (1994). Interval number and fuzzy number linear programming. *Fuzzy Sets and Systems*, *66*(3), 301–306. DOI: 10.1016/0165-0114(94)90097-3

Sharma, J., & Swarup, K. (1977). Indefinite quadratic programming and transportation technique.

Simić, V., Milovanović, B., Pantelić, S., Pamučar, D., & Tirkolaee, E. B. (2023). Sustainable route selection of petroleum transportation using a type-2 neutrosophic number based ITARA-EDAS model. *Information Sciences*, *622*, 732–754. DOI: 10.1016/j.ins.2022.11.105

Singh, A., Arora, R., & Arora, S. (2022). Bilevel transportation problem in neutrosophic environment. *Computational & Applied Mathematics*, *41*(1), 1–25. DOI: 10.1007/s40314-021-01711-3

Singh, A., Arora, R., & Arora, S. (2023). A Novel Fully Interval-Valued Intuitionistic Fuzzy Multi-objective Indefinite Quadratic Transportation Problem with an Application to Cost and Wastage Management in the Food Industry. In *Fuzzy Optimization, Decision-making and Operations Research: Theory and Applications* (pp. 87–110). Springer International Publishing. DOI: 10.1007/978-3-031-35668-1_5

Singh, A., Kumar, A., & Appadoo, S. S. (2017). Modified approach for optimization of real life transportation problem in neutrosophic environment. *Mathematical Problems in Engineering, 2017*, 1–9. Advance online publication. DOI: 10.1155/2017/2139791

Smarandache, F. (1998). *A Unifying Field in Logics. Neutrosophy: Neutrosophic Probability. Set and Logic*. American Research Press.

Smarandache, F. (2013). *Introduction to neutrosophic measure, neutrosophic integral, and neutrosophic probability*. Sitech and Education Publisher.

Smarandache, F. (2014). *Introduction to neutrosophic statistics*. Sitech and Education Publishing.

Stancu-Minasian, I. M. (1997). Fractional transportation problem. In *Fractional Programming: Theory, Methods and Applications* (pp. 336–364). Springer Netherlands. DOI: 10.1007/978-94-009-0035-6_11

Tapia, J. F. D. (2021). Evaluating negative emissions technologies using neutrosophic data envelopment analysis. *Journal of Cleaner Production, 286*, 125494. DOI: 10.1016/j.jclepro.2020.125494

Tu, A., Ye, J., & Wang, B. (2021). Neutrosophic Number Optimization Models and Their Application in the Practical Production Process. *Journal of Mathematics, 2021*, 1–8. DOI: 10.1155/2021/6668711

Veeramani, C., Edalatpanah, S. A., & Sharanya, S. (2021). Solving the multi-objective fractional transportation problem through the neutrosophic goal programming approach. *Discrete Dynamics in Nature and Society, 2021*, 1–17. DOI: 10.1155/2021/7308042

Verma, R., Biswal, M. P., & Biswas, A. (1997). Fuzzy programming technique to solve multi-objective transportation problems with some non-linear membership functions. *Fuzzy Sets and Systems, 91*(1), 37–43. DOI: 10.1016/S0165-0114(96)00148-0

Wang, H., Smarandache, F., Zhang, Y., & Sunderraman, R. (2010). Single valued neutrosophic sets. *Infinite study, 12*, 20110.

Wang, Q., Huang, Y., Kong, S., Ma, X., Liu, Y., Das, S. K., & Edalatpanah, S. A. (2021). A Novel Method for Solving Multiobjective Linear Programming Problems with Triangular Neutrosophic Numbers. *Journal of Mathematics, 2021*, 1–8. DOI: 10.1155/2021/6631762

Williams, A. C. (1963). A stochastic transportation problem. *Operations Research, 11*(5), 759–770. DOI: 10.1287/opre.11.5.759

Xie, F., Butt, M. M., Li, Z., & Zhu, L. (2017). An upper bound on the minimal total cost of the transportation problem with varying demands and supplies. *Omega, 68*, 105–118. DOI: 10.1016/j.omega.2016.06.007

Yang, L., & Liu, L. (2007). Fuzzy fixed charge solid transportation problem and algorithm. *Applied Soft Computing, 7*(3), 879–889. DOI: 10.1016/j.asoc.2005.11.011

Ye, J. (2016). Multiple-attribute group decision-making method under a neutrosophic number environment. *Journal of Intelligent Systems, 25*(3), 377–386. DOI: 10.1515/jisys-2014-0149

Zadeh, L. A. (1965). Fuzzy sets. *Information and Control, 8*(3), 338–353. DOI: 10.1016/S0019-9958(65)90241-X

APPENDIX

Preliminary Definitions

Definition 1 Fuzzy set (Zadeh, 1965) Let X be the universe of discourse. A fuzzy set in X is a set of two elements defined as where is the membership function such that

Definition 2 Intuitionistic fuzzy set (Atanassov, 1986) Let X be the universe of discourse. An IFS in X is a set of three elements defined as where , are the membership and non-membership functions such that

Definition 3 Neutrosophic Set (Smarandache, 1998) Let X be the universe of discourse. A neutrosophic set in X is a set of four elements defined as

$$T'= \{\langle x, \lambda_{T''}(x), \mu_{T''}(x), \psi_{T''}(x)\rangle : x \in X\}$$

where $\lambda_{T''}(x), \mu_{T''}(x), \psi_{T''}(x) \subseteq]^{-}0, 1^{+}[$are the membership and non-membership functions such that

$$^{-}0 \leq \inf\lambda_{T''}(X) + \inf\mu_{T''}(X) + \inf\psi_{T''} \leq \sup\lambda_{T''}(X) + \sup\mu_{T''}(X) + \sup\psi_{T''} \leq 3^{+}\forall x \in X.$$

Definition 4 Single-valued neutrosophic sets (Wang et al., 2010) A single-valued neutrosophic set over X is defined as the set where , such that + .

Definition 5 Neutrosophic Number (Smarandache 1998, 2013, 2014) A NN is of the form with determinate part b and indeterminate part . Here, and denotes indeterminacy.

Definition 6 Interval number (Moore, 1979; Alefeld & Herzberger, 1983) An interval number on R is denoted as where and are the left and right limit of the interval number A on R.

Remark 1 For any interval , Centre of and Half-width of It can be observed that .

Definition 7 Ranking of intervals (Ishibuchi and Tanaka, 1990; Das et al., 1999): Let and be two given intervals. Then, the order relation is defined as

Example 1: For and if then B becomes an interval number of the form [$7 + 3.0, 7 + 3.(0.15)] = [7, 7.45]$. For sure, $B \geq 7$ which means a possibility for number B to be a little bigger than 7.

Operational Relations of NNs

Let, $B_1 = u + u'I$ and $B_2 = v + v'I$ be two NNs. Then, their arithmetic operations are given by

(1) $B_1 + B_2 = u + v + (u' + v')I$;
(2) $B_1 - B_2 = u - v + (u' - v')I$;
(3) $B_1 \times B_2 = uv + (uv' + vu' + u'v')I$. In particular, when B_1=0 and B_2=I, we get the equation $0\times I=0$;
(4) $B_1^2 = (u + u'I)^2 = u^2 + (2uu' + u'^2)I$. In particular, when B_1=I, we get $I^2=I$.

(5) $\frac{B_1}{B_2} = \frac{u + u'I}{v + v'I} = \frac{u}{v} - \frac{vu' - uv'}{v(v + v')} I$ for $v\neq 0$ and $v \neq -v'$.

Example 2: Let us take two NNs, B_1=4+5I and B_2=3+2I. Then, the following values can be calculated using the above operational laws.

(1) B_1+B_2=7+7I;
(2) B_1–B_2=1+3I;
(3) B_1×B_2=12+33I;
(4) $B_1^2 = (4 + 5I)^2 = 16 + 65I$;

(5) $\frac{B_1}{B_2} = \frac{4 + 5I}{3 + 2I} = \frac{4}{3} - \frac{7}{15} I = 1.33 + 0.46I$.

Chapter 11
Neutrosophic Inventory Model With Coordinated Rework Stations and Distribution Centers With Ant Colony Optimization

M. Renee Miriam
Madurai Kamaraj University, India

Nivetha Martin
Arul Anandar College (Autonomous), Karumathur, India

Akbar Rezaei
Payame Noor University, Iran

Seyyed Ahmad Edalatpanah
https://orcid.org/0000-0001-9349-5695
Ayandegan Institute of Higher Education, Iran

ABSTRACT

In the realm of supply chain management, achieving optimal inventory control and efficient rework processes within distribution centers are critical for maintaining competitiveness and customer satisfaction. This chapter introduces an approach that integrates a neutrosophic inventory model (NIM) with coordinated rework stations and distribution centers, bolstered by ant colony optimization (ACO). Coordinated rework stations within distribution centers aim to minimize cycle times and improve product quality. ACO, inspired by the foraging behavior of real ants, is employed to

DOI: 10.4018/979-8-3693-3204-7.ch011

adaptively optimize the dynamic and uncertain nature of supply chain and rework processes. Simulation studies validate the effectiveness of the proposed model, showcasing its ability to improve overall supply chain performance, reduce costs, and enhance resilience to uncertainties. This research contributes to advancing supply chain methodologies by offering a holistic approach to inventory and rework optimization through neutrosophic logic and swarm intelligence.

1. INTRODUCTION

In the dynamic landscape of supply chain management, the effective coordination of inventory systems, rework stations, and distribution centers plays a pivotal role in optimizing operational efficiency and minimizing costs. Traditional inventory models often struggle to adapt to the uncertainties and complexities inherent in modern supply chains. To address these challenges, the integration of neutrosophic logic—a mathematical framework for handling indeterminacy, vagueness, and inconsistency—provides a promising avenue for modeling and decision-making in uncertain environments. This research focuses on the development and application of a Neutrosophic Inventory Model that goes beyond conventional approaches by incorporating coordinated rework stations and distribution centers. Rework stations are strategically positioned within the supply chain to address defects and imperfections in products, while distribution centers play a crucial role in the timely and efficient delivery of goods. The synergy between these elements is harnessed through the implementation of Ant Colony Optimization (ACO), a nature-inspired algorithm that mimics the foraging behavior of ants to find optimal solutions in complex problem spaces.

Coordinated rework stations play a pivotal role in enhancing the overall quality and reliability of products within a supply chain. In traditional inventory models, defects and imperfections in products are often treated as inevitable disruptions, leading to increased costs and potential customer dissatisfaction. The introduction of coordinated rework stations represents a paradigm shift by strategically placing checkpoints within the supply chain to address and rectify defects promptly. This proactive approach not only minimizes the negative impact of defective items on inventory levels but also contributes to a significant reduction in waste and rework costs. By integrating rework stations into the inventory model, companies can ensure that defective products are identified and corrected efficiently, leading to improved product quality and customer satisfaction.

Distribution centers serve as the linchpin in the supply chain, playing a crucial role in the timely and efficient movement of goods from manufacturers to end consumers. The strategic placement and effective management of distribution centers contribute to streamlined logistics, reduced lead times, and overall cost optimization. These centers act as hubs for consolidating and redistributing products, allowing for economies of scale in transportation and warehousing. Additionally, distribution centers enhance responsiveness to fluctuating demand patterns by strategically positioning inventory closer to the end-users. The significance of distribution centers extends beyond mere storage and transportation; they serve as key nodes in the supply chain network, facilitating agility and adaptability.

The neutrosophic framework allows the representation of uncertainties associated with demand fluctuations, lead times, and other variables, providing a more realistic and robust model for decision-makers. By introducing coordinated rework stations into the inventory model, the research aims to enhance the overall quality of products within the supply chain, mitigating the impact of defects on inventory levels and customer satisfaction. Furthermore, the utilization of Ant Colony Optimization contributes to the optimization of routing and scheduling decisions in distribution networks. The adaptive and decentralized nature of ACO enables the system to dynamically respond to changes in demand patterns and supply chain disruptions, thereby improving the overall responsiveness of the system. By leveraging advanced technologies and optimization algorithms, such as Ant Colony Optimization, distribution centers can dynamically adjust their operations to meet evolving market demands, ensuring a resilient and responsive supply chain.

This research work, combining neutrosophic logic, coordinated rework stations, and Ant Colony Optimization, seeks to advance the state-of-the-art in supply chain optimization and also contribute significantly to the improvement of product quality, cost reduction, and overall supply chain efficiency. The findings of this study are anticipated to offer valuable insights into designing resilient and adaptive inventory systems capable of navigating the intricacies of contemporary supply chains by reflecting a holistic approach to address the complexities and uncertainties inherent in contemporary supply chains.

The remaining contents of the paper are organized as follows, section 2 presents the literature of Neutrosophic supply chain models, section 3 describes the framework of the proposed model, section 4 validates the model with a simulated numerical example and discusses the model under different cases, section 5 comprises the inferences and observations obtained from the model, section 6 sketches the industrial applications of the model and the final section concludes the work.

2. LITERATURE REVIEW

This section presents the state of art of development of supply chain inventory models in Neutrosophic environment. The existing research gaps and significant contribution of this research work are also presented in this section. Ahmad et al presented a modified neutrosophic fuzzy optimization model for closed-loop supply chain management under uncertainty (Ahmad et al., 2020). Fallah and Nozari proposed a neutrosophic mathematical programming model for the optimization of multi-objective sustainable biomass supply chain network design (Fallah & Nozari, 2021). Szmelter-Jarosz et al expanded on this foundation by proposing a neutrosophic fuzzy optimization model tailored for optimal sustainable closed-loop supply chain networks, particularly in the context of the challenges posed by the COVID-19 pandemic (Szmelter-Jarosz et al., 2021). Nabeeh et al introduced a neutrosophic evaluation model for blockchain technology in supply chain management (Nabeeh et al., 2022). Ahmad introduced an interactive neutrosophic optimization technique for multi-objective programming problems, specifically applied to pharmaceutical supply chain management (Ahmad, 2022).

Aytekin et al applied a neutrosophic framework to evaluate factors influencing the performance and sustainability of supply chains in the textile industry (Aytekin et al., 2022). Ismail et al contributed to the literature with a neutrosophic approach for multi-factor analysis, focusing on uncertainty and sustainability in supply chain performance (Ismail et al., 2023). Khan et al extended the application of neutrosophic theory to risk assessment in the halal supply chain under uncertain conditions (Khan et al., 2023). Dohale et al analyzed enablers of circular supply chains in the Indian apparel industry, employing the Neutrosophic-ISM method (Dohale et al., 2023). Abdel Mouty et al explored the role of the Internet of Things and Industry 4.0 in healthcare supply chains using neutrosophic sets (AbdelMouty et al., 2023). Mohamed et al employed a neutrosophic model to examine challenges faced by manufacturing businesses in adopting green supply chain practices (Mohamed et al., 2023). Farid and Riaz have contributed insights into single-valued neutrosophic dynamic aggregation information with time sequence preference, specifically for IoT technology in supply chain management (Farid & Riaz, 2023). Deb and Islam applied neutrosophic interval-valued goal programming to a supply chain inventory model for deteriorating items with time-dependent demand (Deb & Islam, 2023). Miriam et al contribute to the evolving landscape of inventory management and product distribution by introducing a rework warehouse inventory model within a neutrosophic environment (Miriam et al., 2023). This study contributes to the ongoing discourse on innovative inventory management practices, particularly in the context of quality preservation and distribution efficiency within a neutrosophic framework. The rework warehouse management model is developed to handle the defective items caused during the locomotion of items from one supply source to several destinations. The model addresses the defective items that are classified into minor and major. Based on this research work, the authors extend the model by classifying the defective items into minor, major and critical. This newly developed model will set stage for developing more generalized neutrosophic based supply chain model to handle defective items of all categories with the integration of rework stations and distribution centers.

Ant-colony optimization algorithm is one of the bio-inspired algorithms employed in model optimization. Yang, Li, and Tan proposed a research framework for solving the shortest path problem in interval valued neutrosophic graphs using the ant colony algorithm, showcasing the adaptability of swarm intelligence to handle uncertainty in graph structures (Yang et al., 2020). Building upon this, Broumi, Raut, and Behera extended the application of the ant colony algorithm to address shortest path problems with triangular neutrosophic arc weights, providing further insights into the robustness of the algorithm in handling more complex and nuanced graph representations (Broumi et al., 2023). In the realm of inventory management, Wang focused on the path optimization of Vendor-Managed Inventory (VMI) for large logistics enterprises, introducing the ant colony algorithm as a solution method (Wang, 2022). This indicates a growing interest in integrating nature-inspired algorithms to enhance logistics efficiency within inventory systems. Dong, Duan, and Li explored joint optimization and visualization of inventory transportation in agricultural logistics using the ant colony algorithm (Dong et al., 2022), showcasing its applicability in addressing specific challenges within agricultural supply chains. Oudouar and Zaoui presented a novel hybrid heuristic based on the ant colony algorithm for solving the multi-product inventory routing problem (Oudouar & Zaoui, 2021).

Based on the literature, the following research gaps are identified.

(i) There is a lack of exploration in neutrosophic-based supply chain models incorporating the concepts of rework stations and distribution centers.
(ii) The classification of defective items into three distinct categories has not been thoroughly discussed in the current research landscape.
(iii) The application of the ant colony optimization algorithm for optimizing inventory models within the context of supply chain management has not been extensively explored.

To bridge the research gaps a neutrosophic based supply chain model with costs associated with rework stations and distribution centers is developed and optimized using ant colony optimization. This is considered as one of the significant contributions of this research work.

3. MODEL DEVELOPMENT

This part develops an inventory model that incorporates warehouse management rework. The model aims to address defective items in the warehouse by identifying various defects, including minor, major, and critical defects. The minor and major defects are dealt with by reworking them in the warehouse, while the critical defects are handled at the service stations that are close to the warehouse zone.

3.1 Notations

z: The lot size acquired from the manufacturing facility.
k: Cost of production per unit
F_T: Fixed acquisition cost
D: The product's demand rate
P: Total percentage of reworkable objects.
P_{R_1}: The percentage of reworkable items in a lot that have minor flaws that can be remedied without the use of tooling.
P_{R_2}: The percentage of reworkable items in a lot that have serious flaws that can be fixed with tooling.
P_{R_3}: The percentage of reworkable items in a lot that have critical flaws and must be sent to the service station.
t_1: Inspection period
t_2: Time until reworked minor faulty goods are received.
t_3: Time until the reworked major faulty items are received.
t_4: Time until the reworked critical faulty items are received.
t_5: Time left to utilize the whole inventory.
Z_1: Inventory amount following the inspection time.
Z_2: Stock level after the division of reworkable objects into minor, major, and crucial categories
Z_3: Stock level immediately before obtaining the revised minor faulty items.
Z_4: Stock level immediately following the arrival of the modified minor faulty items.
Z_5: Stock level immediately before obtaining the major faulty goods that were reworked.
Z_6: Stock level immediately following the arrival of the major faulty items that were reworked.
Z_7: Stock level immediately before obtaining the reworked critical defective item.
Z_8: Stock level immediately upon the arrival of the reworked critical defective item.
S_P: The selling price per unit of superior goods

H_c: Holding cost per unit
r: Inspection rate
W_1: Minor faulty items rework rate
W_2: Major faulty item rework rate
W_3: Critical faulty item rework rate
T: Length of cycle

3.2 Assumptions

1) There are enough good items to meet demand during inspection period t_1 and until time t_2 is up.
2) The goods available during period t_3—aside from the major and critical items—are adequate to meet demand.
3) During the period t_4, the goods that are not the critical item are adequate to meet the demand.

3.3 Determination of Profit per Cycle

The total net revenue and total profit per cycle is determined by

$$t_1 = \frac{z}{r}.$$

$$t_2 = \frac{zP_{R_1}}{W_1}.$$

$$t_3 = \frac{zP_{R_2}}{W_2}.$$

$$t_4 = \frac{zP_{R_3}}{W_3}$$

$$t_5 = \frac{Y_8}{D}.$$

$$Z_1 = \left(1 - \frac{D}{r}\right)z.$$

$$Z_2 = \left(1 - P - \frac{D}{r}\right)z.$$

$$Z_3 = \left(1 - P - \frac{D}{r} - \frac{DP_{R_1}}{W_1}\right)z.$$

$$Z_4 = \left(1 - P_{R_2} - P_{R_3} - \frac{D}{r} - \frac{DP_{R_1}}{W_1}\right)z$$

$$Z_5 = \left(1 - P_{R_2} - P_{R_3} - \frac{D}{r} - \frac{DP_{R_1}}{W_1} - \frac{DP_{R_2}}{W_2}\right)z.$$

$$Z_6 = \left(1 - P_{R_3} - \frac{D}{r} - \frac{DP_{R_1}}{W_1} - \frac{DP_{R_2}}{W_2}\right)z.$$

$$Z_7 = \left(1 - P_{R_3} - \frac{D}{r} - \frac{DP_{R_1}}{W_1} - \frac{DP_{R_2}}{W_2} - \frac{DP_{R_3}}{W_3}\right)z.$$

$$Z_8 = \left(1 - \frac{D}{r} - \frac{DP_{R_1}}{W_1} - \frac{DP_{R_2}}{W_2} - \frac{DP_{R_3}}{W_3}\right)z.$$

Acquisition cost = $F_T + kz$
Appraisal Cost = A_C
Total Rework cost = $R_1 P_{R_1} z + R_2 P_{R_2} z + R_3 P_{R_3} z + C_S$
Packaging Cost = P_C
Green Disposal Cost = G_C

Transportation cost to the three Distribution Centers= *DC1 + DC2 + DC3*

Holding Cost =

$$\frac{z^2}{2D} - \frac{Pz^2P_{R_1}}{L_1} - \frac{P_{R_2}{}^2z^2}{L_2} - \frac{z^2P_{R_3}{}^2}{L_3} - \frac{z^2P_{R_2}P_{R_3}}{L_2}$$

$$T = t_1 + t_2 + t_3 + t_4$$

$$T = \frac{z}{D}$$

The Total Cost = Acquisition Cost + Appraisal Cost + Holding Cost + Total Rework Cost + Packaging Cost + Green Disposal Cost + Transportation cost to Distribution Centers

$$\text{Total Cost} = F_T + pz + A_c + H_c\left[\frac{z^2}{2D} - \frac{Pz^2P_{R_1}}{L_1} - \frac{P_{R_2}{}^2z^2}{L_2} - \frac{z^2P_{R_3}{}^2}{L_3} - \frac{z^2P_{R_2}P_{R_3}}{L_2}\right]$$
$$+R_1P_{R_1}z + R_2P_{R_2}z + R_3P_{R_3}z + C_S + P_C + G_C + DC1 + DC2 + DC3$$

$$\text{The Total Average Cost} = \frac{F_TD}{z} + pD + \frac{A_cD}{z} + R_1P_{R_1}D + R_2P_{R_2}D + R_3P_{R_3}D$$
$$+\frac{H_CD}{z} * \left[\frac{z^2}{2D} - \frac{Pz^2P_{R_1}}{L_1} - \frac{P_{R_2}{}^2z^2}{L_2} - \frac{z^2P_{R_3}{}^2}{L_3} - \frac{z^2P_{R_2}P_{R_3}}{L_2}\right] + \frac{C_SD}{z} + \frac{P_CD}{z} + \frac{G_CD}{z}$$
$$+ \frac{DC_1D}{z} + \frac{DC_2D}{z} + \frac{DC_3D}{z}$$

3.4 Problem Description

The lot is received at time $t = 0$. At time t_1, the inspection process is carried out. During this phase, the faulty materials are classified as minor, major, or critical. There are three different rework rates for faulty items: W_1 for minor defects, W_2 for severe defects, and W_3 for critical defects.

The rework of the defective items begins at time t_1. At the end of time t_2, the minor defective items are totally reworked and added to the inventory. At the conclusion of time t_3, the major defective items are totally reworked and added to the inventory. At the conclusion of time t_4, the critical defective items are totally reworked and added to the inventory. During the time t_5, the whole stockpile is cleared out. T denotes the total length of the cycle.

The inventory level is Z_1 at the conclusion of the inspection period. Z_2 is the inventory level following the removal of significant, minor, and critical defective items. Z_3 is the inventory level that existed right before the revised minor defective items were added. Z_4 is the inventory level following the addition of revised items with minor flaws. Z_5 is the inventory level right before major defective items that have been reworked are added. Z_6 is the inventory level following the addition of revised items with serious flaws. Z_7 is the inventory level that existed immediately prior to the addition of reworked critical defective items. Z_8 is the inventory level following the addition of revised items with critical flaws. Figure 1 represents the overall system.

Figure 1. Representation of the System

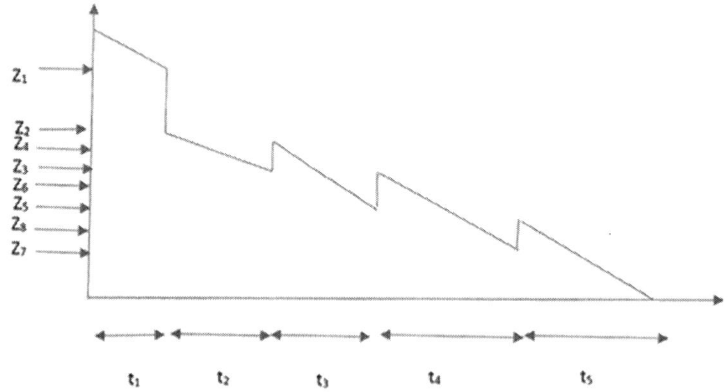

4. MODEL VALIDATION

The developed model is optimized using Ant Colony Optimization and then in Neutrosophic model implemented using MATLAB Software.

4.1 Ant Colony Optimization

The term "ant colony optimization" refers to a collection of optimization approaches based on how the colonies of ants might function. Artificial 'ants', such as simulation agents, discover optimal solutions by navigating a parameter space that represents all potential solutions. Real ants leave behind pheromones that lead one another to resources as they explore their environment. In a similar fashion, the simulated 'ants' record their locations as well as the degree of quality of their solutions, allowing more ants to discover better answers in later simulation rounds.

The steps involving in finding the optimal solution using the ant colony optimization are:

1) Establishing the goal function.
    - Define the objective function that needs to be minimized or maximized.
    - Minimize or Maximize $f(x)$ subject to constraints, where x is the solution vector.
2) Setting the initial pheromone levels.
    - Initialize the pheromone levels on the edges(solution) in the solution space.
    - Mathematical expression:

$$\tau_{ij}^{(0)} = \text{constant},$$

where $\tau_{ij}^{(0)}$ is the initial pheromone level on edge $i\,j$ at iteration 0.

3) Setting the values to zero.
    - Initialize values related to the objective function and ant solutions.
    - Mathematical solutions:

$f_{best}=\infty$ (or $-\infty$ for minimization or maximization, respectively)
Ant solutions = 0

4) Analyzing the Ant solutions' objective function.

- Evaluating the objective function for each ant's solution.
- Mathematical expression:

$f(\text{Ant}_i)$ for each ant solution Ant_i

5) Adjusting the pheromone levels.
   - Update pheromone levels based on the quality of solutions.
   - Mathematical expression:

$$\Delta\tau_{ij} = \frac{1}{f(Ant_i)}$$

$$\tau_{ij}^{(k+1)} = (1-\rho) \bullet \tau_{ij}^{(k)} + \rho \bullet \Delta_{ij}$$

Here, ρ is the pheromone evaporation rate.

6) Pheromone deposit on Optimal Solution:
   - Apply additional pheromone deposit on the edge corresponding to the best solution found.
   - Mathematical expression:

$$\Delta\tau_{best} = \frac{Q}{f_{best}},$$

where Q is a constant representing the amount of pheromone deposited.

7) Pheromone Decay:
   - Update pheromone levels with decay to stimulate the natural decay of pheromones over time.
   - Mathematical expression:

$$\tau_{ij}^{(k+1)} = \left(1-\rho_d\right) \bullet \tau_{ij}^{(k)},$$

where ρ_d is the pheromone decay rate.

8) Optimal Solution Determination:
   - After several iterations, determine the optimal solution based on the accumulated pheromone levels.
   - Mathematical expression:

Optimal Solution = arg min $f(x)$

4.2 Fundamentals of Neutrosophic Sets

4.2.1 Neutrosophic Sets

A truth-membership function, an indeterminacy-membership function, and a falsity-membership function, each of which is defined from $X \rightarrow [0,1]$, each separately characterise a neutrosophic set.

4.2.2 Arithmetic Operations of Neutrosophic Sets

Let $X = \ < (u_1, v_1, r_1, s_1): \ \tau_A, \ \epsilon_A, \ \mu_A >$ and $Y = \ < (u_2, v_2, r_2, s_2): \ \tau_B, \ \epsilon_B, \ \mu_B >$ be two single valued neutrosophic numbers and σ not equal to 0, then

1) $X + Y = \ < u_1 + u_2, \ v_1 + v_2, r_1 + r_2, s_1 + s_2): \tau_A \wedge \tau_B, \mu_A \vee \mu_B, \varepsilon_A \vee \varepsilon_B >$
2) $X - Y = \ < (u_1 - s_2, v_1 - r_2, \ r_1 - v_2, s_1 - u_2): \tau_A \wedge \tau_B, \mu_A \vee \mu_B, \varepsilon_A \vee \varepsilon_B >$
3) $XY = \ < \left(u_1 u_2, \ v_1 v_2, r_1 r_2, s_1 s_2\right): \tau_A \wedge \tau_B, \mu_A \vee \mu_B, \varepsilon_A \vee \varepsilon_B > \left(s_1 > 0, s_2 > 0\right)$

$= \ < \left(u_1 s_2, \ v_1 r_2, r_1 v_2, s_1 v_2\right): \tau_A \wedge \tau_B, \mu_A \vee \mu_B, \varepsilon_A \vee \varepsilon_B > \left(s_1 < 0, s_2 > 0\right)$

$= \ < s_1 s_2, r_1 r_2, v_1 v_2, u_1 u_2: \tau_A \wedge \tau_B, \mu_A \vee \mu_B, \varepsilon_A \vee \varepsilon_B > \left(s_1 < 0, s_2 < 0\right)$

4) $X/Y = \ < \left(u_1/s_2, \ v_1/r_2, r_1/v_2, s_1/v_2\right): \tau_A \wedge \tau_B, \mu_A \vee \mu_B, \varepsilon_A \vee \varepsilon_B > \left(s_1 > 0, s_2 > 0\right)$

$= \ < s_1/s_2, r_{1/} r_2, v_1/v_2, u_1/u_2: \tau_A \wedge \tau_B, \mu_A \vee \mu_B, \varepsilon_A \vee \varepsilon_B > \left(s_1 < 0, s_2 > 0\right)$

$= \ < \left(s_2/u_1, \ r_2/v_1, v_2/r_1, v_2/s_1\right): \tau_A \wedge \tau_B, \mu_A \vee \mu_B, \varepsilon_A \vee \varepsilon_B > \left(s_1 < 0, s_2 < 0\right)$

5) $\sigma X = \ < \left(\sigma u_1,\ \sigma u_2, \sigma u_3,\ \sigma u_4\right): \tau_A,\ \mu_A,\ \epsilon_A \ > \quad (\sigma > 0)$

$= \ < \left(\sigma s_1,\ \sigma r_1,\ \sigma v_1, \sigma u_1\right): \tau_A,\ \mu_A,\ \epsilon_A \ > \quad (\sigma < 0)$

6) $X^{-1} = \ < \left(\frac{1}{\delta_1}, \frac{1}{\gamma_1}, \frac{1}{\beta_1}, \frac{1}{u_1}\right): \tau_B,\ \mu_B,\ \epsilon_B \ > X \neq 0$

4.2.3 Defuzzification of Neutrosophic Set

The respective score value $A(X)$ by defuzzifying a single value trapezoidal neutrosophic number $X \ = \ < (u, v, r, s): \tau, \mu, \epsilon \ >$ is given by $A(X) \ = \ \frac{1}{12}\left[u_1 + v_1 + r_1 + s_1\right] \times (2 + \tau - \mu - \epsilon)$

4.3 Numerical Example (Crisp)

Consider the following scenario: Manufactured products are damaged during transportation due to sudden climate disruption and inappropriate packing. The items cannot be returned because the production center is located so far away, yet the product must be provided due to high demand. The products have arrived in the rework warehouse. The model parameters with crisp values are listed below in Table 1 and Table 2.

Table 1. Model parameters with crisp values

F_T	D	k	R_1	P_{R_1}	R_2	P_{R_2}	R_3	P_{R_3}	A_C	H_C	W_1	W_2
1000	6000	250	10	0.05	20	0.02	20	0.03	10	5	20000	5000

Table 2. Model parameters with crisp values

W_3	P_C	P	S_P	$DC1$	$DC2$	$DC3$	C_S	G_C	r
1000	20	0.07	300	20	40	60	10	40	200000

The concave nature of the Total Average Cost is represented in Figure 2 with the help of the model parameters.

Figure 2. Concave nature of the total average cost

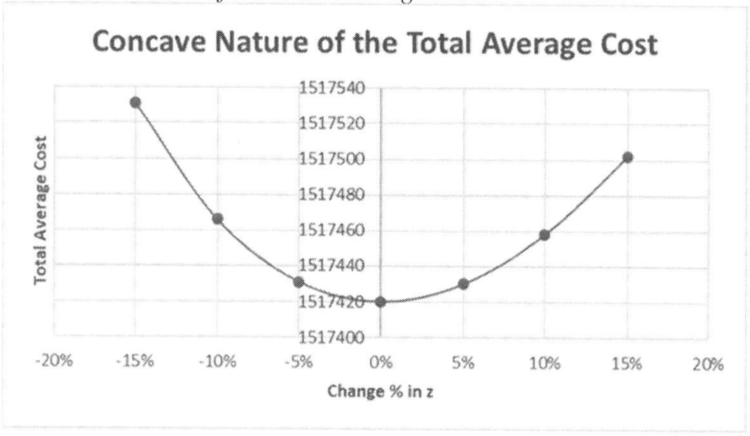

Solution Obtained using Ant Colony Optimization is
Best Solution (z): 1710.19
Total Profit: 2.8258e+05
Objective Function Value: 1517420

4.4 Numerical Example (Neutrosophic)

Let us consider the same example as mentioned in Section 4.1 with neutrosophic parametric values.

$F_T = (1100,1150,1200,1250)$: 0.9 0.3 0.2; $F_T = 940$

$D = (9200, 9500, 9800, 10100)$; 0.8 0.3 0.2; $D = 7398.333$

$k = (283.433,300,314,321)$; 0.7 0.3 0.4; $k = 203.0722$

$R_1 = (10.5,11,11.5,12)$; 0.5 0.2 0.3; $R_1 = 7.5$

P_{R_1}

$= (0.0475,0.05,0.0525,0.055)$; 0.6 0.3 0.1;

P_{R_1}

$= 0.03758$

$R_2 = (21,22,23,24); 0.8\ 0.1\ 0.3; R_2 = 19.6$

P_{R_2}

$= (0.019,0.02,0.021,0.022); 0.7\ 0.2\ 0.3;$

P_{R_2}

$= 0.01503$

$R_3 = (19,20,21,22); 0.9\ 0.1\ 0.3; R_3 = 17.0833$

P_{R_3}

$= (0.0285,0.03,0.0315,0.033); 0.8\ 0.4\ 0.1;$

P_{R_3}

$= 0.023575$

$P_C = (20,21,22,23); 0.8\ 0.2\ 0.3; P_C = 16.4833$

$G_C = (38,40,42,44); 0.8\ 0.1\ 0.1; G_C = 35.533$

$H_C = (4.75,5,5.25,5.5); 0.8\ 0.2\ 0.4; H_C = 3.75833$

$A_C = (9.5,10,10.5,11); 0.7\ 0.2\ 0.3; I = 7.5166$

$P = (0.0665,0.07,0.0735,0.077); 0.7\ 0.2\ 0.2; P = 0.0550$

$Sp = (285,300,315,330); 0.8\ 0.1\ 0.1; Sp = 266.5$

$DC1 = (19,20,21,22); 0.8\ 0.1\ 0.2; DC1 = 17.0833$

$DC2 = (38,40,42,44); 0.7\ 0.1\ 0.2; DC2 = 32.8$

$DC3 = (57,60,63,66); 0.6\ 0.3\ 0.2; DC3 = 43.05$

$C_S = (9.5,10,10.5,11); 0.8\ 0.1\ 0.2; C_S = 8.5410$

Solution Obtained using Neutrosophic Optimization is $z = 2094.038$
Total Average Cost = 1517420
Total Profit = 454235.9

5. DISCUSSIONS

5.1 Sensitivity Analysis

Sensitivity of the parameters is determined. The change in the Optimal lot size, Optimal Total Average Cost and Optimal Total Profit with respect to different parameters $DC1$, $DC2$, $DC3$, F_T, H_C, k, G_C, C_S, P_C, D and S_P for crisp and neutrosophic values is calculated.

1) Analysis is performed by varying the variable $DC1$ by changing 5%, 10% and 15% of its value increasingly and decreasingly.

The change in Total Profit with respect to the variable $DC1$(Crisp) is depicted in Figure 3.

Figure 3. Change in total profit with respect to DC1 (crisp)

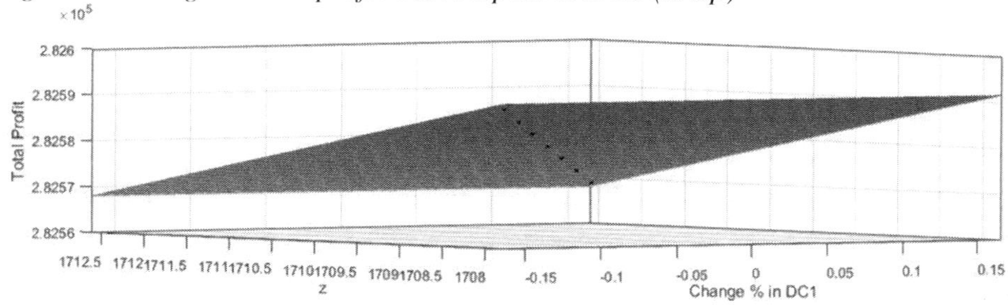

The change in Total Profit with respect to the variable $DC1$(Neutrosophic) is depicted in Figure 4.

Figure 4. Change in total profit with respect to DC1 (neutrosophic)

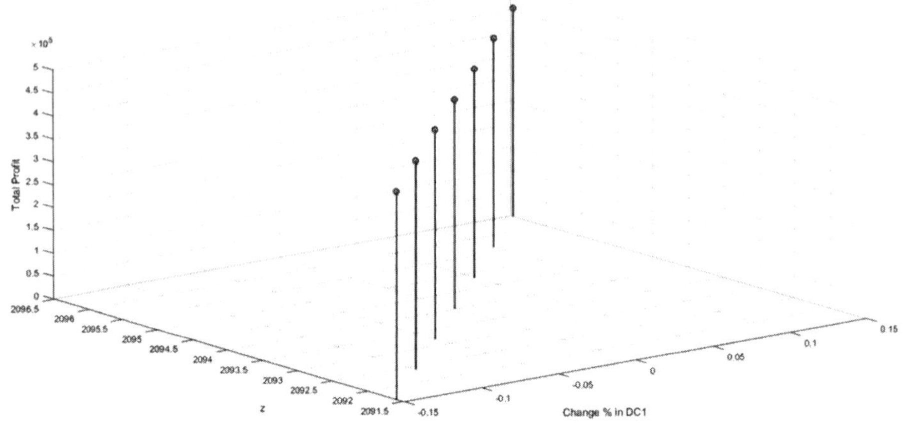

Implication: A positive change in the transportation cost to the distribution center 1 implies a slight increase in the lot size, slight increase in TAC and decrease in Total Profit per cycle. *DC1* is a less sensitive parameter.

2) Analysis is performed by varying the variable *DC2* by changing 5%, 10% and 15% of its value increasingly and decreasingly.

The change in Total Profit with respect to the variable *DC2*(Crisp) is depicted in Figure 5.

Figure 5. Change in total profit with respect to DC2 (crisp)

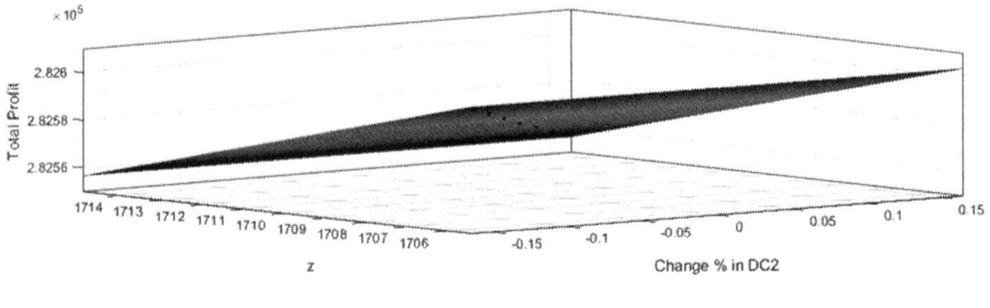

The change in Total Profit with respect to the variable *DC2*((Neutrosophic) is depicted in Figure 6.

Figure 6. Change in total profit with respect to DC2 (neutrosophic)

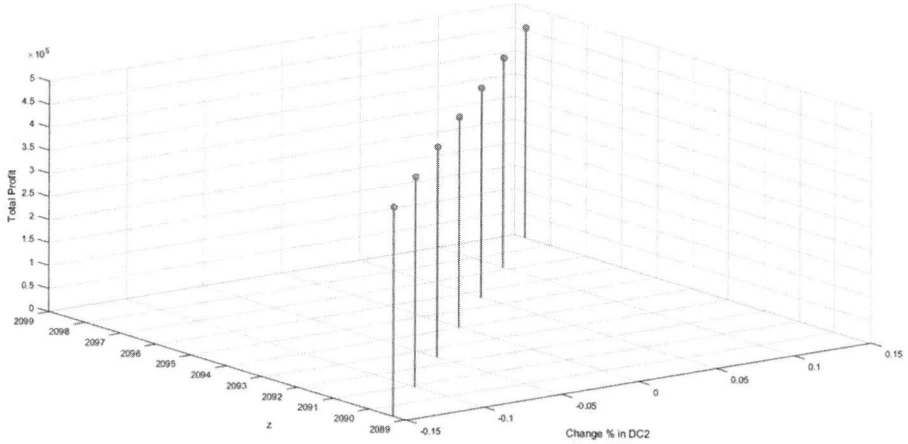

Implication: A positive change in the transportation cost to the distribution center 2 implies a slight increase in lot size, slight increase in TAC and decrease in Total Profit per cycle. *DC2* is a less sensitive parameter.

3) Analysis is performed by varying the variable *DC3* by changing 5%, 10% and 15% of its value increasingly and decreasingly.

The change in Total Profit with respect to the variable *DC3* (Crisp) is depicted in Figure 7.

Figure 7. Change in total profit with respect to DC3 (crisp)

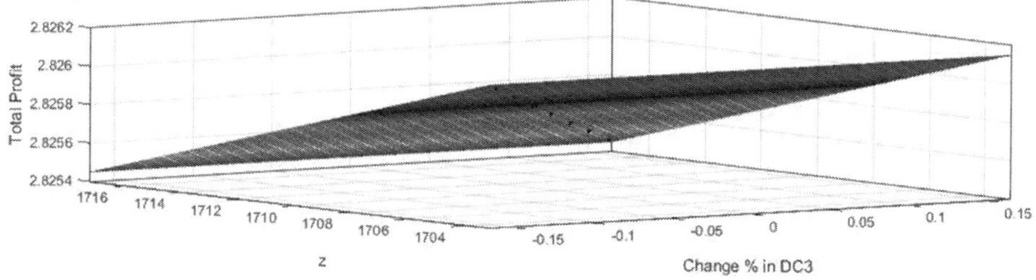

The change in Total Profit with respect to the variable *DC3* (Neutrosophic) is depicted in Figure 8.

Figure 8. Change in total profit with respect to DC3 (neutrosophic)

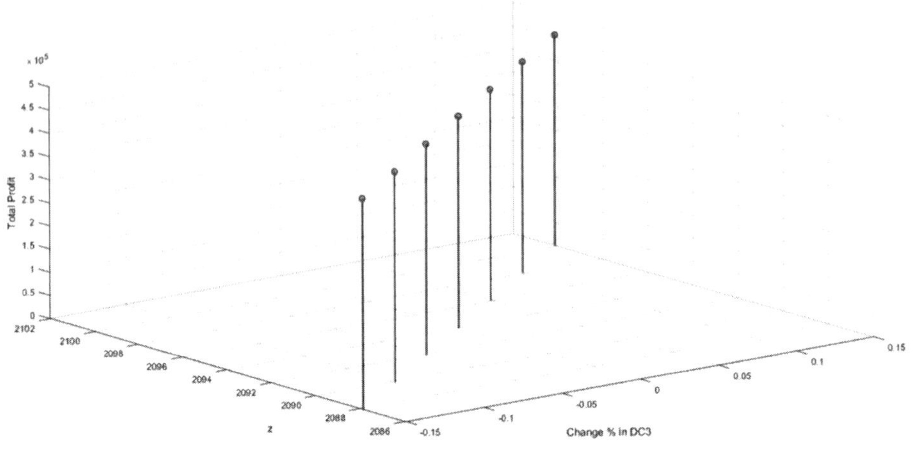

Implication: A positive change in the transportation cost to the distribution center 3 implies a slight increase in lot size, slight increase in TAC and decrease in Total Profit per cycle. *DC3* is a less sensitive parameter.

4) Analysis is performed by varying the variable F_T by changing 5%, 10% and 15% of its value increasingly and decreasingly.

The change in Total Profit with respect to the variable F_T is depicted in Figure 9.

Figure 9. Change in total profit with respect to F_T (crisp)

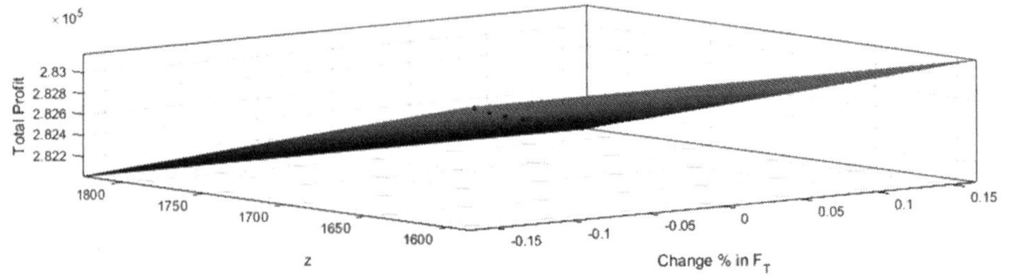

The change in Total Profit with respect to the variable F_T is depicted in Figure 10.

Figure 10. Change in total profit with respect to F_T (neutrosophic)

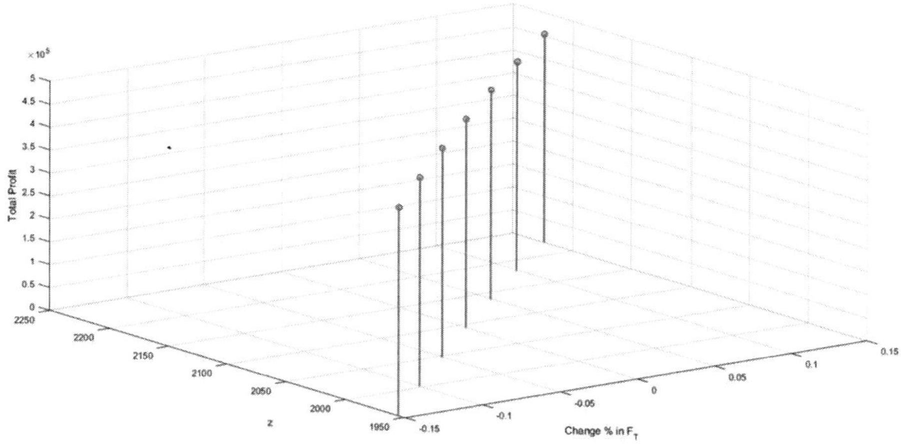

Implication: A positive change in the fixed acquisition cost implies huge increase in lot size, significant increase in TAC and a significant decrease in Total Profit per cycle. F_T is a moderately sensitive parameter.

5) Analysis is performed by varying the variable H_C by changing 5%, 10% and 15% of its value increasingly and decreasingly.

The change in Total Profit with respect to the variable H_C is depicted in Figure 11.

Figure 11. Change in total profit with respect to H_C (crisp)

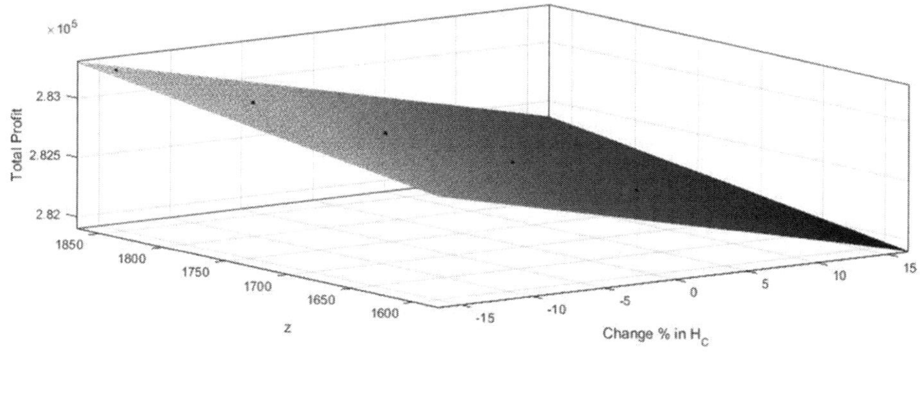

The change in Total Profit with respect to the variable H_C is depicted in Figure 12.

Figure 12. Change in total profit with respect to H_C (neutrosophic)

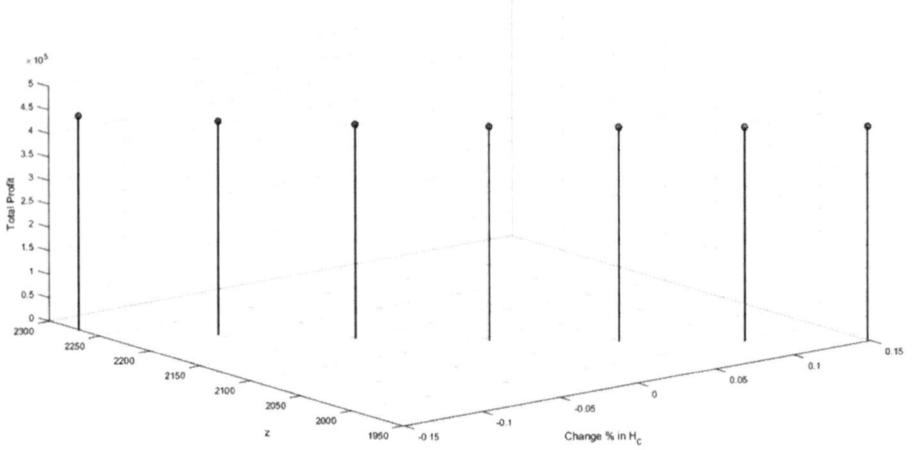

Implications: A positive change in the holding cost per unit implies a huge decrease in lot size, a significant increase in TAC and a considerable significant in Total Profit per cycle. H_C is a highly sensitive parameter.

6) Analysis is performed by varying the variable k by changing 5%, 10% and 15% of its value increasingly and decreasingly.

The change in Total Profit with respect to the variable *k* is depicted in Figure 13.

Figure 13. Change in total profit with respect to k (crisp)

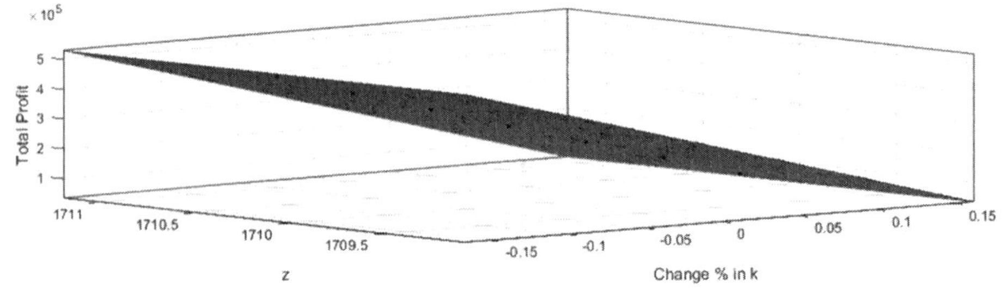

The change in Total Profit with respect to the variable *k* is depicted in Figure 14.

Figure 14. Change in total profit with respect to k (neutrosophic)

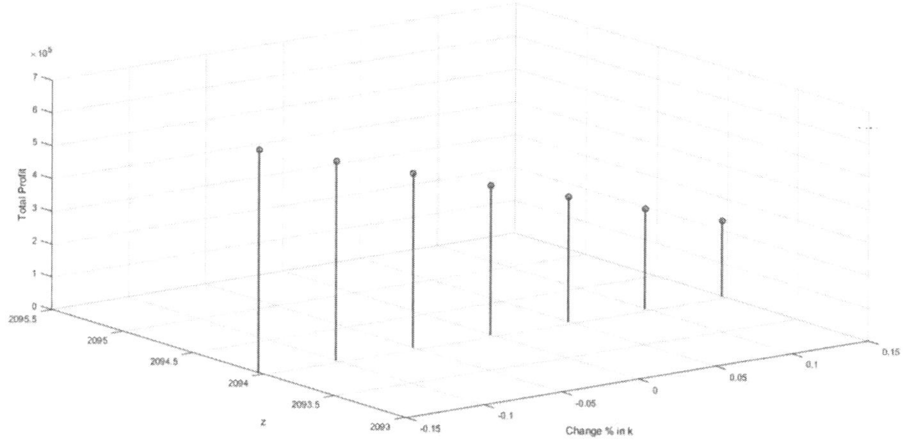

Implication: A positive change in cost of production per unit implies a constant lot size, huge increase in TAC and decrease in Total Profit per cycle. k is a moderately sensitive parameter.

7) Analysis is performed by varying the variable Gc by changing 5%, 10% and 15% of its value increasingly and decreasingly.

The change in Total Profit with respect to the variable Gc is depicted in Figure 15.

Figure 15. Change in total profit with respect to G_C (crisp)

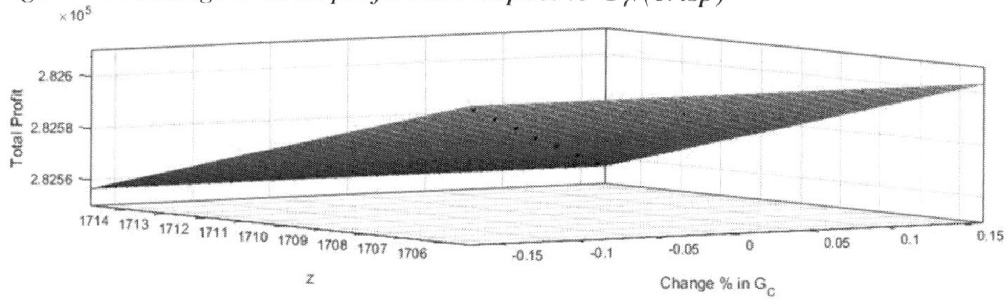

The change in Total Profit with respect to the variable Gc is depicted in Figure 16.

Figure 16. Change in total profit with respect to G_C (neutrosophic)

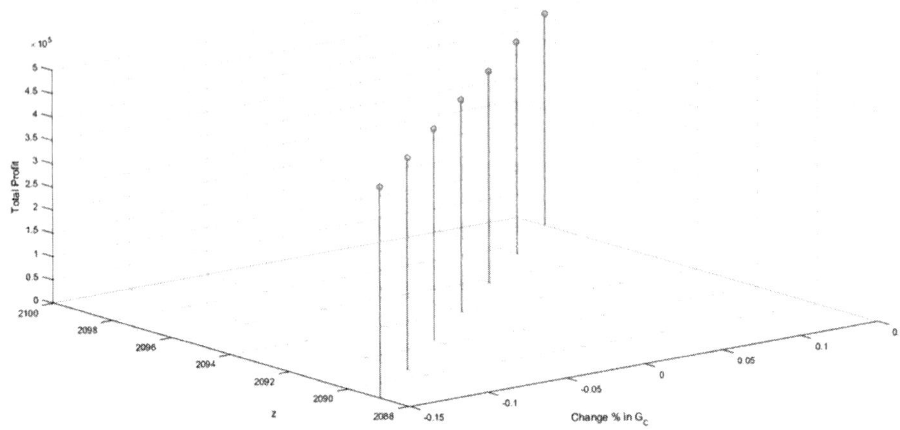

Implication: A positive change in the green disposal cost implies a slight increase in lot size, slight increase in TAC and a decrease in Total Profit per cycle. *Gc* is a less sensitive parameter.

8) Analysis is performed by varying the variable *Cs* by changing 5%, 10% and 15% of its value increasingly and decreasingly.

The change in Total Profit with respect to the variable *Cs* is depicted in Figure 17.

Figure 17. Change in total profit with respect to C_S (crisp)

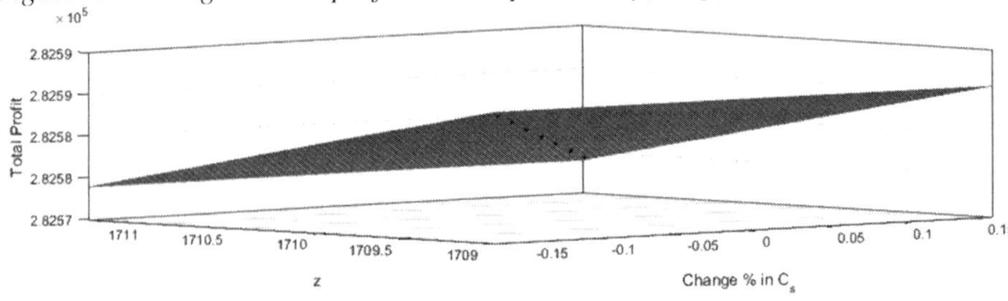

The change in Total Profit with respect to the variable *Cs* is depicted in Figure 18.

Figure 18. Change in total profit with respect to C_S (neutrosophic)

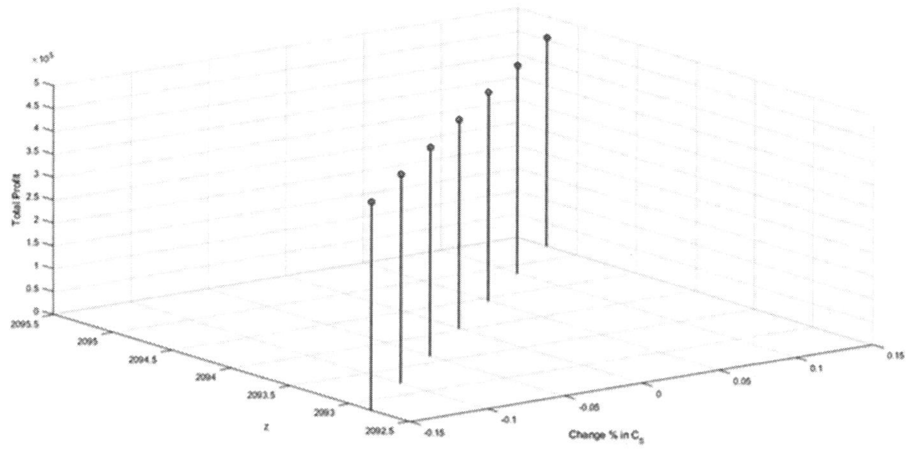

Implication: A positive change in the cost of transportation to the service center implies a slight increase in lot size, slight increase in TAC slight and a decrease in Total Profit per cycle. C_S is a less sensitive parameter.

9) Analysis is performed by varying the variable *Pc* (Crisp) by changing 5%, 10% and 15% of its value increasingly and decreasingly.

The change in Total Profit with respect to the variable *Pc* is depicted in Figure 19.

Figure 19. Change in total profit with respect to P_C (crisp)

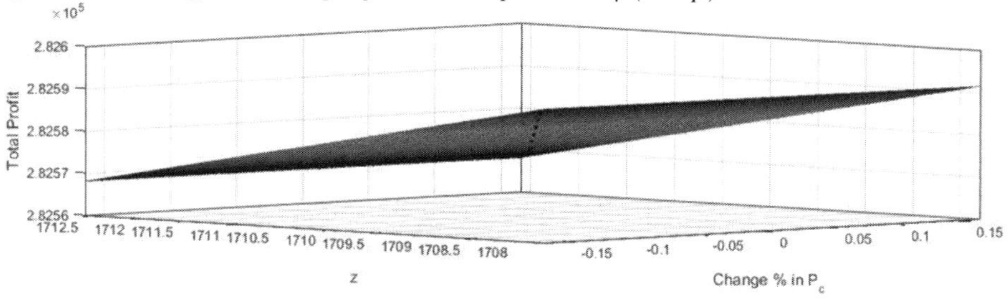

The change in Total Profit with respect to the variable *Pc* is depicted in Figure 20.

Figure 20. Change in total profit with respect to P_C (neutrosophic)

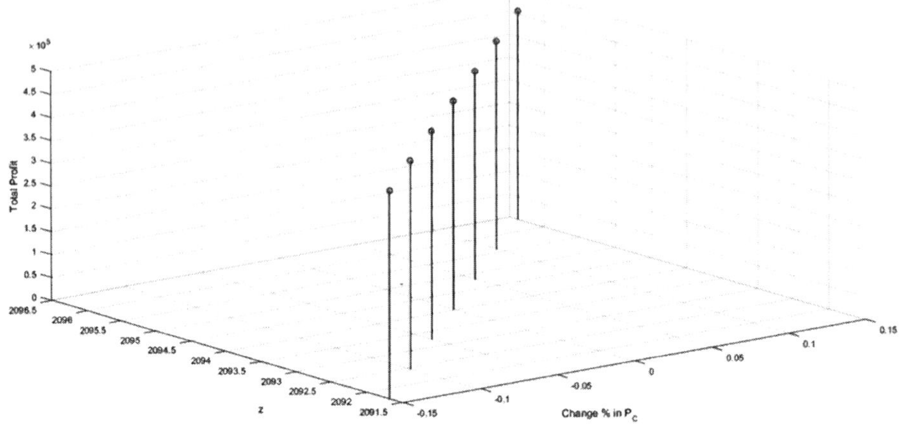

Implication: A positive change in the packaging cost implies a slight increase in lot size, slight increase in TAC and a decrease in Total Profit per cycle. The parameter P_C is less sensitive.

10) Analysis is performed by varying the variable *D* by changing 5%, 10% and 15% of its value increasingly and decreasingly.

Change in Total Profit with respect to the variable *D* is depicted in Figure 21.

Figure 21. Change in total profit with respect to D (crisp)

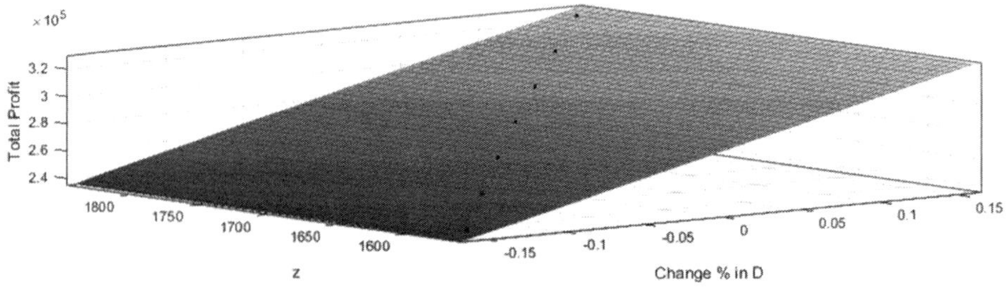

The change in Total Profit with respect to the variable *D* is depicted in Figure 22.

Figure 22. Change in total profit with respect to D (neutrosophic)

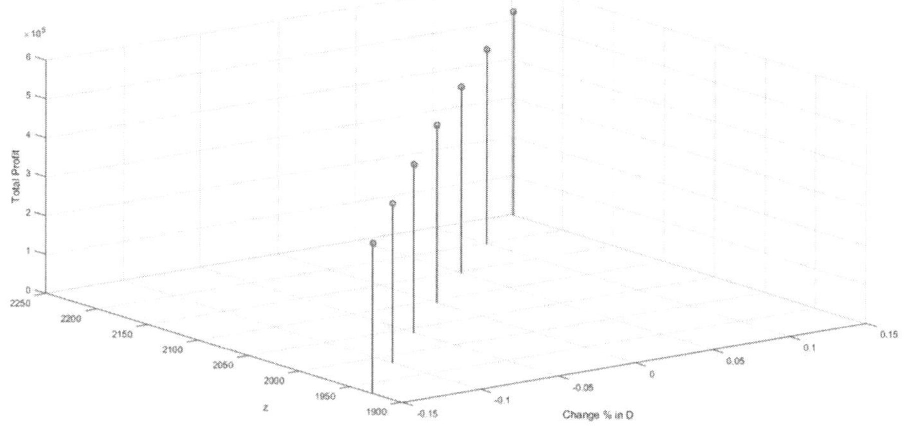

Implication: A positive change in the demand rate implies a huge increase in lot size, huge increase in TAC and an increase in Total Profit per cycle. Parameter D is highly sensitive.

11) Analysis is performed by varying the variable S_P (Crisp) by changing 5%, 10% and 15% of its value increasingly and decreasingly.

The change in Total Profit with respect to the variable S_P (Crisp) is depicted in Figure 23.

Figure 23. Change in total profit with respect to S_p (crisp)

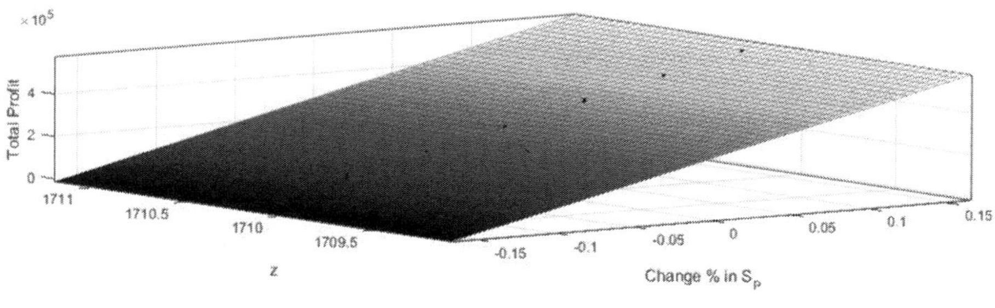

The change in Total Profit with respect to the variable S_P (Neutrosophic) is depicted in Figure 24.

Figure 24. Change in total profit with respect to S_p(neutrosophic)

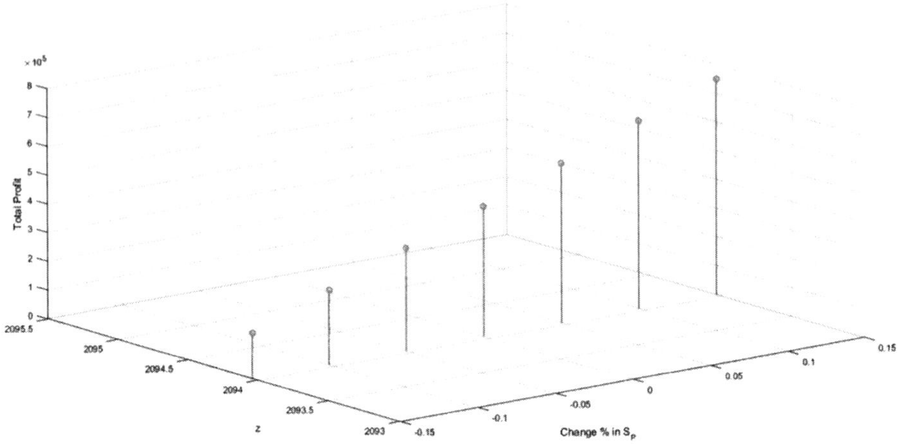

Implication: A positive change in the unit selling price implies a constant lot size, constant TAC and a huge increase in Total Profit.

The result obtained by performing sensitivity analysis by varying the parameter *DC1*(Crisp) is provided in Table 3.

Table 3. Analysis of change in DC1 (crisp)

Change % in *DC1* (Crisp)	*DC1*	*z*	TAC	TP
-15%	17	1708.051	1517410	282590.4
-10%	18	1708.764	1517413	282586.9
-5%	19	1709.477	1517417	282583.4
0	20	1710.19	1517420	282579.9
5%	21	1710.902	1517424	282576.4
10%	22	1711.614	1517427	282572.9
15%	23	1712.326	1517431	282569.4

The result obtained by performing sensitivity analysis by varying the parameter DC1(Neutrosophic) is provided in Table 4.

Table 4. Analysis of change in DC1(neutrosophic)

Change % in *DC1* (Neutrosophic)	*DC1*	*z*	**TAC**	**TP**
-15%	14.52081	2091.6	1517411	454245
-10%	15.37497	2092.413	1517414	454241.9
-5%	16.22914	2093.226	1517417	454238.9
0	17.0833	2094.038	1517420	454235.9
5%	17.93747	2094.85	1517423	454232.9
10%	18.79163	2095.662	1517426	454229.9
15%	19.6458	2096.474	1517429	454226.8

The result obtained by performing sensitivity analysis by varying the parameter *DC2*(Crisp) is provided in Table 5.

Table 5. Analysis of change in DC2 (crisp)

Change % in *DC2* (Crisp)	*DC2*	*z*	**TAC**	**TP**
-15%	34	1705.909	1517399	282601
-10%	36	1707.337	1517406	282593.9
-5%	38	1708.764	1517413	282586.9
0	40	1710.19	1517420	282579.9
5%	42	1711.614	1517427	282572.9
10%	44	1713.038	1517434	282565.9
15%	46	1714.46	1517441	282558.9

The result obtained by performing sensitivity analysis by varying the parameter *DC2*(Neutrosophic) is provided in Table 6.

Table 6. Analysis of change in DC2(neutrosophic)

Change % in *DC2* (Neutrosophic)	*DC2*	*z*	**TAC**	**TP**
-15%	27.88	2089.354	1517402	454253.3
-10%	29.52	2090.917	1517408	454247.5
-5%	31.16	2092.478	1517414	454241.7
0	32.8	2094.038	1517420	454235.9
5%	34.44	2095.597	1517426	454230.1
10%	36.08	2097.155	1517431	454224.3
15%	37.72	2098.712	1517437	454218.5

The result obtained by performing sensitivity analysis by varying the parameter *DC3*(Crisp) is provided in Table 7.

Table 7. Analysis of change in DC3 (crisp)

Change % in *DC3* (Crisp)	*DC3*	*z*	**TAC**	**TP**
-15%	51	1703.764	1517388	282611.5
-10%	54	1705.909	1517399	282601
-5%	57	1708.051	1517410	282590.4
0	60	1710.19	1517420	282579.9
5%	63	1712.326	1517431	282569.4
10%	66	1714.46	1517441	282558.9
15%	69	1716.591	1517452	282548.4

The result obtained by performing sensitivity analysis by varying the parameter *DC3*(Neutrosophic) is provided in Table 8.

Table 8. Analysis of change in DC3(neutrosophic)

Change % in *DC3* (Neutrosophic)	*DC3*	*z*	**TAC**	**TP**
-15%	36.5925	2087.888	1517397	454258.7
-10%	38.745	2089.94	1517405	454251.1
-5%	40.8975	2091.99	1517412	454243.5
0	43.05	2094.038	1517420	454235.9
5%	45.2025	2096.084	1517427	454228.3
10%	47.355	2098.128	1517435	454220.7
15%	49.5075	2100.17	1517443	454213.1

The result obtained by performing sensitivity analysis by varying the parameter F_T (Crisp) is provided in Table 9.

Table 9. Analysis of change in F_T (crisp)

Change % in F_T(Crisp)	F_T	*z*	**TAC**	**TP**
-15%	850	1599.736	1516876	283123.7
-10%	900	1637.382	1517062	282938.4
-5%	950	1674.182	1517243	282757.2

continued on following page

Table 9. Continued

Change % in F_T (Crisp)	F_T	z	TAC	TP
0	1000	1710.19	1517420	282579.9
5%	1050	1745.455	1517594	282406.3
10%	1100	1780.022	1517764	282236.1
15%	1150	1813.93	1517931	282069.1

The result obtained by performing sensitivity analysis by varying the parameter F_T (Neutrosophic) is provided in Table 10.

Table 10. Analysis of change in F_T(neutrosophic)

Change % in F_T (Neutrosophic)	F_T	z	TAC	TP
-15%	799	1955.36	1516905	454751.1
-10%	846	2002.654	1517080	454575.4
-5%	893	2048.856	1517252	454403.8
0	940	2094.038	1517420	454235.9
5%	987	2138.267	1517584	454071.6
10%	1034	2181.598	1517745	453910.6
15%	1081	2224.086	1517903	453752.7

The result obtained by performing sensitivity analysis by varying the parameter H_C (Crisp) is provided in Table 11.

Table 11. Analysis of change in H_C (crisp)

Change % in Hc (Crisp)	Hc	z	TAC	TP
-15%	4.25	1854.961	1516763	283237
-10%	4.5	1802.698	1516988	283012
-5%	4.75	1754.618	1517207	282793.1
0	5	1710.19	1517420	282579.9
5%	5.25	1668.974	1517628	282371.9
10%	5.5	1630.602	1517831	282168.9
15%	5.75	1594.76	1518030	281970.4

The result obtained by performing sensitivity analysis by varying the parameter H_C (Neutrosophic) is provided in Table 12.

Table 12. Analysis of change in H_C(neutrosophic)

Change % in *Hc* (Neutrosophic)	*Hc*	*z*	**TAC**	**TP**
-15%	3.194581	2271.303	1516813	454843.1
-10%	3.382497	2207.31	1517021	454635.1
-5%	3.570414	2148.438	1517223	454432.9
0	3.75833	2094.038	1517420	454235.9
5%	3.946247	2043.572	1517612	454043.8
10%	4.134163	1996.587	1517800	453856.2
15%	4.32208	1952.701	1517983	453672.8

The result obtained by performing sensitivity analysis by varying the parameter *k*(Crisp) is provided in Table 13.

Table 13. Analysis of change in k (crisp)

Change % in *k* (Crisp)	*k*	*z*	**TAC**	**TP**
-15%	212.5	1710.19	1292420	507579.9
-10%	225	1710.19	1367420	432579.9
-5%	237.5	1710.19	1442420	357579.9
0	250	1710.19	1517420	282579.9
5%	262.5	1710.19	1592420	207579.9
10%	275	1710.19	1667420	132579.9
15%	287.5	1710.19	1742420	57579.88

The result obtained by performing sensitivity analysis by varying the parameter *k*(Neutrosophic) is provided in Table 14.

Table 14. Analysis of change in k(neutrosophic)

Change % in *k* (Neutrosophic)	*k*	*z*	**TAC**	**TP**
-15%	172.6114	2094.038	1292060	679595.3
-10%	182.765	2094.038	1367180	604475.5
-5%	192.9186	2094.038	1442300	529355.7
0	203.0722	2094.038	1517420	454235.9
5%	213.2258	2094.038	1592540	379116.1
10%	223.3794	2094.038	1667659	303996.3
15%	233.533	2094.038	1742779	228876.5

The result obtained by performing sensitivity analysis by varying the parameter G_C (Crisp) is provided in Table 15.

Table 15. Analysis of change in G_C(crisp)

Change % in *Gc* (Crisp)	*Gc*	*z*	TAC	TP
-15%	34	1705.909	1517399	282601
-10%	36	1707.337	1517406	282593.9
-5%	38	1708.764	1517413	282586.9
0	40	1710.19	1517420	282579.9
5%	42	1711.614	1517427	282572.9
10%	44	1713.038	1517434	282565.9
15%	46	1714.46	1517441	282558.9

The result obtained by performing sensitivity analysis by varying the parameter G_C (Neutrosophic) is provided in Table 16.

Table 16. Analysis of change in G_C(neutrosophic)

Change % in *Gc* (Neutrosophic)	*Gc*	*z*	TAC	TP
-15%	30.20305	2088.964	1517401	454254.7
-10%	31.9797	2090.657	1517407	454248.5
-5%	33.75635	2092.348	1517414	454242.2
0%	35.533	2094.038	1517420	454235.9
5%	37.30965	2095.727	1517426	454229.6
10%	39.0863	2097.415	1517432	454223.3
15%	40.86295	2099.101	1517439	454217.1

The result obtained by performing sensitivity analysis by varying the parameter C_S (Crisp) is provided in Table 17.

Table 17. Analysis of change in C_S(crisp)

Change % in *Cs*(Crisp)	*Cs*	*z*	TAC	TP
-15%	8.5	1709.12	1517415	282585.1
-10%	9	1709.477	1517417	282583.4
-5%	9.5	1709.833	1517418	282581.6

continued on following page

Table 17. Continued

Change % in *Cs*(Crisp)	*Cs*	*z*	TAC	TP
0	10	1710.19	1517420	282579.9
5%	10.5	1710.546	1517422	282578.1
10%	11	1710.902	1517424	282576.4
15%	11.5	1711.258	1517425	282574.6

The result obtained by performing sensitivity analysis by varying the parameter C_S (Neutrosophic) is provided in Table 18.

Table 18. Analysis of change in C_S(neutrosophic)

Change % in *Cs* (Neutrosophic)	*Cs*	*z*	TAC	TP
-15%	7.25985	2092.82	1517415	454240.4
-10%	7.6869	2093.226	1517417	454238.9
-5%	8.11395	2093.632	1517418	454237.4
0%	8.541	2094.038	1517420	454235.9
5%	8.96805	2094.444	1517421	454234.4
10%	9.3951	2094.85	1517423	454232.9
15%	9.82215	2095.256	1517424	454231.4

The result obtained by performing sensitivity analysis by varying the parameter P_C (Crisp) is provided in Table 19.

Table 19. Analysis of change in P_C(crisp)

Change % in *Pc* (Crisp)	*Pc*	*z*	TAC	TP
-15%	17	1708.051	1517410	282590.4
-10%	18	1708.764	1517413	282586.9
-5%	19	1709.477	1517417	282583.4
0	20	1710.19	1517420	282579.9
5%	21	1710.902	1517424	282576.4
10%	22	1711.614	1517427	282572.9
15%	23	1712.326	1517431	282569.4

The result obtained by performing sensitivity analysis by varying the parameter P_C (Neutrosophic) is provided in Table 20.

Table 20. Analysis of change in P_C(neutrosophic)

Change % in *Pc* (Neutrosophic)	*Pc*	*z*	TAC	TP
-15%	14.01081	2091.686	1517411	454244.6
-10%	14.83497	2092.47	1517414	454241.7
-5%	15.65914	2093.254	1517417	454238.8
0%	16.4833	2094.038	1517420	454235.9
5%	17.30747	2094.822	1517423	454233
10%	18.13163	2095.605	1517426	454230.1
15%	18.9558	2096.388	1517429	454227.2

The result obtained by performing sensitivity analysis by varying the parameter *D*(Crisp) is provided in Table 21.

Table 21. Analysis of change in D(crisp)

Change % in *D* (Crisp)	*D*	*z*	TAC	TP
-15%	5100	1574.883	1290422	239578
-10%	5400	1621.169	1366094	253905.8
-5%	5700	1666.24	1441760	268239.9
0	6000	1710.19	1517420	282579.9
5%	6300	1753.104	1593075	296925.3
10%	6600	1795.057	1668724	311275.8
15%	6900	1836.115	1744369	325631

The result obtained by performing sensitivity analysis by varying the parameter *D*(Neutrosophic) is provided in Table 22.

Table 22. Analysis of change in D(neutrosophic)

Change % in *D* (Neutrosophic)	*D*	*z*	TAC	TP
-15%	6288.583	1928.93	1290373	385534.5
-10%	6658.5	1985.428	1366061	408429.3
-5%	7028.416	2040.424	1441743	431329.9
0	7398.333	2094.038	1517420	454235.9
5%	7768.25	2146.373	1593092	477146.9
10%	8138.166	2197.521	1668759	500062.5
15%	8508.083	2247.563	1744422	522982.4

The result obtained by performing sensitivity analysis by varying the parameter S_p (Crisp) is provided in Table 23.

Table 23. Analysis of change in S_p(crisp)

Change % in S_p (Crisp)	S_p	z	TAC	TP
-15%	255	1710.19	1517420	12579.88
-10%	270	1710.19	1517420	102579.9
-5%	285	1710.19	1517420	192579.9
0	300	1710.19	1517420	282579.9
5%	315	1710.19	1517420	372579.9
10%	330	1710.19	1517420	462579.9
15%	345	1710.19	1517420	552579.9

The result obtained by performing sensitivity analysis by varying the parameter S_p (Neutrosophic) is provided in Table 24.

Table 24. Analysis of change in S_p(neutrosophic)

Change % in S_p (Neutrosophic)	S_p	z	TAC	TP
-15%	226.525	2094.038	1517420	158487.5
-10%	239.85	2094.038	1517420	257070.3
-5%	253.175	2094.038	1517420	355653.1
0	266.5	2094.038	1517420	454235.9
5%	279.825	2094.038	1517420	552818.7
10%	293.15	2094.038	1517420	651401.5
15%	306.475	2094.038	1517420	749984.3

6. INDUSTRIAL IMPLICATIONS

This research work presents a multifaceted solution with broad applicability across diverse industrial domains. The production domain comprising various sectors, in manufacturing, the model's integration of coordinated rework stations ensures stringent quality control by identifying and rectifying defects promptly, minimizing waste, and optimizing production processes. Distribution centers, strategically positioned within the supply chain, find application in sectors ranging from retail to e-commerce, enhancing logistics efficiency, reducing lead times, and meeting dynamic consumer demands. Furthermore, the utilization of Ant Colony Optimization in the distribution network scheduling ensures adaptive and optimal routing, particularly beneficial in industries with intricate and time-sensitive supply chain networks. This innovative model holds promise in addressing the challenges of uncertainty, variability, and quality management, making it a valuable tool for industries seeking resilient, responsive, and cost-effective supply chain solutions.

CONCLUSION

This research work represents a cutting-edge approach to supply chain management that addresses the complexities and uncertainties pervasive in modern industrial settings. By integrating neutrosophic logic, coordinated rework stations, and Ant Colony Optimization, this model offers a comprehensive solution for optimizing inventory systems, improving product quality, and enhancing distribution network efficiency. The coordinated rework stations proactively manage defects, reducing waste and ensuring higher product quality, while distribution centers strategically positioned in the supply chain enhance logistics and responsiveness to market demands. The application of Ant Colony Optimization optimizes routing and scheduling decisions, adding a layer of adaptability to the system. This interdisciplinary framework has significant implications across various industries, promising increased operational efficiency, reduced costs, and improved customer satisfaction. As industries continue to grapple with dynamic market conditions, this innovative model provides a forward-looking strategy for achieving resilience and agility in supply chain management.

REFERENCES

AbdelMouty, A. M., Abdel-Monem, A., Aal, S. I. A., & Ismail, M. M. (2023). Analysis the Role of the Internet of Things and Industry 4.0 in Healthcare Supply Chain Using Neutrosophic Sets. *Neutrosophic Systems with Applications*, *4*, 33–42.

Ahmad, F. (2022). Interactive neutrosophic optimization technique for multiobjective programming problems: An application to pharmaceutical supply chain management. *Annals of Operations Research*, *311*(2), 551–585.

Ahmad, F., Adhami, A. Y., & Smarandache, F. (2020). Modified neutrosophic fuzzy optimization model for optimal closed-loop supply chain management under uncertainty. In *Optimization theory based on neutrosophic and plithogenic sets* (pp. 343–403). Academic Press.

Aytekin, A., Okoth, B. O., Korucuk, S., Karamaşa, Ç., & Tirkolaee, E. B. (2022). A neutrosophic approach to evaluate the factors affecting performance and theory of sustainable supply chain management: Application to textile industry. *Management Decision*, *61*(2), 506–529.

Broumi, S., Raut, P. K., & Behera, S. P. (2023). Solving shortest path problems using an ant colony algorithm with triangular neutrosophic arc weights. *International Journal of Neutrosophic Science*, *20*(4), 128–137.

Deb, S. C., & Islam, S. (2023). Application of Neutrosophic Interval valued Goal Programming to a Supply Chain Inventory Model for Deteriorating Items with Time Dependent Demand. *Neutrosophic Sets and Systems*, *53*(1), 35.

Dohale, V., Ambilkar, P., Kumar, A., Mangla, S. K., & Bilolikar, V. (2023). Analyzing the enablers of circular supply chain using Neutrosophic-ISM method: Lessons from the Indian apparel industry. *International Journal of Logistics Management*, *34*(3), 611–643.

Dong, B., Duan, M., & Li, Y. (2022). Exploration of Joint Optimization and Visualization of Inventory Transportation in Agricultural Logistics Based on Ant Colony Algorithm. *Computational Intelligence and Neuroscience*. PMID: 35755759

Fallah, M., & Nozari, H. (2021). Neutrosophic mathematical programming for optimization of multi-objective sustainable biomass supply chain network design. *Computer Modeling in Engineering & Sciences*, *129*(2), 927–951.

Farid, H. M. A., & Riaz, M. (2023). Single-valued neutrosophic dynamic aggregation information with time sequence preference for IoT technology in supply chain management. *Engineering Applications of Artificial Intelligence*, *126*, 106940.

Ismail, M. M., Ibrahim, M. M., & Zaki, S. (2023). A Neutrosophic Approach for Multi-Factor Analysis of Uncertainty and Sustainability of Supply Chain Performance. *Neutrosophic Sets and Systems*, *58*(1), 16.

Khan, S., Haleem, A., & Khan, M. I. (2023). A risk assessment framework using neutrosophic theory for the halal supply chain under an uncertain environment. *Arab Gulf Journal of Scientific Research.*

Mohamed, Z., Ismail, M. M., & Abd El-Gawad, A. F. (2023). Neutrosophic Model to Examine the Challenges Faced by Manufacturing Businesses in Adopting Green Supply Chain Practices and to Provide Potential Solutions. *Neutrosophic Systems With Applications*, *3*, 45–52.

Nabeeh, N. A., Mohamed, M., Abdel-Monem, A., Abdel-Basset, M., Sallam, K. M., El-Abd, M., & Wagdy, A. (2022, July). A Neutrosophic Evaluation Model for Blockchain Technology in Supply Chain Management. In *2022 IEEE International Conference on Fuzzy Systems (FUZZ-IEEE)* (pp. 1-8). IEEE.

Oudouar, F., & Zaoui, E. M. (2021, November). A novel hybrid heuristic based on ant colony algorithm for solving multi-product inventory routing problem. In *International Conference on Advanced Technologies for Humanity* (pp. 519-529). Cham: Springer International Publishing.

Renee Miriam, M., Martin, N., Aleeswari, A., & Broumi, S. (2023). Rework Warehouse Inventory Model for Product Distribution with Quality Conservation in Neutrosophic Environment. *International Journal of Neutrosophic Science*, *21*(2), 177–195.

Szmelter-Jarosz, A., Ghahremani-Nahr, J., & Nozari, H. (2021). A neutrosophic fuzzy optimisation model for optimal sustainable closed-loop supply chain network during COVID-19. *Journal of Risk and Financial Management*, *14*(11), 519.

Wang, Y. (2022). Inventory Path Optimization of VMI Large Logistics Enterprises Based on Ant Colony Algorithm. *Mobile Information Systems*, 2022.

Yang, L., Li, D., & Tan, R. (2020). Research on the shortest path solution method of interval valued neutrosophic graphs based on the ant colony algorithm. *IEEE Access : Practical Innovations, Open Solutions*, *8*, 88717–88728.

Chapter 12
A Probabilistic Approach for Renewable Energy Alternative Selection Through Correlation–Based Neutrosophic TOPSIS Approach

Biplab Sinha Mahapatra
https://orcid.org/0000-0002-9651-3083
Haldia Institute of Technology, India

Mihir Baran Bera
https://orcid.org/0000-0002-2677-6101
Haldia Institute of Technology, India

Manoj Kumar Mondal
https://orcid.org/0009-0009-1034-3211
Haldia Institute of Technology, India

Pinaki Pratim Acharjya
https://orcid.org/0000-0002-0305-2661
Haldia Institute of Technology, India

ABSTRACT

The selection of renewable energy alternatives becomes critical due to several quantitative and qualitative factors describing the criteria. Often these criteria cannot be appropriately quantified due to linguistic assessment. Hence, selecting

DOI: 10.4018/979-8-3693-3204-7.ch012

renewable energy alternatives becomes a complex, uncertain multi-criteria decision-making (MCDM) problem. Several soft set-based TOPSIS approaches are used to solve uncertain MCDM problems. In this chapter, the authors introduce a neutrosophic soft set-based TOPSIS approach to solve MCDM problems in an uncertain situation. The TOPSIS methods calculate the relative measure of distance between its positive ideal solution (PIS) and negative ideal solution (NIS). This distance is often measured by Euclidean or Hamming techniques which create ambiguity and computational complexities. The authors introduce a correlation-based TOPSIS approach to solve an MCDM problem to avoid this ambiguity and computational hazards. A numerical discussion of renewable energy alternative selection is given to establish the proposed approach.

INTRODUCTION

Decision-making is essential to human life since it lays out all the options based on the decision-maker's (DM) judgment data and selects the most advantageous option. DMs used precise numbers to assess early social development. When MCDM problems get increasingly complex, experts cannot provide precise numerical estimates of solutions. The complexity and vagueness of human decisions demonstrate the inadequacy of crisp set theory. Fuzzy set (FS) (Zadeh, 1965) theory allows experts to indicate their level of satisfaction with a member's performance within a unit interval. FSs provide the basis for questionable assessments but cannot quantify non-membership grade. Intuitionistic fuzzy sets (IFSs) (Atanassov, 1986) have both satisfaction and discontent, making them preferable to conventional FS. The IFS represents a greater degree of ambiguity in decision-making since they provide information on the benefits and drawbacks of the various options. Atanassov's approach has limitations since its total support is capped at one. Molodtsov (1999) introduced soft set, which was an important development to consider uncertainty. Later Smarandache introduced several extensions of soft set such as HyperSoft Set (Smarandache, 2018), IndetermSoft Set, IndetermHyperSoft Set (Smarandache, 2022), and TreeSoft Set (Smarandache, 2023). These advanced soft sets are important tools to deal with uncertainty and vague information. Smarandache proposed (Smarandache, 1999; Smarandache, 2004) single-valued neutrosophic set (SVNS), which is a generalization of the IFS, is capable of handling inaccurate information and is based on three values (truth, indeterminacy, and falsity). The truth, indeterminacy and falsity degrees are independent of each other. So, the SVNSs are an effective, human-focused way to capture and model deception and uncertainty in intuitive decision-making processes.

The Technique for Order Preference by Similarity to Ideal Solution (TOPSIS) was introduced in the early 1980s by Hwang &Yoon (1981). TOPSIS selects the option with the highest distance from the NIS and the lowest distance from the PIS. This technique ranks alternatives based on their distances from two reference locations, known as ideal solutions. There are numerous applications of the TOPSIS approach in uncertain, complex MCDM problems. The conventional TOPSIS method considers only distance measures, not similarity or probability. Chen (2000) proposed fuzzy TOPSIS using triangular fuzzy numbers. Sahin et al. (2016) extended the TOPSIS approach by SVNS for supplier selection. Biswas et al. (2016) solved TOPSIS by SVN Euclidean distance measure. Abdel-Basset et al. (2019) employed neutrosophic TOPSIS technique for selecting medical devices. Karasan&Kaya(2020b) proposed a neutrosophic TOPSIS approach for evaluating different network controllers and relays for aerial vehicles. Hezamet al. (2021) applied neutrosophic TOPSIS to rank COVID-19 vaccine alternatives. Nafei et al. (2021) defined a score function on fuzzy NSs to solve hotel site selection problems by the TOPSIS approach. Pouresmaeil et al. (2022) developed a score function for solving interval neutrosophic MCDM problems by the TOPSIS approach. Ridvan et al. (2021) developed a novel TOPSIS that optimizes the distance, similarity, and magnitude closeness coefficients in a neutrosophic environment. Mollaoglu et al. (2022) identified alternate fuel sources for ship investment choices using SVNS based TOPSIS approach. Based on the Dice and Jaccard vector measures, Ozlu&Karaaslan (2022) created similarity measures in TOPSIS under SVNS type-2 information.

MCDM Solved by Different Methods

Evaluating and selecting the most suitable alternatives is one of the most difficult challenges for a DM. This process is known as MCDM problem. MCDM methods are used when several alternatives have conflicting and non-comparable choice criteria. MCDM approach is a well-known tool for solving complex real-life problems due to its capacity to appraise numerous options using various judgment criteria. Popular MCDM strategies for decision-making include:

Simple Additive Weighting Method (SAW)

SAW is a classic approach for evaluating alternatives that are perhaps the most prevalent and widely used method. In this procedure, each alternative's performance is calculated by multiplying its rating on each criterion by its weight and summing these products' overall criteria. Many studies have employed this straightforward, easy-to-understand strategy (Chen et al., 1992). Recent SAW applications include predictive torque control of induction motors (Muddineni et al., 2017) and laboratory preference for COVID-19 testing (Gul, 2021).

Complex Proportional Assessment (COPRAS)

This method provides non-dominant significance and utility solutions; hence, alternatives are progressively organized and assessed. The COPRAS technique is a practical MCDM approach that evaluates options using a ratio based on two metrics. This method has been used in practical MCDM applications like choosing a type 2 diabetes medication (Rani et al., 2020a), healthcare evaluation (Mishra et al., 2020b) and selection of desalination technology (Mishra et al., 2022a).

Weighted Aggregated Sum Product Assessment (WASPAS)

Zavadskas et al. (2012) proposed the model WASPAS, which is a hybridization of the SAW model. The WASPAS method was used in certain recent studies such as reservoir flood control management policy (Mishra& Rani, 2018), multi-criteria physician selection problem (Rani et al., 2020b) and location selection for medical waste disposal (Mishra& Rani, 2021).

Technique for Order Preference by Similarity to Ideal Solution (TOPSIS)

TOPSIS was introduced by Hwang & Yoon (1981) as an effective approach for solving multi-attribute decision-making (MADM) or multi-criteria decision-making (MCDM) problems. It was further developed by Hwang et al. (1993). This technique ranks alternatives based on their distances from two reference locations, known as ideal solutions. Opricovic and Tzeng (2004) note that the TOPSIS technique does not evaluate the significance of these distances. Researchers are familiar with this MCDM approach. Current TOPSIS techniques can be found in the following studies: Evaluation of China's Shaanxi province's resource-based cities' economic transition and upgrading (Chen et al., 2018) and cloud computing (Khorsandan-dRamezanpour, 2020).

VIseKriterijumskaOptimizacija I KompromisnoResenje (VIKOR)

The VIKOR method is a reference based MCDM approach for resolving non-comparable and conflicting criterion choice problems. The method uses an ideal solution as a reference point and compares alternatives to it using the Manhattan and

Chebyshev distances. Opricovic and Tzeng (2004) describe in their work that the significant distinctions between the VIKOR and TOPSIS methods. This approach has been implemented in practical MCDM problems as follows: location section of an airport (SennarogluandCelebi, 2018), risk evaluation of construction projects (Wang et al., 2018), renewable energy technology (Rani et al., 2019) and personal selection problem (Krishankumar et al., 2020).

Evaluation based on Distance from Average Solution (EDAS)

Keshavarz Ghorabaee et al. (2015) introduced the EDAS in the MCDM approach. EDAS method estimates alternatives' desirability by their distance from an average solution. Like other MCDM techniques, EDAS can tackle non-comparable and conflicting criteria. EDAS is used to solve a variety of real-world MCDM issues, including choosing a green supplier (Zhang et al., 2019), evaluation of health-care waste disposal (Mishra et al., 2020a), investment problem (Hanand Wei, 2020) and selecting sustainable reverse logistics providers (Mishra et al., 2022b).

Multi-objective Optimization on the Basis of Ratio Analysis (MOORA)

MOORA was proposed initially by Brauers and Zavadskas in (2006). MOORA and COPRAS methods are quite similar. However, MOORA utilizes the vector for the initial decision matrix rather than the sum normalisation. Recent applications are MOORA can be found in EV charging station selection (Raniand Mishra,2021), regional green development level evaluation (Luoand Lin, 2022), reverse logistics provider choice of EV (Yang et al., 2022) and green supplier selection (Gai et al., 2022).

Combinative Distance Based Assessment (CODAS)

CODAS technique is based on the Euclidean and taxicab distances from the NIS, which is the worst criteria value. CODAS generally employ the Euclidean distance. If the two solutions are not comparable according to the Euclidean distance (for example, they are similar), the taxicab distance is employed as an alternate measurement. However most current CODAS method applications are financial management performance analysis (He et al., 2019), MNC financial models (Zhou et al., 2020), renewable energy options (Deveci et al., 2020), vehicle shredding facility location selection (Simic et al., 2021a) and transit investment taxation (Simic et al., 2021b).

Elimination and Choice Expressing Reality (ELECTRE)

The ELECTRE method is the most extensively used outranking approach. Researchers spent much time adapting the ELECTRE approach to fuzzy decision contexts. Some recent realistic applications are cellular phone service providers' performance (Mishra et al. (2020c), decision-making on the most suitable renewable energy sources (Karacsanand Kahraman, 2020b), analyzing failure modes and impacts for risk assessment (Akram et al., 2021) and selection of the best possible biomedical material (Kiricsci et al., 2022).

The scope of MCDM approaches is broader than the methodologies mentioned above. But these procedures are reviewed since they significantly impact the solution to the MCDM problem.

PRELIMINARY CONCEPT ON SOFT SET

The majority of parameters in real-world circumstances are imprecise, which is another word for flawed or wrong. The SVNS may explain the truth, hesitation, and falsehood of the provided parameters to get over this imprecision.

Definition 1:*Soft Set (*Molodtsov,1999)*: Let the power set of the universal set X be P(X), and S be the set of attributes and a mapping is defined as F: S P(X). Then, a Soft Set over X is defined as (F, X).*

Definition 2: *IndetermSoft Set (Smarandache,2022): Let P(H) be the power set of a non-empty subset H of the universal set X. Let the set of attributes be A. Then the function F*: $A \rightarrow P(H)$ be an IndetermSoft Set if at least one of the following properties satisfy:

i) Hesitancy exists in A.
ii) Hesitancy exists in H or P(H) .
iii) *Hesitancy exists in F such that* ∃ at least one $u \in A$, such that $F(u)=$ indeterminate (missing information, incomplete, contradictory, or not unique).

Several sources can't give clear or complete information about the sets $A, H, P(H)$ or the function F. IndetermSoft Set is an extension of classical (determinate) Soft Set, which deals with incomplete information or data.

Smarandache has extended the Soft Set to the Hyper Soft set by converting function F into a multi-attribute function. He has also introduced the hybrids of Crisp, Fuzzy, Intuitionistic Fuzzy, Neutrosophic, various fuzzy modifications, and Plithogenic Hyper Soft Set.

Definition 3: *HyperSoft Set(Smarandache, 2018):Let P(X) be the power set of X, the universal set. Let* $a_1, a_2, \ldots, a_n$ for $n \geq 1$ be 'n' distinct attributes, whose corresponding attribute values are respectively the sets $A_1, A_2, \ldots, A_n$, with $A_i \cap A_j = \phi$, for $i \neq j$, and $i, j \in \{1, 2, \ldots, n\}$. Then the pair $(F, A_1 \times A_2 \times \ldots \times A_n)$, where $F: A_1 \times A_2 \times \ldots \times A_n \rightarrow P(X)$, is called a HyperSoft Set over.

Definition 4: *IndetermHyperSoft Set(Smarandache,2022): Let P(H) the power set of H, a non-empty subset of the universal set X. Let* $a_1, a_2, \ldots, a_n$ for $n \geq 1$, be 'n' distinct attributes, whose corresponding attribute-values are respectively the sets $A_1, A_2, \ldots, A_n$, with $A_i \cap A_j = \phi$, for $i \neq j$, and $i, j \in \{1, 2, \ldots, n\}$. Then the pair $(F, A_1 \times A_2 \times \ldots \times A_n)$, where, $F: A_1 \times A_2 \times \ldots \times A_n \rightarrow P(X)$ is called an IndetermHyperSoft Set over X if at least one of the bellow occurs:

i) $A_1, A_2, \ldots, A_n$ has some indeterminacy.
ii) the sets H or P(H) have some indeterminacy.
iii) *there exist at least one n-ordered pair*$(k_1, k_2, \ldots, k_n) \in A_1 \times A_2 \times \ldots \times A_n$ such that $F(k_1, k_2, \ldots, k_n)$= indeterminate (missing information, incomplete, contradictory, or not unique).

Definition 5: *Tree soft Set(Smarandache,2023):Let P(H) the power set of H, a non-empty subset of the universal set X. Let* $A = \{A_1, A_2, \ldots, A_n\}$, for integer $n \geq 1$ be a set of attributes (parameters, factors, etc.), where $A_1, A_2, \ldots, A_n$ are considered attributes of first level (owing to their one-digit labels).

Each attribute A_i, $1 \leq i \leq n$, is formed by sub-attributes:

$$A_1 = \{A_{1,1}, A_{1,2}, \cdots\}$$

$$A_2 = \{A_{2,1}, A_{2,2}, \cdots\}$$

....

$$A_n = \{A_{n,1}, A_{n,2}, \cdots\}$$

where the above $A_{i,j}$ are sub-attributes (or attributes of second level) (owing to their two-digit labels).

Again, each sub-attribute $A_{i,j}$ is formed by sub-sub-attributes (or attributes of third level): $A_{i,j,k}$, and so on, as much refinement as needed into each application, up to sub-sub-…-sub-attributes (or attributes of m-level (or having m digits into the indexes): $A_{i1,i2,\cdots im}$.

Therefore, a graph-tree is formed, that we denote as Tree(A), whose root is A (considered of level zero), then nodes of level 1, level 2, up to level m.

We call leaves of the graph-tree, all terminal nodes (nodes that have no descendants). Then the Tree Soft Set is:

F: $P(Tree(A))\rightarrow P(H)$

Tree(A) is the set of all nodes and leaves (from level 1 to level m) of the graph-tree, and P(Tree(A)) is the power set of the *Tree(A)*.

All node sets of the Tree Soft Set of level m are:

$$Tree(A) = \left\{ A_{i_1}; i_1 = 1, 2, \cdots \right\}$$

*The nodes of level 1 form the first set, the nodes of level 2 form the second set, the nodes of level 3 form the third set, and so on until the nodes of level m create the final set. A Tree Soft Set with two levels (m=*2) graph-tree is called Multi Soft Set [Alkhazaleh et al. (2010)].

Definition 6: *Single-valued neutrosophic set (SNVS) (Smarandache,2004):*

Suppose R is the set of the universe of discourse. An SVNS (S) of a single-valued independent variable (x) is defined by $S = \left\{ \langle x; [\alpha_S(x), \beta_S(x), \gamma_S(x)] \rangle : x \in R \right\}$, where $\alpha_S(x)$, $\beta_S(x)$, $\gamma_S(x)$ stand for the concepts of truth, hesitation, and membership functions of falsehood, respectively. Here, $\alpha_S(x)$: $R\rightarrow[0,1]$ is the truth membership function, $\beta_S(x)$: $R\rightarrow[0,1]$ is the hesitation membership function, and the falsity membership function is $\gamma_S(x)$: $R\rightarrow[0,1]$ with $0\leq \alpha_S+\beta_S+\gamma_S \leq 3$.

Definition 7: *Membership and non-membership of SVNS (Liu and Wang, 2014):*

Let $\alpha_P, \beta_P, \gamma_P \in [0,1]$ and $r_1, r_2, r_3 \in R$ such that $r_1 \leq r_2 \leq r_3$. Then a single-valued tri-angular neutrosophic fuzzy number is $P = \langle (r_1, r_2, r_3); \alpha_P, \beta_P, \gamma_P \rangle$ is a special NS on the real line set R, *whose truth, hesitation and falsity membership functions are given as follows:*

$$\alpha_p(x) = \begin{cases} \alpha_P \dfrac{(x - r_1)}{(r_2 - r_1)}, & if \quad r_1 \leq x \leq r_2 \\ \alpha_P, & if \quad x = r_2 \\ \alpha_P \dfrac{(r_3 - x)}{(r_3 - r_2)}, & if \quad r_2 \leq x \leq r_3 \\ 0 & Otherwise \end{cases}$$

$$\beta_P(x)=\begin{cases}\beta_P\dfrac{(r_2-x)+\beta_P(x-r_1)}{(r_2-r_1)}, & if \quad r_1\le x\le r_2\\ \beta_P, & if \quad x=r_2\\ \beta_P\dfrac{(x-r_2)+\beta_P(r_3-x)}{(r_3-r_2)}, & if \quad r_2\le x\le r_3\\ 1 & Otherwise\end{cases}$$

and

$$\gamma_P(x)=\begin{cases}\gamma_P\dfrac{(r_2-x)+\gamma_P(x-r_1)}{(r_2-r_1)}, & if \quad r_1\le x\le r_2\\ \gamma_P, & if \quad x=r_2\\ \gamma_P\dfrac{(x-r_2)+\gamma_P(r_3-x)}{(r_3-r_2)}, & if \quad r_2\le x\le r_3\\ 1 & Otherwise\end{cases}$$

A diagram of SNVS is given in figure 1

Figure 1. A diagram of SNVS

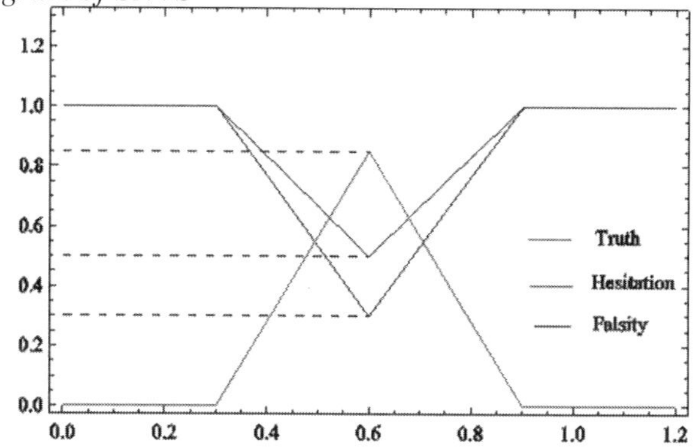

CONCEPT OF TOPSIS METHOD

One of the key techniques in the MCDM field is TOPSIS (Hwang et al., 1992; Yoon & Hwang, 1995), which has gained enormous popularity in applications and as the basis for many method developments. The core tenet of TOPSIS is that the optimum answer is the one that is closest to the PIS but furthest from the NIS. By using a global index that is based on how far away from the ideal solutions each alternative is, options are rated. The TOPSIS needs discrete attribute weight values to be produced before any calculations can be performed. The modified TOPSIS (Deng et al., 2000), which is likewise based on the TOPSIS approach, has gained popularity due to its innovative use of the methodology for objective weight elicitation that is based on Shannon's (Shannon, 2001) entropy theory. The weight in TOPSIS represents the decision maker's proportional preference for the criterion. However, in modified TOPSIS, attribute weights are elicited based on their entropy-measured significance. When the attribute weight is to be considered, each of these algorithms use the same Euclidean distance metric. Due to the strong similarities between these two approaches' mathematical frameworks and their suitability for solving the same class of MCDM issues, decision-makers choose any of them.

Let us consider a MCDM problem with 'm' distinct alternatives defined as $E=\{E_1,E_2,\ldots,E_m\}$ and 'n' criteria $\psi=\{\psi_1,\psi_2,\ldots,\psi_n\}$. Let $\sigma=(\sigma_1,\sigma_2,\ldots,\sigma_n)$ be the respective weights for the 'n' criteria $\psi=\{\psi_1,\psi_2,\ldots,\psi_n\}$where $\sigma_j\in[0,1]\forall j=1,2,\ldots,n$, where $\sum_{j=1}^{n}\sigma_j = 1$. Let ψ_B and ψ_C represent the benefit and cost criteria collection, respectively. Where $\psi = \psi_B \cup \psi_C$ and $\psi_B \cap \psi_C = \varphi$. The decision matrix $E = (E_{ij})_{m\times n}$ represents the performance ratings of the criteria for each alternative given in equation (1).

$$E = \begin{array}{c} \\ E_1 \\ E_2 \\ \vdots \\ E_m \end{array} \begin{array}{c} \begin{array}{ccccc} \psi_1 & \psi_2 & \psi_3 & \cdots & \psi_n \end{array} \\ \begin{bmatrix} E_{11} & E_{12} & E_{13} & \cdots & E_{1n} \\ E_{21} & E_{22} & E_{23} & \cdots & E_{2n} \\ \vdots & \vdots & \vdots & \cdots & \vdots \\ E_{m1} & E_{m2} & E_{m3} & \cdots & E_{mn} \end{bmatrix} \end{array} \tag{1}$$

Here, the stages of the TOPSIS methodology are demonstrated.

Step 1: **Create a calculation for the normalized performance ratings:**

To generate normalized performance ratings from Eq. (1), vector normalization is used. Each performance rating E_{ij} is divided by its norm, in this approach. This is given by the relation:

$$E_{ij}^{N} = \frac{E_{ij}}{\sqrt{\sum_{i=1}^{m} E_{ij}^{2}}} \tag{2}$$

By using dimensionless units, this translation procedure makes it simpler to compare different qualities.

Step 2: **Combining ratings and weight:**

The weighted normalized decision matrix (F) is calculated combining the performance rating with the corresponding weights of the criteria.

$$F = [f_{ij}]_{m \times n}$$

and

$$f_{ij} = \sigma_j * E_{ij} ; i = 1,2,...,m; j = 1,2,...,n \tag{3}$$

Step 3: **Finding PIS and NIS:**

Let E^+ and E^- represents the PIS and NIS set respectively, which will be determined from the weighted normalized matrix F and defined by $E^+ = (f_1^+, f_2^+, f_3^+, ... f_n^+)$ and $E^- = (f_1^-, f_2^-, f_3^-, ... f_n^-)$, where

$$f_j^+ = \begin{cases} \max\limits_{1 \le i \le m} [f_{ij}], where \quad \psi_j \in \psi_B \\ \min\limits_{1 \le i \le m} [f_{ij}], where \quad \psi_j \in \psi_C \end{cases} \tag{4}$$

and

$$f_j^- = \begin{cases} \min\limits_{1 \le i \le m} [f_{ij}], where \quad \psi_j \in \psi_B \\ \max\limits_{1 \le i \le m} [f_{ij}], where \quad \psi_j \in \psi_C \end{cases} \tag{5}$$

Step 4: **Discover the separation values:**

The Euclidean distance theory is used to figure out the separation measure, that determines the distance between each option grade and both the PIS and the NIS. In equations (6) and (7), the processes for deriving PIS and NIS separation are depicted, respectively.

$$D_j^+ = \sqrt{\sum_{j=1}^{n} (f_{ij} - f_j^+)^2}; \; i = 1,2,...,m \tag{6}$$

And

$$D_j^- = \sqrt{\sum_{j=1}^{n} (f_{ij} - f_j^-)^2}; \; i = 1,2,...,m \tag{7}$$

For finding the separation measure, later on the modification has been done in the equation no. (6) and (7). The modified equations then rewritten as follows:

$$D_i^+ = \sqrt{\sum_{j=1}^{n} \sigma_j (f_{ij} - f_j^+)^2}; \; i = 1,2,...,m \tag{8}$$

And

$$D_i^- = \sqrt{\sum_{j=1}^{n} \sigma_j (f_{ij} - f_j^-)^2}; \; i = 1, 2, ..., m \tag{9}$$

Step 5: **Generate an aggregate preference score:**

Now the aggregate preference score (Z_i) for each option is (E_i) obtained by the relation shown in equation (10).

$$Z_i = \frac{D_i^+}{D_i^+ + D_i^-} \tag{10}$$

Following the above steps, in any MCDM problem, we can find the rank of the alternatives, based on the higher performance values of (Z_i).

PROPOSED NEUTROSOPHIC TOPSIS TECHNIQUE

The neutrosophic TOPSIS approach for resolving the MCDM issue with uncertainty is demonstrated in the present section. Then, the expert provides an SVNS rating for a criterion of an alternative. The SVNS rating is then used to create a neutrosophic decision matrix. The PIS, the NIS, the type I and II correlation measures, the Index function, and their attributes are all defined using the decision matrix.

Let us consider m distinct alternatives defined as $E = \{E_1, E_2, \ldots, E_m\}$ and n criteria $\psi = \{\psi_1, \psi_2, \ldots, \psi_n\}$. Let $\sigma = (\sigma_1, \sigma_2, \ldots, \sigma_n$ be the respective weights for the 'n' criteria $\psi = \{\psi_1, \psi_2, \ldots, \psi_n\}$ where $\sigma_j \in [0,1] \forall j = 1,2,\ldots,n$ and $\sum_{j=1}^{n} \sigma_j = 1$ Let ψ_B and ψ_C represent the benefit and cost criteria collection, respectively. Where $\psi = \psi_B \cup \psi_C$ and $\psi_B \cap \psi_C = \varphi$. Each entry in the decision matrix $E = (E_{ij})_{m \times n}$ are SVNSs as shown below:

$$E = \begin{array}{c} \\ E_1 \\ E_2 \\ \vdots \\ E_m \end{array} \begin{array}{c} \begin{array}{ccccc} \psi_1 & \psi_2 & \psi_3 & \cdots & \psi_n \end{array} \\ \begin{bmatrix} (\alpha_{11}, \beta_{11}, \gamma_{11}) & (\alpha_{12}, \beta_{12}, \gamma_{12}) & (\alpha_{13}, \beta_{13}, \gamma_{13}) & \cdots & (\alpha_{1n}, \beta_{1n}, \gamma_{1n}) \\ (\alpha_{21}, \beta_{21}, \gamma_{21}) & (\alpha_{22}, \beta_{22}, \gamma_{22}) & (\alpha_{23}, \beta_{23}, \gamma_{23}) & \cdots & (\alpha_{2n}, \beta_{2n}, \gamma_{2n}) \\ \vdots & \vdots & \vdots & \cdots & \vdots \\ (\alpha_{m1}, \beta_{m1}, \gamma_{m1}) & (\alpha_{m2}, \beta_{m2}, \gamma_{m2}) & (\alpha_{m3}, \beta_{m3}, \gamma_{m3}) & \cdots & (\alpha_{mn}, \beta_{mn}, \gamma_{mn}) \end{bmatrix} \end{array} \tag{11}$$

The element $E_{ij} = (\alpha_{ij}, \beta_{ij}, \gamma_{ij})$ denote the ij^{th} rating of SVNS equation (11) by the expert. The characteristic Z_i corresponding to the alternative E_i in the ith row is represented as

$$Z_i = \{(\psi_1; E_{i1}), (\psi_2; E_{i2}), ..., (\psi_n; E_{in})\}$$
$$= \{(\psi_1; \alpha_{i1}; \beta_{i1}; \gamma_{i1}), (\psi_2; \alpha_{i2}; \beta_{i2}; \gamma_{i2}), ..., (\psi_n; \alpha_{in}; \beta_{in}; \gamma_{in})\}$$

Definition 8: *Positive ideal solution (PIS) and Negative ideal solution (NIS):* Let E_+ and E_- denote the PIS and NIS represented as

$$E_+ = \{(\psi_1; E_{+1}), (\psi_2; E_{+2}), ..., (\psi_n; E_{+n})\} \text{ and}$$

$$E_- = \{(\psi_1; E_{-1}), (\psi_2; E_{-2}), ..., (\psi_n; E_{-n})\}$$

where E_{+j} *and* E_{-j} are defined as $E_{+j} = (\alpha_{+j}, \beta_{+j}, \gamma_{+j})$ where

$$(\alpha_{+j}, \beta_{+j}, \gamma_{+j}) = \begin{cases} \left(\max_{1 \le i \le m} [\alpha_{ij}], \min_{1 \le i \le m} [\beta_{ij}], \min_{1 \le i \le m} [\gamma_{ij}]\right), \text{ if } \psi_j \in \psi_B \\ \left(\min_{1 \le i \le m} [\alpha_{ij}], \max_{1 \le i \le m} [\beta_{ij}], \max_{1 \le i \le m} [\gamma_{ij}]\right), \text{ if } \psi_j \in \psi_C \end{cases} \tag{12}$$

And $E_{-j} = (\alpha_{-j}, \beta_{-j}, \gamma_{-j})$, where

$$(\alpha_{-j}, \beta_{-j}, \gamma_{-j}) = \begin{cases} \left(\min_{1 \le i \le m} [\alpha_{ij}], \max_{1 \le i \le m} [\beta_{ij}], \max_{1 \le i \le m} [\gamma_{ij}]\right), \text{ if } \psi_j \in \psi_B \\ \left(\max_{1 \le i \le m} [\alpha_{ij}], \min_{1 \le i \le m} [\beta_{ij}], \min_{1 \le i \le m} [\gamma_{ij}]\right), \text{ if } \psi_j \in \psi_C \end{cases} \tag{13}$$

Definition 9: *Weighted neutrosophic correlation coefficient [Lin et al. (2019)]:*

Let E_s, E_t *be the two neutrosophic characteristics in the neutrosophic decision-making matrix* $E, \sigma = (\sigma_1, \sigma_2, ..., \sigma_n)$, $\sum_{j=1}^{n} \sigma_j = 1$ *be the weight vector related with the criteria and* $\overline{\alpha}_j = \frac{\sum_{i=1}^{m} \alpha_{ij}}{m}$, $\overline{\beta}_j = \frac{\sum_{i=1}^{m} \beta_{ij}}{m}$, $\overline{\gamma}_j = \frac{\sum_{i=1}^{m} \gamma_{ij}}{m}$. *Then, the weighted neutrosophic correlation coefficient between* E_s, E_t *is defined as*

$$\theta^\sigma(E_s, E_t) = \frac{1}{3}\left[\theta_\alpha{}^\sigma(E_s, E_t) + \theta_\beta{}^\sigma(E_s, E_t) + \theta_\gamma{}^\sigma(E_s, E_t)\right] \tag{14}$$

Where

$$\theta_\alpha{}^\sigma(E_s, E_t) = \frac{\sum_{j=1}^{n} \sigma_j \left[\alpha_{sj}^2 - \overline{\alpha}_j^2\right] \cdot \left[\alpha_{tj}^2 - \overline{\alpha}_j^2\right]}{\sqrt{\sum_{j=1}^{n} \sigma_j \left[\alpha_{sj}^2 - \overline{\alpha}_j^2\right]} \sqrt{\sum_{j=1}^{n} \sigma_j \left[\alpha_{tj}^2 - \overline{\alpha}_j^2\right]}} \tag{15}$$

$$\theta_\beta{}^\sigma(E_s, E_t) = \frac{\sum_{j=1}^{n} \sigma_j \left[\beta_{sj}^2 - \bar{\beta}_j^2\right] \cdot \left[\beta_{tj}^2 - \bar{\beta}_j^2\right]}{\sqrt{\sum_{j=1}^{n} \sigma_j \left[\beta_{sj}^2 - \bar{\beta}_j^2\right]} \sqrt{\sum_{j=1}^{n} \sigma_j \left[\beta_{tj}^2 - \bar{\beta}_j^2\right]}} \tag{16}$$

$$\theta_\gamma{}^\sigma(E_s, E_t) = \frac{\sum_{j=1}^{n} \sigma_j \left[\gamma_{sj}^2 - \bar{\gamma}_j^2\right] \cdot \left[\gamma_{tj}^2 - \bar{\gamma}_j^2\right]}{\sqrt{\sum_{j=1}^{n} \sigma_j \left[\gamma_{sj}^2 - \bar{\gamma}_j^2\right]} \sqrt{\sum_{j=1}^{n} \sigma_j \left[\gamma_{tj}^2 - \bar{\gamma}_j^2\right]}} \tag{17}$$

The denominator in the formulas above is expected to be non-zero, just like in the un-weighted preceding instance.

Definition 10: *Weighted Type I and Type II closeness measures: Let $\theta^\sigma(E_i, E_+)$ and $\theta^\sigma(E_i, E_-)$ represent the neutrosophic correlation coefficients for PIS and NIS respectively for the alternative E_i. Also, let $R_I{}^\sigma(E_i), R_{II}{}^\sigma(E_i)$ represent the weighted type I and type II closeness measure and where*

$$R_I{}^\sigma(E_i) = \frac{1 - \theta^\sigma(E_i, E_-)}{2 - \theta^\sigma(E_i, E_+) - \theta^\sigma(E_i, E_-)} \tag{18}$$

And

$$R_{II}{}^\sigma(E_i) = \frac{1 + \theta^\sigma(E_i, E_-)}{2 + \theta^\sigma(E_i, E_+) + \theta^\sigma(E_i, E_-)} \tag{19}$$

It is assumed that the denominator of $R_I{}^\sigma(E_i)$ and $R_{II}{}^\sigma(E_i)$ are not zero.

There is a positive or negative relationship between E_i, E_+ if $\theta^\sigma(E_i, E_+)$ are positive or negative, respectively. Better selection of E_i is indicated by a larger value of $\theta^\sigma(E_i, E_+)$. The worst choice of E_i is represented by a larger value of $\theta^\sigma(E_i, E_-)$. The type I closeness measure $R_I{}^\sigma(E_i)$ and type II closeness measure $R_{II}{}^\sigma(E_i)$ assist the DM in selecting a better alternative.

Definition 11: *Weighted index value: Let $R_I{}^\sigma(E_i), R_{II}{}^\sigma(E_i)$ denote the weighted type I and type II closeness measures for the alternative E_i respectively and let ε denote a closeness parameter, where $0 \leq \varepsilon \leq 1$. The neutrosophic correlation-based weighted index is $I^\sigma(E_i) = \varepsilon R_I{}^\sigma(E_i) + (1 - \varepsilon) R_{II}{}^\sigma(E_i)$. The type I and type II closeness measurements impact the parameter ε. The larger value of ε indicates that neutrosophic correlation closeness $I^\sigma(E_i)$ would focus on $R_I{}^\sigma(E_i)$ and smaller value of ε indicates type II focus on $R_{II}{}^\sigma(E_i)$.*

Figure 2.The flow chart of the algorithm

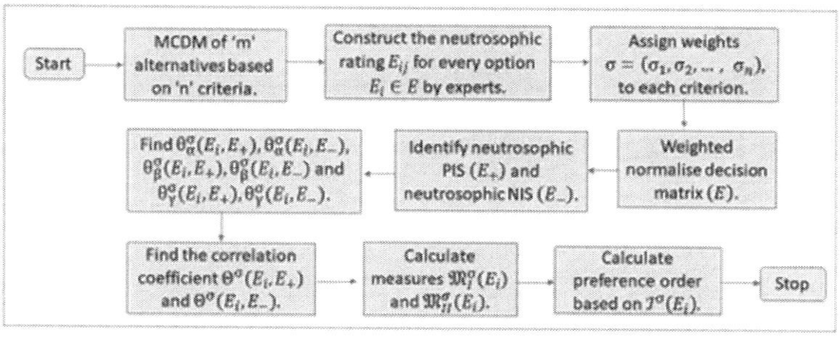

DESCRIPTION OF RENEWABLE ENERGY: ALTERNATIVES AND CRITERIA

Renewable energy is defined as energy produced by renewable natural resources, such as wind,sunlight, water, and geothermal heat. It is seen as an alternative to fossil fuels (coal, oil, and natural gas), which are non-renewable energy sources, because it has a substantially smaller environmental effect and provides sustainability over the long term. Renewable energy is derived from natural resources that regenerate faster than they are depleted. All around us, there are several renewable energy sources. Several popular forms of renewable energy include wind energy, solar energy, hydropower, biomass energy, geothermal energy, tidal energy. Injurious pollutants including sulphur dioxide, nitrogen oxides, and particulates are not released by renewable energy sources. When fossil fuels are utilized to generate energy, harmful greenhouse gases such as carbon dioxide are released. The combustion of fossil fuels produces more emissions than the generation of power from renewable sources. The transition from fossil fuels to renewable energy is critical to resolving the climate crisis. Renewable energies are now more inexpensive in most countries and employ three times as many people as fossil fuels. All nations have access to renewable energy sources; however, their potential has not yet been completely realized. In view of the International Renewable Energy Agency (IRENA), renewable energy sources can and ought to provide 90% of all electricity in the future.

Nowadays, the cheapest source of energy in most of the globe is renewable energy. Renewable energy sources contribute to lowering carbon footprints and preventing climate change by emitting little or no greenhouse gas emissions. It is becoming increasingly interesting due to declining costs. By 2030, cheap power from renewable sources may account for 65% of the world's total electrical supply. The World Health Organization (WHO) guesses that roughly 99% of people worldwide breathe air that is dangerous to their health. A few widely used renewable energy sources are hydropower, solar energy, geothermal energy, ocean energy, wind energy, bio-energy, etc.

Renewable Energy Alternatives

Now we select which one is the desire for the renewable energy source though we may use any kind of renewable energy source. The following are four alternatives that we consider here, and we shall provide a rank according to our aspiration.

Solar energy: Solar energy is a form of renewable energy that uses the sun's energy to produce heat or electricity. It is a safe and environmentally friendly form of energy that does not cause harmful emissions or have an impact on global warming. There are two main ways to transform solar energy into useful types of energy: photovoltaic (PV) systems and solar thermal systems. In turn, this increases energy security and independence for nations and people by reducing dependence on fossil fuels. Solar energy is the most abundant source of energy, and it may even be utilized under gloomy situations. Photovoltaic (PV) panels or solar radiation concentrators are used in solar energy systems to convert sunlight into electrical energy. Because of dramatic price drops in solar panel manufacture, solar panels are now not only affordable, but often the cheapest source of electricity.

Wind energy: A renewable energy source known as wind energy uses the wind's energy to produce electricity. It is a clean energy source that does not emit greenhouse gases or other air pollutants when in use. Although wind energy has been utilized for centuries, the effectiveness and scale of wind power generation have been greatly increased by the development of contemporary wind turbines, also known as wind generators or windmills. Wind is a plentiful and sustainable resource; therefore it won't run out over time. The fuel (wind) is free once a wind turbine is installed; hence running expenses are negligible for the duration of the turbine.In wind energy, large wind turbines deployed on land or at sea harness the kinetic energy of moving air. Wind energy has a bigger technological potential than the entire quantity of electricity generated globally. Although there are powerful winds in many places across the world, the finest places to produce wind power are frequently in remote areas.

Hydro-energy: Hydro-energy, commonly referred to as hydropower or hydroelectric power, is a sustainable energy source that harnesses the energy of moving or falling water to produce electricity. It is among the most traditional and popular types of renewable energy. A river or a reservoir built by a dam serves as the typical water sources for hydroelectric power plants. Water diversion systems divert a portion of a river's flow while avoiding the construction of a dam, whereas dams are built to regulate the flow of water and create a reservoir. The energy of water flowing from higher elevations to lower ones is captured by hydropower. Hydropower is presently the most important source of renewable energy in the electrical sector.

Bioenergy: Bioenergy is defined as power produced by biomass, or organic material originating from plants and animals. Due to the fact that it makes use of freshly produced plant material, which may be regenerated through natural processes, it is a type of renewable energy. Combustion, fermentation, and the creation of biofuels from organic material are just a few of the processes that can create bioenergy. Agricultural crops (like corn, sugarcane, and soybeans), forestry waste, energy crops (like switchgrass and miscanthus), organic waste (like animal manure and food scraps), and specialised energy crops grown for bioenergy production are just a few of the many sources of biomass that can be used to make bioenergy. A range of organic resources referred to as biomass is used to create bioenergy, including wood, charcoal, dung, and other manures for the generation of heat and electricity, as well as agricultural crops for the creation of liquid biofuels. The majority of biomass is utilized in rural regions for space heating, lighting, and cooking.

Criteria of the Renewable Energy Alternatives

There are some criteria for every alternative depending on which we can select the one. Setting up plants is necessary to obtain energy. Depending on the type of installation and the particular requirements of the project, installation costs might vary widely. Costs can vary based on a number of variables, including location, degree of difficulty, and state of the market. The price of installing a renewable energy system can vary greatly depending on several variables, including the technology used, the scale of the system, the location, the available subsidies, and the requirements of the particular project. The installation cost can be considered as one criterion. Maintain the plant; there is some maintenance cost that we must pay attention to. The individual asset or item that needs maintenance will determine the maintenance expenses. The cost of maintaining a renewable energy facility can differ based on a number of variables, including the technology used, the size of the plant, and the location. A few general maintenance cost factors are equipment maintenance, labor costs, spare parts and materials, monitoring and control systems etc. Therefore, maintenance cost is another criterion. A crucial component of setup is the plant area. Choose the kind of energy plant you wish to build, such as a biomass, solar, wind, hydroelectric, or geothermal one. Each energy source has unique specifications and is appropriate for various geographical locations. Hence the area needed to build up the plant is another criterion. The preceding three criteria need a minimum value; hence, they are considered non-beneficial. What is the expected lifespan of any plant? We can increase our revenue if the plant lives a long time. The longevity of a project plant can vary significantly based on a number of variables, including the kind of plant, the operating environment, the quantity of maintenance and care it receives, and the general quality of its construction. Several variables, including the kind of energy plant, its capacity, location, operating costs, government incentives, and market circumstances, can have a substantial impact on the financial return from an energy plant. The various energy facilities produce varied financial returns. As a result, a plant's lifespan (lifetime) and its financial return (profit) are used as criteria. The maximum value is the desire for both lifetime and profit; thus, both are viewed as beneficial criteria.

RANKING OF RENEWABLE ENERGY ALTERNATIVES: A NUMERICAL ELASTRATION

In the present section, we construct the MCDM problem of ranking the renewable energy with four alternatives, which we already explain in the previous section 6.1, namely: solar (E_1), wind (E_2), hydro (E_3) and biomass (E_4). Each of which has five criteria (discoursed in section 6.2): installation cost (ψ_1), maintenance cost (ψ_2), plant area (ψ_3), plant's life span (ψ_4) and profit (ψ_5). To get the neutrosophic rating of the criteria, an expert in the specified field rate the criteria by a four-point neutrosophic rating as given in table 1.

Table 1. Four-point Linguistic scale with neutrosophic rating

Linguistic Variable	**Low (L)**	**Medium (M)**	**High (H)**	**Very High (VH)**
Membership function	(0.1, 0.3, 0.8)	(0.4, 0.25, 0.7)	(0.7, 0.15, 0.4)	(0.9, 0.2, 0.1)

The weight associated with the criteria is very much essential. According to the DM's knowledge, the significance of the criteria is listed as follows. Since the installation cost for any plant is one of the prior criteria, more importance is given to that as 0.35. Profits from the project should be as desirable as the capital invested in the plant. Profit can thus be credited for another significant factor and a weight of 0.3. Maintenance cost can be considered less desirable; hence the weight of this criterion is taken as 0.15. The other two criteria, the life span of the project and the area needed for the project, are given equal importance as they are equally essential in the plant, and tolerance is given 0.1. So, the weight vector is considered as $\sigma=$ (0.35, 0.15, 0.10, 0.10, 0.30). The neutrosophic rating for each criterion according to each alternative is given in table 2.

Table 2. Linguistic term-based decision matrix

Criteria→ / Alternatives↓	ψ_1	ψ_2	ψ_3	ψ_4	ψ_5
E_1	L	L	H	H	M
E_2	VH	H	VH	L	H
E_3	H	M	M	VH	L
E_4	M	VH	L	M	VH

The weighted normalized SVNS decision matrix is build up by four alternatives and five criteria as follows:

$$E = \begin{bmatrix} (0.03,0.23,0.25) & (0.01,0.10,0.11) & (0.06,0.03,0.04) & (0.06,0.03,0.04) & (0.10,0.16,0.18) \\ (0.26,0.15,0.03) & (0.09,0.05,0.05) & (0.07,0.04,0.01) & (0.01,0.06,0.07) & (0.17,0.10,0.11) \\ (0.20,0.11,0.12) & (0.05,0.08,0.09) & (0.03,0.05,0.06) & (0.07,0.04,0.01) & (0.02,0.19,0.21) \\ (0.12,0.19,0.21) & (0.11,0.06,0.01) & (0.01,0.06,0.07) & (0.03,0.05,0.06) & (0.22,0.13,0.03) \end{bmatrix}$$

The neutrosophicPIS and the neutrosophic NIS are given in table 3.

Table 3. Neutrosophic PIS and Neutrosophic NIS

Neutrosophic ideal solution	Components	1	2	3	4	5
PIS	Membership	0.03	0.01	0.01	0.07	0.22
	Hesitant	0.23	0.10	0.06	0.03	0.10
	Non-membership	0.25	0.11	0.07	0.01	0.03
NIS	Membership	0.26	0.11	0.07	0.01	0.02
	Hesitant	0.11	0.05	0.03	0.06	0.19
	Non-membership	0.03	0.01	0.01	0.07	0.21

An approximation of a choice may be made by the PIS and NIS. The best option is shown by the PIS's highest membership function, which shows that it is. The worst option would be chosen if the non-membership of the NIS was at its lowest level. The next table 4 gives the weighted correlation, type I and type II closeness measures and index value to rank the alternatives.

Table 4. Weighted correlation coefficient, type I and type II closeness measures, Index Values and Ranking

Options	$\theta^{\sigma}(E_i,E_+)$	$\theta^{\sigma}(E_i,E_-)$	$R_I^{\sigma}(E_i)$	$R_{II}^{\sigma}(E_i)$	$I^{\sigma}(E_i)$	Ranking
E_1	0.61	-0.58	1.64	0.41	1.0227	II
E_2	-0.41	0.36	0.61	1.43	1.0219	III
E_3	-0.84	0.93	0.07	1.78	0.9253	IV
E_4	0.95	-0.82	2.09	0.16	1.1213	I

Therefore, the ranking of the renewable energy alternatives is as follows: $E_4 \succ E_1 \succ E_2 \succ E_3$. Figure 2 shows the ranking of the alternatives along with their index values.

Figure 3. Ranking of the renewable energy alternatives

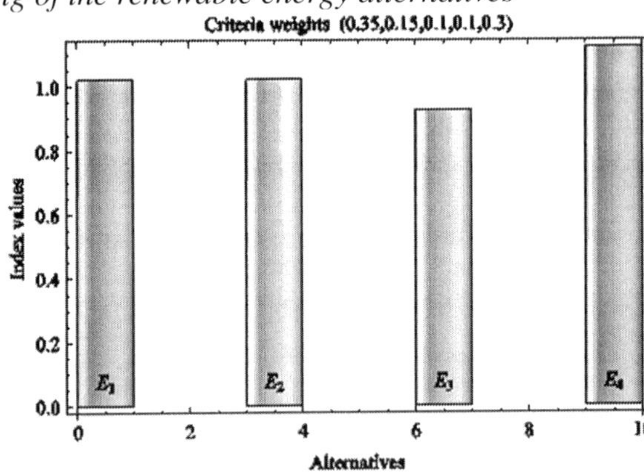

The change of ranking for different closeness parameter ε is given in figure 3.

Figure 4. Ranking of the alternatives for different closeness parameter ε

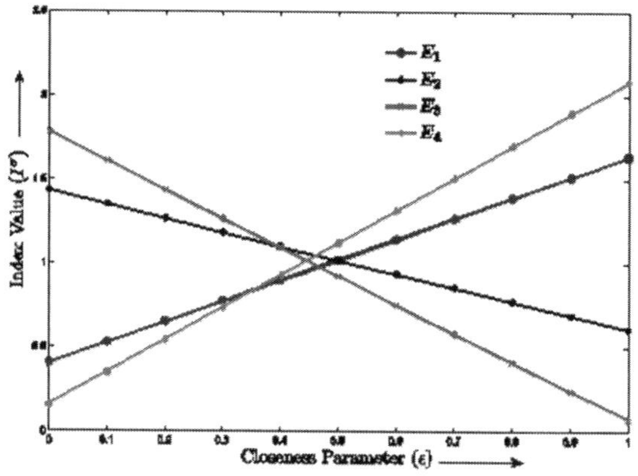

Figure 3 clears that when the closeness parameter is near zero or one, the rankings of the alternatives are significantly different. The differences in the index values of the options are noticeably small when the closeness parameter is close to 0.5. If the DM assigns greater weight to type I or type II proximity measures, however, the ranking of the options is determined by a significantly more significant margin.

CONCLUSION

In the recent decade, fuzzy TOPSIS has been used to handle several MCDM problems. In this fuzzy TOPSIS technique, Euclidean and Hamming distance measures compute the distance between PIS and NIS. It is still being determined which distance measures should be utilized to calculate the distance. This research solves an uncertain MCDM issue using the neutrosophic TOPSIS method. We skip the distance measure technique in the TOPSIS method, using the correlation measure instead. The correlation measure is reliable for determining the relationships between criteria by PIS and NIS. The suggested method is effectively implemented to decide the alternative renewable energy choices based on its conflicting criteria. The application demonstrates that the suggested method for solving an uncertain MCDM issue is simple. This method may determine the MCDM for health and safety management, environmental and construction management, energy management,military applications, and industrial engineering.By using advanced soft sets, one may expand the TOPSIS technique for handling uncertain MCDM problems.

REFERENCES

Abdel-Basset, M., Manogaran, G., Gamal, A., & Smarandache, F. (2019). A group decision making framework based on neutrosophic topsis approach for smart medical device selection. *Journal of Medical Systems*, *43*(2), 1–13. DOI: 10.1007/s10916-019-1156-1 PMID: 30627801

Akram, M., Luqman, A., & Alcantud, J. C. R. (2021). Risk evaluation in failure modes and effects analysis: Hybrid TOPSIS and ELECTRE I solutions with Pythagorean fuzzy information. *Neural Computing & Applications*, *33*(11), 5675–5703. DOI: 10.1007/s00521-020-05350-3

Alkhazaleh, S., Salleh, A. R., Hassan, N., & Ahmad, A. G. (2010, November). Multisoft Sets. In *Proc. 2nd International Conference on Mathematical Sciences* (pp. 910-917).

Atanassov, K. T. (1986). Intuitionistic fuzzy sets. *Fuzzy Sets and Systems*, *20*(1), 87–96. DOI: 10.1016/S0165-0114(86)80034-3

Biswas, P., Pramanik, S., & Giri, B. C. (2016). Topsismethod for multi-attribute-groupdecision-makingundersingle-valuedneutrosophicenvironment. *Neural Computing & Applications*, *27*(3), 727–737. DOI: 10.1007/s00521-015-1891-2

Brauers, W. K., & Zavadskas, E. K. (2006). The MOORA method and its application to privatization in a transition economy. *Control and Cybernetics*, *35*(2), 445–469.

Chen, C. T. (2000). Extensions of the topsis for group decision-making under fuzzy environment. *Fuzzy Sets and Systems*, *114*(1), 1–9. DOI: 10.1016/S0165-0114(97)00377-1

Chen, S. J., Hwang, C. L., Chen, S. J., & Hwang, C. L. (1992). *Fuzzy multiple attribute decision making methods*. Springer Berlin Heidelberg. DOI: 10.1007/978-3-642-46768-4

Chen, W., Shen, Y., & Wang, Y. (2018). Evaluation of economic transformation and upgrading of resource-based cities in Shaanxi province based on an improved TOPSIS method. *Sustainable Cities and Society*, *37*, 232–240. DOI: 10.1016/j.scs.2017.11.019

Deng, H., Yeh, C. H., & Willis, R. J. (2000). Inter-company comparison using modified TOPSIS with objective weights. *Computers & Operations Research*, *27*(10), 963–973. DOI: 10.1016/S0305-0548(99)00069-6

Deveci, K., Cin, R., & Kağızman, A. (2020). A modified interval valued intuitionistic fuzzy CODAS method and its application to multi-criteria selection among renewable energy alternatives in Turkey. *Applied Soft Computing*, *96*, 106660. DOI: 10.1016/j.asoc.2020.106660

Gai, L., Liu, H. C., Wang, Y., & Xing, Y. (2023). Green supplier selection and order allocation using linguistic Z-numbers MULTIMOORA method and bi-objective non-linear programming. *Fuzzy Optimization and Decision Making*, *22*(2), 267–288. DOI: 10.1007/s10700-022-09392-1

Gul, S. (2021). Fermatean fuzzy set extensions of SAW, ARAS, and VIKOR with applications in COVID-19 testing laboratory selection problem. *Expert Systems: International Journal of Knowledge Engineering and Neural Networks*, *38*(8), e12769. DOI: 10.1111/exsy.12769 PMID: 34511690

Han, L., & Wei, C. (2020). An extended EDAS method for multicriteria decision-making based on multivalued neutrosophic sets. *Complexity*, *2020*, 1–9. DOI: 10.1155/2020/7578507

He, T., Wei, G., Wei, C., & Wang, J. (2019). CODAS method for Pythagorean 2-tuple linguistic multiple attribute group decision making. *IEEE Access*. IEEE.

Hezam, I. M., Nayeem, M. K., Foul, A., & Alrasheedi, A. F. (2021). Covid-19 vaccine: A neutrosophic mcdm approach for determining the priority groups. *Results in Physics*, *20*, 103654. DOI: 10.1016/j.rinp.2020.103654 PMID: 33520620

Hwang, C. L., Lai, Y. J., & Liu, T. Y. (1993). A new approach for multiple objective decision making. *Computers & Operations Research*, *20*(8), 889–899. DOI: 10.1016/0305-0548(93)90109-V

Hwang, C. L., & Yoon, K. (1981). Methods for multiple attribute decision making. *Multiple attribute decision making: methods and applications a state-of-the-art survey*, 58-191.

Hwang, F. P., Chen, S. J., & Hwang, C. L. (1992). *Fuzzy multiple attribute decision making: Methods and applications*. Springer Berlin/Heidelberg.

Karaşan, A., & Kahraman, C. (2020a). Selection of the most appropriate renewable energy alternatives by using a novel interval-valued neutrosophic ELECTRE I method. *Informatica (Vilnius)*, *31*(2), 225–248. DOI: 10.15388/20-INFOR388

Karasan, A., & Kaya, I. (2020b). Neutrosophic TOPSIS method for technology evaluation of unmanned aerial vehicles (UAVs). In *Intelligent and Fuzzy Techniques in Big Data Analytics and Decision Making:Proceedings of the INFUS 2019 Conference* (pp. 665-673). Springer International Publishing.

Keshavarz Ghorabaee, M., Zavadskas, E. K., Olfat, L., & Turskis, Z. (2015). Multi-criteria inventory classification using a new method of evaluation based on distance from average solution (EDAS). *Informatica (Vilnius)*, *26*(3), 435–451. DOI: 10.15388/Informatica.2015.57

Khorsand, R., & Ramezanpour, M. (2020). An energy-efficient task-scheduling algorithm based on a multi-criteria decision-making method in cloud computing. *International Journal of Communication Systems*, *33*(9), e4379. DOI: 10.1002/dac.4379

Kirişci, M., Demir, I., & Şimşek, N. (2022). Fermatean fuzzy ELECTRE multi-criteria group decision-making and most suitable biomedical material selection. *Artificial Intelligence in Medicine*, *127*, 102278. DOI: 10.1016/j.artmed.2022.102278 PMID: 35430046

Krishankumar, R., Premaladha, J., Ravichandran, K. S., Sekar, K. R., Manikandan, R., & Gao, X. Z. (2020). A novel extension to VIKOR method under intuitionistic fuzzy context for solving personnel selection problem. *Soft Computing*, *24*(2), 1063–1081. DOI: 10.1007/s00500-019-03943-2

Lin, Y. L., Ho, L. H., Yeh, S. L., & Chen, T. Y. (2019). A pythagoreanfuzzytopsis method based on novel correlation measures and its application to multiple criteria decision analysis of inpatient stroke rehabilitation. *International Journal of Computational Intelligence Systems*, *12*(1), 410–425. DOI: 10.2991/ijcis.2018.125905657

Liu, P., & Wang, Y. (2014). Multiple attribute decision-making method based on single-valued neutrosophic normalized weighted bonferroni mean. *Neural Computing & Applications*, *25*(7), 2001–2010. DOI: 10.1007/s00521-014-1688-8

Luo, S., & Liu, J. (2022). An innovative index system and HFFS-MULTIMOORA method based group decision-making framework for regional green development level evaluation. *Expert Systems with Applications*, *189*, 116090. DOI: 10.1016/j.eswa.2021.116090

Mishra, A. R., Liu, P., & Rani, P. (2022). COPRAS method based on interval-valued hesitant Fermatean fuzzy sets and its application in selecting desalination technology. *Applied Soft Computing*, *119*, 108570. DOI: 10.1016/j.asoc.2022.108570

Mishra, A. R., Mardani, A., Rani, P., & Zavadskas, E. K. (2020a). A novel EDAS approach on intuitionistic fuzzy set for assessment of health-care waste disposal technology using new parametric divergence measures. *Journal of Cleaner Production*, *272*, 122807. DOI: 10.1016/j.jclepro.2020.122807

Mishra, A. R., & Rani, P. (2018). Interval-valued intuitionistic fuzzy WASPAS method: Application in reservoir flood control management policy. *Group Decision and Negotiation*, *27*(6), 1047–1078. DOI: 10.1007/s10726-018-9593-7

Mishra, A. R., & Rani, P. (2021). Multi-criteria healthcare waste disposal location selection based on Fermatean fuzzy WASPAS method. *Complex & Intelligent Systems*, *7*(5), 2469–2484. DOI: 10.1007/s40747-021-00407-9 PMID: 34777968

Mishra, A. R., Rani, P., Mardani, A., Pardasani, K. R., Govindan, K., & Alrasheedi, M. (2020b). Healthcare evaluation in hazardous waste recycling using novel interval-valued intuitionistic fuzzy information based on complex proportional assessment method. *Computers & Industrial Engineering*, *139*, 106140. DOI: 10.1016/j.cie.2019.106140

Mishra, A. R., Rani, P., & Pandey, K. (2022). Fermatean fuzzy CRITIC-EDAS approach for the selection of sustainable third-party reverse logistics providers using improved generalized score function. *Journal of Ambient Intelligence and Humanized Computing*, *13*(1), 1–17. DOI: 10.1007/s12652-021-02902-w PMID: 33584868

Mishra, A. R., Singh, R. K., & Motwani, D. (2020c). Intuitionistic fuzzy divergence measure-based ELECTRE method for performance of cellular mobile telephone service providers. *Neural Computing & Applications*, *32*(8), 3901–3921. DOI: 10.1007/s00521-018-3716-6

Mollaoglu, M., Bucak, U., Demirel, H., & Balin, A. (2022). Evaluationofvarious fuelalternativesintermsofsustainabilityfortheshipinvestmentdecisionusingsinglevalued neutrosophic numbers with topsis methods. *Proceedings of the Institution of Mechanical Engineers, PartM:Journal of Engineering for the Maritime Environment*, *237*(1), 215–226.

Molodtsov, D. (1999). Soft set theory—First results. *Computers & Mathematics with Applications (Oxford, England)*, *37*(4-5), 19–31. DOI: 10.1016/S0898-1221(99)00056-5

Muddineni, V. P., Sandepudi, S. R., & Bonala, A. K. (2017). Improved weighting factor selection for predictive torque control of induction motor drive based on a simple additive weighting method. *Electric Power Components and Systems*, *45*(13), 1450–1462. DOI: 10.1080/15325008.2017.1347215

Nafei, A. H., Javadpour, A., Nasseri, H., & Yuan, W. (2021). Optimizedscore function and its application in group multiattribute decision making based on fuzzy neutrosophic sets. *International Journal of Intelligent Systems*, *36*(12), 7522–7543. DOI: 10.1002/int.22597

Opricovic, S., & Tzeng, G. H. (2004). Compromise solution by MCDM methods: A comparative analysis of VIKOR and TOPSIS. *European Journal of Operational Research, 156*(2), 445–455. DOI: 10.1016/S0377-2217(03)00020-1

Ozlu, S., & Karaaslan, F. (2022). Hybrid similarity measures of single-valued neutrosophic type-2 fuzzy sets and their application to MCDM based on TOPSIS. *Soft Computing, 26*(9), 4059–4080. DOI: 10.1007/s00500-022-06824-3

Pouresmaeil, H., Khorram, E., & Shivanian, E. (2022). A parametric scoring function and the associated method for interval neutrosophic multi-criteria decision-making. *Evolving Systems, 13*(2), 347–359. DOI: 10.1007/s12530-021-09394-1

Rani, P., & Mishra, A. R. (2021). Fermatean fuzzy Einstein aggregation operators-based MULTIMOORA method for electric vehicle charging station selection. *Expert Systems with Applications, 182*, 115267. DOI: 10.1016/j.eswa.2021.115267

Rani, P., Mishra, A. R., & Mardani, A. (2020a). An extended Pythagorean fuzzy complex proportional assessment approach with new entropy and score function: Application in pharmacological therapy selection for type 2 diabetes. *Applied Soft Computing, 94*, 106441. DOI: 10.1016/j.asoc.2020.106441

Rani, P., Mishra, A. R., & Pardasani, K. R. (2020b). A novel WASPAS approach for multi-criteria physician selection problem with intuitionistic fuzzy type-2 sets. *Soft Computing, 24*(3), 2355–2367. DOI: 10.1007/s00500-019-04065-5

Rani, P., Mishra, A. R., Pardasani, K. R., Mardani, A., Liao, H., & Streimikiene, D. (2019). A novel VIKOR approach based on entropy and divergence measures of Pythagorean fuzzy sets to evaluate renewable energy technologies in India. *Journal of Cleaner Production, 238*, 117936. DOI: 10.1016/j.jclepro.2019.117936

Ridvan, S., Aslan, F., & Gokec, D. K. (2021). A single-valued neutrosophic multi-criteria group decision approach with DPL-TOPSIS method based on optimization. *International Journal of Intelligent Systems, 36*(7), 3339–3366. DOI: 10.1002/int.22418

Sahin, R., & Yigider, M. (2016). A multi-criteria neutrosophic group decision making method based topsis for supplier selection. *Applied Mathematics & Information Sciences, 10*(5), 1–10. DOI: 10.18576/amis/100525

Sennaroglu, B., & Celebi, G. V. (2018). A military airport location selection by AHP integrated PROMETHEE and VIKOR methods. *Transportation Research Part D, Transport and Environment, 59*, 160–173. DOI: 10.1016/j.trd.2017.12.022

Shannon, C. E. (2001). A mathematical theory of communication. *Mobile Computing and Communications Review, 5*(1), 3–55. DOI: 10.1145/584091.584093

Simic, V., Gokasar, I., Deveci, M., & Isik, M. (2021a). Fermatean fuzzy group decision-making based CODAS approach for taxation of public transit investments. *IEEE Transactions on Engineering Management.*

Simic, V., Karagoz, S., Deveci, M., & Aydin, N. (2021b). Picture fuzzy extension of the CODAS method for multi-criteria vehicle shredding facility location. *Expert Systems with Applications*, *175*, 114644. DOI: 10.1016/j.eswa.2021.114644

Smarandache, F. (1999). A unifying field in logics: Neutrosophic logic. In *Philosophy* (pp. 1–141). American Research Press.

Smarandache, F. (2004). A geometric interpretation of the neutrosophic set-A generalization of the intuitionistic fuzzy set. *arXiv preprint math/0404520.*

Smarandache, F. (2005). Generalization of the intuitionistic fuzzy logic to the neutrosophic fuzzy set. *International Journal of Pure and Applied Mathematics*, *24*(3), 287–297.

Smarandache, F. (2018). Extension of soft set to hypersoft set, and then to plithogenichypersoft set. *Neutrosophic Sets and Systems, 22*(1), 168-170.

Smarandache, F. (2023). New Types of Soft Sets: HyperSoft Set, IndetermSoft Set, IndetermHyperSoft Set, and TreeSoft Set. *International Journal of Neutrosophic Science*, *20*(4), 58–64. DOI: 10.54216/IJNS.200404

Smarandache, F., Abdel-Basset, M., & Broumi, S. (2022). Neutrosophic Sets and Systems, Vol. 50, 2022. *Neutrosophic Sets and Systems*, *50*(1), 40.

Wang, L., Zhang, H. Y., Wang, J. Q., & Li, L. (2018). Picture fuzzy normalized projection-based VIKOR method for the risk evaluation of construction project. *Applied Soft Computing*, *64*, 216–226. DOI: 10.1016/j.asoc.2017.12.014

Yang, C., Wang, Q., Pan, M., Hu, J., Peng, W., Zhang, J., & Zhang, L. (2022). A linguistic Pythagorean hesitant fuzzy MULTIMOORA method for third-party reverse logistics provider selection of electric vehicle power battery recycling. *Expert Systems with Applications*, *198*, 116808. DOI: 10.1016/j.eswa.2022.116808

Yoon, K. P., & Hwang, C. L. (1995). *Multiple attribute decision making: an introduction*. Sage publications. DOI: 10.4135/9781412985161

Zadeh, L. A. (1965). Fuzzy sets. *Information and Control*, *8*(3), 338–353. DOI: 10.1016/S0019-9958(65)90241-X

Zavadskas, E. K., Turskis, Z., Antucheviciene, J., & Zakarevicius, A. (2012). Optimization of weighted aggregated sum product assessment. *Elektronika ir Elektrotechnika*, *122*(6), 3–6. DOI: 10.5755/j01.eee.122.6.1810

Zhang, S., Gao, H., Wei, G., Wei, Y., & Wei, C. (2019). Evaluation based on distance from average solution method for multiple criteria group decision making under picture 2-tuple linguistic environment. *Mathematics*, *7*(3), 243. DOI: 10.3390/math7030243

Chapter 13
A Proposed Neutrosophic Probability Model for Normalized DifferenceVegetation Index Using Remote Sensing:
Model Building on Climate

Shan E. Fatima
Government College University, Lahore, Pakistan

Hina Khan
Government College University, Lahore, Pakistan

Kanwal Javaid
https://orcid.org/0000-0003-3143-8868
Governemnt College University, Lahore, Pakistan

ABSTRACT

In this research work, a new neutrosophic probability model named a neutrosophic Topp-Leone exponentiated generalized exponential distribution is proposed to model the normalized difference vegetation index (NDVI) of Pakistan in the year 2022. The MODIS-Merra-2 satellite was used in this research, which gave spatial and spectral information on NDVI in the Earth. NDVI maps were produced from the acquired data, and NDVI sensors computed the normalized difference vegetation

DOI: 10.4018/979-8-3693-3204-7.ch013

index. Mathematical characteristics of the proposed model of NDVI are derived, including central tendency measures, dispersion estimators, inferential statistics, moment generating function, order statistics, reliability measures, quantile-based statistical measures, Rényi entropy, and graphical representations of the density function are also presented. The maximum likelihood technique is used to estimate the parameters for the proposed probability model.

INTRODUCTION

A mathematical method known as fuzzy logic enables the representation and processing of hazy or inaccurate data. It is based on the idea of fuzzy sets, which generalize crisp sets in the sense that they permit items to have degrees of membership ranging from 0 to 1.

The degree to which an element belongs to a fuzzy set is represented by membership functions, which are given to variables and propositions in fuzzy logic. Depending on the nature of the problem, the membership functions can be defined using a variety of shapes, including triangular, trapezoidal, or Gaussian curves. Fuzzy logic is enhanced by the concept of neutrality or indeterminacy in the neutrosophicsense. Neutrosophic logic includes indeterminate truth values in addition to true and false values to deal with circumstances in which the truth or falsity of a statement cannot be decided or is only partially known.

A new division of philosophyisintroduced as neutrosophic statistics, which is executed as a generalization of fuzzy logic(Smarandache, 2000), (Rivieccio, 2008). This type of statistics is used for uncertain environments(Kashihara, 1996). It gained significance in utilizing the ambivalent phenomenon and estimating values in intervals instead of single values(Wang et al., 2005), (Smarandache & Pramanik, 2016).The set of neutrosophic variables is defined in three components indeterminate (I), valid (T), and false (F) (Smarandache, 2003), (Salama, 2018) uses these three components to improve grayscale images(Salama, Smarandache, & Eisa, 2014) also (Salama et al., 2022)has developed "Neutrosophic Closed Set and Neutrosophic Continuous Functions" in application to GIS and topology(Vasantha Kandasamy, 2006),(Salama, Smarandache, & Kroumov, 2014), (Salama & Alblowi, 2012),(Smarandache, 2014).A new set theory based on the neutrosophic random variable is derived and named it neutrosophic set theory(Salama & Alblowi, 2012). "Neutrosophic crisp neighbourhoods system for the neutrosophic crisp points" has been established for local neutrosophic functions (Salama, Smarandache, et al.), (Smarandache, 1999). Programming in the most potent, easily assessable software, EXCEL, has been developed to analyze neutrosophic data sets (Salama, El-Ghareeb, et al.). The effect of neutrosophic random variables in decision trees is a significant area in classical decision-making(Abdel-Basset et al., 2019), (Khan et al., 2018)to reach the best decision with more general and accurate information(Olgun & Hatip), (Pramanik & Mondal, 2015). This technique is used as comparison purpose with other different classical methods to reach most precise decision(Basha et al., 2018), (Zeema & Christopher, 2019), (Khalifa et al., 2022), (Jain et al., 2020). Development of Neutrosophic Data envelopment (DAE) methods to analyze the performance organizations whether private and public (Mao et al.). Neutrosophic Erlang queuing service model is derived tomodify its parameters for better data handling (Zeina, 2020).

The development of neutrosophic distributions, along with neutrosophic statistical techniques, played an essential role in analyzing the data efficiently. Neutrosophic lognormal distribution has been developed, and environmental data on nitrogen oxide has been utilized to provide better estimates(Khan et al., 2021). Neutrosophic Kumaraswamy, Pareto, and Rayleigh distributions have been derived (Ahsan-ul-Haq, 2022), (Aslam et al., 2019),(Aslam, 2020). Neutrosophic one-way and two-wayANOVA is produced to deal with uncertainty in data sets available in the market(Miari et al., 2022).NANOVA is used to check the hostility level among university students(Aslam, 2019). F test is modified with the assumption that the two counts of the climatology of the U.S.followPoisson distribution, and the data set is based on uncertain events of climate(Aslam, 2021b). Neutrosophic Analysis of the mean test was developed and compared its efficiency with classical analysis of the mean using the data of wind power data (Aslam, 2021a).Neutrosophic beta distribution, along with somemathematical characteristics and application on flood peak data sets of Wheaton River Canada, is analyzed(Sherwani et al.). Binomial, Normal,and Weibulldistributions are also modified in the neutrosophic version as available data are vague and imprecise (Sherwani et al., 2021), (Alhasan & Smarandache), (Patro & Smarandache). Poisson, Exponential, and Uniform distribution are also generalized under the umbrella of Neutrosophic logic (Alhabib et al., 2018).

Climate change is a big challenge for scientists and researchers. To deal with such changes,which occur every second and affect our environment,new inventions need to be studied andtheir effects on human life(Khan et al., 2021), (Aslam, 2021a, 2021b). Wind speed data from Pakistan is analyzed using newly proposed neutrosophic measures of skewness and kurtosis(Aslam, 2021c)

The rest of this study is organized asthe development of the newly proposed NTLEG-E distribution for NDVI of Pakistan, major essential characteristics of the newly proposed model, analysis of NDVI through mapping, and the last section containingthe conclusion of the article.

NEUTROSOPHIC TOPP-LEONE EXPONENTIATED GENERALIZED-EXPONENTIAL DISTRIBUTION

In this section, the Probability Density and Distribution Function ofNTLEG-E distributions is derived to model NDVI of Pakistan. Thefour-parameterNTLEG-E distribution's CDF is given by:

$$F_{NTLEG-G}(x) = \left[1 - \{1 - \{1 - e^{-\lambda_N a_N x}\}^{b_N}\}^2\right]^{\alpha_N} \quad (2.1)$$

The PDF of NTLEG-E is as follows:

$$f_{NTLEG-G}(x) = 2\alpha_N a_N b_N \lambda_N e^{-\lambda_N a_N x}(1 - e^{-\lambda_N a_N x})^{b_N - 1}$$

$$\left[1 - \{1 - e^{-\lambda_N a_N x}\}^{b_N}\right]\left[1 - [1 - \{1 - e^{-\lambda_N a_N x}\}^{b_N}]^2\right]^{(\alpha_N - 1)}, a_N, b_N, \alpha_N, \lambda_N, x > 0 \quad (2.2)$$

Where 'a' is the location parameter, 'b' and 'λ' are scale parameters, and 'c' is the shape parameter. The following figure shows the shapes of PDF on different parametric values of NTLEG-E distribution.

The parametric effect on PDF of NTLEG-E distribution is as follows:

Figure 1. Effect of parameters on PDF of NTLEG-E distribution

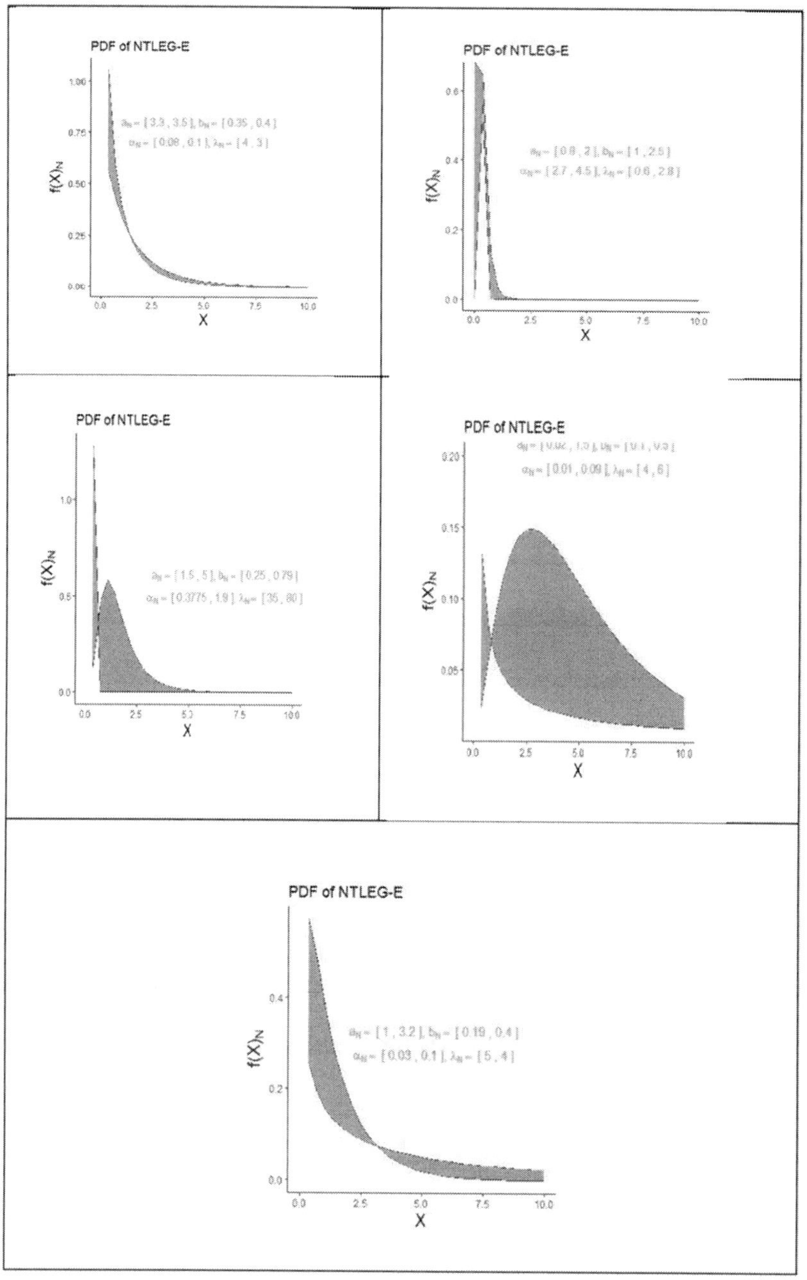

The following table shows the sub-models of NTLEG-E distribution. All the special cases of NTLEG-E probability models are new neutrosophic versions of the developed distributions.

Table 1. New probability models (special cases) of NTLEG-E distribution

Parametric Values	Sub-Models	New Models
a=1	$F(x) = [\{(1 - e^{-\lambda_N x})^{b_N}\}^2]^{\alpha_N}$	**NTLEEx**
b=1	$F(x) = (1 - e^{-2\lambda_N a_N x})^{\alpha_N}$	**NTLEx(a)**
a=b=1	$F(x) = (1 - e^{-2\lambda_N x})^{\alpha_N}$	**NTLEx**
b=α=a=1	$F(x) = 1 - e^{-2\lambda_N x}$	**N.Weibull (β=2)**
a=b=1	$F(x) = (1 - e^{-2\lambda_N x})^{\alpha_N}$	**NEW (β=2)**

Reliability Measures of NTLEG-E Distribution

In this section, essentialproperties in the reliability theory of NTLEG-E distribution areobtained along with graphs of some characteristics.

Survival Rate Function

If $X \sim NTLEGE(a_N, b_N, \alpha_N, \lambda_N)$, the survival function of NTLEG-E can be obtained by using the following formulae(Nzei et al., 2020):

$$S(x, \varnothing) = \overline{F}(x, \varnothing) = 1 - F(x, \varnothing) \tag{2.3}$$

By using (2.1) in (2.3), the survival function of TLEG-E distribution is expressed as:

$$S(x) = 1 - [1 - \{1 - \{1 - e^{-\lambda_N a_N x}\}^{b_N}\}^2]^{\alpha_N} \tag{2.4}$$

Hazard Rate Function

The instantaneous hazard ratefunction of a random variable '*X*' is defined as the ratio of the PDF and survival rate function. The hazard rate is also known as the instantaneous failure rate(Nzei et al., 2020).

$$h(x, \varnothing) = \frac{f(x, \varnothing)}{s(x, \varnothing)} s(x, \varnothing) > 0$$

Where

$$S(x, \varnothing) = 1 - F(x, \varnothing)$$

So,

$$h(x) = \frac{f(x, \varnothing)}{1 - F(x, \varnothing)} \tag{2.5}$$

Substituting equations (2.1) and (2.2) in the above equation, the failure rate for the NTLEG-E distribution can be given as:

$$h(x) = \frac{2\alpha_N a_N b_N \lambda_N e^{-\lambda_N a_N x}\left(1 - e^{-\lambda_N a_N x}\right)^{b_N - 1}\left[1 - \{1 - e^{-\lambda_N a_N x}\}^{b_N}\right]\left[1 - [1 - \{1 - e^{-\lambda_N a_N x}\}^{b_N}]^2\right]^{(\alpha_N - 1)}}{1 - \left[1 - \{1 - \{1 - e^{-\lambda_N a_N x}\}^{b_N}\}^2\right]^{\alpha_N}} \tag{2.6}$$

Reverse Hazard Rate Function

The reverse hazard function is derived by using the following formula(Nzei et al., 2020):

$$\widehat{R}\left(x\right) = \frac{f(x)}{1 - S(x)} \tag{2.7}$$

By substituting (2.3) and (2.2) in the above formula, the reverse hazard rate function of NTLEG-Eis as follows:

$$\widehat{R}\left(x\right) = \frac{2\alpha_N a_N b_N \lambda_N e^{-\lambda_N a_N x}\left(1 - e^{-\lambda_N a_N x}\right)^{b_N - 1}\left[1 - \{1 - e^{-\lambda_N a_N x}\}^{b_N}\right]\left[1 - [1 - \{1 - e^{-\lambda_N a_N x}\}^{b_N}]^2\right]^{(\alpha_N - 1)}}{\left[1 - \{1 - \{1 - e^{-\lambda_N a_N x}\}^{b_N}\}^2\right]^{\alpha_N}} \tag{2.8}$$

Cumulative Hazard Rate Function

The cumulativehazard rate function is obtained by utilizing the following formula(Nzei et al., 2020):

$$\omega(x) = -ln(s(x)) \tag{2.9}$$

By substituting (2.3) in (2.9), the cumulative hazard function of NTLEG-E is as follows:

$$\omega(x) = -ln\left(1 - \left[1 - \{1 - \{1 - e^{-\lambda_N a_N x}\}^{b_N}\}^2\right]^{\alpha_N}\right)$$

Odds Function

The odds function is the ratio of the distribution Function and survival rate function of the distribution. The general formulae for the odds function is given as(Nzei et al., 2020):

$$O_{TLEG-E} = \frac{F(x)}{S(x)} \tag{2.10}$$

By substituting (2.1) and (2.4) in the expression mentioned above of the odds function, the odds function of TLEG-E is as follows:

$$O_{TLEGE} = \frac{\left[1-\{1-\{1-e^{-\lambda_N a_N x}\}^{b_N}\}^2\right]^{a_N}}{1-\left[1-\{1-\{1-e^{-\lambda_N a_N x}\}^{b_N}\}^2\right]^{a_N}}$$

Quantile Based Characteristics

Quantile-based functions are commonly used in reliability theory. Quantile functionis helpful in such situations as they may not need the life-testing tryouts to last until the consumption of all subjects(Parzen, 1979).NTLEG-E distribution is used to derive quantile-based measures.

Quantiles of NTLEG-E Distribution

The quantile function is a very significant feature of any probability modelas it is used to generate data. The quantile function of NTLEG-E when $F(x)=u$ is as follows(Gilchrist, 2000):

$$x = -\frac{ln\left[1-\{1-\sqrt{1-u^{\frac{1}{a_N}}}\}^{\frac{1}{b_N}}\right]}{a_N\lambda_N} \tag{2.11}$$

The quantile function is used to find the different quantiles. For example, for different values of u first quartile, third quartile, median, and 90th percentiles are obtained.

Table 2. Quantiles of NTLEG-E Distribution

$Q_1 = -\frac{ln\left[1 - \left\{1 - \sqrt{1 - (0.25)^{\frac{1}{d_N}}}\right\}^{\frac{1}{b_N}}\right]}{a_N \lambda_N}$
$\text{Median} = -\frac{ln\left[1 - \left\{1 - \sqrt{1 - (0.5)^{\frac{1}{d_N}}}\right\}^{\frac{1}{b_N}}\right]}{a_N \lambda_N}$
$Q_3 = -\frac{ln\left[1 - \left\{1 - \sqrt{1 - (0.75)^{\frac{1}{d_N}}}\right\}^{\frac{1}{b_N}}\right]}{a_N \lambda_N}$
$P_{90} = -\frac{ln\left[1 - \left\{1 - \sqrt{1 - (0.9)^{\frac{1}{d_N}}}\right\}^{\frac{1}{b_N}}\right]}{a_N \lambda_N}$

Semi-Inter Quartile Range

The semi-inter-quartile range is half of the inter-quartile which is as follows(Gilchrist, 2000):

$$Q.D = \frac{Q_3 - Q_1}{2}$$

Using formulae of the third and first quartile, the expression of quartile deviation for the NTLEG-E probability model is as follows:

$$Q.D = ln\frac{\left[1 - \left\{1 - \sqrt{1 - (0.25)^{\frac{1}{d_N}}}\right\}^{\frac{1}{b_N}}\right]}{a_N \lambda_N \left[1 - \left\{1 - \sqrt{1 - (0.75)^{\frac{1}{d_N}}}\right\}^{\frac{1}{b_N}}\right]} \tag{2.12}$$

Quantile Hazard Function

Conferring to Gilchrist (2000), the QF's as the hazard quantile function is as follows:

$$H(u) = [(1 - u)q(u)]^{-1} \tag{2.13}$$

Where $q(u)$is a quantile density function, and 'u' isa random number. The quantile hazard function of NTLEG-E distribution using equation (2.13) is:

$$H(u) = \frac{-b_N\sqrt{1 - u^{1/\alpha_N}\alpha_N x\left(1 - \sqrt{1 - u^{1/\alpha_N}}\right)\left(1 - \left(1 - \sqrt{1 - u^{1/\alpha_N}}\right)^{1/b_N}\right)a_N\lambda_N}}{(1 - u)\left(1 - \sqrt{1 - u^{1/\alpha_N}}\right)^{1/b_N} u^{1/\alpha_N}}$$

Reverse Hazard Quantile Function

The general Reverse hazard quantile function is introduced byNair and Sankaran (2009) which is expressed as:

$$A(u) = (uq(u))^{-1} \tag{2.14}$$

By substituting the quantile density function in equation (2.14), the reverse hazard quantile function is given below:

$$A(u) = \frac{-b_N\sqrt{1 - u^{1/\alpha_N}\alpha_N x\left(1 - \sqrt{1 - u^{1/\alpha_N}}\right)\left(1 - \left(1 - \sqrt{1 - u^{1/\alpha_N}}\right)^{1/b_N}\right)a_N\lambda_N}}{u\left(1 - \sqrt{1 - u^{1/\alpha_N}}\right)^{1/b_N} u^{1/\alpha_N}} \tag{2.15}$$

Qausi Range(Inter-p Range)

The lower-p deviation ld(p) and upper-p deviation ud(p) can be used for the inter-P range, which is also called quasi-range. It is a spread function.Quasi-range is defined as(Gilchrist, 2000):

$$ipr(p) = \widetilde{Q}(q) - \widetilde{Q}(p) = ld(p) + ud\left(p\right)\frac{r-1}{n} < p \leq \frac{r}{n}\, and\, p$$

$$= \left(1 - q\right) \tag{2.16}$$

where $\widetilde{Q}\left(p\right)$ is empirical quartile function. By substituting ld(p) and ud(p), the ipr(p) for TLEG-E distribution is as follows:

$$ipr(p) = -\frac{1}{a_N\lambda_N}\left[ln\left\{\left[1 - \left\{1 - \sqrt{1 - (0.75)^{\frac{1}{\alpha_N}}}\right\}^{\frac{1}{b_N}}\right]\left[1 - \left\{1 - \sqrt{1 - (0.25)^{\frac{1}{\alpha_N}}}\right\}^{\frac{1}{b_N}}\right]\right\}\right] \tag{2.17}$$

Mode of NTLEG-E Distribution

The probability density function of NTLEG-E distribution is used to obtain the formulae of mode(A TALAAT, 2018).

$$f_{TLEG-G}(x) = 2\alpha_N a_N b_N \lambda_N e^{-\lambda_N a_N x}(1 - e^{-\lambda_N a_N x})^{b_N - 1}[1 - \{1 - e^{-\lambda_N a_N x}\}^{b_N}]$$

$$\left[1-\left[1-\{1-e^{-\lambda_N a_N x}\}^{b_N}\right]^2\right]^{(\alpha_N-1)},x>0$$

Taking natural log on both sides of the equation, then differentiate both sides with respect to x, put $f'(x) = 0$. Since $f(x)>0$, then the mode can be found numerically by solving the following equation

$$=-\lambda_N a_N+\frac{(b_N-1)\lambda_N a_N e^{-\lambda_N a_N x}}{(1-e^{-\lambda_N a_N x})}-\frac{b_N\{1-e^{-\lambda_N a_N x}\}^{b_N-1}\lambda_N a_N e^{-\lambda_N a_N x}}{\left[1-\{1-e^{-\lambda_N a_N x}\}^{b_N}\right]}$$

$$+\frac{2(\alpha_N-1)\lambda_N a_N b_N e^{-\lambda_N a_N x}(1-\{1-e^{-\lambda_N a_N x}\}^{b_N})\{1-e^{-\lambda_N a_N x}\}^{b_N-1}}{\left[1-\left[1-\{1-e^{-\lambda_N a_N x}\}^{b_N}\right]^2\right]}=0$$

Moments of TLEG-E Distribution

Moments play anessential role in the measurement of the mean, variance, kurtosis, and skewness. The general expression to define the raw moments of a random variable$X$ is as follows(Nzei et al., 2020)

$$E(x^r)=\int_0^\infty x^r f(x;a_N,b_N,\alpha_N,\lambda_N)dx$$

By the definition of complete gamma function, we obtained

$$E(x^r)=\frac{2\alpha_N a_N b_N\lambda_N\sum_{p,q,r=0}^{\infty}(-1)^{p+q+r}\binom{\alpha-1}{p}\binom{2p+1}{q}\binom{b(q+1)-1}{r}}{\left(-\lambda_N a_N(r+1)\right)^{r+1}}\Gamma r+1$$

Skewness and Kurtosis

Kurtosis and skewness are fundamental properties of any distribution.These measures accurately define the shape and peak of available data. The two estimates of TLEG-E distribution can be derived through the following formulae of Bowley (B) skewness(Nzei et al., 2020),Galton (G) skewness coefficient by Gilchrist (2000), and Moors (M) kurtosis(Nzei et al., 2020):

$$G=\frac{QR}{IQR}$$

$$B=\frac{Q\left(\frac{3}{4}\right)+Q\left(\frac{1}{4}\right)-2Q\left(\frac{1}{2}\right)}{Q\left(\frac{3}{4}\right)-Q\left(\frac{1}{4}\right)}$$

and

$$B = \frac{Q\left(\frac{7}{8}\right) - Q\left(\frac{3}{8}\right) - Q\left(\frac{5}{8}\right) + Q\left(\frac{1}{8}\right)}{Q\left(\frac{3}{4}\right) - Q\left(\frac{1}{4}\right)}$$

Moment Generating Function (MGF)

The moment generating a function of the NTLEG-E distribution can be derived by using the following general formula of MGF(Nzei et al., 2020)

$$M_X(t) = E(e^{tx}) = \int_0^{\infty} e^{tx}.f(x; a_N, b_N, \alpha_N, \lambda_N)\, dx$$

By considering the expression of $E(x^r)$ we obtain,

$$M_x(t) = \sum_{s=0}^{\infty} \frac{(t)^s}{s!} \frac{2\alpha_N a_N b_N \lambda_N \sum_{p,q,r=0}^{\infty} (-1)^{p+q+r} \binom{\alpha - 1}{p} \binom{2p+1}{q} \binom{b(q+1)-1}{r}}{\left(-\lambda_N a_N (r+1)\right)^{r+1}} \lceil r+1$$

Renyi Entropy

It is a measure of the uncertainty measure of a probability model and the randomness of the data. Renyi entropy if $X \sim f(x)$ is defined as(Nzei et al., 2020):

$$I_R(v) = \frac{1}{1-v} log\left[\int_0^{\infty} [f(x)]^v dx\right]$$

If $X \sim NTLEG - E(x; a_N, b_N, \alpha_N, \lambda_N)$ then $I_R(v)$ is:

$$\frac{1}{1-v} log\left[\int_0^{\infty} [f(x)]^v dx\right]$$

$$= \frac{1}{1-v} log\left[\frac{\left(2\alpha_N a_N b_N \lambda_N\right)^v}{\lambda_N a_N (r+v)}\right] \sum_{p,q,r=0}^{\infty} (-1)^{p+q+r} \binom{v(\alpha-1)}{p} \binom{2p+v}{q} \binom{bq+(b-1)}{r}$$

Maximum Likelihood Estimates

Suppose $x_1, x_2, x_3, \ldots, x_n$ be a random sample from NTLEG-E distribution,then the log-likelihood function becomes(Mahdavi & Kundu, 2017):

$$L = \prod_{i=1}^{n} f(x)$$

$$L = \prod_{i=1}^{n} 2\alpha_N a_N b_N \lambda_N e^{-\lambda_N a_N x} (1 - e^{-\lambda_N a_N x})^{b_N - 1}$$

$$\left[1 - \{1 - e^{-\lambda_N a_N x}\}^{b_N}\right] \left[1 - [1 - \{1 - e^{-\lambda_N a_N x}\}^{b_N}]^2\right]^{(\alpha_N - 1)}$$

The four normal equations for $\hat{a}, \hat{b}, \hat{\lambda}$ *and* $\hat{\alpha}$ are as follows to estimate the parametric values of the TLEG-E distribution.

$$\frac{\partial lnL}{\partial(\hat{\alpha}_N)} = \frac{n}{\alpha_N} + \sum ln\left[1 - \{1 - (1 - e^{-\lambda_N a_N x})^{b_N}\}^2\right] = 0$$

$$\frac{\partial lnL}{\partial(\hat{b}_N)} = \frac{n}{b_N} + \sum ln(1 - e^{-\lambda_N a_N x}) + \sum ln\left[\frac{(1 - e^{-\lambda_N a_N x})}{1 - (1 - e^{-\lambda_N a_N x})^{b_N}}\right]$$

$$+(\alpha_N - 1)\frac{\sum ln\left[2\{-(1 - e^{-\lambda_N a_N x})^{b_N}\}\right]}{\left[1 - 1 - \{(1 - e^{-\lambda_N a_N x})^{b_N}\}^2\right]} = 0$$

$$\frac{\partial lnL}{\partial(\hat{a}_N)} = \frac{n}{a_N} - \sum(\lambda_N x) + (b_N - 1)\frac{\sum ln\left[-e^{-\lambda_N a_N x}(-\lambda_N x)\right]}{(1 - e^{-\lambda a x})}$$

$$+\frac{\sum ln\left[-(1 - e^{-\lambda_N a_N x})^{b_N}\lambda_N x e^{-\lambda_N a_N x}\right]}{\{(1 - e^{-\lambda a x})^{b}\}}$$

$$+(\alpha_N - 1)\frac{\sum ln\left[-\{1 - (1 - e^{-\lambda_N a_N x})\}^2\{\lambda_N x(1 - e^{-\lambda_N a_N x})(-e^{-\lambda_N a_N x})\}\right]}{\left[1 - \{1 - (1 - e^{-\lambda_N a_N x})^{b_N}\}^2\right]}$$

$$\frac{\partial lnL}{\partial(\hat{\lambda}_N)} = \frac{n}{\lambda_N} + \sum(a_N x) + (b_N - 1)\frac{\sum ln\left(a_N x e^{-\lambda_N a_N x}\right)}{(1 - e^{-\lambda_N a_N x})} + \frac{\sum ln\left[-(1 - e^{-\lambda_N a_N x})a_N x e^{-\lambda_N a_N x}\right]}{\{1 - (1 - e^{-\lambda_N a_N x})^{b_N}\}}$$

$$+(\alpha_N - 1)\frac{\sum\left[\{1 - (1 - e^{-\lambda_N a_N x})^{b_N}\}^2(1 - e^{-\lambda_N a_N x})^{b_N} a_N x e^{-\lambda_N a_N x}\right]}{\left[1 - \{1 - (1 - e^{-\lambda_N a_N x})^{b_N}\}^2\right]} = 0$$

The equation, as mentioned above, can be solvedsimultaneously. The Newton-Raphson method may be employed for the Standard algorithm to derive the final expression of these non-linear equations for parametric values. Since it is a simple two-dimensional optimization delinquent, getting some preliminary guesses is also not challenging.

STUDY AREA

The terrain in Pakistan includes a complex mixture of plains, deserts, forests, and plateaus, stretching from the southern Indian Ocean coast to the northern mountains of the Karakoram, Hindukush, and Himalayan regions. Pakistan's Sindh and Punjab provinces are located on the northwest corner of the Indian plate. In contrast, Baluchistan and most of Khyber Pakhtunkhwa are located within the Eurasian plate, primarily made up of the Iranian Plateau.

India borders Pakistan in the east, Afghanistan borders Pakistan in the northwest, Iran borders Pakistan in the west, and China borders Pakistan in the northeast, as seen in Figures 3.1(a) and (b). The location's latitude and longitude are 30.3753° N, and 69.3451° W. Pakistan has an overall area of 803,936 km^2, as seen in Figure 2 (c). Pakisțan is located in the temperate zone, directly above the tropic of cancer. There are both tropical and temperate climates. While the Punjab province experiences plentiful rainfall, arid conditions exist in the coastal south, characterised by a monsoon season with sufficient rainfall and, a dry season with less rainfall, and significant changes between temperature extremes at specific sites. In various country regions, annual rainfall ranges from less than 10 inches to over 150 inches. Yet, these generalisations should not mask the noticeable variations in specific places.

Figure 2. Maps showing the location and outline of the study area. A) shows the world, B) shows the study area location in South Asia, C) shows the study area, Pakistan.

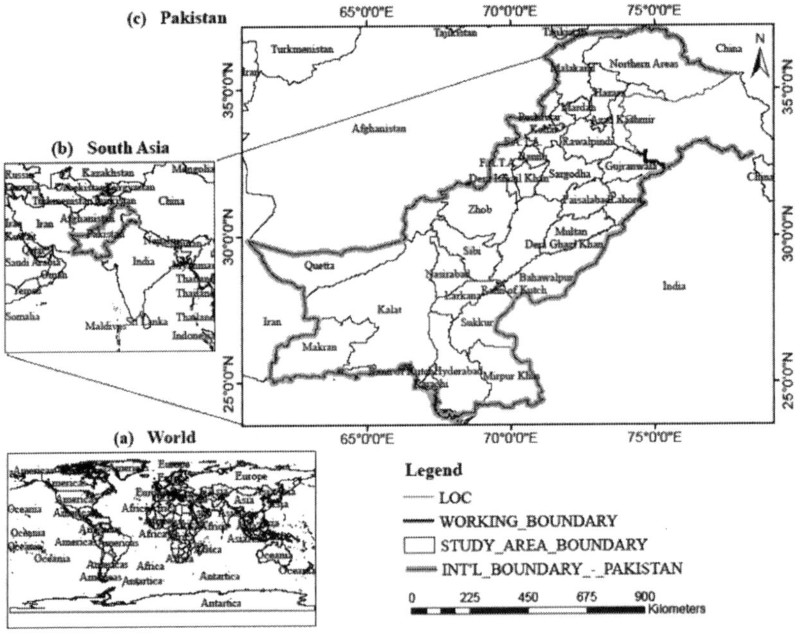

Acquisition of Data

The National Aeronautics and Space Administration (NASA)-World View provides the data. Data are gathered between January 2022 to December 2022. The Moderate Resolution ImagingSpectroradiometer (MODIS)-Merra-2 satellite was used, which gave spatial and spectral information on NDVI in the Earth of cell size 250m. NDVI sensors compute normalised difference vegetation index. The normalised difference between leaf area and canopy chlorophyll content, or NDVI, is estimated and is commonly used to track leaf senescence in the autumn and spring green-up.

Spatial and Temporal Patterns of NDVI

Spatial presentation of NDVI through maps are presented in this section to observe the difference in vegetation index over the year of 2022 in Pakistan.Satellite data is used to create the Normalised Difference Vegetation Index (NDVI) grids and maps. The information gives a broad picture of the state and dynamics of the vegetation throughout Pakistan and quantifies the amount of living greenery.

Figure 3. Maps of NDVI of Pakistan

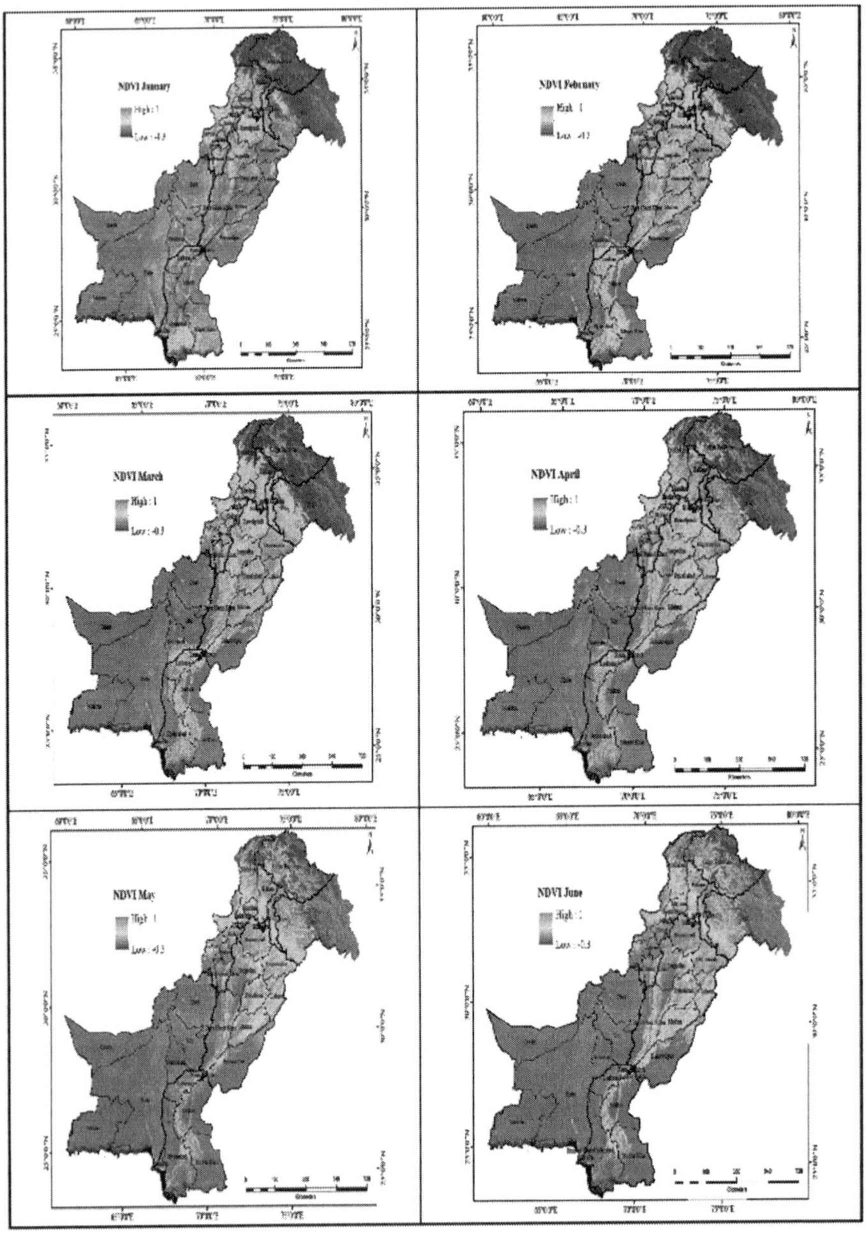

Figure 4. Maps of NDVI of Pakistan

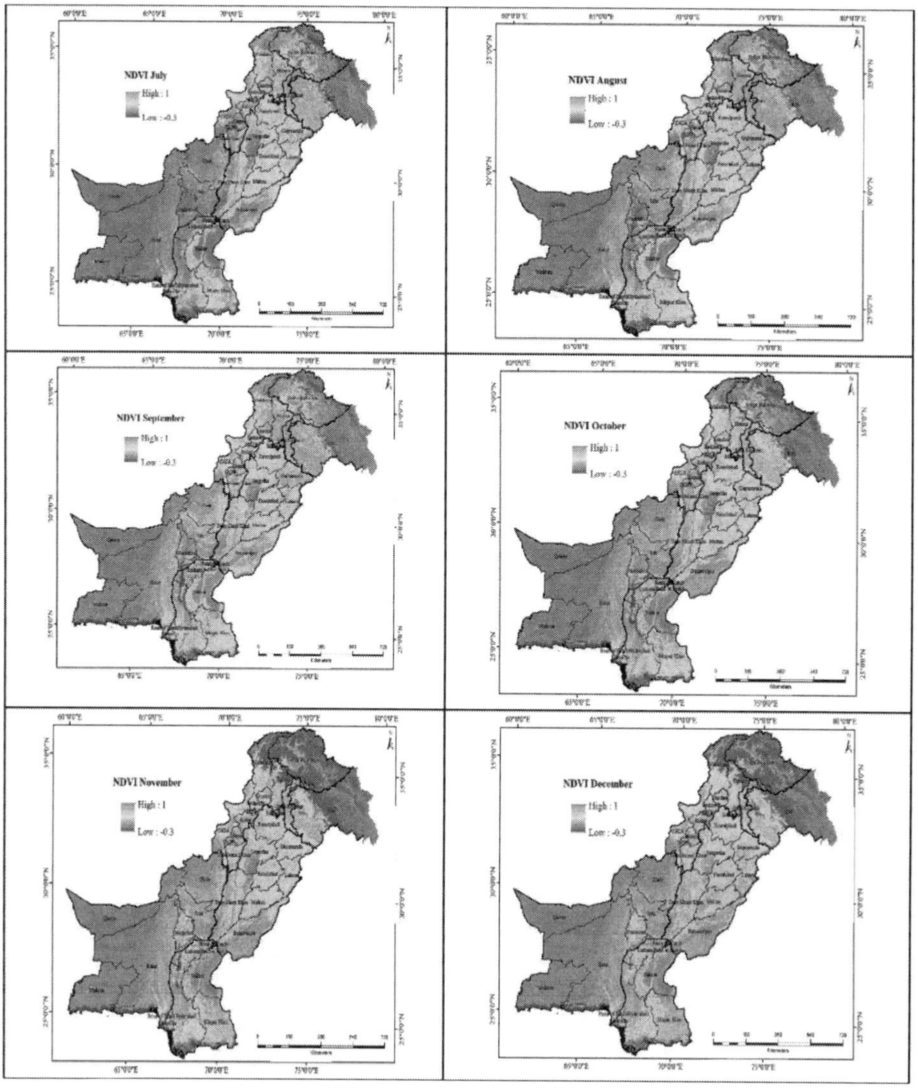

The MODIS-Merra-2 satellite was used in this research, which gave spatial and spectral information on NDVI in the Earth. NDVI maps were produced from the acquired data, andNDVI sensors computed the normalized difference vegetation index. The normalized difference between leaf area and canopy chlorophyll content, or NDVI, is estimated and is commonly used to track leaf senescence in the autumn and spring green-up.NDVI values generally fall between -1.0 and 1.0, with negative values denoting clouds and water, positive values close to zero denoting bare soil, and higher positive values of NDVI ranging from 0.1 to 0.5 representing sparse vegetation to 0.6 and above representing thick green vegetation.In Pakistan, the assessment of seasonal variability in NDVI was evaluated with the help of MISR analysis, which was performed on the collected data by NASA for the year 2022. NDVI is frequently used in forestry, ecology, and agriculture to track the health and growth of vegetation as well as to pinpoint damaged or stressed areas. Moreover, vegetation types can be mapped and classified, and variations in vegetation cover over time can be identified using NDVI readings. In this research, the changes in NDVI were tracked down to observe the density of plants in Pakistan over 2022.

It is observed that from Hyderabad to Malakand, the NDVI value seems to be 1, which shows high greenery from January to March. The trend changes from April onward till June, and maps show low greenery in the south of Pakistan. The green areas in Dera Ghazi Khan to Malakand shrinks in the month of April, whereas from Kashmir to Malakand, it shrinks in the month of May. In the month of May, low greenery levels can also be seen fromHyderabad to Islamabad. However July onward, it is observed that the region of Balochistan and nearby areas become lighter red, and the southeast area of Pakistan becomes green again. So, over the year 2022, it is observed that northeast areas of Pakistan were greener as compared to the end of the year.

CONCLUSION

A new proposed novel continuous probability model with four parameterscalled Topp-Leone Exponentiated Generalized--Exponential (TLEG-E) is developed to address the need for modeling NDVI effectively. Special cases of the proposed model, explicit expressions of some important mathematical characteristics,i.e., ordinary moments, moment generating function, reliability measure, quantile function-based measures for fatigue of components, Rényi entropy, and order statisticsof new proposed distribution are derived. ML method is employed for estimating the new model parameters for efficient modeling.Mapping of NDVI data of Pakistan is also considered in order to observe its pattern in 2022. It is observed that the green area in Pakistan shrinks from January to May; however, from July onwards, green areas can be seen, which might be due to the plantation of new trees or seasonal vegetation in those areas.

REFERENCES

Abdel-Basset, M., Smarandache, F., & Ye, J. (2019). Special issue on "Applications of neutrosophic theory in decision making-recent advances and future trends". *Complex & Intelligent Systems*, *5*(4), 363–364. DOI: 10.1007/s40747-019-00127-1

Ahsan-ul-Haq, M. (2022). Neutrosophic Kumaraswamy Distribution with Engineering Application. *Neutrosophic Sets and Systems*, *49*(1), 17.

Alhabib, R., Ranna, M. M., Farah, H., & Salama, A. (2018). Some neutrosophic probability distributions. *Neutrosophic Sets and Systems*, *22*, 30–38.

Alhasan, K. F. H., & Smarandache, F. *Neutrosophic Weibull distribution and neutrosophic family Weibull distribution*. Infinite Study.

Aslam, M. (2019). Neutrosophic analysis of variance: Application to university students. *Complex & Intelligent Systems*, *5*(4), 403–407. DOI: 10.1007/s40747-019-0107-2

Aslam, M. (2021a). Analyzing wind power data using analysis of means under neutrosophic statistics. *Soft Computing*, *25*(10), 7087–7093. DOI: 10.1007/s00500-021-05661-0

Aslam, M. (2021b). Neutrosophic statistical test for counts in climatology. *Scientific Reports*, *11*(1), 1–5. DOI: 10.1038/s41598-021-97344-x PMID: 34497328

Aslam, M. (2021c). A study on skewness and kurtosis estimators of wind speed distribution under indeterminacy. *Theoretical and Applied Climatology*, *143*(3), 1227–1234. DOI: 10.1007/s00704-020-03509-5

Aslam, M., Khan, N., & Al-Marshadi, A. H. (2019). Design of variable sampling plan for pareto distribution using neutrosophic statistical interval method. *Symmetry*, *11*(1), 80. DOI: 10.3390/sym11010080

Aslam, M. A. (2020). Neutrosophic Rayleigh distribution with some basic properties and application. In *Neutrosophic Sets in Decision Analysis and Operations Research* (pp. 119–128). IGI Global. DOI: 10.4018/978-1-7998-2555-5.ch006

Basha, S. H., Tharwat, A., Ahmed, K., & Hassanien, A. E. (2018). A predictive model for seminal quality using neutrosophic rule-based classification system. International conference on advanced intelligent systems and informatics

Gilchrist, W. (2000). *Statistical modelling with quantile functions*. CRC Press. DOI: 10.1201/9781420035919

Jain, A., Pal Nandi, B., Gupta, C., & Tayal, D. K. (2020). Senti-NSetPSO: Large-sized document-level sentiment analysis using Neutrosophic Set and particle swarm optimization. *Soft Computing*, *24*(1), 3–15. DOI: 10.1007/s00500-019-04209-7

Kashihara, K. (1996). *Comments and topics on Smarandache notions and problems*. Infinite Study.

Khalifa, N. E., Loey, M., Chakrabortty, R. K., & Taha, M. H. N. (2022). Within the Protection of COVID-19 Spreading: A Face Mask Detection Model Based on the Neutrosophic RGB with Deep Transfer Learning. *Neutrosophic Sets and Systems*, *50*, 320–335.

Khan, M., Son, L. H., Ali, M., Chau, H. T. M., Na, N. T. N., & Smarandache, F. (2018). Systematic review of decision making algorithms in extended neutrosophic sets. *Symmetry*, *10*(8), 314. DOI: 10.3390/sym10080314

Khan, Z., Amin, A., Khan, S. A., & Gulistan, M. (2021). Statistical Development of the Neutrosophic Lognormal Model with Application to Environmental Data. *Neutrosophic Sets and Systems*, *47*(1), 1.

Mahdavi, A., & Kundu, D. (2017). A new method for generating distributions with an application to exponential distribution. *Communications in Statistics. Theory and Methods*, *46*(13), 6543–6557. DOI: 10.1080/03610926.2015.1130839

Mao, X., Guoxi, Z., Fallah, M., & Edalatpanah, S. (2020). A neutrosophic-based approach in data envelopment analysis with undesirable outputs. *Mathematical Problems in Engineering*, *2020*, 1–8. DOI: 10.1155/2020/7626102

Miari, M., Anan, M. T., & Zeina, M. B. (2022). Neutrosophic two way ANOVA. *International Journal of Neutrosophic Science*, *18*(3), 73–83. DOI: 10.54216/IJNS.180306

Nair, N. U., & Sankaran, P. (2009). Quantile-based reliability analysis. *Communications in Statistics. Theory and Methods*, *38*(2), 222–232. DOI: 10.1080/03610920802187430

Nzei, L. C., Eghwerido, J. T., & Ekhosuehi, N. (2020). Topp-Leone Gompertz Distribution: Properties and Applications. *Journal of Data Science : JDS*, *18*(4), 782–794. DOI: 10.6339/JDS.202010_18(4).0012

Olgun, N., & Hatip, A. *The effect of the neutrosophic logic on the decision tree* (Vol. 7).

Parzen, E. (1979). Density quantile estimation approach to statistical data modelling. In *Smoothing Techniques for Curve Estimation* (pp. 155–180). Springer. DOI: 10.1007/BFb0098495

Patro, S., & Smarandache, F. *The neutrosophic statistical distribution, more problems, more solutions*. Infinite Study.

Pramanik, S., & Mondal, K. (2015). Interval neutrosophic multi-attribute decision-making based on grey relational analysis. *Neutrosophic Sets and Systems*, *9*(1), 13–22.

Rivieccio, U. (2008). Neutrosophic logics: Prospects and problems. *Fuzzy Sets and Systems*, *159*(14), 1860–1868. DOI: 10.1016/j.fss.2007.11.011

Salama, A. (2018). Neutrosophic approach to grayscale images domain.

Salama, A., & Alblowi, S. (2012). Neutrosophic set and neutrosophic topological spaces. *IOSR Journal of Mathematics (IOSR-JM)*, *3*(4).

Salama, A., El-Ghareeb, H. A., Manie, A. M., & Smarandache, F. Introduction to develop some software programs for dealing with neutrosophic sets.

Salama, A., Smarandache, F., & Alblowi, S. *New neutrosophic crisp topological concepts*. Infinite Study.

Salama, A., Smarandache, F., & Eisa, M. (2014). *Introduction to image processing via neutrosophic techniques*. Infinite Study.

Salama, A., Smarandache, F., & Kromov, V. (2022). Neutrosophic closed set and neutrosophic continuous functions. *Collected Papers. Volume IX: On Neutrosophic Theory and Its Applications in Algebra*, 25.

Salama, A., Smarandache, F., & Kroumov, V. (2014). *Neutrosophic crisp sets & neutrosophic crisp topological spaces*. Infinite Study.

Sherwani, R. A. K., Aslam, M., Raza, M. A., Farooq, M., Abid, M., & Tahir, M. (2021). Neutrosophic Normal Probability Distribution—A Spine of Parametric Neutrosophic Statistical Tests: Properties and Applications. In *Neutrosophic Operational Research* (pp. 153-169). Springer.

Sherwani, R. A. K., Naeem, M., Aslam, M., Raza, M., Abid, M., & Abbas, S. *Neutrosophic beta distribution with properties and applications*. Infinite Study.

Smarandache, F. (1999). A unifying field in Logics: Neutrosophic Logic. In *Philosophy* (pp. 1–141). American Research Press.

Smarandache, F. (2000). Neutrosophy. *arXiv preprint math/0010099*.

Smarandache, F. (2003). Definiton of neutrosophic logic-a generalization of the intuitionistic fuzzy logic. *EUSFLAT Conf.*

Smarandache, F. (2014). *Neutrosophic Theory and Its Applications, Vol. I: Collected Papers*. Infinite Study.

Smarandache, F., & Pramanik, S. (2016). *New trends in neutrosophic theory and applications* (Vol. 1). Infinite Study.

Vasantha Kandasamy, W. (2006). Smarandache Neutrosophic algebraic structures. *arXiv Mathematics e-prints*, math/0603708.

Wang, H., Smarandache, F., Sunderraman, R., & Zhang, Y.-Q. (2005). *interval neutrosophic sets and logic: theory and applications in computing: Theory and applications in computing* (Vol. 5). Infinite Study.

Zeema, J. L., & Christopher, D. F. X. (2019). Evolving optimized neutrosophic C means clustering using behavioral inspiration of artificial bacterial foraging (ONCMC-ABF) in the prediction of Dyslexia. *Journal of King Saud University. Computer and Information Sciences*.

Zeina, M. B. (2020). Erlang Service Queueing Model with Neutrosophic Parameters. *International Journal of Neutrosophic Science*, *6*(2), 106–112. DOI: 10.54216/IJNS.060202

Chapter 14
Development of Some New Hybrid Structures of Hypersoft Set With Possibility-Degree Settings

Atiqe Ur Rahman
https://orcid.org/0000-0001-6320-9221
University of Management and Technology, Lahore, Pakistan

Florentin Smarandache
https://orcid.org/0000-0002-5560-5926
University of New Mexico, USA

Muhammad Saeed
https://orcid.org/0000-0002-7284-6908
University of Management and Technology, Lahore, Pakistan

Khuram Ali Khan
https://orcid.org/0000-0002-3468-2295
University of Sargodha, Pakistan

ABSTRACT

The concept of a hypersoft membership function is introduced in the extension of a soft set known as a hypersoft set, permitting it to handle complicated and uncertain information in a more powerful and flexible manner. Many academics have already become fascinated with this new area of study, leading to the development of a number of hybrid structures. This chapter develops some new hybrid hypersoft

DOI: 10.4018/979-8-3693-3204-7.ch014

set structures by taking into account multiple fuzzy set-like settings and possibility degree-based settings collectively. Additionally, numerical examples are included to clarify the concept of these structures. Researchers can utilize this work to better understand and apply a variety of mathematical ideas.

1. INTRODUCTION

A data analyst must manage many types of uncertainty when examining information-based data. It is unquestionably a difficult task in every way. Ideas (Zadeh, 1965) (Atanassov, 1986) (Samarandache, 1999) are considered trustworthy regarding the handling of such uncertainties and impreciseness. However, it is observed that these ideas are unable to manage situations where attributes/parameters have to be considered. Therefore, Molodtsov (1999) put forward the idea of the soft set (SOS) that is meant to equip the fuzzy set-like structures with parameterization mode. In SOS, the classical belonging mapping is replaced with approximate mapping that considers the set of parameters as its domain and the power set of initial space of alternatives as its co-domain. In other words, the SOS provides a single argument domain to approximate the alternatives. However, the SOS itself is inadequate with the settings which demand a multi-argument domain for the approximation of alternatives. Smarandache (2018) addressed this issue by putting forward the idea of the hypersoft set (HypSOS) that employs MaaM to provide a multi-argument domain. This domain is obtained by taking the product of attribute-valued sub-classes. Saeed *et al.* (2022) characterized various fundamental concepts and notions of HypSOS to enhance its applicability in other branches of study. Yolcu and Öztürk (2021), Debnath (2021), Ihsan, Rahman, & Saeed (2021), Khan, Gulistan, and Wahab (2021), and Kamacı and Saqlain (2021) developed the hybrid structures of HypSOS with fuzzy set-like environments and discussed their applications. Yolcu, Smarandache, and Öztürk (2021) developed the hybrid structure of HypSOS with an intuitionistic fuzzy set and discussed its applications. Recently, some new types of SOS and HypSOS have been introduced (Smarandache, 2022a) (Smarandache, 2022b) (Samrandache, 2022c) (Smarandache, 2023) (Ihsan *et al.*, 2022) to tackle various decision-making situations.

Zadeh (1978) discussed the possibility theory as a basis for the fuzzy set. In such a theory, a possibility grade is used that is meant to assess the acceptance level of any approximation. Several authors (Alkhazaleh, Salleh, & Hassan, 2011) (Bashir, Salleh, & Alkhazaleh, 2012) (Alhazaymeh & Hassan, 2012) (Zhang & Shu, 2014) (Alhazaymeh & Hassan, 2013) (Karaaslan, 2017) (Jia-hua, Zhang, He, 2019) (Khalil, 2019) (Karaaslan, 2016) (Bashir & Salleh, 2012) (Karaaslan, 2016) have already used possibility setting with soft set-like environments to develop various possibility

SOS-like structures. Recently such settings have been employed with HypSOS-like environments to introduce new structures, (Rahman *et al.,* 2022) (Rahman *et al.,* 2022) (Rahman, Saeed, & Garg, 2022) (Rahman *et al.,* 2023) (Zhao *et al.,* 2023) (Al-Hijjawi & Alkhazaleh, 2023).

The core aim of this chapter is to introduce hybrid set structures of hypersoft set with possibility degree settings which may help the decision makers to furnish their acceptance levels in terms of different fuzzy set-like membership grades, like intuitionistic fuzzy membership grades, Pythagorean fuzzy membership grades, neutrosophic membership grades, etc. Consequently, many real-world decision-making problems relating to such situations may be tackled.

This chapter basically is aimed to extend the above-mentioned possibility SOS-like and HypSOS-like structures and then to introduce different types of possibility HypSOS-like structures.

Table 1. Notations and abbreviations

Notations	Full Name	Notations	Full Name
$\mathfrak{A}^{F}$	Collection of fuzzy subsets	$\mathfrak{A}^{IVIF}$	Collection of interval-valued Intuitionistic fuzzy subsets
$\mathfrak{A}^{IF}$	Collection of intuitionistic fuzzy subsets	$\mathfrak{A}^{IVPF}$	Collection of interval-valued picture fuzzy subsets
$\mathfrak{A}^{PyF}$	Collection of Pythagorean fuzzy subsets	$\mathfrak{A}^{IVsvN}$	Collection of interval-valued sv-neutrosophic subsets
$\mathfrak{A}^{PF}$	Collection of picture fuzzy subsets	$\mathfrak{A}^{IVF}$	Collection of interval-valued fuzzy subsets
$\mathfrak{A}^{svN}$	Collection of sv-neutrosophic subsets	**MaaM**	Multi-argument approximate mapping

2. PRELIMINARIES

This part presents some essential terms for proper understanding of the main results. The notation $2^{\widehat{U}}$ represents the power set of $\widehat{U}$ (initial space of objects).

Definition 2.1: Soft Set (Molodtsov, 1999)

A SOS A is the collection of object $(\Psi_A, \widehat{E})$ characterized by an approximate mapping $\Psi_A : \widehat{E} \rightarrow 2^{\widehat{U}}$ and defined as $A = \left\{ \left(\Psi_A(e), e \right) : e \in \widehat{E} \wedge \Psi_A(e) \subseteq 2^{\widehat{U}} \right\}$ where $\Psi_A(e)$ is e-approximate element of A corresponding to attribute e and $\widehat{E}$ is the set of distinct attributes.

Definition 2.2: Hypersoft Set (Smarandache, 2018)

A HypSS H is the collection of object $(\Psi_H, \widehat{\Theta})$ characterized by an approximate mapping $\Psi_H: \widehat{\Theta} \to 2^{\widehat{U}}$ and defined as $H = \left\{ \left(\Psi_H(\hat{\theta}), \hat{\theta} \right) : \hat{\theta} \in \widehat{\Theta} \wedge \Psi_H(\hat{\theta}) \subseteq 2^{\widehat{U}} \right\}$ where $\Psi_H(\hat{\theta})$ is $\hat{\theta}-$ approximate element of H corresponding to attribute-valued tuple $\wp$ and $\widehat{\Theta} = \widehat{\Theta}_1 \times \widehat{\Theta}_2 \times \ldots \times \widehat{\Theta}_n$. The sets $\widehat{\Theta}_1, \widehat{\Theta}_2, \ldots, \widehat{\Theta}_n$ are attribute-valued non-overlapping sets.

3. HYBRID STRUCTURES OF HYPERSOFT SETS

In this section, some new hybrid structures of hypersoft set are discussed with illustrative examples.

3.1 Hybrid Structures of Fuzzy Hypersoft Sets With Possibility Settings

Possibility degree is actually meant for the acceptance level provided by decision makers for the approximation of alternatives based on some parametric tuples. In accessible literature, Rahman *et al.* (2023), Zhao *et al.* (2023) and Al-Hijjawi & Alkhazaleh (2023) developed hybrid structures of hypersoft set by considering possibility degree in terms of fuzzy membership grades. However, in this section, possibility degree is considered in terms of other fuzzy set-like membership grades, e.g., intuitionistic fuzzy membership grades, neutrosophic membership grades, etc. It is meant that acceptance level may be provided in terms of intuitionistic fuzzy membership grades, neutrosophic membership grades, etc. therefore, by considering such cases, we have developed these hybrid set structures for hypersoft set with possibility degree-based settings.

The set $\widehat{\Theta} = \widehat{\Theta}_1 \times \widehat{\Theta}_2 \times \ldots \times \widehat{\Theta}_n$ is the product of attribute-valued disjoint sets $\widehat{\Theta}_i$, $i = 1,2,3,\ldots,n$ with respect to n different attributes in $\widehat{E}$.

Definition 3.1.1: Possibility Fuzzy Hypersoft Set of Type-1

A possibility fuzzy hypersoft set of type-1 $_1$ is defined as

$_1=$

$$\left\{ \left(\hat{\theta}, \left\langle \frac{\hat{u}}{\psi_F(\hat{\theta})(\hat{u})}, \mu_F(\hat{\theta}) \right\rangle \right) : \hat{u} \in \widehat{U} \wedge \hat{\theta} \in \widehat{\Theta} \right\}$$

where $\psi_F(\hat{\theta})(\hat{u})$, $\mu_F(\hat{\theta}) \subseteq \mathfrak{A}^F$ and $\psi_F(\hat{\theta})(\hat{u})$, $\mu_F(\hat{\theta}) \in [0,1]$.

Example 3.1.1: Let $\widehat{U} = \{\hat{u}_1, \hat{u}_2, \hat{u}_3, \hat{u}_4\}$ consisting of four models of air coolers and $\widehat{\Theta} = \{\hat{\theta}_1, \hat{\theta}_2, \hat{\theta}_3, \hat{\theta}_4\}$ corresponding to attribute-valued sets

$\widehat{\Theta}_1 = \{800, 1000\}, \widehat{\Theta}_2 = \{180, 230\}$,

$\widehat{\Theta}_3 = \{55\}$, and $\widehat{\Theta}_4 = \{200\}$ with respect to attributes $\wp_1$= water tank capacity in liters, $\wp_2$= voltage, $\wp_3$= air pressure in feet, and $\wp_4$= weight in kilograms respectively then possibility fuzzy hypersoft set of type-1 $_1$ can be constructed as

$$\mathcal{E}_1 = \left\{ \begin{array}{l} \left(\hat{\theta}_1, \left\{\left\langle \frac{\hat{u}_1}{0.21}, 0.31 \right\rangle, \left\langle \frac{\hat{u}_2}{0.31}, 0.41 \right\rangle, \left\langle \frac{\hat{u}_3}{0.41}, 0.51 \right\rangle, \left\langle \frac{\hat{u}_4}{0.51}, 0.61 \right\rangle \right\}\right), \\ \left(\hat{\theta}_2, \left\{\left\langle \frac{\hat{u}_1}{0.22}, 0.32 \right\rangle, \left\langle \frac{\hat{u}_2}{0.32}, 0.42 \right\rangle, \left\langle \frac{\hat{u}_3}{0.42}, 0.52 \right\rangle, \left\langle \frac{\hat{u}_4}{0.52}, 0.62 \right\rangle \right\}\right), \\ \left(\hat{\theta}_3, \left\{\left\langle \frac{\hat{u}_1}{0.23}, 0.33 \right\rangle, \left\langle \frac{\hat{u}_2}{0.33}, 0.43 \right\rangle, \left\langle \frac{\hat{u}_3}{0.43}, 0.53 \right\rangle, \left\langle \frac{\hat{u}_4}{0.53}, 0.63 \right\rangle \right\}\right), \\ \left(\hat{\theta}_4, \left\{\left\langle \frac{\hat{u}_1}{0.24}, 0.34 \right\rangle, \left\langle \frac{\hat{u}_2}{0.34}, 0.44 \right\rangle, \left\langle \frac{\hat{u}_3}{0.44}, 0.54 \right\rangle, \left\langle \frac{\hat{u}_4}{0.54}, 0.64 \right\rangle \right\}\right), \end{array} \right\}$$

Definition 3.1.2: Possibility Fuzzy Hypersoft Set of Type-2

A possibility fuzzy hypersoft set of type-2 $_2$ is defined as

$_1$=

$$\left\{ \left(\hat{\theta}, \left\langle \frac{\hat{u}}{\psi_F(\hat{\theta})(\hat{u})}, \mu_{IF}(\hat{\theta}) \right\rangle \right) : \hat{u} \in \widehat{U} \wedge \hat{\theta} \in \widehat{\Theta} \right\}$$

where

$\psi_F(\hat{\theta})(\hat{u}) \subseteq \mathfrak{A}^F, \mu_{IF}(\hat{\theta}) \subseteq \mathfrak{A}^{IF}$

and

$\psi_F(\hat{\theta})(\hat{u}) \in [0,1], \mu_{IF}(\hat{\theta}) = \langle T_\mu(\hat{\theta}), F_\mu(\hat{\theta}) \rangle$

with

$T_{\mu}(\hat{\theta}),F_{\mu}(\hat{\theta})\in [0,1]$

and

$0 \leq T_{\mu}(\hat{\theta})+F_{\mu}(\hat{\theta})\leq 1$

.

Example 3.1.2: Considering the assumptions from Example 3.1.1, we can construct possibility fuzzy hypersoft set of type-2 $_2$ can be constructed as

$$\mathfrak{E}_2 = \left\{ \begin{array}{l} \left(\hat{\theta}_1,\left\{\left\langle\frac{\hat{u}_1}{0.21},\langle 0.31,0.21\rangle\right\rangle,\left\langle\frac{\hat{u}_2}{0.31},\langle 0.41,0.31\rangle\right\rangle,\left\langle\frac{\hat{u}_3}{0.41},\langle 0.51,0.41\rangle\right\rangle,\left\langle\frac{\hat{u}_4}{0.51},\langle 0.61,0.21\rangle\right\rangle\right\}\right), \\ \left(\hat{\theta}_2,\left\{\left\langle\frac{\hat{u}_1}{0.22},\langle 0.32,0.22\rangle\right\rangle,\left\langle\frac{\hat{u}_2}{0.32},\langle 0.42,0.32\rangle\right\rangle,\left\langle\frac{\hat{u}_3}{0.42},\langle 0.52,0.42\rangle\right\rangle,\left\langle\frac{\hat{u}_4}{0.52},\langle 0.62,0.22\rangle\right\rangle\right\}\right), \\ \left(\hat{\theta}_3,\left\{\left\langle\frac{\hat{u}_1}{0.23},\langle 0.33,0.23\rangle\right\rangle,\left\langle\frac{\hat{u}_2}{0.33},\langle 0.43,0.33\rangle\right\rangle,\left\langle\frac{\hat{u}_3}{0.43},\langle 0.53,0.43\rangle\right\rangle,\left\langle\frac{\hat{u}_4}{0.53},\langle 0.63,0.23\rangle\right\rangle\right\}\right), \\ \left(\hat{\theta}_4,\left\{\left\langle\frac{\hat{u}_1}{0.24},\langle 0.34,0.24\rangle\right\rangle,\left\langle\frac{\hat{u}_2}{0.34},\langle 0.44,0.34\rangle\right\rangle,\left\langle\frac{\hat{u}_3}{0.44},\langle 0.54,0.44\rangle\right\rangle,\left\langle\frac{\hat{u}_4}{0.54},\langle 0.64,0.24\rangle\right\rangle\right\}\right) \end{array} \right\}$$

Definition 3.1.3: Possibility Fuzzy Hypersoft Set of Type-3

A possibility fuzzy hypersoft set of type-3 $_3$ is defined as

$_3=$

$$\left\{\left(\hat{\theta},\left\langle\frac{\hat{u}}{\psi_F(\hat{\theta})(\hat{u})},\mu_{PyF}(\hat{\theta})\right\rangle\right):\hat{u}\in\hat{U}\wedge\hat{\theta}\in\hat{\Theta}\right\}$$

where

$\psi_F(\hat{\theta})(\hat{u})\subseteq \mathfrak{A}^F,\mu_{PyF}(\hat{\theta})\subseteq \mathfrak{A}^{PyF}$

and

$\psi_F(\hat{\theta})(\hat{u}) \in [0,1], \mu_{PyF}(\hat{\theta}) = \langle T_{\mu}(\hat{\theta}), F_{\mu}(\hat{\theta}) \rangle$

with

$T_{\mu}(\hat{\theta}), F_{\mu}(\hat{\theta}) \in [0,1]$

and

$0 \leq T_{\mu}^2(\hat{\theta}) + F_{\mu}^2(\hat{\theta}) \leq 1$

.

Example 3.1.3: Considering the assumptions from Example 3.1.1, we can construct possibility fuzzy hypersoft set of type-3 $_3$ can be constructed as

$$\mathfrak{E}_3 = \left\{ \begin{array}{l} \left(\hat{\theta}_1, \left\{ \left\langle \frac{\hat{u}_1}{0.21}, \langle 0.5, 0.6 \rangle \right\rangle, \left\langle \frac{\hat{u}_2}{0.31}, \langle 0.7, 0.5 \rangle \right\rangle, \left\langle \frac{\hat{u}_3}{0.41}, \langle 0.5, 0.8 \rangle \right\rangle, \left\langle \frac{\hat{u}_4}{0.51}, \langle 0.6, 0.5 \rangle \right\rangle \right\} \right), \\ \left(\hat{\theta}_2, \left\{ \left\langle \frac{\hat{u}_1}{0.22}, \langle 0.9, 0.4 \rangle \right\rangle, \left\langle \frac{\hat{u}_2}{0.32}, \langle 0.4, 0.8 \rangle \right\rangle, \left\langle \frac{\hat{u}_3}{0.42}, \langle 0.4, 0.7 \rangle \right\rangle, \left\langle \frac{\hat{u}_4}{0.52}, \langle 0.5, 0.8 \rangle \right\rangle \right\} \right), \\ \left(\hat{\theta}_3, \left\{ \left\langle \frac{\hat{u}_1}{0.23}, \langle 0.7, 0.5 \rangle \right\rangle, \left\langle \frac{\hat{u}_2}{0.33}, \langle 0.8, 0.5 \rangle \right\rangle, \left\langle \frac{\hat{u}_3}{0.43}, \langle 0.6, 0.5 \rangle \right\rangle, \left\langle \frac{\hat{u}_4}{0.53}, \langle 0.5, 0.6 \rangle \right\rangle \right\} \right), \\ \left(\hat{\theta}_4, \left\{ \left\langle \frac{\hat{u}_1}{0.24}, \langle 0.4, 0.9 \rangle \right\rangle, \left\langle \frac{\hat{u}_2}{0.34}, \langle 0.8, 0.4 \rangle \right\rangle, \left\langle \frac{\hat{u}_3}{0.44}, \langle 0.7, 0.4 \rangle \right\rangle, \left\langle \frac{\hat{u}_4}{0.54}, \langle 0.8, 0.5 \rangle \right\rangle \right\} \right), \end{array} \right\}$$

Definition 3.1.4: Possibility Fuzzy Hypersoft Set of Type-4

A possibility fuzzy hypersoft set of type-4 $_4$ is defined as

$_4 =$

$$\left\{ \left(\hat{\theta}, \left\langle \frac{\hat{u}}{\psi_F(\hat{\theta})(\hat{u})}, \mu_{PF}(\hat{\theta}) \right\rangle \right) : \hat{u} \in \hat{U} \wedge \hat{\theta} \in \hat{\Theta} \right\}$$

where

$$\psi_F(\hat{\theta})(\hat{u}) \subseteq \mathfrak{A}^F, \mu_{PF}(\hat{\theta}) \subseteq \mathfrak{A}^{PF}$$

and

$$\psi_F(\hat{\theta})(\hat{u}) \in [0,1], \mu_{PF}(\hat{\theta}) = \langle T_\mu(\hat{\theta}), I_\mu(\hat{\theta}), F_\mu(\hat{\theta}) \rangle$$

with

$$T_\mu(\hat{\theta}), I_\mu(\hat{\theta}), F_\mu(\hat{\theta}) \in [0,1]$$

and

$$0 \leq T_\mu(\hat{\theta}) + I_\mu(\hat{\theta}) + F_\mu(\hat{\theta}) \leq 1$$

.

Example 3.1.4: Considering the assumptions from Example 3.1.1, we can construct possibility fuzzy hypersoft set of type-4 $_4$ can be constructed as

$$\mathcal{E}_4 = \left\{ \begin{array}{l} \left(\hat{\theta}_1, \left\{ \left\langle \frac{\hat{u}_1}{.2}, \langle .5,.1,.2 \rangle \right\rangle, \left\langle \frac{\hat{u}_2}{.3}, \langle .2,.5,.2 \rangle \right\rangle, \left\langle \frac{\hat{u}_3}{.4}, \langle .4,.2,.3 \rangle \right\rangle, \left\langle \frac{\hat{u}_4}{.5}, \langle .6,.1,.2 \rangle \right\rangle \right\} \right), \\ \left(\hat{\theta}_2, \left\{ \left\langle \frac{\hat{u}_1}{.3}, \langle .1,.4,.4 \rangle \right\rangle, \left\langle \frac{\hat{u}_2}{.4}, \langle .3,.4,.1 \rangle \right\rangle, \left\langle \frac{\hat{u}_3}{.5}, \langle .2,.3,.3 \rangle \right\rangle, \left\langle \frac{\hat{u}_4}{.6}, \langle .2,.4,.2 \rangle \right\rangle \right\} \right), \\ \left(\hat{\theta}_3, \left\{ \left\langle \frac{\hat{u}_1}{.4}, \langle .3,.5,.1 \rangle \right\rangle, \left\langle \frac{\hat{u}_2}{.5}, \langle .2,.5,.2 \rangle \right\rangle, \left\langle \frac{\hat{u}_3}{.6}, \langle .6,.2,.1 \rangle \right\rangle, \left\langle \frac{\hat{u}_4}{.7}, \langle .5,.2,.2 \rangle \right\rangle \right\} \right), \\ \left(\hat{\theta}_4, \left\{ \left\langle \frac{\hat{u}_1}{.5}, \langle .4,.1,.1 \rangle \right\rangle, \left\langle \frac{\hat{u}_2}{.6}, \langle .2,.2,.2 \rangle \right\rangle, \left\langle \frac{\hat{u}_3}{.7}, \langle .7,.1,.1 \rangle \right\rangle, \left\langle \frac{\hat{u}_4}{.8}, \langle .4,.2,.1 \rangle \right\rangle \right\} \right), \end{array} \right\}$$

Definition 3.1.5: Possibility Fuzzy Hypersoft Set of Type-5

A possibility fuzzy hypersoft set of type-5 $_5$ is defined as

$_5$=

$$\left\{\left(\hat{\theta}, \left\langle \frac{\hat{u}}{\psi_F(\hat{\theta})(\hat{u})}, \mu_{svN}(\hat{\theta}) \right\rangle\right) : \hat{u} \in \widehat{U} \wedge \hat{\theta} \in \widehat{\Theta}\right\}$$

where

$$\psi_F(\hat{\theta})(\hat{u}) \subseteq \mathfrak{A}^F, \mu_{svN}(\hat{\theta}) \subseteq \mathfrak{A}^{svN}$$

and

$$\psi_F(\hat{\theta})(\hat{u}) \in [0,1], \mu_{svN}(\hat{\theta}) = \langle T_\mu(\hat{\theta}), I_\mu(\hat{\theta}), F_\mu(\hat{\theta}) \rangle$$

with

$$T_\mu(\hat{\theta}), I_\mu(\hat{\theta}), F_\mu(\hat{\theta}) \in [0,1]$$

and

$$0 \leq T_\mu(\hat{\theta}) + I_\mu(\hat{\theta}) + F_\mu(\hat{\theta}) \leq 3$$

.

Example 3.1.5: Considering the assumptions from Example 3.1.1, we can construct possibility fuzzy hypersoft set of type-5 $_5$ can be constructed as

$$\mathfrak{E}_5 = \left\{\begin{array}{l}
\left(\hat{\theta}_1, \left\{\left\langle \frac{\hat{u}_1}{.2}, \langle .5,.6,.6 \rangle \right\rangle, \left\langle \frac{\hat{u}_2}{.3}, \langle .6,.7,.7 \rangle \right\rangle, \left\langle \frac{\hat{u}_3}{.4}, \langle .7,.8,.8 \rangle \right\rangle, \left\langle \frac{\hat{u}_4}{.5}, \langle .8,.9,.9 \rangle \right\rangle\right\}\right), \\
\left(\hat{\theta}_2, \left\{\left\langle \frac{\hat{u}_1}{.3}, \langle .6,.5,.5 \rangle \right\rangle, \left\langle \frac{\hat{u}_2}{.4}, \langle .7,.6,.6 \rangle \right\rangle, \left\langle \frac{\hat{u}_3}{.5}, \langle .8,.7,.7 \rangle \right\rangle, \left\langle \frac{\hat{u}_4}{.6}, \langle .9,.8,.8 \rangle \right\rangle\right\}\right), \\
\left(\hat{\theta}_3, \left\{\left\langle \frac{\hat{u}_1}{.4}, \langle .7,.6,.7 \rangle \right\rangle, \left\langle \frac{\hat{u}_2}{.5}, \langle .8,.7,.8 \rangle \right\rangle, \left\langle \frac{\hat{u}_3}{.6}, \langle .9,.8,.9 \rangle \right\rangle, \left\langle \frac{\hat{u}_4}{.7}, \langle .6,.5,.9 \rangle \right\rangle\right\}\right), \\
\left(\hat{\theta}_4, \left\{\left\langle \frac{\hat{u}_1}{.5}, \langle .8,.7,.6 \rangle \right\rangle, \left\langle \frac{\hat{u}_2}{.6}, \langle .9,.6,.5 \rangle \right\rangle, \left\langle \frac{\hat{u}_3}{.7}, \langle .6,.7,.5 \rangle \right\rangle, \left\langle \frac{\hat{u}_4}{.8}, \langle .6,.8,.8 \rangle \right\rangle\right\}\right),
\end{array}\right\}$$

Definition 3.1.6: Possibility Fuzzy Hypersoft Set of Type-6

A possibility fuzzy hypersoft set of type-6 $_6$ is defined as

$$_6 =$$

$$\left\{ \left(\hat{\theta}, \left\langle \frac{\hat{u}}{\psi_F(\hat{\theta})(\hat{u})}, \mu_{IVF}(\hat{\theta}) \right\rangle \right) : \hat{u} \in \widehat{U} \wedge \hat{\theta} \in \widehat{\Theta} \right\}$$

where

$$\psi_F(\hat{\theta})(\hat{u}) \subseteq \mathfrak{A}^F, \mu_{IVF}(\hat{\theta}) \subseteq \mathfrak{A}^{IVF}$$

and

$$\psi_F(\hat{\theta})(\hat{u}) \in [0,1], \mu_{IVF}(\hat{\theta}) = \left[L_\mu(\hat{\theta}), U_\mu(\hat{\theta}) \right]$$

with $L_\mu(\hat{\theta}), U_\mu(\hat{\theta}) \in [0,1]$.

Example 3.1.6: Considering the assumptions from Example 3.1.1, we can construct possibility fuzzy hypersoft set of type-6 $_6$ can be constructed as

$$\xi_6 =$$

$$\left\{ \begin{array}{l} \left(\hat{\theta}_1, \left\{ \left\langle \frac{\hat{u}_1}{0.21}, [0.21, 0.31] \right\rangle, \left\langle \frac{\hat{u}_2}{0.31}, [0.31, 0.41] \right\rangle, \left\langle \frac{\hat{u}_3}{0.41}, [0.41, 0.51] \right\rangle, \left\langle \frac{\hat{u}_4}{0.51}, [0.21, 0.61] \right\rangle \right\} \right), \\ \left(\hat{\theta}_2, \left\{ \left\langle \frac{\hat{u}_1}{0.22}, [0.22, 0.32] \right\rangle, \left\langle \frac{\hat{u}_2}{0.32}, [0.32, 0.42] \right\rangle, \left\langle \frac{\hat{u}_3}{0.42}, [0.42, 0.52] \right\rangle, \left\langle \frac{\hat{u}_4}{0.52}, [0.22, 0.62] \right\rangle \right\} \right), \\ \left(\hat{\theta}_3, \left\{ \left\langle \frac{\hat{u}_1}{0.23}, [0.23, 0.33] \right\rangle, \left\langle \frac{\hat{u}_2}{0.33}, [0.33, 0.43] \right\rangle, \left\langle \frac{\hat{u}_3}{0.43}, [0.43, 0.53] \right\rangle, \left\langle \frac{\hat{u}_4}{0.53}, [0.23, 0.63] \right\rangle \right\} \right), \\ \left(\hat{\theta}_4, \left\{ \left\langle \frac{\hat{u}_1}{0.24}, [0.24, 0.34] \right\rangle, \left\langle \frac{\hat{u}_2}{0.34}, [0.34, 0.44] \right\rangle, \left\langle \frac{\hat{u}_3}{0.44}, [0.44, 0.54] \right\rangle, \left\langle \frac{\hat{u}_4}{0.54}, [0.24, 0.64] \right\rangle \right\} \right) \end{array} \right\}$$

3.2 Hybrid Structures of Intuitionistic Fuzzy Hypersoft Sets With Possibility Settings

Definition 3.2.1: Possibility Intuitionistic Fuzzy Hypersoft Set of Type-1

A possibility intuitionistic fuzzy hypersoft set of type-1 λ_1 is defined as

$$\lambda_1 = \left\{ \left(\hat{\theta}, \left\langle \frac{\hat{u}}{\psi_{IF}(\hat{\theta})(\hat{u})}, \mu_F(\hat{\theta}) \right\rangle \right) : \hat{u} \in \hat{U} \wedge \hat{\theta} \in \hat{\Theta} \right\}$$

where

$$\psi_{IF}(\hat{\theta})(\hat{u}) = \langle T_{\psi}(\hat{\theta})(\hat{u}), F_{\psi}(\hat{\theta})(\hat{u}) \rangle \subseteq \mathfrak{A}^{IF},\ \mu_F(\hat{\theta}) \subseteq \mathfrak{A}^{F}$$

and

$$T_{\psi}(\hat{\theta})(\hat{u}), F_{\psi}(\hat{\theta})(\hat{u}),\ \mu_F(\hat{\theta}) \in [0,1]$$

such that $0 \leq T_{\psi}(\hat{\theta})(\hat{u}) + F_{\psi}(\hat{\theta})(\hat{u}) \leq 1$.

Example 3.2.1: Considering the assumptions from Example 3.1.1, we can construct possibility intuitionistic fuzzy hypersoft set of type-1 λ_1 can be constructed as

$$\lambda_1 = \left\{ \begin{array}{l} \left(\hat{\theta}_1, \left\{ \left\langle \frac{\hat{u}_1}{\langle .2,.5 \rangle}, 0.31 \right\rangle, \left\langle \frac{\hat{u}_2}{\langle .3,.6 \rangle}, 0.41 \right\rangle, \left\langle \frac{\hat{u}_3}{\langle .4,.4 \rangle}, 0.51 \right\rangle, \left\langle \frac{\hat{u}_4}{\langle .5,.2 \rangle}, 0.61 \right\rangle \right\} \right), \\ \left(\hat{\theta}_2, \left\{ \left\langle \frac{\hat{u}_1}{\langle .2,.3 \rangle}, 0.32 \right\rangle, \left\langle \frac{\hat{u}_2}{\langle .3,.4 \rangle}, 0.42 \right\rangle, \left\langle \frac{\hat{u}_3}{\langle .4,.2 \rangle}, 0.52 \right\rangle, \left\langle \frac{\hat{u}_4}{\langle .5,.1 \rangle}, 0.62 \right\rangle \right\} \right), \\ \left(\hat{\theta}_3, \left\{ \left\langle \frac{\hat{u}_1}{\langle .2,.5 \rangle}, 0.33 \right\rangle, \left\langle \frac{\hat{u}_2}{\langle .3,.3 \rangle}, 0.43 \right\rangle, \left\langle \frac{\hat{u}_3}{\langle .4,.1 \rangle}, 0.53 \right\rangle, \left\langle \frac{\hat{u}_4}{\langle .5,.2 \rangle}, 0.63 \right\rangle \right\} \right), \\ \left(\hat{\theta}_4, \left\{ \left\langle \frac{\hat{u}_1}{\langle .2,.2 \rangle}, 0.34 \right\rangle, \left\langle \frac{\hat{u}_2}{\langle .3,.5 \rangle}, 0.44 \right\rangle, \left\langle \frac{\hat{u}_3}{\langle .4,.5 \rangle}, 0.54 \right\rangle, \left\langle \frac{\hat{u}_4}{\langle .5,.3 \rangle}, 0.64 \right\rangle \right\} \right), \end{array} \right\}$$

Definition 3.2.2: Possibility Intuitionistic Fuzzy Hypersoft Set of Type-2

A possibility intuitionistic fuzzy hypersoft set of type-2 λ_2 is defined as

$$\lambda_2 = \left\{ \left(\hat{\theta}, \left\langle \frac{\hat{u}}{\psi_{IF}(\hat{\theta})(\hat{u})}, \mu_{IF}(\hat{\theta}) \right\rangle \right) : \hat{u} \in \hat{U} \wedge \hat{\theta} \in \hat{\Theta} \right\}$$

where

$$\psi_{IF}(\hat{\theta})(\hat{u}) = \langle T_{\psi}(\hat{\theta})(\hat{u}), F_{\psi}(\hat{\theta})(\hat{u}) \rangle \subseteq \mathfrak{A}^{IF}, \mu_{IF}(\hat{\theta}) \subseteq \mathfrak{A}^{IF}$$

and

$$T_{\psi}(\hat{\theta})(\hat{u}), F_{\psi}(\hat{\theta})(\hat{u}) \in [0,1], \mu_{IF}(\hat{\theta}) = \langle T_{\mu}(\hat{\theta}), F_{\mu}(\hat{\theta}) \rangle, T_{\mu}(\hat{\theta}), F_{\mu}(\hat{\theta}) \in [0,1]$$

with

$$0 \leq T_{\psi}(\hat{\theta})(\hat{u}) + F_{\psi}(\hat{\theta})(\hat{u}) \leq 1$$

and

$$0 \leq T_{\mu}(\hat{\theta}) + F_{\mu}(\hat{\theta}) \leq 1$$

.

Example 3.2.2: Considering the assumptions from Example 3.1.1, we can construct possibility intuitionistic fuzzy hypersoft set of type-2 λ_2 can be constructed as

$$\lambda_2 = \left\{ \begin{array}{l} \left(\hat{\theta}_1, \left\{ \left\langle \frac{\hat{u}_1}{\langle .2, .4 \rangle}, \langle 0.31, 0.21 \rangle \right\rangle, \left\langle \frac{\hat{u}_2}{\langle .3, .4 \rangle}, \langle 0.41, 0.31 \rangle \right\rangle, \left\langle \frac{\hat{u}_3}{\langle .4, .5 \rangle}, \langle 0.51, 0.41 \rangle \right\rangle, \left\langle \frac{\hat{u}_4}{\langle .5, .2 \rangle}, \langle 0.61, 0.21 \rangle \right\rangle \right\} \right), \\ \left(\hat{\theta}_2, \left\{ \left\langle \frac{\hat{u}_1}{\langle .2, .3 \rangle}, \langle 0.32, 0.22 \rangle \right\rangle, \left\langle \frac{\hat{u}_2}{\langle .3, .2 \rangle}, \langle 0.42, 0.32 \rangle \right\rangle, \left\langle \frac{\hat{u}_3}{\langle .4, .4 \rangle}, \langle 0.52, 0.42 \rangle \right\rangle, \left\langle \frac{\hat{u}_4}{\langle .5, .1 \rangle}, \langle 0.62, 0.22 \rangle \right\rangle \right\} \right), \\ \left(\hat{\theta}_3, \left\{ \left\langle \frac{\hat{u}_1}{\langle .1, .5 \rangle}, \langle 0.33, 0.23 \rangle \right\rangle, \left\langle \frac{\hat{u}_2}{\langle .1, .4 \rangle}, \langle 0.43, 0.33 \rangle \right\rangle, \left\langle \frac{\hat{u}_3}{\langle .5, .1 \rangle}, \langle 0.53, 0.43 \rangle \right\rangle, \left\langle \frac{\hat{u}_4}{\langle .4, .2 \rangle}, \langle 0.63, 0.23 \rangle \right\rangle \right\} \right), \\ \left(\hat{\theta}_4, \left\{ \left\langle \frac{\hat{u}_1}{\langle .3, .3 \rangle}, \langle 0.34, 0.24 \rangle \right\rangle, \left\langle \frac{\hat{u}_2}{\langle .3, .5 \rangle}, \langle 0.44, 0.34 \rangle \right\rangle, \left\langle \frac{\hat{u}_3}{\langle .4, .5 \rangle}, \langle 0.54, 0.44 \rangle \right\rangle, \left\langle \frac{\hat{u}_4}{\langle .5, .3 \rangle}, \langle 0.64, 0.24 \rangle \right\rangle \right\} \right), \end{array} \right\}$$

Definition 3.2.3: Possibility Intuitionistic Fuzzy Hypersoft Set of Type-3

A possibility intuitionistic fuzzy hypersoft set of type-3 λ_3 is defined as

$$\lambda_3 = \left\{ \left(\hat{\theta}, \left\langle \frac{\hat{u}}{\psi_{IF}(\hat{\theta})(\hat{u})}, \mu_{PyF}(\hat{\theta}) \right\rangle \right) : \hat{u} \in \hat{U} \wedge \hat{\theta} \in \hat{\Theta} \right\}$$

where

$$\psi_{IF}(\hat{\theta})(\hat{u}) = \langle T_{\psi}(\hat{\theta})(\hat{u}), F_{\psi}(\hat{\theta})(\hat{u}) \rangle \subseteq \mathfrak{A}^{IF}, \mu_{PyF}(\hat{\theta}) \subseteq \mathfrak{A}^{PyF}$$

and

$$T_{\psi}(\hat{\theta})(\hat{u}), F_{\psi}(\hat{\theta})(\hat{u}) \in [0,1]$$

with

$$0 \leq T_{\psi}(\hat{\theta})(\hat{u}) + F_{\psi}(\hat{\theta})(\hat{u}) \leq 1$$

and

$$\mu_{PyF}(\hat{\theta}) = \langle T_{\mu}(\hat{\theta}), F_{\mu}(\hat{\theta}) \rangle$$

with

$T_\mu(\hat{\theta}), F_\mu(\hat{\theta}) \in [0,1]$

and

$0 \leq T_\mu^2(\hat{\theta}) + F_\mu^2(\hat{\theta}) \leq 1$

.

Example 3.2.3: Considering the assumptions from Example 3.1.1, we can construct possibility intuitionistic fuzzy hypersoft set of type-3 λ_3 can be constructed as

$$\lambda_3 = \left\{ \begin{array}{l} \left(\hat{\theta}_1, \left\{\left\langle \frac{\hat{u}_1}{\langle .2,.1\rangle}, \langle 0.5, 0.6\rangle \right\rangle, \left\langle \frac{\hat{u}_2}{\langle .2,.3\rangle}, \langle 0.7, 0.5\rangle \right\rangle, \left\langle \frac{\hat{u}_3}{\langle .2,.4\rangle}, \langle 0.5, 0.8\rangle \right\rangle, \left\langle \frac{\hat{u}_4}{\langle .2,.5\rangle}, \langle 0.6, 0.5\rangle \right\rangle \right\}\right), \\ \left(\hat{\theta}_2, \left\{\left\langle \frac{\hat{u}_1}{\langle .3,.1\rangle}, \langle 0.9, 0.4\rangle \right\rangle, \left\langle \frac{\hat{u}_2}{\langle .3,.2\rangle}, \langle 0.4, 0.8\rangle \right\rangle, \left\langle \frac{\hat{u}_3}{\langle .3,.4\rangle}, \langle 0.4, 0.7\rangle \right\rangle, \left\langle \frac{\hat{u}_4}{\langle .3,.5\rangle}, \langle 0.5, 0.8\rangle \right\rangle \right\}\right), \\ \left(\hat{\theta}_3, \left\{\left\langle \frac{\hat{u}_1}{\langle .4,.1\rangle}, \langle 0.7, 0.5\rangle \right\rangle, \left\langle \frac{\hat{u}_2}{\langle .4,.2\rangle}, \langle 0.8, 0.5\rangle \right\rangle, \left\langle \frac{\hat{u}_3}{\langle .4,.3\rangle}, \langle 0.6, 0.5\rangle \right\rangle, \left\langle \frac{\hat{u}_4}{\langle .4,.5\rangle}, \langle 0.5, 0.6\rangle \right\rangle \right\}\right), \\ \left(\hat{\theta}_4, \left\{\left\langle \frac{\hat{u}_1}{\langle .5,.1\rangle}, \langle 0.4, 0.9\rangle \right\rangle, \left\langle \frac{\hat{u}_2}{\langle .5,.2\rangle}, \langle 0.8, 0.4\rangle \right\rangle, \left\langle \frac{\hat{u}_3}{\langle .5,.3\rangle}, \langle 0.7, 0.4\rangle \right\rangle, \left\langle \frac{\hat{u}_4}{\langle .5,.4\rangle}, \langle 0.8, 0.5\rangle \right\rangle \right\}\right), \end{array} \right\}$$

Definition 3.2.4: Possibility Intuitionistic Fuzzy Hypersoft Set of Type-4

A possibility intuitionistic fuzzy hypersoft set of type-4 λ_4 is defined as

$$\lambda_4 = \left\{ \left(\hat{\theta}, \left\langle \frac{\hat{u}}{\psi_{IF}(\hat{\theta})(\hat{u})}, \mu_{PF}(\hat{\theta}) \right\rangle \right) : \hat{u} \in \widehat{U} \wedge \hat{\theta} \in \widehat{\Theta} \right\}$$

where

$\psi_{IF}(\hat{\theta})(\hat{u}) = \langle T_\psi(\hat{\theta})(\hat{u}), F_\psi(\hat{\theta})(\hat{u}) \rangle \subseteq \mathfrak{A}^{IF}$,

and

$T_{\psi}(\hat{\theta})(\hat{u}), F_{\psi}(\hat{\theta})(\hat{u}) \in [0,1]$

with $0 \leq T_{\psi}(\hat{\theta})(\hat{u}) + F_{\psi}(\hat{\theta})(\hat{u}) \leq 1$. Similarly

$\mu_{PF}(\hat{\theta}) \subseteq \mathfrak{A}^{PF}$

,

$\mu_{PF}(\hat{\theta}) = \langle T_{\mu}(\hat{\theta}), I_{\mu}(\hat{\theta}), F_{\mu}(\hat{\theta})\rangle$

with

$T_{\mu}(\hat{\theta}), I_{\mu}(\hat{\theta}), F_{\mu}(\hat{\theta}) \in [0,1]$

and $0 \leq T_{\mu}(\hat{\theta}) + I_{\mu}(\hat{\theta}) + F_{\mu}(\hat{\theta}) \leq 1$.

Example 3.2.4: Considering the assumptions from Example 3.1.1, we can construct possibility intuitionistic fuzzy hypersoft set of type-4 λ_4 can be constructed as

$$\lambda_4 = \left\{ \begin{array}{l} \left(\hat{\theta}_1, \left\{\left\langle \frac{\hat{u}_1}{\langle .2,.1\rangle}, \langle .5,.1,.2\rangle \right\rangle, \left\langle \frac{\hat{u}_2}{\langle .2,.3\rangle}, \langle .2,.5,.2\rangle \right\rangle, \left\langle \frac{\hat{u}_3}{\langle .2,.4\rangle}, \langle .4,.2,.3\rangle \right\rangle, \left\langle \frac{\hat{u}_4}{\langle .2,.5\rangle}, \langle .6,.1,.2\rangle \right\rangle\right\}\right), \\ \left(\hat{\theta}_2, \left\{\left\langle \frac{\hat{u}_1}{\langle .3,.1\rangle}, \langle .1,.4,.4\rangle \right\rangle, \left\langle \frac{\hat{u}_2}{\langle .3,.2\rangle}, \langle .3,.4,.1\rangle \right\rangle, \left\langle \frac{\hat{u}_3}{\langle .3,.4\rangle}, \langle .2,.3,.3\rangle \right\rangle, \left\langle \frac{\hat{u}_4}{\langle .3,.5\rangle}, \langle .2,.4,.2\rangle \right\rangle\right\}\right), \\ \left(\hat{\theta}_3, \left\{\left\langle \frac{\hat{u}_1}{\langle .4,.1\rangle}, \langle .3,.5,.1\rangle \right\rangle, \left\langle \frac{\hat{u}_2}{\langle .4,.2\rangle}, \langle .2,.5,.2\rangle \right\rangle, \left\langle \frac{\hat{u}_3}{\langle .4,.3\rangle}, \langle .6,.2,.1\rangle \right\rangle, \left\langle \frac{\hat{u}_4}{\langle .4,.5\rangle}, \langle .5,.2,.2\rangle \right\rangle\right\}\right), \\ \left(\hat{\theta}_4, \left\{\left\langle \frac{\hat{u}_1}{\langle .5,.1\rangle}, \langle .4,.1,.1\rangle \right\rangle, \left\langle \frac{\hat{u}_2}{\langle .5,.2\rangle}, \langle .2,.2,.2\rangle \right\rangle, \left\langle \frac{\hat{u}_3}{\langle .5,.3\rangle}, \langle .7,.1,.1\rangle \right\rangle, \left\langle \frac{\hat{u}_4}{\langle .5,.4\rangle}, \langle .4,.2,.1\rangle \right\rangle\right\}\right), \end{array} \right\}$$

Definition 3.2.5: Possibility Intuitionistic Fuzzy Hypersoft Set of Type-5

A possibility intuitionistic fuzzy hypersoft set of type-5 λ_5 is defined as

$$\lambda_5 = \left\{ \left(\hat{\theta}, \left\langle \frac{\hat{u}}{\psi_{IF}(\hat{\theta})(\hat{u})}, \mu_{svN}(\hat{\theta}) \right\rangle \right) : \hat{u} \in \hat{U} \wedge \hat{\theta} \in \hat{\Theta} \right\}$$

where

$\psi_{IF}(\hat{\theta})(\hat{u}) = \langle T_{\psi}(\hat{\theta})(\hat{u}), F_{\psi}(\hat{\theta})(\hat{u})\rangle \subseteq \mathfrak{A}^{IF}$,

and

$T_{\psi}(\hat{\theta})(\hat{u}), F_{\psi}(\hat{\theta})(\hat{u}) \in [0,1]$

with $0 \leq T_{\psi}(\hat{\theta})(\hat{u}) + F_{\psi}(\hat{\theta})(\hat{u}) \leq 1$. Similarly

$\mu_{svN}(\hat{\theta}) \subseteq \mathfrak{A}^{svN}$

,

$\mu_{svN}(\hat{\theta}) = \langle T_{\mu}(\hat{\theta}), I_{\mu}(\hat{\theta}), F_{\mu}(\hat{\theta})\rangle$

with $T_{\mu}(\hat{\theta}), I_{\mu}(\hat{\theta}), F_{\mu}(\hat{\theta}) \in [0,1]$ and $0 \leq T_{\mu}(\hat{\theta}) + I_{\mu}(\hat{\theta}) + F_{\mu}(\hat{\theta}) \leq 3$.

Example 3.2.5: Considering the assumptions from Example 3.1.1, we can construct possibility intuitionistic fuzzy hypersoft set of type-5 λ_5 can be constructed as

$$\lambda_5 = \left\{ \begin{array}{l} \left(\hat{\theta}_1, \left\{\left\langle \frac{\hat{u}_1}{\langle .2,.1\rangle}, \langle .5,.6,.6\rangle \right\rangle, \left\langle \frac{\hat{u}_2}{\langle .2,.3\rangle}, \langle .6,.7,.7\rangle \right\rangle, \left\langle \frac{\hat{u}_3}{\langle .2,.4\rangle}, \langle .7,.8,.8\rangle \right\rangle, \left\langle \frac{\hat{u}_4}{\langle .2,.5\rangle}, \langle .8,.9,.9\rangle \right\rangle \right\}\right), \\ \left(\hat{\theta}_2, \left\{\left\langle \frac{\hat{u}_1}{\langle .3,.1\rangle}, \langle .6,.5,.5\rangle \right\rangle, \left\langle \frac{\hat{u}_2}{\langle .3,.2\rangle}, \langle .7,.6,.6\rangle \right\rangle, \left\langle \frac{\hat{u}_3}{\langle .3,.4\rangle}, \langle .8,.7,.7\rangle \right\rangle, \left\langle \frac{\hat{u}_4}{\langle .3,.5\rangle}, \langle .9,.8,.8\rangle \right\rangle \right\}\right), \\ \left(\hat{\theta}_3, \left\{\left\langle \frac{\hat{u}_1}{\langle .4,.1\rangle}, \langle .7,.6,.7\rangle \right\rangle, \left\langle \frac{\hat{u}_2}{\langle .4,.2\rangle}, \langle .8,.7,.8\rangle \right\rangle, \left\langle \frac{\hat{u}_3}{\langle .4,.3\rangle}, \langle .9,.8,.9\rangle \right\rangle, \left\langle \frac{\hat{u}_4}{\langle .4,.5\rangle}, \langle .6,.5,.9\rangle \right\rangle \right\}\right), \\ \left(\hat{\theta}_4, \left\{\left\langle \frac{\hat{u}_1}{\langle .5,.1\rangle}, \langle .8,.7,.6\rangle \right\rangle, \left\langle \frac{\hat{u}_2}{\langle .5,.2\rangle}, \langle .9,.6,.5\rangle \right\rangle, \left\langle \frac{\hat{u}_3}{\langle .5,.3\rangle}, \langle .6,.7,.5\rangle \right\rangle, \left\langle \frac{\hat{u}_4}{\langle .5,.4\rangle}, \langle .6,.8,.8\rangle \right\rangle \right\}\right), \end{array} \right\}$$

Definition 3.2.6: Possibility Intuitionistic Fuzzy Hypersoft Set of Type-6

A possibility intuitionistic fuzzy hypersoft set of type-6 λ_6 is defined as

$$\lambda_6 = \left\{ \left(\hat{\theta}, \left\langle \frac{\hat{u}}{\psi_{IF}(\hat{\theta})(\hat{u})}, \mu_{IVF}(\hat{\theta}) \right\rangle \right) : \hat{u} \in \widehat{U} \wedge \hat{\theta} \in \widehat{\Theta} \right\}$$

where

$$\psi_{IF}(\hat{\theta})(\hat{u}) = \langle T_{\psi}(\hat{\theta})(\hat{u}), F_{\psi}(\hat{\theta})(\hat{u}) \rangle \subseteq \mathfrak{A}^{IF}$$

and

$$T_{\psi}(\hat{\theta})(\hat{u}), F_{\psi}(\hat{\theta})(\hat{u}) \in [0,1]$$

such that $0 \leq T_{\psi}(\hat{\theta})(\hat{u}) + F_{\psi}(\hat{\theta})(\hat{u}) \leq 1$.

Similarly $\mu_{IVF}(\hat{\theta}) \subseteq \mathfrak{A}^{IVF}$ with $\mu_{IVF}(\hat{\theta}) = \left[L_{\mu}(\hat{\theta}), U_{\mu}(\hat{\theta}) \right]$ and $L_{\mu}(\hat{\theta}), U_{\mu}(\hat{\theta}) \in [0,1]$.

Example 3.2.6: Considering the assumptions from Example 3.1.1, we can construct possibility intuitionistic fuzzy hypersoft set of type-6 λ_6 can be constructed as

$$\lambda_6 = \left\{ \begin{array}{l} \left(\hat{\theta}_1, \left\{ \left\langle \frac{\hat{u}_1}{\langle .2, .1 \rangle}, [0.21, 0.31] \right\rangle, \left\langle \frac{\hat{u}_2}{\langle .2, .3 \rangle}, [0.31, 0.41] \right\rangle, \left\langle \frac{\hat{u}_3}{\langle .2, .4 \rangle}, [0.41, 0.51] \right\rangle, \left\langle \frac{\hat{u}_4}{\langle .2, .5 \rangle}, [0.21, 0.61] \right\rangle \right\} \right), \\ \left(\hat{\theta}_2, \left\{ \left\langle \frac{\hat{u}_1}{\langle .3, .1 \rangle}, [0.22, 0.32] \right\rangle, \left\langle \frac{\hat{u}_2}{\langle .3, .2 \rangle}, [0.32, 0.42] \right\rangle, \left\langle \frac{\hat{u}_3}{\langle .3, .4 \rangle}, [0.42, 0.52] \right\rangle, \left\langle \frac{\hat{u}_4}{\langle .3, .5 \rangle}, [0.22, 0.62] \right\rangle \right\} \right), \\ \left(\hat{\theta}_3, \left\{ \left\langle \frac{\hat{u}_1}{\langle .4, .1 \rangle}, [0.23, 0.33] \right\rangle, \left\langle \frac{\hat{u}_2}{\langle .4, .2 \rangle}, [0.33, 0.43] \right\rangle, \left\langle \frac{\hat{u}_3}{\langle .4, .3 \rangle}, [0.43, 0.53] \right\rangle, \left\langle \frac{\hat{u}_4}{\langle .4, .5 \rangle}, [0.23, 0.63] \right\rangle \right\} \right), \\ \left(\hat{\theta}_4, \left\{ \left\langle \frac{\hat{u}_1}{\langle .5, .1 \rangle}, [0.24, 0.34] \right\rangle, \left\langle \frac{\hat{u}_2}{\langle .5, .2 \rangle}, [0.34, 0.44] \right\rangle, \left\langle \frac{\hat{u}_3}{\langle .5, .3 \rangle}, [0.44, 0.54] \right\rangle, \left\langle \frac{\hat{u}_4}{\langle .5, .5 \rangle}, [0.24, 0.64] \right\rangle \right\} \right), \end{array} \right\}$$

3.3 Hybrid Structures of Pythagorean Fuzzy Hypersoft Sets With Possibility Settings

Definition 3.3.1: Possibility Pythagorean Fuzzy Hypersoft Set of Type-1

A possibility Pythagorean fuzzy hypersoft set of type-1 η_1 is defined as

$$\eta_1 = \left\{ \left(\hat{\theta}, \left\langle \frac{\hat{u}}{\psi_{PyF}(\hat{\theta})(\hat{u})}, \mu_F(\hat{\theta}) \right\rangle \right) : \hat{u} \in \hat{U} \wedge \hat{\theta} \in \hat{\Theta} \right\}$$

where

$$\psi_{PyF}(\hat{\theta})(\hat{u}) = \langle T_{\psi}(\hat{\theta})(\hat{u}), F_{\psi}(\hat{\theta})(\hat{u}) \rangle \subseteq \mathfrak{A}^{PyF}$$

and

$$T_{\psi}(\hat{\theta})(\hat{u}), F_{\psi}(\hat{\theta})(\hat{u}) \in [0,1]$$

such that $0 \leq T_{\psi}^2(\hat{\theta})(\hat{u}) + F_{\psi}^2(\hat{\theta})(\hat{u}) \leq 1$.
Similarly $\mu_F(\hat{\theta}) \subseteq \mathfrak{A}^F$ with $\mu_F(\hat{\theta}) \in [0,1]$.

Example 3.3.1: Considering the assumptions from Example 3.1.1, we can construct possibility Pythagorean fuzzy hypersoft set of type-1 η_1 can be constructed as

$$\eta_1 = \left\{ \begin{array}{l} \left(\hat{\theta}_1, \left\{ \left\langle \frac{\hat{u}_1}{\langle .5,.6 \rangle}, 0.31 \right\rangle, \left\langle \frac{\hat{u}_2}{\langle .5,.7 \rangle}, 0.41 \right\rangle, \left\langle \frac{\hat{u}_3}{\langle .5,.8 \rangle}, 0.51 \right\rangle, \left\langle \frac{\hat{u}_4}{\langle .5,.8 \rangle}, 0.61 \right\rangle \right\} \right), \\ \left(\hat{\theta}_2, \left\{ \left\langle \frac{\hat{u}_1}{\langle .6,.5 \rangle}, 0.32 \right\rangle, \left\langle \frac{\hat{u}_2}{\langle .6,.6 \rangle}, 0.42 \right\rangle, \left\langle \frac{\hat{u}_3}{\langle .6,.7 \rangle}, 0.52 \right\rangle, \left\langle \frac{\hat{u}_4}{\langle .6,.7 \rangle}, 0.62 \right\rangle \right\} \right), \\ \left(\hat{\theta}_3, \left\{ \left\langle \frac{\hat{u}_1}{\langle .7,.5 \rangle}, 0.33 \right\rangle, \left\langle \frac{\hat{u}_2}{\langle .7,.6 \rangle}, 0.43 \right\rangle, \left\langle \frac{\hat{u}_3}{\langle .7,.7 \rangle}, 0.53 \right\rangle, \left\langle \frac{\hat{u}_4}{\langle .7,.4 \rangle}, 0.63 \right\rangle \right\} \right), \\ \left(\hat{\theta}_4, \left\{ \left\langle \frac{\hat{u}_1}{\langle .8,.5 \rangle}, 0.34 \right\rangle, \left\langle \frac{\hat{u}_2}{\langle .8,.5 \rangle}, 0.44 \right\rangle, \left\langle \frac{\hat{u}_3}{\langle .8,.4 \rangle}, 0.54 \right\rangle, \left\langle \frac{\hat{u}_4}{\langle .8,.4 \rangle}, 0.64 \right\rangle \right\} \right), \end{array} \right\}$$

Definition 3.3.2: Possibility Pythagorean Fuzzy Hypersoft Set of Type-2

A possibility Pythagorean fuzzy hypersoft set of type-2 η_2 is defined as

$$\eta_2 = \left\{ \left(\hat{\theta}, \left\langle \frac{\hat{u}}{\psi_{PyF}(\hat{\theta})(\hat{u})}, \mu_{IF}(\hat{\theta}) \right\rangle \right) : \hat{u} \in \widehat{U} \wedge \hat{\theta} \in \widehat{\Theta} \right\}$$

where

$$\psi_{PyF}(\hat{\theta})(\hat{u}) = \langle T_{\psi}(\hat{\theta})(\hat{u}), F_{\psi}(\hat{\theta})(\hat{u}) \rangle \subseteq \mathfrak{A}^{PyF}, \mu_{IF}(\hat{\theta}) \subseteq \mathfrak{A}^{IF}$$

and

$$T_{\psi}(\hat{\theta})(\hat{u}), F_{\psi}(\hat{\theta})(\hat{u}) \in [0,1], \mu_{IF}(\hat{\theta}) = \langle T_{\mu}(\hat{\theta}), F_{\mu}(\hat{\theta}) \rangle, T_{\mu}(\hat{\theta}), F_{\mu}(\hat{\theta}) \in [0,1]$$

with

$$0 \leq T_{\psi}^2(\hat{\theta})(\hat{u}) + F_{\psi}^2(\hat{\theta})(\hat{u}) \leq 1$$

and

$$0 \leq T_{\mu}(\hat{\theta}) + F_{\mu}(\hat{\theta}) \leq 1$$

.

Example 3.3.2: Considering the assumptions from Example 3.1.1, we can construct possibility Pythagorean fuzzy hypersoft set of type-2 η_2 can be constructed as

$$\eta_2 = \left\{ \begin{array}{l} \left(\hat{\theta}_1, \left\{ \left\langle \frac{\hat{u}_1}{\langle .5, .6 \rangle}, \langle 0.31, 0.21 \rangle \right\rangle, \left\langle \frac{\hat{u}_2}{\langle .5, .7 \rangle}, \langle 0.41, 0.31 \rangle \right\rangle, \left\langle \frac{\hat{u}_3}{\langle .5, .8 \rangle}, \langle 0.51, 0.41 \rangle \right\rangle, \left\langle \frac{\hat{u}_4}{\langle .5, .8 \rangle}, \langle 0.61, 0.21 \rangle \right\rangle \right\} \right), \\ \left(\hat{\theta}_2, \left\{ \left\langle \frac{\hat{u}_1}{\langle .6, .5 \rangle}, \langle 0.32, 0.22 \rangle \right\rangle, \left\langle \frac{\hat{u}_2}{\langle .6, .6 \rangle}, \langle 0.42, 0.32 \rangle \right\rangle, \left\langle \frac{\hat{u}_3}{\langle .6, .7 \rangle}, \langle 0.52, 0.42 \rangle \right\rangle, \left\langle \frac{\hat{u}_4}{\langle .6, .7 \rangle}, \langle 0.62, 0.22 \rangle \right\rangle \right\} \right), \\ \left(\hat{\theta}_3, \left\{ \left\langle \frac{\hat{u}_1}{\langle .7, .5 \rangle}, \langle 0.33, 0.23 \rangle \right\rangle, \left\langle \frac{\hat{u}_2}{\langle .7, .6 \rangle}, \langle 0.43, 0.33 \rangle \right\rangle, \left\langle \frac{\hat{u}_3}{\langle .7, .7 \rangle}, \langle 0.53, 0.43 \rangle \right\rangle, \left\langle \frac{\hat{u}_4}{\langle .7, .4 \rangle}, \langle 0.63, 0.23 \rangle \right\rangle \right\} \right), \\ \left(\hat{\theta}_4, \left\{ \left\langle \frac{\hat{u}_1}{\langle .8, .5 \rangle}, \langle 0.34, 0.24 \rangle \right\rangle, \left\langle \frac{\hat{u}_2}{\langle .8, .5 \rangle}, \langle 0.44, 0.34 \rangle \right\rangle, \left\langle \frac{\hat{u}_3}{\langle .8, .4 \rangle}, \langle 0.54, 0.44 \rangle \right\rangle, \left\langle \frac{\hat{u}_4}{\langle .8, .4 \rangle}, \langle 0.64, 0.24 \rangle \right\rangle \right\} \right), \end{array} \right\}$$

Definition 3.3.3: Possibility Pythagorean Fuzzy Hypersoft Set of Type-3

A possibility Pythagorean fuzzy hypersoft set of type-3 η_3 is defined as

$$\eta_3 = \left\{ \left(\hat{\theta}, \left\langle \frac{\hat{u}}{\psi_{PyF}(\hat{\theta})(\hat{u})}, \mu_{PyF}(\hat{\theta}) \right\rangle \right) : \hat{u} \in \hat{U} \wedge \hat{\theta} \in \widehat{\Theta} \right\}$$

where

$$\psi_{PyF}(\hat{\theta})(\hat{u}) = \langle T_{\psi}(\hat{\theta})(\hat{u}), F_{\psi}(\hat{\theta})(\hat{u}) \rangle \subseteq \mathfrak{A}^{PyF}, \mu_{PyF}(\hat{\theta}) \subseteq \mathfrak{A}^{PyF}$$

and

$$T_{\psi}(\hat{\theta})(\hat{u}), F_{\psi}(\hat{\theta})(\hat{u}) \in [0, 1]$$

with

$$0 \leq T_{\psi}^2(\hat{\theta})(\hat{u}) + F_{\psi}^2(\hat{\theta})(\hat{u}) \leq 1$$

and

$$\mu_{PyF}(\hat{\theta}) = \langle T_{\mu}(\hat{\theta}), F_{\mu}(\hat{\theta}) \rangle$$

with

$T_{\mu}(\hat{\theta}), F_{\mu}(\hat{\theta}) \in [0,1]$

and .

$0 \leq T_{\mu}^{2}(\hat{\theta}) + F_{\mu}^{2}(\hat{\theta}) \leq 1$

.

Example 3.3.3: Considering the assumptions from Example 3.1.1, we can construct possibility Pythagorean fuzzy hypersoft set of type-3 η_3 can be constructed as

$$\eta_3 = \left\{ \begin{array}{l} \left(\hat{\theta}_1, \left\{ \left\langle \frac{\hat{u}_1}{\langle .5,.6 \rangle}, \langle 0.5, 0.6 \rangle \right\rangle, \left\langle \frac{\hat{u}_2}{\langle .5,.7 \rangle}, \langle 0.7, 0.5 \rangle \right\rangle, \left\langle \frac{\hat{u}_3}{\langle .5,.8 \rangle}, \langle 0.5, 0.8 \rangle \right\rangle, \left\langle \frac{\hat{u}_4}{\langle .5,.8 \rangle}, \langle 0.6, 0.5 \rangle \right\rangle \right\} \right), \\ \left(\hat{\theta}_2, \left\{ \left\langle \frac{\hat{u}_1}{\langle .6,.5 \rangle}, \langle 0.9, 0.4 \rangle \right\rangle, \left\langle \frac{\hat{u}_2}{\langle .6,.6 \rangle}, \langle 0.4, 0.8 \rangle \right\rangle, \left\langle \frac{\hat{u}_3}{\langle .6,.7 \rangle}, \langle 0.4, 0.7 \rangle \right\rangle, \left\langle \frac{\hat{u}_4}{\langle .6,.7 \rangle}, \langle 0.5, 0.8 \rangle \right\rangle \right\} \right), \\ \left(\hat{\theta}_3, \left\{ \left\langle \frac{\hat{u}_1}{\langle .7,.4 \rangle}, \langle 0.7, 0.5 \rangle \right\rangle, \left\langle \frac{\hat{u}_2}{\langle .7,.5 \rangle}, \langle 0.8, 0.5 \rangle \right\rangle, \left\langle \frac{\hat{u}_3}{\langle .7,.6 \rangle}, \langle 0.6, 0.5 \rangle \right\rangle, \left\langle \frac{\hat{u}_4}{\langle .7,.7 \rangle}, \langle 0.5, 0.6 \rangle \right\rangle \right\} \right), \\ \left(\hat{\theta}_4, \left\{ \left\langle \frac{\hat{u}_1}{\langle .8,.4 \rangle}, \langle 0.4, 0.9 \rangle \right\rangle, \left\langle \frac{\hat{u}_2}{\langle .8,.5 \rangle}, \langle 0.8, 0.4 \rangle \right\rangle, \left\langle \frac{\hat{u}_3}{\langle .8,.5 \rangle}, \langle 0.7, 0.4 \rangle \right\rangle, \left\langle \frac{\hat{u}_4}{\langle .8,.4 \rangle}, \langle 0.8, 0.5 \rangle \right\rangle \right\} \right), \end{array} \right\}$$

Definition 3.3.4: Possibility Pythagorean Fuzzy Hypersoft Set of Type-4

A possibility Pythagorean fuzzy hypersoft set of type-4 η_4 is defined as

$$\eta_4 = \left\{ \left(\hat{\theta}, \left\langle \frac{\hat{u}}{\psi_{PyF}(\hat{\theta})(\hat{u})}, \mu_{PF}(\hat{\theta}) \right\rangle \right) : \hat{u} \in \widehat{U} \wedge \hat{\theta} \in \widehat{\Theta} \right\}$$

where

$\psi_{PyF}(\hat{\theta})(\hat{u}) = \langle T_{\psi}(\hat{\theta})(\hat{u}), F_{\psi}(\hat{\theta})(\hat{u}) \rangle \subseteq \mathfrak{A}^{PyF}$,

and

$T_{\psi}(\hat{\theta})(\hat{u}), F_{\psi}(\hat{\theta})(\hat{u}) \in [0,1]$

with

$0 \leq T_{\psi}^{2}(\hat{\theta})(\hat{u}) + F_{\psi}^{2}(\hat{\theta})(\hat{u}) \leq 1$

.

Similarly

$\mu_{PF}(\hat{\theta}) \subseteq \mathfrak{A}^{PF}$

,

$\mu_{PF}(\hat{\theta}) = \langle T_{\mu}(\hat{\theta}), I_{\mu}(\hat{\theta}), F_{\mu}(\hat{\theta}) \rangle$

with

$T_{\mu}(\hat{\theta}), I_{\mu}(\hat{\theta}), F_{\mu}(\hat{\theta}) \in [0,1]$

and

$0 \leq T_{\mu}(\hat{\theta}) + I_{\mu}(\hat{\theta}) + F_{\mu}(\hat{\theta}) \leq 1$

.

Example 3.3.4: Considering the assumptions from Example 3.1.1, we can construct possibility Pythagorean fuzzy hypersoft set of type-4 η_4 can be constructed as

$$\eta_4 = \left\{ \begin{array}{l} \left(\hat{\theta}_1, \left\{ \left\langle \frac{\hat{u}_1}{\langle .5,.6 \rangle}, \langle .5,.1,.2 \rangle \right\rangle, \left\langle \frac{\hat{u}_2}{\langle .5,.7 \rangle}, \langle .2,.5,.2 \rangle \right\rangle, \left\langle \frac{\hat{u}_3}{\langle .5,.8 \rangle}, \langle .4,.2,.3 \rangle \right\rangle, \left\langle \frac{\hat{u}_4}{\langle .5,.8 \rangle}, \langle .6,.1,.2 \rangle \right\rangle \right\} \right), \\ \left(\hat{\theta}_2, \left\{ \left\langle \frac{\hat{u}_1}{\langle .6,.5 \rangle}, \langle .1,.4,.4 \rangle \right\rangle, \left\langle \frac{\hat{u}_2}{\langle .6,.6 \rangle}, \langle .3,.4,.1 \rangle \right\rangle, \left\langle \frac{\hat{u}_3}{\langle .6,.7 \rangle}, \langle .2,.3,.3 \rangle \right\rangle, \left\langle \frac{\hat{u}_4}{\langle .6,.7 \rangle}, \langle .2,.4,.2 \rangle \right\rangle \right\} \right), \\ \left(\hat{\theta}_3, \left\{ \left\langle \frac{\hat{u}_1}{\langle .7,.4 \rangle}, \langle .3,.5,.1 \rangle \right\rangle, \left\langle \frac{\hat{u}_2}{\langle .7,.5 \rangle}, \langle .2,.5,.2 \rangle \right\rangle, \left\langle \frac{\hat{u}_3}{\langle .7,.6 \rangle}, \langle .6,.2,.1 \rangle \right\rangle, \left\langle \frac{\hat{u}_4}{\langle .7,.7 \rangle}, \langle .5,.2,.2 \rangle \right\rangle \right\} \right), \\ \left(\hat{\theta}_4, \left\{ \left\langle \frac{\hat{u}_1}{\langle .8,.4 \rangle}, \langle .4,.1,.1 \rangle \right\rangle, \left\langle \frac{\hat{u}_2}{\langle .8,.5 \rangle}, \langle .2,.2,.2 \rangle \right\rangle, \left\langle \frac{\hat{u}_3}{\langle .8,.5 \rangle}, \langle .7,.1,.1 \rangle \right\rangle, \left\langle \frac{\hat{u}_4}{\langle .8,.4 \rangle}, \langle .4,.2,.1 \rangle \right\rangle \right\} \right), \end{array} \right\}$$

Definition 3.3.5: Possibility Pythagorean Fuzzy Hypersoft Set of Type-5

A possibility Pythagorean fuzzy hypersoft set of type-5 η_5 is defined as

$$\eta_5 = \left\{ \left(\hat{\theta}, \left\langle \frac{\hat{u}}{\psi_{PyF}(\hat{\theta})(\hat{u})}, \mu_{svN}(\hat{\theta}) \right\rangle \right) : \hat{u} \in \widehat{U} \wedge \hat{\theta} \in \widehat{\Theta} \right\}$$

where

$$\psi_{PyF}(\hat{\theta})(\hat{u}) = \langle T_\psi(\hat{\theta})(\hat{u}), F_\psi(\hat{\theta})(\hat{u}) \rangle \subseteq \mathfrak{A}^{PyF},$$

and

$$T_\psi(\hat{\theta})(\hat{u}), F_\psi(\hat{\theta})(\hat{u}) \in [0,1]$$

with $0 \leq T_\psi^2(\hat{\theta})(\hat{u}) + F_\psi^2(\hat{\theta})(\hat{u}) \leq 1$.

Similarly

$$\mu_{svN}(\hat{\theta}) \subseteq \mathfrak{A}^{svN}$$

,

$$\mu_{svN}(\hat{\theta}) = \langle T_\mu(\hat{\theta}), I_\mu(\hat{\theta}), F_\mu(\hat{\theta}) \rangle$$

with

$$T_{\mu}(\hat{\theta}), I_{\mu}(\hat{\theta}), F_{\mu}(\hat{\theta}) \in [0,1]$$

and

$$0 \leq T_{\mu}(\hat{\theta}) + I_{\mu}(\hat{\theta}) + F_{\mu}(\hat{\theta}) \leq 3$$

.

Example 3.3.5: Considering the assumptions from Example 3.1.1, we can construct possibility Pythagorean fuzzy hypersoft set of type-5 η_5 can be constructed as

$$\eta_5 = \left\{ \begin{array}{l} \left(\hat{\theta}_1, \left\{ \left\langle \frac{\hat{u}_1}{\langle .5,.6 \rangle}, \langle .5,.6,.6 \rangle \right\rangle, \left\langle \frac{\hat{u}_2}{\langle .5,.7 \rangle}, \langle .6,.7,.7 \rangle \right\rangle, \left\langle \frac{\hat{u}_3}{\langle .5,.8 \rangle}, \langle .7,.8,.8 \rangle \right\rangle, \left\langle \frac{\hat{u}_4}{\langle .5,.8 \rangle}, \langle .8,.9,.9 \rangle \right\rangle \right\} \right), \\ \left(\hat{\theta}_2, \left\{ \left\langle \frac{\hat{u}_1}{\langle .6,.5 \rangle}, \langle .6,.5,.5 \rangle \right\rangle, \left\langle \frac{\hat{u}_2}{\langle .6,.6 \rangle}, \langle .7,.6,.6 \rangle \right\rangle, \left\langle \frac{\hat{u}_3}{\langle .6,.7 \rangle}, \langle .8,.7,.7 \rangle \right\rangle, \left\langle \frac{\hat{u}_4}{\langle .6,.7 \rangle}, \langle .9,.8,.8 \rangle \right\rangle \right\} \right), \\ \left(\hat{\theta}_3, \left\{ \left\langle \frac{\hat{u}_1}{\langle .7,.4 \rangle}, \langle .7,.6,.7 \rangle \right\rangle, \left\langle \frac{\hat{u}_2}{\langle .7,.5 \rangle}, \langle .8,.7,.8 \rangle \right\rangle, \left\langle \frac{\hat{u}_3}{\langle .7,.6 \rangle}, \langle .9,.8,.9 \rangle \right\rangle, \left\langle \frac{\hat{u}_4}{\langle .7,.7 \rangle}, \langle .6,.5,.9 \rangle \right\rangle \right\} \right), \\ \left(\hat{\theta}_4, \left\{ \left\langle \frac{\hat{u}_1}{\langle .8,.4 \rangle}, \langle .8,.7,.6 \rangle \right\rangle, \left\langle \frac{\hat{u}_2}{\langle .8,.4 \rangle}, \langle .9,.6,.5 \rangle \right\rangle, \left\langle \frac{\hat{u}_3}{\langle .8,.5 \rangle}, \langle .6,.7,.5 \rangle \right\rangle, \left\langle \frac{\hat{u}_4}{\langle .8,.5 \rangle}, \langle .6,.8,.8 \rangle \right\rangle \right\} \right) \end{array} \right\}$$

Definition 3.3.6: Possibility Pythagorean Fuzzy Hypersoft Set of Type-6

A possibility Pythagorean fuzzy hypersoft set of type-6 η_6 is defined as

$$\eta_6 = \left\{ \left(\hat{\theta}, \left\langle \frac{\hat{u}}{\psi_{PyF}(\hat{\theta})(\hat{u})}, \mu_{IVF}(\hat{\theta}) \right\rangle \right) : \hat{u} \in \hat{U} \wedge \hat{\theta} \in \hat{\Theta} \right\}$$

where

$$\psi_{PyF}(\hat{\theta})(\hat{u}) = \langle T_{\psi}(\hat{\theta})(\hat{u}), F_{\psi}(\hat{\theta})(\hat{u}) \rangle \subseteq \mathfrak{A}^{PyF}$$

and

$$T_\psi(\hat{\theta})(\hat{u}), F_\psi(\hat{\theta})(\hat{u}) \in [0,1]$$

such that $0 \leq T^2_\psi(\hat{\theta})(\hat{u}) + F^2_\psi(\hat{\theta})(\hat{u}) \leq 1$.

Similarly $\mu_{IVF}(\hat{\theta}) \subseteq \mathfrak{A}^{IVF}$ with $\mu_{IVF}(\hat{\theta}) = \left[L_\mu(\hat{\theta}), U_\mu(\hat{\theta})\right]$ and $L_\mu(\hat{\theta}), U_\mu(\hat{\theta}) \in [0,1]$.

Example 3.3.6: Considering the assumptions from Example 3.1.1, we can construct possibility Pythagorean fuzzy hypersoft set of type-6 η_6 can be constructed as

$$\eta_6 = \left\{ \begin{array}{l} \left(\hat{\theta}_1, \left\{\left\langle \frac{\hat{u}_1}{\langle .5, .6\rangle}, [0.21, 0.31]\right\rangle, \left\langle \frac{\hat{u}_2}{\langle .5, .7\rangle}, [0.31, 0.41]\right\rangle, \left\langle \frac{\hat{u}_3}{\langle .5, .8\rangle}, [0.41, 0.51]\right\rangle, \left\langle \frac{\hat{u}_4}{\langle .5, .8\rangle}, [0.21, 0.61]\right\rangle\right\}\right), \\ \left(\hat{\theta}_2, \left\{\left\langle \frac{\hat{u}_1}{\langle .6, .5\rangle}, [0.22, 0.32]\right\rangle, \left\langle \frac{\hat{u}_2}{\langle .6, .6\rangle}, [0.32, 0.42]\right\rangle, \left\langle \frac{\hat{u}_3}{\langle .6, .7\rangle}, [0.42, 0.52]\right\rangle, \left\langle \frac{\hat{u}_4}{\langle .6, .7\rangle}, [0.22, 0.62]\right\rangle\right\}\right), \\ \left(\hat{\theta}_3, \left\{\left\langle \frac{\hat{u}_1}{\langle .7, .4\rangle}, [0.23, 0.33]\right\rangle, \left\langle \frac{\hat{u}_2}{\langle .7, .5\rangle}, [0.33, 0.43]\right\rangle, \left\langle \frac{\hat{u}_3}{\langle .7, .6\rangle}, [0.43, 0.53]\right\rangle, \left\langle \frac{\hat{u}_4}{\langle .7, .7\rangle}, [0.23, 0.63]\right\rangle\right\}\right), \\ \left(\hat{\theta}_4, \left\{\left\langle \frac{\hat{u}_1}{\langle .8, .4\rangle}, [0.24, 0.34]\right\rangle, \left\langle \frac{\hat{u}_2}{\langle .8, .5\rangle}, [0.34, 0.44]\right\rangle, \left\langle \frac{\hat{u}_3}{\langle .8, .4\rangle}, [0.44, 0.54]\right\rangle, \left\langle \frac{\hat{u}_4}{\langle .8, .5\rangle}, [0.24, 0.64]\right\rangle\right\}\right), \end{array} \right\}$$

3.4 Hybrid Structures of Picture Fuzzy Hypersoft Sets With Possibility Settings

Definition 3.4.1: Possibility Picture Fuzzy Hypersoft Set of Type-1

A possibility picture fuzzy hypersoft set of type-1 δ_1 is defined as

$$\delta_1 = \left\{ \left(\hat{\theta}, \left\langle \frac{\hat{u}}{\psi_{PF}(\hat{\theta})(\hat{u})}, \mu_F(\hat{\theta}) \right\rangle\right) : \hat{u} \in \widehat{U} \wedge \hat{\theta} \in \widehat{\Theta} \right\}$$

where

$$\psi_{PF}(\hat{\theta})(\hat{u}) = \langle T_\psi(\hat{\theta})(\hat{u}), I_\psi(\hat{\theta})(\hat{u}), F_\psi(\hat{\theta})(\hat{u})\rangle \subseteq \mathfrak{A}^{PF}$$

and

$$T_\psi(\hat{\theta})(\hat{u}), I_\psi(\hat{\theta})(\hat{u}), F_\psi(\hat{\theta})(\hat{u}) \in [0,1]$$

such that

$0 \leq T_{\psi}(\hat{\theta})(\hat{u}) + I_{\psi}(\hat{\theta})(\hat{u}) + F_{\psi}(\hat{\theta})(\hat{u}) \leq 1$

.

Similarly

$\mu_F(\hat{\theta}) \subseteq \mathfrak{A}^F$

with

$\mu_F(\hat{\theta}) \in [0,1]$

.

Example 3.4.1: Considering the assumptions from Example 3.1.1, we can construct possibility picture fuzzy hypersoft set of type-1 λ_1 can be constructed as

$$\lambda_1 = \left\{ \begin{array}{l} \left(\hat{\theta}_1, \left\{ \left\langle \frac{\hat{u}_1}{\langle .2,.5,.1 \rangle}, .3 \right\rangle, \left\langle \frac{\hat{u}_2}{\langle .3,.5,.1 \rangle}, .4 \right\rangle, \left\langle \frac{\hat{u}_3}{\langle .4,.4,.1 \rangle}, .5 \right\rangle, \left\langle \frac{\hat{u}_4}{\langle .5,.2,.1 \rangle}, .6 \right\rangle \right\} \right), \\ \left(\hat{\theta}_2, \left\{ \left\langle \frac{\hat{u}_1}{\langle .2,.3,.2 \rangle}, .2 \right\rangle, \left\langle \frac{\hat{u}_2}{\langle .3,.4,.2 \rangle}, .3 \right\rangle, \left\langle \frac{\hat{u}_3}{\langle .4,.2,.2 \rangle}, .4 \right\rangle, \left\langle \frac{\hat{u}_4}{\langle .5,.1,.2 \rangle}, .5 \right\rangle \right\} \right), \\ \left(\hat{\theta}_3, \left\{ \left\langle \frac{\hat{u}_1}{\langle .2,.5,.3 \rangle}, .1 \right\rangle, \left\langle \frac{\hat{u}_2}{\langle .3,.3,.3 \rangle}, .2 \right\rangle, \left\langle \frac{\hat{u}_3}{\langle .4,.1,.3 \rangle}, .3 \right\rangle, \left\langle \frac{\hat{u}_4}{\langle .5,.2,.2 \rangle}, .4 \right\rangle \right\} \right), \\ \left(\hat{\theta}_4, \left\{ \left\langle \frac{\hat{u}_1}{\langle .2,.2,.4 \rangle}, .5 \right\rangle, \left\langle \frac{\hat{u}_2}{\langle .3,.5,.1 \rangle}, .6 \right\rangle, \left\langle \frac{\hat{u}_3}{\langle .4,.4,.1 \rangle}, .7 \right\rangle, \left\langle \frac{\hat{u}_4}{\langle .5,.3,.1 \rangle}, .8 \right\rangle \right\} \right), \end{array} \right\}$$

Definition 3.4.2: Possibility Picture Fuzzy Hypersoft Set of Type-2

A possibility picture fuzzy hypersoft set of type-2 δ_2 is defined as

$$\delta_2 = \left\{ \left(\hat{\theta}, \left\langle \frac{\hat{u}}{\psi_{PF}(\hat{\theta})(\hat{u})}, \mu_{IF}(\hat{\theta}) \right\rangle \right) : \hat{u} \in \widehat{U} \wedge \hat{\theta} \in \widehat{\Theta} \right\}$$

where

$$\psi_{PF}(\hat{\theta})(\hat{u}) = \langle T_{\psi}(\hat{\theta})(\hat{u}), I_{\psi}(\hat{\theta})(\hat{u}), F_{\psi}(\hat{\theta})(\hat{u})\rangle \subseteq \mathfrak{A}^{PF}$$

and

$$T_{\psi}(\hat{\theta})(\hat{u}), I_{\psi}(\hat{\theta})(\hat{u}), F_{\psi}(\hat{\theta})(\hat{u}) \in [0,1],$$

with $0 \leq T_{\psi}(\hat{\theta})(\hat{u}) + I_{\psi}(\hat{\theta})(\hat{u}) + F_{\psi}(\hat{\theta})(\hat{u}) \leq 1$. Similarly

$$\mu_{IF}(\hat{\theta}) \subseteq \mathfrak{A}^{IF}, \mu_{IF}(\hat{\theta}) = \langle T_{\mu}(\hat{\theta}), F_{\mu}(\hat{\theta})\rangle$$

and

$$T_{\mu}(\hat{\theta}), F_{\mu}(\hat{\theta}) \in [0,1]$$

with

$$0 \leq T_{\mu}(\hat{\theta}) + F_{\mu}(\hat{\theta}) \leq 1$$

.

Example 3.4.2: Considering the assumptions from Example 3.1.1, we can construct possibility picture fuzzy hypersoft set of type-2 δ_2 can be constructed as

$$\delta_2 = \left\{ \begin{array}{l} \left(\hat{\theta}_1, \left\{ \left\langle \frac{\hat{u}_1}{\langle .2, .4, .1\rangle}, \langle .3, .2\rangle \right\rangle, \left\langle \frac{\hat{u}_2}{\langle .3, .4, .1\rangle}, \langle .4, .3\rangle \right\rangle, \left\langle \frac{\hat{u}_3}{\langle .4, .4, .1\rangle}, \langle .5, .4\rangle \right\rangle, \left\langle \frac{\hat{u}_4}{\langle .5, .2, .1\rangle}, \langle .6, .2\rangle \right\rangle \right\} \right), \\ \left(\hat{\theta}_2, \left\{ \left\langle \frac{\hat{u}_1}{\langle .2, .3, .2\rangle}, \langle .3, .1\rangle \right\rangle, \left\langle \frac{\hat{u}_2}{\langle .3, .2, .2\rangle}, \langle .4, .2\rangle \right\rangle, \left\langle \frac{\hat{u}_3}{\langle .4, .4, .1\rangle}, \langle .5, .3\rangle \right\rangle, \left\langle \frac{\hat{u}_4}{\langle .5, .1, .2\rangle}, \langle .6, .1\rangle \right\rangle \right\} \right), \\ \left(\hat{\theta}_3, \left\{ \left\langle \frac{\hat{u}_1}{\langle .1, .5, .3\rangle}, \langle .3, .3\rangle \right\rangle, \left\langle \frac{\hat{u}_2}{\langle .1, .4, .3\rangle}, \langle .3, .4\rangle \right\rangle, \left\langle \frac{\hat{u}_3}{\langle .5, .1, .3\rangle}, \langle .3, .5\rangle \right\rangle, \left\langle \frac{\hat{u}_4}{\langle .4, .2, .3\rangle}, \langle .3, .6\rangle \right\rangle \right\} \right), \\ \left(\hat{\theta}_4, \left\{ \left\langle \frac{\hat{u}_1}{\langle .3, .3, .3\rangle}, \langle .3, .4\rangle \right\rangle, \left\langle \frac{\hat{u}_2}{\langle .3, .5, .2\rangle}, \langle .3, .5\rangle \right\rangle, \left\langle \frac{\hat{u}_3}{\langle .4, .4, .2\rangle}, \langle .3, .6\rangle \right\rangle, \left\langle \frac{\hat{u}_4}{\langle .3, .3, .2\rangle}, \langle .3, .3\rangle \right\rangle \right\} \right), \end{array} \right\}$$

Definition 3.4.3: Possibility Picture Fuzzy Hypersoft Set of Type-3

A possibility picture fuzzy hypersoft set of type-3 δ_3 is defined as

$$\delta_3 = \left\{ \left(\hat{\theta}, \left\langle \frac{\hat{u}}{\psi_{PF}(\hat{\theta})(\hat{u})}, \mu_{PyF}(\hat{\theta}) \right\rangle \right) : \hat{u} \in \widehat{U} \wedge \hat{\theta} \in \widehat{\Theta} \right\}$$

where

$$\psi_{PF}(\hat{\theta})(\hat{u}) = \langle T_\psi(\hat{\theta})(\hat{u}), I_\psi(\hat{\theta})(\hat{u}), F_\psi(\hat{\theta})(\hat{u}) \rangle \subseteq \mathfrak{A}^{PF}$$

and

$$T_\psi(\hat{\theta})(\hat{u}), I_\psi(\hat{\theta})(\hat{u}), F_\psi(\hat{\theta})(\hat{u}) \in [0,1]$$

with $0 \leq T_\psi(\hat{\theta})(\hat{u}) + I_\psi(\hat{\theta})(\hat{u}) + F_\psi(\hat{\theta})(\hat{u}) \leq 1$. Similarly

$$\mu_{PyF}(\hat{\theta}) \subseteq \mathfrak{A}^{PyF}$$

and

$$\mu_{PyF}(\hat{\theta}) = \langle T_\mu(\hat{\theta}), F_\mu(\hat{\theta}) \rangle$$

with $T_\mu(\hat{\theta}), F_\mu(\hat{\theta}) \in [0,1]$ and $0 \leq T_\mu^2(\hat{\theta}) + F_\mu^2(\hat{\theta}) \leq 1$.

Example 3.4.3: Considering the assumptions from Example 3.1.1, we can construct possibility picture fuzzy hypersoft set of type-3 δ_3 can be constructed as

$$\lambda_3 = \left\{ \begin{array}{l} \left(\hat{\theta}_1, \left\{ \left\langle \frac{\hat{u}_1}{\langle .2,.1,.1\rangle}, \langle .5,.6\rangle \right\rangle, \left\langle \frac{\hat{u}_2}{\langle .2,.3,.1\rangle}, \langle .7,.5\rangle \right\rangle, \left\langle \frac{\hat{u}_3}{\langle .2,.4,.1\rangle}, \langle .5,.8\rangle \right\rangle, \left\langle \frac{\hat{u}_4}{\langle .2,.5,.1\rangle}, \langle .6,.5\rangle \right\rangle \right\} \right), \\ \left(\hat{\theta}_2, \left\{ \left\langle \frac{\hat{u}_1}{\langle .3,.1,.2\rangle}, \langle .9,.4\rangle \right\rangle, \left\langle \frac{\hat{u}_2}{\langle .3,.2,.2\rangle}, \langle .4,.8\rangle \right\rangle, \left\langle \frac{\hat{u}_3}{\langle .3,.4,.2\rangle}, \langle .4,.7\rangle \right\rangle, \left\langle \frac{\hat{u}_4}{\langle .3,.5,.1\rangle}, \langle .5,.8\rangle \right\rangle \right\} \right), \\ \left(\hat{\theta}_3, \left\{ \left\langle \frac{\hat{u}_1}{\langle .4,.1,.3\rangle}, \langle .7,.5\rangle \right\rangle, \left\langle \frac{\hat{u}_2}{\langle .4,.2,.3\rangle}, \langle .8,.5\rangle \right\rangle, \left\langle \frac{\hat{u}_3}{\langle .4,.3,.1\rangle}, \langle .6,.5\rangle \right\rangle, \left\langle \frac{\hat{u}_4}{\langle .4,.4,.1\rangle}, \langle .5,.6\rangle \right\rangle \right\} \right), \\ \left(\hat{\theta}_4, \left\{ \left\langle \frac{\hat{u}_1}{\langle .5,.1,.2\rangle}, \langle .4,.9\rangle \right\rangle, \left\langle \frac{\hat{u}_2}{\langle .5,.2,.1\rangle}, \langle .8,.4\rangle \right\rangle, \left\langle \frac{\hat{u}_3}{\langle .5,.3,.1\rangle}, \langle .7,.4\rangle \right\rangle, \left\langle \frac{\hat{u}_4}{\langle .5,.2,.1\rangle}, \langle .8,.5\rangle \right\rangle \right\} \right), \end{array} \right\}$$

Definition 3.4.4: Possibility Picture Fuzzy Hypersoft Set of Type-4

A possibility picture fuzzy hypersoft set of type-4 δ_4 is defined as

$$\delta_4 = \left\{ \left(\hat{\theta}, \left\langle \frac{\hat{u}}{\psi_{PF}(\hat{\theta})(\hat{u})}, \mu_{PF}(\hat{\theta}) \right\rangle \right) : \hat{u} \in \widehat{U} \wedge \hat{\theta} \in \widehat{\Theta} \right\}$$

where

$$\psi_{PF}(\hat{\theta})(\hat{u}) = \langle T_\psi(\hat{\theta})(\hat{u}), I_\psi(\hat{\theta})(\hat{u}), F_\psi(\hat{\theta})(\hat{u}) \rangle \subseteq \mathfrak{A}^{PF},$$

and

$$T_\psi(\hat{\theta})(\hat{u}), I_\psi(\hat{\theta})(\hat{u}), F_\psi(\hat{\theta})(\hat{u}) \in [0,1]$$

with

$$0 \leq T_\psi(\hat{\theta})(\hat{u}) + I_\psi(\hat{\theta})(\hat{u}) + F_\psi(\hat{\theta})(\hat{u}) \leq 1$$

. Similarly

$$\mu_{PF}(\hat{\theta}) \subseteq \mathfrak{A}^{PF}$$

,

$$\mu_{PF}(\hat{\theta}) = \langle T_{\mu}(\hat{\theta}), I_{\mu}(\hat{\theta}), F_{\mu}(\hat{\theta}) \rangle$$

with

$$T_{\mu}(\hat{\theta}), I_{\mu}(\hat{\theta}), F_{\mu}(\hat{\theta}) \in [0,1]$$

and

$$0 \leq T_{\mu}(\hat{\theta}) + I_{\mu}(\hat{\theta}) + F_{\mu}(\hat{\theta}) \leq 1$$

.

Example 3.4.4: Considering the assumptions from Example 3.1.1, we can construct possibility picture fuzzy hypersoft set of type-4 δ_4 can be constructed as

$$\delta_4 = \left\{ \begin{array}{l} \left(\hat{\theta}_1, \left\{ \left\langle \frac{\hat{u}_1}{\langle .2,.1,.1 \rangle}, \langle .5,.1,.2 \rangle \right\rangle, \left\langle \frac{\hat{u}_2}{\langle .2,.3,.1 \rangle}, \langle .2,.5,.2 \rangle \right\rangle, \left\langle \frac{\hat{u}_3}{\langle .2,.4,.1 \rangle}, \langle .4,.2,.3 \rangle \right\rangle, \left\langle \frac{\hat{u}_4}{\langle .2,.5,.1 \rangle}, \langle .6,.1,.2 \rangle \right\rangle \right\} \right), \\ \left(\hat{\theta}_2, \left\{ \left\langle \frac{\hat{u}_1}{\langle .3,.1,.2 \rangle}, \langle .1,.4,.4 \rangle \right\rangle, \left\langle \frac{\hat{u}_2}{\langle .3,.2,.2 \rangle}, \langle .3,.4,.1 \rangle \right\rangle, \left\langle \frac{\hat{u}_3}{\langle .3,.4,.2 \rangle}, \langle .2,.3,.3 \rangle \right\rangle, \left\langle \frac{\hat{u}_4}{\langle .3,.5,.1 \rangle}, \langle .2,.4,.2 \rangle \right\rangle \right\} \right), \\ \left(\hat{\theta}_3, \left\{ \left\langle \frac{\hat{u}_1}{\langle .4,.1,.2 \rangle}, \langle .3,.5,.1 \rangle \right\rangle, \left\langle \frac{\hat{u}_2}{\langle .4,.2,.1 \rangle}, \langle .2,.5,.2 \rangle \right\rangle, \left\langle \frac{\hat{u}_3}{\langle .4,.3,.1 \rangle}, \langle .6,.2,.1 \rangle \right\rangle, \left\langle \frac{\hat{u}_4}{\langle .4,.4,.1 \rangle}, \langle .5,.2,.2 \rangle \right\rangle \right\} \right), \\ \left(\hat{\theta}_4, \left\{ \left\langle \frac{\hat{u}_1}{\langle .5,.1,.2 \rangle}, \langle .4,.1,.1 \rangle \right\rangle, \left\langle \frac{\hat{u}_2}{\langle .5,.2,.1 \rangle}, \langle .2,.2,.2 \rangle \right\rangle, \left\langle \frac{\hat{u}_3}{\langle .5,.3,.1 \rangle}, \langle .7,.1,.1 \rangle \right\rangle, \left\langle \frac{\hat{u}_4}{\langle .2,.4,.1 \rangle}, \langle .4,.2,.1 \rangle \right\rangle \right\} \right), \end{array} \right\}$$

Definition 3.4.5: Possibility Picture Fuzzy Hypersoft Set of Type-5

A possibility picture fuzzy hypersoft set of type-5 δ_5 is defined as

$$\delta_5 = \left\{ \left(\hat{\theta}, \left\langle \frac{\hat{u}}{\psi_{PF}(\hat{\theta})(\hat{u})}, \mu_{svN}(\hat{\theta}) \right\rangle \right) : \hat{u} \in \widehat{U} \wedge \hat{\theta} \in \widehat{\Theta} \right\}$$

where

$$\psi_{PF}(\hat{\theta})(\hat{u}) = \langle T_{\psi}(\hat{\theta})(\hat{u}), I_{\psi}(\hat{\theta})(\hat{u}), F_{\psi}(\hat{\theta})(\hat{u}) \rangle \subseteq \mathfrak{A}^{PF},$$

and

$$T_{\psi}(\widehat{\theta})(\hat{u}), I_{\psi}(\widehat{\theta})(\hat{u}), F_{\psi}(\widehat{\theta})(\hat{u}) \in [0,1]$$

with $0 \leq T_{\psi}(\widehat{\theta})(\hat{u}) + I_{\psi}(\widehat{\theta})(\hat{u}) + F_{\psi}(\widehat{\theta})(\hat{u}) \leq 1$. Similarly $\mu_{svN}(\widehat{\theta}) \subseteq \mathfrak{A}^{svN}$,

$$\mu_{svN}(\widehat{\theta}) = \langle T_{\mu}(\widehat{\theta}), I_{\mu}(\widehat{\theta}), F_{\mu}(\widehat{\theta}) \rangle$$

with $T_{\mu}(\widehat{\theta}), I_{\mu}(\widehat{\theta}), F_{\mu}(\widehat{\theta}) \in [0,1]$ and $0 \leq T_{\mu}(\widehat{\theta}) + I_{\mu}(\widehat{\theta}) + F_{\mu}(\widehat{\theta}) \leq 3$.

Example 3.4.5: Considering the assumptions from Example 3.1.1, we can construct possibility picture fuzzy hypersoft set of type-5 δ_5 can be constructed as

$$\delta_5 = \left\{ \begin{array}{l} \left(\hat{\theta}_1, \left\{ \left\langle \frac{\hat{u}_1}{\langle .2,.1,.1 \rangle}, \langle .5,.6,.6 \rangle \right\rangle, \left\langle \frac{\hat{u}_2}{\langle .2,.3,.1 \rangle}, \langle .6,.7,.7 \rangle \right\rangle, \left\langle \frac{\hat{u}_3}{\langle .2,.4,.1 \rangle}, \langle .7,.8,.8 \rangle \right\rangle, \left\langle \frac{\hat{u}_4}{\langle .2,.5,.1 \rangle}, \langle .8,.9,.9 \rangle \right\rangle \right\} \right), \\ \left(\hat{\theta}_2, \left\{ \left\langle \frac{\hat{u}_1}{\langle .3,.1,.2 \rangle}, \langle .6,.5,.5 \rangle \right\rangle, \left\langle \frac{\hat{u}_2}{\langle .3,.2,.2 \rangle}, \langle .7,.6,.6 \rangle \right\rangle, \left\langle \frac{\hat{u}_3}{\langle .3,.4,.2 \rangle}, \langle .8,.7,.7 \rangle \right\rangle, \left\langle \frac{\hat{u}_4}{\langle .3,.5,.1 \rangle}, \langle .9,.8,.8 \rangle \right\rangle \right\} \right), \\ \left(\hat{\theta}_3, \left\{ \left\langle \frac{\hat{u}_1}{\langle .4,.1,.3 \rangle}, \langle .7,.6,.7 \rangle \right\rangle, \left\langle \frac{\hat{u}_2}{\langle .4,.2,.3 \rangle}, \langle .8,.7,.8 \rangle \right\rangle, \left\langle \frac{\hat{u}_3}{\langle .4,.3,.2 \rangle}, \langle .9,.8,.9 \rangle \right\rangle, \left\langle \frac{\hat{u}_4}{\langle .4,.4,.1 \rangle}, \langle .6,.5,.9 \rangle \right\rangle \right\} \right), \\ \left(\hat{\theta}_4, \left\{ \left\langle \frac{\hat{u}_1}{\langle .5,.1,.2 \rangle}, \langle .8,.7,.6 \rangle \right\rangle, \left\langle \frac{\hat{u}_2}{\langle .5,.2,.1 \rangle}, \langle .9,.6,.5 \rangle \right\rangle, \left\langle \frac{\hat{u}_3}{\langle .5,.3,.1 \rangle}, \langle .6,.7,.5 \rangle \right\rangle, \left\langle \frac{\hat{u}_4}{\langle .3,.4,.1 \rangle}, \langle .6,.8,.8 \rangle \right\rangle \right\} \right), \end{array} \right\}$$

Definition 3.4.6: Possibility Picture Fuzzy Hypersoft Set of Type-6

A possibility picture fuzzy hypersoft set of type-6 δ_6 is defined as

$$\delta_6 = \left\{ \left(\hat{\theta}, \left\langle \frac{\hat{u}}{\psi_{PF}(\widehat{\theta})(\hat{u})}, \mu_{IVF}(\widehat{\theta}) \right\rangle \right) : \hat{u} \in \widehat{U} \wedge \hat{\theta} \in \widehat{\Theta} \right\}$$

where

$$\psi_{PF}(\widehat{\theta})(\hat{u}) = \langle T_{\psi}(\widehat{\theta})(\hat{u}), I_{\psi}(\widehat{\theta})(\hat{u}), F_{\psi}(\widehat{\theta})(\hat{u}) \rangle \subseteq \mathfrak{A}^{PF}$$

and

$T_\psi(\hat{\theta})(\hat{u}), I_\psi(\hat{\theta})(\hat{u}), F_\psi(\hat{\theta})(\hat{u}) \in [0,1]$

such that $0 \leq T_\psi(\hat{\theta})(\hat{u}) + I_\psi(\hat{\theta})(\hat{u}) + F_\psi(\hat{\theta})(\hat{u}) \leq 1$. Similarly

$\mu_{IVF}(\hat{\theta}) \subseteq \mathfrak{A}^{IVF}$

with

$\mu_{IVF}(\hat{\theta}) = \left[L_\mu(\hat{\theta}), U_\mu(\hat{\theta})\right]$

and

$L_\mu(\hat{\theta}), U_\mu(\hat{\theta}) \in [0,1]$

.

Example 3.4.6: Considering the assumptions from Example 3.1.1, we can construct possibility picture fuzzy hypersoft set of type-6 δ_6 can be constructed as

$$\delta_6 = \left\{ \begin{array}{l} \left(\hat{\theta}_1, \left\{ \left\langle \frac{\hat{u}_1}{\langle .2,.1,.1 \rangle}, [.2,.3] \right\rangle, \left\langle \frac{\hat{u}_2}{\langle .2,.3,.1 \rangle}, [.3,.4] \right\rangle, \left\langle \frac{\hat{u}_3}{\langle .2,.4,.1 \rangle}, [.4,.5] \right\rangle, \left\langle \frac{\hat{u}_4}{\langle .2,.5,.1 \rangle}, [.5,.6] \right\rangle \right\} \right), \\ \left(\hat{\theta}_2, \left\{ \left\langle \frac{\hat{u}_1}{\langle .3,.1,.2 \rangle}, [.6,.7] \right\rangle, \left\langle \frac{\hat{u}_2}{\langle .3,.2,.2 \rangle}, [.7,.8] \right\rangle, \left\langle \frac{\hat{u}_3}{\langle .3,.4,.2 \rangle}, [.8,.9] \right\rangle, \left\langle \frac{\hat{u}_4}{\langle .3,.5,.1 \rangle}, [.2,.4] \right\rangle \right\} \right), \\ \left(\hat{\theta}_3, \left\{ \left\langle \frac{\hat{u}_1}{\langle .4,.1,.2 \rangle}, [.3,.5] \right\rangle, \left\langle \frac{\hat{u}_2}{\langle .4,.2,.2 \rangle}, [.4,.6] \right\rangle, \left\langle \frac{\hat{u}_3}{\langle .4,.3,.1 \rangle}, [.5,.7] \right\rangle, \left\langle \frac{\hat{u}_4}{\langle .4,.4,.1 \rangle}, [.6,.8] \right\rangle \right\} \right), \\ \left(\hat{\theta}_4, \left\{ \left\langle \frac{\hat{u}_1}{\langle .5,.1,.1 \rangle}, [.7,.9] \right\rangle, \left\langle \frac{\hat{u}_2}{\langle .5,.2,.1 \rangle}, [.2,.5] \right\rangle, \left\langle \frac{\hat{u}_3}{\langle .5,.3,.1 \rangle}, [.3,.6] \right\rangle, \left\langle \frac{\hat{u}_4}{\langle .2,.4,.1 \rangle}, [.4,.7] \right\rangle \right\} \right), \end{array} \right\}$$

3.5 Hybrid Structures of sv-Neutrosophic Hypersoft Sets With Possibility Settings

Definition 3.5.1: Possibility sv-Neutrosophic Fuzzy Hypersoft Set of Type-1

A possibility sv-neutrosophic hypersoft set of type-1 ϖ_1 is defined as

$$\varpi_1 = \left\{ \left(\hat{\theta}, \left\langle \frac{\hat{u}}{\psi_{svN}(\hat{\theta})(\hat{u})}, \mu_F(\hat{\theta}) \right\rangle \right) : \hat{u} \in \widehat{U} \wedge \hat{\theta} \in \widehat{\Theta} \right\}$$

where

$$\psi_{svN}(\hat{\theta})(\hat{u}) = \langle T_\psi(\hat{\theta})(\hat{u}), I_\psi(\hat{\theta})(\hat{u}), F_\psi(\hat{\theta})(\hat{u}) \rangle \subseteq \mathfrak{A}^{svN}$$

and

$$T_\psi(\hat{\theta})(\hat{u}), I_\psi(\hat{\theta})(\hat{u}), F_\psi(\hat{\theta})(\hat{u}) \in [0,1]$$

such that $0 \leq T_\psi(\hat{\theta})(\hat{u}) + I_\psi(\hat{\theta})(\hat{u}) + F_\psi(\hat{\theta})(\hat{u}) \leq 3$. Similarly

$$\mu_F(\hat{\theta}) \subseteq \mathfrak{A}^F$$

with

$$\mu_F(\hat{\theta}) \in [0,1]$$

.

Example 3.5.1: Considering the assumptions from Example 3.1.1, we can construct possibility sv-neutrosophic hypersoft set of type-1 ϖ_1 can be constructed as

$$\varpi_1 = \left\{ \begin{array}{l} \left(\hat{\theta}_1, \left\{ \left\langle \frac{\hat{u}_1}{\langle .8,.8,.9 \rangle}, .3 \right\rangle, \left\langle \frac{\hat{u}_2}{\langle .8,.8,.8 \rangle}, .4 \right\rangle, \left\langle \frac{\hat{u}_3}{\langle .8,.8,.7 \rangle}, .5 \right\rangle, \left\langle \frac{\hat{u}_4}{\langle .8,.8,.6 \rangle}, .6 \right\rangle \right\} \right), \\ \left(\hat{\theta}_2, \left\{ \left\langle \frac{\hat{u}_1}{\langle .7,.7,.9 \rangle}, .2 \right\rangle, \left\langle \frac{\hat{u}_2}{\langle .7,.7,.8 \rangle}, .3 \right\rangle, \left\langle \frac{\hat{u}_3}{\langle .7,.7,.7 \rangle}, .4 \right\rangle, \left\langle \frac{\hat{u}_4}{\langle .7,.7,.6 \rangle}, .5 \right\rangle \right\} \right), \\ \left(\hat{\theta}_3, \left\{ \left\langle \frac{\hat{u}_1}{\langle .6,.6,.9 \rangle}, .1 \right\rangle, \left\langle \frac{\hat{u}_2}{\langle .6,.6,.8 \rangle}, .2 \right\rangle, \left\langle \frac{\hat{u}_3}{\langle .6,.6,.7 \rangle}, .3 \right\rangle, \left\langle \frac{\hat{u}_4}{\langle .6,.9,.6 \rangle}, .4 \right\rangle \right\} \right), \\ \left(\hat{\theta}_4, \left\{ \left\langle \frac{\hat{u}_1}{\langle .9,.9,.8 \rangle}, .5 \right\rangle, \left\langle \frac{\hat{u}_2}{\langle .9,.9,.7 \rangle}, .6 \right\rangle, \left\langle \frac{\hat{u}_3}{\langle .9,.9,.6 \rangle}, .7 \right\rangle, \left\langle \frac{\hat{u}_4}{\langle .9,.9,.5 \rangle}, .8 \right\rangle \right\} \right), \end{array} \right\}$$

Definition 3.5.2: Possibility sv-Neutrosophic Hypersoft Set of Type-2

A possibility sv-neutrosophic hypersoft set of type-2 ϖ_2 is defined as

$$\varpi_2 = \left\{ \left(\hat{\theta}, \left\langle \frac{\hat{u}}{\psi_{svN}(\hat{\theta})(\hat{u})}, \mu_{IF}(\hat{\theta}) \right\rangle \right) : \hat{u} \in \widehat{U} \wedge \hat{\theta} \in \widehat{\Theta} \right\}$$

where

$$\psi_{svN}(\hat{\theta})(\hat{u}) = \langle T_{\psi}(\hat{\theta})(\hat{u}), I_{\psi}(\hat{\theta})(\hat{u}), F_{\psi}(\hat{\theta})(\hat{u}) \rangle \subseteq \mathfrak{A}^{svN}$$

and

$$T_{\psi}(\hat{\theta})(\hat{u}), I_{\psi}(\hat{\theta})(\hat{u}), F_{\psi}(\hat{\theta})(\hat{u}) \in [0,1],$$

with $0 \leq T_{\psi}(\hat{\theta})(\hat{u}) + I_{\psi}(\hat{\theta})(\hat{u}) + F_{\psi}(\hat{\theta})(\hat{u}) \leq 3$. Similarly

$$\mu_{IF}(\hat{\theta}) \subseteq \mathfrak{A}^{IF}, \mu_{IF}(\hat{\theta}) = \langle T_{\mu}(\hat{\theta}), F_{\mu}(\hat{\theta}) \rangle$$

and

$$T_{\mu}(\hat{\theta}), F_{\mu}(\hat{\theta}) \in [0,1]$$

with

$$0 \leq T_{\mu}(\hat{\theta}) + F_{\mu}(\hat{\theta}) \leq 1$$

.

Example 3.5.2: Considering the assumptions from Example 3.1.1, we can construct possibility sv-neutrosophic hypersoft set of type-2 ϖ_2 can be constructed as

$$\varpi_2 = \left\{ \begin{array}{l} \left(\hat{\theta}_1, \left\{ \left\langle \frac{\hat{u}_1}{\langle .8,.8,.9 \rangle}, \langle .3,.2 \rangle \right\rangle, \left\langle \frac{\hat{u}_2}{\langle .8,.8,.8 \rangle}, \langle .4,.3 \rangle \right\rangle, \left\langle \frac{\hat{u}_3}{\langle .8,.8,.7 \rangle}, \langle .5,.4 \rangle \right\rangle, \left\langle \frac{\hat{u}_4}{\langle .8,.8,.6 \rangle}, \langle .6,.2 \rangle \right\rangle \right\} \right), \\ \left(\hat{\theta}_2, \left\{ \left\langle \frac{\hat{u}_1}{\langle .7,.7,.9 \rangle}, \langle .3,.1 \rangle \right\rangle, \left\langle \frac{\hat{u}_2}{\langle .7,.7,.8 \rangle}, \langle .4,.2 \rangle \right\rangle, \left\langle \frac{\hat{u}_3}{\langle .7,.7,.7 \rangle}, \langle .5,.3 \rangle \right\rangle, \left\langle \frac{\hat{u}_4}{\langle .7,.7,.6 \rangle}, \langle .6,.1 \rangle \right\rangle \right\} \right), \\ \left(\hat{\theta}_3, \left\{ \left\langle \frac{\hat{u}_1}{\langle .6,.6,.9 \rangle}, \langle .3,.3 \rangle \right\rangle, \left\langle \frac{\hat{u}_2}{\langle .6,.6,.8 \rangle}, \langle .3,.4 \rangle \right\rangle, \left\langle \frac{\hat{u}_3}{\langle .6,.6,.7 \rangle}, \langle .3,.5 \rangle \right\rangle, \left\langle \frac{\hat{u}_4}{\langle .6,.9,.6 \rangle}, \langle .3,.6 \rangle \right\rangle \right\} \right), \\ \left(\hat{\theta}_4, \left\{ \left\langle \frac{\hat{u}_1}{\langle .9,.9,.8 \rangle}, \langle .3,.4 \rangle \right\rangle, \left\langle \frac{\hat{u}_2}{\langle .9,.9,.7 \rangle}, \langle .3,.5 \rangle \right\rangle, \left\langle \frac{\hat{u}_3}{\langle .9,.9,.6 \rangle}, \langle .3,.6 \rangle \right\rangle, \left\langle \frac{\hat{u}_4}{\langle .9,.9,.5 \rangle}, \langle .3,.3 \rangle \right\rangle \right\} \right), \end{array} \right\}$$

Definition 3.5.3: Possibility sv-Neutrosophic Hypersoft Set of Type-3

A possibility sv-neutrosophic hypersoft set of type-3 ϖ_3 is defined as

$$\varpi_3 = \left\{ \left(\hat{\theta}, \left\langle \frac{\hat{u}}{\psi_{svN}(\hat{\theta})(\hat{u})}, \mu_{PyF}(\hat{\theta}) \right\rangle \right) : \hat{u} \in \hat{U} \wedge \hat{\theta} \in \hat{\Theta} \right\}$$

where

$$\psi_{svN}(\hat{\theta})(\hat{u}) = \langle T_\psi(\hat{\theta})(\hat{u}), I_\psi(\hat{\theta})(\hat{u}), F_\psi(\hat{\theta})(\hat{u}) \rangle \subseteq \mathfrak{A}^{svN}$$

and

$$T_\psi(\hat{\theta})(\hat{u}), I_\psi(\hat{\theta})(\hat{u}), F_\psi(\hat{\theta})(\hat{u}) \in [0,1]$$

with $0 \leq T_\psi(\hat{\theta})(\hat{u}) + I_\psi(\hat{\theta})(\hat{u}) + F_\psi(\hat{\theta})(\hat{u}) \leq 3$. Similarly

$$\mu_{PyF}(\hat{\theta}) \subseteq \mathfrak{A}^{PyF}$$

and

$$\mu_{PyF}(\hat{\theta}) = \langle T_\mu(\hat{\theta}), F_\mu(\hat{\theta}) \rangle$$

with $T_\mu(\hat{\theta}), F_\mu(\hat{\theta}) \in [0,1]$ and $0 \le T_\mu^2(\hat{\theta}) + F_\mu^2(\hat{\theta}) \le 1$.

Example 3.5.3: Considering the assumptions from Example 3.1.1, we can construct possibility sv-neutrosophic hypersoft set of type-3 ϖ_3 can be constructed as

$$\varpi_3 = \left\{ \begin{array}{l} \left(\hat{\theta}_1, \left\{ \left\langle \frac{\hat{u}_1}{\langle .8,.8,.9 \rangle}, \langle .5,.6 \rangle \right\rangle, \left\langle \frac{\hat{u}_2}{\langle .8,.8,.8 \rangle}, \langle .7,.5 \rangle \right\rangle, \left\langle \frac{\hat{u}_3}{\langle .8,.8,.7 \rangle}, \langle .5,.8 \rangle \right\rangle, \left\langle \frac{\hat{u}_4}{\langle .8,.8,.6 \rangle}, \langle .6,.5 \rangle \right\rangle \right\} \right), \\ \left(\hat{\theta}_2, \left\{ \left\langle \frac{\hat{u}_1}{\langle .7,.7,.9 \rangle}, \langle .9,.4 \rangle \right\rangle, \left\langle \frac{\hat{u}_2}{\langle .7,.7,.8 \rangle}, \langle .4,.8 \rangle \right\rangle, \left\langle \frac{\hat{u}_3}{\langle .7,.7,.7 \rangle}, \langle .4,.7 \rangle \right\rangle, \left\langle \frac{\hat{u}_4}{\langle .7,.7,.6 \rangle}, \langle .5,.8 \rangle \right\rangle \right\} \right), \\ \left(\hat{\theta}_3, \left\{ \left\langle \frac{\hat{u}_1}{\langle .6,.6,.9 \rangle}, \langle .7,.5 \rangle \right\rangle, \left\langle \frac{\hat{u}_2}{\langle .6,.6,.8 \rangle}, \langle .8,.5 \rangle \right\rangle, \left\langle \frac{\hat{u}_3}{\langle .6,.6,.7 \rangle}, \langle .6,.5 \rangle \right\rangle, \left\langle \frac{\hat{u}_4}{\langle .6,.9,.6 \rangle}, \langle .5,.6 \rangle \right\rangle \right\} \right), \\ \left(\hat{\theta}_4, \left\{ \left\langle \frac{\hat{u}_1}{\langle .9,.9,.8 \rangle}, \langle .4,.9 \rangle \right\rangle, \left\langle \frac{\hat{u}_2}{\langle .9,.9,.7 \rangle}, \langle .8,.4 \rangle \right\rangle, \left\langle \frac{\hat{u}_3}{\langle .9,.9,.6 \rangle}, \langle .7,.4 \rangle \right\rangle, \left\langle \frac{\hat{u}_4}{\langle .9,.9,.5 \rangle}, \langle .8,.5 \rangle \right\rangle \right\} \right), \end{array} \right\}$$

Definition 3.5.4: Possibility sv-Neutrosophic Hypersoft Set of Type-4

A possibility sv-neutrosophic hypersoft set of type-4 ϖ_4 is defined as

$$\varpi_4 = \left\{ \left(\hat{\theta}, \left\langle \frac{\hat{u}}{\psi_{svN}(\hat{\theta})(\hat{u})}, \mu_{PF}(\hat{\theta}) \right\rangle \right) : \hat{u} \in \hat{U} \wedge \hat{\theta} \in \hat{\Theta} \right\}$$

where

$$\psi_{svN}(\hat{\theta})(\hat{u}) = \langle T_\psi(\hat{\theta})(\hat{u}), I_\psi(\hat{\theta})(\hat{u}), F_\psi(\hat{\theta})(\hat{u}) \rangle \subseteq \mathfrak{A}^{svN}$$

and

$$T_\psi(\hat{\theta})(\hat{u}), I_\psi(\hat{\theta})(\hat{u}), F_\psi(\hat{\theta})(\hat{u}) \in [0,1]$$

with $0 \le T_\psi(\hat{\theta})(\hat{u}) + I_\psi(\hat{\theta})(\hat{u}) + F_\psi(\hat{\theta})(\hat{u}) \le 3$. Similarly

$$\mu_{PF}(\hat{\theta}) \subseteq \mathfrak{A}^{PF}$$

,

$$\mu_{PF}(\hat{\theta}) = \langle T_{\mu}(\hat{\theta}), I_{\mu}(\hat{\theta}), F_{\mu}(\hat{\theta}) \rangle$$

with $T_{\mu}(\hat{\theta}), I_{\mu}(\hat{\theta}), F_{\mu}(\hat{\theta}) \in [0,1]$ and $0 \leq T_{\mu}(\hat{\theta}) + I_{\mu}(\hat{\theta}) + F_{\mu}(\hat{\theta}) \leq 1$.

Example 3.5.4: Considering the assumptions from Example 3.1.1, we can construct possibility sv-neutrosophic hypersoft set of type-4 ϖ_4 can be constructed as

$$\varpi_4 = \left\{ \begin{array}{l} \left(\hat{\theta}_1, \left\{ \left\langle \frac{\hat{u}_1}{\langle .8,.8,.9 \rangle}, \langle .5,.1,.2 \rangle \right\rangle, \left\langle \frac{\hat{u}_2}{\langle .8,.8,.8 \rangle}, \langle .2,.5,.2 \rangle \right\rangle, \left\langle \frac{\hat{u}_3}{\langle .8,.8,.7 \rangle}, \langle .4,.2,.3 \rangle \right\rangle, \left\langle \frac{\hat{u}_4}{\langle .8,.8,.6 \rangle}, \langle .6,.1,.2 \rangle \right\rangle \right\} \right), \\ \left(\hat{\theta}_2, \left\{ \left\langle \frac{\hat{u}_1}{\langle .7,.7,.9 \rangle}, \langle .1,.4,.4 \rangle \right\rangle, \left\langle \frac{\hat{u}_2}{\langle .7,.7,.8 \rangle}, \langle .3,.4,.1 \rangle \right\rangle, \left\langle \frac{\hat{u}_3}{\langle .7,.7,.7 \rangle}, \langle .2,.3,.3 \rangle \right\rangle, \left\langle \frac{\hat{u}_4}{\langle .7,.7,.6 \rangle}, \langle .2,.4,.2 \rangle \right\rangle \right\} \right), \\ \left(\hat{\theta}_3, \left\{ \left\langle \frac{\hat{u}_1}{\langle .6,.6,.9 \rangle}, \langle .3,.5,.1 \rangle \right\rangle, \left\langle \frac{\hat{u}_2}{\langle .6,.6,.8 \rangle}, \langle .2,.5,.2 \rangle \right\rangle, \left\langle \frac{\hat{u}_3}{\langle .6,.6,.7 \rangle}, \langle .6,.2,.1 \rangle \right\rangle, \left\langle \frac{\hat{u}_4}{\langle .6,.9,.6 \rangle}, \langle .5,.2,.2 \rangle \right\rangle \right\} \right), \\ \left(\hat{\theta}_4, \left\{ \left\langle \frac{\hat{u}_1}{\langle .9,.9,.8 \rangle}, \langle .4,.1,.1 \rangle \right\rangle, \left\langle \frac{\hat{u}_2}{\langle .9,.9,.7 \rangle}, \langle .2,.2,.2 \rangle \right\rangle, \left\langle \frac{\hat{u}_3}{\langle .9,.9,.6 \rangle}, \langle .7,.1,.1 \rangle \right\rangle, \left\langle \frac{\hat{u}_4}{\langle .9,.9,.5 \rangle}, \langle .4,.2,.1 \rangle \right\rangle \right\} \right), \end{array} \right\}$$

Definition 3.5.5: Possibility sv-Neutrosophic Hypersoft Set of Type-5

A possibility sv-neutrosophic hypersoft set of type-5 ϖ_5 is defined as

$$\varpi_5 = \left\{ \left(\hat{\theta}, \left\langle \frac{\hat{u}}{\psi_{svN}(\hat{\theta})(\hat{u})}, \mu_{svN}(\hat{\theta}) \right\rangle \right) : \hat{u} \in \hat{U} \wedge \hat{\theta} \in \hat{\Theta} \right\}$$

where

$$\psi_{svN}(\hat{\theta})(\hat{u}) = \langle T_{\psi}(\hat{\theta})(\hat{u}), I_{\psi}(\hat{\theta})(\hat{u}), F_{\psi}(\hat{\theta})(\hat{u}) \rangle \subseteq \mathfrak{A}^{svN}$$

and

$$T_{\psi}(\hat{\theta})(\hat{u}), I_{\psi}(\hat{\theta})(\hat{u}), F_{\psi}(\hat{\theta})(\hat{u}) \in [0,1]$$

with $0 \leq T_{\psi}(\hat{\theta})(\hat{u}) + I_{\psi}(\hat{\theta})(\hat{u}) + F_{\psi}(\hat{\theta})(\hat{u}) \leq 3$. Similarly

$$\mu_{svN}(\hat{\theta}) \subseteq \mathfrak{A}^{svN}$$

,

$$\mu_{svN}(\hat{\theta}) = \langle T_{\mu}(\hat{\theta}), I_{\mu}(\hat{\theta}), F_{\mu}(\hat{\theta})\rangle$$

with $T_{\mu}(\hat{\theta}), I_{\mu}(\hat{\theta}), F_{\mu}(\hat{\theta}) \in [0,1]$ and $0 \leq T_{\mu}(\hat{\theta}) + I_{\mu}(\hat{\theta}) + F_{\mu}(\hat{\theta}) \leq 3$.

Example 3.5.5: Considering the assumptions from Example 3.1.1, we can construct possibility sv-neutrosophic hypersoft set of type-5 ϖ_5 can be constructed as

$$\varpi_5 = \left\{ \begin{array}{l} \left(\hat{\theta}_1, \left\{ \left\langle \frac{\hat{u}_1}{\langle .8,.8,.9\rangle}, \langle .5,.6,.6\rangle \right\rangle, \left\langle \frac{\hat{u}_2}{\langle .8,.8,.8\rangle}, \langle .6,.7,.7\rangle \right\rangle, \left\langle \frac{\hat{u}_3}{\langle .8,.8,.7\rangle}, \langle .7,.8,.8\rangle \right\rangle, \left\langle \frac{\hat{u}_4}{\langle .8,.8,.6\rangle}, \langle .8,.9,.9\rangle \right\rangle \right\} \right), \\ \left(\hat{\theta}_2, \left\{ \left\langle \frac{\hat{u}_1}{\langle .7,.7,.9\rangle}, \langle .6,.5,.5\rangle \right\rangle, \left\langle \frac{\hat{u}_2}{\langle .7,.7,.8\rangle}, \langle .7,.6,.6\rangle \right\rangle, \left\langle \frac{\hat{u}_3}{\langle .7,.7,.7\rangle}, \langle .8,.7,.7\rangle \right\rangle, \left\langle \frac{\hat{u}_4}{\langle .7,.7,.6\rangle}, \langle .9,.8,.8\rangle \right\rangle \right\} \right), \\ \left(\hat{\theta}_3, \left\{ \left\langle \frac{\hat{u}_1}{\langle .6,.6,.9\rangle}, \langle .7,.6,.7\rangle \right\rangle, \left\langle \frac{\hat{u}_2}{\langle .6,.6,.8\rangle}, \langle .8,.7,.8\rangle \right\rangle, \left\langle \frac{\hat{u}_3}{\langle .6,.6,.7\rangle}, \langle .9,.8,.9\rangle \right\rangle, \left\langle \frac{\hat{u}_4}{\langle .6,.9,.6\rangle}, \langle .6,.5,.9\rangle \right\rangle \right\} \right), \\ \left(\hat{\theta}_4, \left\{ \left\langle \frac{\hat{u}_1}{\langle .9,.9,.8\rangle}, \langle .8,.7,.6\rangle \right\rangle, \left\langle \frac{\hat{u}_2}{\langle .9,.9,.7\rangle}, \langle .9,.6,.5\rangle \right\rangle, \left\langle \frac{\hat{u}_3}{\langle .9,.9,.6\rangle}, \langle .6,.7,.5\rangle \right\rangle, \left\langle \frac{\hat{u}_4}{\langle .9,.9,.5\rangle}, \langle .6,.8,.8\rangle \right\rangle \right\} \right), \end{array} \right\}$$

Definition 3.5.6: Possibility sv-Neutrosophic Hypersoft Set of Type-6

A possibility sv-neutrosophic hypersoft set of type-6 ϖ_6 is defined as

$$\delta_6 = \left\{ \left(\hat{\theta}, \left\langle \frac{\hat{u}}{\psi_{PF}(\hat{\theta})(\hat{u})}, \mu_{IVF}(\hat{\theta}) \right\rangle \right) : \hat{u} \in \widehat{U} \wedge \hat{\theta} \in \widehat{\Theta} \right\}$$

where

$$\psi_{svN}(\hat{\theta})(\hat{u}) = \langle T_{\psi}(\hat{\theta})(\hat{u}), I_{\psi}(\hat{\theta})(\hat{u}), F_{\psi}(\hat{\theta})(\hat{u})\rangle \subseteq \mathfrak{A}^{svN}$$

and

$$T_{\psi}(\hat{\theta})(\hat{u}), I_{\psi}(\hat{\theta})(\hat{u}), F_{\psi}(\hat{\theta})(\hat{u}) \in [0,1]$$

such that $0 \leq T_{\psi}(\hat{\theta})(\hat{u}) + I_{\psi}(\hat{\theta})(\hat{u}) + F_{\psi}(\hat{\theta})(\hat{u}) \leq 3$. Similarly

$$\mu_{IVF}(\hat{\theta}) \subseteq \mathfrak{A}^{IVF}$$

with

$$\mu_{IVF}(\hat{\theta}) = \left[L_{\mu}(\hat{\theta}), U_{\mu}(\hat{\theta})\right]$$

and

$$L_{\mu}(\hat{\theta}), U_{\mu}(\hat{\theta}) \in [0,1]$$

.

Example 3.5.6: Considering the assumptions from Example 3.1.1, we can construct possibility sv-neutrosophic hypersoft set of type-6 ϖ_6 can be constructed as

$$\varpi_6 = \left\{ \begin{array}{l} \left(\hat{\theta}_1, \left\{\left\langle \frac{\hat{u}_1}{\langle .8,.8,.9\rangle}, [.2,.3]\right\rangle, \left\langle \frac{\hat{u}_2}{\langle .8,.8,.8\rangle}, [.3,.4]\right\rangle, \left\langle \frac{\hat{u}_3}{\langle .8,.8,.7\rangle}, [.4,.5]\right\rangle, \left\langle \frac{\hat{u}_4}{\langle .8,.8,.6\rangle}, [.5,.6]\right\rangle\right\}\right), \\ \left(\hat{\theta}_2, \left\{\left\langle \frac{\hat{u}_1}{\langle .7,.7,.9\rangle}, [.6,.7]\right\rangle, \left\langle \frac{\hat{u}_2}{\langle .7,.7,.8\rangle}, [.7,.8]\right\rangle, \left\langle \frac{\hat{u}_3}{\langle .7,.7,.7\rangle}, [.8,.9]\right\rangle, \left\langle \frac{\hat{u}_4}{\langle .7,.7,.6\rangle}, [.2,.4]\right\rangle\right\}\right), \\ \left(\hat{\theta}_3, \left\{\left\langle \frac{\hat{u}_1}{\langle .6,.6,.9\rangle}, [.3,.5]\right\rangle, \left\langle \frac{\hat{u}_2}{\langle .6,.6,.8\rangle}, [.4,.6]\right\rangle, \left\langle \frac{\hat{u}_3}{\langle .6,.6,.7\rangle}, [.5,.7]\right\rangle, \left\langle \frac{\hat{u}_4}{\langle .6,.9,.6\rangle}, [.6,.8]\right\rangle\right\}\right), \\ \left(\hat{\theta}_4, \left\{\left\langle \frac{\hat{u}_1}{\langle .9,.9,.8\rangle}, [.7,.9]\right\rangle, \left\langle \frac{\hat{u}_2}{\langle .9,.9,.7\rangle}, [.2,.5]\right\rangle, \left\langle \frac{\hat{u}_3}{\langle .9,.9,.6\rangle}, [.3,.6]\right\rangle, \left\langle \frac{\hat{u}_4}{\langle .9,.9,.5\rangle}, [.4,.7]\right\rangle\right\}\right), \end{array} \right\}$$

3.6 Hybrid Structures of Interval-valued Fuzzy Hypersoft Sets With Possibility Settings

Definition 3.6.1: Possibility Interval-valued Fuzzy Hypersoft Set of Type-1

A possibility interval-valued fuzzy hypersoft set of type-1 π_1 is defined as

$$\pi_1 = \left\{ \left(\hat{\theta}, \left\langle \frac{\hat{u}}{\psi_{IVF}(\hat{\theta})(\hat{u})}, \mu_F(\hat{\theta})\right\rangle\right) : \hat{u} \in \widehat{U} \wedge \hat{\theta} \in \widehat{\Theta} \right\}$$

where

$$\psi_{IVF}(\hat{\theta})(\hat{u}) = \left[L_{\psi}(\hat{\theta})(\hat{u}), U_{\psi}(\hat{\theta})(\hat{u})\right] \subseteq \mathfrak{A}^{IVF}$$

and

$$L_{\psi}(\hat{\theta})(\hat{u}), U_{\psi}(\hat{\theta})(\hat{u}) \in [0,1]$$

.

Similarly $\mu_F(\hat{\theta}) \subseteq \mathfrak{A}^F$ with $\mu_F(\hat{\theta}) \in [0,1]$.

Example 3.6.1: Considering the assumptions from Example 3.1.1, we can construct possibility interval-valued fuzzy hypersoft set of type-1 π_1 can be constructed as

$$\pi_1 = \left\{ \begin{array}{l} \left(\hat{\theta}_1, \left\{\left\langle \frac{\hat{u}_1}{[.2,.3]}, 0.31 \right\rangle, \left\langle \frac{\hat{u}_2}{[.2,.4]}, 0.41 \right\rangle, \left\langle \frac{\hat{u}_3}{[.2,.5]}, 0.51 \right\rangle, \left\langle \frac{\hat{u}_4}{[.2,.6]}, 0.61 \right\rangle \right\}\right), \\ \left(\hat{\theta}_2, \left\{\left\langle \frac{\hat{u}_1}{[.3,.4]}, 0.32 \right\rangle, \left\langle \frac{\hat{u}_2}{[.3,.5]}, 0.42 \right\rangle, \left\langle \frac{\hat{u}_3}{[.3,.6]}, 0.52 \right\rangle, \left\langle \frac{\hat{u}_4}{[.3,.7]}, 0.62 \right\rangle \right\}\right), \\ \left(\hat{\theta}_3, \left\{\left\langle \frac{\hat{u}_1}{[.4,.5]}, 0.33 \right\rangle, \left\langle \frac{\hat{u}_2}{[.4,.6]}, 0.43 \right\rangle, \left\langle \frac{\hat{u}_3}{[.4,.7]}, 0.53 \right\rangle, \left\langle \frac{\hat{u}_4}{[.4,.8]}, 0.63 \right\rangle \right\}\right), \\ \left(\hat{\theta}_4, \left\{\left\langle \frac{\hat{u}_1}{[.5,.6]}, 0.34 \right\rangle, \left\langle \frac{\hat{u}_2}{[.5,.7]}, 0.44 \right\rangle, \left\langle \frac{\hat{u}_3}{[.5,.8]}, 0.54 \right\rangle, \left\langle \frac{\hat{u}_4}{[.5,.9]}, 0.64 \right\rangle \right\}\right), \end{array} \right\}$$

Definition 3.6.2: Possibility Interval-valued Fuzzy Hypersoft Set of Type-2

A possibility interval-valued fuzzy hypersoft set of type-2 π_2 is defined as

$$\pi_2 = \left\{ \left(\hat{\theta}, \left\langle \frac{\hat{u}}{\psi_{IVF}(\hat{\theta})(\hat{u})}, \mu_{IF}(\hat{\theta}) \right\rangle \right) : \hat{u} \in \hat{U} \wedge \hat{\theta} \in \hat{\Theta} \right\}$$

where

$$\psi_{IVF}(\hat{\theta})(\hat{u}) = \left[L_{\psi}(\hat{\theta})(\hat{u}), U_{\psi}(\hat{\theta})(\hat{u})\right] \subseteq \mathfrak{A}^{IVF}$$

and

$L_\psi(\hat{\theta})(\hat{u}), U_\psi(\hat{\theta})(\hat{u}) \in [0,1]$

.

Similarly $\mu_{IF}(\hat{\theta}) \subseteq \mathfrak{A}^{IF}$ with $\mu_{IF}(\hat{\theta}) = \langle T_\mu(\hat{\theta}), F_\mu(\hat{\theta})\rangle, T_\mu(\hat{\theta}), F_\mu(\hat{\theta}) \in [0,1]$ such that

$0 \leq T_\mu(\hat{\theta}) + F_\mu(\hat{\theta}) \leq 1$

.

Example 3.6.2: Considering the assumptions from Example 3.1.1, we can construct possibility interval-valued fuzzy hypersoft set of type-2 π_2 can be constructed as

$$\pi_2 = \left\{ \begin{array}{l} \left(\hat{\theta}_1, \left\{\left\langle \frac{\hat{u}_1}{[.2,.3]}, \langle .31,.21\rangle \right\rangle, \left\langle \frac{\hat{u}_2}{[.2,.4]}, \langle .41,.31\rangle \right\rangle, \left\langle \frac{\hat{u}_3}{[.2,.5]}, \langle .51,.41\rangle \right\rangle, \left\langle \frac{\hat{u}_4}{[.2,.6]}, \langle .61,.21\rangle \right\rangle \right\}\right), \\ \left(\hat{\theta}_2, \left\{\left\langle \frac{\hat{u}_1}{[.3,.4]}, \langle .32,.22\rangle \right\rangle, \left\langle \frac{\hat{u}_2}{[.3,.5]}, \langle .42,.32\rangle \right\rangle, \left\langle \frac{\hat{u}_3}{[.3,.6]}, \langle .52,.42\rangle \right\rangle, \left\langle \frac{\hat{u}_4}{[.3,.7]}, \langle .62,.22\rangle \right\rangle \right\}\right), \\ \left(\hat{\theta}_3, \left\{\left\langle \frac{\hat{u}_1}{[.4,.5]}, \langle .33,.23\rangle \right\rangle, \left\langle \frac{\hat{u}_2}{[.4,.6]}, \langle .43,.33\rangle \right\rangle, \left\langle \frac{\hat{u}_3}{[.4,.7]}, \langle .53,.43\rangle \right\rangle, \left\langle \frac{\hat{u}_4}{[.4,.8]}, \langle .63,.23\rangle \right\rangle \right\}\right), \\ \left(\hat{\theta}_4, \left\{\left\langle \frac{\hat{u}_1}{[.5,.6]}, \langle .34,.24\rangle \right\rangle, \left\langle \frac{\hat{u}_2}{[.5,.7]}, \langle .44,.34\rangle \right\rangle, \left\langle \frac{\hat{u}_3}{[.5,.8]}, \langle .54,.44\rangle \right\rangle, \left\langle \frac{\hat{u}_4}{[.5,.9]}, \langle .64,.24\rangle \right\rangle \right\}\right), \end{array} \right\}$$

Definition 3.6.3: Possibility Interval-valued Fuzzy Hypersoft Set of Type-3

A possibility interval-valued fuzzy hypersoft set of type-3 π_3 is defined as

$$\pi_3 = \left\{ \left(\hat{\theta}, \left\langle \frac{\hat{u}}{\psi_{IVF}(\hat{\theta})(\hat{u})}, \mu_{PyF}(\hat{\theta}) \right\rangle \right) : \hat{u} \in \widehat{U} \wedge \hat{\theta} \in \widehat{\Theta} \right\}$$

where

$\psi_{IVF}(\hat{\theta})(\hat{u}) = \left[L_\psi(\hat{\theta})(\hat{u}), U_\psi(\hat{\theta})(\hat{u})\right] \subseteq \mathfrak{A}^{IVF}$

and

$T_{\psi}(\hat{\theta})(\hat{u}), F_{\psi}(\hat{\theta})(\hat{u}) \in [0,1]$

.

Similarly $\mu_{PyF}(\hat{\theta}) \subseteq \mathfrak{A}^{PyF}$ with $\mu_{PyF}(\hat{\theta}) = \langle T_{\mu}(\hat{\theta}), F_{\mu}(\hat{\theta})\rangle$,

$T_{\mu}(\hat{\theta}), F_{\mu}(\hat{\theta}) \in [0,1]$

and

$0 \leq T_{\mu}^2(\hat{\theta}) + F_{\mu}^2(\hat{\theta}) \leq 1$

.

Example 3.6.3: Considering the assumptions from Example 3.1.1, we can construct possibility interval-valued fuzzy hypersoft set of type-3 π_3 can be constructed as

$$\pi_3 = \left\{ \begin{array}{l} \left(\hat{\theta}_1, \left\{\left\langle \frac{\hat{u}_1}{[.2,.3]}, \langle .5,.6\rangle \right\rangle, \left\langle \frac{\hat{u}_2}{[.2,.4]}, \langle .7,.5\rangle \right\rangle, \left\langle \frac{\hat{u}_3}{[.2,.5]}, \langle .5,.8\rangle \right\rangle, \left\langle \frac{\hat{u}_4}{[.2,.6]}, \langle .6,.5\rangle \right\rangle \right\}\right), \\ \left(\hat{\theta}_2, \left\{\left\langle \frac{\hat{u}_1}{[.3,.4]}, \langle .9,.4\rangle \right\rangle, \left\langle \frac{\hat{u}_2}{[.3,.5]}, \langle .4,.8\rangle \right\rangle, \left\langle \frac{\hat{u}_3}{[.3,.6]}, \langle .4,.7\rangle \right\rangle, \left\langle \frac{\hat{u}_4}{[.3,.7]}, \langle .5,.8\rangle \right\rangle \right\}\right), \\ \left(\hat{\theta}_3, \left\{\left\langle \frac{\hat{u}_1}{[.4,.5]}, \langle .7,.5\rangle \right\rangle, \left\langle \frac{\hat{u}_2}{[.4,.6]}, \langle .8,.5\rangle \right\rangle, \left\langle \frac{\hat{u}_3}{[.4,.7]}, \langle .6,.5\rangle \right\rangle, \left\langle \frac{\hat{u}_4}{[.4,.8]}, \langle .5,.6\rangle \right\rangle \right\}\right), \\ \left(\hat{\theta}_4, \left\{\left\langle \frac{\hat{u}_1}{[.5,.6]}, \langle .4,.9\rangle \right\rangle, \left\langle \frac{\hat{u}_2}{[.5,.7]}, \langle .8,.4\rangle \right\rangle, \left\langle \frac{\hat{u}_3}{[.5,.8]}, \langle .7,.4\rangle \right\rangle, \left\langle \frac{\hat{u}_4}{[.5,.9]}, \langle .8,.5\rangle \right\rangle \right\}\right), \end{array} \right\}$$

Definition 3.6.4: Possibility Interval-Valued Fuzzy Hypersoft Set of Type-4

A possibility interval-valued fuzzy hypersoft set of type-4 π_4 is defined as

$$\pi_4 = \left\{ \left(\hat{\theta}, \left\langle \frac{\hat{u}}{\psi_{IVF}(\hat{\theta})(\hat{u})}, \mu_{PF}(\hat{\theta}) \right\rangle \right) : \hat{u} \in \widehat{U} \wedge \hat{\theta} \in \widehat{\Theta} \right\}$$

where

$$\psi_{IVF}(\hat{\theta})(\hat{u}) = \left[L_{\psi}(\hat{\theta})(\hat{u}), U_{\psi}(\hat{\theta})(\hat{u})\right] \subseteq \mathfrak{A}^{IVF},$$

and

$$L_{\psi}(\hat{\theta})(\hat{u}), U_{\psi}(\hat{\theta})(\hat{u}) \in [0,1]$$

.

Similarly $\mu_{PF}(\hat{\theta}) \subseteq \mathfrak{A}^{PF}$, $\mu_{PF}(\hat{\theta}) = \langle T_{\mu}(\hat{\theta}), I_{\mu}(\hat{\theta}), F_{\mu}(\hat{\theta})\rangle$ with $T_{\mu}(\hat{\theta}), I_{\mu}(\hat{\theta}), F_{\mu}(\hat{\theta}) \in [0,1]$ and

$$0 \leq T_{\mu}(\hat{\theta}) + I_{\mu}(\hat{\theta}) + F_{\mu}(\hat{\theta}) \leq 1$$

.

Example 3.6.4: Considering the assumptions from Example 3.1.1, we can construct possibility interval-valued fuzzy hypersoft set of type-4 π_4 can be constructed as

$$\pi_4 = \left\{ \begin{array}{l} \left(\hat{\theta}_1, \left\{\left\langle \frac{\hat{u}_1}{[.2,.3]}, \langle .5,.1,.2\rangle \right\rangle, \left\langle \frac{\hat{u}_2}{[.2,.4]}, \langle .2,.5,.2\rangle \right\rangle, \left\langle \frac{\hat{u}_3}{[.2,.5]}, \langle .4,.2,.3\rangle \right\rangle, \left\langle \frac{\hat{u}_4}{[.2,.6]}, \langle .6,.1,.2\rangle \right\rangle \right\}\right), \\ \left(\hat{\theta}_2, \left\{\left\langle \frac{\hat{u}_1}{[.3,.4]}, \langle .1,.4,.4\rangle \right\rangle, \left\langle \frac{\hat{u}_2}{[.3,.5]}, \langle .3,.4,.1\rangle \right\rangle, \left\langle \frac{\hat{u}_3}{[.3,.6]}, \langle .2,.3,.3\rangle \right\rangle, \left\langle \frac{\hat{u}_4}{[.3,.7]}, \langle .2,.4,.2\rangle \right\rangle \right\}\right), \\ \left(\hat{\theta}_3, \left\{\left\langle \frac{\hat{u}_1}{[.4,.5]}, \langle .3,.5,.1\rangle \right\rangle, \left\langle \frac{\hat{u}_2}{[.4,.6]}, \langle .2,.5,.2\rangle \right\rangle, \left\langle \frac{\hat{u}_3}{[.4,.7]}, \langle .6,.2,.1\rangle \right\rangle, \left\langle \frac{\hat{u}_4}{[.4,.8]}, \langle .5,.2,.2\rangle \right\rangle \right\}\right), \\ \left(\hat{\theta}_4, \left\{\left\langle \frac{\hat{u}_1}{[.5,.6]}, \langle .4,.1,.1\rangle \right\rangle, \left\langle \frac{\hat{u}_2}{[.5,.7]}, \langle .2,.2,.2\rangle \right\rangle, \left\langle \frac{\hat{u}_3}{[.5,.8]}, \langle .7,.1,.1\rangle \right\rangle, \left\langle \frac{\hat{u}_4}{[.5,.9]}, \langle .4,.2,.1\rangle \right\rangle \right\}\right), \end{array} \right\}$$

Definition 3.6.5: Possibility Interval-Valued Fuzzy Hypersoft Set of Type-5

A possibility interval-valued fuzzy hypersoft set of type-5 π_5 is defined as

$$\pi_5 = \left\{ \left(\hat{\theta}, \left\langle \frac{\hat{u}}{\psi_{IVF}(\hat{\theta})(\hat{u})}, \mu_{svN}(\hat{\theta}) \right\rangle\right) : \hat{u} \in \widehat{U} \wedge \hat{\theta} \in \widehat{\Theta} \right\}$$

where

$$\psi_{IVF}(\hat{\theta})(\hat{u}) = \left[L_{\psi}(\hat{\theta})(\hat{u}), U_{\psi}(\hat{\theta})(\hat{u})\right] \subseteq \mathfrak{A}^{IVF},$$

and

$$L_{\psi}(\hat{\theta})(\hat{u}), U_{\psi}(\hat{\theta})(\hat{u}) \in [0,1]$$

.

Similarly $\mu_{svN}(\hat{\theta}) \subseteq \mathfrak{A}^{svN}$, $\mu_{svN}(\hat{\theta}) = \langle T_{\mu}(\hat{\theta}), I_{\mu}(\hat{\theta}), F_{\mu}(\hat{\theta})\rangle$ with

$$T_{\mu}(\hat{\theta}), I_{\mu}(\hat{\theta}), F_{\mu}(\hat{\theta}) \in [0,1]$$

and

$$0 \leq T_{\mu}(\hat{\theta}) + I_{\mu}(\hat{\theta}) + F_{\mu}(\hat{\theta}) \leq 3$$

.

Example 3.6.5: Considering the assumptions from Example 3.1.1, we can construct possibility interval-valued fuzzy hypersoft set of type-5 π_5 can be constructed as

$$\pi_5 = \left\{ \begin{array}{l} \left(\hat{\theta}_1, \left\{\left\langle \frac{\hat{u}_1}{[.2,.3]}, \langle .5,.6,.6\rangle \right\rangle, \left\langle \frac{\hat{u}_2}{[.2,.4]}, \langle .6,.7,.7\rangle \right\rangle, \left\langle \frac{\hat{u}_3}{[.2,.5]}, \langle .7,.8,.8\rangle \right\rangle, \left\langle \frac{\hat{u}_4}{[.2,.6]}, \langle .8,.9,.9\rangle \right\rangle \right\}\right), \\ \left(\hat{\theta}_2, \left\{\left\langle \frac{\hat{u}_1}{[.3,.4]}, \langle .6,.5,.5\rangle \right\rangle, \left\langle \frac{\hat{u}_2}{[.3,.5]}, \langle .7,.6,.6\rangle \right\rangle, \left\langle \frac{\hat{u}_3}{[.3,.6]}, \langle .8,.7,.7\rangle \right\rangle, \left\langle \frac{\hat{u}_4}{[.3,.7]}, \langle .9,.8,.8\rangle \right\rangle \right\}\right), \\ \left(\hat{\theta}_3, \left\{\left\langle \frac{\hat{u}_1}{[.4,.5]}, \langle .7,.6,.7\rangle \right\rangle, \left\langle \frac{\hat{u}_2}{[.4,.6]}, \langle .8,.7,.8\rangle \right\rangle, \left\langle \frac{\hat{u}_3}{[.4,.7]}, \langle .9,.8,.9\rangle \right\rangle, \left\langle \frac{\hat{u}_4}{[.4,.8]}, \langle .6,.5,.9\rangle \right\rangle \right\}\right), \\ \left(\hat{\theta}_4, \left\{\left\langle \frac{\hat{u}_1}{[.5,.6]}, \langle .8,.7,.6\rangle \right\rangle, \left\langle \frac{\hat{u}_2}{[.5,.7]}, \langle .9,.6,.5\rangle \right\rangle, \left\langle \frac{\hat{u}_3}{[.5,.8]}, \langle .6,.7,.5\rangle \right\rangle, \left\langle \frac{\hat{u}_4}{[.5,.9]}, \langle .6,.8,.8\rangle \right\rangle \right\}\right), \end{array} \right\}$$

Definition 3.6.6: Possibility Interval-Valued Fuzzy Hypersoft Set of Type-6

A possibility interval-valued fuzzy hypersoft set of type-6 π_6 is defined as

$$\pi_6 = \left\{ \left(\hat{\theta}, \left\langle \frac{\hat{u}}{\psi_{IVF}(\hat{\theta})(\hat{u})}, \mu_{IVF}(\hat{\theta}) \right\rangle\right) : \hat{u} \in \hat{U} \wedge \hat{\theta} \in \widehat{\Theta} \right\}$$

where

$$\psi_{IVF}(\hat{\theta})(\hat{u}) = \left[L_{\psi}(\hat{\theta})(\hat{u}), U_{\psi}(\hat{\theta})(\hat{u})\right] \subseteq \mathfrak{A}^{IVF}$$

and

$$T_{\psi}(\hat{\theta})(\hat{u}), F_{\psi}(\hat{\theta})(\hat{u}) \in [0,1]$$

.

Similarly $\mu_{IVF}(\hat{\theta}) \subseteq \mathfrak{A}^{IVF}$ with $\mu_{IVF}(\hat{\theta}) = \left[L_{\mu}(\hat{\theta}), U_{\mu}(\hat{\theta})\right]$ and $L_{\mu}(\hat{\theta}), U_{\mu}(\hat{\theta}) \in [0,1]$.

Example 3.6.6: Considering the assumptions from Example 3.1.1, we can construct possibility interval-valued fuzzy hypersoft set of type-6 π_6 can be constructed as

$$\pi_6 = \left\{ \begin{array}{l} \left(\hat{\theta}_1, \left\{ \left\langle \frac{\hat{u}_1}{[.2,.3]}, [.21,.31] \right\rangle, \left\langle \frac{\hat{u}_2}{[.2,.4]}, [.31,.41] \right\rangle, \left\langle \frac{\hat{u}_3}{[.2,.5]}, [.41,.51] \right\rangle, \left\langle \frac{\hat{u}_4}{[.2,.6]}, [.21,.61] \right\rangle \right\} \right), \\ \left(\hat{\theta}_2, \left\{ \left\langle \frac{\hat{u}_1}{[.3,.4]}, [.22,.32] \right\rangle, \left\langle \frac{\hat{u}_2}{[.3,.5]}, [.32,.42] \right\rangle, \left\langle \frac{\hat{u}_3}{[.3,.6]}, [.42,.52] \right\rangle, \left\langle \frac{\hat{u}_4}{[.3,.7]}, [.22,.62] \right\rangle \right\} \right), \\ \left(\hat{\theta}_3, \left\{ \left\langle \frac{\hat{u}_1}{[.4,.5]}, [.23,.33] \right\rangle, \left\langle \frac{\hat{u}_2}{[.4,.6]}, [.33,.43] \right\rangle, \left\langle \frac{\hat{u}_3}{[.4,.7]}, [.43,.53] \right\rangle, \left\langle \frac{\hat{u}_4}{[.4,.8]}, [.23,.63] \right\rangle \right\} \right), \\ \left(\hat{\theta}_4, \left\{ \left\langle \frac{\hat{u}_1}{[.5,.6]}, [.24,.34] \right\rangle, \left\langle \frac{\hat{u}_2}{[.5,.7]}, [.34,.44] \right\rangle, \left\langle \frac{\hat{u}_3}{[.5,.8]}, [.44,.54] \right\rangle, \left\langle \frac{\hat{u}_4}{[.5,.9]}, [.24,.64] \right\rangle \right\} \right), \end{array} \right\}$$

4. CONCLUSION

The expansion of a soft set known as a hypersoft set introduces the idea of a hypersoft membership function, enabling it to handle complex and uncertain information in a more powerful and flexible way. In this chapter, several settings that are similar to fuzzy sets are considered, along with settings based on possibility degree. Numerical examples are also provided to help explain the principle behind these structures. This work can be used by researchers to comprehend and apply a range of mathematical concepts. The proposed structures can be used for developing various algebraic and topological structures.

ACKNOWLEDGMENT

This research received no specific grant from any funding agency in the public, commercial, or not-for-profit sectors.

REFERENCES

Al-Hijjawi, S., & Alkhazaleh, S. (2023). Possibility Neutrosophic Hypersoft Set (PNHSS). *Neutrosophic Sets and Systems*, *53*, 117–129.

Alhazaymeh, K., & Hassan, N. (2012). Possibility vague soft set and its application in decision making. *International Journal of Pure and Applied Mathematics*, *77*(4), 549–563.

Alhazaymeh, K., & Hassan, N. (2013). Possibility interval-valued vague soft set. *Applied Mathematical Sciences (Ruse)*, *7*(140), 6989–6994. DOI: 10.12988/ams.2013.310576

Alkhazaleh, S., Salleh, A. R., & Hassan, N. (2011). Possibility fuzzy soft set. *Advances in Decision Sciences*, 2011.

Atanassov, K. (1986). Intuitionistic fuzzy sets. *Fuzzy Sets and Systems*, *20*(1), 87–96. DOI: 10.1016/S0165-0114(86)80034-3

Bashir, M., & Salleh, A. R. (2012). Possibility fuzzy soft expert set. *Ozean Journal of Applied Sciences*, *12*, 208–211.

Bashir, M., Salleh, A. R., & Alkhazaleh, S. (2012). Possibility intuitionistic fuzzy soft set. *Advances in Decision Sciences*, 2012.

Debnath, S. (2021). Fuzzy hypersoft sets and its weightage operator for decision making. *Journal of Fuzzy Extension and Applications*, *2*(2), 163–170.

Ihsan, M., Rahman, A. U., & Saeed, M. (2021). Fuzzy Hypersoft Expert Set with Application in Decision Making for the Best Selection of Product. *Neutrosophic Sets and Systems*, *46*, 318–336.

Ihsan, M., Saeed, M., Rahman, A. U., & Smarandache, F. (2022). Multi-Attribute Decision Support Model Based on Bijective Hypersoft Expert Set. *Punjab University Journal of Mathematics*, *54*(1), 55–73. DOI: 10.52280/pujm.2022.540105

Jia-hua, D., Zhang, H., & He, Y. (2019). Possibility Pythagorean fuzzy soft set and its application. *Journal of Intelligent & Fuzzy Systems*, *36*(1), 413–421. DOI: 10.3233/JIFS-181649

Kamacı, H., & Saqlain, M. (2021). n-ary Fuzzy Hypersoft Expert Sets. *Neutrosophic Sets and Systems*, *43*, 180–211.

Karaaslan, F. (2016a). Correlation coefficient between possibility neutrosophic soft sets. *Math. Sci. Lett*, *5*(1), 71–74. DOI: 10.18576/msl/050109

Karaaslan, F. (2016b). Similarity measure between possibility neutrosophic soft sets and its applications. *University Politehnica of Bucharest Scientific Bulletin-Series A-Applied Mathematics and Physics*, *78*(3), 155–162.

Karaaslan, F. (2017). Possibility neutrosophic soft sets and PNS-decision making method. *Applied Soft Computing*, *54*, 403–414. DOI: 10.1016/j.asoc.2016.07.013

Khalil, A. M., Li, S. G., Li, H. X., & Ma, S. Q. (2019). Possibility m-polar fuzzy soft sets and its application in decision-making problems. *Journal of Intelligent & Fuzzy Systems*, *37*(1), 929–940. DOI: 10.3233/JIFS-181769

Khan, S., Gulistan, M., & Wahab, H. A. (2021). Development of the structure of q-Rung Orthopair Fuzzy Hypersoft Set with basic Operations. *Punjab University Journal of Mathematics*, *53*(12), 881–892. DOI: 10.52280/pujm.2021.531204

Molodtsov, D. (1999). Soft set theory—First results. *Computers & Mathematics with Applications (Oxford, England)*, *37*(4-5), 19–31. DOI: 10.1016/S0898-1221(99)00056-5

Rahman, A. U., Saeed, M., & Abd El-Wahed Khalifa, H. (2022). Multi-attribute decision-making based on aggregations and similarity measures of neutrosophic hypersoft sets with possibility setting. *Journal of Experimental & Theoretical Artificial Intelligence*, 1–26.

Rahman, A. U., Saeed, M., & Garg, H. (2022). An innovative decisive framework for optimized agri-automobile evaluation and HRM pattern recognition via possibility fuzzy hypersoft setting. *Advances in Mechanical Engineering*, *14*(10). DOI: 10.1177/16878132221132146

Rahman, A. U., Saeed, M., Khalifa, H. A. E. W., & Afifi, W. A. (2022). Decision making algorithmic techniques based on aggregation operations and similarity measures of possibility intuitionistic fuzzy hypersoft sets. *AIMS Mathematics*, *7*(3), 3866–3895. DOI: 10.3934/math.2022214

Rahman, A. U., Saeed, M., Mohammed, M. A., Abdulkareem, K. H., Nedoma, J., & Martinek, R. (2023). Fppsv-NHSS: Fuzzy parameterized possibility single valued neutrosophic hypersoft set to site selection for solid waste management. *Applied Soft Computing*, *140*, 110273. DOI: 10.1016/j.asoc.2023.110273

Rahman, A. U., Saeed, M., Mohammed, M. A., Krishnamoorthy, S., Kadry, S., & Eid, F. (2022). An integrated algorithmic MADM approach for heart diseases' diagnosis based on neutrosophic hypersoft set with possibility degree-based setting. *Life (Chicago, Ill.)*, *12*(5), 729. DOI: 10.3390/life12050729 PMID: 35629396

Saeed, M., Rahman, A. U., Ahsan, M., & Smarandache, F. (2022). Theory of Hypersoft Sets: Axiomatic Properties, Aggregation Operations, Relations, Functions and Matrices. *Neutrosophic Sets and Systems*, *51*, 744–765.

Smarandache, F. (1999). A unifying field in Logics: Neutrosophic Logic. In *Philosophy* (pp. 1-141). American Research Press.

Smarandache, F. (2018). Extension of Soft Set to Hypersoft Set, and then to Plithogenic Hypersoft Set. *Neutrosophic Sets and Systems*, *22*, 168–170.

Smarandache, F. (2022a). Introduction to the IndetermSoft Set and IndetermHyperSoft Set. *Neutrosophic Sets and Systems*, *50*, 629–650.

Smarandache, F. (2022b). Soft set product extended to hypersoft set and indetermsoft set cartesian product extended to indetermhypersoft set. *Journal of Fuzzy Extension and Applications, 3*(4), 313-316.

Smarandache, F. (2022c). Practical Applications of IndetermSoft Set and IndetermHyperSoft Set and Introduction to TreeSoft Set as an extension of the MultiSoft Set. *Neutrosophic Sets and Systems*, *51*, 939–947.

Smarandache, F. (2023). New Types of Soft Sets" HyperSoft Set, IndetermSoft Set, IndetermHyperSoft Set, and TreeSoft Set": An Improved Version. *Neutrosophic Systems with Applications*, *8*, 35–41. DOI: 10.61356/j.nswa.2023.41

Yolcu, A., & Öztürk, T. Y. (2021). Fuzzy hypersoft sets and it's application to decision-making. In Smarandache, F., Saeed, M., Abdel-Baset, M., & Saqlain, M. (Eds.), *Theory and Application of Hypersoft Set* (pp. 50–64). Pons Publishing House.

Yolcu, A., Smarandache, F., & Öztürk, T. Y. (2021). Intuitionistic fuzzy hypersoft sets. *Communications Faculty of Sciences University of Ankara Series A1 Mathematics and Statistics*, *70*(1), 443–455. DOI: 10.31801/cfsuasmas.788329

Zadeh, L. A. (1965). Fuzzy sets. *Information and Control*, *8*(3), 338–353. DOI: 10.1016/S0019-9958(65)90241-X

Zadeh, L. A. (1978). Fuzzy sets as a basis for a theory of possibility. *Fuzzy Sets and Systems*, *1*(1), 3–28. DOI: 10.1016/0165-0114(78)90029-5

Zhang, H. D., & Shu, L. (2014). Possibility multi-fuzzy soft set and its application in decision making. *Journal of Intelligent & Fuzzy Systems*, *27*(4), 2115–2125. DOI: 10.3233/IFS-141176

Zhao, J., Li, B., Rahman, A. U., & Saeed, M. (2023). An intelligent multiple-criteria decision-making approach based on sv-neutrosophic hypersoft set with possibility degree setting for investment selection. *Management Decision*, *61*(2), 472–485. DOI: 10.1108/MD-04-2022-0462

Chapter 15

MCDM Using Normalized Weighted Bonferroni Mean Operator in Fermatean Neutrosophic Environment

A. Revathy
https://orcid.org/0000-0003-1382-7317
Sri Eshwar College of Engineering, India

V. Inthumathi
NGM College, Pollachi, India

S. Krishnaprakash
https://orcid.org/0000-0003-3415-4457
Sri Krishna College of Engineering and Technology, India

S. Gomathi
Sri GVG Vishalakshi College for Women, India

N. Akiladevi
Sri Eshwar College of Engineering, India

ABSTRACT

Fermatean fuzzy set (FFS) is a comprehensive form of intuitionistic fuzzy set (IFS) which has wide range for truth values (TV) and false values (FV). The neutrosophic set (NS) can quantify the indeterminacy of fuzzy characteristics of the dataset

DOI: 10.4018/979-8-3693-3204-7.ch015

beyond the TV and FV independently. Fermatean neutrosophic set (FNS) set is an effective tool to handle the uncertainty in multi criteria decision making (MCDM) since it incorporates the important aspects of NS as well as FFS. For MCDM, the Bonferroni mean operator, which addresses the interdependencies between attributes, is a beneficial tool in certain circumstances because of its easy accessibility and stability. In this chapter Fermatean neutrosophic normalized weighted Bonferroni mean (FNNWBM) operator is presented and their features are examined. The Fermatean neutrosophic numbers (FNN) are aggregated using FNNWBM operator and the alternatives are ranked by the FNN extended score function in MCDM. The effectiveness of the proposed operator is checked through the obtained results' simulation and comparison analysis with other existing methods.

INTRODUCTION

In contemporary decision-making scenarios, the ability to handle uncertainty and imprecision has become increasingly critical. Traditional fuzzy set theories have been instrumental in capturing vagueness within data; however, they often fall short in addressing the full spectrum of uncertainty present in complex decision contexts. Senapati and Yager (2020) presents an in-depth exploration of Fermatean fuzzy sets, aiming to elucidate their theoretical underpinnings and practical implications. Fermatean fuzzy sets have emerged as a promising extension, offering a broader range of truth values (TV) and false values (FV) to better encapsulate uncertainty. In addition to Fermatean fuzzy sets, the paper likely touches upon the significance of neutrosophic sets, introduced by Smarandache (1998) in capturing the inherent indeterminacy and ambiguity present in real-world data. Neutrosophic sets extend the traditional fuzzy set paradigm by incorporating neutral elements, thereby offering a more comprehensive representation of uncertainty. Gonul (2022) introduced Fermatean fuzzy topological spaces and Broumi (2023) discussed Fermatean neutrosophic matrices and their basic operations. Palanikumar et al. (2022) proposed neutrosophic Fermatean fuzzy soft with aggregation operators based on VIKOR and TOPSIS method for MCGDM. Furthermore, Yager (1988) may delve into the Bonferroni mean operator, which plays a crucial role in aggregating information and addressing interdependencies between attributes in multi-criteria decision-making contexts. By considering the minimum values among a set of attributes, the Bonferroni mean operator offers a robust mechanism for capturing conservative estimates and ensuring the integrity of decision outcomes. Revathy (2022) used Fermatean fuzzy normalised Bonferroni mean operator in MCDM. Kishorekumar (2022), and Ma et al (2024) used Interval-valued picture fuzzy geometric Bonferroni mean aggregation operators and interval-valued Fermatean fuzzy bonferroni mean operators for MCDM. Kanchana et al. (2024) have done comprehensive analysis of neutrosophic Bonferroni mean operator. In light of these advancements, the exploration of Fermatean fuzzy sets, neutrosophic sets, and the Bonferroni mean operator represents a significant contribution to the field of decision science. By synthesizing these concepts, researchers aim to develop more comprehensive frameworks for decision-making under uncertainty, thereby enhancing the robustness and reliability of decision outcomes across various domains.

Research Gap and Motivation

Despite the advancements in multi-criteria decision making (MCDM) techniques, there remains a gap in effectively handling uncertainty beyond truth values (TV) and false values (FV) in decision-making processes. Traditional fuzzy set theories and even some neutrosophic set approaches may not fully capture the complexity of uncertainty present in real-world decision-making scenarios, especially when dealing with multi-dimensional and interdependent criteria.

The motivation behind this research lies in the necessity to develop more comprehensive tools for MCDM that can accommodate a wider range of uncertainty beyond the binary framework of truth and false values. Real-world decision-making often involves ambiguous, imprecise, and conflicting information, which cannot be adequately represented by conventional methods. By integrating Fermatean fuzzy sets and neutrosophic sets, this research aims to address these limitations and provide decision-makers with more robust and flexible tools for handling uncertainty in MCDM.

Contribution

The primary contribution of this research lies in the development and exploration of the Fermatean neutrosophic normalized weighted Bonferroni mean operator (FNNWBMO) as a novel approach for aggregating Fermatean neutrosophic numbers in MCDM. By combining aspects of Fermatean fuzzy sets, neutrosophic sets, and the Bonferroni mean operator, the proposed approach offers a more nuanced and comprehensive framework for decision-making under uncertainty. The research contributes to advancing the field of MCDM by providing decision-makers with a more powerful and flexible toolset for handling complex decision scenarios. Additionally, the evaluation and comparison of the proposed approach against existing methods help validate its effectiveness and identify areas for further improvement and refinement in future research endeavors.

PRELIMINARIES

Definition 1: Fermatean Fuzzy Set

A set $F_\gamma = \{\langle \gamma, \mu(\gamma), \nu(\gamma) : \gamma \in \Gamma \rangle\}$ in the universe of discourse X is called Fermatean fuzzy set (FFS) if $0 \le \mu(\gamma)^3 + \upsilon(\gamma)^3 \le 1$ where $\mu(\gamma):\Gamma \to [0,1]$, $\nu(\gamma):\Gamma \to [0,1]$ and $\pi = \sqrt[3]{1 - \mu(\gamma)^3 - \nu(\gamma)^3}$ are TV, FV and indeterminacy of $\gamma \in \Gamma$.

Definition 2: Neutrosophic Set

A set $F_\gamma = \{\langle \gamma, \mu(\gamma), \pi(\gamma), \nu(\gamma) : \gamma \in \Gamma \rangle\}$ in the universe of discourse X is called Neutrosophic set (NS) if $0 \le \mu(\gamma) + \pi(\gamma) + \upsilon(\gamma) \le 3$ where $\mu(\gamma):\Gamma \to [0,1]$, $\pi(\gamma):\Gamma \to [0,1]$, $\nu(\gamma):\Gamma \to [0,1]$ are TV, FV and indeterminacy of $\gamma \in \Gamma$.

Definition 3: Fermatean Neutrosophic Set

A set $F_\gamma = \{\langle \gamma, \mu(\gamma), \pi(\gamma), \nu(\gamma) : \gamma \in \Gamma \rangle\}$ in the universe of discourse X is called Fermatean neutrosophic set (FNS) in the universe of discourse Γ if $\mu(\gamma):\Gamma \to [0,1]$, $\pi(\gamma):\Gamma \to [0,1]$, $\nu(\gamma):\Gamma \to [0,1]$ that represent TV, FV and indeterminacy of $\gamma \in \Gamma$ respectively satisfy the condition $0 \le \mu(\gamma)^3 + \upsilon(\gamma)^3 \le 1$ and $0 \le \pi(\gamma) \le 1$. Therefore for FNS, $0 \le \mu(\gamma)^3 + \pi(\gamma)^3 + \upsilon(\gamma)^3 \le 2$ for all $\gamma \in \Gamma$. The component $(\mu_\gamma, \pi_\gamma, \nu_\gamma)$ of FNS is called Fermatean neutrosophic number (FNN).

Definition 4

For a FNN, $F_\gamma = (\mu_\gamma, \pi_\gamma, \nu_\gamma)$ we define

(i) The complement as $F_\gamma^c = (\nu_\gamma, 1 - \pi_\gamma, \mu_\gamma)$.
(ii) An extended score value as $S(F_\gamma) = \mu_\gamma^3(1 + \pi_\gamma^3(1 - \mu_\gamma^3 - \nu_\gamma^3))$.
(iii) The accuracy function as $A_c(F_\gamma) = \mu_\gamma^3 + \pi_\gamma^3 + \nu_\gamma^3$.

Definition 5

For a FNN, $F_{\gamma_1} = (\mu_{\gamma_1}, \pi_{\gamma_1}, \nu_{\gamma_1})$ and $F_{\gamma_2} = (\mu_{\gamma_2}, \pi_{\gamma_2}, \nu_{\gamma_2})$ be any two FNSs and let $\delta > 0$. Then the operations are defined as follows

(i) $F_{\gamma_1} \cup F_{\gamma_2} = (\max(\mu_{\gamma_1}, \mu_{\gamma_2}), \min(\pi_{\gamma_1}, \pi_{\gamma_2}), \min(\nu_{\gamma_1}, \nu_{\gamma_2}))$
(ii) $F_{\gamma_1} \cap F_{\gamma_2} = (\min(\mu_{\gamma_1}, \mu_{\gamma_2}), \max(\pi_{\gamma_1}, \pi_{\gamma_2}), \max(\nu_{\gamma_1}, \nu_{\gamma_2}))$
(iii) $F_{\gamma_1} \oplus F_{\gamma_2} = \left(\sqrt[3]{\mu_{\gamma_1}^3 + \mu_{\gamma_2}^3 - \mu_{\gamma_1}^3 \mu_{\gamma_2}^3}, \pi_{\gamma_1}\pi_{\gamma_2}, \nu_{\gamma_1}\nu_{\gamma_2}\right)$
(iv) $F_{\gamma_1} \otimes F_{\gamma_2} = \left(\mu_{\gamma_1}\mu_{\gamma_2}, \sqrt[3]{\pi_{\gamma_1}^3 + \pi_{\gamma_2}^3 - \pi_{\gamma_1}^3 \pi_{\gamma_2}^3}, \sqrt[3]{\nu_{\gamma_1}^3 + \nu_{\gamma_2}^3 - \nu_{\gamma_1}^3 \nu_{\gamma_2}^3}\right)$

(v) $\delta F_{\gamma} = \left(\sqrt[3]{1-\left(1-\mu_{\gamma}^{3}\right)^{\delta}}, \pi_{\gamma}^{\delta}, \upsilon_{\gamma}^{\delta}\right)$

(vi) $F_{\gamma}^{\delta} = \left(\mu_{\gamma}^{\delta}, \sqrt[3]{1-\left(1-\pi_{\gamma}^{3}\right)^{\delta}}, \sqrt[3]{1-\left(1-\nu_{\gamma}^{3}\right)^{\delta}}\right)$

Definition 6

Bonferroni mean operator

Let $\rho,\sigma \geq 0$, and $\alpha_1,\alpha_2,\alpha_3,\ldots,\alpha_k$ be non-negative numbers. Then Bonferroni mean (BM) is defined as

$$BM^{\rho,\sigma}(\alpha_1, \alpha_2, \alpha_3, \ldots, \alpha_k) = \left(\frac{1}{k(k-1)}\sum_{\substack{\psi,\eta=1 \\ \psi\neq\eta}}^{k} \alpha_{\psi}^{\rho}\alpha_{\eta}^{\sigma}\right)^{\frac{1}{\rho+\sigma}}$$

FERMATEAN NEUTROSOPHIC NORMALIZED WEIGHTED BONFERRONI MEAN OPERATOR (FNNWBMO)

In this section we define FNNWBM. Also its properties are examined.

Definition 7

Let $w = (w_1, w_2, \ldots, w_k)^T$ be the weight vector of FNNs, $fn_i = \left(\mu_{fn_i}, \pi_{fn_i}, \nu_{fn_i}\right), i=1,2,3,\ldots,k,$ where w_i is the importance degree of fn_i with $w_i \geq 0$ and

$$\sum_{i=1}^{k} w_i = 1.$$

The Fermatean Neutrosophic Normalized Weighted Bonferroni Mean Operator (FNNWBM) denoted by

$$FNNWBM^{\alpha,\beta}(fn_1, fn_2, \cdots, fn_k) = \left(\bigoplus_{\substack{\psi,\eta=1 \\ \psi\neq\eta}}^{k} \frac{w_{\psi}w_{\eta}}{1-w_{\psi}}\left(fn_{\psi}{}^{\alpha} \otimes fn_{\eta}{}^{\beta}\right)\right)^{\frac{1}{\alpha+\beta}} \tag{1}$$

Theorem 1

For $\psi,\eta > 0$ and a collection of FNNs $fn_i = \left(\mu_{fn_i}, \pi_{fn_i}, \nu_{fn_i}\right), i=1,2,3,\ldots,k,$ the aggregation FNNWBM is also a FNN and it is of the form

$$FNNWBM^{\alpha,\beta}(fn_1, fn_2, \cdots, fn_k) = \left(\sqrt[3]{\left(1 - \prod_{\substack{\psi,\eta=1 \\ \psi\neq\eta}}^{k}\left(1 - \mu_{fn_{\psi}}^{3\alpha}\mu_{fn_{\eta}}^{3\beta}\right)^{\frac{\varpi_{\psi}\varpi_{\eta}}{1-\varpi_{\psi}}}\right)^{\frac{1}{\alpha+\beta}}},\right.$$

$$\sqrt[3]{1-\left(1-\prod_{\substack{\psi,\eta=1\\ \psi\neq\eta}}^{k}\left(1-(1-\pi_{fn_\psi}^{3})^{\alpha}(1-\pi_{fn_\eta}^{3})^{\beta}\right)^{\frac{\varpi_\psi\varpi_\eta}{1-\varpi_\psi}}\right)^{\frac{1}{\alpha+\beta}}},$$

$$\left.\sqrt[3]{1-\left(1-\prod_{\substack{\psi,\eta=1\\ \psi\neq\eta}}^{k}\left(1-(1-\nu_{fn_\psi}^{3})^{\alpha}(1-\nu_{fn_\eta}^{3})^{\beta}\right)^{\frac{\varpi_\psi\varpi_\eta}{1-\varpi_\psi}}\right)^{\frac{1}{\alpha+\beta}}}\right) \quad (2)$$

Proof. By Definition 6,

$$fn_\psi^\alpha = \left(\mu_{fn_\psi}^\alpha, \sqrt[3]{1-(1-\pi_{fn_\psi}^3)^\alpha}, \sqrt[3]{1-(1-\nu_{fn_\psi}^3)^\alpha}\right)$$

and

$$fn_\eta^\beta = \left(\mu_{fn_\eta}^\beta, \sqrt[3]{1-(1-\pi_{fn_\eta}^3)^\beta}, \sqrt[3]{1-(1-\nu_{fn_\eta}^3)^\beta}\right)$$

gives

$$fn_\psi^\alpha \otimes fn_\eta^\beta = \left(\mu_{fn_\psi}^\alpha \mu_{fn_\eta}^\beta, \sqrt[3]{1-(1-\pi_{fn_\psi}^3)^\alpha(1-\pi_{fn_\eta}^3)^\beta}, \sqrt[3]{1-(1-\nu_{fn_\psi}^3)^\alpha(1-\nu_{fn_\eta}^3)^\beta}\right)$$

First we shall prove that

$$\bigoplus_{\substack{\psi,\eta=1\\ \psi\neq\eta}}^{k} \frac{\varpi_\psi\varpi_\eta}{1-\varpi_\psi}\left(fn_\psi{}^\alpha \otimes fn_\eta{}^\beta\right) =$$

$$\left(\sqrt[3]{1-\prod_{\substack{\psi,\eta=1\\ \psi\neq\eta}}^{k}\left(1-\mu_{fn_\psi}^{3\alpha}\mu_{fn_\eta}^{3\beta}\right)^{\frac{\varpi_\psi\varpi_\eta}{1-\varpi_\psi}}}, \sqrt[3]{\prod_{\substack{\psi,\eta=1\\ \psi\neq\eta}}^{k}\left(1-(1-\pi_{fn_\psi}^3)^\alpha(1-\pi_{fn_\eta}^3)^\beta\right)^{\frac{\varpi_\psi\varpi_\eta}{1-\varpi_\psi}}},\right.$$

$$\left.\sqrt[3]{\prod_{\substack{\psi,\eta=1\\ \psi\neq\eta}}^{k}\left(1-(1-\nu_{fn_\psi}^3)^\alpha(1-\nu_{fn_\eta}^3)^\beta\right)^{\frac{\varpi_\psi\varpi_\eta}{1-\varpi_\psi}}}\right) \quad (3)$$

Using mathematical induction.
When $k = 2$,

$$\bigoplus_{\substack{\psi,\eta=1\\ \psi\neq\eta}}^{2}\frac{\varpi_\psi\varpi_\eta}{1-\varpi_\psi}\left(fn_\psi{}^{\alpha}\otimes fn_\eta{}^{\beta}\right) = \frac{\varpi_1\varpi_2}{1-\varpi_1}\left(fn_1{}^{\alpha}\otimes fn_2{}^{\beta}\right) \oplus \frac{\varpi_2\varpi_1}{1-\varpi_2}\left(fn_2{}^{\alpha}\otimes fn_1{}^{\beta}\right)$$

$$= \left(\sqrt[3]{1-\prod_{\substack{\psi,\eta=1\\ \psi\neq\eta}}^{2}\left(1-\mu_{fn_\psi}^{3\alpha}\mu_{fn_\eta}^{3\beta}\right)^{\frac{\varpi_\psi\varpi_\eta}{1-\varpi_\psi}}}, \sqrt[3]{\prod_{\substack{\psi,\eta=1\\ \psi\neq\eta}}^{2}\left(1-(1-\pi_{fn_\psi}^{3})^{\alpha}(1-\pi_{fn_\eta}^{3})^{\beta}\right)^{\frac{\varpi_\psi\varpi_\eta}{1-\varpi_\psi}}},\right.$$

$$\left.\sqrt[3]{\prod_{\substack{\psi,\eta=1\\ \psi\neq\eta}}^{2}\left(1-(1-\nu_{fn_\psi}^{3})^{\alpha}(1-\nu_{fn_\eta}^{3})^{\beta}\right)^{\frac{\varpi_\psi\varpi_\eta}{1-\varpi_\psi}}}\right)$$

The Equation (3) is true for $k = 2$.
Let us assume that it holds for $k = n$.

$$\bigoplus_{\substack{\psi,\eta=1\\ \psi\neq\eta}}^{n}\frac{\varpi_\psi\varpi_\eta}{1-\varpi_\psi}\left(fn_\psi{}^{\alpha}\otimes fn_\eta{}^{\beta}\right) = \left(\sqrt[3]{1-\prod_{\substack{\psi,\eta=1\\ \psi\neq\eta}}^{n}\left(1-\mu_{fn_\psi}^{3\alpha}\mu_{fn_\eta}^{3\beta}\right)^{\frac{\varpi_\psi\varpi_\eta}{1-\varpi_\psi}}},\right.$$

$$\left.\sqrt[3]{\prod_{\substack{\psi,\eta=1\\ \psi\neq\eta}}^{n}\left(1-(1-\pi_{fn_\psi}^{3})^{\alpha}(1-\pi_{fn_\eta}^{3})^{\beta}\right)^{\frac{\varpi_\psi\varpi_\eta}{1-\varpi_\psi}}}, \sqrt[3]{\prod_{\substack{\psi,\eta=1\\ \psi\neq\eta}}^{n}\left(1-(1-\nu_{fn_\psi}^{3})^{\alpha}(1-\nu_{fn_\eta}^{3})^{\beta}\right)^{\frac{\varpi_\psi\varpi_\eta}{1-\varpi_\psi}}}\right) \quad (4)$$

Then for $k = n+ 1$.

$$\bigoplus_{\substack{\psi,\eta=1\\ \psi\neq\eta}}^{n+1}\frac{\varpi_\psi\varpi_\eta}{1-\varpi_\psi}\left(fn_i{}^{\alpha}\otimes fn_j{}^{\beta}\right) = \left(\bigoplus_{\substack{\psi,\eta=1\\ \psi\neq\eta}}^{n}\frac{\varpi_\psi\varpi_\eta}{1-\varpi_\psi}\left(fn_\psi{}^{\alpha}\otimes fn_\eta{}^{\beta}\right)\right)$$

$$\oplus \left(\bigoplus_{\psi=1}^{n}\frac{\varpi_i\varpi_{n+1}}{1-\varpi_\psi}\left(fn_\psi{}^{\alpha}\otimes fn_{n+1}{}^{\beta}\right)\right) \oplus \left(\bigoplus_{\eta=1}^{n}\frac{\varpi_{n+1}\varpi_j}{1-\varpi_{n+1}}\left(fn_{n+1}{}^{\alpha}\otimes fn_\eta{}^{\beta}\right)\right) \quad (5)$$

By Definition (6) we have

$$\left(\bigoplus_{\psi=1}^{n}\frac{\varpi_\psi\varpi_{n+1}}{1-\varpi_\psi}\left(fn_\psi{}^{\alpha}\otimes fn_{n+1}{}^{\beta}\right)\right) = \left(\sqrt[3]{1-\prod_{\psi=1}^{n}\left(1-\mu_{fn_\psi}^{3\alpha}\mu_{fn_{n+1}}^{3\beta}\right)^{\frac{\varpi_\psi\varpi_{n+1}}{1-\varpi_\psi}}},\right.$$

$$\left.\sqrt[3]{\prod_{\psi=1}^{n}\left(1-(1-\pi_{fn_\psi}^{3})^{\rho}(1-\pi_{fn_{n+1}}^{3})^{\sigma}\right)^{\frac{\varpi_\psi\varpi_{n+1}}{1-\varpi_\psi}}}, \sqrt[3]{\prod_{\psi=1}^{n}\left(1-(1-\nu_{fn_\psi}^{3})^{\alpha}(1-\nu_{fn_{n+1}}^{3})^{\beta}\right)^{\frac{\varpi_\psi\varpi_{n+1}}{1-\varpi_\psi}}}\right) \quad (6)$$

and

$$\left(\bigoplus_{\eta=1}^{n}\frac{\varpi_{n+1}\varpi_\eta}{1-\varpi_{n+1}}\left(fn_{n+1}{}^{\alpha}\otimes fn_\eta{}^{\beta}\right)\right) = \left(\sqrt[3]{1-\prod_{\psi=1}^{n}\left(1-\mu_{fn_{n+1}}^{3\alpha}\,\theta_{fn_\eta}^{3\beta}\right)^{\frac{\varpi_{n+1}\varpi_\eta}{1-\varpi_{n+1}}}},\right.$$

$$\sqrt[3]{\prod_{\psi=1}^{n}\left(1-(1-\pi_{fn_{n+1}}^3)^\alpha(1-\pi_{fn_\eta}^3)^\beta\right)^{\frac{\varpi_{n+1}\varpi_\eta}{1-\varpi_{n+1}}}},\sqrt[3]{\prod_{\psi=1}^{n}\left(1-(1-\nu_{fn_{n+1}}^3)^\alpha(1-\nu_{fn_\eta}^3)^\beta\right)^{\frac{\varpi_{n+1}\varpi_\eta}{1-\varpi_{n+1}}}}\Bigg) \quad (7)$$

Then from Equations (4) to (7), we get

$$\bigoplus_{\substack{\psi,\eta=1\\ \psi\neq\eta}}^{n+1}\frac{\varpi_\psi\varpi_\eta}{1-\varpi_\psi}\left(fn_\psi{}^\alpha\otimes fn_\eta{}^\beta\right)=\Bigg(\sqrt[3]{1-\prod_{\substack{\psi,\eta=1\\ \psi\neq\eta}}^{n}\left(1-\mu_{fn_\psi}^{3\alpha}\mu_{fn_{n+1}}^{3\beta}\right)^{\frac{\varpi_\psi\varpi_\eta}{1-\varpi_\psi}}},$$

$$\sqrt[3]{\prod_{\substack{\psi,\eta=1\\ \psi\neq\eta}}^{n}\left(1-(1-\pi_{fn_\psi}^3)^\alpha(1-\pi_{fn_{n+1}}^3)^\beta\right)^{\frac{\varpi_\psi\varpi_\eta}{1-\varpi_\psi}}},\sqrt[3]{\prod_{\substack{\psi,\eta=1\\ \psi\neq\eta}}^{n}\left(1-(1-\nu_{fn_\psi}^3)^\alpha(1-\nu_{fn_{n+1}}^3)^\sigma\right)^{\frac{\varpi_\psi\varpi_\eta}{1-\varpi_\psi}}}\Bigg)$$

$$\oplus\Bigg(\sqrt[3]{1-\prod_{\psi=1}^{n}\left(1-\mu_{fn_\psi}^{3\alpha}\mu_{fn_{n+1}}^{3\beta}\right)^{\frac{\varpi_\psi\varpi_{n+1}}{1-\varpi_\psi}}},\sqrt[3]{\prod_{\psi=1}^{n}\left(1-(1-\pi_{fn_\psi}^3)^\alpha(1-\pi_{fn_{n+1}}^3)^\beta\right)^{\frac{\varpi_\psi\varpi_{n+1}}{1-\varpi_\psi}}},$$

$$\sqrt[3]{\prod_{\psi=1}^{n}\left(1-(1-\nu_{fn_\psi}^3)^\alpha(1-\nu_{fn_{n+1}}^3)^\beta\right)^{\frac{\varpi_\psi\varpi_{n+1}}{1-\varpi_\psi}}}\Bigg)\oplus\Bigg(\sqrt[3]{1-\prod_{\eta=1}^{n}\left(1-\mu_{fn_{n+1}}^{3\alpha}\mu_{fn_j}^{3\beta}\right)^{\frac{\varpi_{n+1}\varpi_\eta}{1-\varpi_{n+1}}}},$$

$$\sqrt[3]{\prod_{\eta=1}^{n}\left(1-(1-\pi_{fn_{n+1}}^3)^\alpha(1-\pi_{fn_\eta}^3)^\beta\right)^{\frac{\varpi_{n+1}\varpi_\eta}{1-\varpi_{n+1}}}},\sqrt[3]{\prod_{\eta=1}^{n}\left(1-(1-\nu_{fn_{n+1}}^3)^\alpha(1-\nu_{fn_\eta}^3)^\beta\right)^{\frac{\varpi_{n+1}\varpi_\eta}{1-\varpi_{n+1}}}}\Bigg)$$

$$=\Bigg(\sqrt[3]{1-\prod_{\substack{\psi,\eta=1\\ \psi\neq\eta}}^{n+1}\left(1-\mu_{fn_\psi}^{3\alpha}\mu_{fn_\eta}^{3\beta}\right)^{\frac{\varpi_\psi\varpi_\eta}{1-\varpi_\psi}}},\sqrt[3]{\prod_{\substack{\psi,\eta=1\\ \psi\neq\eta}}^{n+1}\left(1-(1-\pi_{fn_\psi}^3)^\alpha(1-\pi_{fn_\eta}^3)^\beta\right)^{\frac{\varpi_\psi\varpi_\eta}{1-\varpi_\psi}}},$$

$$\sqrt[3]{\prod_{\substack{\psi,\eta=1\\ \psi\neq\eta}}^{n+1}\left(1-(1-\nu_{fn_\psi}^3)^\alpha(1-\nu_{fn_\eta}^3)^\beta\right)^{\frac{\varpi_\psi\varpi_\eta}{1-\varpi_\psi}}}\Bigg)$$

Thus the Equation (3) is true for $k=n+1$ and hence for every k. Hence by Equation (1) we get Equation (2).

Since

$$0\leq\sqrt[3]{\left(1-\prod_{\substack{\psi,\eta=1\\ \psi\neq\eta}}^{k}\left(1-\mu_{fn_\psi}^{3\alpha}\mu_{fn_\eta}^{3\beta}\right)^{\frac{\varpi_\psi\varpi_\eta}{1-\varpi_\psi}}\right)^{\frac{1}{\alpha+\beta}}}\leq 1,$$

$$0\leq\sqrt[3]{1-\left(1-\prod_{\substack{\psi,\eta=1\\ \psi\neq\eta}}^{k}\left(1-(1-\pi_{fn_\psi}^3)^\alpha(1-\pi_{fn_\eta}^3)^\beta\right)^{\frac{\varpi_\psi\varpi_\eta}{1-\varpi_\psi}}\right)^{\frac{1}{\alpha+\beta}}}\leq 1$$

And

$$0\leq\sqrt[3]{1-\left(1-\prod_{\substack{\psi,\eta=1\\ \psi\neq\eta}}^{k}\left(1-(1-\nu_{fn_\psi}^3)^\alpha(1-\nu_{fn_\eta}^3)^\beta\right)^{\frac{\varpi_\psi\varpi_\eta}{1-\varpi_\psi}}\right)^{\frac{1}{\alpha+\beta}}}\leq 1$$

We have

$$\left(\sqrt[3]{\left(1 - \prod_{\substack{\psi,\eta=1 \\ \psi\neq\eta}}^{k}\left(1 - \mu_{fn_\psi}^{3\alpha}\mu_{fn_\eta}^{3\beta}\right)^{\frac{\varpi_\psi\varpi_\eta}{1-\varpi_\psi}}\right)^{\frac{1}{\alpha+\beta}}}\right)^3$$

$$+ \left(\sqrt[3]{1 - \left(1 - \prod_{\substack{\psi,\eta=1 \\ \psi\neq\eta}}^{k}\left(1 - (1 - \nu_{fn_\psi}^{3})^{\alpha}(1 - \nu_{fn_\eta}^{3})^{\beta}\right)^{\frac{\varpi_\psi\varpi_\eta}{1-\varpi_\psi}}\right)^{\frac{1}{\alpha+\beta}}}\right)^3$$

$$= \left(1 - \prod_{\substack{\psi,\eta=1 \\ \psi\neq\eta}}^{k}\left(1 - \mu_{fn_\psi}^{3\alpha}\mu_{fn_\eta}^{3\beta}\right)^{\frac{\varpi_\psi\varpi_\eta}{1-\varpi_\psi}}\right)^{\frac{1}{\alpha+\beta}}$$

$$+ 1 - \left(1 - \prod_{\substack{\psi,\eta=1 \\ \psi\neq\eta}}^{k}\left(1 - (1 - \nu_{fn_\psi}^{3})^{\alpha}(1 - \nu_{fn_\eta}^{3})^{\beta}\right)^{\frac{\varpi_\psi\varpi_\eta}{1-\varpi_\psi}}\right)^{\frac{1}{\alpha+\beta}}$$

$$\leq \left(1 - \prod_{\substack{\psi,\eta=1 \\ \psi\neq\eta}}^{k}\left(1 - (1 - (1 - \nu_{fn_\psi}^{3})^{\alpha}(1 - \nu_{fn_\eta}^{3})^{\beta}\right)^{\frac{\varpi_\psi\varpi_\eta}{1-\varpi_\psi}}\right)^{\frac{1}{\alpha+\beta}}$$

$$+1 - \left(1 - \prod_{\substack{\psi,\eta=1 \\ \psi\neq\eta}}^{k}\left(1 - (1 - \eta_{fn_\psi}^{2})^{\alpha}(1 - \eta_{fn_\eta}^{2})^{\beta}\right)^{\frac{\varpi_\psi\varpi_\eta}{1-\varpi_\psi}}\right)^{\frac{1}{\alpha+\beta}} = 1$$

and

$$0 \leq \left(\sqrt[3]{\left(1 - \prod_{\substack{\psi,\eta=1 \\ \psi\neq\eta}}^{k}\left(1 - \mu_{fn_\psi}^{3\alpha}\mu_{fn_\eta}^{3\beta}\right)^{\frac{\varpi_\psi\varpi_\eta}{1-\varpi_\psi}}\right)^{\frac{1}{\alpha+\beta}}}\right)^3$$

$$+ \left(\sqrt[3]{1 - \left(1 - \prod_{\substack{\psi,\eta=1 \\ \psi\neq\eta}}^{k}\left(1 - (1 - \pi_{fn_\psi}^{3})^{\alpha}(1 - \pi_{fn_\eta}^{3})^{\beta}\right)^{\frac{\varpi_\psi\varpi_\eta}{1-\varpi_\psi}}\right)^{\frac{1}{\alpha+\beta}}}\right)^3$$

$$+\left(\sqrt[3]{1 - \left(1 - \prod_{\substack{\psi,\eta=1 \\ \psi\neq\eta}}^{k}\left(1 - (1 - \nu_{fn_\psi}^{3})^{\alpha}(1 - \nu_{fn_\eta}^{3})^{\beta}\right)^{\frac{\varpi_\psi\varpi_\eta}{1-\varpi_\psi}}\right)^{\frac{1}{\alpha+\beta}}}\right)^3 \leq 2 \qquad (8)$$

Which fulfill the proof of the theorem.

Also FNNWBM has the following properties.

Property 1: (Idempotency) If the FNNs, $fn_i = fn = \left(\mu_{fn}, \pi_{fn}, \nu_{fn}\right)$, $i=1,2,\ldots,k$, then

$$FNNWBM^{\alpha,\beta}(fn_1, fn_2, \cdots, fn_k) = fn$$

Proof. By definition 7,

$$FNNWBM^{\alpha,\beta}(fn, fn, \cdots, fn) = \left(\bigoplus_{\substack{\psi,\eta=1 \\ \psi\neq\eta}}^{k} \frac{w_\psi w_\eta}{1 - w_\psi} (fn^\alpha \otimes fn^\beta) \right)^{\frac{1}{\alpha+\beta}}$$

$$= \left((fn^\alpha \otimes fn^\beta) \sum_{\substack{\psi,\eta=1 \\ \psi\neq\eta}}^{k} \frac{w_\psi w_\eta}{1 - w_\psi} \right)^{\frac{1}{\alpha+\beta}} = fn \left(\sum_{\psi=1}^{k} w_\psi \sum_{\substack{\eta=1 \\ \psi\neq\eta}}^{k} \frac{w_\eta}{1 - w_\psi} \right)^{\frac{1}{\alpha+\beta}} = fn$$

Which completes the proof.

Property 2: (Monotonicity) If $fn_{ai} = (\mu_{fn_{ai}}, \pi_{fn_{ai}}, \nu_{fn_{ai}}), i=1,2,\ldots,k$ and $fn_{bi} = (\mu_{fn_{bi}}, \pi_{fn_{bi}}, \nu_{fn_{bi}}), i=1,2,\ldots,k$ are two collection of FFNs with $\mu_{fn_{ai}} \leq \mu_{fn_{bi}}$, $\pi_{fn_{ai}} \geq \pi_{fn_{bi}}$, $\nu_{fn_{ai}} \geq \nu_{fn_{bi}}$ then

$$FNNWBM^{\alpha,\beta}(fn_{a1}, fn_{a2}, \cdots, fn_{ak}) \leq FNNWBM^{\alpha,\beta}(fn_{b1}, fn_{b2}, \cdots, fn_{bk}) \quad (9)$$

Proof: $\mu_{fn_{ai}} \leq \mu_{fn_{bi}}$ implies $\mu_{fn_{a\psi}}^{3\alpha} \mu_{fn_{a\eta}}^{3\beta} \leq \mu_{fn_{b\psi}}^{3\alpha} \mu_{fn_{b\eta}}^{3\beta}$

$$\sqrt[3]{\prod_{\substack{\psi,\eta=1 \\ \psi\neq\eta}}^{k} \left(1 - \mu_{fn_{a\psi}}^{3\alpha} \mu_{fn_{a\eta}}^{3\beta}\right)^{\frac{\varpi_\psi \varpi_\eta}{1-\varpi_\psi}}} \geq \sqrt[3]{\prod_{\substack{\psi,\eta=1 \\ \psi\neq\eta}}^{k} \left(1 - \mu_{fn_{b\psi}}^{3\alpha} \mu_{fn_{b\eta}}^{3\beta}\right)^{\frac{\varpi_\psi \varpi_\eta}{1-\varpi_\psi}}}$$

$$\sqrt[3]{1 - \prod_{\substack{\psi,\eta=1 \\ \psi\neq\eta}}^{k} \left(1 - \mu_{fn_{a\psi}}^{3\alpha} \mu_{fn_{a\eta}}^{3\beta}\right)^{\frac{\varpi_\psi \varpi_\eta}{1-\varpi_\psi}}} \leq \sqrt[3]{1 - \prod_{\substack{\psi,\eta=1 \\ \psi\neq\eta}}^{k} \left(1 - \mu_{fn_{b\psi}}^{3\alpha} \mu_{fn_{b\eta}}^{3\beta}\right)^{\frac{\varpi_\psi \varpi_\eta}{1-\varpi_\psi}}}$$

$$\sqrt[3]{\left(1 - \prod_{\substack{\psi,\eta=1 \\ \psi\neq\eta}}^{k} \left(1 - \mu_{fn_{a\psi}}^{3\alpha} \mu_{fn_{a\eta}}^{3\beta}\right)^{\frac{\varpi_\psi \varpi_\eta}{1-\varpi_\psi}}\right)^{\frac{1}{\alpha+\beta}}} \leq \sqrt[3]{\left(1 - \prod_{\substack{\psi,\eta=1 \\ \psi\neq\eta}}^{k} \left(1 - \mu_{fn_{b\psi}}^{3\alpha} \mu_{fn_{b\eta}}^{3\beta}\right)^{\frac{\varpi_\psi \varpi_\eta}{1-\varpi_\psi}}\right)^{\frac{1}{\alpha+\beta}}}$$

Since $\pi_{fn_{ai}} \geq \pi_{fn_{bi}}$

$$\sqrt[3]{\prod_{\substack{\psi,\eta=1 \\ \psi\neq\eta}}^{k} \left(1 - (1 - \pi_{fn_{a\psi}}^{3})^{\alpha} (1 - \pi_{fn_{a\eta}}^{3})^{\beta}\right)^{\frac{\varpi_\psi \varpi_\eta}{1-\varpi_\psi}}} \geq \sqrt[3]{\prod_{\substack{\psi,\eta=1 \\ \psi\neq\eta}}^{k} \left(1 - (1 - \pi_{fn_{b\psi}}^{3})^{\alpha} (1 - \pi_{fn_{b\eta}}^{3})^{\beta}\right)^{\frac{\varpi_\psi \varpi_\eta}{1-\varpi_\psi}}}$$

$$\sqrt[3]{\left(1 - \prod_{\substack{\psi,\eta=1 \\ \psi\neq\eta}}^{k} \left(1 - (1 - \pi_{fn_{a\psi}}^{3})^{\alpha} (1 - \pi_{fn_{a\eta}}^{3})^{\beta}\right)^{\frac{\varpi_\psi \varpi_\eta}{1-\varpi_\psi}}\right)^{\frac{1}{\alpha+\beta}}}$$

$$\leq \sqrt[3]{\left(1 - \prod_{\substack{\psi,\eta=1 \\ \psi\neq\eta}}^{k} \left(1 - (1 - \pi_{fn_{b\psi}}^{3})^{\alpha} (1 - \pi_{fn_{b\eta}}^{3})^{\beta}\right)^{\frac{\varpi_\psi \varpi_\eta}{1-\varpi_\psi}}\right)^{\frac{1}{\alpha+\beta}}}$$

$$\sqrt[3]{1-\left(1-\prod_{\substack{\psi,\eta=1\\ \psi\neq\eta}}^{k}\left(1-\left(1-\pi_{fn_{a\psi}}^{3}\right)^{\alpha}\left(1-\pi_{fn_{a\eta}}^{3}\right)^{\beta}\right)^{\frac{\varpi_{\psi}\varpi_{\eta}}{1-\varpi_{\psi}}}\right)^{\frac{1}{\alpha+\beta}}}$$

$$\geq\sqrt[3]{1-\left(1-\prod_{\substack{\psi,\eta=1\\ \psi\neq\eta}}^{k}\left(1-\left(1-\pi_{fn_{b\psi}}^{3}\right)^{\alpha}\left(1-\pi_{fn_{b\eta}}^{3}\right)^{\beta}\right)^{\frac{\varpi_{\psi}\varpi_{\eta}}{1-\varpi_{\psi}}}\right)^{\frac{1}{\alpha+\beta}}}$$

Since $\nu_{fn_{ai}}\geq\nu_{fn_{bi}}$

$$\sqrt[3]{\prod_{\substack{\psi,\eta=1\\ \psi\neq\eta}}^{k}\left(1-\left(1-\nu_{fn_{a\psi}}^{3}\right)^{\alpha}\left(1-\nu_{fn_{a\eta}}^{3}\right)^{\beta}\right)^{\frac{\varpi_{\psi}\varpi_{\eta}}{1-\varpi_{\psi}}}}\geq\sqrt[3]{\prod_{\substack{\psi,\eta=1\\ \psi\neq\eta}}^{k}\left(1-\left(1-\nu_{fn_{b\psi}}^{3}\right)^{\alpha}\left(1-\nu_{fn_{b\eta}}^{3}\right)^{\beta}\right)^{\frac{\varpi_{\psi}\varpi_{\eta}}{1-\varpi_{\psi}}}}$$

$$\sqrt[3]{\left(1-\prod_{\substack{\psi,\eta=1\\ \psi\neq\eta}}^{k}\left(1-\left(1-\nu_{fn_{a\psi}}^{3}\right)^{\alpha}\left(1-\nu_{fn_{a\eta}}^{3}\right)^{\beta}\right)^{\frac{\varpi_{\psi}\varpi_{\eta}}{1-\varpi_{\psi}}}\right)^{\frac{1}{\alpha+\beta}}}$$

$$\leq\sqrt[3]{\left(1-\prod_{\substack{\psi,\eta=1\\ \psi\neq\eta}}^{k}\left(1-\left(1-\nu_{fn_{b\psi}}^{3}\right)^{\alpha}\left(1-\nu_{fn_{b\eta}}^{3}\right)^{\beta}\right)^{\frac{\varpi_{\psi}\varpi_{\eta}}{1-\varpi_{\psi}}}\right)^{\frac{1}{\alpha+\beta}}}$$

$$\sqrt[3]{1-\left(1-\prod_{\substack{\psi,\eta=1\\ \psi\neq\eta}}^{k}\left(1-\left(1-\nu_{fn_{a\psi}}^{3}\right)^{\alpha}\left(1-\nu_{fn_{a\eta}}^{3}\right)^{\beta}\right)^{\frac{\varpi_{\psi}\varpi_{\eta}}{1-\varpi_{\psi}}}\right)^{\frac{1}{\alpha+\beta}}}$$

$$\geq\sqrt[3]{1-\left(1-\prod_{\substack{\psi,\eta=1\\ \psi\neq\eta}}^{k}\left(1-\left(1-\nu_{fn_{b\psi}}^{3}\right)^{\alpha}\left(1-\nu_{fn_{b\eta}}^{3}\right)^{\beta}\right)^{\frac{\varpi_{\psi}\varpi_{\eta}}{1-\varpi_{\psi}}}\right)^{\frac{1}{\alpha+\beta}}}$$

Therefore

$$\left(\sqrt[3]{\left(1-\prod_{\substack{\psi,\eta=1\\ \psi\neq\eta}}^{k}\left(1-\mu_{fn_{a\psi}}^{3\alpha}\mu_{fn_{a\eta}}^{3\beta}\right)^{\frac{\varpi_{\psi}\varpi_{\eta}}{1-\varpi_{\psi}}}\right)^{\frac{1}{\alpha+\beta}}}\right.$$

$$-\sqrt[3]{1-\left(1-\prod_{\substack{\psi,\eta=1\\ \psi\neq\eta}}^{k}\left(1-\left(1-\pi_{fn_{a\psi}}^{3}\right)^{\alpha}\left(1-\pi_{fn_{a\eta}}^{3}\right)^{\beta}\right)^{\frac{\varpi_{\psi}\varpi_{\eta}}{1-\varpi_{\psi}}}\right)^{\frac{1}{\alpha+\beta}}}$$

$$-\left(\sqrt[3]{\left(1-\prod_{\substack{\psi,\eta=1\\ \psi\neq\eta}}^{k}\left(1-\mu_{fn_{b\psi}}^{3\alpha}\mu_{fn_{b\eta}}^{3\beta}\right)^{\frac{\varpi_{\psi}\varpi_{\eta}}{1-\varpi_{\psi}}}\right)^{\frac{1}{\alpha+\beta}}}\right.$$

$$-\sqrt[3]{1-\left(1-\prod_{\substack{\psi,\eta=1\\ \psi\neq\eta}}^{k}\left(1-\left(1-\pi_{fn_{b\psi}}^{3}\right)^{\alpha}\left(1-\pi_{fn_{b\eta}}^{3}\right)^{\beta}\right)^{\frac{\varpi_{\psi}\varpi_{\eta}}{1-\varpi_{\psi}}}\right)^{\frac{1}{\alpha+\beta}}}$$

$$\sqrt[3]{1-\left(1-\prod_{\substack{\psi,\eta=1\\ \psi\neq\eta}}^{k}\left(1-(1-\nu_{fn_{a\psi}}^{3})^{\alpha}(1-\nu_{fn_{a\eta}}^{3})^{\beta}\right)^{\frac{\varpi_\psi\varpi_\eta}{1-\varpi_\psi}}\right)^{\frac{1}{\alpha+\beta}}}$$

$$\leq -\sqrt[3]{1-\left(1-\prod_{\substack{\psi,\eta=1\\ \psi\neq\eta}}^{k}\left(1-(1-\nu_{fn_{b\psi}}^{3})^{\alpha}(1-\nu_{fn_{b\eta}}^{3})^{\beta}\right)^{\frac{\varpi_\psi\varpi_\eta}{1-\varpi_\psi}}\right)^{\frac{1}{\alpha+\beta}}}$$

Property 3: (Commutativity) Consider the set of FNNs $fn_i = (\mu_{fn_i}, \pi_{fn_i}, \nu_{fn_i})$, i=1,2,…,k. If any one permutation of $fn_1, fn_2, \ldots, fn_k \dot{fn}_1, \dot{fn}_2, \ldots, \dot{fn}_k$ then

$$\mathrm{FNNWBM}^{\alpha,\beta}(fn_1, fn_2, \cdots, fn_k) = \mathrm{FNNWBM}^{\alpha,\beta}(\dot{fn}_1, \dot{fn}_2, \cdots \dot{fn}_k) \tag{10}$$

Proof:

Since $\dot{fn}_1, \dot{fn}_2, \cdots \dot{fn}_k$ is any one permutation of $fn_1, fn_2, \ldots, fn_k$

$$FNNWBM^{\alpha,\beta}(fn_1, fn_2, \cdots, fn_k) = \left(\bigoplus_{\substack{\psi,\eta=1\\ \psi\neq\eta}}^{k} \frac{w_\psi w_\eta}{1-w_\psi}\left(fn_\psi^{\alpha} \otimes fn_\eta^{\beta}\right)\right)^{\frac{1}{\alpha+\beta}}$$

$$= \left(\bigoplus_{\substack{\psi,\eta=1\\ \psi\neq\eta}}^{k} \frac{w_\psi w_\eta}{1-w_\psi}\left(\dot{fn}_\psi^{\alpha} \otimes \dot{fn}_\eta^{\beta}\right)\right)^{\frac{1}{\alpha+\beta}} = \mathrm{FNNWBM}^{\alpha,\beta}(\dot{fn}_1, \dot{fn}_2, \cdots \dot{fn}_k)$$

Property 4: (Boundedness) Consider $fn_i = (\mu_{fn_i}, \pi_{fn_i}, \nu_{fn_i})$, i=1,2,3,…,k as a set of FNNs, and let $fn^- = (\min_i\{\mu_{fn_i}\}, \max_i\{\pi_{fn_i}\}, \max_i\{\nu_{fn_i}\})$

$$fn^+ = (\max_i\{\mu_{fn_i}\}, \min_i\{\pi_{fn_i}\}, \min_i\{\nu_{fn_i}\})$$

Then

$$fn^- \leq FNNWBM^{\alpha,\beta}(fn_1, fn_2, \cdots, fn_k) \leq fn^+. \tag{11}$$

Proof:

$$\min_i\{\mu_{fn_i}\} \leq \mu_{fn_i} \leq \max_i\{\mu_{fn_i}\},$$

$$\max_i\{\pi_{fn_i}\} \leq \pi_{fn_i} \leq \min_i\{\pi_{fn_i}\},$$

$$\max_i\{\lambda_{fn_i}\} \leq \lambda_{fn_i} \leq \min_i\{\lambda_{fn_i}\},$$

we have $fn^{-} \leq fn$ and $fn \leq fn^{+}$. By monotonicity of *FNNWBM* operator

$$FNNWBM^{\alpha,\beta}(fn^{-},fn^{-},\cdots,fn^{-}) \leq FNNWBM^{\alpha,\beta}(fn_1,fn_2,\cdots,fn_k)$$

$$FNNWBM^{\alpha,\beta}(fn_1,fn_2,\cdots,fn_k) \leq FNNWBM^{\alpha,\beta}(fn^{+},fn^{+},\cdots,fn^{+})$$

Therefore

$$FNNWBM^{\alpha,\beta}(fn^{-},fn^{-},\cdots,fn^{-}) \leq FNNWBM^{\alpha,\beta}(fn_1,fn_2,\cdots,fn_k) \leq FNNWBM^{\alpha,\beta}(fn^{+},fn^{+},\cdots,fn^{+})$$

Then by idempotency $fn^{-} \leq FNNWBM^{\alpha,\beta}(fn_1,fn_2,\cdots,fn_k) \leq fn^{+}$.

MULTI CRITERIA DECISION MAKING USING FERMATEAN NEUTROSOPHIC NORMALIZED WEIGHTED BONFERRONI MEAN OPERATOR

Algorithm for integrating Fermatean Neutrosophic Sets in the selection of Kabaddi players:

Defining Selection Criteria

Identify the key criteria for assessing Kabaddi player performance. These criteria could include agility, strength, speed, tactical awareness, teamwork, and overall game understanding.

Recruitment of Scouts: Choose scouts based on their expertise, experience, and understanding of Kabaddi dynamics. Ensure that scouts are familiar with the selection criteria and possess the ability to assess players effectively.

Scouting Process: Each scout independently evaluates Kabaddi players across the predefined criteria. Scouts may use qualitative assessments or numerical scales to denote varying degrees of proficiency in each criterion.

Conversion of Qualitative Assessments to Numerical Representations: If qualitative assessments are used, convert them to numerical representations using Fermatean neutrosophic numbers or other appropriate techniques.

Aggregation of Assessments: Integrate assessments from multiple scouts and criteria to compute aggregate scores for each Kabaddi player. Consider the weight assigned to each criterion and scout's judgment in the aggregation process.

Ranking and Analysis: Analyze the aggregate scores to rank Kabaddi players based on their performance. Identify strengths, weaknesses, and potential areas for improvement across different players and skill sets.

Interpretation and Recommendations: Interpret the evaluation results and provide actionable recommendations for player selection or training programs. Highlight standout performers and areas for focused development based on the analysis of assessment data.

Monitor the effectiveness of the player selection process and make adjustments as necessary to ensure the team's competitiveness and success.

Application of Fermatean Neutrosophic Normalized Weighted Bonferroni Mean Operator in Selection of Kabaddi Players

The integration of Fermatean Neutrosophic Sets in the selection of Kabaddi players revolutionizes the approach to talent identification and team building. In this dynamic framework, scouts (S1, S2, S3) are tasked with objectively evaluating the performance of Kabaddi players (P1-P6) across a range of predefined criteria: agility (C1), strength (C2), speed (C3), tactical awareness (C4), teamwork (C5), and overall game understanding (C6). Each criterion serves as a pivotal indicator of a player's suitability for competitive Kabaddi.

Criteria for Selecting Kabaddi Players

C1-Agility: measures a player's ability to change direction swiftly, evade defenders, and execute dynamic maneuvers on the Kabaddi court.
C2-Strength: evaluates a player's physical prowess, including muscular endurance, power, and resilience during physical confrontations.
C3-Speed: assesses a player's quickness and acceleration, crucial for both offensive raids and defensive maneuvers in Kabaddi.
C4-Tactical Awareness: delves into a player's strategic acumen, decision-making abilities, and adaptability to evolving game situations.
C5-Teamwork: encompasses a player's capacity to collaborate effectively with teammates, communicate on the court, and execute coordinated plays.
C6-Overall Game Understanding: serves as a holistic gauge, considering a player's comprehension of Kabaddi rules, nuances, and positional responsibilities.

Throughout the scouting process, scouts (S1, S2, S3) leverage their expertise and insights to conduct a thorough and impartial assessment of Kabaddi players. They meticulously evaluate players across diverse criteria, capturing the intricacies of their performances with precision and objectivity. By employing Fermatean Neutrosophic Sets, the scouts navigate the evaluation landscape adeptly, weighing the significance of each criterion and synthesizing assessments to form comprehensive player profiles.

Recruitment of Scouts

The selection of scouts is a pivotal aspect of the player selection process, ensuring expertise, objectivity, and credibility in player assessments. Several factors guide the recruitment of scouts:

Expertise: Scouts are chosen based on their deep understanding of Kabaddi dynamics, including game strategies, player roles, and tactical nuances. This ensures that assessments are conducted with insight and proficiency.
Experience: Experienced scouts bring invaluable knowledge and perspective to the selection process, having observed and analyzed countless Kabaddi matches and player performances.
Objectivity: It is essential to select scouts who can maintain objectivity and impartiality in player evaluations. They should assess players based on performance metrics and criteria, free from personal biases or affiliations.
Credibility: Scouts must possess credibility and integrity in their assessments, adhering to ethical standards and professional conduct throughout the scouting process.
Communication Skills: Effective communication skills are essential for scouts to articulate their observations, provide constructive feedback, and collaborate with coaching staff and team management.

By recruiting scouts who embody these qualities, the player selection process ensures fairness, accuracy, and transparency in talent identification and team formation.

The integration of Fermatean Neutrosophic Sets in the selection of Kabaddi players heralds a new era of data-driven decision-making and talent development. By leveraging advanced evaluation techniques and objective criteria, teams can identify and nurture top talent, enhancing their competitive edge and performance on the Kabaddi court. Moreover, the player selection process fosters transparency, accountability, and trust within the Kabaddi community, ensuring that players are evaluated fairly and equitably based on their abilities and potential.

Furthermore, the integration of Fermatean Neutrosophic Sets promotes continuous improvement and innovation in player scouting and team selection methodologies. By analyzing assessment data, identifying emerging trends, and incorporating stakeholder feedback, teams can refine their selection criteria, enhance evaluation techniques, and adapt to evolving Kabaddi dynamics. This iterative process of improvement enables teams to stay ahead of the curve, maximize player potential, and achieve sustained success in competitive Kabaddi tournaments and leagues.

The decision makers will give opinion on the players in terms of linguistic term. The FNN corresponding to the FN linguistic term is given in Table 1.

Table 1. Linguistic terms with corresponding fermatean neutrosophic numbers

Linguistic Terms	T	F	I
Extremely good (EG)	1	0	0
Very Very good (VVG)	0.95	0.37	0.32
Very good (VG)	0.85	0.39	0.34
Good (G)	0.76	0.37	0.32
Medium good (MG)	0.64	0.49	0.45
Medium (M)	0.56	0.57	0.52
Medium bad (MB)	0.43	0.69	0.65
Bad (B)	0.38	0.77	0.72
Very bad (VB)	0.28	0.89	0.85
Very very bad (VVB)	0.19	0.99	0.94
Extremely bad (EB)	0	1	0.95

Linguistic term of each players corresponding to the six criteria evaluated by three scouts (S1, S2, S3) are displayed in Table 2.

Table 2. Scouts' decisions in linguistic terms

Scout	Player	C1	C2	C3	C4	C5	C6
S1	P1	EG	M	VVG	VVB	MG	B
	P2	VG	G	MB	MG	VG	G
	P3	M	VVG	EG	B	G	VVB
	P4	MB	G	MG	EG	VG	M
	P5	VVG	B	MB	VVB	EG	G
	P6	VG	MG	G	M	MB	EG
	P7	G	VVB	VVG	MB	VG	B

continued on following page

Table 2. Continued

Scout	Player	C1	C2	C3	C4	C5	C6
S2	P1	MB	M	VG	G	MG	VG
	P2	EG	G	MG	B	VVG	VVB
	P3	M	G	VG	MB	G	MG
	P4	G	B	EG	VVB	EG	VVG
	P5	MB	G	M	VG	VG	MG
	P6	B	VG	EG	VVG	EG	G
	P7	G	VVB	MG	EG	M	MG
S3	P1	VG	VVG	G	MG	VG	B
	P2	MB	G	M	VG	EG	MG
	P3	EG	VG	VVG	B	G	MB
	P4	M	MB	G	VG	MG	EG
	P5	VVB	B	EG	EG	VVG	VG
	P6	VG	G	M	VVB	MB	MG
	P7	B	VG	VVG	G	EG	M

The linguistic terms are replaced by FNN with help of Table 1 and are displayed in Table 3.

Table 3. Fermatean neutrosophic values for linguistic terms

Scout	Player	C1	C2	C3	C4	C5	C6
S1	P1	(1,0,1)	(0.56, 0.57, 0.52)	(0.95, 0.37, 0.32)	(0.19, 0.99, 0.94)	(0.64, 0.37, 0.32)	(0.38, 0.77, 0.72)
	P2	(0.85, 0.39, 0.34)	(0.76, 0.37, 0.32)	(0.43, 0.69, 0.65)	(0.64, 0.37, 0.32)	(0.85, 0.39, 0.34)	(0.76, 0.37, 0.32)
	P3	(0.56, 0.57, 0.52)	(0.95, 0.37, 0.32)	(1,0,1)	(0.38, 0.77, 0.72)	(0.76, 0.37, 0.32)	(0.19, 0.99, 0.94)
	P4	(0.43, 0.69, 0.65)	(0.76, 0.37, 0.32)	(0.64, 0.37, 0.32)	(1,0,1)	(0.85, 0.39, 0.34)	(0.56, 0.57, 0.52)
	P5	(0.95, 0.37, 0.32)	(0.38, 0.77, 0.72)	(0.43, 0.69, 0.65)	(0.19, 0.99, 0.94)	(1,0,1)	(0.76, 0.37, 0.32)
	P6	(0.85, 0.39, 0.34)	(0.64, 0.37, 0.32)	(0.76, 0.37, 0.32)	(0.56, 0.57, 0.52)	(0.43, 0.69, 0.65)	(1,0,1)
	P7	(0.76, 0.37, 0.32)	(0.19, 0.99, 0.94)	(0.95, 0.37, 0.32)	(0.43, 0.69, 0.65)	(0.85, 0.39, 0.34)	(0.38, 0.77, 0.72)
S2	P1	(0.43, 0.69, 0.65)	(0.56, 0.57, 0.52)	(0.85, 0.39, 0.34)	(0.76, 0.37, 0.32)	(0.64, 0.37, 0.32)	(0.85, 0.39, 0.34)
	P2	(1,0,1)	(0.76, 0.37, 0.32)	(0.64, 0.37, 0.32)	(0.38, 0.77, 0.72)	(0.95, 0.37, 0.32)	(0.19, 0.99, 0.94)
	P3	(0.56, 0.57, 0.52)	(0.76, 0.37, 0.32)	(0.85, 0.39, 0.34)	(0.43, 0.69, 0.65)	(0.76, 0.37, 0.32)	(0.64, 0.37, 0.32)
	P4	(0.76, 0.37, 0.32)	(0.38, 0.77, 0.72)	(1,0,1)	(0.19, 0.99, 0.94)	(1,0,1)	(0.95, 0.37, 0.32)
	P5	(0.43, 0.69, 0.65)	(0.76, 0.37, 0.32)	(0.56, 0.57, 0.52)	(0.85, 0.39, 0.34)	(0.85, 0.39, 0.34)	(0.64, 0.37, 0.32)
	P6	(0.38, 0.77, 0.72)	(0.85, 0.39, 0.34)	(1,0,1)	(0.95, 0.37, 0.32)	(1,0,1)	(0.76, 0.37, 0.32)
	P7	(0.76, 0.37, 0.32)	(0.19, 0.99, 0.94)	(0.64, 0.37, 0.32)	(1,0,1)	(0.56, 0.57, 0.52)	(0.64, 0.37, 0.32)
S3	P1	(0.85, 0.39, 0.34)	(0.95, 0.37, 0.32)	(0.76, 0.37, 0.32)	(0.64, 0.37, 0.32)	(0.85, 0.39, 0.34)	(0.38, 0.77, 0.72)
	P2	(0.43, 0.69, 0.65)	(0.76, 0.37, 0.32)	(0.56, 0.57, 0.52)	(0.85, 0.39, 0.34)	(1,0,1)	(0.64, 0.37, 0.32)
	P3	(1,0,1)	(0.85, 0.39, 0.34)	(0.95, 0.37, 0.32)	(0.38, 0.77, 0.72)	(0.76, 0.37, 0.32)	(0.43, 0.69, 0.65)
	P4	(0.56, 0.57, 0.52)	(0.43, 0.69, 0.65)	(0.76, 0.37, 0.32)	(0.85, 0.39, 0.34)	(0.64, 0.37, 0.32)	(1,0,1)
	P5	(0.19, 0.99, 0.94)	(0.38, 0.77, 0.72)	(1,0,1)	(1,0,1)	(0.95, 0.37, 0.32)	(0.85, 0.39, 0.34)
	P6	(0.85, 0.39, 0.34)	(0.76, 0.37, 0.32)	(0.56, 0.57, 0.52)	(0.19, 0.99, 0.94)	(0.43, 0.69, 0.65)	(0.64, 0.37, 0.32)
	P7	(0.38, 0.77, 0.72)	(0.85, 0.39, 0.34)	(0.95, 0.37, 0.32)	(0.76, 0.37, 0.32)	(1,0,1)	(0.56, 0.57, 0.52)

Since there are three scout members for decision making, the average of all the scout members for each player against each criteria are evaluated and it is displayed in Table 4.

Table 4. Fermatean neutrosophic averaging values

	C1	C2	C3	C4	C5	C6
P1	(0.76, 0.36, 0.33)	(0.69, 0.50, 0.45)	(0.85, 0.38, 0.33)	(0.53, 0.62, 0.57)	(0.76, 0.33, 0.30)	(0.59, 0.51, 0.48)
P2	(0.76, 0.36, 0.33)	(0.76, 0.37, 0.32)	(0.54, 0.58, 0.54)	(0.67, 0.42, 0.39)	(0.98, 0.12, 0.11)	(0.53, 0.62, 0.57)
P3	(0.71, 0.38, 0.35)	(0.85, 0.38, 0.33)	(0.93, 0.25, 0.22)	(0.40, 0.74, 0.70)	(0.76, 0.37, 0.32)	(0.42, 0.72, 0.68)
P4	(0.58, 0.54, 0.50)	(0.52, 0.61, 0.56)	(0.80, 0.29, 0.26)	(0.73, 0.33, 0.31)	(0.88, 0.16, 0.15)	(0.84, 0.31, 0.28)
P5	(0.52, 0.68, 0.64)	(0.51, 0.64, 0.59)	(0.66, 0.42, 0.39)	(0.73, 0.33, 0.31)	(0.98, 0.12, 0.11)	(0.80, 0.29, 0.26)
P6	(0.69, 0.52, 0.47)	(0.75, 0.42, 0.37)	(0.77, 0.31, 0.28)	(0.57, 0.64, 0.59)	(0.62, 0.46, 0.43)	(0.80, 0.29, 0.26)
P7	(0.63, 0.50, 0.45)	(0.41, 0.79, 0.74)	(0.85, 0.41, 0.36)	(0.73, 0.35, 0.32)	(0.85, 0.19, 0.17)	(0.53, 0.61, 0.56)

Criterion weights plays an important role in decision making process. The criterion weight of all the six criteria (C1, C2, C3, C4, C5, C6) is (0.2, 0.15, 0.15, 0.2, 0.2, 0.1). The FNN of each player is aggregated through FNNWBM. Here we taken the values of α and β equally as 1. For each player the six criteria are aggregated using FNNWBM given in below Equation (12).

$$FNNWBM^{\alpha,\beta}(fn_1, fn_2, fn_3, fn_4, fn_5, fn_6)$$

$$= \left(\begin{array}{l} \sqrt[3]{\left(1 - \prod_{\substack{\psi,\eta=1 \\ \psi\neq\eta}}^{6} \left(1 - \mu_{fn_\psi}^{3\alpha} \mu_{fn_\eta}^{3\beta}\right)^{\frac{\varpi_\psi \varpi_\eta}{1-\varpi_\psi}}\right)^{\frac{1}{2}}}, \\ \sqrt[3]{1 - \left(1 - \prod_{\substack{\psi,\eta=1 \\ \psi\neq\eta}}^{6} \left(1 - (1 - \pi_{fn_\psi}^{3})^{\alpha} (1 - \pi_{fn_\eta}^{3})^{\beta}\right)^{\frac{\varpi_\psi \varpi_\eta}{1-\varpi_\psi}}\right)^{\frac{1}{2}}}, \\ \sqrt[3]{1 - \left(1 - \prod_{\substack{\psi,\eta=1 \\ \psi\neq\eta}}^{6} \left(1 - (1 - \nu_{fn_\psi}^{3})^{\alpha} (1 - \nu_{fn_\eta}^{3})^{\beta}\right)^{\frac{\varpi_\psi \varpi_\eta}{1-\varpi_\psi}}\right)^{\frac{1}{2}}} \end{array} \right) \quad (12)$$

The FNNBW for α =1, β =1is calculated by using the above Equation (12). The FNN extended score function defined in Definition (4) is used to find the score value of players and it is displayed in Table 5.

Table 5. Ranking of players

Players	FNNBM(α =1, β =1)	Score value	Ranking of the players
P1	(0.64, 0.62, 0.59)	0.289	5
P2	(0.68, 0.59, 0.56)	0.341	1
P3	(0.66, 0.64, 0.23)	0.287	6
P4	(0.67, 0.57, 0.28)	0.301	3
P5	(0.66, 0.62, 0.39)	0.297	4
P6	(0.70, 0.63, 0.30)	0.340	2
P7	(0.64, 0.64, 0.19)	0.263	7

Thus we have P2 > P6 > P4 > P5> P1> P3 > P7.

COMPARISON ANALYSIS

The decisions obtained by using different values of α and β and also using other existing methods are tabulated.

Table 6. Comparison of ranking

α and β values		comparison of the rank of the alternatives						
α	**β**	**P1**	**P2**	**P3**	**P4**	**P5**	**P6**	**P7**
α =1	β =1	5	1	6	3	4	2	7
α =0.5	β =0.5	6	1	5	3	4	2	7
α =2	β =2	6	2	5	3	4	1	7
α =3	β =3	5	2	6	3	4	1	7
Existing Methods								
Kano Method		6	2	5	3	4	1	7
FNBM		6	2	5	3	4	1	7
FNWBM		6	2	5	3	4	1	7

Figure 1. Ranking of players

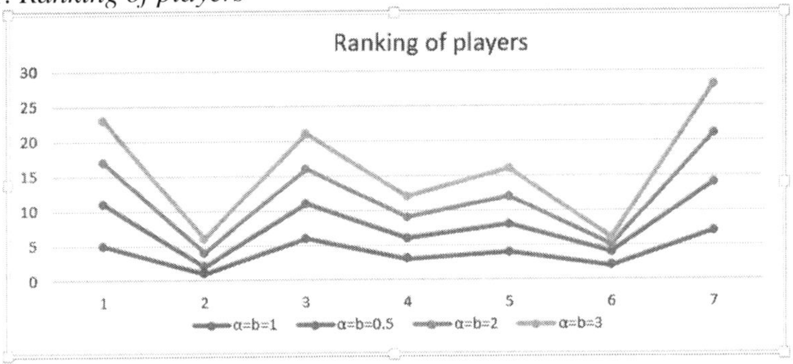

The rank of the players for various values of α and β is displayed in the chart below.

The comparison displayed above illustrates how adjusting values of α and β affect the ranking of alternatives. Thus, the ranking of the alternative is influenced by the interrelationship of the criteria, which is the most important component of FNNWBM approach. When generating a decision based on the interrelation of the criteria, the values of α and β can be selected suitably. In the circumstance only the positive of the criteria is considered for decision making process we can set β as zero.

Limitations

Fermatean neutrosophic sets could be desirable in specific circumstances where the membership and non-membership has wide range and also the indeterminacy occurs. They may not be appropriate for all situations, depending on the unique characteristics at such case alternative mathematical frameworks may be more pertinent. Despite these limitations, Fermatean neutrosophic sets offer an essential basis for articulating and debating ambiguity and indeterminacy in a variety of contexts. Since the suggested FNNWBM can only consider the interrelationships between two criteria but are unsuccessful to capture the interrelationships among three or more than three criteria. To overcome these barriers and improve Fermatean neutrosophic sets efficiency in applications in real life, additional research and development are required.

CONCLUSION AND FUTURE WORK

In the neutrosophic circumstances, the traditional Bonferroni mean operator and possibility degree have been improved to more effectively organize and model the uncertainties and indeterminacy inside multi-attribute decision analysis. The neutrosophic Bonferroni operator can integrate ambiguous, undecided, and uncertain opinions or reviews gathered from numerous decision-makers. In this study we have introduced FNNWBM operator and their properties are investigated. By our proposed FNNWBM operator several types of challenging multiple attribute decision-making problems involving uncertain, inconsistent information and high range of TV and FV can be handled. In this study FNNWBM operator is applied in selecting the best option of players among the available options and its effectiveness is ensured by the comparison analysis. In future We will examine the integration of artificial intelligence and machine learning approaches with the FNNWBM operator. It may also be used in data fusion, decision support systems, and uncertainty modeling, leveraging artificial intelligence skills for better problem-solving.

REFERENCES

Broumi, S. (2023). Fermatean Neutrosophic Matrices and Their Basic Operations. *Neutrosophic Sets and Systems*, *58*(1), 35.

Gonul Bilgin, N., Pamučar, D., & Riaz, M. (2022). Fermatean neutrosophic topological spaces and an application of neutrosophic kano method. *Symmetry*, *14*(11), 2442. DOI: 10.3390/sym14112442

Kanchana, A., Nagarajan, D., Broumi, S., & Prastyo, D. D. (2024). A Comprehensive Analysis of Neutrosophic Bonferroni Mean Operator. *Neutrosophic Sets and Systems*, *64*, 57–76.

Kishorekumar, M., Karpagadevi, M., Mariappan, R., Krishnaprakash, S., & Revathy, A. (2023, February). Interval-valued picture fuzzy geometric Bonferroni mean aggregation operators in multiple attributes. In *2023 Fifth International Conference on Electrical, Computer and Communication Technologies (ICECCT)* (pp. 1-8). IEEE.

Ma, X., Sun, H., Qin, H., Wang, Y., & Zheng, Y. (2024). A novel multi-attribute decision making method based on interval-valued fermatean fuzzy bonferroni mean operators. *Journal of Intelligent & Fuzzy Systems*, 1-21.

Palanikumar, M., Iampan, A., & Broumi, S. (2022). MCGDM based on VIKOR and TOPSIS proposes neutrsophic Fermatean fuzzy soft with aggregation operators. *International Journal of Neutrosophic Science*, *19*(3), 85–94. DOI: 10.54216/IJNS.190308

Revathy, A., Inthumathi, V., Krishnaprakash, S., & Kishorekumar, M. (2023). Fermatean fuzzy normalised Bonferroni mean operator in multi criteria decision making on selection of electric bike. In *2023 Fifth International Conference on Electrical, Computer and Communication Technologies (ICECCT)* (pp. 1-7). IEEE.

Senapati, T., & Yager, R. R. (2020). Fermatean fuzzy sets. *Journal of Ambient Intelligence and Humanized Computing*, *11*(2), 663–674. DOI: 10.1007/s12652-019-01377-0

Smarandache, F. (1998). Neutrosophy: neutrosophic probability, set, and logic: analytic synthesis & synthetic analysis.

Yager, R. R. (1988). On ordered weighted averaging aggregation operators in multicriteria decisionmaking. *IEEE Transactions on Systems, Man, and Cybernetics*, *18*(1), 183–190. DOI: 10.1109/21.87068

Chapter 16
Neutrosophic Optimization and Its Uncertainty Quantification

Srinivasan Vijayabalaji
University College of Engineering, Anna University, Panruti, India

Parthasarathy Balaji
Measi Academy of Architecture, India

Gunalan Venkadesh
Krishnasamy College of Engineering and Technology, Cuddalore, India

ABSTRACT

The purpose of this chapter is to introduce the critical path problem in generalized Fermatean neutrosophic set. An algorithm to find the critical path using generalized Fermatean neutrosophic set is also provided with an example. The chapter is conducted by comparing this work with the existing algorithm to validate the new work. Furthermore, the authors intend to expand the Fermatean neutrosophic set to include previously defined score, accuracy, and certainty functions. In operations research, they want to present an algorithm for detecting the critical path using the Fermatean neutrosophic set and compare the results. They intend to compare the currently available traditional critical path method to the Fermatean neutrosophic set, along with suitable circumstances.

DOI: 10.4018/979-8-3693-3204-7.ch016

INTRODUCTION

Fuzzy sets, an influential framework in uncertainty theory, was introduced by Zadeh (Zadeh, 1965) in 1965. Researchers consider fuzzy sets to be extraordinarily beneficial for the real life application of innovative ideas in several mathematical domains. Operation research (Paneerselvam, R. 2008) provides management with a scientific foundation upon which to formulate effective and expeditious solutions to problems. It attempts to mitigate the risks associated with decision-making by imposing specific conditions. In addition to economic considerations, every individual should take into account all other factors when making decisions, so that their solution is applicable in all respects. Among those that rely on scientific principles to inform decision making, decision theory and mathematical or quantitative measures are the most significant. The individual employs mathematical or quantitative methods to arrive at decisions, as stated previously. A quantitative approach to management problems necessitates the deliberate, logical, systematic, scientific definition, analysis and resolution of decision problems. Such methods should be grounded in data, facts, information, and logic, rather than being predicated on predetermined rules. In the contemporary business environment, individuals are confronted with intricate systems that involve interdependent factors and a criterion for effective system performance that is similarly complex. In such situations, traditional approaches to decision-making are deemed woefully inadequate. The significance of scientific methods in facilitating optimal decision-making cannot be denied. Operations research examines a system and using logical analysis and analytical techniques. The parameters of these constraints are identified, quantified, and relationships are investigated to the greatest extent feasible. From this information, alternative decisions are derived. Operations research rationalizes decision-making by providing a clear picture of potential repercussions. Quantitative methods encompass an extremely comprehensive scope. They are implemented in the problem-solving processes of diverse organizations, including for-profit and non-profit sectors, government entities, and service units. They are applicable to a wide range of issues, including but not limited to making decisions regarding plant location, replacement issues, inventory control, production scheduling, return on investment analysis, portfolio selection, and marketing research. Quantitative methods encompass two critical elements: the accessibility of well-organized models and problem-solving techniques; and a disposition towards scientific inquiry aimed at acquiring further understanding in the realm of organizational management. As a result, the quantitative approach has been implemented not only in conventional business but also in addressing social issues, public policy concerns, national and international issues, and interpersonal issues. In reality, quantitative techniques are applicable to decision-making in every aspect of life and are not restricted to any particular field.

Two factors have contributed to the development of the quantitative approach to management problems. Ongoing research endeavours are concentrated on the exploration and making tools that works better and methodologies to address various types of decision problems. In this pursuit of exploring uncharted territories, attempts have been made to broaden the scope and practical utility of existing techniques. The invention of Linear Programming (LP) and the Simplex approach in 1947 by American Mathematician George B. Dontzig facilitated the advancement of novel methodologies and applications through the collaboration and efforts of academic and industrial stakeholders. An extensive array of Operations research consultants is readily accessible to address a variety of challenges. Similarly, the Operation Research Society of India (1959) assists in the resolution of numerous issues in India. Presently, Operations research methodologies are incorporated into high school curricula. Operations research is indispensable to every organization, especially within the framework of the decision-making process. The need to make decisions is a consequence of our shared existence in a world characterized by diverse aspirations, requirements, and objectives, for which resources can be scarce at times. Each of us competes to utilize these resources in order to achieve our objectives. The decision-making process is divided into two distinct phases. Initially, one is developing the aims and objectives, listing the environmental limitations, and locating and assessing the options make up the first stage. The subsequent phase entails determining the most advantageous course of action given a collection of constraints. The authors (Krishnaprabha et al., 2022) implemented an innovative ranking strategy and the allocation table method (ATM) to tackle an intuitionistic fuzzy transportation problem with interval values. Additionally, they introduced an improved version of the ATM. The researchers (Vijayabalaji & Balaji, 2015) proposed the concept of a rough matrix theory. Using the principles of rough matrix theory, they offer a method for making decisions based on multiple criteria in the game of cricket. The incorporation of Multiple Criteria Decision Making (MCDM), Rough matrices and the Assignment model by the researchers (Vijayabalaji & Balaji, 2020) has inspired the concept of selecting the best 11 performers in all three formats. Samarandache (Samarandache, 2020) introduced the indeterminacy function, true membership and false membership functions to the new concept of the Neutrosophic set, which has proven to be extremely useful for researchers venturing into uncharted territory in order to manage ambiguous and inconsistent situations. In order to applied in authentic scientific methods and technical domains. The concept of the multi-valued neutrosophic set (MVNs) was developed to address situations when the exact degree of truth, indeterminacy, and falsehood of some statements cannot be accurately represented using traditional methods, but rather require the consideration of several interval values. In addition, the author (Samarandache, 2020) collaborates with neutosophic triplets to establish the accuracy and certainty functions and the single-valued neutrosophic number(SVNn) for determining the score function. In the article (Broumi S. et al., 2017) utilized single-valued neutrosophic graphs with score, accuracy and assurance functions to solve the shortest path problem. In this extension, the researchers (Broumi S. et al., 2017) utilize a neutrosophic set with a trapezoidal number to determine the shortest path to a graph, in addition to their work with interval-valued neutrosophic numbers. A remarkable and fundamental connectional optimization problem, the SPP manifests itself in numerous scientific and engineering domains, such as transportation, highway systems as well as additional technology. The SPP issue in an identified network finds the least possible path length weight to connect the two specified locations. In addition to time and value, the weights designated to the edges of a given network may also represent other essential aspects of existence. The caller is expected to possess knowledge of the parameters (such as length and duration) that differentiate various nodes in the conventional shortest route problem. In practice, however, parameter information across distinct nodes is perpetually ambiguous. Several approaches have been devised to determine the shortest path (SP) in fuzzy sets spanning distinct kinds of input files.

In a fuzzy environment, in the article (Prasanta Kumar R. et al., 2023) calculated the fuzzy shortest path utilizing a multivalued neutrosophic number. On the other hand, in real-world scenarios, many forms of ambiguity are typically brought about by malfunctions, incomplete data, or other variables like bad weather or traffic. In these instances, a fuzzy number is used because it can be difficult to ascertain the precise optimal path in the available networks. The weights assigned to the edges in real-world situations include shortest path scheduling, transportation, and vehicle green routing problems don't always have to be known because of variable elements like the state of the weather and traffic. In similar situations, experts stress handling unpredictability resulting from the uncertainties of the SPP by applying ideas of probability.

In 1994, in the research article (Nasution, 1994) clarified his views by looking at fuzzy interactive subtraction and pointing out that fuzzy numbers can only have a physical interpretation for their non-negative parts. The critical route technique also includes network computation, which consists of a sequence of project operations that determines the minimal required project duration. They used to identify the critical route based on three characteristics for each event: the earliest starting time, the latest finishing time, and the slack of every event. The crucial path is the path that connects the start and finish events with zero slack. They built a network to aid in the management of real project implementation. The technique relies on the ease access of deterministic time durations for every task. This criteria is tough to meet in a real time scenario in which numerous actions will be carried out for the first time. In most cases, there is only a vague sense of activity periods, which must then be estimated subjectively.

In their study, the authors (Mazlum M. et al, 2015) investigated the use of standard PERT and CPM project management approaches, as well as fuzzy PERT and fuzzy CPM, which are utilised in fuzzy project management, to develop and plan an online internet branch project. The study's findings will be analysed at the conclusion. The study uses specific and ambiguous activity periods from three distinct firms. CPM and PERT optimisation are analysed at specific activity times, as well as when triangular fuzzy numbers are used for fuzzy data.

The researchers (Liu, D., & Hu, C., 2020) proposed an evolving critical path method for the scheduling of projects based on a generalised fuzzy similarity paradigm. The discussion revolved around the use of the vector integration similarity method for fuzzy node labelling. Additionally, the focus was on determining the time duration and critical path of project network problems under fuzzy conditions. Furthermore, calculating greatly diminishes the amount of information, including the loss of original data resulting from defuzzification. The uncertain project network, which comprises multiple time frame factors as well as intricate reactions, is successfully managed to determine the critical path and project duration. Furthermore, contractors are increasingly recognising the necessity of planning and scheduling to enable successful project execution and business benefits such as timely completion, competitive bids, and customisation. In the development and execution of projects, the critical path method and programme evaluation review technique (PERT), both traditional network planning and control technologies, are regularly and widely utilised to tackle critical path concerns of networks. In contrast, the implementation of CPM necessitates a distinct duration for each activity and a precedence relationship between them, whereas conventional PERT relies on rudimentary assumptions, such as the utilisation of beta and normal distributions, to streamline the model. In lieu of relying on stochastic assumptions, historical data and human judgement are necessary to ascertain a suitable distribution for task durations. Complete and accurate statistical data are unattainable due to the unpredictability of project, the uncertainty of the execution environment and activity realisations. As a result, it is challenging to maintain the assumption that is founded on multiple probability distributions. Significant research findings have been acquired, leading to the conclusion that in this particular case, project scheduling based on fuzzy theory is far more efficient than project scheduling based on probability theory.

In their study, (Dorfeshan, Y., & Mousavi, S. M., 2019). introduced a novel approach called the TOPSIS-COPRAS methodology, which incorporates Pythagorean fuzzy sets and takes into account the expert weights for solving the project critical path problem. The critical path is being determined in a Pythagorean fuzzy set environment using the COPRAS and TOPSIS methods to facilitate decision-making.

Uncertainty significantly influences decision-making challenges, which encompass defective and insufficient knowledge pertaining to real-world circumstances. In order for decision-making methods to account for all possible uncertainty, Pythagorean fuzzy sets(PFSs) are implemented. PFSs offer a broader space to account for the grades of agreement, disagreement, and hesitancy when compared to intuitionistic fuzzy sets (IFSs). In addition, they address the unpredictability task in the real-world projects and establish CPM by taking efficient criteria, including cost, time, quality, safety and risks into account. To accomplish this, they expand a novel group decision methodology that utilises complex proportional assessment (COPRAS) methods as outlined in PFSs and techniques for order of preference by similarity to ideal solution (TOPSIS). In order to provide greater precision regarding the relative importance of each expert, a recently adjusted iteration of the suggested methodology is executed. In conclusion, they deliberated on a case study extracted from the literature that concerned the workflow schema of a marble processing plant project. This case study was presented to further exemplify the potential of the proposed methodology.

The researchers (Nowpada, Ravi Shankar & Saradhi, B., 2011) proposed a fuzzy critical path approach for Interval-Valued activity networks. Their discussion revolved around intervals, namely the earliest times, latest times, and float times. The approach involves identifying the most recent times by eliminating negative interval times, determining the criticality degree for floating time, and identifying the critical path based on the highest criticality degree. The researchers (Dubois, D., & Prade, H.,1988c). assigned varying levels of significance to each action on a critical path for a randomly selected group of activities. The researchers (Dubois et al., 2003) expanded their concept of a fuzzy arithmetic operations model to calculate the most recent time for each action in a project network. The researchers (Hapke, M., & Slowinski, R. 1996) determine the first time at which each activity in a project network can begin by employing fuzzy arithmetic operations. The researchers (Jin-Shing Yao, & Feng-Tse Lin. 2000) utilised the signed distance ranking approach for fuzzy numbers to identify the important path in a fuzzy project network. The authors (Shyi-Ming Chen & Tao-Hsing Chang, 2001) employed the defuzzification technique to identify potential essential paths in a fuzzy project network. The researchers (Chanas, S., & Zieliński, P. 2001) propose that the duration of each action can be represented as a precise value, a range of values, or a fuzzy number. The authors (Shyi-Ming Chen, & Tao-Hsing Chang. 2001) introduced a crucial degree approach to address the management of completion time and all activities in a project network.

Presently, we apply the aforementioned concepts in operations research to the critical path method's finding the critical path problem by utilizing neutrosophic triplet functions and comparing the results. Determining the critical path of a given network with arc length weights determined by a single-valued neutrosophic set is the principal objective of this study. We discover the optimal solution by utilizing the Neutrophic Fermatean set as a tool to determine the critical path through the applicability of the score, accuracy and certainty functions.

In the second Section of this paper delves into essential definitions pertaining to single - valued neutrosophic (SVN) sets, which are neutrosophic sets. Regarding neutrosophic data, we describe a method for determining critical path with connected edges in the third section. In section four we illustrate how the suggested approach resolved a practical case. The comparison study between the existing critical path methods (CPM) with our result is detailed in Section 4.3. Section 5 presents the findings and recommendations for future research.

PRELIMINARIES

Definition 1 (Raut, Prasanta Kumar et al., 2023)

A neutrosophic set N in a universal set X is defined as N= $\{x, (T_N(x), I_N(x), F_N(x)): x \in X\}$, where $T_N(x)$, $I_N(x)$, $F_N(x)$ are degree of true membership, indeterminacy and false memberships, with the restriction of $0^- \leq T_N(x)+I_N(x)+F_N(x) \leq 3^+$.

Definition 2 (Raut, Prasanta Kumar et al., 2023)

A Fermatean neutrosophic set N in universal set X is represented by $\{x, (T_N(x), I_N(x), F_N(x)): x \in X\}$, where $0 \leq (T_N(x))^3 + (F_N(x))^3 \leq 1$, with the restriction $0 \leq (I_N(x))^3 \leq 1$ and $0 \leq (T_N(x))^3 + (I_N(x))^3 + (F_N(x))^3 \leq 2$.

Definition 3 (Raut, Prasanta Kumar et al., 2023)

Assume that $N_1 = \{T_{N_1}, I_{N_1}, F_{N_1}\}$ and $N_2 = \{T_{N_2}, I_{N_2}, F_{N_2}\}$ are two neutrosophic sets with the neutrosophic number with multiple values, then the functions of Single valued neutrosophic numbers are given below:

i) $N_1 + N_2 = \{T_{N_1} + T_{N_2} - T_{N_1}T_{N_2}, I_{N_1}I_{N_2}, F_{N_1}F_{N_2}\}$

ii) $N_1 \times N_2 = \{T_{N_1}T_{N_2}, I_{N_1} + I_{N_2} - I_{N_1}I_{N_2}, F_{N_1} + F_{N_2} - F_{N_1}F_{N_2}\}$

iii) $\delta N_1 = \{1 - (1 - T_{N_1})^{\delta}, (I_{N_1})^{\delta}, (F_{N_1})^{\delta}\}$

iv) $N_1^{\delta} = \{(T_{N_1})^{\delta}, 1 - (1 - I_{N_1})^{\delta}, 1 - (1 - F_{N_1})^{\delta}\}$ with $\delta > 0$.

Definition 4 (Raut, Prasanta Kumar et al., 2023)

If $N_1 = \{T_{N_1}, I_{N_1}, F_{N_1}\}$ be a Neutrosophic number of single value. Then, the score function are defined as the value $SN_1 = \frac{2 + T_{N_1} - I_{N_1} - F_{N_1}}{3}$.

The accuracy function takes the values as $a(N_1) = \{T_{N_1} - F_{N_1}\}$

The certainty function is defined as $c(N_1) = T_{N_1}$.

GENERALIZED FERMATEAN NEUTROSOPHIC SET

Definition 1

We extend our notion in the Fermatean neutrosophic set from the closed interval bound from 0 to 2 to 0 to 3, in accordance with the concept presented in [15]. It is provided below:

In universal set X, a generalisation of Fermatean neutrosophic set N is designated by $\{x, (T_N(x), I_N(x), F_N(x)): x \in X\}$, where the closed interval between 0 and 1 contains the sum of the cubes of membership degree and false membership degrees.

$$0 \leq (T_N(x))^3 + (F_N(x))^3 \leq 1.$$

Moreover, the closed interval 0 and 1 contains the cube of indeterminacy degree.

$$0 \leq (I_N(x))^3 \leq 1,$$

therefore the sum of cubes of all three membership degrees lies between the closed interval 0 and 3. This is symbolised by

$$0 \leq (T_N(x))^3 + (I_N(x))^3 + (F_N(x))^3 \leq 3.$$

Algorithm to Find the Critical Path by Using Neutrosophic Number

Step 1: Select a network analysis problem on operation research to find critical path.

Step 2: Obtain fuzzy numbers by converting all of the real values.

Step 3: To establish a fuzzy number with a true membership function.

$$T = \frac{n-0}{n}, \text{ Here 'n' represents } n = \text{Max}\begin{cases} W_{ij} \; if\, i \neq j\; i = 1,2,..\, j = 2,3\ldots \\ 0,\, if\, i = j \end{cases}$$

Where W_{ij} - weight of the ith node to j^{th} node of the network.

Step 4: From step 3 we can find False membership function using the formula $F=1-T$. Further let us assume the indeterminacy number be any arbitrary number I between 0 and 1.

Step 5: Choose any node as the source and any node as the destination (except the source) of the specified network, then discover all pathways that connect the source and destination nodes.

Step 6: Apply the fuzzy number, accuracy, and indeterminacy functions to transform every conceivable edge value of a discrete valued neutrosophic number (SVN) into a single valued neutrosophic number (SVN).

Step 7: Once all edge values have been obtained, a path's average must be determined.

Step 8: Once all edge values have been obtained, a path's average must be determined.

CRITICAL PATH PROBLEMS AND ITS SOLUTION USING GENERALIZED FERMATEAN NEUTROSOPHIC NUMBER

Problem Description With Tabular Form

A online medicine delivery company delivers his products to the different regions in all over the country the regions are consider as node and the delivery distance consider as an acticity (in terms of Kms).

Table 1. Activity and delivery distance

Acitivity	1-2	1-3	1-4	2-5	3-2	3-4	4-5
Delivery distance (Kms)	11	5	7	11	6	5	9

Step 1: Construct a network analysis problem on operation research to find critical path.

Table 2. Activity and edge weight

Acitivity	Edge Weight (W_{ij})
1 - 2	11
1 - 3	5
1 - 4	7
2 - 5	11
3 - 2	6
3 - 4	5
4 - 5	9

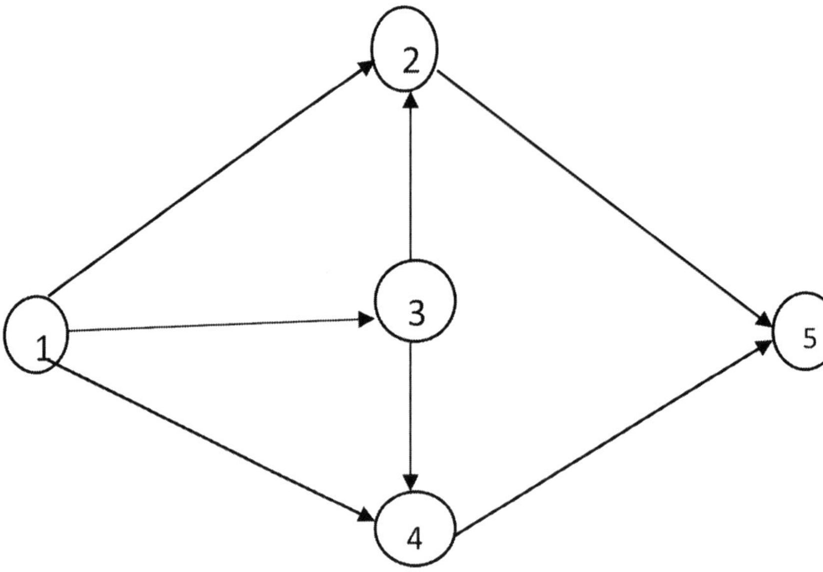

Apply step 2 & 3 to find

$$n = \text{Max}\begin{cases} W_{ij} \; if\; i \neq j \; i = 1,2,..\, j = 2,3\ldots \\ 0,\; if\; i = j \end{cases}$$

n = Max{ 11, 5, 7, 11, 6, 5, 9 } = 11.

Table 3. Fuzzy number

Activity	Edge Weight (W_{ij})	Fuzzy number $T = \frac{n-0}{n}$
1 - 2	11	1
1 - 3	5	0.45
1 - 4	7	0.63
2 - 5	11	1
3 - 2	6	0.54
3 - 4	5	0.45
4 - 5	9	0.81

Step 4: Let us find the Indeterminacy function and false membership function from true membership values.

Table 4. Indeterminacy function and false membership function

Activity	T	I	F = 1 - T
1-2	1	0.8	0
1-3	0.45	0.6	0.55
1-4	0.63	0.5	0.37
2-5	1	0.3	0
3-2	0.54	0	0.46
3-4	0.45	0.2	0.55
4-5	0.81	0.4	0.19

Step 5: Tabulate SVN values

Table 5. Single valued neutrosophic number

Activity	SVN
1 - 2	[1, 0.8, 0]
1 - 3	[0.45, 0.6, 0.55]
1 - 4	[0.63, 0.5, 0.37]
2 - 5	[1, 0.3, 0]
3 - 2	[0.54, 0, 0.46]
3 - 4	[0.45, 0.2, 0.55]
4 - 5	[0.81, 0.4, 0.19]

Step 6: Now by the definition 4

Table 6. Score function

Activity	Score function
1-2	0.73
1-3	0.43
1-4	0.59
2-5	0.9
3-2	0.69
3-4	0.57
4-5	0.74

Step 7: Find all possible paths from source to destination and also find distance of the respective path

Table 7. Distance of the path

Possible critical paths	Distance of the path
1-2-5	0.82
1-3-2-5	0.68
1-3-4-5	**0.58**
1-4-5	0.67

Step 8: The minimum distance of the path value is 0.58 and critical path of the problem is 1-3-4-5.

Comparative Analysis

We evaluate our approach in comparison to the critical path method that is currently in use in (Paneerselvam, R., 2008).

Table 8. Comparative analysis

Comparison	By Classical Critical path method (Paneerselvam, R., 2008).	Our method by using Neutrosophic
Critical paths	1-3-2-5	**1-3-4-5**

Problem Description by Tabular Form With Dummy Activity

A two-wheeler company has made the decision to update and renovate one of its satellite locations. After the renovations are finished, some of the current office equipment will be disposed of and the rest will be returned to the branch. A handful of carefully chosen contractors are invited to submit tenders. Everything related to the remodelling project, with the exception of the initial removal and replacement of the outdated equipment, would fall under the purview of the contractors. The following list includes the main project components, their durations, and the components that come just before them.

Table 9. Description

Activity	Description	Duration (Weeks)	Immediate Predecessors
A	Create new facilities	14	-
B	Invite contractors to submit bids.	4	A
C	Choose the contractor	2	B
D	Coordinate specifics with the chosen contractor.	1	C
E	Determine which machinery will be used.	2	A
F	Establish equipment storage	3	E
G	Manage the disposal of additional equipment	2	E
H	Acquire fresh instruments	4	E
I	Accept the shipment of new equipment	3	H, L
J	Renovations are performed.	12	K
K	Eliminate obsolete equipment for storage	4	D, F, G
L	Clearing following the contractor's completion	2	J
M	Equipment that has been used for storage	2	H, L

Construct the network analysis problem on operation research to find critical path.

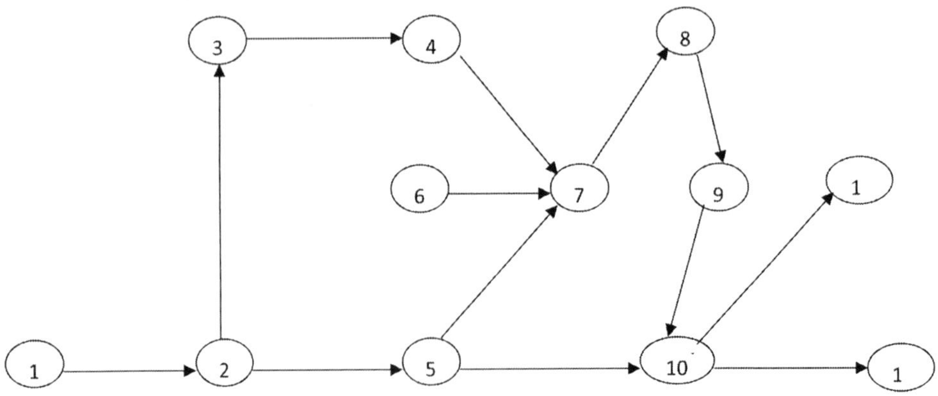

Table 10. Duration of weeks

Activity	Duration of weeks Weight (W_{ij})
1-2	14
2-3	4
2-5	2
3-4	2
4-7	1
5-7	2
5-10	4
6-7	3
7-8	4
8-9	12
9-10	2
10-11	2
10-12	3

Step 1: There are two sources and two destinations which lead to find critical path might be challenging. So we introduce the dummy activities between 5 to 6 and 11 to 12.

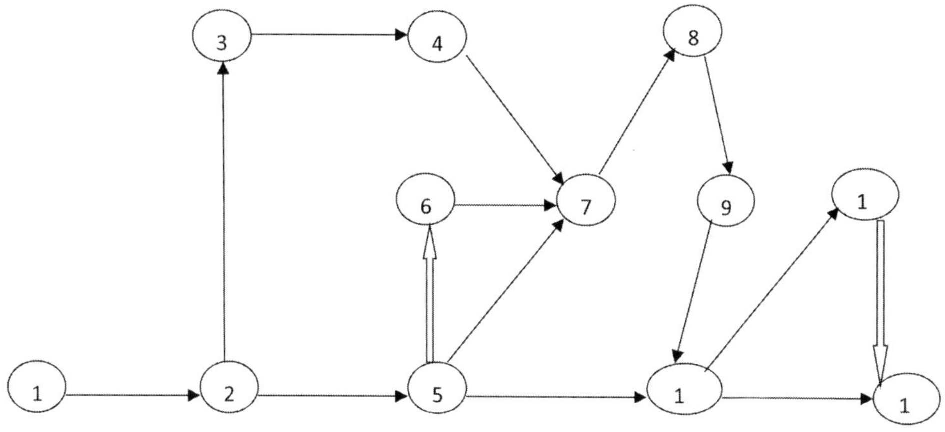

Table 11. Introducing dummy activity with zero weight

Activity	Duration of weeks Weight (W_{ij})
1-2	14
2-3	4
2-5	2
3-4	2
4-7	1
5-7	2
5-6	**0**
5-10	4
6-7	3
7-8	4
8-9	12
9-10	2
10-11	2
10-12	3
11-12	**0**

Apply Step 2 & 3: To find

$$n = \text{Max}\begin{cases} W_{ij} \; if\, i \neq j \; i = 1,2,..j = 2,3\ldots \\ 0, \; if\, i = j \end{cases}$$

n = Max{ 14, 4, 2, 2, 1, 2, 0, 4, 3, 4, 12, 2, 2, 3, 0 } = 14

Table 12. Fuzzy number

Activity	Duration of weeks Weight (W_{ij})	Fuzzy number $T = \frac{n-0}{n}$
1-2	14	1
2-3	4	0.29
2-5	2	0.14
3-4	2	0.14
4-7	1	0.07
5-7	2	0.14
5-6	0	0
5-10	4	0.29
6-7	3	0.21
7-8	4	0.29
8-9	12	0.86
9-10	2	0.14
10-11	2	0.14
10-12	3	0.21
11-12	0	0

Step 4: Let us find the Indeterminacy function and false membership function from true membership values.

Table 13. Indeterminacy function and false membership function

Activity	T	I	F = 1 - T
1-2	1	0.4	0
2-3	0.29	0.7	0.71
2-5	0.14	0.5	0.86
3-4	0.14	0.3	0.86
4-7	0.07	0.2	0.93
5-7	0.14	0.7	0.86
5-6	0	0.8	1
5-10	0.29	0.3	0.71
6-7	0.21	0.4	0.79
7-8	0.29	0.9	0.71
8-9	0.86	0.6	0.14
9-10	0.14	0.2	0.86

continued on following page

Table 13. Continued

Activity	T	I	F = 1 - T
10-11	0.14	0	0.86
10-12	0.21	0.8	0.79
11-12	0	0.3	1

Step 5: Tabulate SVN values

Table 14. Single valued neutrosophic number

Activity	SVN
1-2	[1, 0.4, 0]
2-3	[0.29, 0.7, 0.71]
2-5	[0.14, 0.5, 0.86]
3-4	[0.14, 0.3, 0.86]
4-7	[0.07, 0.2, 0.93]
5-7	[0.14, 0.7, 0.86]
5-6	[0, 0.8, 1]
5-10	[0.29, 0.3, 0.71]
6-7	[0.21, 0.4, 0.79]
7-8	[0.29, 0.9, 0.71]
8-9	[0.86, 0.6, 0.14]
9-10	[0.14, 0.2, 0.86]
10-11	[0.14, 0, 0.86
10-12	[0.21, 0.8, 0.79]
11-12	[0, 0.3, 1]

Step 6: Now by the definition 4

Table 15. Score function

Activity	SVN
1-2	0.87
2-3	0.29
2-5	0.26
3-4	0.33

continued on following page

Table 15. Continued

Activity	SVN
4-7	0.31
5-7	0.19
5-6	0.07
5-10	0.43
6-7	0.34
7-8	0.23
8-9	0.71
9-10	0.36
10-11	0.43
10-12	0.21
11-12	0.23

Step 7: Find all possible paths from source to destination and also find distance of the respective path

Table 16. Distance of the path

Possible critical paths	Distance of the path
1-2-3-4-7-8-9-10-11-12	0.42
1-2-5-7-8-9-10-11-12	0.41
1-2-5-6-7-8-9-10-11-12	0.39
1-2-5-10-11-12	0.44
1-2-5-10-12	0.44
1-2-3-4-7-8-9-10-12	0.41
1-2-5-7-8-9-10-12	0.40
1-2-5-6-7-8-9-10-12	**0.38**

Step 8: The minimum distance of the path value is 0.38 and critical path of the problem is 1-2-5-6-7-8-9-10-12.

Comparative Analysis

We evaluate our approach in comparison to the current critical route method.

Table 17. Comparative analysis

Comparison	By Classical Critical path method (Sharma, J K., 2016).	Our method by using Fermatean Neutrsophic set
Critical path	1-2-3-4-7-8-9-10-12	**1-2-5-6-7-8-9-10-12**

CONCLUSION

In this research article, the Fermatean neutrosophic set is expanded to incorporate previously established score, accuracy, and certainty functions. In the field of operations research, we proposed an algorithm that would discover the critical path by making use of the Fermatean neutrosophic set and comparing our findings. With existing classical critical path method and Fermatean Neutrosophic set with suitable examples. First example in terms of simple ordinary method in easy manner and another example by introducing dummy activities to find critical path. In this chapter we state that the new way of finding critical path is possible by using Fermatean neutrosophic set. In addition, we are attempting to incorporate this Fermatean neutrosophic set into the programme evaluation review technique (PERT) in order to locate the most effective solution to the network problem.

ACKNOWLEDGMENT

The authors would like to express their whole hearted thanks to the reviewers for their valuable comments which helped us a lot to re-write the chapter in the present form.

REFERENCES

Broumi, S., Bakali, A., Talea, M., & Smarandache, F. (2019). Computation of shortest path problem in a network with sv-triangular neutrosophic numbers. *Uluslararası Yönetim Bilişim Sistemleri Ve Bilgisayar Bilimleri Dergisi*, *3*(2), 41–51. DOI: 10.33461/uybisbbd.588290

Broumi, S., Talea, M., Bakali, A., Smarandache, F., & Kishore Kumar, P. K. (2017). Shortest path problem on single valued neutrosophic graphs. *2017 International Symposium on Networks, Computers and Communications (ISNCC)*. IEEE. DOI: 10.1109/ISNCC.2017.8071993

Chanas, S., & Zieliński, P. (2001). Critical path analysis in the network with fuzzy activity times. *Fuzzy Sets and Systems*, *122*(2), 195–204. DOI: 10.1016/S0165-0114(00)00076-2

Chen, S.-M., & Chang, T.-H. (2001). Finding multiple possible critical paths using fuzzy PERT. *IEEE Transactions on Systems, Man, and Cybernetics. Part B, Cybernetics*, *31*(6), 930–937. DOI: 10.1109/3477.969496 PMID: 18244858

Chen, S. P. (2007). Analysis of critical paths in a project network with fuzzy activity times. *European Journal of Operational Research*, *183*(1), 442–459. DOI: 10.1016/j.ejor.2006.06.053

Dorfeshan, Y., & Mousavi, S. M. (2019). A group TOPSIS-COPRAS methodology with Pythagorean fuzzy sets considering weights of experts for project critical path problem. *Journal of Intelligent & Fuzzy Systems*, *36*(2), 1375–1387. DOI: 10.3233/JIFS-172252

Dubois, D., Fargier, H., & Galvagnon, V. (2003). On latest starting times and floats in activity networks with ill-known durations. *European Journal of Operational Research*, *147*(2), 266–280. DOI: 10.1016/S0377-2217(02)00560-X

Dubois, D., & Prade, H. (1988c). Default reasoning and possibility theory. *Artificial Intelligence*, *35*(2), 243–257. DOI: 10.1016/0004-3702(88)90014-8

Hapke, M., & Slowinski, R. (1996). Fuzzy priority heuristics for project scheduling. *Fuzzy Sets and Systems*, *83*(3), 291–299. DOI: 10.1016/0165-0114(95)00338-X

Krishna Prabha, S., Hema, P., Balaji, P., & Kalaiselvi, B. (2022). Interval Value Intuitionistic Fuzzy Transportation Problem Via Modified ATM. *Journal of Emerging Technologies and Innovative Research*, *4*, 633–643.

Liu, D., & Hu, C. (2020). A dynamic critical path method for project scheduling based on a generalised fuzzy similarity. *The Journal of the Operational Research Society*, *72*(2), 458–470. DOI: 10.1080/01605682.2019.1671150

Mazlum, M., & Güneri, A. F. (2015). CPM, PERT and Project Management with Fuzzy Logic Technique and Implementation on a Business. *Procedia: Social and Behavioral Sciences*, *210*, 348–357. DOI: 10.1016/j.sbspro.2015.11.378

Nasution, S. (1994). Fuzzy critical path method. *IEEE Transactions on Systems, Man, and Cybernetics*, *24*(1), 48–57. DOI: 10.1109/21.259685

Nowpada & Saradhi. (2011). Fuzzy critical path method in interval valued activity networks. *International Journal of Pure and Applied Sciences and Technology*, *3*, 72-79.

Paneerselvam, R. (2008). *Operations research* (3rd ed.). PHI Learning Private Ltd.

Raut, P. K., Behera, S. P., Broumi, S., & Mishra, D. (2023). Calculation of Fuzzy shortest path problem using Multi-valued Neutrosophic number under fuzzy environment. *Neutrosophic Sets and Systems*, *57*. https://digitalrepository.unm.edu/nss_journal/vol57/iss1/24

Sharma, J. K. (2016). Operations Research Theory and Applications. Am imprint of Laxmi Publications.

Smarandache, F. (2020). The Score, Accuracy, and Certainty Functions determine a Total Order on the Set of Neutrosophic Triplets (T, I, F). *Neutrosophic Sets and Systems*,*38*, 1. https://digitalrepository.unm.edu/nss_journal/vol38/iss1/1

Vijayabalaji, S., & Balaji, P. (2015). MCDM method in cricket by rough matrix theory. *International Journal of Mathematical Analysis*, *9*, 869–875. DOI: 10.12988/ijma.2015.5380

Vijayabalaji, S., & Balaji, P. (2020). Best'11 strategy in cricket using MCDM, rough matrix and assignment model. *Journal of Intelligent & Fuzzy Systems*, *39*(5), 7431–7447. DOI: 10.3233/JIFS-200784

Yao, J.-S., & Lin, F.-T. (2000). Fuzzy critical path method based on signed distance ranking of fuzzy numbers. *IEEE Transactions on Systems, Man, and Cybernetics. Part A, Systems and Humans*, *30*(1), 76–82. DOI: 10.1109/3468.823483

Zadeh, L. (1965). Fuzzy sets. *Information and Control*, *8*(3), 338–353. DOI: 10.1016/S0019-9958(65)90241-X

Chapter 17

Interval Valued Neutrosophic Information System and Its Applications to Decision Making

V. Lakshmana Gomathi Nayagam
National Institute of Technology, Tiruchirappalli, India

Daniel P.
http://orcid.org/0009-0001-2181-6362
St. Xavier's College, India

Bharanidharan R.
National Institute of Technology, Tiruchirappalli, India

ABSTRACT

Smarandache introduced neutrosophic sets, interval-valued / n-valued refined neutrosophic sets due to their utility in various research fields. Multi criteria decision making problem in which evaluations of the alternatives based on criteria are given in the form of single-valued / interval-valued neutrosophic triplets (SVNT / IVNT) is an important area in neutrosophic research. Nayagam, et al have developed total ordering method to rank SVNT / IVNT. Few dominance-based ranking techniques have also been developed in the recent years to identify the best alternative in an information system. In this article, single valued and interval valued neutrosophic information systems (SVNIS and IVNIS) are introduced by defining generalized weighted dominance relation on objects and generalized weighted dominance degree

DOI: 10.4018/979-8-3693-3204-7.ch017

of objects. Further, IVNIS is compared with Intuitionistic fuzzy interval information system (IFIIS) by an example. Further, algorithms to identify optimal objects from SVNIS / IVNIS are developed and are illustrated through numerical examples.

1. INTRODUCTION

Zadeh introduced fuzzy sets (Zadeh, 1965) to deal impreciseness and ambiguity which occurs in a real-life situation where classical set theory is not capable to handle those situations. Fuzzy sets assign a membership grade between 0 to 1 to each element of its universe of discourse which generalizes the classical set. After the introduction of fuzzy sets, Atanassov introduced intuitionistic fuzzy sets (Atanassov K., 1986) which is a generalization of fuzzy sets. In literature, intuitionistic fuzzy sets are used in many research applications, since it is not only involving membership value but also involves non- membership value which is more logical in most of the scenarios. In intuitionistic fuzzy sets, membership and non-membership degree ranges from 0 to 1 with the condition that sum of those degrees should not exceed the value 1. Many MCDM problems based on intuitionistic fuzzy sets have been analysed in the literature (Nayagam & Sivaraman, 2011) (Sivaraman, Nayagam, & Ponalagusamy, 2014) (Nayagam, Jeevaraj, & Sivaraman, 2016) and total ordering algorithm were developed for various types of fuzzy numbers (Nayagam, Jeevaraj, & Sivaraman, 2016) (Velu & Ramalingam, 2023) (Nayagam, Jeevaraj, & Dhanasekaran, 2016)There are many generalizations of fuzzy sets available in the literature like picture fuzzy sets, hesitant fuzzy sets, Pythagorean fuzzy sets, circular intuitionistic fuzzy sets.

Later Smarandache introduced neutrosophic sets (Smarandache, 2004) as a generalization of intuitionistic fuzzy sets. In neutrosophic sets membership degree, non-membership degree and indeterminacy degree all included and all of them are ranges between 0 to 1. Neutrosophic sets are widely used because of its utility in many research domains in which MCDM is very significant because of its utility. In MCDM problems, the goal is to find the best alternative from the given set of alternatives with respect to the given criteria (Geetha, VLG, & Ponalagusamy, 2013) (Nayagam, Ponnialagan, & Jeevaraj, 2020). When a rating of an alternatives with respect to the criteria is given in the form of neutrosophic triplets in a MCDM problem, a total ordering algorithm is needed to rank the neutrosophic triplets. To derive total ordering on neutrosophic triplets, Smarandache developed total ordering on single valued neutrosophic triplets and developed a ranking order in interval valued neutrosophic triplets in (Smarandache, 2020). Later, Lakshmana Gomathi Nayagam, et al. developed total ordering on single valued and interval valued neutrosophic triplets in (V Lakshmana Gomathi Nayagam, 2023). Further, n valued refined neutrosophic sets are also developed in the literature (Smarandache, 2014) and total ordering has been derived on n valued refined neutrosophic sets in the literature (Nayagam & Bharanidharan, 2023). Thus, neutrosophic sets are often used in multi criteria decision making problems. Now, we are going to develop another decision-making research area which is called information system based on the IVNT.

Generally, in fuzzy information system, objects are evaluated by a set of values in the unit interval or a set of linguistic terms. In an information system, comparison of data plays a crucial role to define a dominance degree. Now in this proposed work, we are going to develop Interval Valued Neutrosophic Information System. This proposed information system is based on the single valued and interval valued neutrosophic triplets (IVNT). First, we propose Single Valued Neutrosophic Information System and a numerical example as a study of its applications in decision making. Then we define Interval Valued Neutrosophic Information System, which is another major research area in neutrosophic theory. A few dominance-based ranking techniques have grown up in the past few years (Zhan, Wang, Carlos, & Zhan, 2022) (Li, 2019), (Gope & Das, 2014) wherein an information system utilizes a dominance degree and the entire dominance degree (Sivaraman, Nayagam, & Ponalagusamy, 2012) to identify the best alternative. Even while the basis of intuitionistic fuzzy sets opened the door to make decisions based on qualitative data in the absence of complete knowledge, it is not applicable to situations where there is indeterminacy in the qualitative data. This leads to the introduction of the IVNIS and its generalizations. In this work, we introduce information system-based on interval valued neutrosophic triplets and then we define generalized weighted dominance relation on objects and generalized weighted dominance degree of objects using ranking methods proposed on IVNT in (Nayagam & Bharanidharan, 2023) (V Lakshmana Gomathi Nayagam, 2023). After defining generalized weighted dominance degree of objects, we study some of its properties and we also inherit a difference between dominance degree on IVNIS and dominance degree on IFIIS (Intuitionistic fuzzy interval information system) by a comparative analysis. Further, an algorithm to identify optimal objects from an IVNIS is derived and numerical examples are illustrated to show how the proposed algorithm identifies the optimal objects from IVNIS.

2. PRELIMINARIES

In this section, we recall some definitions from the literature which are required to understand further sections.

Definition 2.1. (Smarandache, 2004) Let $M=\{(T,I,F)|T,I,F\in[0,1],0\leq T+I+F\leq 3\}$ be a collection of all single valued neutrosophic triplets (SVNT), where T,I and F represent degree of membership, degree of indeterminacy and degree of non-membership respectively.

Definition 2.2. (Nayagam & Bharanidharan, 2023) A single valued neutrosophic triplet membership score S^{+}: $M\rightarrow[0,1]$ is defined as

$$S^{+}(T,I,F)=\frac{2+(T-F)(2-I)-I}{4}.$$

Definition 2.3. (Nayagam & Bharanidharan, 2023) A single valued neutrosophic triplet non-membership score S^{-}: $M\rightarrow[0,1]$ is defined as

$$S^{-}(T,I,F)=\frac{2+(F-T)(2-I)-I}{4}.$$

Definition 2.4. (Nayagam & Bharanidharan, 2023) A single valued neutrosophic triplet average score C: $M\rightarrow[0,1]$ is defined as

$$C=\frac{T+F}{2}.$$

Definition 2.5. (Smarandache, 2004) Let

$$M=\{([T^L,T^U],[I^L,I^U],[F^L,F^U])\,|\,T^L,T^U,I^L,I^U,F^L,F^U\in[0,1],0\leq T^U+I^U+F^U\leq 3\}$$

be a collection of all interval valued neutrosophic triplets (IVNT) with the conditions that

$$T^L\leq T^U,F^L\leq F^U,I^L\leq I^U.$$

Definition 2.6. (Nayagam & Bharanidharan, 2023) An interval valued neutrosophic triplet membership score S^{+}: $M\rightarrow[0,1]$ is defined as

$$S^{+}(T,I,F)=\frac{8+(T^L+T^U-F^L-F^U)(4-I^L-I^U)-2(I^L+I^U)}{16}.$$

Definition 2.7. (Nayagam & Bharanidharan, 2023) An interval valued neutrosophic triplet non-membership score S^{-}: $M \rightarrow [0,1]$ is defined as

$$S^{-}(T,I,F) = \frac{8 + (F^L + F^U - T^L - T^U)(4 - I^L - I^U) - 2(I^L + I^U)}{16}.$$

Definition 2.8. (Nayagam & Bharanidharan, 2023)An interval valued neutrosophic triplet average score C: $M \rightarrow [0,1]$ is defined as

$$C(T,I,F) = \frac{(T^L + T^U + F^L + F^U)}{4}.$$

Definition 2.9. (Nayagam & Bharanidharan, 2023)An interval valued neutrosophic triplet positive range score $S'^{+}: M \rightarrow [0,1]$ is defined as

$$S'^{+}(T,I,F) = \frac{8 + (T^U - T^L - F^U + F^L)(4 + I^L - I^U) - 2(I^U - I^L)}{12}.$$

Definition 2.10. (Nayagam & Bharanidharan, 2023) An interval valued neutrosophic triplet negative range score $S'^{-}: M \rightarrow [0,1]$ is defined as

$$S'^{-}(T,I,F) = \frac{8 + (F^U - F^L - T^U + T^L)(4 + I^L - I^U) - 2(I^U - I^L)}{12}.$$

Definition 2.11. (Nayagam & Bharanidharan, 2023) An interval valued neutrosophic triplet average range score C': $M \rightarrow [0,1]$ is defined as

$$C'(T,I,F) = \frac{(T^U - T^L + F^U - F^L)}{4}.$$

Definition 2.12. (Sivaraman, Nayagam, & Ponalagusamy, Intuitionistic Fuzzy Interval Information System, 2012) A quadruple $IS = (M, A_t, W_d, \tau)$ is said to be an information system if $M(\neq\phi)$ is a finite set and $A_t(\neq\phi)$ is finite collection of attributes, $W_d = \cup_{a \in A_t} W_a$ where W_a is a domain of attribute $a \in A_t$ and τ: $M \times A_t \rightarrow W_d$ is a total / information function such that $\tau(m,a) \in W_a$.

3. SINGLE VALUED NEUTROSOPHIC INFORMATION SYSTEM (SVNIS)

In this section, single valued neutrosophic information system is developed from ordinary information system based on single valued neutrosophic dominance relation and single valued neutrosophic entire dominance function.

3.1 Dominance Relation on Single Valued Neutrosophic Sets

Definition 3.1. If W_d is a set of single valued neutrosophic triplets, then an information system $IS=(M,A_t,W_d,\tau)$ is called single valued neutrosophic information system (SVNIS). Since $\tau(m,a)\in W_d$, $\tau(m,a)=(T_a^m,I_a^m,F_a^m)$, where $T_a^m,I_a^m,F_a^m \in [0,1]$ with $0 \le T_a^m+I_a^m+F_a^m \le 3$.

Definition 3.2. Let $a\in A_t$ be a criterion and $p,q\in M$. Now we define $p>_a q$ as

i. $p>_a q$ if $S^+(\tau(p,a))> S^+(\tau(q,a))$,
ii. $p>_a q$ if $S^+(\tau(p,a))=S^+(\tau(q,a))$ and $S^-(\tau(p,a))< S^-(\tau(q,a))$,
iii. $p>_a q$ if $S^+(\tau(p,a))=S^+(\tau(q,a))$, $S^-(\tau(p,a))=S^-(\tau(q,a))$ and $C(\tau(p,a))> C(\tau(q,a))$,
iv. $p=_a q$ if $S^+(\tau(p,a))=S^+(\tau(q,a))$, $S^-(\tau(p,a))=S^-(\tau(q,a))$ and $C(\tau(p,a))=C(\tau(q,a))$.

In the above, $p>_a q$ indicates p is better than q with respect to criterion a and $p=_a q$ indicates that p is equally good as q with respect to criterion a and $S^+(\tau(p,a))$, $S^-(\tau(p,a))$ and $C(\tau(p,a))$ are considered as in definitions 2.2 – 2.4 for $\tau(p,a)=(T_a^p, I_a^p, F_a^p)\in W_d$, where $T_a^p,I_a^p,F_a^p \in [0,1]$.

Definition 3.3. Let $SVNIS=(M,A_t,W_d,\tau)$ and $A \subseteq A_t$. Now we define $G_A(p,q)$ as a collection of all attributes in A such that $p>_a q$ i.e., $(G_A(p,q)=\{a \in A|p>_a q\})$. Further we define $E_A(p,q)$ as a collection of all attributes in A such that $p=_a q$ i.e., $(E_A(p,q)=\{a \in A|p=_a q\})$.

Definition 3.4. Let $SVNIS=(M,A_t,W_d,\tau)$ and $A \subseteq A_t$. Let w_a be the weight for $a\in A$ based on the preference of a. The weighted single valued neutrosophic dominance relation $SR_A^*:M\times M \rightarrow [0,1]$ is defined as

$$SR_A^*(p,q)=\sum_{a\in G_A(p,q)} w_a+\frac{1}{2}\sum_{a\in E_A(p,q)} w_a$$

such that $\sum_{a\in A} w_a = SR_A^*(p,q)+SR_A^*(q,p)=1$.

Definition 3.5. Let $SVNIS=(M,A_t,W_d,\tau)$ and $A \subseteq A_t$. The generalized weighted single valued neutrosophic dominance relation SR_A: $M\times M\rightarrow[0,1]$ defined as

$$SR_A(p,q)=\sum_{a\in G_A(p,q)} w_a\lambda_a(p,q)+\frac{1}{2}\sum_{a\in E_A(p,q)} w_a$$

where,

$\lambda_a(p,q)=$

$$\begin{cases}\delta_1(S^+(p)-S^+(q)), & if \quad S^+(p)> S^+(q)\\ \delta_2(S^-(q)-S^-(p)), & if \quad S^+(p)= S^+(q), S^-(p)< S^-(q)\\ \delta_3(C(p)-C(q)), & if \quad S^+(p)= S^+(q), S^-(p)= S^-(q), C(p)> C(q)\end{cases}$$

with $\delta_1>\delta_2>\delta_3>0$ and $\delta_1+\delta_2+\delta_3=1$.

In the above, $\lambda_a(p,q)$ for $p>_a q$ with respect to a measure the generalized dominance based on the preferences δ_i of score functions.

Definition 3.6. Let $SVNIS=(M,A_t,W_d,\tau)$ and $A \subseteq A_t$. Then single valued neutrosophic entire dominance degree of each object $p_i \in M$ is defined as $SR_A(p_i)=\frac{1}{|M|}\sum_{j=1}^{|M|} SR_A\left(p_i,p_j\right)$.

Theorem 3.1. Let $SVNIS=(M,A_t,W_d,\tau)$ and $A \subseteq A_t$ then the generalized weighted interval valued neutrosophic dominance relation SR_A: $M\times M\rightarrow[0,1]$ satisfies the following properties,

1. $0 \leq SR_A(p,q)\leq 1$.
2. $SR_A(p,p)=\frac{1}{2}$.
3. $SR_A(p,q)+SR_A(q,p)< 1$.
4. $SR_A(p,r)\geq SR_A(q,r)\, if p\geq_a q for all a \in A$.
5. $SR_A(p,r)< SR_A(q,r)\, if p<_a q for all a \in A$.

Proof.

1. Since $\sum_{a\in A} w_a$ and $0 \leq w_a \leq 1, 0 \leq \sum_{a\in G_A(p,q)} w_A + \frac{1}{2}\sum_{a\in E_A(p,q)} w_a \leq 1$. Now, from $0 \leq \lambda_a(p,q)\leq 1$,

$$0 \leq \sum_{a\in G_A(p,q)} w_a\lambda_a(p,q)+\frac{1}{2}\sum_{a\in E_A(p,q)} w_a \leq 1.$$

Hence $0 \leq SR_A(p,q)\leq 1$.

Now, $SR_A(p,p)=\sum_{a\in G_A(p,p)} w_a\lambda_a(p,p)+\frac{1}{2}\sum_{a\in E_A(p,p)} w_a$. Here $G_A(p,p)=\phi$ and $E_A(p,p)=A$. Hence $SR_A(p,p)=0+\frac{1}{2}\sum_{a\in A} w_a = \frac{1}{2},[\because \sum_{a\in A} w_a = 1]$.

2. Clearly,

$$SR_A(p,q)+SR_A(q,p)=\sum_{a\in G_A(p,q)} w_a\lambda_a(p,q)+\frac{1}{2}\sum_{a\in E_A(p,q)} w_a+\sum_{a\in G_A(q,p)} w_a\lambda_a(q,p)$$

$$+\frac{1}{2}\sum_{a\in E_A(q,p)} w_a=\sum_{a\in G_A(p,q)} w_a\lambda_a(p,q)+\sum_{a\in E_A(p,q)} w_a$$

$$+\sum_{a\in G_A(q,p)} w_a\lambda_a(q,p)< 1,[\because \sum_{a\in A} w_a = 1 and 0 < \lambda_a < 1].$$

3. Let $w_{a_p}(w_{a_q})$ be the weights for a with respect to $p>_{a_p} r(q>_{a_q} r)$ if

$a_p \in G_A(p,r)(a_q \in G_A(q,r))$

and w_a is the weight for a with respect to $p=_{a_p} r(q=_{a_q} r)$ if $a_p \in E_A(p,r)(a_q \in E_A(q,r))$.

$$SR_A(p,r)=\sum_{a_p\in G_A(p,r)} w_{a_p}\lambda_{a_p}(p,r)+\frac{1}{2}\sum_{a_p\in E_A(p,r)} w_{a_p}$$

$$SR_A(q,r)=\sum_{a_q\in G_A(q,r)} w_{a_q}\lambda_{a_p}(q,r)+\frac{1}{2}\sum_{a_q\in E_A(q,r)} w_{a_q}$$

By definition of $E_A(p,r)$ and $E_A(q,r)$, $\sum_{a_p\in E_A(p,r)} w_{a_p} = \sum_{a_q\in E_A(q,r)} w_{a_q}$. By given $p>_a q$ for all $a\in A$ and $q>_{a_q} r$ implies $p>_{a_p} r$. Hence $G_A(q,r)\subseteq G_A(p,r)$. This implies $w_{a_p} \geq w_{a_q}$ and $\lambda_{a_p}(p,r)\geq \lambda_{a_q}(q,r)$. Hence $SR_A(p,r)\geq SR_A(q,r)$.

4. Proof similar to 4.

3.2 Algorithm for Ranking Objects in Single Valued Neutrosophic Information System

Let (M,A_t,W_d,τ) be a single valued neutrosophic information system with the information function $\tau(m,a)=(T_a^m,I_a^m,F_a^m)$, where $m \in M, a \in A \subseteq A_t$.

Step 1: Determine $S^+(T_a^m,I_a^m,F_a^m)$, $S^-(T_a^m,I_a^m,F_a^m)$ and $C(T_a^m,I_a^m,F_a^m)$ for all $a\in A$, $m\in M$ by using definitions 2.2-2.4.
Step 2: Find $G_A(p,q)$ and $E_A(p,q)$ by using definition 3.3 for all $p,q\in M$.
Step 3: Compute the generalized weighted single valued neutrosophic dominance relation $SR_A(p,q)$ by using definition 3.5.

Step 4: Compute the single valued neutrosophic entire dominance degree $SR_A(p_i)$ for each $p_i \in M$ by using definition 3.6.
Step 5: The objects are ranked using entire dominance degree $SR_A(p_i)$
Step 6: The object p_i which has the highest value of $SR_A(p_i)$ is selected as the best one.

3.3 Numerical Example

Consider an investor interested in making a company investment problem in (V Lakshmana Gomathi Nayagam, 2023). The investor selects one out of five companies based on factors such as risk(a_{t_1}), raw material availability(a_{t_2}), availability of labor(a_{t_3}), market demand (a_{t_4}) and production quantity(a_{t_5}). Let us consider those five companies as automobile company (m_1), food manufacturing (m_2), electrical manufacturing (m_3), oil (m_4), and pharmaceutical (m_5). The single valued neutrosophic triplets for each company with respect to each criterion are displayed in SVNIS in Table 1.

Table 1. Single valued neutrosophic information system

	a_{t_1}	a_{t_2}	a_{t_3}	a_{t_4}	a_{t_5}
m_1	(0.7,0.4,0.3)	(0.27,0.4,0.7)	(0.5.0.2.0.3)	(0.5,0.9,0.4)	(0.1,0.3,0.8)
m_2	(0.5,0.8,0.2)	(0.15,0.36,0.78)	(0.9,0,0.2)	(0.6,0.96,0.45)	(0.3,0.5,0.75)
m_3	(0.9,0.1,0.1)	(0.3,0.6,0.9)	(0.4,0.2,0.2)	(0.72,0.85,0.3)	(0.3,0.45,0.87)
m_4	(0.8,0.6,0.3)	(0.1,0.8,0.2)	(1,0.5,0)	(0.57,0.8,0.35)	(0.1,0.8,0.6)
m_5	(0.65,0.2,0.8)	(0.2,0.45,0.65)	(0.7,0.4,0.6)	(0.4,0.7,0.6)	(0.7,0.2,0.3)

In the above SVNIS, $M = \{m_1, m_2, m_3, m_4, m_5\}$ are objects and
$A = \left\{ a_{t_1}, a_{t_2}, a_{t_3}, a_{t_4}, a_{t_5} \right\}$ are attributes.
Step 1. Employing the definitions 2.2-2.4, S^+,S^- and C are determined, and these are combined in Table 2, 3 and 4 respectively for $(T^{m_j}_{a_{t_i}}, I^{m_j}_{a_{t_i}}, F^{m_j}_{a_{t_i}})$ where i,j=1,2,3,4,5.

Table 2. The values of $S^+(T^{m_j}_{a}, I^{m_j}_{a}, F^{m_j}_{a})$ for i,j=1,2,3,4,5

	a_{t_1}	a_{t_2}	a_{t_3}	a_{t_4}	a_{t_5}
m_1	0.56	0.23	0.54	0.30	0.13
m_2	0.39	0.15	0.85	0.3	0.21
m_3	0.86	0.14	0.54	0.41	0.17
m_4	0.53	0.27	0.75	0.37	0.15
m_5	0.38	0.21	0.44	0.26	0.63

Table 3. The values of $S^-(T^{m_j}_a, I^{m_j}_a, F^{m_j}_a)$ for i,j=1,2,3,4,5

	a_{t_1}	a_{t_2}	a_{t_3}	a_{t_4}	a_{t_5}
m_1	0.24	0.57	0.36	0.25	0.72
m_2	0.21	0.67	0.15	0.22	0.54
m_3	0.1	0.56	0.36	0.17	0.61
m_4	0.18	0.33	0	0.23	0.45
m_5	0.52	0.56	0.36	0.39	0.27

Table 4. The values of $C(T^{m_j}_a, I^{m_j}_a, F^{m_j}_a)$ for i,j=1,2,3,4,5

	a_{t_1}	a_{t_2}	a_{t_3}	a_{t_4}	a_{t_5}
m_1	0.5	0.49	0.4	0.45	0.45
m_2	0.35	0.47	0.55	0.53	0.53
m_3	0.5	0.6	0.3	0.51	0.59
m_4	0.55	0.15	0.5	0.46	0.35
m_5	0.73	0.43	0.65	0.5	0.5

Step 2. Let us consider companies m_1 and m_3 for $G_{a_{t_i}}(m_1, m_3)$ and $E_{a_{t_i}}(m_1, m_3)$. By definition 3.3, $G_A(m_1,m_3) = \{\alpha \in A | m_1 >_a m_3\}$ and $E_A(m_1,m_3) = \{\alpha \in A | m_1 =_a m_3\}$. Now,

$m_3 >_{a_{t_1}} m_1$ since $S^+(\tau(m_3, a_{t_1})) = 0.86 > 0.56 = S^+(\tau(m_1, a_{t_1}))$,

$m_1 >_{a_{t_2}} m_3$ since $S^+(\tau(m_1, a_{t_2})) = 0.23 > 0.14 = S^+(\tau(m_3, a_{t_2}))$,

$m_1 >_{a_{t_3}} m_3$ since $S^+(\tau(m_1, a_{t_3})) = 0.54 = S^+(\tau(m_3, a_{t_3}))$,

$S^-(\tau(m_1, a_{t_3})) = 0.36 = S^-(\tau(m_3, a_{t_3}))$ and

$C(\tau(m_1, a_{t_3})) = 0.4 > 0.3 = C(\tau(m_3, a_{t_3}))$.

$m_3 >_{a_{t_4}} m_1$

since $S^+(\tau(m_3, a_{t_4})) = 0.41 > 0.30 = S^+(\tau(m_1, a_{t_4}))$,

$m_3 >_{a_{t_5}} m_1$

since $S^+(\tau(m_3, a_{t_5})) = 0.17 > 0.13 = S^+(\tau(m_1, a_{t_5}))$,

Hence

$$G_A(m_1,m_3)=\left\{a_{t_2},a_{t_3}\right\},G_A(m_3,m_1)=\left\{a_{t_1},a_{t_4},a_{t_5}\right\}$$

and $E_A(m_1,m_3)=\Phi = E_A(m_3,m_1)$.

Similarly, we can find $G_A(m_i,m_j)$ and $E_A(m_i,m_j)$ for all i,j=1,2,3,4,5 and the values are tabulated in Table 5.

Table 5. The set $G_A(m_i,m_j)$ for i,j=1,2,3,4,5 and $i \neq j$

	m_1	m_2	m_3	m_4	m_5
m_1	-	$\{a_{t_1},a_{t_2}\}$	$\{a_{t_2},a_{t_3}\}$	$\{a_{t_1}\}$	$\{a_{t_1},a_{t_2},a_{t_3},a_{t_4}\}$
m_2	$\{a_{t_3},a_{t_4},a_{t_5}\}$	-	$\{a_{t_2},a_{t_3},a_{t_5}\}$	$\{a_{t_3},a_{t_5}\}$	$\{a_{t_1},a_{t_3},a_{t_4}\}$
m_3	$\{a_{t_1},a_{t_4},a_{t_5}\}$	$\{a_{t_1},a_{t_4}\}$	-	$\{a_{t_1},a_{t_4},a_{t_5}\}$	$\{a_{t_1},a_{t_3},a_{t_4}\}$
m_4	$\{a_{t_2},a_{t_3},a_{t_4},a_{t_5}\}$	$\{a_{t_1},a_{t_2},a_{t_4}\}$	$\{a_{t_2},a_{t_3}\}$	-	$\{a_{t_1},a_{t_2},a_{t_3},a_{t_4}\}$
m_5	$\{a_{t_5}\}$	$\{a_{t_2},a_{t_5}\}$	$\{a_{t_2},a_{t_5}\}$	$\{a_{t_5}\}$	-

Step 3. Now, a_{t_2} and a_{t_3} have different weights because of their preference levels, even though m_1 is preferable to m_3 with respect to a_{t_2} and a_{t_3}. Consider δ_1=0.5, δ_2=0.3, δ_3=0.2 as definition 3.5.

$$SR_A(m_1,m_3)=\sum_{a\in G_A(m_1,m_3)} w_a \lambda_a(m_1,m_3)+\frac{1}{2}\sum_{a\in E_A(m_1,m_3)} w_a$$

$$=w_{a_{t_2}}\lambda_{a_{t_2}}(m_1,m_3)+w_{a_{t_3}}\lambda_{a_{t_3}}(m_1,m_3)+0$$

$$=w_{a_{t_2}}\delta_1(S^{+}(m_1)-S^{+}(m_3))+w_{a_{t_3}}\delta_3(C(m_1)-C(m_3))$$

$$=0.045\,w_{a_{t_2}}+0.02\,w_{a_{t_3}}$$

Choose $w_{a_{t_2}}$ and $w_{a_{t_3}}$ with the condition $\sum_{i=1}^{n} w_{a_{t_i}} = 1$ and $w_{a_{t_2}} > w_{a_{t_3}}$. Now, a_{t_1}, a_{t_4} and a_{t_5} have different weights because of their preference levels, even though m_3 is preferable to m_1 with respect to a_{t_1}, a_{t_4} and a_{t_5}.

$$SR_A(m_3,m_1)=0.15\,w_{a_{t_1}}+0.055\,w_{a_{t_4}}+0.02\,w_{a_{t_5}}$$

Choose w_{a_i}, w_{a_i} and w_{a_i} with the condition $\sum_{i=1}^{5} w_{a_i} = 1$ and $w_{a_i} > w_{a_i} > w_{a_i}$. Let us consider $w_{a_i} = 0.4, w_{a_i} = 0.16, w_{a_i} = 0.14, w_{a_i} = 0.2$ and $w_{a_i} = 0.1$. Hence, $SR_A(m_1,m_3)=0.01$ and $SR_A(m_3,m_1)=0.073$. Similarly, we can find $SR_A(m_i,m_j)$ for $i,j=1,2,3,4,5$ and it is presented in Table 6.

Table 6. The values of $SR_A(m_i,m_j)$ for $i,j=1,2,3,4,5$

	m_1	m_2	m_3	m_4	m_5
m_1	0.5	0.069	0.01	0.002	0.03
m_2	0.069	0.5	0.034	0.014	0.065
m_3	0.073	0.103	0.5	0.062	0.091
m_4	0.053	0.038	0.037	0.5	0.053
m_5	0.10	0.0678	0.073	0.096	0.5

Step 4. Now the single valued neutrosophic entire dominance degree $SR_A(m_i)$ of each object is found by definition 3.6. For example,

$$SR_A(m_1)=\frac{1}{5}(SR_A(m_1,m_1)$$
$$+SR_A(m_1,m_2)+SR_A(m_1,m_3)+SR_A(m_1,m_4)+SR_A(m_1,m_5))=0.11298$$

and $SR_A(m_i)$ is given in Table 7.

Table 7. The values of $SR_A(m_i)$ fori,j=1,2,3,4,5

	m_1	m_2	m_3	m_4	m_5
SR_A	0.11298	0.13644	0.16567	0.136028	0.16726

Step 5. By single valued neutrosophic entire dominance degree $SR_A(m_i)$, we can conclude that $m_5>m_3>m_2>m_4>m_1$.

Thus, by step 5, m_5 is a selected as the best object from the SVNIS.

4. INTERVAL VALUED NEUTROSOPHIC INFORMATION SYSTEM (IVNIS)

In this section, interval valued neutrosophic information system is developed as a generalization of SVNIS. Further, interval valued neutrosophic dominance relation and interval valued neutrosophic entire dominance relation are introduced.

4.1 Dominance Relation on Interval Valued Neutrosophic Set

Definition 4.1. If W_d is a set of interval-valued neutrosophic triplets, then an information system $IS=(M,A_t,W_d,\tau)$ is called interval-valued neutrosophic information system (IVNIS). Since $\tau(m,a)\in W_d$,

$$\tau(m,a) = ([T_a^L, T_a^U],[I_a^L, I_a^U],[F_a^L, F_a^U]),$$

where $T_a^L, T_a^U, I_a^L, I_a^U, F_a^L, F_a^U \in [0,1]$ with $T_a^L < T_a^U, I_a^L < I_a^U, F_a^L < F_a^U$.

Definition 4.2. Let $a\in A_t$ be a criterion and $p,q\in M$. Now we define $p>_a q$ as

i. $p>_a q$ if $S^+(\tau(p,a)) > S^+(\tau(q,a))$,
ii. $p>_a q$ if $S^+(\tau(p,a)) = S^+(\tau(q,a))$ and $S^-(\tau(p,a)) < S^-(\tau(q,a))$,
iii. $p>_a q$ if $S^+(\tau(p,a)) = S^+(\tau(q,a)), S^-(\tau(p,a)) = S^-(\tau(q,a))$ and $C(\tau(p,a)) > C(\tau(q,a))$,
iv. $p>_a q$ if $S^+(\tau(p,a)) = S^+(\tau(q,a)), S^-(\tau(p,a)) = S^-(\tau(q,a)), C(\tau(p,a)) = C(\tau(q,a))$

and $S'^+(\tau(p,a)) > S'^+(\tau(q,a))$.

v. $p>_a q$ if $S^+(\tau(p,a)) = S^+(\tau(q,a)), S^-(\tau(p,a)) = S^-(\tau(q,a)), C(\tau(p,a)) = C(\tau(q,a))$,

$S'^+(\tau(p,a)) = S'^+(\tau(q,a))$

and $S'^-(\tau(p,a)) < S'^-(\tau(q,a))$,

vi. $p>_a q$ if $S^+(\tau(p,a)) = S^+(\tau(q,a)), S^-(\tau(p,a)) = S^-(\tau(q,a)), C(\tau(p,a)) = C(\tau(q,a))$,

$S'^+(\tau(p,a)) = S'^+(\tau(q,a)), S'^-(\tau(p,a)) = S'^-(\tau(q,a))$

and $C'(\tau(p,a)) > C'(\tau(q,a))$,

vii. $p=_a q$ if $S^+(\tau(p,a)) = S^+(\tau(q,a)), S^-(\tau(p,a)) = S^-(\tau(q,a)), C(\tau(p,a)) = C(\tau(q,a))$,

$S'^+(\tau(p,a)) = S'^+(\tau(q,a)), S'^-(\tau(p,a)) = S'^-(\tau(q,a))$

and $C'(\tau(p,a)) = C'(\tau(q,a))$.

In the above, $p >_a q$ indicates p is better than q with respect to criterion a and $p =_a q$ indicates that p is equally good as q with respect to criterion a and

$S^{+}(\tau(p,a)), S^{-}(\tau(p,a)) C(\tau(p,a)), S'^{+}(\tau(p,a)), S'^{-}(\tau(p,a))$ and $C'(\tau(p,a))$

are considered as in definitions 2.6 through 2.11 for $\tau(p,a) = ([T_a^L, T_a^U],[I_a^L, I_a^U],[F_a^L, F_a^U])$, where $T_a^L, T_a^U, I_a^L, I_a^U, F_a^L, F_a^U \in [0,1]$ with $T_a^L < T_a^U, I_a^L < I_a^U, F_a^L < F_a^U$.

Definition 4.3. Let $IVNIS = (M, A_t, W_d, \tau)$ and $A \subseteq A_t$. Now we define $G_A(p,q)$ as a collection of all attributes in A such that $p >_a q$ i.e., ($G_A(p,q) = \{a \in A | p >_a q\}$). Further we define $E_A(p,q)$ as a collection of all attributes in A such that $p =_a q$ i.e., ($E_A(p,q) = \{a \in A | p =_a q\}$).

Definition 4.4. Let $IVNIS = (M, A_t, W_d, \tau)$ and $A \subseteq A_t$. Let w_a be the weight for $a \in A$ based on the preference of a. The weighted interval valued neutrosophic dominance relation $SR_A^*: M \times M \rightarrow [0,1]$ is defined as

$$SR_A^*(p,q) = \sum_{a \in G_A(p,q)} w_a + \frac{1}{2} \sum_{a \in E_A(p,q)} w_a$$

such that $\sum_{a \in A} w_a = SR_A^*(p,q) + SR_A^*(q,p) = 1$.

Definition 4.5. Let $IVNIS = (M, A_t, W_d, \tau)$ and $A \subseteq A_t$. The generalized weighted interval valued neutrosophic dominance relation SR_A: $M \times M \rightarrow [0,1]$ defined as

$$SR_A(p,q) = \sum_{a \in G_A(p,q)} w_a \lambda_a(p,q) + \frac{1}{2} \sum_{a \in E_A(p,q)} w_a$$

where,

$$\lambda_a(p,q) = \begin{cases} \delta_1(S^{+}(p) - S^{+}(q)), \text{ if } S^{+}(p) > S^{+}(q) \\ \delta_2(S^{-}(q) - S^{-}(p)), \text{ if } S^{+}(p) = S^{+}(q), S^{-}(p) < S^{-}(q) \\ \delta_3(C(p) - C(q)), \text{ if } S^{+}(p) = S^{+}(q), S^{-}(p) = S^{-}(q), C(p) > C(q) \\ \delta_4(S'^{+}(p) - S'^{+}(q)), \text{ if } S^{+}(p) = S^{+}(q), S^{-}(p) = S^{-}(q), \\ \quad C(p) = C(q), S'^{+}(p) > S'^{+}(q) \\ \delta_5(S'^{-}(q) - S'^{-}(p)), \text{ if } S^{+}(p) = S^{+}(q), S^{-}(p) = S^{-}(q), C(p) = C(q), \\ \quad S'^{+}(p) = S'^{+}(q), S'^{-}(p) > S'^{-}(q) \\ \delta_6(C'(p) - C'(q)), \text{if } S^{+}(p) = S^{+}(q), S^{-}(p) = S^{-}(q), C(p) = C(q), \\ \quad S'^{+}(p) = S'^{+}(q), S'^{-}(p) = S'^{-}(q), C'(p) > C'(q) \end{cases}$$

with $\delta_1>\delta_2>\delta_3>\delta_4>\delta_5>\delta_6>0$ and $\delta_1+\delta_2+\delta_3+\delta_4+\delta_5+\delta_6=1$.

In the above, $\lambda_a(p,q)$ for $p>_a q$ with respect to a measures the generalized dominance based on the preferences δ_i of score functions.

Definition 4.6. Let $IVNIS=(M,A_t,W_d,\tau)$ and $A\subseteq A_t$. Then interval valued neutrosophic entire dominance degree of each object $p_i\in M$ is defined as $SR_A(p_i)=\frac{1}{|M|}\sum_{j=1}^{|M|} SR_A(p_i,p_j)$.

4.2 Algorithm for Ranking Objects in Interval Valued Neutrosophic Information System

Let (M,A_t,W_d,τ) be an interval valued neutrosophic information system with the information function $\tau(m,a)=([T_a^L,T_a^U],[I_a^L,I_a^U],[F_a^L,F_a^U])$, where $m\in M, a\in A\subseteq A_t$.

Step 1: Determine

$$S^{+}([T_a^L,T_a^U],[I_a^L,I_a^U],[F_a^L,F_a^U]),$$

$$S^{-}([T_a^L,T_a^U],[I_a^L,I_a^U],[F_a^L,F_a^U]),$$

$$C([T_a^L,T_a^U],[I_a^L,I_a^U],[F_a^L,F_a^U]),$$

$$S'^{+}([T_a^L,T_a^U],[I_a^L,I_a^U],[F_a^L,F_a^U]),$$

$$S'^{-}([T_a^L,T_a^U],[I_a^L,I_a^U],[F_a^L,F_a^U])$$

and

$$C'([T_a^L,T_a^U],[I_a^L,I_a^U],[F_a^L,F_a^U])$$

for all $a\in A$, $m\in M$ by using definitions 2.6-2.11.

Step 2: Find $G_A(p,q)$ and $E_A(p,q)$ by using definition 4.3 for all $p,q \in M$.

Step 3: Compute the generalized weighted interval valued neutrosophic dominance Relation $SR_A(p,q)$ by using definition 4.5

Step 4: Compute the interval valued neutrosophic entire dominance degree $SR_A(p_i)$ for Each $p_i \in M$ by using definition 4.6

Step 5: The objects are ranked using entire dominance degree $SR_A(p_i)$

Step 6: The object p_i which has the highest value of $SR_A(p_i)$ is selected as the best one.

4.3 Numerical Example

Consider an investor interested in making a company investment problem in (Aiwua, Jianguoa, & Hongjunb, 2015) . The investor selects one out of four companies based on factors such as earning estimate analysis(a_{t_1}) growth analysis (a_{t_2})and environmental impact analysis (a_{t_3}). Let us consider those four companies as bookshop (m_1), chemical plant (m_2), supermarket (m_3) and food company (m_4). The interval valued neutrosophic triplets for each company with respect to each criterion are displayed in IVNIS in Table 8.

Table 8. Interval valued neutrosophic information system

	a_{t_1}	a_{t_2}	a_{t_3}
m_1	([0.4,0.5],[0.2,0.3],[0.3,0.4])	([0.4,0.6],[0.1,0.3],[0.2,0.4])	([0.4,0.5],[0.2,0.3],[0.7,0.9])
m_2	([0.6,0.7],[0.1,0.2],[0.2,0.3])	([0.6,0.7],[0.1,0.2],[0.2,0.3])	([0.8,0.9],[0.3,0.5],[0.3,0.6])
m_3	([0.4,0.6],[0.2,0.3],[0.3,0.4])	([0.5,0.6],[0.2,0.3],[0.3,0.4])	([0.7,0.9],[0.2,0.4],[0.4,0.5])
m_4	([0.7,0.8],[0.0,0.1],[0.1,0.2])	([0.6,0.7],[0.1,0.2],[0.1,0.3])	([0.8,0.9],[0.3,0.4],[0.6,0.7])

In the above IVNIS, $M= \{m_1,m_2,m_3,m_4\}$ are objects and $A = \left\{ a_{t_1}, a_{t_2}, a_{t_3} \right\}$ are attributes.

Step 1. Employing the definitions 2.6-2.11, S^+, S^-, C, S'^+, S'^- and C' are determined, and these are combined in Table 9, 10, 11, 12, 13 and 14 respectively for

$$([T^L_{a_{t_j}}, T^U_{a_{t_j}}],[I^L_{a_{t_j}}, I^U_{a_{t_j}}],[F^L_{a_{t_j}}, F^U_{a_{t_j}}])_{m_i}$$

where i=1,2,3,4 and j=1,2,3.

Table 9. The values of $S^{+}([T^{L}_{a}, T^{U}_{a}],[I^{L}_{a}, I^{U}_{a}],[F^{L}_{a}, F^{U}_{a}])$ *where i=1,2,3,4 and j=1,2,3*

	a_{t_1}	a_{t_2}	a_{t_3}
m_1	0.4813	0.5400	0.5906
m_2	0.6475	0.6475	0.2400
m_3	0.5031	0.5250	0.2844
m_4	0.7800	0.6702	0.3300

Table 10. The values of $S^{+}([T^{L}_{a}, T^{U}_{a}],[I^{L}_{a}, I^{U}_{a}],[F^{L}_{a}, F^{U}_{a}])$ *where i=1,2,3,4 and j=1,2,3*

	a_{t_1}	a_{t_2}	a_{t_3}
m_1	0.3938	0.3600	0.2844
m_2	0.2775	0.2775	0.5600
m_3	0.3719	0.3500	0.5906
m_4	0.1950	0.2544	0.4950

Table 11. The values of $C([T^{L}_{a}, T^{U}_{a}],[I^{L}_{a}, I^{U}_{a}],[F^{L}_{a}, F^{U}_{a}])$ *where i=1,2,3,4 and j=1,2,3*

	a_{t_1}	a_{t_2}	a_{t_3}
m_1	0.4000	0.4000	0.6250
m_2	0.4500	0.4500	0.6500
m_3	0.4000	0.4500	0.6250
m_4	0.4500	0.4250	0.7500

Table 12. The values of $S'^{+}([T^{L}_{a}, T^{U}_{a}],[I^{L}_{a}, I^{U}_{a}],[F^{L}_{a}, F^{U}_{a}])$ *where i=1,2,3,4 and j=1,2,3*

	a_{t_1}	a_{t_2}	a_{t_3}
m_1	0.6500	0.6333	0.6825
m_2	0.6500	0.6500	0.6967
m_3	0.6825	0.6500	0.6175
m_4	0.6500	0.6175	0.6500

Table 13. The values of $S'^{-}([T_a^L, T_a^U],[I_a^L, I_a^U],[F_a^L, F_a^U])$ where i=1,2,3,4 and j=1,2,3

	a_{t_1}	a_{t_2}	a_{t_3}
m_1	0.6500	0.6333	0.6175
m_2	0.6500	0.6500	0.5700
m_3	0.6175	0.6500	0.6825
m_4	0.6500	0.6825	0.6500

Table 14. The values of $C'([T_a^L, T_a^U],[I_a^L, I_a^U],[F_a^L, F_a^U])$ where i=1,2,3,4 and j=1,2,3

	a_{t_1}	a_{t_2}	a_{t_3}
m_1	0.0500	0.1000	0.0750
m_2	0.0500	0.0500	0.1000
m_3	0.0750	0.0500	0.0750
m_4	0.0500	0.0750	0.0500

Step 2. We can obtain $G_A(m_i,m_j)$ and $E_A(m_i,m_j)$ by definition 4.3 as follows,

Table 15. The set $G_A(m_i,m_j)$ for i,j=1,2,3,4 and p≠q

	m_1	m_2	m_3	m_4
m_1	-	$\{a_{t_3}\}$	$\{a_{t_2},a_{t_3}\}$	$\{a_{t_3}\}$
m_2	$\{a_{t_1},a_{t_2}\}$	-	$\{a_{t_1},a_{t_2}\}$	ϕ
m_3	$\{a_{t_1}\}$	$\{a_{t_3}\}$	-	ϕ
m_4	$\{a_{t_1},a_{t_2}\}$	$\{a_{t_1},a_{t_2},a_{t_3}\}$	$\{a_{t_1},a_{t_2},a_{t_3}\}$	-

Clearly $E_A(m_i,m_j)=\phi$ for $i,j=1,2,3,4$ and $i\neq j$.

Step 3. By using definition 4.5, we can acquire $SR_A(m_i,m_j)$ for $i,j=1,2,3,4$. Now $m_i >_a m_j$ or $m_j >_a m_i$ is determined by the initial score S^+ alone. We may thus consider $\delta_1 \cong 1$ without losing generality. Hence

$$SR_A(m_i,m_j) = \sum_{a\in G_A(m_i,m_j)} w_a \lambda_a(m_i,m_j) + \frac{1}{2}\sum_{a\in E_A(m_i,m_j)} w_a$$

$$= \sum_{a\in G_A(m_i,m_j)} w_a \delta_1 (S^+(m_i) - S^+(m_j)) + 0$$

$$= \sum_{a\in G_A(m_i,m_j)} w_a (S^+(m_i) - S^+(m_j))$$

Choose $w_{a_{t_x}}, w_{a_{t_y}}$ and $w_{a_{t_z}}$ with the condition $w_{a_{t_x}} + w_{a_{t_y}} + w_{a_{t_z}} = 1$ and $w_{a_{t_x}} > w_{a_{t_y}} > w_{a_{t_z}}$ where $x,y,z \in \{1,2,3\}$ and x,y,z are mutually distinct. Let us consider $w_{a_{t_x}} = 0.5, w_{a_{t_y}} = 0.3$ and $w_{a_{t_z}} = 0.2$. In this way $SR_A(m_i,m_j)$ determined and presented in Table 16.

Table 16. $SR_A(m_i,m_j)$values for i,j=1,2,3,4

	m_1	m_2	m_3	m_4
m_1	0.50	0.175	0.157	0.078
m_2	0.073	0.50	0.111	0.00
m_3	0.018	0.008	0.50	0.00
m_4	0.176	0.101	0.192	0.50

Step 4. Now the interval valued neutrosophic entire dominance degree $SR_A(m_i)$ of each object is found by definition 4.6. For example,

$$SR_A(m_1) = \frac{1}{4}(SR_A(m_1,m_1) + SR_A(m_1,m_2) + SR_A(m_1,m_3) + SR_A(m_1,m_4)) = 0.2275$$

and $SR_A(m_i)$ is given in Table 17.

Table 17. $SR_A(m_i)$ values for i=1,2,3,4

	m_1	m_2	m_3	m_4
SR_A	0.2275	0.1710	0.1315	0.2423

Step 5. By interval valued neutrosophic entire dominance degree $SR_A(m_i)$ we can conclude that $m_4 > m_1 > m_2 > m_3$.

So, by step 5, m_4> is a selected as the best object from the IVNIS.

5. COMPARATIVE STUDY WITH INTUITIONISTIC FUZZY INTERVAL INFORMATION SYSTEM

In this section, we are going to compare our interval valued neutrosophic information system with existing intuitionistic fuzzy interval information system (Sivaraman, Nayagam, & Ponalagusamy, Intuitionistic Fuzzy Interval Information System, 2012). Since every interval valued intuitionistic fuzzy number can be viewed as an interval valued neutrosophic triplets, we can view an intuitionistic fuzzy interval information system by our proposed interval valued neutrosophic information system as follows.

Let us consider the following intuitionistic fuzzy interval information system.

Table 18. Intuitionistic fuzzy interval information system

	a_{t_1}	a_{t_2}	a_{t_3}
m_1	([0.2,0.3],[0.25,0.4])	([0.2,0.3],[0.25,0.4])	([0.2,0.3],[0.25,0.4])
m_2	([0.1,0.4],[0.2,0.5])	([0.1,0.4],[0.2,0.5])	([0.1,0.4],[0.2,0.5])
m_3	([0.1,0.2],[0.09,0.25])	([0.01,0.394],[0.2,0.3])	([0,0.1],[0,0.09])
m_4	([0,0.2],[0.1,0.1])	([0.01,0.01],[0,0])	([0.024,0.035],[0.02,0.02])

For Table 18, we apply score function $S([a,b],[c,d]) = \frac{a + b - d(1 - b) - c(1 - a)}{2}$ to each interval valued intuitionistic fuzzy number. Thus, we get following Table 19.

Table 19. Score value of Table 18

	a_{t_1}	a_{t_2}	a_{t_3}
m_1	0.01	0.01	0.01
m_2	0.01	0.01	0.01
m_3	0.01	0.01	0.01
m_4	0.01	0.01	0.01

From the above table, we note that when we apply the ranking algorithm for Intuitionistic fuzzy interval information system given in (Sivaraman, Nayagam, & Ponalagusamy, Intuitionistic Fuzzy Interval Information System, 2012), we get the ranking as $m_1=m_2=m_3=m_4$. So, we are unable to find the best alternative. This shortcoming appears, since the ranking method in (Sivaraman, Nayagam, & Ponalagusamy, Intuitionistic Fuzzy Interval Information System, 2012) lacks total ordering.

Now, for the same problem, we apply our ranking method. First, we transform the Intuitionistic fuzzy interval information system into Interval valued neutrosophic information system as shown in Table 20. Interval valued neutrosophic information system

	a_{t_1}	a_{t_2}	a_{t_3}
m_1	([.2,.3],[.3,.55],[.25,.4])	([.2,.3],[.3,.55],[.25,.4])	([.2,.3],[.3,.55],[.25,.4])
m_2	([.1,.4],[.1,.7],[.2,.5])	([.1,.4].[.1,.7],[.2,.5])	([.1,.4],[.1,.7],[.2,.5])
m_3	([.1,.2],[.55,.81],[.09,.25])	([.01,.394],[.31,.79],[.2,.3])	([0,.1],[.81,1],[0,.09])
m_4	([0,.2],[.7,.9],[.1,.1])	([.01,.01],[.99,.99],[0,0])	([.024,.04],[.95,.96],[.02,.02]])

Table 21. Score value of Table 20

	a_{t_1}	a_{t_2}	a_{t_3}
m_1	0.37	0.37	0.37
m_2	0.36	0.36	0.36
m_3	0.32	0.35	0.28
m_4	0.3	0.26	0.26

The set $G_A(m_i,m_j)$ for i,j=1,2,3,4 and $p \neq q$ is given in Table 22.

Table 22. The set $G_A(m_i,m_j)$ for i,j=1,2,3,4 and $p \neq q$

	m_1	m_2	m_3	m_4
m_1	-	$\{a_{t_1},a_{t_2},a_{t_3}\}$	$\{a_{t_1},a_{t_2},a_{t_3}\}$	$\{a_{t_1},a_{t_2},a_{t_3}\}$
m_2	ϕ	-	$\{a_{t_1},a_{t_2},a_{t_3}\}$	$\{a_{t_1},a_{t_2},a_{t_3}\}$
m_3	ϕ	ϕ	-	$\{a_{t_1},a_{t_2},a_{t_3}\}$
m_4	ϕ	ϕ	ϕ	-

Clearly $E_A(m_i,m_j)=\phi$ for i,j=1,2,3,4 and $i \neq j$

Step 3. By using definition 4.5, we can acquire $SR_A(m_i,m_j)$ for i,j=1,2,3,4. Now $m_i >_a m_j$ or $m_j >_a m_i$ is determined by the initial score S^+ alone. We may thus consider $\delta_1 \cong 1$ without losing generality. Hence

$$SR_A(m_i,m_j)=\sum_{a\in G_A(m_i,m_j)} w_a \lambda_a(m_i,m_j)+\frac{1}{2}\sum_{a\in E_A(m_i,m_j)} w_a$$
$$=\sum_{a\in G_A(m_i,m_j)} w_a \delta_1(S^+(m_i)-S^+(m_j))+0$$
$$=\sum_{a\in G_A(m_i,m_j)} w_a(S^+(m_i)-S^+(m_j))$$

Choose $w_{a_{t_1}}, w_{a_{t_2}}$ and $w_{a_{t_3}}$ with the condition $w_{a_{t_1}}+w_{a_{t_2}}+w_{a_{t_3}}=1$ and $w_{a_{t_x}} > w_{a_{t_y}} > w_{a_{t_z}}$ where $x,y,z \in \{1,2,3\}$ and x,y,z are mutually distinct. Let us consider $w_{a_{t_1}} = 0.5, w_{a_{t_2}} = 0.3$ and $w_{a_{t_3}} = 0.2$. In this way $SR_A(m_i,m_j)$ determined and presented in Table 23.

Table 23. $SR_A(m_i,m_j)$ values for i,j=1,2,3,4

	m_1	m_2	m_3	m_4
m_1	0.50	0.01	0.049	0.09
m_2	0	0.50	0.039	0.08
m_3	0	0.039	0.50	0.041
m_4	0	0	0	0.50

Now the interval valued neutrosophic entire dominance degree $SR_A(m_i)$ of each object is found by definition 4.6. For example,

$$SR_A(m_1)=\frac{1}{4}(SR_A(m_1,m_1)+SR_A(m_1,m_2)+SR_A(m_1,m_3)+SR_A(m_1,m_4))=0.1623$$

and $SR_A(m_i)$ is given in Table 24.

Table 24. $SR_A(m_i)$ values for i=1,2,3,4

	m_1	m_2	m_3	m_4	
SR_A	0.1623	0.155	0.145	0.125	

By interval valued neutrosophic entire dominance degree $SR_A(m_i)$ we can conclude that $m_1>m_2>m_3>m_4$.

Hence m_1 is selected as the best object from the IVNIS.

Now from this section we can conclude that our proposed IVNIS is more efficient than IFIIS, since our proposed method ranks the alternatives whereas IFIIS method failed to rank them. For any given fuzzy information system our method will rank those alternatives effectively, since our method possesses a set of score functions which inherits total ordering. Note that, this is possible for all the cases where a given fuzzy information system is converted into a neutrosophic information system.

6. CONCLUSION

Decision and the selection of an object from a set of available objects based on predetermined criteria is a crucial problem in many domains, especially in education, medicine, engineering, etc., This work has begun with a study of single valued neutrosophic information system and introduced a ranking method from SVNIS based on a generalized weighted single valued dominance relation and dominance degree. Subsequently, the interval valued neutrosophic information system has been examined, and a mechanism for ranking from the IVNIS according to the generalized weighted interval valued dominance relation and dominance degree has been presented and studied by numerical example. Through comparative analysis, it has been confirmed that the proposed information system is better than the intuitionistic fuzzy interval-valued information system. In near future, n valued refined neutrosophic information system will be developed by defining n valued refined neutrosophic dominance relation and dominance degree.

Funding: The third author would like to acknowledge Council of Scientific and Industrial Research (CSIR-HRDG) India, grant number 09/895(0014)/2019-EMR-I for funding.

REFERENCES

Abdel-Basset, M., & Mohamed, M. (2021). Multicriteria group decision making based on neutrosophic analytic hierarchy process Suggested modifications. *Neutrosophic Sets and Systems, 43.*

Aiwua, Z., Jianguoa, D., & Hongjunb, G. (2015). Interval valued neutrosophic sets and multi-attribute decision-making based on generalized weighted aggregation operator. *Journal of Intelligent & Fuzzy Systems*, *29*, 2697–2706.

Al-Sharqi, , A. A.-Q. (2021). Interval-valued complex neutrosophic soft set and its applications in decision-making. *Neutrosophic Sets and Systems*, *40*, 149–168.

Atanassov, K. (1986). Intuitionistic fuzzy sets. *Fuzzy Sets and Systems*, *20*, 87–96.

Atanassov, K. T., & Gargov, G. (1989). Interval valued intuitionistic fuzzy sets. *Fuzzy Sets and Systems*, *31*, 343–349.

Bhat, A. S. (2021). A novel score and accuracy function for neutrosophic sets and their real- world applications to multi-criteria decision-making process. *Neutrosophic Sets and Systems*, *41*, 168–197.

Colhon, M., Tilea, M., González-Marcos, A., Reşceanu, A., Smarandache, F., & Navaridas-Nalda, F. (2023). A Neutrosophic Decision-Making Model for Determining Young People's Active Engagement. *International Journal of Information Technology & Decision Making.*

Geetha, S., VLG, N., & Ponalagusamy, R. (2013). Multi-criteria interval valued intuitionistic fuzzy decision making using a new score function. *KIM 2013 knowledge and information management conference.*

Gope, S., & Das, S. (2014). Fuzzy Dominance Matrix and its Application in Decision Making Problems. *International Journal of Soft Computing and Engineering, 4.*

Kalyan Mondal, S. P. (2021). NN-TOPSIS strategy for MADM in neutrosophic number setting. *Neutrosophic sets and system, 47*, 66-92.

Lakshmana Gomathi Nayagam, , B. R. (2023). A Total Order on Single Valued and Interval Valued Neutrosophic Triplets. *Neutrosophic Sets and Systems*, *55*, 20.

Li, J. W.-h. (2019). Multi-criteria decision-making method based on dominance degree and BWM with probabilistic hesitant fuzzy information. *International Journal of Machine Learning and Cybernetics.*

Nayagam, V., Ponnialagan, D., & Jeevaraj, S. (2020). Similarity measure on incomplete imprecise interval information and its applications. *Neural Computing & Applications*, *32*, 3749–3761. DOI: 10.1007/s00521-019-04277-8

Nayagam, V. L., Jeevaraj, & Sivaraman, G. (2016). Total ordering for intuitionistic fuzzy numbers. *Complexity*, *21*, 54–66. DOI: 10.1002/cplx.21783

Nayagam, V. L., & Bharanidharan, R. (2023). A Total Ordering on n - Valued Refined Neutrosophic Sets using Dictionary Ranking based on Total ordering on n - Valued Neutrosophic Tuplets. *Neutrosophic Sets and Systems, 58.*

Nayagam, V. L., Jeevaraj, S., & Dhanasekaran, P. (2016). A linear ordering on the class of Trapezoidal intuitionistic fuzzy numbers. *Expert Systems with Applications*, *60*, 269–279. DOI: 10.1016/j.eswa.2

Nayagam, V. L., & Sivaraman, G. (2011). Ranking of interval-valued intuitionistic fuzzy sets. *Applied Soft Computing*, *11*, 3368–3372. DOI: 10.1016/j.asoc.2011.01.008

Shio Gai Quek, H. G. (2023). VIKOR and TOPSIS framework with a truthful-distance measure for the (t, s)-regulated interval-valued neutrosophic soft set. *Soft Computing*.

Sivaraman, G., Nayagam, V. L., & Ponalagusamy, R. (2012)., Intuitionistic Fuzzy Interval Information System. *International Journal of Computer Theory and Engineering, 4.*

Sivaraman, G., Nayagam, V. L., & Ponalagusamy, R. (2014). A complete ranking of incomplete interval information. *Expert Systems with Applications*, *41*, 1947–1954. DOI: 10.1016/j.eswa

Smarandache, F. (2004). Neutrosophic set a generalization of the intuitionistic fuzzy set. *International Journal of Pure and Applied Mathematics*.

Smarandache, F. (2014). Refined neutrosophic logic and its application to physics.

Smarandache, F. (2020). The score, accuracy, and certainty functions determine a total order on the set of neutrosophic triplets (t, i, f). *Neutrosophic Sets and Systems*, *38*, 1–14.

Smarandache, F., Deli, I., & Said, B. (2014). N - valued interval neutrosophic sets and their application in medical diagnosis. *Critical Review*.

Sushil. (2018). Interpretive Multi-Criteria Valuation of Flexibility Initiatives on Direct Value Chain. *Benchmarking An International Journal, 25*, 3720-3742.

Velu, L., & Ramalingam, B. (2023). Total Ordering on Generalized 'n' Gonal Linear Fuzzy Numbers. *Int J Comput Intell Syst*, *16*, 23. DOI: 10.1007/s44196-022-00180-8

Zadeh, L. (1965). Fuzzy sets. *Information and Control*, 338–353.

Zhan, J., Wang, W., Carlos, A. R., & Zhan, J. (2022). A three-way decision approach with prospect-regret theory via fuzzy set pair dominance degrees for incomplete information systems. *Information Sciences*, *617*, 310–330.

Chapter 18

On Multi-Criteria Job Sequencing Decision-Making Problem via Fermatean Pentapartitioned Neutrosophic Set

R. Subha
Nirmala College for Women, Coimbatore, India

K. Mokana
Nirmala College for Women, Coimbatore, India

ABSTRACT

Every strategy for making decisions involves some degree of uncertainty. Numerous theories exist to tackle this uncertainty in real-world models. The word "scheduling" appears so often in our daily lives. We can organize and arrange tasks in the right order with the aid of scheduling. It is highly appropriate for handling production management since it helps to boost earnings and shorten the total amount of time needed to do the assigned tasks. In the natural sequencing problem, which is a specific kind of scheduling problem, a timetable is entirely determined by the order in which the jobs are completed. One significant use of operations research is job sequencing, which identifies the best order for tasks in order to reduce the overall amount of time that has passed. This chapter solves a machine sequencing problem with suitable numerical applications using a Feramatean pentapartitioned neutrosophic set technique. The findings give room for more investigation into the problems this study identified and future research directions.

DOI: 10.4018/979-8-3693-3204-7.ch018

INTRODUCTION

Making decisions is one of the most important things we do as humans. We frequently find ourselves in circumstances where we must choose between several options and criteria. Real-world scenarios are unpredictable and frequently involve ambiguity and imprecision. Zadeh (1965) created fuzzy theory, which is one of the most significant instruments for addressing the ambiguity that permeates our everyday life. A membership function T in a fuzzy set is used to determine an element's degree of membership in the discourse universe with respect to the unit interval [0, 1]. Given that the fuzzy set only included the membership function, Atanassov (1986) believed it to be incomplete. In 1986, he presented the idea of the intuitive fuzzy set, which comprises both membership and non-membership functions. The idea of the neutrosophic set, which includes an element's truth membership, falsity membership, and intermediate membership, was first presented by Smarandache (1999). The Pythgorean fuzzy set was first introduced by Yager R.R (1981), and it is currently used to tackle a lot of science and engineering challenges. Using MCDM approaches, Yager (1981), Zhang, and Xu (2014) applied the Pythagorean membership grades. PFS was used to MADM models by Peng and Yang. The application of Pythagorean interval-valued fuzzy maclaurin symmetric mean operators to MADM issues was studied by Weital (2018), Wei, and Li (2019). Garg used the idea of linear programming to apply PFN to models of decision-making. Li and Lu (2019) talked about the PFS similarity distance measure and its uses. PFS transportation models were used by Kumar et al. (2018).

When we encounter scenarios where multiple jobs need to be completed within a certain time frame, we have a work sequencing challenge. One of the most significant uses of operations research is the solution of sequencing problems. Finding the best order for the jobs on the machines to complete in order to minimize the overall elapsed time is the primary goal of the sequencing problem. Johnson's (1954) contributions to the field of production scheduling were among the best and most well-known. He minimizes both the overall idle time of the machines and the total elapsed time of all the jobs in his method. A general approach for the N jobs and M machines sequencing problems was given by Smith and Dudek (1967). Fuzzy sets, which were created by Zadeh (1965), have been widely used in recent years to solve uncertainty-related real-world issues. The literature on fuzzy sequencing contains a wide range of applications.

In daily life, the term "scheduling" is used so frequently. Actually, what the average person uses are "schedules," not scheduling. Take a bus or a class schedule, for instance. A timetable often indicates the dates on which events are expected to occur as well as a strategy for the timing of certain tasks. Thus, scheduling aids in the planning and sequential arrangement of tasks. Dealing with production management is highly appropriate because it helps to maximize profit and reduce the overall time needed to complete the assigned tasks. A pure sequencing problem is a type of stated scheduling problem where a timetable is entirely determined by the sequence in which the jobs are completed. The best order of occupations to minimize total elapsed time is determined through job sequencing, a crucial application of operations research. Stated differently, the primary goal of the sequencing problem is to determine the order in which various machines do tasks in order to maximize a given performance, such as the lowest cost, the shortest total elapsed time, the highest profit, and the completion of work by the deadline. These performances all need meticulous preparation and focus on the little things.

The remainder of the document is structured as follows: Part 2 includes the preliminary information and fundamental definitions. In Section 3, the Fermatean Pentapartitioned Neutrosophic Sets Sequencing Problem is formulated mathematically. This study presents a new approach to use Fermatean Pentapartitioned Neutrosophic Numbers (FPNN) to address the task sequencing problem. The best sequence to minimize the sequence's total elapsed time is determined using the job-sequencing technique. In order to solve the task sequencing problem in the Fermatean pentapartitioned neutrosophic environment, the score function is listed as one of the arithmetic operations of FPN. To illustrate the idea, an algorithm is first created to solve the problem and applied to a numerical example. The conclusion is presented in the end.

PRELIMINARIES

Senapati and Yager (2020) provide some basic definitions of FFSs. Senapati and Yager (2019) and Sahoo (2021) are supplied in a modified version in this part I.

Fermatean Fuzzy Set (FFS): Sahoo L. (2017)

A Fermatean fuzzy set (FFS) on a widely wide-spread set X is written as: $F = \{(x, \phi(x),\psi(x): x\in X)\}$. Such that $\phi(x)$: $X \rightarrow [0, 1]$ and $\psi(x)$: $X\rightarrow [0,1]$ with $0\leq (\phi(x))^3+ (\psi(x))^3 \leq 1, \forall x\in X$. Also, $\phi(x)$ and $\psi(x)$ are expressed as the degree of membership and non-membership of $x \in X$ in F.

The degree of indeterminacy for any FFS is expressed by

$$\xi(x) = \sqrt[3]{1 - (\phi(x))^3+ (\psi(x))^3}.$$

The set $F = \{(x, \phi(x),\psi(x): x\in X)\}$ is denoted as $F = (\phi,\psi)$ due to explicitness.

Definition: Subha R., K.Mohana (2023)

Let X be a universe. An object of the shape A Fermatean pentapartitioned neutrosophic set (FPN) A on X is A = {< x, $T_{A,}$ C_{A}, K_{A}, U_{A}, $F_{A,}$) >: x € X}

$$(T_A)^3 + (C_A)^3 + (K_A)^3 + (U_A)^3 + (F_A)^3 \leq 3$$

In this case, the truth membership is denoted by $T_A(x)$.
The contradiction membership is denoted by $C_A(x)$.
The ignorance membership is denoted by U_A (x).
The unknown membership is denoted by $K_A(x)$.
The false membership is denoted by $F_A(x)$.

Elementary Operations on FPNS: Subha R., K.Mohana (2023)

Considering the three FPNS, we have F = $(T_A, C_A, K_A, U_A, F_A)$,

$$F1 = \left(T_{A_1}, C_{A_1}, K_{A_1}, U_{A_1}\ F_{A_1}\right)$$

and $F_2 = (T_{A_2}, C_{A_2}, K_{A_2}, U_{A_2}, F_{A_2})$ on the common set X and λ>0. The following is an expression for the elementary operations on the FPNS:

Addition:

$$F_1 \oplus F_2 =$$

$$\Big(\sqrt[3]{\left(T_{A_1}\right)^3 + \left(T_{A_2}\right)^3 - \left(T_{A_1}\right)^3\left(T_{A_2}\right)^3}, \sqrt[3]{\left(C_{A_1}\right)^3 + \left(C_{A_2}\right)^3 - \left(C_{A_1}\right)^3\left(C_{A_2}\right)^3}$$

$$\sqrt[3]{\left(K_{A_1}\right)^3 + \left(K_{A_2}\right)^3 - \left(K_{A_1}\right)^3\left(K_{A_2}\right)^3}, \sqrt[3]{\left(U_{A_1}\right)^3 + \left(U_{A_2}\right)^3 - \left(U_{A_1}\right)^3\left(U_{A_2}\right)^3},$$

$$F_{A_1} \cdot F_{A_2}\Big)$$

Multiplication:

$$F1 \otimes F2 = \Big(T_{A_1} \cdot T_{A_2},$$

$$\sqrt[3]{\left(C_{A_1}\right)^3 + \left(C_{A_2}\right)^3 - \left(C_{A_1}\right)^3\left(C_{A_2}\right)^3}, \sqrt[3]{\left(K_{A_1}\right)^3 + \left(K_{A_2}\right)^3 - \left(K_{A_1}\right)^3\left(K_{A_2}\right)^3}$$

$$\sqrt[3]{\left(U_{A_1}\right)^3 + \left(U_{A_2}\right)^3 - \left(U_{A_1}\right)^3\left(U_{A_2}\right)^3}, \sqrt[3]{\left(F_{A_1}\right)^3 + \left(F_{A_2}\right)^3 - \left(F_{A_1}\right)^3\left(F_{A_2}\right)^3}\Big)$$

Scalar Multiplication:

$\lambda \odot F =$

$$\left(\sqrt[3]{1-(1-T_A{}^3)^{\lambda}}, \sqrt[3]{1-(1-C_A{}^3)^{\lambda}}, \sqrt[3]{1-(1-K_A{}^3)^{\lambda}}, \sqrt[3]{1-(1-U_A{}^3)^{\lambda}}, F_A{}^{\lambda}\right)$$

Exponent:

$$F^{\lambda} = \left(T_A{}^{\lambda}, \sqrt[3]{1-(1-C_A{}^3)^{\lambda}}, \sqrt[3]{1-(1-K_A{}^3)^{\lambda}}, \sqrt[3]{1-(1-U_A{}^3)^{\lambda}}, \sqrt[3]{1-(1-F_A{}^3)^{\lambda}}\right)$$

Union:

$F_1 \cup F_2 = (\max(T_{A,}\ T_B), \max(C_{A,}\ C_B), \min(K_A, K_B)_, \min(U_A, U_B), \min(F_A, F_B))$

Intersection:

$F_1 \cap F_2 = (\min(T_{A,}\ T_B), \min(C_{A,}\ C_B), \max(K_A, K_B)_, \max(U_A, U_B), \max(F_A, F_B))$

Complement:

$F^c = (F_A, U_A, K_A, C_A, T_A)$

Fermatean Fuzzy Score Function: Sahoo L. (2017)

Let us consider what needs to be taken into account for an FFS F = (ϕ, ψ). The score function for F is then represented by the following symbol:

$$S^*(F) = \frac{1}{2}(1+\phi^3-\psi^3).(\mathrm{Min}(\phi, \psi)).$$

Ranking of Fermatean Fuzzy Sets: Sahoo L. (2017)

Let us assume two FFSs, $F_1 = (\phi_1, \psi_1)$ and $F_2 = (\phi_2, \psi_2)$, the corresponding ranking laws of F_1 and F_2 are described in the following way:

Case 1: when $S^*(F_1) \geq S^*(F_2)$ with $A^*(F_1) > A^*(F_2)$ iff $F_1 > F_2$

Case 2: when $S^*(F_1) \leq S^*(F_2)$ with $A^*(F_1) < A^*(F_2)$ iff $F_1 < F_2$

Case 3: when $S^*(F_1) = S^*(F_2)$ with $A^*(F_1) < A^*(F_2)$ iff $F_1 = F_2$

Proposed Fermatean Pentapartitioned Neutrosophic Score Function: Subha R. & K. Mohana (2023)

In this section, we constructed a ranking mechanism for organizing the FPNS in circumstances involving selection.

Fermatean Pentapartitioned Neutrosophic Score Function: Subha R. & K. Mohana (2023)

Consider the following FPNS F = (T_A, C_A, K_A, U_A, F_A). Hence the following represents the score function for

$$F{:}^*(F) = \frac{1}{5}\begin{pmatrix} (1+\boldsymbol{T}_A^{\,3}-\boldsymbol{F}_A^{\,3}),\ (1+\boldsymbol{C}_A^{\,3}-\boldsymbol{F}_A^{\,3}), \\ (1+\boldsymbol{K}_A^{\,3}-\boldsymbol{F}_A^{\,3}),\ (1+\boldsymbol{U}_A^{\,3}-\boldsymbol{F}_A^{\,3}.\ Min\,(\boldsymbol{T}_A, \boldsymbol{F}_A) \end{pmatrix} \tag{1}$$

Remark.
Consider an FPNS F = (T_A, C_A, K_A, U_A, F_A), then S*(F) ∈ [0, 1].

Ranking of Fermatean Pentapartitioned Neutrosophic Sets (RFPNS): Subha R., K.Mohana (2023)

Consider these two FPNSs:

$$F_1 = \left(\mathbf{T}_{A_1}, \mathbf{C}_{A_1}, \mathbf{K}_{A_1}, \mathbf{U}_{A_1}\ \mathbf{F}_{A_1}\right)\ and\ F_2 = \left(\mathbf{T}_{A_2}, \mathbf{C}_{A_2}, \mathbf{K}_{A_2}, \mathbf{U}_{A_2}, \mathbf{F}_{A_2}\right)$$

and the ranking laws for F_1 and F_2 are described in the following way:
Case 1: when S* (F_1) ≥ S* (F_2) with A* (F_1) > A* (F_2) iff $F_1 > F_2$
Case 2: when S* (F_1) ≤ S* (F_2) with A* (F_1) < A* (F_2) iff $F_1 < F_2$
Case 3: when S* (F_1) = S* (F_2) with A* (F_1) < A* (F_2) iff $F_1 = F_2$ ranking laws.

Example:

Let F_1 = (0.5, 0.2, 0.5, 0.3, 0.7), F_2 = (0.2, 0.4, 0.5, 0.4, 0.6) be two FPNSs, then we have

$$S^*(F) = \tfrac{1}{5}\left((1+\boldsymbol{T}_A^{\,3}-\boldsymbol{F}_A^{\,3}),\ (1+\boldsymbol{C}_A^{\,3}-\boldsymbol{F}_A^{\,3}),\ (1+\boldsymbol{K}_A^{\,3}-\boldsymbol{F}_A^{\,3}),\ (1+\boldsymbol{U}_A^{\,3}-\boldsymbol{F}_A^{\,3})\right)$$

Min (T_A, F_A).

$$S^*(\boldsymbol{F}_1) = \tfrac{1}{5}(1+0.5^3-0.7^3),\ (1+0.2^3-0.7^3),\ (1+0.5^3-0.7^3),\ \left(1+0.3^3-0.7^3\right)$$

Min (0.5, 0.7) = 0.0278157807

And

$$S*(F_2) = \frac{1}{5}(1 + 0.2^3 - 0.6^3), (1 + 0.4^3 - 0.6^3), (1 + 0.5^3 - 0.6^3), (1 + 0.4^3 - 0.6^3)$$

Min (0.2, 0.6) = 0.0207081242

Hence $S*(F_1) \geq S*(F_2)$ with $A*(F_1) > A*(F_2)$ iff $F_1 > F_2$

JOB SEQUENCING PROBLEM

Consider 'n' jobs (1, 2, 3....n) to be processed at each machine A, B, C, ... Each job is processed in a specific order through the equipment. The time that each job should take on each machine is known. The objective is to identify a sequence among (n!)^m number of potential sequences, combinations, or orders for processing the jobs so that the total elapsed time for all the jobs is minimal.

In mathematical perspective, let us consider

A_t= time for job t on Machine A
B_t= time for job t on Machine B, etc.

T stands for total lapsed time, or the amount of time from the beginning of the first job to the end of the last one. Next, the task is to ascertain a series of jobs $i_1, i_2, i_3, \ldots, i_n$ for every machine, whereas $(i_1, i_2, i_3, \ldots, i_n)$ the permutation of the numbers will minimize T.

Terminology and Notations

This paper will utilize the following keywords and notations:

1) **Number of Machines:** This is the total number of service facilities that a project must transit through in order to be finished.
2) **Processing Order:** This refers to the sequence in which the machines finish all of the tasks. If every job is to be processed on machine A, machine B, and machine C in that sequence, the processing order is ABC. The sequence in which different machines are needed to finish the task is indicated.

3) **Processing Time:** The amount of time needed for a task on every computer is known as the processing time. It refers to the amount of time needed for each task on each machine. The notation will show how long each job takes the machine to process (i = 1, 2, 3, … n); j = 1, 2, 3,... m).
4) **Idle Time on a Machine:** This represents the amount of time that a machine is idle throughout the entire amount of elapsed time. The idle time of machine j between the ending of the $(i-1)^{th}$ job and the beginning of the ith job is indicated by the notation X_{ij}.
5) **Total Elapsed Time:** The amount of time from the beginning of processing the first job on the first machine to the end of the final job on the last machine is referred to as the total elapsed time. It is the sum of each machine's processing and idle times. That is the amount of time that separates beginning the first job and finishing the last one. If idle time is present, it is also included in this. The letter T will be used to represent it.
6) **No Passing Rule:** The order in which jobs are to be processed on two separate machines is indicated by this rule. For instance, each work must go to machine 'A' first and then machine 'B' if n jobs are to be processed in the sequence AB on two machines, A and B. According to this rule, passing is prohibited. That is, each machine's jobs are kept in the same order. This rule indicates that each task will go to machine 'A' first and then machine B if each of the n jobs is to be processed through two machines, A and B, in the order AB.

Main Premises

1. Multiple operations cannot be processed by a single machine at once.
2. Once an operation begins, it must be carried out through to the end.
3. A job is an entity that may not be handled by multiple machines at once, despite the fact that it consists of numerous distinct elements.
4. Before starting any other operation that needs to happen, each one needs to be finished.
5. The sequence in which procedures are completed has no bearing on the processing time intervals.
6. Every kind of machine is singular.
7. A project is completed as quickly as feasible, based on the specifications of the order.
8. Before the time frame under review starts, every job is known and prepared to be processed.
9. There is very little time needed to switch jobs between machines.

Solution of Sequencing Problem

Currently, there are answers for the following cases:

1) Assume that there are 'n' jobs and that machines A and B process the jobs in sequence AB.
2) Assume that n jobs are processed in the ABC order by three machines A, B, and C.
3) M machines and two jobs. The order in which the machines process each work is specified and need not be the same for both jobs.
4) Issues with m machines and n jobs.

An Algorithm to solve Job Sequencing using Fermatean Pentapartitioned Neutrosophic Numbers

Our sequencing issues come in a variety of forms. The goal of every problem is to reduce the overall amount of time spent and the amount of time spent waiting for the jobs. Let's now examine the scenario where two machines process n jobs each. Using this approach, the sequencing problem can be expressed as follows since the machine processing time is treated as a Fermatean pentapartitioned Neutrosophic number:

Table 1. Two Machines Process N Jobs

Jobs → Machines↓	J_1	J_2	....	J_n
M_1	t_{11}	t_{12}	....	t_{1n}
M_2	t_{21}	t_{22}	.....	t_{2n}

Here t_{ij} is the Fermatean pentapartitioned neutrosophic number, which indicates the time duration captured by the ith job on the jth machine.

Algorithm for Solving Problem

Johnson developed an iterative process that may be summarized as follows for figuring out the best sequence for an n-job 2-machine sequencing problem:

1. Start
2. Determine the scoring function for each and every processing time t_{ij}.
3. Find the traditional sequencing value by using the score function formula for all processing times t_{ij}.

4. Examine the $A_i's$ and $B_i's$ for i = 1, 2 …., n and determine min[A_i,B_i].
5. (i) If this minimum is A_k for some i = k, perform the kth job first of all.
   (ii) If this minimum is B_r for some i = r, perform the rth job last of all.
6. (i) In case of a tie for minima A_k=B_r, proceed with processing the kth job initially and the rth job last.
   (ii) If the tie for the minimum occurs among the $A_i's$, select the job corresponding to the minimum of $B_i's$ and process it first of all.
   (iii) If the tie for minimum occurs among the $B_i's$, select the job corresponding to the minimum of $A_i's$ and process it in the last. Then go to the next step.
7. Once all of the jobs have been assigned, cross out the ones that have already been assigned and repeat steps 2 through 4 to arrange the jobs next to each other or next to last.
8. Determine the total amount of time that has passed and the idle time for both Machines.

Numerical Example

Example 3.5.1

The two machines A and B must be used for each of the five jobs in the order AB. The amount of time needed to complete each job's operations is recorded as FPNS. The following are the processing times:

Table 2. Input Type-1 FPNSP

Jobs → Machines↓	J_1	J_2	J_3	J_4	J_5
M_1	(0.4,0.3,0.6,0.5,0.7)	(0.5,0.4,0.6,0.7,0.4)	(0.8,0.6,0.2,0.1,0.3)	(0.6,0.4,0.7,0.8,0.3)	(0.9,0.7,0.8,0.4,0.1)
M_2	(0.4,0.6,0.3,0.5,0.2)	(0.7,0.6,0.4,0.2,0.3)	(0.6,0.3,0.4,0.2,0.4)	(0.7,0.3,0.2,0.4,0.3)	(0.8,0.6,0.3,0.2,0.1)

Find the five jobs that should be completed in the best order to reduce the elapsed time T. Determine how much time the machines were idle overall at this time.

Answer:

Step 1: Find traditional table values using the above equation (1) and the score function formula. Then, the traditional table becomes

Table 3. Traditional Sequencing Problem

Jobs → Machines↓	J_1	J_2	J_3	J_4	J_5
M_1	0.0532	0.3479	0.3441	0.5898	0.9987
M_2	0.3757	0.4127	0.1964	0.2846	0.4896

Step 2: Examine the minimum value that occurs, whether in machine M_1 or M_2. Here, the smallest value, 0.0532 occurs in job 1, which is in job J_1 in the first machine M_1 Job sequences occur first of all. So the job sequences in this table are

Table 4.

J_1				

Step 3: The diminished table is

Table 5. Diminished Table

Jobs → Machines↓	J_2	J_3	J_4	J_5
M_1	0.3479	0.3441	0.5898	0.9987
M_2	0.4127	0.1964	0.2846	0.4896

Step 4: Next, examine the minimum value that occurs whether the machine is M_1 or M_2. Here, the smallest value is 0.1964, which occurs in job 3, which is in J_3 in the second machine M_3 Job sequences occur last of all. So the job sequences in this table are

Table 6.

J_1				J_3

Step 5: The diminished table is

Table 7.

Jobs → Machines↓	J_2	J_4	J_5
M_1	0.3479	0.5898	0.9987
M_2	0.4127	0.2846	0.4896

Step 6: Next, examine minimum value that occurs whether the machine M_1 or M_2. Here, the smallest value is 0.2846, occurs in job 4, which is in J_4 in the second machine M_2 Job sequences occur last of all. So the job sequences in this table are

Table 8.

J_1			J_4	J_3

Step 7: The diminished table is

Table 9.

Jobs → Machines↓	J_2	J_5
M_1	0.3479	0.9987
M_2	0.4127	0.4896

Step 8: Next, examine the minimum value that occurs, whether in machine M_1 or M_2. Here, the smallest value is 0.3479, which occurs in job 2, which is in J_2 in the first machine M_1 Job sequences occur first of all. So the job sequences in this table are

Table 10.

J_1	J_2		J_4	J_3

Step 9: Finally, the optimal sequence will become

Table 11.

J_1	J_2	J_5	J_4	J_3

Step 10: To find the total elapsed time

Furthermore, using the individual processing time provided in the issue statement, it is also possible to determine the lowest elapsed time corresponding to the optimal sequencing. The information is provided in Table 12:

Table 12. Total Elapsed Time

Job sequence	Machine A		Machine B	
	Time in	Time out	Time in	Time out
J_1	0	0.0532	0.0532	0.4289
J_2	0.0532	0.4011	0.4289	0.8416
J_5	0.4011	1.3998	1.3998	1.8894
J_4	1.3998	1.9896	1.9896	2.2742
J_3	1.9896	2.3337	2.3337	2.5301

Consequently, 2.5301 hours is the minimum time, or the amount of time needed to complete tasks 1 through 3. The machine "A" idles for 0.1931 hours (from 2.337 to 2.5301 hours) throughout this period, while the machine "B" idles for only 0.7711 hours (from 1.759 to 2.5301 hours).

As a result, 2.5301 hours is the least total elapsed time.

Idle time of machine M_1 = 2.530 - 2.3337 = 0.1963 hours

Idle time of machine M_2 = 2.530 - 1.759 = 0.771 hours

Processing 'n' Jobs Through 'm' Machines

Let each of the n jobs be processed through m machines, say $M_1, M_2, M_3, \ldots, M_m$ in order $M_1, M_2, M_3, \ldots, M_m$, and T_{ij} denote the time taken by the ith machine to complete the jth job. The step-by-step procedure for obtaining an optimal sequence is as follows:

1. Start
2. Determine the scoring function for each and every processing time t_{ij}.
3. Find the traditional sequencing value by using the above equation (1) and the FPN score function formula for all processing times t_{ij}.
4. Determine (i) $\operatorname{Min}_j \boldsymbol{T}_{1j}$, (ii) $\operatorname{Min}_j \boldsymbol{T}_{mj}$, and (iii) $\operatorname{Max}_j \boldsymbol{T}_{2j}, \boldsymbol{T}_{3j} \ldots \boldsymbol{T}_{(m-1)j}$ for j = 1, 2, 3 … n.
5. Then check whether (i) $\operatorname{Min}_j \boldsymbol{T}_{1j} \geq \operatorname{Max}_j \boldsymbol{T}_{2j}, \boldsymbol{T}_{3j} \ldots \boldsymbol{T}_{(m-1)j}$ or (ii) $\operatorname{Min}_j \boldsymbol{T}_{mj} \geq \operatorname{Max}_j \boldsymbol{T}_{2j}, \boldsymbol{T}_{3j} \ldots \boldsymbol{T}_{(m-1)j}$
6. If the inequalities of steps 4 and 5 suggest that are not satisfied, this technique fails. Hence, go to the next step.

7. Taking into account two hypothetical machines, G and H, convert the m-machine problem to a 2-machine problem, in order that $T_{Gj} = T_{1j} + T_{2j} + \ldots + T_{(m-1)j}$ and $T_{Hj} = T_{2j} + T_{3j} + \ldots + T_{mj}$. Now use the optimal sequence algorithm to find the best order for the jobs to be completed using e-machines.
8. Apart from the prerequisites specified in step 4 and 5, if $T_{2j} + T_{3j} + \ldots + T_{(m-1)j} = C$ (a fixed positive constant) for all j = 1, 2. …., n, then use the optimal sequence algorithm to find the optimal sequence for n tasks and two machines, M_1 and M_m in the order M_1M_m.

Table 13.

Machine	Job				
	1	2	3	...	n
M_1	A_1	A_2	A_3	...	A_n
M_2	B_1	B_2	B_3	...	B_n
M_3	C_1	C_2	C_3	...	C_n
...	...	...	...	...	...
M_m	...	...	...	...	....

Example

Every one of the four jobs needs to pass through all six machines M_i.

The processing timings for i = 1, 2, 3, 4, 5, 6 in the sequence $M_1, M_2, \ldots, M_6$ are provided. The amount of time needed to complete each job's operations is recorded as FPNS.

Table 14. Input Fermatean Pentapartitioned Neutrosophic Number

Machines	Jobs			
	1	2	3	4
M_1	(0.6,0.3,0.4,0.2,0.4)	(0.5,0.4,0.6,0.7,0.4)	(0.7,0.6,0.4,0.9,0.3)	(0.7,0.6,0.4,0.2,0.3)
M_2	(0.4,0.3,0.6,0.5,0.7)	(0.7,0.6,0.4,0.2,0.3)	(0.9,0.7,0.8,0.4,0.1)	(0.6,0.3,0.4,0.2,0.4)
M_3	(0.1,0.9,0.6,0.5,0.7)	(0.2,0.8,0.6,0.7,0.4)	(0.9,0.7,0.8,0.4,0.1)	(0.1,0.9,0.7,0.8,0.3)
M_4	(0.1,0.7,0.2,0.6,0.8)	(0.4,0.7,0.3,0.6,0.2)	(0.3,0.8,0.3,0.6,0.2)	(0.7,0.1,0.2,0.4,0.3)
M_5	(0.1,0.3,0.4,0.6,0.8)	(0.8,0.6,0.3,0.2,0.1)	(0.6,0.3,0.4,0.2,0.4)	(0.9,0.7,0.2,0.4,0.1)
M_6	(0.7,0.6,0.4,0.2,0.3)	(0.5,0.4,0.6,0.7,0.4)	(0.7,0.6,0.4,0.2,0.3)	(0.8,0.6,0.2,0.1,0.3)

Find the correct sequence in which to do these 4 jobs in order to minimize the total elapsed time T.

Solution:

Step 1: Find out the traditional sequencing table values using the above equation (1) and the score function formula. Then, the traditional table becomes

Table 15. Traditional Sequencing Problem

Machines	Job			
	1	2	3	4
M_1	0.0838	0.1251	0.1653	0.0955
M_2	0.0269	0.0955	0.0745	0.0838
M_3	0.0125	0.0403	0.0745	0.0648
M_4	0.0028	0.0694	0.0755	0.0782
M_5	0.00196	0.0379	0.0838	0.0496
M_6	0.0955	0.1392	0.0955	0.1012

Step 2: In this table, find

$$\underset{j}{\text{Min}}\, \boldsymbol{T}_{1j} = 0.0838, \underset{j}{\text{Min}}\, \boldsymbol{T}_{6j} = 0.0955 \textbf{ and } \underset{j}{\text{Max}}\, \boldsymbol{T}_{2j}, \boldsymbol{T}_{3j}, \boldsymbol{T}_{4j}, \boldsymbol{T}_{5j} = 0.0955$$

Step 3: Then check whether

(i) $\underset{j}{\text{Min}}\, \boldsymbol{T}_{1j} \geq \underset{j}{\text{Max}}\, \boldsymbol{T}_{2j}, \boldsymbol{T}_{3j}, \boldsymbol{T}_{4j}, \boldsymbol{T}_{5j}$ or
(ii) $\underset{j}{\text{Min}}\, \boldsymbol{T}_{6j} \geq \underset{j}{\text{Max}}\, \boldsymbol{T}_{2j}, \boldsymbol{T}_{3j}, \boldsymbol{T}_{4j}, \boldsymbol{T}_{5j}$ are satisfied.

Step 4: Next, the problem can be transformed into a 4-job and 2-machine problem.

Table 16. Converted 4-jobs 2-machines

Machine	Job			
	1	2	3	4
G	0.12796	0.3682	0.4736	0.3719
H	0.13966	0.3823	0.4038	0.3776

Step 5: Next, examine the minimum value that occurs, whether the machine is G or H. Here the smallest value of 0.12796 occurs in the job 1 in first machine G job sequence occur first of all. So the job sequence in this table is

Table 17.

1			

Step 6: The diminished table is

Table 18.

Machine	Job		
	2	3	4
G	0.3682	0.4736	0.3719
H	0.3823	0.4038	0.3776

Step 7: Next, examine the minimum value occurs, whether the machine is G or H. Here, the smallest value of 0.3682 occurs in the job 2 in first machine G job sequence occur first of all. So the job sequence in this table is

Table 19.

1	2		

Step 8: The diminished table is

Table 20.

Machine	Job	
	3	4
G	0.4736	0.3719
H	0.4038	0.3776

Step 9: Next, examine the minimum value occurs, whether the machine is G or H. Here, the smallest value of 0.3719 occurs in the job 4 in first machine G job sequence occur first of all. So the job sequence in this table is

Table 21.

1	2	4	

Step 10: The optimal sequence becomes

Table 22.

1	2	4	3

Step 11: To calculate the total elapsed time:

Table 23. Total Elapsed Time

Job sequence	Machine M_1		Machine M_2		Machine M_3	
	Time in	Time out	Time in	Time out	Time in	Time out
1	0	0.0838	0.0838	0.1107	0.1107	0.1232
3	0.0838	0.2089	0.2089	0.3044	0.3044	0.3447
2	0.2089	0.3044	0.3044	0.3882	0.3882	0.453
4	0.3044	0.4697	0.4697	0.5442	0.5442	0.6187

Table 24. Total Elapsed Time Cont.

Job sequence	Machine M_4		Machine M_5		Machine M_6	
	Time in	Time out	Time in	Time out	Time in	Time out
1	0.1232	0.126	0.126	0.12796	0.12796	0.22346
3	0.3447	0.4141	0.4141	0.452	0.452	0.5912
2	0.453	0.5312	0.5312	0.5808	0.5912	0.6924
4	0.6187	0.6942	0.6942	0.778	0.778	0.8735

Hence, the total elapsed time is 0.8735 hours.
Idle time of Machine M_1 = 0.4038 hours
Idle time of Machine M_2 = 0.5928 hours
Idle time of Machine M_3 = 0.6814 hours
Idle time of Machine M_4 = 0.6476 hours
Idle time of Machine M_5 = 0.70024 hours
Idle time of Machine M_6 = 0.4421 hours.

Processing 'n' Jobs Through Three Machines

The problem can be explained as follows:

1) There are just three machines involved: A, B, and C.
2) Every job is completed in the specified ABC order.

3) Transfer of jobs is unpermitted unless the order is strictly followed on each machine.
4) The table below provides the processing timeframes, either exact or estimated.

Table 25.

Machine	Job				
	1	**2**	**3**	**...**	**n**
M_1	A_1	A_2	A_3	...	A_n
M_2	B_1	B_2	B_3	...	B_n
M_3	C_1	C_2	C_3	...	C_n

Optimal Solution: Ideal Resolution: As of yet, there is no universal process for determining the best sequence in this situation. On the other hand, Johnson's (1954) earlier approach can be expanded to include the unique situation in which one or both of the subsequent circumstances apply:

The following is a step-by-step process for getting an optimal sequence:

1. Start
2. Determine the scoring function for each and every processing time t_{ij}.
3. Find the traditional sequencing value by using the above equation (1) and the FPN score function formula for all processing times t_{ij}.
4. The minimum time on machine A$\geq$ The maximum time on machine B.
5. The minimum time on machine C$\geq$ The maximum time on machine B.
6. If the inequalities of steps 4 and 5 suggest that are not satisfied, this technique fails. Hence, go to the next step.
7. Taking into account two hypothetical machines, G and H, convert the m-machine problem to a 2-machine problem, in order that $T_{Gj} = T_{1j} + T_{2j}$ and $T_{Hj} = T_{2j} + T_{3j}$. Now use the optimal sequence algorithm to find the best order for the jobs to be completed using e-machines.
8. Apart from the prerequisites specified in steps 4 and 5, if $T_{2j} + T_{3j} + \ldots + T_{(m-1)j} = C$ (a fixed positive constant) for all j = 1, 2. …., n, then use the optimal sequence algorithm to find the optimal sequence for n tasks and two machines, M_1 and M_3 in the order M_1M_3.
9. We must determine the total elapsed time and the idle time of the machines after determining the optimal sequence.
10. End.

Example

Each of the five jobs needs to pass through machines A, B, and C in the ABC order. The amount of time needed to complete each job's operations is recorded as FPNS. The processing times are listed in the following table.

.

Table 26.

Machine	Job				
	1	2	3	4	5
M_1	(0.8,0.6,0.2,0.1,0.3)	(0.3,0.8,0.3,0.6,0.2)	(0.6,0.3,0.4,0.2,0.4)	(0.5,0.4,0.6,0.7,0.4)	(0.7,0.6,0.4,0.9,0.3)
M_2	(0.4,0.3,0.6,0.5,0.7)	(0.4,0.6,0.3,0.5,0.2)	(0.7,0.3,0.4,0.6,0.8)	(0.8,0.6,0.3,0.2,0.1)	(0.7,0.1,0.2,0.4,0.3)
M_3	(0.7,0.6,0.4,0.2,0.3)	(0.6,0.3,0.4,0.2,0.4)	(0.7,0.3,0.2,0.4,0.3)	(0.6,0.4,0.7,0.8,0.3)	(0.7,0.6,0.4,0.2,0.3)

Find the best sequence for five jobs that will reduce the total amount of time (T) needed to finish them all.

Solution:

Table 27. Traditional Sequencing Problem

Machine	Job				
	1	2	3	4	5
M_1	0.1012	0.1653	0.0838	0.0782	0.1392
M_2	0.0269	0.0581	0.0356	0.0379	0.0755
M_3	0.0955	0.0838	0.0803	0.1446	0.0955

Step 2: In this table, find

$$\underset{j}{\text{Min}}\, T_{1j} = 0.0782,\ \underset{j}{\text{Min}}\, t_{3j} = 0.0803 \text{ and } \underset{j}{\text{Max}}\, T_{2j} = 0.0755$$

Step 3: Proceed to verify if

(i) $\underset{j}{\text{Min}}\, T_{1j} \geq \underset{j}{\text{Max}}\, T_{2j}$ alternatively,
(ii) $\underset{j}{\text{Min}}\, T_{3j} \geq \underset{j}{\text{Max}}\, T_{2j}$ are satisfied.

Step 4: Next, the problem can be changed into a 5-job and 2-machine problem.

Table 28. converted 5-jobs 2-machines

Machine	Job				
	1	2	3	4	5
G	0.1281	0.2234	0.1194	0.1161	0.2147
H	0.1224	0.1419	0.1159	0.1825	0.171

Step 5: Examine the minimum value that occurs whether the machine G or H. Here the smallest value of 0.1159 occurs in job 3 in first machine H. Job sequences occur last of all. So the job sequence in this table is

Table 29.

				3

Step 6: The reduced table is

Table 30.

Machine	Job			
	1	2	4	5
G	0.1281	0.2234	0.1161	0.2147
H	0.1224	0.1419	0.1825	0.171

Step 7: Next, examine the minimum value that occurs, whether in machine G or H. Here, the smallest value is 0.1161, which occurs in job 4 in the first machine G job sequence. So the job sequence in this table is

Table 31.

4				3

Step 7: The reduced table is

Table 32.

Machine	Job		
	1	2	5
G	0.1281	0.2234	0.2147
H	0.1224	0.1419	0.171

Step 8: Next, examine the minimum value that occurs, whether in machine G or H. Here, the smallest value is 0.1224, which occurs in job 1 in the first machine H job sequence last of all. So the job sequence in this table is

Table 33.

4			1	3

Step 9: The reduced table is

Table 34.

Machine	Job	
	2	5
G	0.2234	0.2147
H	0.1419	0.171

Step 10: Next, examine the minimum value that occurs, whether the machine is G or H. Here, the smallest value is 0.1419, which occurs in job 2 in the first machine H job sequence, last of all. So the job sequence in this table is

Table 35.

4		2	1	3

Step 11: The optimal sequence becomes

Table 36.

4	5	2	1	3

Step 12: To calculate the total elapsed time:

Table 37. Total Elapsed Time

Job sequence	Machine M_1		Machine M_2		Machine M_3	
	Time in	Time out	Time in	Time out	Time in	Time out
1	0	0.1012	0.1012	0.1281	0.1281	0.2236
3	0.1012	0.2665	0.2665	0.3246	0.3246	0.4084
2	0.2665	0.3503	0.3503	0.3859	0.4084	0.4887
4	0.3503	0.4285	0.4285	0.4664	0.4887	0.6333
5	0.4285	0.5677	0.5677	0.6432	0.6432	0.7387

Hence, the total elapsed time is 0.7387 hours
Idle time of Machine $M_1 = 0.171$ hours
Idle time of Machine $M_2 = 0.5047$ hours
Idle time of Machine $M_3 = 0.239$ hours

CONCLUSION

The problem of uncertainty in fermatean pentapartitioned neutrosophic sequencing problems is addressed with a proposed algorithm. To find the best job sequence that minimizes the total elapsed time of the fermatean pentapartitioned neutrosophic job sequencing problem, the existing operations on fermatean pentapartitioned neutrosophic numbers and score function are utilized. The approach described here might be applied in the future to resolve practical issues.

REFERENCES

Antony Crispin Sweety, A., & Janshi, R. (2021). Fermatean Neutrosophic Sets. *IJARCCE, 10*(6), 24–27.

Atanassov, K. (1986). Intuitionistic Fuzzy Sets. *Fuzzy Sets and Systems, 20*(1), 87–96. DOI: 10.1016/S0165-0114(86)80034-3

Hong, T., & Chuang, T. (1999). New triangular fuzzy Johnson algorithm. *Computers & Industrial Engineering, 36*(1), 179–200. DOI: 10.1016/S0360-8352(99)00008-X

Johnson, S. M. (1954). Optimal Two and Three Stage Production schedules with setup Times Included. *Naval Research Logistics Quarterly, 1*(1), 61–68. DOI: 10.1002/nav.3800010110

Kumar, P. S. (2018). PSK method for solving intuitionistic fuzzy solid transportation problems. *International Journal of Fuzzy System Applications, 7*(4), 62–99. DOI: 10.4018/IJFSA.2018100104

Li, X. & Lu M, (2019). Some Novel Similarity and distance and Measures of Pythagorean Fuzzy Sets and their Applications. *J. Intel, Fuzzy Syst. 37*, 1781-1799.

McCahon, C. S., & Lee, E. S. (1990). Job Sequencing with fuzzy processing times, computers &. *Mathematics for Applications, 19*(7), 31–41.

Peng, X., & Yang, Y. (2016). Pythagorean Fuzzy Choquet Integral Based Mabac Method for Multiple attribute Group Decision Making. *International Journal of Intelligent Systems, 31*(10), 989–1020. DOI: 10.1002/int.21814

Radha, R., & Stanis Arul Mary, A. (2021). Pentapartitioned Neutrosophic Pythagorean set. *IRJASH, 3*(2S), 62–82. DOI: 10.47392/irjash.2021.041

. (2020). Rama malik, Surapati Pramanik (2020), Pentapartitioned Neutrosophic set and its properties. *Neutrosophic Sets and Systems, 36*, 184–192.

Sahoo, L. (2017). Solving Job Sequencing problem with fuzzy processing times. *IJARIIE, 3*(4), 3326–3329.

Selvakumari, K., & Santhi, S. (2018). An Approach for solving Fuzzy Sequencing Problems with octagonal Fuzzy numbers using Robust Ranking Techniques. *Int. Jour. Of Mathematics Trends and Technology, 56*(3), 148–152. DOI: 10.14445/22315373/IJMTT-V56P521

Smarandache, F. (1999). A Unifying Field in Logics. Neutrosophy: Neutrosophic Probability, Set and Logic. American research Press, Rehoboth.

Smith, R. D., & Dudek, R. D. (1967). A General Algorithm for solution of the N-jobs, M-Machine sequencing problem of the Flow Shop. *Operations Research, 21*(1).

Subha, R., & Mohana, K. (2023). Fermatean pentapartitioned neutrosophic sets. *Indian Journal of Natural Sciences, 14*.

Wei, G., & Lu, M. (2018). Pythagorean Fuzzy Maclaurin Symmetric Mean Operators in Multiple Attribute decision Making. *IEEE Access : Practical Innovations, Open Solutions*, *6*, 7866–7884. DOI: 10.1109/ACCESS.2018.2877725

Wei, G. (2019). The Novel Generalized Exponential *Entropy for intuitionistic Fuzzy Sets and Interval Valued Intuitionistic,* 2339.

Yager, R. R. (1981). A procedure for ordering fuzzy subsets of the unit interval. *Information Sciences*, *24*(2), 143–161. DOI: 10.1016/0020-0255(81)90017-7

Zadeh, L. (1965). Fuzzy sets. *Information and Control*, *8*(3), 87–96. DOI: 10.1016/S0019-9958(65)90241-X

Zhang, X., & Xu, Z. (2014). Extensioon of TOPSIS to Multicriteria decision Making with Pythagorean Fuzzy Sets. *International Journal of Intelligent Systems*, *29*(12), 1061–1078. DOI: 10.1002/int.21676

Zhang, Z., Yang, J. Ye. Y., Hu, Y., & Zhang, Q. (2012). A type of score function on intuitionistic fuzzy set with double parameters and its application to pattern recognition and medical diagnosis. *Procedia Engineering*, *29*, 4336–4342. DOI: 10.1016/j.proeng.2012.01.667

Chapter 19
An Innovative Approach to Group Decision-Making Based on Weighted Hypersoft Expert System

Ajoy Kanti Das
https://orcid.org/0000-0002-9326-1677
Tripura University, India

Rakhal Das
ICFAI University, India

Rupak Datta
https://orcid.org/0000-0002-2469-6925
Tripura University, India

Carlos Granados
Universidad de Antioquia, Colombia

ABSTRACT

The aim of this chapter is to explore the concept of weighted hypersoft expert sets (WHSES) and analyze their core properties, extending from hypersoft expert sets (HSES), also to introduce innovative notions like agree-WHSES, disagree-WHSES, and weighted scores, enhancing understanding of decision consensus in WHSES frameworks. This proposed decision-making method (DMM) is adaptable and suited for real-life decision-making problems (DMPs), demonstrated with practical

DOI: 10.4018/979-8-3693-3204-7.ch019

examples. Furthermore, authors conducted a comparative analysis to validate the effectiveness of our approach in advancing decision science methodologies.

INTRODUCTION

Molodtsov (1999) introduced Soft set theory (SST) as a foundational and highly effective mathematical tool for addressing complexity, vague definitions, and unknown elements in various domains. Unlike other uncertainty modelling methods, SST offers flexibility in describing elements, allowing researchers to choose parameters according to their specific needs. This flexibility significantly simplifies DMPs and enhances decision-making processes, especially when dealing with partial knowledge or uncertain information. While there are alternative mathematical frameworks for handling uncertainties, such as game theory, probability theory, operations analysis, rough set theory, fuzzy set theory, and interval-valued fuzzy set theory, each has its limitations and challenges. One of the key drawbacks of these theories is the lack of parameterization, which restricts their applicability in solving practical problems, particularly in economic, environmental, and social contexts. In contrast, SST excels in overcoming these limitations and provides a clear and effective approach to addressing uncertainties. Molodtsov (1999) pioneered the development of SST, laying down its foundational principles and successfully applying them across various fields. These applications include but are not limited to functions' smoothness analysis, operations analysis, game theory, Riemann integrations, and probability theory. Molodtsov's (Molodtsov, 1999) definition of a soft set over a nonempty universe Ω_H is represented as a pair (Ψ, P), where Ψ is a function defined as Ψ: P $\rightarrow$ P(Ω_H), with P representing a set of parameters and P(Ω_H) denoting the power set of Ω_H. This formalization provides a structured framework for utilizing soft sets in modelling uncertain information and making informed decisions in diverse scenarios.

Maji et al. (2003) made significant contributions to soft set theory (SST) by introducing various new concepts such as subsets, complements, unions, and intersections, along with their applications in DMPs. This work marked the first usage of SST to address practical decision-making problems (Maji et al. 2002). Since then, researchers have delved into the broader properties and applications of SST. Maji et al. (2001) introduced fuzzy soft sets as a foundation for the theory of fuzzy soft sets, which has since found applications in various fields such as decision-making, image processing, and data analysis. Chen et al. (2005) presented the idea of parameterization reduction for soft sets focusing on reducing the complexity of soft sets by parameterization, leading to more efficient and real-life applications in various fields such as decision-making, pattern recognition, and data analysis. This research significantly contributes to advancing the understanding and utilization of soft sets theory, enhancing their effectiveness in solving real-world problems. Alcantud and Santos-García (2017) introduced a novel criterion for soft set based decision-making problems under incomplete information, offering a valuable perspective for handling uncertainties. Similarly, Alcantud and Muñoz Torrecillas (2018) pioneered the discussion on intertemporal-FSS selection, addressing temporal aspects in fuzzy set selection processes. Mukherjee and Das (2018) explored the application of Einstein operations to fuzzy soft multisets, focusing on their implications for decision-making processes. By incorporating these operations, the study enhances the analytical capabilities of fuzzy soft multisets in handling complex decision scenarios, providing valuable insights for practical applications in various domains. Akram and Adeel (2019) presented the TOPSIS approach for multi-attribute group decision making based on interval-valued hesitant-fuzzy-N-soft environments, showcasing the versatility of SST in complex decision models. Building on this, Akram et al. (2019) introduced decision-making technique based on hesitant N-soft sets, introducing HF-N-soft sets and an innovative idea with applications in DMPs. Abdulkareem et al. (2020) introduced a new multiperspective benchmarking framework for selecting image dehazing intelligent techniques. Their approach, based on the BWM and group VIKOR models, offers a comprehensive method for evaluating and choosing the most effective image dehazing algorithms. Petchimuthu et al. (2020) studied the mean operators and generalized products of fuzzy soft matrices, applying them to MCGDM problems. Chen et al. (2020) proposed a decision-making model based on generalized vague N-soft sets, expanding the scope of SST applications in collaborative decision scenarios. In 2021, Dalkılıç and Demirtaş (2021) developed the concept of bipolar fuzzy soft D-metric spaces, exploring further nuances in fuzzy set theory and SST. Bhardwaj and Sharma (2021) proposed a novel uncertainty measure using fuzzy SST and demonstrated its effectiveness in decision-making problems, highlighting the theory's adaptability in handling complex decision scenarios. Das and Granados (2021) presented an innovative DMM using fuzzy soft sets. Their methodology leverages weighted average ratings, offering a sophisticated framework to navigate complex decision scenarios effectively within group settings. Abdulkareem et al. (2021) proposed a novel standardization and selection framework for real-time image dehazing algorithms in multi-foggy scenes. Their approach combines fuzzy Delphi and hybrid multi-criteria decision analysis methods, providing a comprehensive methodology for selecting optimal image dehazing algorithms. This research contributes significantly to the field of image processing by addressing the challenges of hazy environments and enhancing the quality of visual data in real-time applications. Paik and Mondal (2021) introduced a distance-similarity method for solving decision-making problems based on fuzzy sets and fuzzy soft sets. This method offers a novel approach to handling uncertainty and ambiguity in decision-making processes, contributing to the advancement of soft computing techniques in practical applications. In another work, Paik and Mondal (2021) explored the representation and application of fuzzy soft sets in a type-2 environment, and delved into the utilization of fuzzy soft sets in handling uncertainty and complexity within type-2 fuzzy environments, contributing to advancements in computational intelligence. In 2021, Dalkılıç (2021) also contributed to SST research by presenting a new approach for addressing DMPs under uncertainty, further enhancing the theory's practical utility. Fatimah and Alcantud (2021) proposed multi-fuzzy N-SS, expanding SST's capabilities in handling multiple sources of fuzziness in decision contexts. Later, Das and Granados (2022) contributed to the theory by introducing FP-IFS multisets and suggesting an adjustable approach based on these multisets. They also defined new operations on FSSs, studying their applications in decision-making contexts. Dalkiliç (2022) defined topology on virtual fuzzy parameterized Fuzzy SST, adding a new dimension to the theory's mathematical foundations. Ihsan et al. (2022) presented an innovative approach using an intelligent fuzzy parameterized multi-criteria decision-support system. Their methodology integrates intuitionistic fuzzy hypersoft expert sets to enhance the evaluation process for automobiles, contributing to advancements in decision-making within the automotive industry. This work reflects a sophisticated application of fuzzy logic and expert systems in tackling complex evaluation tasks, potentially improving efficiency and accuracy in automobile assessment processes. Recently, Das, Granados, and Bhattacharya (2022) introduced novel operations on fuzzy soft sets, along with their applications in decision-making processes, which contribute to advancing the understanding and utilization of fuzzy soft set theory in various fields, particularly in handling uncertainty and complexity in decision-making scenarios. In 2023, Dalkılıç and Demirtaş (2023) introduced an algorithm for managing COVID-19 outbreaks using SST, showcasing its relevance in addressing contemporary challenges. Darvish Falehi and Torkaman (2023) defined the optimal fractional order interval type-2 fuzzy controller for dynamic voltage restorers, showcasing SST's applications in control systems and engineering. Das and Granados (2023) introduced the notion of IFP-intuitionistic multi fuzzy N-SST, along with induced IFP-hesitant N-SST, broadening SST's scope in handling various types of uncertainties in decision-making processes. Rathnasabapathy and Palanisami (2023) presented a theoretical development of an improved cosine similarity measure tailored for interval-valued intuitionistic fuzzy sets. Their work addresses the need for a more refined similarity measure to accurately assess the similarity between interval-valued intuitionistic fuzzy sets. Similarly, Das, et al. (2023) developed a novel weighted hesitant bipolar-valued fuzzy soft set framework aimed at improving decision-making processes across diverse fields. Their model incorporates weighted ratings to capture nuanced preferences and uncertainties, contributing significantly to analytical robustness and accuracy in practical applications.

Alkhazaleh and Salleh (2011) introduced the concept of Soft Expert Sets (SES), merging Soft Set Theory (SST) with expert systems to underscore the significance of considering all experts' perspectives in DMPs. They highlighted the importance of incorporating diverse viewpoints to enhance decision accuracy and reliability. Building on this, Alkhazaleh and Salleh (2014) extended SESs to fuzzy SESs, exploring their applications and effectiveness in handling uncertainty and vagueness in decision processes. Al-Quran and Hassan (2016) proposed the theory of neutrosophic vague SES, further expanding the applicability of SESs to scenarios involving indeterminacy and unknown factors. Bashir and Salleh (2012) introduced the concept of possibility fuzzy soft expert sets, which combine the flexibility of fuzzy sets with the structural advantages of soft sets to handle uncertainty and imprecision in decision-making processes. This hybrid approach aims to improve the accuracy and reliability of expert systems by incorporating degrees of possibility, thereby addressing the limitations of conventional fuzzy and soft sets. Similarly, in another note, Bashir and Salleh (2012) proposed the fuzzy parameterized soft expert set, a novel approach that extends traditional soft set theory by incorporating fuzzy parameters. This method enhances the capacity to deal with vagueness and uncertainty in expert systems, providing a more nuanced and adaptable framework for various applications. On the other hand, Hazaymeh et al. (2012) focused on generalized fuzzy SESs, contributing significantly to the advancement of fuzzy-based decision models. Their research deepened the understanding of how fuzzy logic can be integrated into expert systems, offering insights into more comprehensive and adaptable decision-making methodologies suitable for complex and dynamic environments. These studies collectively enrich the field of decision sciences by offering diverse perspectives and methodologies to tackle real-world decision challenges. In their work, Alhazaymeh and Hassan (2014) contributed to the field of soft computing by introducing the concept of generalized vague SESs. This extension of vague soft sets incorporates expert opinions, enhancing the capability to handle uncertainty and vagueness in decision-making contexts, which is crucial in various domains such as decision support systems and pattern recognition. In another work, Alhazaymeh and Hassan (2014) focused on mapping operations within the context of Generalized Vague Soft Expert Sets. This work contributes to the advancement of soft computing techniques by addressing mapping principles specifically tailored to handle uncertainty and expert opinions, which are crucial in decision-making and computational intelligence domains. They also demonstrated the practical application of generalized vague soft expert sets in decision-making contexts (Alhazaymeh and Hassan, 2014). Their work showcases the effectiveness of this framework in handling uncertainty and incorporating expert opinions, providing valuable insights into enhancing decision-making processes across various domains. The application of Generalized Vague Soft Expert Sets offers a flexible and adaptable approach to decision support systems, contributing to advancements in computational intelligence and decision science. Sahin et al. (2015) delved into neutrosophic SESs, investigating their properties and potential applications in decision contexts. Pramanik et al. (2015) investigated the TOPSIS method for Single-Valued Neutrosophic SES-based MCDM, offering insights into effective decision strategies. Al-Qudah and Hassan (2017) focused on bipolar fuzzy SESs and demonstrated their utility in addressing bipolarity and ambiguity in decision scenarios. They also explored the application of complex multi-fuzzy SESs (Al-Qudah and Hassan, 2017), providing advanced tools for modeling complex decision environments. Hassan et al. (2018) introduced Q-neutrosophic SESs, studying their characteristics and applicability in decision-making under uncertainty. Ulucay et al. (2018) extended the scope by examining generalized neutrosophic SESs within the framework of multi-criteria decision making (MCDM), showcasing their versatility in handling diverse decision criteria. Abu Qamar et al. (2018) delved into generalized Q-neutrosophic SESs, providing techniques for decision-making under uncertainty and ambiguity.

Liu and Zhang (2017) introduced an extended MCDM method based on neutrosophic hesitant fuzzy information. Their approach aims to enhance decision-making processes by incorporating neutrosophic hesitant fuzzy information, allowing for the effective handling of uncertainties and hesitancies. In another work by Liu and Zhang (2017), they proposed a MCDM method based on neutrosophic hesitant fuzzy Heronian mean aggregation operators. This method leverages Heronian mean aggregation operators within the neutrosophic hesitant fuzzy framework to improve decision outcomes. By integrating these operators, the model can better capture the nuances and uncertainties present in decision-making situations, leading to more informed and accurate decisions. On a related note, Abdel-Basset et al. (2019) proposed an integrated approach combining neutrosophic ANP and VIKOR for sustainable supplier selection. Also, Abdel-Basset et al. (2019) utilized a TOPSIS technique tailored for group DMPs under type-2 neutrosophic numbers. This integrated methodology provided a robust framework for complex decision scenarios, particularly in supplier selection processes requiring sustainability considerations and dealing with uncertainty. Abdel-Baset and colleagues (Abdel-Basset et al., 2019) pioneered a distinctive neutrosophic approach for assessing green supply chain management practices. Abdel-Basset et al. (2020) also extended to grouping decision-making models for heart disease diagnostics using neutrosophic sets. By leveraging neutrosophic logic, which deals with indeterminacy and uncertainty, they provided a robust framework for evaluating complex systems in both environmental sustainability and healthcare domains. On the other hand, Abdel-Basset and collaborators (Abdel-Basset et al., 2020) proposed two innovative frameworks: a bipolar neutrosophic MCDM model for professional selection and an intelligent medical decision support model leveraging soft computing techniques. The bipolar neutrosophic MCDM framework offers a nuanced approach to decision-making, considering both positive and negative aspects, which is particularly useful in professional selection processes where diverse criteria need to be evaluated. Additionally, their (Abdel-Basset et al., 2020) intelligent medical decision support model, based on soft computing, harnesses the power of computational intelligence to assist healthcare professionals in making informed decisions. These frameworks contribute significantly to improving decision-making processes in complex domains such as personnel selection and healthcare, showcasing the potential of advanced computational techniques in real-world applications. Granados et al. (2022) introduced continuous neutrosophic distributions characterized by neutrosophic parameters derived from neutrosophic random variables. This innovative approach expands the scope of neutrosophic theory, offering new insights into nonlinear analysis and its practical applications. Later, Granados et al. (2023) worked on weighted neutrosophic soft multiset theory. This research delves into novel methodologies for decision-making processes, showcasing the theory's efficacy and relevance in operational research contexts, and offering valuable insights for practical implementations.

Adam and Hassan (2016) introduced multi-Q-fuzzy SESs, which combine fuzzy logic with expert systems to provide a robust framework for addressing various decision scenarios. Their work expanded the range of tools available for decision-making, offering improved flexibility and accuracy in handling complex and uncertain information. This innovation has significant implications for industries and fields where precise decision-making is critical, such as finance, healthcare, and engineering. Al-Quran and Hassan (2016) delved into the study of fuzzy parameterized single-value neutrosophic SESs within DMPs. Neutrosophic logic, with its ability to handle indeterminacy and uncertainty, was applied within the framework of expert systems, providing a nuanced approach to decision modeling. In 2018, Smarandache (2018) introduced the hypersoft set (HSS) as a novel structure to overcome the limitations of dealing with disjoint attribute-valued sets for distinct attributes. HSS improves upon traditional soft sets by incorporating multi-attribute valued functions, allowing for a more comprehensive representation of complex data relationships. This innovative approach enables a finer granularity in capturing information, making it particularly valuable for decision-making processes that involve diverse and interconnected attributes. By replacing single attribute-valued functions with multi-attribute ones, HSS enhances the accuracy and relevance of data analysis, leading to more informed and effective decisions in real-life scenarios. In 2020, Saeed et al. (2020) expanded upon the concept of HSSs, delving into fundamental aspects like subsets, complements, aggregation operators, relations, matrices, and mappings within the HSS framework. Their work provided a deeper understanding of the operational mechanics and theoretical underpinnings of HSS, enhancing its applicability in diverse decision-making contexts. Additionally, Abbas et al. (2020)

contributed by defining new operations and presenting results related to HSSs and hypersoft points. Their research extended the theoretical foundations of HSSs, paving the way for more sophisticated applications and methodologies in decision sciences and related fields. Collectively, these studies have advanced the field of HSSs, enriching its toolkit and expanding its potential for addressing complex data analysis and decision-making challenges. Rahman et al. (2020) provided knowledge of complex HSS in 2020 and built HSS hybrids with complex neutrosophic sets, complex fuzzy sets, and complex intuitionistic fuzzy sets. They also talked about subsets, equal sets, null sets, and absolute sets, as well as theoretic operations like complement, union, and intersection. Rahman et al. (2020) explored the concepts of convexity and concavity within HSSs and provided visual representations along with specific instances. Their work contributed to a better understanding of the geometric properties of HSS, offering insights into the structural characteristics that influence decision-making processes. Mohammed et al. (2020) introduced a benchmarking methodology aimed at selecting the most effective COVID-19 diagnostic tool. Their approach incorporated TOPSIS (Technique for Order of Preference by Similarity to Ideal Solution) and entropy methodologies to provide a comprehensive evaluation framework, considering multiple criteria and uncertainties inherent in the diagnostic tool selection process. More recently, Ihsan et al. (2021) proposed the concept of a HSES, which extends the successful ideas of HSSs and soft expert systems (SES). They developed a decision-making model based on HSES, offering a generalized framework tailored for solving real-life DMPs. This innovative approach integrates the versatility of HSSs with expert systems, enhancing the accuracy and applicability of decision-making processes across diverse domains. Recently, Muhammad Zulqarnain et al. (2022) proposed Einstein-ordered weighted aggregation operators for Pythagorean fuzzy hypersoft sets and applied them to solve MCDM problems to enhance decision-making processes by incorporating advanced aggregation techniques tailored for handling Pythagorean fuzzy hypersoft sets efficiently. Musa et al. (2023) introduced N-Hypersoft Sets as an innovative extension of HSSs in their work published in Symmetry. This extension contributes to enhancing the versatility and applicability of HSSs in decision-making and computational intelligence domains.

The study introduces a novel concept called WHSES and thoroughly examines its fundamental properties. It also defines innovative notions such as agree-WHSES, disagree-WHSES, agree weighted score, and disagree weighted score. These concepts are then used to propose an innovative and adaptable DMM for addressing practical DMPs based on both HSES and WHSES. One key aspect of the proposed DMM is the introduction of the weighted score function, which enhances stability and feasibility compared to traditional score functions, such as those used in the Ihsan-DMM (decision-making model). This weighted score function takes into account both agreement and disagreement among experts or decision criteria, providing a more nuanced and reliable assessment mechanism. This concept enables decision-makers to evaluate alternatives based on relative stability rankings, contributing to a more informed and robust decision outcome. Moreover, the DMM's adjustability is enhanced through the use of weighted scores (agree weighted score-disagree weighted score) instead of simple choice values. This adjustability factor allows decision-makers to fine-tune the model according to specific requirements or changing conditions, making the DMM highly adaptable and responsive to dynamic decision scenarios. To validate the effectiveness of the proposed DMM, a real-life illustration is provided, demonstrating its applicability and validity in practical decision-making contexts. Additionally, a comparison analysis with existing methods highlights the uniqueness and advantages of the proposed DMM, emphasizing its stability, adjustability, and overall feasibility in solving complex decision problems.

The chapter structure is outlined as follows:

Section 1: Introduction. The introduction section sets the stage for the entire chapter by providing an overview of the key concepts and topics that will be discussed. It aims to introduce the reader to the fundamental ideas and motivations behind the research presented in the chapter. Additionally, the introduction may outline the objectives, scope, and structure of the chapter, providing a roadmap for what the reader can expect to learn and explore throughout the document.

Section 2: Essential Concepts and Conclusions of SS, SES, HSES, Agree-HSES, and Disagree-HSES. This section presents fundamental concepts and conclusions related to Subset Structures SS, SES, HSES, Agree-HSES, and Disagree-HSES. These concepts form the basis for later discussions and analyses.

Section 3: Introduction to WHSES and Fundamental Properties. In this section, a new idea of WHSES is presented, and its fundamental properties are examined in detail. Additionally, novel notions such as Agree-WHSES, Disagree-WHSES, and Weighted Score (agree weighted score-disagree weighted score) are defined, laying the groundwork for an innovative and adaptable decision-making model for addressing HSES and WHSES based DMPs.

Section 4: Practical Illustration Demonstration. A practical illustration is presented in this section to demonstrate the validity and effectiveness of the proposed technique. This example showcases how the advanced DMM can be applied in practical decision-making scenarios.

Section 5: Comparison Analysis with Existing Method. In Section 5, a comparison analysis is conducted to evaluate the proposed technique against existing methods. This analysis provides insights into the advantages and limitations of the proposed approach compared to traditional methods.

Section 6: Conclusion and Future Work. The chapter concludes in Section 6, summarizing the key findings and contributions of the research. Future work and potential areas for further exploration or improvement are also discussed, providing a roadmap for ongoing research and development in this area.

PRELIMINARIES

Let's consider Ω_H represents the initial universe and the power set of Ω_H is presented by $P(\Omega_H)$.

Definition 2.1 (Molodtsov, 1999) A soft set over a nonempty universe Ω_H is s defined as a pair (ψ, T), where ψ is a function represented by ψ: T→ $P(\Omega_H)$. The mapping ψ assigns each element from the set T to a subset of Ω_H, allowing for the representation of uncertainty or vagueness in the elements of T through various subsets of Ω_H.

Definition 2.2 (Alkhazaleh and Salleh, 2011) Assume that Q be a set of specialists (operators), R is a set of conclusions, and $S=P\times Q\times R$ with $T \subseteq S$, P is a collection of parameters. An SES on Ω_H is a set with a structure (Ψ_S, T) where Ψ_S: $T\rightarrow P(\Omega_H)$ is a mapping.

In this definition, Ψ maps the combination of a specialist q with a conclusion r and a parameter p to a value between 0 and 1, representing the degree of expertise or confidence associated with that particular combination.

Definition 2.3 (Ihsan et al., 2021) A pair (Ψ_H, T) is known as an HSES over Ω_H, where Ψ_H: $T\rightarrow P(\Omega_H)$, such that

(i) $T \subseteq S=P\times Q\times R$
(ii) $P=P_1\times P_2\times P_3\times\ldots\times P_n$ where $P_1,P_2,P_3,\ldots,P_n$ are disjoint attribute sets corresponding to n distinct attributes $p_1,p_2,p_3,\ldots,p_n$
(iii) $Q=\{Q_1,Q_2,Q_3,\ldots,Q_m\}$ be a set of specialists (operators).
(iv) R= {0-disagree, 1-agree} is a collection of conclusions.

Example 2.4 Let $\Omega_H=\{\rho_1,\rho_2,\rho_3,\rho_4,\rho_5,\rho_6,\rho_7,\rho_8,\rho_9,\rho_{10}\}$ be the set of universe and $P_1=\{p_{11},p_{12},p_{13}\}$, $P_2=\{p_{21},p_{22}\}$, $P_3=\{p_{31}\}$ be the disjoint sets of parameters. Let $Q=\{Q_1,Q_2,Q_3\}$ and $R=\{1,0\}$. Then

$$P=P_1\times P_2\times P_3=\left\{\begin{matrix}\sigma_1=(p_{11},p_{21},p_{31}),\sigma_2=(p_{12},p_{21},p_{31}),\sigma_3=(p_{13},p_{21},p_{31}),\\ \sigma_4=(p_{11},p_{22},p_{31}),\sigma_5=(p_{12},p_{22},p_{31}),\sigma_6=(p_{13},p_{22},p_{31}),\end{matrix}\right\},$$

and

$$S = P \times Q \times R = \left\{ \begin{matrix} (\sigma_1, Q_1, 1), (\sigma_1, Q_2, 1), (\sigma_1, Q_3, 1), (\sigma_1, Q_1, 0), (\sigma_1, Q_2, 0), (\sigma_1, Q_3, 0), \\ (\sigma_2, Q_1, 1), (\sigma_2, Q_2, 1), (\sigma_2, Q_3, 1), (\sigma_2, Q_1, 0), (\sigma_2, Q_2, 0), (\sigma_2, Q_3, 0), \\ (\sigma_3, Q_1, 1), (\sigma_3, Q_2, 1), (\sigma_3, Q_3, 1), (\sigma_3, Q_1, 0), (\sigma_3, Q_2, 0), (\sigma_3, Q_3, 0), \\ (\sigma_4, Q_1, 1), (\sigma_4, Q_2, 1), (\sigma_4, Q_3, 1), (\sigma_4, Q_1, 0), (\sigma_4, Q_2, 0), (\sigma_4, Q_3, 0), \\ (\sigma_5, Q_1, 1), (\sigma_5, Q_2, 1), (\sigma_5, Q_3, 1), (\sigma_5, Q_1, 0), (\sigma_5, Q_2, 0), (\sigma_5, Q_3, 0), \\ (\sigma_6, Q_1, 1), (\sigma_6, Q_2, 1), (\sigma_6, Q_3, 1), (\sigma_6, Q_1, 0), (\sigma_6, Q_2, 0), (\sigma_6, Q_3, 0) \end{matrix} \right\},$$

let

$$T = \left\{ \begin{matrix} t_1 = (\sigma_1, Q_1, 1),\ t_2 = (\sigma_1, Q_2, 1),\ t_3 = (\sigma_1, Q_3, 1),\ t_4 = (\sigma_2, Q_1, 1), \\ t_5 = (\sigma_2, Q_2, 1),\ t_6 = (\sigma_2, Q_3, 1),\ t_7 = (\sigma_4, Q_1, 1),\ t_8 = (\sigma_4, Q_2, 1), \\ t_9 = (\sigma_4, Q_3, 1),\ t_{10} = (\sigma_6, Q_1, 1),\ t_{11} = (\sigma_6, Q_2, 1),\ t_{12} = (\sigma_6, Q_3, 1), \\ t'_1 = (\sigma_1, Q_1, 0),\ t'_2 = (\sigma_1, Q_2, 0),\ t'_3 = (\sigma_1, Q_3, 0),\ t'_4 = (\sigma_2, Q_1, 0), \\ t'_4 = (\sigma_2, Q_2, 0),\ t'_6 = (\sigma_2, Q_3, 0),\ t'_7 = (\sigma_4, Q_1, 0),\ t'_8 = (\sigma_4, Q_2, 0), \\ t'_9 = (\sigma_4, Q_3, 0),\ t'_{10} = (\sigma_6, Q_1, 0),\ t'_{11} = (\sigma_6, Q_2, 0),\ t'_{12} = (\sigma_6, Q_3, 0) \end{matrix} \right\}.$$

Then an HSES $(\Psi_{H'}, T)$ can be written as follows:

$$(\Psi_H, T) = \left\{ \begin{matrix} (t_1, \{\rho_1, \rho_2, \rho_5, \rho_7, \rho_9, \rho_{10}\}),\ (t_2, \{\rho_2, \rho_3, \rho_6, \rho_7, \rho_{10}\}),\ (t_3, \{\rho_1, \rho_3, \rho_4, \rho_7\}), \\ (t_4, \{\rho_1, \rho_4, \rho_5, \rho_6, \rho_8\}),\ (t_5, \{\rho_2, \rho_4, \rho_6, \rho_8\}),\ (t_6, \{\rho_2, \rho_3, \rho_5, \rho_9\}), \\ (t_7, \{\rho_1, \rho_3, \rho_4, \rho_7, \rho_8, \rho_{10}\}),\ (t_8, \{\rho_3, \rho_4, \rho_6, \rho_7, \rho_8, \rho_{10}\}), (t_9, \{\rho_1, \rho_2, \rho_5, \rho_9, \rho_{10}\}), \\ (t_{10}, \{\rho_2, \rho_4, \rho_6, \rho_7, \rho_9, \rho_{10}\}),\ (t_{11}, \{\rho_3, \rho_5, \rho_8, \rho_9\}),\ (t_{12}, \{\rho_1, \rho_5, \rho_6, \rho_8, \rho_9\}), \\ (t'_1, \{\rho_3, \rho_4, \rho_6, \rho_8\}), (t'_2, \{\rho_1, \rho_4, \rho_5, \rho_8, \rho_9\}),\ (t'_3, \{\rho_2, \rho_5, \rho_6, \rho_8, \rho_9, \rho_{10}\}), \\ (t'_4, \{\rho_2, \rho_3, \rho_7, \rho_9, \rho_{10}\}),\ (t'_5, \{\rho_1, \rho_3, \rho_5, \rho_7, \rho_9, \rho_{10}\}), (t'_6, \{\rho_1, \rho_4, \rho_6, \rho_7, \rho_8, \rho_{10}\}), \\ (t'_7, \{\rho_2, \rho_5, \rho_6, \rho_9\}),\ (t'_8, \{\rho_1, \rho_2, \rho_5, \rho_9\}),\ (t'_9, \{\rho_3, \rho_4, \rho_6, \rho_7, \rho_8\}), \\ (t'_{10}, \{\rho_1, \rho_3, \rho_5, \rho_8\}),\ (t'_{11}, \{\rho_1, \rho_2, \rho_4, \rho_6, \rho_7, \rho_{10}\}),\ (t'_{12}, \{\rho_2, \rho_3, \rho_4, \rho_7, \rho_{10}\}) \end{matrix} \right\}.$$

Definition 2.5 (Ihsan et al., 2021) An HSES $(\Psi^1_{H'}, T_1)$ is said to be a subset of $(\Psi^2_{H'}, T_2)$ on $\Omega_{H'}$, if

(i) $T_1 \subseteq T_2$,
(ii) $\forall t \in T_1$, $\Psi^1_{H}(t) \subseteq \Psi^2_{H}(t)$, and written as $(\Psi^1_{H'}, T_1) \tilde{\subseteq} (\Psi^2_{H'}, T_2)$.

Definition 2.6 (Ihsan et al., 2021) An agree-HSES $(\Psi_{H'}, T)_{Ag}$ on $\Omega_{H'}$ is an HSE subset of $(\Psi_{H'}, T)$ and is characterized as $(\Psi_{H}, T)_{Ag} = \{\Psi_{Ag}(t) : t \in P \times Q \times \{1\}\}$.

Definition 2.7 (Ihsan et al., 2021) A disagree-HSES $(\Psi_{H'}, T)_{Dag}$ on $\Omega_{H'}$ is an HSE subset of $(\Psi_{H'}, T)$ and is characterized as $(\Psi_{H}, T)_{Dag} = \{\Psi_{Dag}(t') : t' \in P \times Q \times \{0\}\}$.

These definitions lay the groundwork for understanding various structures and mappings used in decision-making frameworks involving sets, subsets, specialists, conclusions, and parameters, particularly in contexts where uncertainty and expertise play crucial roles.

Ihsan-Method

Ihsan et al. (2021) introduced a DMM based on HSES, offering a generalized framework tailored for solving real-life DMPs. This innovative approach integrates the versatility of HSSs with expert systems, enhancing the accuracy and applicability of decision-making processes across diverse domains. The steps of the method are listed below:

Algorithm 1 (Ihsan et al., 2021)

Step1. Enter the HSES (Ψ_{H}, T). This step involves inputting the relevant data or information into the decision-making model. The HSES (Ψ_{H}, T) is a key component of the model and contains subsets related to specialists, conclusions, parameters, and mappings that represent degrees of expertise or confidence.
Step2. Calculate the agree-HSES and disagree-HSES for (Ψ_{H}, T). In this step, the algorithm calculates two subsets of the HSES: the agree-HSES and the disagree-HSES. The agree-HSES contains elements that are in agreement according to certain criteria, while the disagree-HSES contains elements that are not in agreement.

Step3. Calculate $D_j = \sum_i a_{ij}$ for the agree-HSES. This value is likely based on certain computations or assessments related to the elements and their degrees of agreement within the agree-HSES subset.

Step4. Compute $F_j = \sum_i d_{ij}$ for the disagree-HSES. This value is likely based on computations or assessments related to the elements and their degrees of disagreement within the disagree-HSES subset.
Step5. Obtain $S_j = D_j - F_j$ for the HSES. This combined score provides a comprehensive assessment of the HSES, considering both agreements and disagreements among specialists in the decision-making process.
Step6. Find k for which the max $S_j = S_k$ for the HSES

This algorithm outlines a systematic approach to decision-making using HSSs, where various calculations and optimizations are performed to arrive at an optimal decision or solution for real-life DMPs.

WEIGHTED HYPERSOFT EXPERT SET AND ITS THEORETICAL ANALYSIS

In this chapter, the symbol Ω_H represents the starting universe. This could represent a set of objects, elements, or entities under consideration in the research domain. The power set of Ω_H is represented by $P(\Omega_H)$, which includes all possible subsets of Ω_H. This is relevant when dealing with set theory and considering various combinations or partitions of elements from Ω_H. Let $P= P_1 \times P_2 \times P_3 \times \ldots \times P_n$ where $P_1, P_2, P_3, \ldots, P_n$ are disjoint attributive sets corresponding to n distinct attributes $p_1, p_2, p_3, \ldots, p_n$. This suggests that the elements of Ω_H can be classified or categorized based on these attributes, and each set P_k corresponds to a specific attribute. Let $Q= \{Q_1, Q_2, Q_3, \ldots, Q_m\}$ be a set of specialists (operators). $R=\{1,0\}$ be a set of conclusions, which could be outcomes, decisions, or results derived from analyzing the elements of Ω_H based on the attributive sets and specialists' input, and let $T \subseteq S=P \times Q \times R$.

Overall, these symbols and notations help establish the foundational concepts and relationships within the research chapter, particularly in the context of decision-making, classification, or analysis involving a starting universe Ω_H, attributive sets, specialists, and conclusions.

Definition 3.1 A weighted hypersoft expert set (WHSES) is defined as a triple $\langle \Psi_H, T, w_H \rangle$ where (Ψ_H, T) is an HSES over Ω_H and $w_H: T \rightarrow [0,1]$ is a weight function that assigns the weight $w_i = w_H(t_i)$ for every $t_i \in T$. If $w_i = w_H(t_i) = 1$, $\forall t_i \in T$, then $\langle \Psi_H, T, w_H \rangle$ will be generated to be a traditional HSES (Ψ_H, T).

In simpler terms, a WHSES is an extension of an HSES with the addition of a weight function that assigns weights to parameters.

Note We denote the collection of all Weighted Hypersoft Expert Sets (WHSESs) over the universe Ω_H as $WHSES(\Omega_H)$.

Example 3.2 Let's assume that the HSES (Ψ_H, T) as in Example 2.4 and now the expert wants to introduce weighted parameters into this HSES. Additionally, let's assume that the expert has set the weights for the parameters in set T provided below:

$$w_1 = w_H(t_1) = 0.72; \quad w_2 = w_H(t_2) = 0.81; \quad w_3 = w_H(t_3) = 0.75; \quad w_4 = w_H(t_4) = 0.65;$$

$$w_5 = w_H(t_5) = 0.85; \quad w_6 = w_H(t_6) = 0.62; \quad w_7 = w_H(t_7) = 0.55; \quad w_8 = w_H(t_8) = 0.78;$$

$$w_9 = w_H(t_9) = 0.89; \quad w_{10} = w_H(t_{10}) = 0.67; \quad w_{11} = w_H(t_{11}) = 0.84; \quad w_{12} = w_H(t_{12}) = 0.87;$$

$$w_{13} = w_H(t'_1) = 0.72; \quad w_{14} = w_H(t'_2) = 0.81; \quad w_{15} = w_H(t'_3) = 0.75; \quad w_{16} = w_H(t'_4) = 0.65;$$

$$w_{17} = w_H(t'_5) = 0.85; \quad w_{18} = w_H(t'_6) = 0.62; \quad w_{19} = w_H(t'_7) = 0.55; \quad w_{20} = w_H(t'_8) = 0.78;$$

$$w_{21} = w_H(t'_9) = 0.89; \quad w_{22} = w_H(t'_{10}) = 0.67; \quad w_{23} = w_H(t'_{11}) = 0.84; \quad w_{24} = w_H(t'_{12}) = 0.87.$$

Then the HSES (Ψ_H, T) is changed into a WHSES $\langle \Psi_H, T, w_H \rangle$ as follows:

$$\langle \Psi_H, T, w_H \rangle = \left\{ \begin{array}{l} ((t_1, 0.72), \{\rho_1, \rho_2, \rho_5, \rho_7, \rho_9, \rho_{10}\}), ((t_2, 0.81), \{\rho_2, \rho_3, \rho_6, \rho_7, \rho_{10}\}), ((t_3, 0.75), \{\rho_1, \rho_3, \rho_4, \rho_7\}), \\ ((t_4, 0.65), \{\rho_1, \rho_4, \rho_5, \rho_6, \rho_8\}), ((t_5, 0.85), \{\rho_2, \rho_4, \rho_6, \rho_8\}), ((t_6, 0.62), \{\rho_2, \rho_3, \rho_5, \rho_9\}), \\ ((t_7, 0.55), \{\rho_1, \rho_3, \rho_4, \rho_7, \rho_8, \rho_{10}\}), ((t_8, 0.78), \{\rho_3, \rho_4, \rho_6, \rho_7, \rho_8, \rho_{10}\}), ((t_9, 0.89), \{\rho_1, \rho_2, \rho_5, \rho_9, \rho_{10}\}), \\ ((t_{10}, 0.67), \{\rho_2, \rho_4, \rho_6, \rho_7, \rho_9, \rho_{10}\}), ((t_{11}, 0.84), \{\rho_3, \rho_5, \rho_8, \rho_9\}), ((t_{12}, 0.87), \{\rho_1, \rho_5, \rho_6, \rho_8, \rho_9\}), \\ ((t'_1, 0.72), \{\rho_3, \rho_4, \rho_6, \rho_8\}), ((t'_2, 0.81), \{\rho_1, \rho_4, \rho_5, \rho_8, \rho_9\}), ((t'_3, 0.75), \{\rho_2, \rho_5, \rho_6, \rho_8, \rho_9, \rho_{10}\}), \\ ((t'_4, 0.65), \{\rho_2, \rho_3, \rho_7, \rho_9, \rho_{10}\}), ((t'_5, 0.85), \{\rho_1, \rho_3, \rho_5, \rho_7, \rho_9, \rho_{10}\}), ((t'_6, 0.62), \{\rho_1, \rho_4, \rho_6, \rho_7, \rho_8, \rho_{10}\}), \\ ((t'_7, 0.55), \{\rho_2, \rho_5, \rho_6, \rho_9\}), ((t'_8, 0.78), \{\rho_1, \rho_2, \rho_5, \rho_9\}), ((t'_9, 0.89), \{\rho_3, \rho_4, \rho_6, \rho_7, \rho_8\}), \\ ((t'_{10}, 0.67), \{\rho_1, \rho_3, \rho_5, \rho_8\}), ((t'_{11}, 0.84), \{\rho_1, \rho_2, \rho_4, \rho_6, \rho_7, \rho_{10}\}), ((t'_{12}, 0.87), \{\rho_2, \rho_3, \rho_4, \rho_7, \rho_{10}\}) \end{array} \right\}.$$

In this WHSES $\langle \Psi_H, T, w_H \rangle$, each element's membership degree is adjusted based on the weighted values calculated using the specified weights for the parameters in T. This transformation allows for a weighted representation of the HSS (Ψ_H, T), which can be useful in decision-making or analysis processes where certain parameters or attributes are given higher importance or priority.

Definition 3.3 (WHSE-subset) For two WHSESs

$\langle \Psi_H, T_1, w_H^1 \rangle, \langle \Phi_H, T_2, w_H^2 \rangle \in WHSES(\Omega_H)$, $\langle \Psi_H, T_1, w_H^1 \rangle$

is said to be a WHSE-subset of $\langle \Phi_H, T_2, w_H^2 \rangle$ if

(i) $T_1 \subseteq T_2$ and $\forall t \in T_1$, $w_H^1(t) \leq w_H^2(t)$;
(ii) $\forall t \in T_1$, $\Psi_H(t) \subseteq \Phi_H(t)$.

We write $\langle \Psi_H, T_1, w_H^1 \rangle \tilde{\subseteq} \langle \Phi_H, T_2, w_H^2 \rangle$.

Example 3.4 Considering two WHSESs $\langle \Psi_H, T_1, w_H^1 \rangle, \langle \Phi_H, T_2, w_H^2 \rangle \in WHSES(\Omega_H)$ as

$$\langle \Psi_H, T_1, w_H^1 \rangle = \left\{ \begin{array}{l} ((t_1, 0.65), \{\rho_5, \rho_7, \rho_9, \rho_{10}\}), \; ((t'_2, 0.76), \{\rho_4, \rho_5, \rho_8, \rho_9\}), \\ ((t_4, 0.61), \{\rho_1, \rho_4, \rho_5, \rho_6, \rho_8\}), \; ((t'_4, 0.64), \{\rho_2, \rho_3, \rho_7, \rho_9, \rho_{10}\}), \\ ((t_6, 0.33), \{\rho_2, \rho_3, \rho_5, \rho_9\}), \; ((t'_6, 0.45), \{\rho_1, \rho_4, \rho_6, \rho_7, \rho_8, \rho_{10}\}) \end{array} \right\},$$

$$\langle \Psi_H, T_1, w_H^1 \rangle = \left\{ \begin{array}{l} ((t_1, 0.65), \{\rho_5, \rho_7, \rho_9, \rho_{10}\}), \; ((t'_2, 0.76), \{\rho_4, \rho_5, \rho_8, \rho_9\}), \\ ((t_4, 0.61), \{\rho_1, \rho_4, \rho_5, \rho_6, \rho_8\}), \; ((t'_4, 0.64), \{\rho_2, \rho_3, \rho_7, \rho_9, \rho_{10}\}), \\ ((t_6, 0.33), \{\rho_2, \rho_3, \rho_5, \rho_9\}), \; ((t'_6, 0.45), \{\rho_1, \rho_4, \rho_6, \rho_7, \rho_8, \rho_{10}\}) \end{array} \right\},$$

Then $\langle \Psi_H, T_1, w_H^1 \rangle \tilde{\subseteq} \langle \Phi_H, T_2, w_H^2 \rangle$.

Definition 3.5 (Equal-sets) Let $\langle \Psi_H, T_1, w_H^1 \rangle, \langle \Phi_H, T_2, w_H^2 \rangle \in WHSES(\Omega_H)$**.** Then

$\langle \Psi_H, T_1, w_H^1 \rangle$

and

$\langle \Phi_H, T_2, w_H^2 \rangle$

are equal sets, denoted by $\langle \Psi_{\mathrm{H}}, T_1, w_H^1 \rangle = \langle \Phi_H, T_2, w_H^2 \rangle$ if and only if

(i) $T_1 = T_2$ and $\forall t \in T_1,\ w_H^1(t) = w_H^2(t)$;
(ii) $\forall t \in T_1,\ \Psi_H(t) = \Phi_H(t)$.

Proposition 3.6 Let $\langle \Psi_{\mathrm{H}}, T_1, w_H^1 \rangle, \langle \Phi_H, T_2, w_H^2 \rangle, \langle \Upsilon_H, T_3, w_H^3 \rangle \in WHSES(\Omega_H)$. Then

(i) $\langle \Psi_{\mathrm{H}}, T_1, w_H^1 \rangle = \langle \Phi_H, T_2, w_H^2 \rangle$ and

$\langle \Phi_H, T_2, w_H^2 \rangle = \langle \Upsilon_H, T_3, w_H^3 \rangle \Rightarrow \langle \Psi_{\mathrm{H}}, T_1, w_H^1 \rangle = \langle \Upsilon_H, T_3, w_H^3 \rangle$

(ii) $\langle \Psi_{\mathrm{H}}, T_1, w_H^1 \rangle \tilde{\subseteq} \langle \Phi_H, T_2, w_H^2 \rangle$ and

$\langle \Phi_H, T_2, w_H^2 \rangle \tilde{\subseteq} \langle \Upsilon_H, T_3, w_H^3 \rangle \Rightarrow \langle \Psi_{\mathrm{H}}, T_1, w_H^1 \rangle \tilde{\subseteq} \langle \Upsilon_H, T_3, w_H^3 \rangle$;

(iii) $\langle \Psi_{\mathrm{H}}, T_1, w_H^1 \rangle \subseteq \langle \Phi_H, T_2, w_H^2 \rangle$ and

$\langle \Phi_H, T_2, w_H^2 \rangle \tilde{\subseteq} \langle \Psi_{\mathrm{H}}, T_1, w_H^1 \rangle \Rightarrow \langle \Psi_{\mathrm{H}}, T_1, w_H^1 \rangle = \langle \Phi_H, T_2, w_H^2 \rangle$.

Definition 3.7 (Complement of a WHSES) The complement of a *WHSES*

$\langle \Psi_{\mathrm{H}}, T, w_H \rangle \in WHSES(\Omega_H)$

can be represented by $\langle \Psi_{\mathrm{H}}, T, w_H \rangle^C$ and defined by

$\langle \Psi_{\mathrm{H}}, T, w_H \rangle^C = \langle \Psi_H^C, T, w_H^C \rangle$, where $\Psi^C_H: T \rightarrow P(\Omega_H)$ and $w^C_H: T \rightarrow [0,1]$ are functions given by

$\Psi^C_H(t) = (\Psi_H(t))^C$ and $w^C_H(t) = 1 - w_H(t)$

Example 3.8 Considering a WHSES $\langle \Psi_H, T, w_H \rangle \in WHSES(\Omega_H)$as

$$\langle \Psi_H, T, w_H \rangle = \left\{ \begin{matrix} ((t_1, 0.65), \{\rho_5, \rho_7, \rho_9, \rho_{10}\}), & ((t_2', 0.76), \{\rho_4, \rho_5, \rho_8, \rho_9\}), \\ ((t_4, 0.61), \{\rho_1, \rho_4, \rho_5, \rho_6, \rho_8\}), & ((t_4', 0.64), \{\rho_2, \rho_3, \rho_7, \rho_9, \rho_{10}\}), \\ ((t_6, 0.33), \{\rho_2, \rho_3, \rho_5, \rho_9\}), & ((t_6', 0.45), \{\rho_1, \rho_4, \rho_6, \rho_7, \rho_8, \rho_{10}\}) \end{matrix} \right\},$$

then

$$\langle \Psi_H, T, w_H \rangle^C = \left\{ \begin{matrix} ((t_1, 0.35), \{\rho_1, \rho_2, \rho_3, \rho_4, \rho_6, \rho_8\}), & ((t_2', 0.24), \{\rho_1, \rho_2, \rho_3, \rho_6, \rho_7, \rho_{10}\}), \\ ((t_4, 0.39), \{\rho_2, \rho_3, \rho_7, \rho_9, \rho_{10}\}), & ((t_4', 0.36), \{\rho_1, \rho_4, \rho_5, \rho_6, \rho_8\}), \\ ((t_6, 0.67), \{\rho_1, \rho_4, \rho_6, \rho_7, \rho_8, \rho_{10}\}), & ((t_6', 0.55), \{\rho_2, \rho_3, \rho_5, \rho_9\}) \end{matrix} \right\}$$

These definitions and examples provide a foundation for understanding operations and relationships within WHSESs, including subset relationships, equality, and complements. The proposition establishes the relationship between equality and subset relationships, while the example illustrates the computation of a complement for a WHSES.

Now, let's understand the operations of the union and the intersection between two WHSESs and then move on to Proposition 3.12.

Definition 3.9 (Union between two WHSESs) The Union between two WHSESs

$\langle \Psi_H, T_1, w^1_H \rangle, \langle \Phi_H, T_2, w^2_H \rangle \in WHSES(\Omega_H)$

is denoted by

$\langle \Psi_H, T_1, w^1_H \rangle \tilde{\cup} \langle \Phi_H, T_2, w^2_H \rangle$

and defined as $\langle \Psi_H, T_1, w^1_H \rangle \tilde{\cup} \langle \Phi_H, T_2, w^2_H \rangle = \langle \Upsilon_H, T_3, w^3_H \rangle$, where

(i) $T_3 = T_1 \cup T_2$

(ii) $\Upsilon_H(t) = \begin{cases} \Psi_H(t), & \forall t \in T_1 - T_2, \\ \Phi_H(t), & \forall t \in T_2 - T_1, \\ \Psi_H(t) \cup \Phi_H(t), & \forall t \in T_1 \cap T_2 \end{cases}$

(iii) $\forall t \in T_3, w_H^3(t) = \max\{w_H^1(t), w_H^2(t)\}$.

Definition 3.10 (Intersection between two WHSESs) The Intersection between two WHSESs $\langle \Psi_{\mathrm{H}}, T_1, w_H^1 \rangle, \langle \Phi_H, T_2, w_H^2 \rangle \in WHSES(\Omega_H)$ is denoted by $\langle \Psi_{\mathrm{H}}, T_1, w_H^1 \rangle \tilde{\cap} \langle \Phi_H, T_2, w_H^2 \rangle$ and defined as $\langle \Psi_{\mathrm{H}}, T_1, w_H^1 \rangle \tilde{\cap} \langle \Phi_H, T_2, w_H^2 \rangle = \langle \Upsilon_H, T_3, w_H^3 \rangle$, where

(i) $T_3 = T_1 \cup T_2$

(ii) $\Upsilon_H(t) = \begin{cases} \Psi_H(t), & \forall t \in T_1 - T_2, \\ \Phi_H(t), & \forall t \in T_2 - T_1, \\ \Psi_H(t) \cap \Phi_H(t), & \forall t \in T_1 \cap T_2 \end{cases}$

(iii) $\forall t \in T_3, w_H^3(t) = \min\{w_H^1(t), w_H^2(t)\}$,

Example 3.11 If we consider two WHSESs $\langle \Psi_{\mathrm{H}}, T_1, w_H^1 \rangle, \langle \Phi_H, T_2, w_H^2 \rangle \in WHSES(\Omega_H)$ as

$$\langle \Psi_{\mathrm{H}}, T_1, w_H^1 \rangle = \left\{ \begin{matrix} ((t_1, 0.65), \{\rho_5, \rho_7, \rho_9\}), & ((t_2', 0.76), \{\rho_5, \rho_8, \rho_9\}), \\ ((t_3, 0.61), \{\rho_1, \rho_4, \rho_5,\}), & ((t_4', 0.64), \{\rho_2, \rho_3, \rho_7\}), \\ ((t_5, 0.33), \{\rho_2, \rho_3, \rho_5\}), & ((t_6', 0.45), \{\rho_1, \rho_8, \rho_{10}\}) \end{matrix} \right\},$$

$$\langle \Phi_{\mathrm{H}}, T_2, w_H^2 \rangle = \left\{ \begin{matrix} ((t_1, 0.61), \{\rho_7, \rho_9, \rho_{10}\}), & ((t_2', 0.71), \{\rho_4, \rho_5, \rho_8\}), \\ ((t_3, 0.68), \{\rho_1, \rho_4, \rho_5\}), & ((t_4', 0.67), \{\rho_2, \rho_7, \rho_9\}), \\ ((t_7, 0.39), \{\rho_3, \rho_5, \rho_9\}), & ((t_8', 0.42), \{\rho_6, \rho_7, \rho_8\}) \end{matrix} \right\}.$$

Then

$$\langle \Psi_{\mathrm{H}}, T_1, w_H^1 \rangle \tilde{\cup} \langle \Phi_H, T_2, w_H^2 \rangle = \left\{ \begin{matrix} ((t_1, 0.65), \{\rho_5, \rho_7, \rho_9, \rho_{10}\}), & ((t_2', 0.76), \{\rho_4, \rho_5, \rho_8, \rho_9\}), \\ ((t_3, 0.68), \{\rho_1, \rho_4, \rho_5\}), & ((t_4', 0.67), \{\rho_2, \rho_3, \rho_7, \rho_9\}), \\ ((t_5, 0.33), \{\rho_2, \rho_3, \rho_5\}), & ((t_6', 0.45), \{\rho_1, \rho_8, \rho_{10}\}), \\ ((t_7, 0.39), \{\rho_3, \rho_5, \rho_9\}), & ((t_8', 0.42), \{\rho_6, \rho_7, \rho_8\}) \end{matrix} \right\}$$

$$\langle \Psi_{\mathrm{H}}, T_1, w_H^1 \rangle \tilde{\cap} \langle \Phi_H, T_2, w_H^2 \rangle = \left\{ \begin{matrix} ((t_1, 0.61), \{\rho_7, \rho_9\}), & ((t_2', 0.71), \{\rho_5, \rho_8\}), \\ ((t_3, 0.61), \{\rho_1, \rho_4, \rho_5\}), & ((t_4', 0.64), \{\rho_2, \rho_7\}), \\ ((t_5, 0.33), \{\rho_2, \rho_3, \rho_5\}), & ((t_6', 0.45), \{\rho_1, \rho_8, \rho_{10}\}), \\ ((t_7, 0.39), \{\rho_3, \rho_5, \rho_9\}), & ((t_8', 0.42), \{\rho_6, \rho_7, \rho_8\}) \end{matrix} \right\}.$$

Proposition 3.12 Let us consider

$$\langle \Psi_H, T_1, w_H^1\rangle,\ \langle \Phi_H, T_2, w_H^2\rangle,\ \langle \Upsilon_H, T_3, w_H^3\rangle\ \in\ WHSES(\Omega_H),$$

then

[i] Associative Laws

$$\langle \Psi_H, T_1, w_H^1\rangle \tilde{\cup} \left(\langle \Phi_H, T_2, w_H^2\rangle \tilde{\cup} \langle \Upsilon_H, T_3, w_H^3\rangle\right)$$
$$= \left(\langle \Psi_H, T_1, w_H^1\rangle \tilde{\cup} \langle \Phi_H, T_2, w_H^2\rangle\right) \tilde{\cup} \langle \Upsilon_H, T_3, w_H^3\rangle$$

$$\langle \Psi_H, T_1, w_H^1\rangle \tilde{\cap} \left(\langle \Phi_H, T_2, w_H^2\rangle \tilde{\cap} \langle \Upsilon_H, T_3, w_H^3\rangle\right)$$
$$= \left(\langle \Psi_H, T_1, w_H^1\rangle \tilde{\cap} \langle \Phi_H, T_2, w_H^2\rangle\right) \tilde{\cap} \langle \Upsilon_H, T_3, w_H^3\rangle$$

These laws describe how the union and intersection operations distribute over each other within WHSESs $\langle \Psi_H, T_1, w_H^1\rangle, \langle \Phi_H, T_2, w_H^2\rangle$, and $\langle \Upsilon_H, T_3, w_H^3\rangle$.

[ii] Distributive Laws

$$\langle \Psi_H, T_1, w_H^1\rangle \tilde{\cap} \left(\langle \Phi_H, T_2, w_H^2\rangle \tilde{\cup} \langle \Upsilon_H, T_3, w_H^3\rangle\right)$$
$$= \left(\langle \Psi_H, T_1, w_H^1\rangle \tilde{\cap} \langle \Phi_H, T_2, w_H^2\rangle\right) \tilde{\cup} \left(\langle \Psi_H, T_1, w_H^1\rangle \tilde{\cap} \langle \Upsilon_H, T_3, w_H^3\rangle\right)$$

$$\langle \Psi_H, T_1, w_H^1\rangle \tilde{\cup} \left(\langle \Phi_H, T_2, w_H^2\rangle \tilde{\cap} \langle \Upsilon_H, T_3, w_H^3\rangle\right)$$
$$= \left(\langle \Psi_H, T_1, w_H^1\rangle \tilde{\cup} \langle \Phi_H, T_2, w_H^2\rangle\right) \tilde{\cap} \left(\langle \Psi_H, T_1, w_H^1\rangle \tilde{\cup} \langle \Upsilon_H, T_3, w_H^3\rangle\right)$$

These laws also emphasize the distribution of union and intersection operations, similar to the associative laws but in a slightly different context or notation.

[iii] De Morgan's Laws

$$\left(\langle \Psi_H, T_1, w_H^1\rangle \tilde{\cap} \langle \Phi_H, T_2, w_H^2\rangle\right)^C = \langle \Psi_H, T_1, w_H^1\rangle^C \tilde{\cup} \langle \Phi_H, T_2, w_H^2\rangle^C$$

$$\left(\langle \Psi_H, T_1, w_H^1\rangle \tilde{\cup} \langle \Phi_H, T_2, w_H^2\rangle\right)^C = \langle \Psi_H, T_1, w_H^1\rangle^C \tilde{\cap} \langle \Phi_H, T_2, w_H^2\rangle^C$$

De Morgan's Laws describe the relationship between the complement of unions and intersections of sets. They assert that the complement of the union of two WHSESs equals the intersection of their complements, while the complement of the intersection of two WHSESs equals the union of their complements.

These laws are fundamental in set theory and are widely used in various mathematical and logical contexts to manipulate sets and derive new relationships or properties. They provide rules for combining and complementing sets that are essential in reasoning and problem-solving involving sets.

Definition 3.13 Let $\langle \Psi_H, T, w_H\rangle \in WHSES(\Omega_H)$. An agree-WHSES $\langle \Psi_H, T, w_H\rangle_{Ag}$ on Ω_H is a WHSE subset of $\langle \Psi_H, T, w_H\rangle$ and is characterized as

$$\langle \Psi_H, T, w_H\rangle_{Ag} = \left\{((t_i, w_H(t_i)), \Psi_H(t_i)) : t_i \in T \cap P \times Q \times \{1\}\right\}$$

and the agree weighted score is defined by $D_j = \sum_i w_H(t_i) \times a_{ij}$ where

$$a_{ij} = 1,\ if\ \rho_j \in \Psi_H(t_i) = 0,\ if\ \rho_j \notin \Psi_H(t_i).$$

Definition 3.14 Let $\langle \Psi_H, T, w_H\rangle \in WHSES(\Omega_H)$. A disagree-WHSES $\langle \Psi_H, T, w_H\rangle_{Dag}$ on Ω_H is a WHSE subset of $\langle \Psi_H, T, w_H\rangle$ and is characterized as

$$\langle \Psi_H, T, w_H\rangle_{Dag} = \left\{((t'_i, w_H(t'_i)), \Psi_H(t'_i)) : t'_i \in T \cap P \times Q \times \{0\}\right\}$$

and the disagree weighted score is defined by $F_j = \sum_i w_H(t'_i) \times d_{ij}$, where

$$d_{ij} = 1,\ if\ \rho_j \in \Psi_H(t'_i) = 0,\ if\ \rho_j \notin \Psi_H(t'_i).$$

These definitions help in characterizing agree-WHSES and disagree-WHSES based on their weighted scores, providing a framework for analyzing WHSESs in decision-making or problem-solving contexts.

Our innovative DMM for addressing DMPs leverages HSESs and WHSESs. The algorithm involves defining the universal set Ω_H and creating HSES with specialist and conclusion sets. Additionally, it incorporates WHSES with a weight function for optimizing decision rules and enhancing the accuracy of the decision-making process. The steps of our proposed decision-making model are outlined below:

Algorithm 2

Step1. Input a nonempty universe $\Omega_H= \{\rho_1,\rho_2,...,\rho_n\}$, a parameter set $T \subseteq P \times Q \times R$ containing relevant parameters for decision-making.

Step2. Input the HSES (Ψ_H, T) over the universe $\Omega_H= \{\rho_1,\rho_2,...,\rho_n\}$, and assign a weight w_H corresponding to (Ψ_H, T), where w_H: $T\rightarrow[0,1]$ is determined by the expert.

Step3. Compute the WHSES $\langle \Psi_H, T, w_H\rangle$with regards to the weight w_H.

Step4. Calculate the agree-WHSES and the disagree-WHSES for $\langle \Psi_H, T, w_H \rangle$in tabular form.

Step5. Calculate agree weighted score $D_j = \sum_i w_H(t_i) \times a_{ij}$ for the agree-WHSES.

Step6. Compute disagree weighted score$F_j = \sum_i w_H(t'_i) \times d_{ij}$ for the disagree-WHSES.

Step7. Obtain weighted score $S_j= D_j-F_j$ for the WHSES

Step8. The best decision is to choose the option ρ_k if the overall weighted score S_k is maximized. This decision is based on maximizing the combined weighted score that reflects both agreement and disagreement aspects.

Step9. Any of ρ_k can be chosen if ρ_k has a lot of values.

Remark 3.15: In the 8th step of our constructed Decision-Making Model (DMM), it is possible to revisit the 2nd step and fine-tune the weights assigned in the WHSES. This adjustment helps in identifying the best optimal choice, especially in situations where there are numerous optimal choices to choose from. By iteratively refining the weights based on desired outcomes or constraints, the model ensures a more precise and tailored decision-making process.

RESULTS AND DISCUSSIONS

In this section, we present a practical decision-making problem to showcase the effectiveness and validity of our proposed DMM. By applying our innovative DMM based on hypersoft expert sets and weighted hypersoft expert sets to this real-world scenario, we aim to demonstrate the model's ability to generate informed and optimal decisions. The case study will highlight the practical applicability and accuracy of our suggested method in tackling complex decision challenges in various domains.

Example 4.1 To address the hiring process for a manufacturing company's vacancy, a real-life DMP is presented. The company has received 10 applications from qualified individuals and aims to finalize the hiring process using a panel of experts who meet specific criteria. This scenario provides a practical context to demonstrate the validity and effectiveness of our proposed advanced machine learning algorithm based on HSES and WHSES. Through this case study, we will illustrate how our method can assist in making informed and optimal hiring decisions.

Step1: Let us consider 10 candidates to form the set of universe $\Omega_H = \{\rho_1, \rho_2, \rho_3, \rho_4, \rho_5, \rho_6, \rho_7, \rho_8, \rho_9, \rho_{10}\}$ and P_1= Qualification= $\{p_{11}, p_{12}, p_{13}\}$, P_2= Skills= $\{p_{21}, p_{22}\}$, P_3= Experience= $\{p_{31}\}$ be the disjoint sets of parameters. Let $Q = \{Q_1, Q_2, Q_3\}$ be a group of experts (board members) for this selection procedure and $R=\{1,0\}$. Then

$$\mathrm{P}=\mathrm{P}_1 \times \mathrm{P}_2 \times \mathrm{P}_3 = \left\{\begin{matrix} \sigma_1 = (p_{11}, p_{21}, p_{31}), \sigma_2 = (p_{12}, p_{21}, p_{31}), \sigma_3 = (p_{13}, p_{21}, p_{31}), \\ \sigma_4 = (p_{11}, p_{22}, p_{31}), \sigma_5 = (p_{12}, p_{22}, p_{31}), \sigma_6 = (p_{13}, p_{22}, p_{31}), \end{matrix}\right\},$$

And

$$\mathrm{S}=\mathrm{P}\times\mathrm{Q}\times\mathrm{R} = \left\{\begin{matrix} (\sigma_1, \mathrm{Q}_1, 1), & (\sigma_1, \mathrm{Q}_2, 1), & (\sigma_1, \mathrm{Q}_3, 1), & (\sigma_1, \mathrm{Q}_1, 0), & (\sigma_1, \mathrm{Q}_2, 0), & (\sigma_1, \mathrm{Q}_3, 0), \\ (\sigma_2, \mathrm{Q}_1, 1), & (\sigma_2, \mathrm{Q}_2, 1), & (\sigma_2, \mathrm{Q}_3, 1), & (\sigma_2, \mathrm{Q}_1, 0), & (\sigma_2, \mathrm{Q}_2, 0), & (\sigma_2, \mathrm{Q}_3, 0), \\ (\sigma_3, \mathrm{Q}_1, 1), & (\sigma_3, \mathrm{Q}_2, 1), & (\sigma_3, \mathrm{Q}_3, 1), & (\sigma_3, \mathrm{Q}_1, 0), & (\sigma_3, \mathrm{Q}_2, 0), & (\sigma_3, \mathrm{Q}_3, 0), \\ (\sigma_4, \mathrm{Q}_1, 1), & (\sigma_4, \mathrm{Q}_2, 1), & (\sigma_4, \mathrm{Q}_3, 1), & (\sigma_4, \mathrm{Q}_1, 0), & (\sigma_4, \mathrm{Q}_2, 0), & (\sigma_4, \mathrm{Q}_3, 0), \\ (\sigma_5, \mathrm{Q}_1, 1), & (\sigma_5, \mathrm{Q}_2, 1), & (\sigma_5, \mathrm{Q}_3, 1), & (\sigma_5, \mathrm{Q}_1, 0), & (\sigma_5, \mathrm{Q}_2, 0), & (\sigma_5, \mathrm{Q}_3, 0), \\ (\sigma_6, \mathrm{Q}_1, 1), & (\sigma_6, \mathrm{Q}_2, 1), & (\sigma_6, \mathrm{Q}_3, 1), & (\sigma_6, \mathrm{Q}_1, 0), & (\sigma_6, \mathrm{Q}_2, 0), & (\sigma_6, \mathrm{Q}_3, 0) \end{matrix}\right\},$$

Let

$$T = \left\{\begin{matrix} t_1 = (\sigma_1, \mathrm{Q}_1, 1), & t_2 = (\sigma_1, \mathrm{Q}_2, 1), & t_3 = (\sigma_1, \mathrm{Q}_3, 1), & t_4 = (\sigma_2, \mathrm{Q}_1, 1), \\ t_5 = (\sigma_2, \mathrm{Q}_2, 1), & t_6 = (\sigma_2, \mathrm{Q}_3, 1), & t_7 = (\sigma_4, \mathrm{Q}_1, 1), & t_8 = (\sigma_4, \mathrm{Q}_2, 1), \\ t_9 = (\sigma_4, \mathrm{Q}_3, 1), & t_{10} = (\sigma_6, \mathrm{Q}_1, 1), & t_{11} = (\sigma_6, \mathrm{Q}_2, 1), & t_{12} = (\sigma_6, \mathrm{Q}_3, 1), \\ t_1' = (\sigma_1, \mathrm{Q}_1, 0), & t_2' = (\sigma_1, \mathrm{Q}_2, 0), & t_3' = (\sigma_1, \mathrm{Q}_3, 0), & t_4' = (\sigma_2, \mathrm{Q}_1, 0), \\ t_4' = (\sigma_2, \mathrm{Q}_2, 0), & t_6' = (\sigma_2, \mathrm{Q}_3, 0), & t_7' = (\sigma_4, \mathrm{Q}_1, 0), & t_8' = (\sigma_4, \mathrm{Q}_2, 0), \\ t_9' = (\sigma_4, \mathrm{Q}_3, 0), & t_{10}' = (\sigma_6, \mathrm{Q}_1, 0), & t_{11}' = (\sigma_6, \mathrm{Q}_2, 0), & t_{12}' = (\sigma_6, \mathrm{Q}_3, 0) \end{matrix}\right\}.$$

Step2: Then an HSES (Ψ_{H}, T) can be written as follows:

$$(\Psi_{H},T)=\left\{\begin{array}{l}(t_1,\{\rho_1,\rho_2,\rho_5,\rho_7,\rho_9,\rho_{10}\}),\ (t_2,\{\rho_2,\rho_3,\rho_6,\rho_7,\rho_{10}\}),\ (t_3,\{\rho_1,\rho_3,\rho_4,\rho_7\}),\\ (t_4,\{\rho_1,\rho_4,\rho_5,\rho_6,\rho_8\}),\ (t_5,\{\rho_2,\rho_4,\rho_6,\rho_8\}),\ (t_6,\{\rho_2,\rho_3,\rho_5,\rho_9\}),\\ (t_7,\{\rho_1,\rho_3,\rho_4,\rho_7,\rho_8,\rho_{10}\}),\ (t_8,\{\rho_3,\rho_4,\rho_6,\rho_7,\rho_8,\rho_{10}\}),(t_9,\{\rho_1,\rho_2,\rho_5,\rho_9,\rho_{10}\}),\\ (t_{10},\{\rho_2,\rho_4,\rho_6,\rho_7,\rho_9,\rho_{10}\}),\ (t_{11},\{\rho_3,\rho_5,\rho_8,\rho_9\}),\ (t_{12},\{\rho_1,\rho_5,\rho_6,\rho_8,\rho_9\}),\\ (t'_1,\{\rho_3,\rho_4,\rho_6,\rho_8\}),(t'_2,\{\rho_1,\rho_4,\rho_5,\rho_8,\rho_9\}),\ (t'_3,\{\rho_2,\rho_5,\rho_6,\rho_8,\rho_9,\rho_{10}\}),\\ (t'_4,\{\rho_2,\rho_3,\rho_7,\rho_9,\rho_{10}\}),\ (t'_5,\{\rho_1,\rho_3,\rho_5,\rho_7,\rho_9,\rho_{10}\}),(t'_6,\{\rho_1,\rho_4,\rho_6,\rho_7,\rho_8,\rho_{10}\}),\\ (t'_7,\{\rho_2,\rho_5,\rho_6,\rho_9\}),\ (t'_8,\{\rho_1,\rho_2,\rho_5,\rho_9\}),\ (t'_9,\{\rho_3,\rho_4,\rho_6,\rho_7,\rho_8\}),\\ (t'_{10},\{\rho_1,\rho_3,\rho_5,\rho_8\}),\ (t'_{11},\{\rho_1,\rho_2,\rho_4,\rho_6,\rho_7,\rho_{10}\}),\ (t'_{12},\{\rho_2,\rho_3,\rho_4,\rho_7,\rho_{10}\})\end{array}\right\}.$$

Assume that the weights for the parameters in set T are provided as follows:

$w_1 = w_H(t_1) = 0.72;\quad w_2 = w_H(t_2) = 0.81;\quad w_3 = w_H(t_3) = 0.75;\quad w_4 = w_H(t_4) = 0.65;$

$w_5 = w_H(t_5) = 0.85;\quad w_6 = w_H(t_6) = 0.62;\quad w_7 = w_H(t_7) = 0.55;\quad w_8 = w_H(t_8) = 0.78;$

$w_9 = w_H(t_9) = 0.89;\quad w_{10} = w_H(t_{10}) = 0.67;\quad w_{11} = w_H(t_{11}) = 0.84;\quad w_{12} = w_H(t_{12}) = 0.87;$

$w_{13} = w_H(t'_1) = 0.72;\quad w_{14} = w_H(t'_2) = 0.81;\quad w_{15} = w_H(t'_3) = 0.75;\quad w_{16} = w_H(t'_4) = 0.65;$

$w_{17} = w_H(t'_5) = 0.85;\quad w_{18} = w_H(t'_6) = 0.62;\quad w_{19} = w_H(t'_7) = 0.55;\quad w_{20} = w_H(t'_8) = 0.78;$

$w_{21} = w_H(t'_9) = 0.89;\quad w_{22} = w_H(t'_{10}) = 0.67;\quad w_{23} = w_H(t'_{11}) = 0.84;\quad w_{24} = w_H(t'_{12}) = 0.87.$

Step3: Then we have a weighted w_H for HSES (Ψ_{H}, T), where w_H: $T \to [0,1]$ and the HSES (Ψ_{H}, T) is changed into a WHSES $\langle \Psi_{H}, T, w_H \rangle$ as follows:

$$\langle\Psi_{H},T,w_H\rangle=\left\{\begin{array}{l}((t_1,0.72),\{\rho_1,\rho_2,\rho_5,\rho_7,\rho_9,\rho_{10}\}),((t_2,0.81),\{\rho_2,\rho_3,\rho_6,\rho_7,\rho_{10}\}),((t_3,0.75),\{\rho_1,\rho_3,\rho_4,\rho_7\}),\\ ((t_4,0.65),\{\rho_1,\rho_4,\rho_5,\rho_6,\rho_8\}),((t_5,0.85),\{\rho_2,\rho_4,\rho_6,\rho_8\}),((t_6,0.62),\{\rho_2,\rho_3,\rho_5,\rho_9\}),\\ ((t_7,0.55),\{\rho_1,\rho_3,\rho_4,\rho_7,\rho_8,\rho_{10}\}),((t_8,0.78),\{\rho_3,\rho_4,\rho_6,\rho_7,\rho_8,\rho_{10}\}),((t_9,0.89),\{\rho_1,\rho_2,\rho_5,\rho_9,\rho_{10}\}),\\ ((t_{10},0.67),\{\rho_2,\rho_4,\rho_6,\rho_7,\rho_9,\rho_{10}\}),((t_{11},0.84),\{\rho_3,\rho_5,\rho_8,\rho_9\}),((t_{12},0.87),\{\rho_1,\rho_5,\rho_6,\rho_8,\rho_9\}),\\ ((t'_1,0.72),\{\rho_3,\rho_4,\rho_6,\rho_8\}),((t'_2,0.81),\{\rho_1,\rho_4,\rho_5,\rho_8,\rho_9\}),((t'_3,0.75),\{\rho_2,\rho_5,\rho_6,\rho_8,\rho_9,\rho_{10}\}),\\ ((t'_4,0.65),\{\rho_2,\rho_3,\rho_7,\rho_9,\rho_{10}\}),((t'_5,0.85),\{\rho_1,\rho_3,\rho_5,\rho_7,\rho_9,\rho_{10}\}),((t'_6,0.62),\{\rho_1,\rho_4,\rho_6,\rho_7,\rho_8,\rho_{10}\}),\\ ((t'_7,0.55),\{\rho_2,\rho_5,\rho_6,\rho_9\}),((t'_8,0.78),\{\rho_1,\rho_2,\rho_5,\rho_9\}),((t'_9,0.89),\{\rho_3,\rho_4,\rho_6,\rho_7,\rho_8\}),\\ ((t'_{10},0.67),\{\rho_1,\rho_3,\rho_5,\rho_8\}),((t'_{11},0.84),\{\rho_1,\rho_2,\rho_4,\rho_6,\rho_7,\rho_{10}\}),((t'_{12},0.87),\{\rho_2,\rho_3,\rho_4,\rho_7,\rho_{10}\})\end{array}\right\}.$$

Step4: Now, we obtain the agree-WHSES and disagree-WHSES for $\langle \Psi_{H}, T, w_H \rangle$ are shown in Tables 1, 2.

Step5: We calculate the agree weighted score $D_j = \sum_i w_H(t_i) \times a_{ij}$ for the agree-WHSES as in Table 1.

Step6: We compute the disagree weighted score $F_j = \sum_i w_H(t'_i) \times d_{ij}$ for the disagree-WHSES as in Table 2.

Step7: Table 3 displays the results of the computations of the weighted score $S_j = D_j - F_j$ for the WHSES $\langle \Psi_{H}, T, w_H \rangle$.

Step8: From Table 3, we have the option $_6$ is the best optimal decision.

Overall, this process involves transforming the initial HSES based on candidate parameters into a weighted HSES (WHSES), computing the agree and disagree weighted scores, and finally determining the best optimal decision based on the computed scores.

Advantages 4.2: Algorithm 2 streamlines the decision-making process by reducing the number of object choices, thus simplifying the decision for leaders. However, it also provides more detailed data, offering valuable insights to aid decision-making. In cases where numerous optimal choices arise in the 8th step, returning to the 2nd step to adjust the weights allows for fine-tuning and optimization of the decision criteria, ensuring a more precise and informed decision-making process. This iterative approach enhances the model's adaptability and effectiveness in handling complex decision scenarios.

Table 1. Agree-WHSES for $\langle \Psi_H, T, w_H \rangle$

	1	2	3	4	5	6	7	8	9	10
t_1, 0.72	1	1	0	0	1	0	1	0	1	1
t_2, 0.81	0	1	1	0	0	1	1	0	0	1
t_3, 0.75	1	0	1	1	0	0	1	0	0	0
t_4, 0.65	1	0	0	1	1	1	0	1	0	0
t_5, 0.85	0	1	0	1	0	1	0	1	0	0
t_6, 0.62	0	1	1	0	1	0	0	0	1	0
t_7, 0.55	1	0	1	1	0	0	1	1	0	1
t_8, 0.78	0	0	1	1	0	1	1	1	0	1
t_9, 0.89	1	1	0	0	1	0	0	0	1	1
t_{10}, 0.67	0	1	0	1	0	1	1	0	1	1
t_{11}, 0.84	0	0	1	0	1	0	0	1	1	0
t_{12}, 0.87	1	0	0	0	1	1	0	1	1	0
$D_j = \sum_i w_H(t_i) \times a_{ij}$	4.43	4.56	4.35	4.25	4.59	4.69	4.28	4.54	4.61	4.42

Table 2. Disagree-WHSES for $\langle \Psi_{H}, T, w_{H} \rangle$

	1	2	3	4	5	6	7	8	9	10
t'_1, 0.72	0	0	1	1	0	1	0	1	0	0
t'_2, 0.81	1	0	0	1	1	0	0	1	1	0
t'_3, 0.75	0	1	0	0	1	1	0	1	1	1
t'_4, 0.65	0	1	1	0	0	0	1	0	1	1
t'_5, 0.85	1	0	1	0	1	0	1	0	1	1
t'_6, 0.62	1	0	0	1	0	1	1	1	0	1
t'_7, 0.55	0	1	0	0	1	1	0	0	1	0
t'_8, 0.78	1	1	0	0	1	0	0	0	1	0
t'_9, 0.89	0	0	1	1	0	1	1	1	0	0
t'_{10}, 0.67	1	0	1	0	1	0	0	1	0	0
t'_{11}, 0.84	1	1	0	1	0	1	1	0	0	1
t'_{12}, 0.87	0	1	1	1	0	0	1	0	0	1
$F_j = \sum_i w_H(t'_i) \times d_{ij}$	4.57	4.44	4.65	4.75	4.41	4.37	4.72	4.46	4.39	4.58

COMPARISON ANALYSES

The Ihsan-method (Ihsan et al., 2021) may lack the necessary robustness and flexibility to effectively handle HSES based DMPs, primarily due to its limited adaptability to complex decision scenarios and its inability to consider the importance of parameter weights in decision outcomes. Below, we aim to demonstrate that the Ihsan-method (Ihsan et al., 2021) does not offer adequate capabilities to solve DMPs based on HSES.

Step1: Let us consider the HSES (Ψ_{H}, T) as in Example 4.1.
Step2: Then, we have the agree-HSES and disagree-HSES for (Ψ_{H}, T) are shown in Tables 4 and 5.

Step3: We calculate $D_j = \sum_i a_{ij}$ for the agree-HSES as in Table 4.

Step4: We calculate $F_j = \sum_i d_{ij}$ for the disagree-HSES as in Table 5.
Step5: Now, we compute the score $S_j = D_j - F_j$ for the agree-HSES table as shown in Table 6.

Step6: Table 6 indicates that all the score values $S_j = D_j - F_j$ are identical, specifically 0. Therefore, in this scenario, it is not possible to determine the best optimal decision.

The method developed in this chapter presents significant advantages and demonstrates robustness in solving HSES based DMPs, as evidenced by Example 4.1. Unlike the approach (Ihsan et al., 2021), our constructed DMM considers the importance of parameter weights, enabling a nuanced evaluation of decision outcomes based on specified criteria. This feature allows decision-makers to assign varying degrees of importance to parameters, enhancing the customization and accuracy of decision processes. Moreover, the adjustable nature of our DMM sets it apart from the non-adjustable DMM (Ihsan et al., 2021). In complex decision scenarios with multiple optimal choices, our DMM's adaptability shines through as decision-makers can revisit and fine-tune weight assignments in the 2nd step, ensuring the selection of the best outcome. This adjustability factor adds a layer of flexibility and optimization capability to our DMM, making it more versatile and capable of handling dynamic decision environments. The absence of these features in the Ihsan-method (Ihsan et al., 2021) highlights its limitations in effectively addressing the complexities and nuances of HSES-based DMPs. By contrast, our constructed DMM excels in stability, feasibility, and adaptability, offering a superior solution for solving HSES-based DMPs and enhancing decision-making processes across various domains.

Table 3. The weighted score $S_j = D_j - F_j$

Ω_H	$D_j = \sum_i a_{ij}$	$F_j = \sum_i d_{ij}$	$S_j = D_j - F_j$
ρ_1	4.43	4.57	-0.14
ρ_2	4.56	4.44	+0.12
ρ_3	4.35	4.65	-0.30
ρ_4	4.25	4.75	-0.50
ρ_5	4.59	4.41	+0.18
ρ_6	**4.63**	**4.37**	**+0.26**
ρ_7	4.28	4.72	-0.44
ρ_8	4.54	4.46	+0.08
ρ_9	4.61	4.39	+0.22
ρ_{10}	4.42	4.58	-0.16

Table 4. Agree-HSES for the HSES (Ψ_H, T)

	1	2	3	4	5	6	7	8	9	10
t_1	1	1	0	0	1	0	1	0	1	1
t_2	0	1	1	0	0	1	1	0	0	1
t_3	1	0	1	1	0	0	1	0	0	0
t_4	1	0	0	1	1	1	0	1	0	0
t_5	0	1	0	1	0	1	0	1	0	0
t_6	0	1	1	0	1	0	0	0	1	0
t_7	1	0	1	1	0	0	1	1	0	1
t_8	0	0	1	1	0	1	1	1	0	1
t_9	1	1	0	0	1	0	0	0	1	1
t_{10}	0	1	0	1	0	1	1	0	1	1
t_{11}	0	0	1	0	1	0	0	1	1	0
t_{12}	1	0	0	0	1	1	0	1	1	0
$D_j = \sum_i a_{ij}$	6	6	6	6	6	6	6	6	6	6

where

$$a_{ij} = 1, \text{ if } \rho_j \in \Psi_H(t_i)$$
$$= 0, \text{ if } \rho_j \notin \Psi_H(t_i).$$

Table 5. Disagree-HSES for the HSES (Ψ_H, T)

	1	2	3	4	5	6	7	8	9	10
t'_1	0	0	1	1	0	1	0	1	0	0
t'_2	1	0	0	1	1	0	0	1	1	0
t'_3	0	1	0	0	1	1	0	1	1	1
t'_4	0	1	1	0	0	0	1	0	1	1
t'_5	1	0	1	0	1	0	1	0	1	1
t'_6	1	0	0	1	0	1	1	1	0	1
t'_7	0	1	0	0	1	1	0	0	1	0
t'_8	1	1	0	0	1	0	0	0	1	0
t'_9	0	0	1	1	0	1	1	1	0	0
t'_{10}	1	0	1	0	1	0	0	1	0	0
t'_{11}	1	1	0	1	0	1	1	0	0	1

continued on following page

Table 5. Continued

	1	2	3	4	5	6	7	8	9	10
t'_{12}	0	1	1	1	0	0	1	0	0	1
$F_j = \sum_i d_{ij}$	6	6	6	6	6	6	6	6	6	6

where

$$d_{ij} = 1, \text{ if } \rho_j \in \Psi_H(t'_i)$$
$$= 0, \text{ if } \rho_j \notin \Psi_H(t'_i)$$

Table 6. The optimal $S_j = D_j - F_j$

$D_j = \sum_i a_{ij}$	$F_j = \sum_i d_{ij}$	$S_j = D_j - F_j$
6	6	0
6	6	0
6	6	0
6	6	0
6	6	0
6	6	0
6	6	0
6	6	0
6	6	0
6	6	0

The Ihsan-method (Ihsan et al., 2021) may not be sufficient to effectively solve DMPs based on HSESs for several reasons:

1. Complexity of HSES: HSES introduces a higher level of complexity compared to traditional decision-making methods. The Ihsan-method may not adequately handle the intricacies and nuances inherent in HSES, such as weighted scores and multiple criteria.
2. Lack of Adaptability: The Ihsan-method may lack the adaptability required to adjust weights dynamically or handle varying degrees of uncertainty and ambiguity in HSES-based DMPs.
3. Limited Decision Rules: The Ihsan-method may rely on predefined decision rules or criteria, which may not be sufficient to capture the diverse range of decision scenarios that arise in HSES-based DMPs.
4. Optimization Challenges: Optimizing decision outcomes in HSES-based DMPs may require iterative adjustments and fine-tuning of parameters, which the Ihsan-method may not explicitly support.

Overall, while the Ihsan-method may be effective for certain decision-making contexts, it may not provide the comprehensive and flexible framework needed to tackle the complexities of HSES-based DMPs.

CONCLUSION

In this chapter, we have studied and extensively explored the idea of WHSES, highlighting its fundamental operations and characteristics. Moreover, we have defined novel concepts such as agree-WHSES, disagree-WHSES, and weighted scores (including agree and disagree weighted scores). These innovative notations have formed the basis for our proposed advanced and adaptable DMM designed to address HSES and WHSES based DMPs. One of the key strengths of our proposed DMM lies in its utilization of the weighted score function, a departure from traditional scoring functions. This shift enhances the stability and practicality of our method, making it more suitable for real-world applications where decisions are often complex and multifaceted. By incorporating a weighted score, our DMM exhibits a high degree of flexibility, allowing it to accommodate a diverse range of decision-making scenarios without the need for rigid value selection. Algorithm 2, derived from our proposed DMM, offers superior adaptability and versatility compared to existing methods. Its inherent flexibility enables seamless application across various domains, including but not limited to computer science, software engineering, and contemporary life conditions. This adaptability opens doors for future research and development, suggesting potential extensions and enhancements to address emerging challenges and evolving complexities in decision-making processes.

Looking ahead, our future study will delve deeper into the broader properties and operations of WHSES, expanding on their applicability and potential impact. Furthermore, we aim to extend our proposed DMM to encompass a wider array of real-life applications, particularly focusing on pattern recognition and medical diagnostics. By doing so, we aspire to contribute significantly to the advancement of decision-making methodologies and their practical utility in diverse and dynamic environments.

ABBREVIATIONS

DMM: Decision-making method
DMP: Decision making problem
HSES: Hypersoft expert set
HSS: Hypersoft set

MAGDM: Multi-attribute group decision making
MCDM: Multi-criteria decision making
SES: Soft expert set
SST: Soft set theory
WHSES: Weighted hypersoft expert set

REFERENCES

Abbas, M., Murtaza, G., & Smarandache, F. (2020). Basic operations on hypersoft sets and hypersoft point. *Neutrosophic Sets and Systems*, *35*, 407–421.

Abdel-Basset, M., Gamal, A., Son, L.H., & Smarandache, F. (2020). A bipolar neutrosophic multi-criteria decision-making framework for professional selection. *Applied Sciences (Basel, Switzerland)*, *10*(4), 1–22.

Abdel-Basset, M., Chang, V., & Gamal, A. (2019). Evaluation of the green supply chain management practices: A novel neutrosophic approach. *Computers in Industry*, *108*, 210–220. DOI: 10.1016/j.compind.2019.02.013

Abdel-Basset, M., Chang, V., Gamal, A., & Smarandache, F. (2019). An integrated neutrosophic ANP and VIKOR method for achieving sustainable supplier selection: A case study in importing field. *Computers in Industry*, *106*, 94–110. DOI: 10.1016/j.compind.2018.12.017

Abdel-Basset, M., Gamal, A., Manogaran, G., & Long, H. V. (2020). A novel group decision-making model based on neutrosophic sets for heart disease diagnosis. *Multimedia Tools and Applications*, *79*(15-16), 9977–10002. DOI: 10.1007/s11042-019-07742-7

Abdel-Basset, M., Manogaran, G., Gamal, A., & Chang, V. (2020). A Novel Intelligent Medical Decision Support Model Based on Soft Computing and IoT. *IEEE Internet of Things Journal*, *7*(5), 4160–4170. DOI: 10.1109/JIOT.2019.2931647

Abdel-Basset, M., Saleh, M., Gamal, A., & Smarandache, F. (2019). An approach of TOPSIS technique for developing supplier selection with group decision making under type-2 neutrosophic number. *Applied Soft Computing*, *77*, 438–452. DOI: 10.1016/j.asoc.2019.01.035

Abdulkareem, K. H., Arbaiy, N., Zaidan, A. A., Zaidan, B. B., Albahri, O. S., Alsalem, M. A., & Salih, M. M. (2020). A novel multiperspective benchmarking framework for selecting image dehazing intelligent algorithms based on BWM and group VIKOR techniques. *International Journal of Information Technology & Decision Making*, *19*(3), 909–957. DOI: 10.1142/S0219622020500169

Abdulkareem, K. H., Arbaiy, N., Zaidan, A. A., Zaidan, B. B., Albahri, O. S., Alsalem, M. A., & Salih, M. M. (2021). A new standardization and selection framework for real-time image dehazing algorithms from multi-foggy scenes based on fuzzy Delphi and hybrid multi-criteria decision analysis methods. *Neural Computing & Applications*, *33*(4), 1029–1054. DOI: 10.1007/s00521-020-05020-4

Abu Qamar, M., & Hassan, N. (2018). Generalized Q-neutrosophic soft expert set for decision under uncertainty. *Symmetry*, *10*(11), 621. DOI: 10.3390/sym10110621

Adam, F., & Hassan, N. (2016). Multi Q-fuzzy soft expert set and its application. *Journal of Intelligent & Fuzzy Systems*, *30*(2), 943–950. DOI: 10.3233/IFS-151816

Akram, M., & Adeel, A. (2019). TOPSIS approach for MAGDM based on interval-valued hesitant fuzzy N-soft environment. *International Journal of Fuzzy Systems*, *21*(3), 993–1009. DOI: 10.1007/s40815-018-0585-1

Akram, M., Adeel, A., & Alcantud, J. C. R. (2019). Group decision-making methods based on hesitant N-soft sets. *Expert Systems with Applications*, *115*, 95–105. DOI: 10.1016/j.eswa.2018.07.060

Al-Qudah, Y., & Hassan, N. (2017). Bipolar fuzzy soft expert set and its application in decision making. *International Journal of Applied Decision Sciences*, *10*(2), 175–191. DOI: 10.1504/IJADS.2017.084310

Al-Qudah, Y., & Hassan, N. (2017). Complex multi-fuzzy soft expert set and its application. *International Journal of Mathematics and Computer Science*, *14*(1), 149–176.

Al-Quran, A., & Hassan, N. (2016). Neutrosophic vague soft expert set theory. *Journal of Intelligent & Fuzzy Systems*, *30*(6), 3691–3702. DOI: 10.3233/IFS-162118

Al-Quran, A., & Hassan, N. (2016). Fuzzy parameterised single valued neutrosophic soft expert set theory and its application in decision making. *International Journal of Applied Decision Science*, *9*(2), 212–227. DOI: 10.1504/IJADS.2016.080121

Alcantud, J. C. R., & Santos-García, G. (2017). A new criterion for soft set-based decision-making problems under incomplete information. *International Journal of Computational Intelligence Systems*, *10*(1), 394–404. DOI: 10.2991/ijcis.2017.10.1.27

Alcantud, J. C. R., & Torrecillas Muñoz, M. J. (2018). Intertemporal choice of fuzzy soft sets. *Symmetry*, *10*(9), 371. DOI: 10.3390/sym10090371

Alhazaymeh, K., & Hassan, N. (2014). Generalized vague soft expert set. *International Journal of Pure and Applied Mathematics*, *93*(3), 351–360.

Alhazaymeh, K., & Hassan, N. (2014). Mapping on generalized vague soft expert set. *International Journal of Pure and Applied Mathematics*, *93*(3), 369–376. DOI: 10.12732/ijpam.v93i3.7

Alhazaymeh, K., & Hassan, N. (2014). Application of generalized vague soft expert set in decision making. *International Journal of Pure and Applied Mathematics*, *93*(3), 361–367. DOI: 10.12732/ijpam.v93i3.6

Alkhazaleh, S., & Salleh, A. R. (2011). Soft Expert Sets. *Advances in Decision Sciences*, *757868*, 1–13.

Alkhazaleh, S., & Salleh, A. R. (2014). Fuzzy soft expert set and its application. *Applied Mathematics*, *5*(9), 1349–1368. DOI: 10.4236/am.2014.59127

Bashir, M., & Salleh, A. R. (2012). Possibility fuzzy soft expert set. *Ozean Journal of Applied Sciences*, *12*, 208–211.

Bashir, M., & Salleh, A.R. (2012). Fuzzy parameterized soft expert set. Abstract and Applied Analysis. DOI: 10.1155/2012/258361

Bhardwaj, N., & Sharma, P. (2021). An advanced uncertainty measure using fuzzy soft sets: Application to decision-making problems. *Big Data Mining and Analytics*, *4*(2), 94–103. DOI: 10.26599/BDMA.2020.9020020

Chen, D., Tsang, E. C. C., Yeung, D. S., & Wang, X. (2005). The parameterization reduction of soft sets and its applications. *Computers & Mathematics with Applications (Oxford, England)*, *49*(5-6), 757–763. DOI: 10.1016/j.camwa.2004.10.036

Chen, Y., Liu, J., Chen, Z., & Zhang, Y. (2020). Group decision-making method based on generalized vague N-soft sets. In *Chinese Control And Decision Conference (CCDC),* (pp. 4010–4015). IEEE. DOI: 10.1109/CCDC49329.2020.9164602

Dalkılıç, O. (2021). A novel approach to soft set theory in decision-making under uncertainty. *International Journal of Computer Mathematics*, *98*(10), 1935–1945. DOI: 10.1080/00207160.2020.1868445

Dalkiliç, O. (2022). On topological structures of virtual fuzzy parametrized fuzzy soft sets. *Complex & Intelligent Systems*, *8*(1), 337–348. DOI: 10.1007/s40747-021-00378-x

Dalkılıç, O., & Demirtaş, N. (2021). Bipolar fuzzy soft D-metric spaces. *Communications Faculty of Sciences University of Ankara Series A1 Mathematics and Statistics*, *70*(1), 64–73. DOI: 10.31801/cfsuasmas.774658

Dalkılıç, O., & Demirtaş, N. (2023). Algorithms for Covid-19 outbreak using soft set theory: Estimation and application. *Soft Computing*, *27*(6), 3203–3211. DOI: 10.1007/s00500-022-07519-5 PMID: 36268457

Darvish Falehi, A., & Torkaman, H. (2023). Optimal fractional order interval type-2 fuzzy controller for upside-down asymmetric multilevel inverter based dynamic voltage restorer to accurately compensate faulty network voltage. *Journal of Ambient Intelligence and Humanized Computing*, *14*(12), 16683–16701. DOI: 10.1007/s12652-023-04673-y

Das, A. K., & Granados, C. (2021). An advanced approach to fuzzy soft group decision-making using weighted average ratings. *SN Computer Science*, *2*(6), 471. DOI: 10.1007/s42979-021-00873-5

Das, A. K., & Granados, C. (2022). A new fuzzy parameterized intuitionistic fuzzy soft multiset theory and group decision-making. *Journal of Current Science and Technology*, *12*, 547–567.

Das, A. K., & Granados, C. (2023). IFP-intuitionistic multi fuzzy N-soft set and its induced IFP-hesitant N-soft set in decision-making. *Journal of Ambient Intelligence and Humanized Computing*, *14*(8), 10143–10152. DOI: 10.1007/s12652-021-03677-w

Das, A. K., Granados, C., & Bhattacharya, J. (2022). Some new operations on fuzzy soft sets and their applications in decision-making. *Songklanakarin Journal of Science and Technology*, *44*(2), 440–449.

Das, A. K., Gupta, N., & Granados, C. (2023). Weighted hesitant bipolar-valued fuzzy soft set in decision-making. *Songklanakarin Journal of Science and Technology*, *45*(6), 681–690.

Fatimah, F., & Alcantud, J. C. R. (2021). The multi-fuzzy N-soft set and its applications to decision-making. *Neural Computing & Applications*, *33*(17), 11437–11446. DOI: 10.1007/s00521-020-05647-3

Granados, C., Das, A.K. & Osu, B. (2023). Weighted neutrosophic soft multiset and its application to decision making. *Yugoslav Journal of Operations Research, 33*(2), 293-308.

Granados, C., Das, A.K. & Das, B. (2022). Some continuous neutrosophic distributions with neutrosophic parameters based on neutrosophic random variables. *Advances in the Theory of Nonlinear Analysis and its Applications, 33*(2) 293-308.

Hassan, N., Ulucay, V., & Sahin, M. (2018). Q-neutrosophic soft expert set and its application in decision making. *International Journal of Fuzzy System Applications*, *7*(4), 37–61. DOI: 10.4018/IJFSA.2018100103

Hazaymeh, A.A., Abdullah, I.B., Balkhi, Z.T., & Ibrahim, R.I. (2012). Generalized fuzzy soft expert set. *Journal of Applied Mathematics*.

Ihsan, M., Rahman, A. U., & Saeed, M. (2021). Hypersoft Expert Set With Application in Decision Making for Recruitment Process. *Neutrosophic Sets and Systems*, *42*, 191–207.

Ihsan, M., Saeed, M., Rahman, A. U., Khalifa, H. A. E.-W., & El-Morsy, S. (2022). An intelligent fuzzy parameterized multi-criteria decision-support system based on intuitionistic fuzzy hypersoft expert set for automobile evaluation. *Advances in Mechanical Engineering*, *14*(7). Advance online publication. DOI: 10.1177/16878132221110005

Liu, P., & Zhang, L. (2017). An extended multiple criteria decision making method based on neutrosophic hesitant fuzzy information. *Journal of Intelligent & Fuzzy Systems*, *32*(6), 4403–4413. DOI: 10.3233/JIFS-16136

Liu, P., & Zhang, L. (2017). Multiple criteria decision-making method based on neutrosophic hesitant fuzzy Heronian mean aggregation operators. *Journal of Intelligent & Fuzzy Systems*, *32*(1), 303–319. DOI: 10.3233/JIFS-151760

Maji, P. K., Biswas, R., & Roy, A. R. (2001). Fuzzy soft sets. *Journal of Fuzzy Mathematics*, *9*(3), 589–602.

Maji, P. K., Biswas, R., & Roy, A. R. (2002). An application of soft sets in decision-making problem. *Computers & Mathematics with Applications (Oxford, England)*, *44*(8–9), 1077–1083. DOI: 10.1016/S0898-1221(02)00216-X

Maji, P. K., Biswas, R., & Roy, A. R. (2003). Soft set theory. *Computers & Mathematics with Applications (Oxford, England)*, *45*(4–5), 555–562. DOI: 10.1016/S0898-1221(03)00016-6

Mohammed, M. A., Abdulkareem, K. H., Al-Waisy, A. S., Mostafa, S. A., Al-Fahdawi, S., Dinar, A. M., Alhakami, W., Baz, A., Al-Mhiqani, M. N., Alhakami, H., Arbaiy, N., Maashi, M. S., Mutlag, A. A., Garcia-Zapirain, B., & De La Torre Diez, I. (2020). Benchmarking methodology for selection of optimal COVID-19 diagnostic model based on entropy and TOPSIS methods. *IEEE Access : Practical Innovations, Open Solutions*, *8*, 99115–99131. DOI: 10.1109/ACCESS.2020.2995597

Molodtsov, D. (1999). Soft set theory-first results. *Computers & Mathematics with Applications (Oxford, England)*, *37*(4–5), 19–31. DOI: 10.1016/S0898-1221(99)00056-5

Mukherjee, A., & Das, A. K. (2018). Einstein operations on fuzzy soft multisets and decision making. *Boletim da Sociedade Paranaense de Matematica*, *40*, 1–10.

Musa, S. Y., Mohammed, R. A., & Asaad, B. A. (2023). N-Hypersoft Sets: An Innovative Extension of Hypersoft Sets and Their Applications. *Symmetry*, *15*(9), 1795. DOI: 10.3390/sym15091795

Paik, B., & Mondal, S. K. (2021). A distance-similarity method to solve fuzzy sets and fuzzy soft sets based decision-making problems. *Soft Computing*, *24*(7), 5217–5229. DOI: 10.1007/s00500-019-04273-z

Paik, B., & Mondal, S. K. (2021). Representation and application of Fuzzy soft sets in type-2 environment. *Complex & Intelligent Systems*, *7*(3), 1597–1617. DOI: 10.1007/s40747-021-00286-0

Petchimuthu, S., Garg, H., Kamacı, H., & Atagün, A. O. (2020). The mean operators and generalized products of fuzzy soft matrices and their applications in MCGDM. *Computational & Applied Mathematics*, *39*(2), 1–32. DOI: 10.1007/s40314-020-1083-2

Pramanik, S., Dey, P. P., & Giri, B. C. (2015). TOPSIS for single valued neutrosophic soft expert set based multi-attribute decision making problems. *Neutrosophic Sets and Systems*, *10*, 88–95.

Rahman, A. U., Saeed, M., & Smarandache, F. (2020). Convex and Concave Hypersoft Sets with Some Properties. *Neutrosophic Sets and Systems*, *38*, 497–508.

Rahman, A. U., Saeed, M., Smarandache, F., & Ahmad, M. R. (2020). Development of Hybrids of Hypersoft Set with Complex Fuzzy Set, Complex Intuitionistic Fuzzy set and Complex Neutrosophic Set. *Neutrosophic Sets and Systems*, *38*, 335–354.

Rathnasabapathy, P., & Palanisami, D. (2023). A theoretical development of improved cosine similarity measure for interval valued intuitionistic fuzzy sets and its applications. *Journal of Ambient Intelligence and Humanized Computing*, *14*(12), 16575–16587. DOI: 10.1007/s12652-022-04019-0 PMID: 35789601

Saeed, M., Ahsan, M., Siddique, M. K., & Ahmad, M. R. (2020). A Study of The Fundamentals of Hypersoft Set Theory. *International Journal of Scientific and Engineering Research*, *11*, 320–329.

Sahin, M., Alkhazaleh, S., & Ulucay, V. (2015). Neutrosophic soft expert sets. *Applied Mathematics*, *6*(1), 116–127. DOI: 10.4236/am.2015.61012

Smarandache, F. (2018). Extension of Soft Set to Hypersoft Set, and then to Plithogenic Hypersoft Set. *Neutrosophic Sets and Systems*, *22*, 168–170.

Ulucay, V., Sahin, M., & Hassan, N. (2018). Generalized neutrosophic soft expert set for multiple-criteria decision-making. *Symmetry*, *10*(10), 437. DOI: 10.3390/sym10100437

Zulqarnain, M., Siddique, I., Ali, R., Awrejcewicz, J., Karamti, H., Grzelczyk, D., Iampan, A., & Asif, M. (2022). Einstein Ordered Weighted Aggregation Operators for Pythagorean Fuzzy Hypersoft Set With Its Application to Solve MCDM Problem. *IEEE Access : Practical Innovations, Open Solutions*, *10*, 95294–95320. DOI: 10.1109/ACCESS.2022.3203717

Chapter 20
Generalized Plithogenic Sets in Multi-Attribute Decision Making

Nivetha Martin
Arul Anandar College (Autonomous), Karumathur, India

R. Priya
Sethu Institute of Technology, India

Florentin Smarandache
http://orcid.org/0000-0002-5560-5926
University of New Mexico, USA

ABSTRACT

The developments in the field of Plithogeny find extensive applications in multi-attribute decision making. This chapter proposes the generalized version of plithogenic sets as an extension of extended plithogenic sets. The generalized plithogenic sets are of 9-tuple form comprising realistic representations of different attribute types. The attributes are categorized into dominant, recessive, and satisfactory in a generalized plithogenic set. A multi-attribute decision making model on material selection with representations of generalized Plithogenic sets is formulated in this chapter. The proposed model is compared with the representations of extended plithogenic sets and basic plithogenic sets. It is observed that the proposed model is more comprehensive in nature. This chapter suggests decision making models based on generalized plithogenic models to the researchers for determining optimal decisions to the real-life problems.

DOI: 10.4018/979-8-3693-3204-7.ch020

1. INTRODUCTION

Smarandache (2018) developed the theory of Plithogeny and constructed the Plithogenic sets. A plithogenic set is more generally of the form of a 5-tuple (P,a,V,d,c) comprising a set P, an attribute a, the set of attribute values V, the degree of appurtenance d and the degree of contradiction c. The degree of appurtenance scales the magnitude of elements of the set agreeing to the dominant attribute value. The degree of contradiction pertains to the attribute values with respect to dominant attribute value. Sudha et al. (2023) introduced the extended version of the Plithogenic sets of the form (P,a,V,d_D,c_D,d_R,c_R) with the inclusion of degrees of appurtenance and contradiction with respect to both recessive attribute value and dominant attribute value. In a decision making circumstance based on general multi-attribute theory, the alternatives are chosen based on several attributes with equal or different weights. However in a Plithogenic based multi-attribute theory, the decision makers have opportunities to choose the alternatives not only based on the attributes but also on the attribute values. For instance if the decision maker assumes 'Cost' as one attribute in choosing the alternatives, the attribute values shall be considered as ' Cheap', 'Budgetary' and 'Expensive'. In any decision making situation, the attribute 'Cost' is considered as non-benefit attribute and henceforth the attribute value 'Cheap' is assumed to be dominant in the basic Plithogenic set. At the same time in the case of extended Plithogenic sets, the attribute value 'Expensive' is assumed to be recessive. The rank of preference and non-preference of the alternatives shall be determined by using basic and extended plithogenic sets respectively.

On other hand, the choice of the attribute values as dominant and recessive are serving as two extremes. The rankings of the alternatives are determined purely on the extreme acceptance of dominant attribute value and extreme denial of the recessive attribute value. Here is the point, where a decision maker gets stuck between the two extreme attribute values having no idea of considering the intermediate attribute values which may serve as satisfactory. A decision maker at certain instance of time, may prefer to choose the attribute value of 'Budgetary' and this reflects that the decision maker is satisfied with this attribute value in comparison with the dominant and recessive attribute values. This also indicates the ranking preference of the decision maker based on a new category of attribute values called the 'Satisfactory attribute value'. The rankings based on extreme attribute values is now generalized with the inclusion of ranking based on satisfactory attribute value. This gives more realistic representations of making decisions on choosing the alternatives based on different sets of attribute values. The decision makers shall obtain three different ranking sets of the alternatives based on dominant attribute values, satisfactory attribute values and recessive attribute values. The rankings of the alternatives based on dominant attribute values presents a clear picture of the first choice set of the alternatives and the rankings based on satisfactory attribute values presents the second choice set of the alternatives to be chosen. The rankings based on recessive attribute values presents the least choice set of alternatives. The inclusive nature of these generalized plithogenic sets has motivated the authors to make an attempt of developing a multi-attribute decision making model. The remaining contents of this chapter are structured as follows, section 2 sketches the state of art of the applications of Plithogenic sets in decision making. Section 3 describes the conceptualization of Generalized Plithogenic sets. Section 4 presents the steps involved in making decisions based on Plithogenic contradictions with Generalized Plithogenic sets. Section 5 applies the proposed concept to make decisions on material selection. Section 6 discusses the results and the last discussion concludes the work with future scope.

2. STATE OF ART OF WORK

This section presents the contributions of the researchers in the field of Plithogeny based multi-attribute decision making. A plithogenic set (Smarandache, 2018) is a more generalized form which facilitates the representations of degree of appurtenance in either of the forms of crisp, fuzzy, intuitionistic, neutrosophic. Sudha et al (Sudha et al., 2023) have presented a comprehensive review of the applications of Plithogenic sets in decision making. It is observed that the plithogeny based decision making is more compatible in drawing optimal decisions in various scenarios of supply chain management (Abdel-Basset & Mohamed, 2020; Ansari & Kant, 2021; Abdel-Basset et al., 2020; Grida et al., 2020), logistics sector (Korucuk et al., 2020; Sudha & Martin, 2023). Supplier Selection (Abdel-Basset et al., 2021), material selection (Martin et al., 2021), performance analysis of manufacturing industries (Abdel-Basset et al., 2020), food processing (Martin, 2022), sustainable supply chain (Wang et al., 2023), livestock management (Sudha et al., 2022), environmental management (Sudha et al., 2023). The conventional to contemporary decision making methods such as AHP (Analytic Hierarchy Process), TOPSIS (Technique for Order Preference by Similarity to an Ideal Solution), VIKOR (Multicriteria Optimization and Compromise Solution), CRITIC (CRiteria Importance Through Intercriteria Correlation), Best-Worst, PIPRECIA (PIvot Pairwise RElative Criteria Importance Assessment), FUCOM (Full Consistency Method), MAIRCA (MultiAtributive Ideal-Real Comparative Analysis), SWARA (Preference Ranking Organization Method for Enrichment Evaluations), MABAC (Multi-Attributive Border Approximation area Comparison), PROMTHEE (Preference Ranking Organization Method for Enrichment Evaluations), MACBETH (Measuring Attractiveness by a Categorical

Based Evaluation Technique) are discussed under Plithogenic environment. On intense analysis the plithogenic results are found to be more optimal.

The Plithogenic based MADM are further extended with different kinds of representations. Tayal et al (Tayal et al., 2023) integrated sentiment analysis with Plithogenic sets in ranking products. Hema et al (Hema et al., 2023) employed Plithogenic Interval Valued Neutrosophic Hyper-soft Sets in decision-making. The Plithogenic representations blended with different kinds of other hypersoft sets, Pythagorean sets, Fermatean sets have initiated the onset of new genre of decision making methods. Extended Plithogenic sets and the concept of the Plithogenic sociogram were introduced by Martin et al. (2022) and later developed further by Sudha et al. (2023) and constructed Plithogenic sociogram which is applied as decision making tool in choosing the optimal alternatives. In a Plithogenic based MADM, the attribute and sub-attribute classification are based on plithogenic representations and the computations are based on plithogenic aggregate operators of union and intersection. A new Plithogenic decision-making method based on contradictions was developed by Martin et al. (2023). This method of decision-making deals with contradictions and have set new standards of qualifying the alternatives based on contradiction degree indicating the attribute satisfaction.

The decision-making methods based on plithogenic approach summarized in Table 1 predominantly depend on classical plithogenic sets and only a very few applications of extended plithogenic sets exist in literature. The MADM methods are familiar only with these conventional plithogenic sets and not with the extended versions. In addition to it the Plithogenic decision making based on contradictions is also not explored. These are identified as major research gaps of Plithogeny based MADM. This has motivated the authors to develop the theory of generalized plithogenic sets which is considered as the novel aspect of this research work. The generalized plithogenic set is the extended versions of the conventional plithogenic sets and extended plithogenic sets. The plithogenic decision making based on contradictions with generalized plithogenic representations proposed in this work will duly bridge the aforementioned gaps and will set new directions of Plithogenic decision making.

Table 1. Literature review of plithogenic decision-making methods

Authors & Year	Plithogenic based methods	Applications
Mohamed et al (2020) (Abdel-Basset & Mohamed, 2020)	Plithogenic TOPSIS-CRITIC	Supply chain risk management
Ansari et al (2021) (Ansari & Kant, 2021)	Plithogenic based neutrosophic analytic hierarchy process	Sustainable supply chain
Abdel-Basset et al [2020] (Abdel-Basset et al., 2020)	Plithogenic Best-Worst method	Supply chain management
Grida et al (2020) (Grida et al., 2020)	Plithogenic Best-Worst method and VIKOR	IoT based Supply chain
Korucuk et al (2020) (Korucuk et al., 2020)	Plithogenic CRITIC	Logistics
Sudha and Nivetha (2023) (Sudha & Martin, 2023)	Plithogenic PIPRECIA	Logistic Selection
Mohamed et al (2020) (Abdel-Basset et al., 2021)	Plithogenic based rough sets	Supplier selection
Nivetha et al (2021) (Martin et al., 2021)	PROMTHEE Plithogenic Pythagorean Hypergraphic Approach	Smart Material selection
Abdel et al (2020) (Abdel-Basset et al., 2020)	Integrated Plithogenic MCDM	Financial performance
Nivetha (2021) (Martin, 2022)	Plithogenic SWARA-TOPSIS	Food Processing
Wang (2023) (Wang et al., 2023)	Plithogenic Probabilistic Linguistic MAGDM	Sustainable supply chain
Sudha and Nivetha (2022) (Sudha et al., 2022)	Plithogenic CRITIC-MAIRCA	Livestock management
Sudha et al (2023) (Sudha et al., 2023)	MACBETH-MAIRCA Plithogenic	Environmental management
Tayal et al (2023) (Tayal et al., 2023)	Plithogenic with sentimental analysis	Ranking of products
Hema et al (2023) (Hema et al., 2023)	Plithogenic Interval Valued Neutrosophic Hypersoft Sets	Personnel selection
Sudha et al (2023) [Sudha et al., 2023)	Extended Plithogenic Sets in Plithogenic Sociogram	Food processing
Nivetha and Smarandache (2023) (Martin et al., 2023)	Plithogenic based contradictions	Supplier selection
Priyadharshini and Irudayam (2023) (Priyadharshini & Irudayam, 2023)	Plithogenic Single Valued Fuzzy Sets	Obesity
Yon et al (2023) (Yon-Delgado et al., 2023)	Neutrosophic Plithogenic AHP	Higher Education

3. PLITHOGENIC GENERALIZED SETS

This section outlines the conceptualization of Plithogenic generalized sets. The Plithogenic sets introduced by Smarandache is of the form (P,a,V,d,c). This is referred as conventional plithogenic set and it wholly deals with attribute and dominant attribute values. This 5-tuple representation is extended into a 7-tuple of the form (P,a,V,d_D,c_D,d_R,c_R) and this kind of representation depends on both dominant and recessive attribute values. These two representations are further extended to generalized plithogenic sets of 9-tuple form. The generalized plithogenic sets are of the form ($P,a,V,d_D,c_D, d_S,c_S, d_R,c_R$).

In this 9-tuple form,

> $d_D,c_{D,}$ refers to the degree of appurtenance and degree of contradiction with respect to the dominant attribute value.
> d_S,c_S, refers to the degree of appurtenance and degree of contradiction with respect to the satisfactory attribute value.
> d_R,c_R, refers to the degree of appurtenance and degree of contradiction with respect to the recessive attribute value.

The generalized plithogenic sets are more accommodative in nature as it facilitates the dealings of dominant, recessive attribute values in addition to the satisfactory attribute value. These kinds of representations are also too flexible as it smoothens the decision making process. The decision making using conventional and extended versions of plithogenic sets are little rigid as it strictly deals with the extremes. The decision makers are constrained with either choosing based on dominant attribute values or rejecting based on the recessive attribute value. The space for choosing the alternatives is limited in the conventional case, however the chances are more in a generalized form of plithogenic sets.

Let us understand the conceptualization of generalized plithogenic sets with an instance. Let us assume a decision-making situation of performance evaluation of managers for promotion and salary increment. Let us consider a set P of 5 managers, i.e. P= {X_1, X_2, X_3, X_4, X_5}. The attributes that are generally considered for promotion are management skills, leadership qualities and soft skills. However, in this case the decision makers decide to specifically consider the competency and mastery of the managers over technical skills and in this case, it is considered to be the attribute. The set of attribute values is taken as basic, intermediate, advances, expert, specialized.

i.e. Attribute value 'a' is Competency in Technical skills

Attribute set V = {basic (a_{11}), intermediate (a_{12}), advanced (a_{13})}

The following Table 2 presents the classification of dominant, satisfactory and recessive attribute values.

Table 2. Classification of Attribute Values

Attribute – Competency in Technical Skills		
Dominant Attribute value	**Satisfactory Attribute value**	**Recessive Attribute value**
Advanced	intermediate	Basic

Let us understand the aforementioned values in the table. The decision makers actually prefer managers who are advanced in technical skills, however they are also convinced and satisfied if the managers are competent with intermediate technical skills. The decision makers do not prefer mangers with basic technical skills to be promoted. In this case the decision makers shall make the selection of the mangers based on dominant and satisfactory attribute values. This shows the opportunities of selecting the managers based on two different categories of attribute values. This shall also be inferred as selection of managers possessing competency of technical skills ranging from intermediate level to advanced level.

3.1 Classification of Generalized Plithogenic Sets

The generalized plithogenic sets shall be classified based on the representations of the degrees of appurtenance.

3.1.1 Crisp Generalized Plithogenic Set

A generalized plithogenic set is crisp if the degrees of appurtenances i.e d_D, d_S, d_R are crisp in nature. i.e.

$$d_D{:}P \times V \rightarrow \{0,1\} \;\&\; ds{:}P \times V \rightarrow \{0,1\} \;\&\; d_R{:}P \times V \rightarrow \{0,1\},$$
$$C_D{:}V \times V \rightarrow [0,1], \& \; C_S{:}V \times V \rightarrow [0,1], \& \; C_R{:}V \times V \rightarrow [0,1]$$

3.1.2 Fuzzy Generalized Plithogenic Set

A generalized plithogenic set is fuzzy if the degrees of appurtenances i.e d_D, d_S, d_R are fuzzy sets. i.e.

$$d_D{:}P \times V \rightarrow [0,1] \& \; ds{:}P \times V \rightarrow [0,1] \& \; d_R{:}P \times V \rightarrow [0,1],$$
$$C_D{:}V \times V \rightarrow [0,1], C_S{:}V \times V \rightarrow [0,1], C_R{:}V \times V \rightarrow [0,1]$$

3.1.3 Intuitionistic Generalized Plithogenic Set

A generalized plithogenic set is intuitionistic if the degrees of appurtenances i.e d_D, d_S, d_R are intuitionistic fuzzy sets.

$$d_D: P \times V \rightarrow [0,1] \times [0,1] \& \; ds: P \times V \rightarrow [0,1] \times [0,1] \& \; d_R: P \times V \rightarrow [0,1] \times [0,1],$$
$$C_D: V \times V \rightarrow [0,1], C_S: V \times V \rightarrow [0,1], C_R: V \times V \rightarrow [0,1]$$

3.1.4 Neutrosophic Generalized Plithogenic Set

A generalized plithogenic set is neutrosophic if the degrees of appurtenances i.e d_D, d_S, d_R are neutrosophic sets. i.e.

$$d_D: P \times V \rightarrow [0,1]^3 \& \; ds: P \times V \rightarrow [0,1]^3 \& \; d_R: P \times V \rightarrow [0,1]^3,$$
$$C_D: V \times V \rightarrow [0,1], C_S: V \times V \rightarrow [0,1], C_R: V \times V \rightarrow [0,1]$$

3.1.5 Linguistic Generalized Plithogenic Set

A generalized plithogenic set is linguistic if the degrees of appurtenances i.e d_D, d_S, d_R are lingual in nature,

$$C_D: V \times V \rightarrow [0,1], C_S: V \times V \rightarrow [0,1], C_R: V \times V \rightarrow [0,1]$$

3.1.6 Mixed Generalized Plithogenic Set

A generalized plithogenic set is mixed if the degrees of appurtenances i.e d_D, d_S, d_R are of mixed type and they need not be uniform in representations.

$$C_D: V \times V \rightarrow [0,1], C_S: V \times V \rightarrow [0,1], C_R: V \times V \rightarrow [0,1]$$

Table 3 presents the illustration of representing the different kinds of generalized Plithogenic sets with respect to the attribute and attribute values presented in Table 1.

Table 3. Illustration of plithogenic generalized set

Dominant Attribute Value				**Satisfactory Attribute value**				**Recessive Attribute value**			
Specialized (a_{14})				*Advanced (a_{13})*				*Basic (a_{11})*			
Contradiction Degree of a_{13} wrt other attribute values				Contradiction Degree of a_{12} wrt other attribute values				Contradiction Degree of a_{11} wrt other attribute values			
	a_{11}	a_{12}	$\mathbf{a_{13}}$		a_{11}	$\mathbf{a_{12}}$	a_{13}		$\mathbf{a_{11}}$	a_{12}	a_{13}
	1/3	2/3	0		1/3	0	2/3		0	1/3	2/3
Crisp appurtenance degree wrt dominant attribute value				*Crisp appurtenance degree wrt satisfactory attribute value*				*Crisp appurtenance degree wrt recessive attribute value*			
	a_{11}	a_{12}	$\mathbf{a_{13}}$		a_{11}	$\mathbf{a_{12}}$	a_{13}		$\mathbf{a_{11}}$	a_{12}	a_{13}
P1	0	0	1	**P1**	0	1	0	**P1**	1	0	0
P2	1	0	0	**P2**	1	0	0	**P2**	0	0	1
P3	0	1	0	**P3**	0	1	0	**P3**	1	0	0
P4	0	0	1	**P4**	0	1	0	**P4**	1	0	0
P5	0	0	1	**P5**	0	1	0	**P5**	1	0	0
Fuzzy appurtenance degree wrt dominant attribute value				*Fuzzy appurtenance degree wrt satisfactory attribute value*				*Fuzzy appurtenance degree wrt recessive attribute value*			
	a_{11}	a_{12}	$\mathbf{a_{13}}$		a_{11}	$\mathbf{a_{12}}$	a_{13}		$\mathbf{a_{11}}$	a_{12}	a_{13}
P1	0.3	0.4	0.8	**P1**	0.5	0.7	0.2	**P1**	0.8	0.3	0.2
P2	0.9	0.2	0.1	**P2**	0.8	0.2	0.1	**P2**	0.2	0.2	0.9
P3	0.1	0.8	0.3	**P3**	0.4	0.7	0.2	**P3**	0.7	0.2	0.1
P4	0.2	0.3	0.8	**P4**	0.4	0.8	0.1	**P4**	0.8	0.2	0.1
P5	0.4	0.5	0.9	**P5**	0.3	0.9	0.1	**P5**	0.9	0.3	0.2
Intuitionistic appurtenance degree wrt dominant attribute value				*Intuitionistic appurtenance degree wrt satisfactory attribute value*				*Intuitionistic appurtenance degree wrt recessive attribute value*			
	a_{11}	a_{12}	$\mathbf{a_{13}}$		a_{11}	$\mathbf{a_{12}}$	a_{13}		$\mathbf{a_{11}}$	a_{12}	a_{13}
P1	(0.2,0.7)	(0.4,0.5)	(0.8,0.1)	**P1**	(0.5,0.4)	(0.7,0.2)	(0.2,0.7)	**P1**	(0.8,0.2)	(0.3,0.6)	(0.2,0.7)
P2	(0.9,0.1)	(0.2,0.7)	(0.1,0.8)	**P2**	(0.8,0.1)	(0.2,0.7)	(0.1,0.7)	**P2**	(0.2,0.7)	(0.2,0.7)	(0.9,0.1)
P3	(0.1,0.8)	(0.8,0.1)	(0.3,0.6)	**P3**	(0.4,0.5)	(0.7,0.1)	(0.2,0.6)	**P3**	(0.7,0.2)	(0.2,0.7)	(0.1,0.8)

continued on following page

Table 3. Continued

Dominant Attribute Value				Satisfactory Attribute value				Recessive Attribute value			
P4	(0.2,0.7)	(0.3,0.6)	(0.8,0.1)	**P4**	(0.4,0.5)	(0.8,0.1)	(0.1,0.7)	**P4**	(0.8,0.2)	(0.2,0.7)	(0.1,0.7)
P5	(0.4,0.6)	(0.5,0.4)	(0.9,0.1)	**P5**	(0.3,0.5)	(0.9,0.1)	(0.1,0.8)	**P5**	(0.9,0.1)	(0.3,0.6)	(0.2,0.7)
Neutrosophic appurtenance degree wrt dominant attribute value				*Neutrosophic appurtenance degree wrt satisfactory attribute value*				*Neutrosophic appurtenance degree wrt recessive attribute value*			
	a_{11}	a_{12}	$\mathbf{a_{13}}$		a_{11}	$\mathbf{a_{12}}$	a_{13}		$\mathbf{a_{11}}$	a_{12}	a_{13}
P1	(0.3,0.1,0.7)	(0.4,0.2,0.7)	(0.8,0.1,0.1)	**P1**	(0.5,0.4,0.1)	(0.7,0.2,0.1)	(0.2,0.1,0.7)	**P1**	(0.8,0.1,0.2)	(0.3,0.1,0.7)	(0.2,0.1,0.8)
P2	(0.9,0.1,0.2)	(0.2,0.3,0.7)	(0.1,0.2,0.8)	**P2**	(0.8,0.10.1)	(0.2,0.1,0.7)	(0.1,0.2,0.7)	**P2**	(0.2,0.1,0.8)	(0.2,0.1,0.7)	(0.9,0.1,0.1)
P3	(0.1,0.1,0.8)	(0.8,0.1,0.1)	(0.3,0.2,0.7)	**P3**	(0.4,0.2,0.5)	(0.7,0.1,0.2)	(0.2,0.2,0.7)	**P3**	(0.7,0.2,0.1)	(0.2,0.1,0.7)	(0.1,0.2,0.7)
P4	(0.2,0.1,0.7)	(0.3,0.1,0.7)	(0.8,0.1,0.2)	**P4**	(0.4,0.2,0.5)	(0.8,0.1,.1)	(0.1,0.2,0.8)	**P4**	(0.8,0.2,0.1)	(0.2,0.1,0.7)	(0.1,0.2,0.6)
P5	(0.4,0.1,0.6)	(0.5,0.1,0.4)	(0.9,0.1,0.1)	**P5**	(0.3,0.1,0.6)	(0.9,0.1,0.1)	(0.1,0.2,0.7)	**P5**	(0.9,0.1,0.1)	(0.3,0.2,0.7)	(0.2,0.1,0.7)

4. METHODOLOGY

This section sketches out the steps involved in the proposed method of Plithogenic decision making based on contradictions with generalized plithogenic representations. The steps proposed in this method are the extension of the plithogenic method based on contradictions which is proposed by Nivetha et al.

Step 1: Construction of Decision-Making Matrix

The decision-making problem is initially defined and the alternatives and attribute are identified. The decision-making matrix is constructed based on the opinion of the experts. The values of entries may be either crisp, fuzzy, intuitionistic, neutrosophic, linguistic or of mixed representations and in general the entries are denoted by M. In this case the criteria are considered as the attributes and the sub-attribute values are assumed to be the attribute values. For each attribute, the attribute values are classified as dominant, satisfactory and recessive. In addition, the attribute is classified as benefit and non-benefit attribute.

$$D_M = \begin{bmatrix} X_{M11} & \cdots & X_{M1n} \\ \vdots & \ddots & \vdots \\ X_{Mm1} & \cdots & X_{Mmn} \end{bmatrix}$$

Step 2: Finding the Attribute Weights

The attribute weights are determined using any of the decision-making methods of computing the attribute weights and at some instances the attribute weights are assumed to be equal.

Step 3: Determining the Contradiction Degree Matrices

Three different contradiction matrices namely C_D, C_S and C_R are determined based on the contradiction degrees of the attribute values or the attribute values with respect to dominant, satisfactory and recessive attribute values respectively.

$$C_D = \begin{bmatrix} C_{D11} & \cdots & C_{D1n} \\ \vdots & \ddots & \vdots \\ C_{Dm1} & \cdots & C_{Dmn} \end{bmatrix}, C_S = \begin{bmatrix} C_{S11} & \cdots & C_{S1n} \\ \vdots & \ddots & \vdots \\ C_{Sm1} & \cdots & C_{Smn} \end{bmatrix}, C_R = \begin{bmatrix} C_{R11} & \cdots & C_{R1n} \\ \vdots & \ddots & \vdots \\ C_{Rm1} & \cdots & C_{Rmn} \end{bmatrix}$$

Step 4: Finding the Weighted Contradiction Matrices

The weighted contradiction matrices WC_D, WC_S and WC_R are calculated for each of the contradiction matrices by multiplying the respective attribute weights.

Step 5: Computing the Cumulative Score Value of Each Alternative Based on the Classification of Attribute Values

The cumulative score values of each alternative with respect to each attribute value classified as dominant, satisfactory and recessive attribute values is calculated. The weighted score values with respect to dominant and satisfactory are subjected to acceptance domain and the weighted score values with respect to recessive attribute value are subjected to rejection domain. Say, for an alternative Ai, if the respective dominant, satisfactory and recessive score values are $S_{Di,}$ $S_{Si,}$ $S_{Ri,}$ then the cumulative score value of the alternative is calculated as $S_{Di} + S_{Si} - S_{Ri.}$

Step 6: Finding the Final Score Values

The final score values of each of the alternative with respect to both benefit and cost attribute say A_{Bj} and A_{Cj} are finally calculated. The difference between the values is determined, say $A_{Bj} - A_{Cj} = A_{Dj}$

Step 7: Ranking of the Alternatives

The alternatives are arranged in order of their respective ADj difference values, with the highest value placed at the top and subsequent values following in descending order.

5. DECISION MAKING ON PLITHOGENIC CONTRADICTIONS WITH GENERALIZED PLITHOGENIC SETS IN MATERIAL SELECTION

This section applies the proposed decision-making method to the decision making problem on material selection. In a construction industry, the materials used for construction possess varied attributes. The quality of the materials is highly concerned as it reflects the brand and hallmark of the industries. The engineers are very keen on selecting more suitable materials based on certain general attributes such as Adaptability, Longevity, Cost, Safety and Environmental Impacts. The attribute or the attribute values for each of the attribute are as follows,

- Adaptability – {low, moderate, high, complete}
- Longevity – {short, moderate, extended, exceptional}
- Cost- {low, moderate, high, very high, prohibitive}
- Safety – {Negligible, very low, low, moderate, high}

- Environmental Impacts – {Negligible, very low, low, moderate, high, very high}

The description of the aforementioned attribute values is presented in the following Table 4.

Table 4. Description of the attribute values

Attributes				
Adaptability (A1)	**Longevity (A2)**	**Cost (A3)**	**Safety (A4)**	**Environmental Impacts (A5)**
Low Adaptability (LA) The material has minimal capacity to change or adjust in response to different conditions.	**Short Longevity (SL)** The material has a very limited lifespan. It exists or functions for only a short duration before becoming obsolete or non-functional.	**Low Cost (LC)** The material is affordable, requiring a minimal financial investment. It is budget-friendly and does not strain financial resources.	**Negligible Safety (NS)** The material is extremely hazardous and potentially life-threatening.	**Negligible Environmental Impact (NEI)** The material has a very minor effect on the environment. It does not significantly alter natural ecosystems or resources.
Moderate Adaptability (MA) The material make moderate adjustments based on varying conditions. It has a reasonable capacity to change, although it may not respond optimally to all situations.	**Moderate Longevity (ML)** The material has a moderate lifespan. It functions effectively for a reasonable amount of time before showing signs of wear and tear, requiring maintenance or replacement.	**Moderate Cost (MC)** The material has a moderate price. It offers a balance between quality and cost, providing good value for the money spent.	**Very Low Safety (VLS)** The material is highly dangerous. There are substantial risks involved, and safety measures are grossly inadequate, making it extremely risky for individuals involved.	**Low Environmental Impact (LEI)** The environmental impact is relatively small. While there is an effect, it is not substantial and can be easily mitigated or managed.
High Adaptability (HA) The material is highly flexible and can readily adjust to a wide range of conditions. It can respond effectively to changes and uncertainties, ensuring optimal performance in diverse situations.	**Extended Longevity (EL)** The material has a long lifespan, significantly exceeding average expectations. It remains functional and efficient for an extended period, requiring minimal maintenance.	**High Cost (HC)** The material is relatively expensive. It demands a significant financial investment and might be considered a luxury or premium option.	**Low Safety (LS)** The material poses significant risks. Safety measures might be lacking or insufficient, leading to a notable level of danger or hazard.	**Moderate Environmental Impact (MEI)** The material has a noticeable impact on the environment. It may affect local ecosystems, resources, or communities to a moderate extent.

continued on following page

Table 4. Continued

Attributes				
Adaptability (A1)	**Longevity (A2)**	**Cost (A3)**	**Safety (A4)**	**Environmental Impacts (A5)**
Complete Adaptability (CA) The material has the highest level of adaptability. It can seamlessly and fully adjust to any given condition, ensuring optimal performance under any circumstance.	**Exceptional Longevity (ExL)** The material has an exceptionally long lifespan. It endures for an extraordinary duration without significant deterioration, making it highly durable and reliable over time.	**Very High Cost (VHC)** The material is exceptionally costly. It is at the top end of the price range and is considered a premium or exclusive option, often beyond the reach of the average consumer.	**Moderate Safety (MS)** The material is reasonably safe. It meets standard safety regulations and guidelines, with moderate risks that are well-controlled.	**High Environmental Impact (HEI)** The environmental impact is considerable and substantial. It significantly affects natural habitats, biodiversity, or ecosystems, leading to significant changes or damage.
		Prohibitive Cost (PC) The cost is so high that it prevents most people from affording the product or service.	**High Safety (HS)** The material is extremely safe. There are minimal risks, and all necessary precautions have been taken to ensure the highest level of safety.	**Very High Environmental Impact (VHEI)** The material has an extremely severe impact on the environment. It causes critical damage to ecosystems, natural resources, or communities, often with long-lasting consequences.

The classification of the attribute values for each of the attribute are presented in Table 5.

Table 5. Classification of the attribute values

Attribute	**Attribute Value**		
	Dominant	**Satisfactory**	**Recessive**
Adaptability	Complete	High	Low
Longevity	Exceptional	Extended	Short
Cost	Low	Moderate	Prohibitive
Safety	High	Moderate	Negligible
Environmental Impact	Negligible	Low	Very High

The contradiction degrees of each of the attribute values with respect to that of the dominant, satisfactory, recessive attribute values are presented as follows in Table 6.

Table 6. Contradiction degrees of the attribute values

		Attribute- Adaptability (Benefit)			
Attribute values	**LA**	**MA**	**HA**	**CA**	
Dominant (CA)	$\frac{3}{4}$	$\frac{2}{4}$	$\frac{1}{4}$	0	
Satisfactory (HA)	$\frac{3}{4}$	$\frac{2}{4}$	0	$\frac{1}{4}$	
Recessive (LA)	0	$\frac{1}{4}$	$\frac{2}{4}$	$\frac{3}{4}$	
		Attribute- Longevity (Benefit)			
Attribute values	**SL**	**ML**	**EL**	**ExL**	
Dominant (ExL)	$\frac{3}{4}$	$\frac{2}{4}$	$\frac{1}{4}$	0	
Satisfactory (EL)	$\frac{3}{4}$	$\frac{2}{4}$	0	$\frac{1}{4}$	
Recessive (SL)	0	$\frac{1}{4}$	$\frac{2}{4}$	$\frac{3}{4}$	
		Attribute- Cost (Non-Benefit)			
Attribute values	**LC**	**MC**	**HC**	**VHC**	**PC**
Dominant (LC)	0	$\frac{1}{5}$	$\frac{2}{5}$	$\frac{3}{5}$	$\frac{4}{5}$
Satisfactory (MC)	$\frac{1}{5}$	0	$\frac{2}{5}$	$\frac{3}{5}$	$\frac{4}{5}$
Recessive (VHC)	$\frac{4}{5}$	$\frac{3}{5}$	$\frac{2}{5}$	0	$\frac{1}{5}$
		Attribute- Safety (Benefit)			
Attribute values	**NS**	**VLS**	**LS**	**MS**	**HS**
Dominant (HS)	$\frac{4}{5}$	$\frac{3}{5}$	$\frac{2}{5}$	$\frac{1}{5}$	0
Satisfactory (MS)	$\frac{4}{5}$	$\frac{3}{5}$	$\frac{2}{5}$	0	$\frac{1}{5}$
Recessive (NS)	0	$\frac{1}{5}$	$\frac{2}{5}$	$\frac{3}{5}$	$\frac{4}{5}$
		Attribute- Environmental (Non-Benefit)			
Attribute values	**NEI**	**LEI**	**MEI**	**HEI**	**VHEI**
Dominant (NEI)	0	$\frac{1}{5}$	$\frac{2}{5}$	$\frac{3}{5}$	$\frac{4}{5}$
Satisfactory (LEI)	$\frac{1}{5}$	0	$\frac{2}{5}$	$\frac{3}{5}$	$\frac{4}{5}$
Recessive (VHEI)	$\frac{4}{5}$	$\frac{3}{5}$	$\frac{2}{5}$	$\frac{1}{5}$	0

Let us consider a decision-making matrix comprising ten alternatives of construction materials and five attribute. Each entry of the matrix is expressed as a linguistic variable pertaining to that of each attribute values as in Table 7.

Table 7. Initial decision-making matrix

	Adaptability	Longevity	Cost	Safety	Environmental Impact
M1	LA	ML	MC	NS	LEI
M2	MA	ML	MC	MS	MEI
M3	MA	SL	PC	LS	HEI
M4	CA	EL	LC	NS	NEI
M5	LA	ExL	HC	HS	VHEI
M6	HA	SL	HC	HS	NEI
M7	HA	ML	VHC	VLS	LEI
M8	CA	ML	VHC	NS	MEI
M9	MA	EL	LC	LS	MEI
M10	LA	EL	LC	LS	NEI

The contradiction matrix with respect to the dominant attribute values of each of the attribute is as follows in Table 8

Table 8. Contradiction matrix w.r.t dominant attribute value

	Adaptability	Longevity	Cost	Safety	Environmental Impact
M1	0.75	0.5	0.2	0.8	0.2
M2	0.5	0.5	0.2	0.2	0.4
M3	0.5	0.75	0.8	0.4	0.6
M4	0	0.5	0	0.8	0
M5	0.75	0	0.4	0	0.8
M6	0.25	0.75	0.4	0	0
M7	0.25	0.5	0.6	0.6	0.2
M8	0	0.5	0.6	0.8	0.4
M9	0.5	0.25	0	0.4	0.4
M10	0.75	0.25	0	0.4	0

The contradiction matrix with respect to the satisfactory attribute values of each of the attribute is as follows in Table 9.

Table 9. Contradiction matrix w.r.t satisfactory attribute value

	Adaptability	Longevity	Cost	Safety	Environmental Impact
M1	0.75	0.5	0	0.8	0
M2	0.5	0.5	0	0	0.4
M3	0.5	0.75	0.8	0.4	0.6
M4	0.25	0	0.2	0.8	0.2
M5	0.75	0.25	0.4	0.2	0.8
M6	0	0.75	0.4	0.2	0.2
M7	0	0.5	0.3	0.6	0
M8	0.25	0.5	0.6	0.8	0.4
M9	0.5	0	0.2	0.4	0.4
M10	0.75	0	0.2	0.4	0.2

The contradiction matrix with respect to the recessive attribute values of each of the attribute is as follows in Table 10

Table 10. Contradiction matrix w.r.t recessive attribute value

	Adaptability	Longevity	Cost	Safety	Environmental Impact
M1	0	0.25	0.6	0	0.6
M2	0.25	0.25	0.6	0.6	0.4
M3	0.25	0	0.2	0.4	0.2
M4	0.75	0.5	0.8	0	0.8
M5	0	0.75	0.4	0.8	0
M6	0.5	0	0.4	0.8	0.8
M7	0.5	0.25	0	0.2	0.6
M8	0.75	0.25	0	0	0.4
M9	0.25	0.5	0.8	0.4	0.4
M10	0	0.5	0.8	0.4	0.8

The cumulative score values of the alternatives with respect to each of the attribute is determined using Step 5 are presented in Table 11.

Table 11. Cumulative score values of the attribute

	Adaptability	Longevity	Cost	Safety	Environmental Impact
M1	1.5	0.75	-0.4	1.6	-0.4
M2	0.75	0.75	-0.4	-0.4	0.4
M3	0.75	1.5	1.4	0.4	1
M4	-0.5	0	-0.6	1.6	-0.6
M5	1.5	-0.5	0.4	-0.6	1.6
M6	-0.25	1.5	0.4	-0.6	-0.6
M7	-0.25	0.75	0.9	1	-0.4
M8	-0.5	0.75	1.2	1.6	0.4
M9	0.75	-0.25	-0.6	0.4	0.4
M10	1.5	-0.25	-0.6	0.4	-0.6

The weighted cumulative score values of the alternatives are as follows in Table 12; in this case the weights are assumed to be equal.

Table 12. Weighted cumulative score values of the attributes

	Adaptability	Longevity	Cost	Safety	Environmental Impact
M1	0.3	0.15	-0.08	0.32	-0.08
M2	0.15	0.15	-0.08	-0.08	0.08
M3	0.15	0.3	0.28	0.08	0.2
M4	-0.1	0	-0.12	0.32	-0.12
M5	0.3	-0.1	0.08	-0.12	0.32
M6	-0.05	0.3	0.08	-0.12	-0.12
M7	-0.05	0.15	0.18	0.2	-0.08
M8	-0.1	0.15	0.24	0.32	0.08
M9	0.15	-0.05	-0.12	0.08	0.08
M10	0.3	-0.05	-0.12	0.08	-0.12

The final score values of the alternatives with respect to both benefit and cost attribute are determined using step 6 and presented in Table 13 and Figure 1 makes a graphical representation of the ranking.

Table 13. Final score values of the alternatives with equal weights

	Score values wrt Benefit attribute	Score values wrt Non-Benefit attribute	Differences	Rank
M1	0.77	-0.16	0.93	1
M2	0.22	0	0.22	4
M3	0.53	0.48	0.05	7
M4	0.22	-0.24	0.46	3
M5	0.08	0.4	-0.32	10
M6	0.13	-0.04	0.17	6
M7	0.3	0.1	0.2	9
M8	0.37	0.32	0.05	7
M9	0.18	-0.04	0.22	4
M10	0.33	-0.24	0.57	2

Figure 1. Score values of the alternatives based on equal weights w.r.t benefit and non-benefit criteria

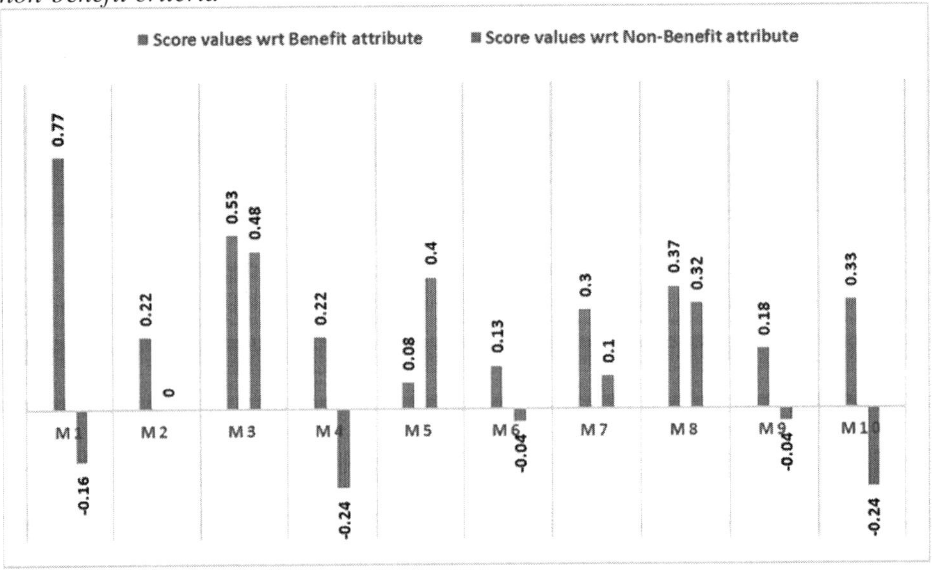

Let us apply the same decision-making method with different attribute weights. Using the methods of AHP, CRITIC and Entropy the attribute weights obtained are presented in the Table 14.

Table 14. Attribute weights using different methods

Methods	Attribute Weights				
	A1	A2	A3	A4	A5
AHP	0.21	0.24	0.11	0.27	0.17
CRITIC	0.18	0.21	0.09	0.28	0.24
Entropy	0.19	0.23	0.12	0.25	0.21

The final score values obtained with different attribute weights are presented in the Table 15 and Figure 2 makes a visual representation of the same.

Table 15. Final score values of the alternatives with different weights

	Score values with Different Attribute Weights		
	AHP	CRITIC	Entropy
M1	1.039	1.0075	0.9895
M2	0.2055	0.1205	0.179
M3	0.3015	0.196	0.2095
M4	0.495	0.556	0.503
M5	-0.283	-0.423	-0.364
M6	0.2037	0.21	0.2255
M7	0.3665	0.4075	0.351
M8	0.307	0.3115	0.2495
M9	0.2035	0.1525	0.173
M10	0.531	0.5275	0.5255

Figure 2. Ranking of the alternatives w.r.t different weights

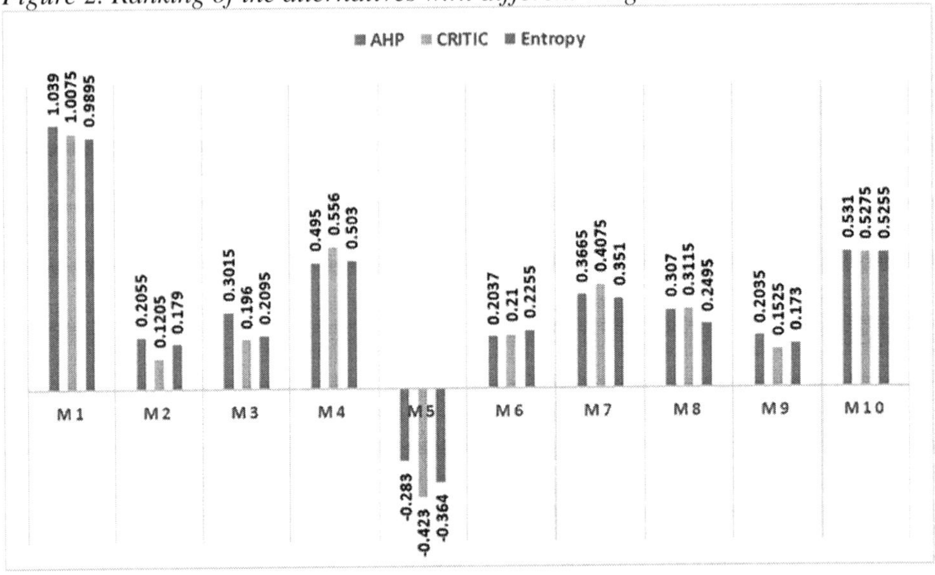

The ranking of the alternatives with respect to the score values in Table 15 is presented in Table 16

Table 16. Ranking of the alternatives with different weights

Alternatives	Ranking with Different Attribute weights		
	AHP	CRITIC	Entropy
M1	1	1	1
M2	7	9	8
M3	6	7	7
M4	3	2	3
M5	10	10	10
M6	8	6	6
M7	4	4	4
M8	5	5	5
M9	9	8	9
M10	2	3	2

The rank correlation coefficient in Table 17 is obtained to determine the consistency between the ranking of the alternatives obtained using the proposed plithogenic based contradiction

Table 17. Correlation coefficient of ranking results

	Equal Weights	AHP	CRITIC	Entropy
Equal weights	1	0.584319	0.559717	0.572018
AHP		1	0.927273	0.963636
CRITIC			1	0.975757576
Entropy				1

It is observed that the ranking results obtained using the method of CRITIC and Entropy are more consistent and hence the ranking of the alternatives based on these two methods are found to be more optimal.

DISCUSSIONS

The ranking results obtained using the proposed plithogenic based decision making method with generalized plithogenic representations are more feasible in comparison with extended plithogenic sets and conventional plithogenic sets. The conventional or the classical plithogenic sets deals with dominant attribute values. The extended Plithogenic sets deal with both dominant and recessive attribute values. However, the generalized plithogenic sets deal with dominant, satisfactory and recessive attribute values. This makes the proposed decision-making method based on plithogenic contradictions with generalized plithogenic representations more comprehensive in nature. The method of CRITIC and Entropy applied to find the attribute weights make the ranking of the alternatives more optimal. The proposed Plithogenic generalized sets provides a lot of space for the decision makers to make choices between the range's values of satisfactory to dominant. The generalized version is more pragmatic and realistic in nature as it does not constrain the decision makers with any limitations in choice making of the alternatives.

CONCLUSION

This chapter has introduced and explored the concept of generalized plithogenic sets as a significant advancement in the field of multi-attribute decision making. The extension of plithogenic sets, particularly in the form of a 9-tuple structure, offers a more comprehensive representation of diverse attribute types, categorizing them into dominant, recessive, and satisfactory. The application of these sets in a material selection model has demonstrated the effectiveness and superiority of the proposed generalized plithogenic sets over both extended plithogenic sets and basic plithogenic sets.

The comparison revealed that the generalized plithogenic sets provide a more nuanced and realistic approach to modelling decision-making scenarios. This contributes to the development of robust decision-making models that can better address the complexities inherent in real-life problems. The findings suggest that the generalized plithogenic sets offer researchers and practitioners a valuable tool for making optimal decisions across various domains.

Future directions in this research area could involve further refinement and enhancement of the generalized plithogenic sets model. Researchers may explore the integration of advanced mathematical techniques, optimization algorithms, or artificial intelligence approaches to improve the accuracy and efficiency of decision-making processes. Additionally, extending the application of generalized plithogenic sets to other domains beyond material selection could open up new avenues for research and practical implementation.

REFERENCES

Abdel-Basset, M., Ding, W., Mohamed, R., & Metawa, N. (2020). An integrated plithogenic MCDM approach for financial performance evaluation of manufacturing industries. *Risk Management*, *22*, 192–218.

Abdel-Basset, M., & Mohamed, R. (2020). A novel plithogenic TOPSIS-CRITIC model for sustainable supply chain risk management. *Journal of Cleaner Production*, *247*, 119586.

Abdel-Basset, M., Mohamed, R., Smarandache, F., & Elhoseny, M. (2021). A new decision-making model based on plithogenic set for supplier selection. *Computers, Materials & Continua*, *66*(3), 2751–2769.

Abdel-Basset, M., Mohamed, R., Zaied, A. E. N. H., Gamal, A., & Smarandache, F. (2020). Solving the supply chain problem using the best-worst method based on a novel Plithogenic model. In *Optimization theory based on neutrosophic and plithogenic sets* (pp. 1–19). Academic Press.

Ansari, Z. N., & Kant, R. (2021). A plithogenic based neutrosophic analytic hierarchy process framework to analyse the barriers hindering adoption of eco-innovation practices in supply chain. *International Journal of Sustainable Engineering*, *14*(6), 1509–1524.

Grida, M., Mohamed, R., & Zaied, , A. N. (2020). A Novel Plithogenic MCDM Framework for Evaluating the Performance of IoTBased Supply Chain. *Neutrosophic Sets and Systems*, *33*, 321–338.

Hema, R., Sudharani, R., & Kavitha, M. (2023). A Novel Approach on Plithogenic Interval Valued Neutrosophic Hyper-soft Sets and its Application in Decision Making. *Indian Journal of Science and Technology*, *16*(32), 2494–2502.

Korucuk, S., Demir, E., Karamasa, C., & Stević, Ž. (2020). Determining the dimensions of the innovation ability in logistics sector by using plithogenic-critic method: An application in Sakarya Province. *International Review (Steubenville, Ohio)*, (1-2), 119–127.

Martin, N. (2022). Plithogenic SWARA-TOPSIS Decision Making on Food Processing Methods with Different Normalization Techniques. *Advances in Decision Making, 69*.

Martin, N., Smarandache, F., & Broumi, S. (2021). PROMTHEE plithogenic pythagorean hypergraphic approach in smart materials selection. *Int J Neutrosophic Sci*, *13*, 52–60.

Martin, N., Smarandache, F., & Priya, R. (2022). Introduction to Plithogenic Sociogram with preference representations by Plithogenic Number. *Journal of Fuzzy Extension and Applications, 3*(1), 96-108.

Martin, N., Smarandache, F., & Sudha, S. (2023). A novel method of decision making based on plithogenic contradictions. *Neutrosophic Systems with Applications, 10*, 12–24.

Priyadharshini, S. P., & Irudayam, F. N. (2023). An analysis of obesity in school children during the pandemic COVID-19 using plithogenic single valued fuzzy sets. *Neutrosophic Systems with Applications*, *9*, 24–28.

Smarandache, F. (2018). Plithogeny, plithogenic set, logic, probability, and statistics. *arXiv preprint arXiv:1808.03948.*

Sudha, S., & Martin, N. (2023, June). Comparative analysis of Plithogenic neutrosophic PIPRECIA over neutrosophic AHP in criteria ordering of logistics selection. In *AIP Conference Proceedings* (*Vol. 2649*, No. 1). AIP Publishing.

Sudha, S., Martin, N., Anand, M. C. J., Palanimani, P. G., Thirunamakkani, T., & Ranjitha, B. (2023). MACBETH-MAIRCA Plithogenic Decision-Making on Feasible Strategies of Extended Producer's Responsibility towards Environmental Sustainability. [IJNS]. *International Journal of Neutrosophic Science*, *22*(2), 114–130.

Sudha, S., Martin, N., & Broumi, S. (2022). Plithogenic CRITIC-MAIRCA Ranking of Feasible Livestock Feeding Stuffs. *International Journal of Neutrosophic Science*, *18*(4), 160–173.

Sudha, S., Martin, N., & Smarandache, F. (2023). State of Art of Plithogeny Multi Criteria Decision Making Methods. *Neutrosophic Sets and Systems*, *56*(1), 27.

Sudha, S., Martin, N., & Smarandache, F. (2023). Applications of Extended Plithogenic Sets in Plithogenic Sociogram. *International Journal of Neutrosophic Science, 20*(4), 8-35.

Tayal, D. K., Yadav, S. K., & Arora, D. (2023). Personalized ranking of products using aspect-based sentiment analysis and Plithogenic sets. *Multimedia Tools and Applications*, *82*(1), 1261–1287.

Wang, P., Lin, Y., Fu, M., & Wang, Z. (2023). VIKOR method for plithogenic probabilistic linguistic MAGDM and application to sustainable supply chain financial risk evaluation. *International Journal of Fuzzy Systems*, *25*(2), 780–793.

Yon-Delgado, J. C., Yon-Delgado, M. R., Aguirre-Baique, N., Gamarra-Salinas, R., & Ponce-Bardales, Z. E. (2023). Neutrosophic Plithogenic AHP Model for Inclusive Higher Education Program Selection. *International Journal of Neutrosophic Science*, *21*(1), 50–0.

Chapter 21
Solving Neutrosophic Minimum Spanning Tree Problem by Least Edge Weight Algorithm

Shayathri Linganathan
Vellore Institute of Techhnology, India

Purusotham Singamsetty
Vellore Institute of Technology, India

ABSTRACT

A minimum spanning tree problem (MST) is a tree that identifies a subset of edges which connects all the vertices of a connected, undirected, and edge-weighted graph with the least total weight. In general, the edge weight can be distance, cost, time, etc. However, in case of any ambiguity of information, the edge weight may not be a deterministic value. A neutrosophic set is a powerful tool for complexity, dealing with imprecise, ambiguous, and inconsistent information in the actual wide world. In neutrosophic MST, the edge weight is represented by a neutrosophic number set. When compared to the fuzzy MST, neutrosophic MST graphs give more accuracy and compatibility in a neutrosophic environment. To solve the neutrosophic MST, least edge weight algorithm is proposed and executed in MATLAB.

DOI: 10.4018/979-8-3693-3204-7.ch021

INTRODUCTION

The minimum spanning tree (MST) problem is a well-known problem in combinatorial optimization as well as in graph theory. A tree is, more generally speaking, a connected network without cycles that extends to all of the graph's vertices. In classical graph theory, almost all minimal spanning tree problem has weights assigned to edges in real numbers but in real-life practical problems the parameters are not naturally precise there exists uncertainty and ambiguity. The MST problem can be resolved using well-known methods like the prim and kruskal algorithms as well as when the edge weights are assigned as a crisps number. In the neutrosophic minimum spanning tree, the crisp numbers are neutrosophic values or consider the neutrosophic number in the neutrosophic set model. This model is a useful tool for solving problems in the real world because it can depict the uncertainty (partial, inconsistent, and indeterminate knowledge) that exists there. Some researchers used the stochastic minimum spanning tree problem, in which the presumption may not hold in practical situations.

In such cases, fuzzy and neutrosophic variables can be used to solve the minimal spanning tree problems. The minimum cost/weight spanning tree problem has wide applications such as processing images, transportation, wireless telecommunication, and cluster analysis. In neutrosophic minimum spanning tree applications used in building cable networks and road network problems.

LITERATURE REVIEW

The concepts of the classic set, fuzzy set Zadeh (1965), interval-valued fuzzy set Turksen (1986), intuitionistic fuzzy set Atanassov (1986), etc., enrich the concept of a neutrosophic set in the standard framework. The neutrosophic set proposed by Smarandache (1998), the author proposes the neutrosophic set to address the practical issues facing in our day-to-day span. The neutrosophic sets attract a lot of attention in order to tackle a range of real-world problems requiring some ambiguity, imprecision, incompleteness, inconsistency, and indeterminacy. Moreover, inside the unit interval [0, 1], the values of the three membership functions are true, independent, and false. The author also discusses the characteristics of single-valued neutrosophic sets.

Triangular fuzzy numbers are very distinct from SVNSs (Single Valued Neutrosophic sets). A development of intuitionistic fuzzy sets, SVNSs was introduced by Wang et al. (2010). SVNSs are a type of set that has applications in actual science and engineering. Generally speaking, Hassan's LC-method can be used to solve the minimum spanning tree (2012).

Zhang et al (2014) investigation 's of interval neutrosophic sets and their use in multi-criteria decision-making problems. Ye (2014) described a technique for determining the minimum spanning tree of a single-valued neutrosophic network with SVNSs as the representation of the vertices. A neutrosophic environment is acceptable given the inconsistent, partial, and ambiguous nature of the information, according to Mandal et al (2016) proposed optimal minimal spanning tree principles. They used the suggested strategy to solve a network problem with various criteria in their work. Impressed by the work of Broumi (2016 &2018) focused on the concept of a minimum spanning tree with neutrosophic least edge weights and also studied the concept of single-valued neutrosophic graphs. Smarandache (1998) defines the neutrosophic concept in the neutrosophic ellipse (see figure-1),

Figure 1. Neutrosophic ellipse to the neutrosophic concept

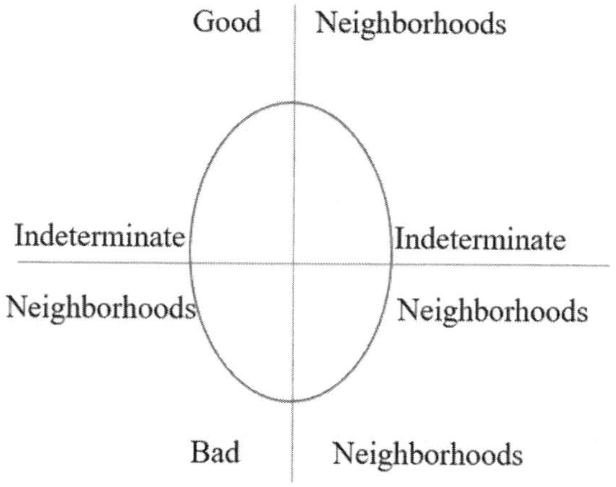

Here by Figure-1 G, B, and I various subsets of Good (G), Bad (B), and Indeterminate (I) are included in]-0,1+ [(i.e., ||-01+||)].

Numerous papers are published on the topic of minimum spanning tree problem in fuzzy environment also some are noted here, Fuzzy logic, a branch of mathematics, utilizes degrees of membership functions instead of rigid true/false categorization, thus adeptly managing uncertainty, ambiguity, and imprecision.

Fuzzy concepts are relatively used in neutrosophic concepts for knowing the background we absorb the technics in fuzzy environment. Itoh and Ishii (1996) was first formulated a MST with fuzzy edge weights using the necessity measure the author constrained in programming based technics. Chang and Lee (1999) introduced the existence ranking index to rank fuzzy edge weights in the spanning tree problem through three approaches. Yama kami et.al (2005) introduced evolutionary algorithm for solving the Minimum spanning tree problem in fuzzy environment. Asady et. al. (2007) ranking fuzzy numbers are used to solve the distance minimization in real world problems some were prove the basis of robustness, low-cost and tractability, hence to overcome all the thinks we use the fuzzy ranking aspect tools.

Ye (2014) created a methodical strategy for tackling the Minimum Spanning Tree Problem (MSTP) in a neutrosophic network/graph, where the characteristics of nodes are articulated in neutrosophic sets, and the connections between nodes represent the differences between these characteristics. Mandal and Basu (2016) have tackled an interesting problem involving multiple criteria represented by edge weights in a network, utilizing neutrosophic sets. Their approach, based on similarity measures, likely aims to assess the resemblance or proximity between different elements in the network, possibly to make decisions or optimize certain objectives. This combination of multiple criteria and neutrosophic sets suggests a nuanced and flexible approach to problem-solving, which could be valuable in complex decision-making scenarios. The idea of a scoring function was first presented by Deli et al. (2015) the score values of two bipolar neutrosophic numbers are the same, they can be differentiated using the accuracy and certainty functions.

By referring to the score function ranking neutrosophic sets with spanning tree problems are analysed in the application of decision-making by Nancy et al. (2016). Smarandache et al. (2016) evaluate the single valued neutrosophic graphs, degree, order and size. Bipolar neutrosophic numbers were used by Mullai et al. (2017) to study the shortest path problem using the least spanning tree algorithm. Kandasamy (2018) a dual-valued version of NMSTP and introduced a clustering technique to categorize data clusters effectively. Adhikary et.al. (2024) introduced the modified prims algorithm to solve minimum spanning tree network using with neutrosophic numbers.

The proposed model constructs the least edge weight Algorithm for the neutrosophic edge weight MST problem. The preferred link matrix (least cost) is built using the distance matrix in the approach that is being provided, and two further stages are also carried out using the node-set matrix. The node-set matrix is used for the final output for neutrosophic edge weight MSP.

PRELIMINARIES

Definition One: Neutrosophic set

A neutrosophic set A in the universal set X then T_A, I_A and F_A is the true, indeterminacy and falsity membership function in Real standard or non-standard subsets of]-0 1+ [respectively.

Definition Two: Single valued numbers:

Let X be a universal set. Single valued neutrosophic set (SVNS) A in X. $T_A(x)$, $I_A(x)$, and $F_A(x) \in [0,1]$ for each point x in X. Where $T_A(x)$ is truth membership, $I_A(x)$ is indeterminacy membership and $F_A(x)$ is a falsity membership.

$$A = \{ < x{:}T_A(x), I_A(x), F_A(x) > x \in X\}$$

Where X is discrete then SVNs A be calculated as,

$$A = \frac{\sum_{i=1}^{n} T(x_i), I(x_i), F(x_i)}{x_i}, \qquad x_i \in X$$

In the neutrosophic set, all three membership functions have satisfied the following condition.

$$0 \leq T(x_i) + I(x_i) + F(x_i) \leq 3$$

Definition Three: Score Functions

Mullai et al. (2017), Nancy et al. (2016) and Zhang et al. (2010) created the score functions and the relationship order between the SVNs to order the neutrosophic universal sets. Where T_A, I_A and F_A be SVNs that represent the truth, Indeterminacy and falsity membership function in A.

PROBLEM DESCRIPTION

Assume T is a spanning tree, which is a connected graph devoid of cycles. Any two vertices in the neutrosophic graph have precisely one path connecting them, and T has n-1 edges (where n is the number of nodes). A neutrosophic minimum spanning tree (MST) for a graph G=(V, E, W), where V- vertices/nodes, E-edges(links) and W- weights assigned to edges of the neutrosophic graph is defined as a subgraph of G that is a neutrosophic tree, contains all vertices of G, and has |V|-1 edges such that all its costs/weights of G is minimum.

The cost and lengths associated with the graph's edges are used in a crisp environment for the MST problem. However, the arc lengths in real-world situations could be erratic or unclear. The choice is made by the decision-maker in light of inadequate or insufficient information due to a lack of supporting data.

In this paper, consider an undirected neutrosophic weighted graph has been considered. Instead of crisp values, the weights of the neutrosophic graph are shown as the neutrosophic edge values. We have developed an algorithm to address this challenge to create the MST.

LEAST EDGE WEIGHT ALGORITHM OF NEUTROSOPHIC UNDIRECTED GRAPH FOR MINIMUM SPANNING TREE PROBLEM

In this section, motivated by the work of Broumi.et.al., (2016), for finding the least cost for the minimum spanning tree, we proposed a new least edge weight algorithm for the given problem.

The aim of the Least edge weight algorithm for neutrosophic adjacency matrix for the given problem. The goal of the approach is to read the distance matrix (neutrosophic weight matrix) of the provided undirected network, build a preferred link matrix (PLM) with a set of least-weight connections, build a node set matrix (NSM), and then draw the MST that does not contain a cycle.

The following steps are mainly used for constructing this neutrosophic least edge weight algorithm,

Steps:

1. Input the neutrosophic adjacency matrix $A = \left[a_{ij}\right]_{n\times n}$ for the given network.
2. We converted the provided neutrosophic matrix into a scoring matrix X_1, X_2 and X_3 by using the score function.
3. For given X_K in step 2 in each column j, we identify the element with the lowest weight (the preferred link), and we set all other elements to zero.
4. Utilizing step 3, create the preferred link matrix (PLM).
5. With the PLM matrix created in step 4, construct the nodes-set matrix (NSM).
6. If there is a group of node pairs, maintain one of them, and delete the node pair with the highest cost after making sure the other node pairs do not create a cycle.
7. Creating the new spanning tree by combining the node pairs from step 5
8. Output the new neutrosophic least edge weight minimum spanning tree.

*   * This algorithm is implemented by using MATLAB.

The MATLAB pseudocode is given,

Input = Score matrix ($n \times n$)

Where i=1, 2, 3, ……. n

Construct S (i,:)

find PLM

Sort S (i,:)

Construct NSM

Index S (i,:)

Output =Draw the minimum spanning tree by using G.

EXAMPLE FOR NEUTROSOPHIC EDGE WEIGHT GRAPH

Consider the neutrosophic graph G = (V, E, W), where V- vertices/nodes, E- edges(links) and W weights assigned for G. Where the example considered in figure-2 as V=6, E=9, W=weights and steps are defined below,

Figure 2. Neutrosophic Graph G

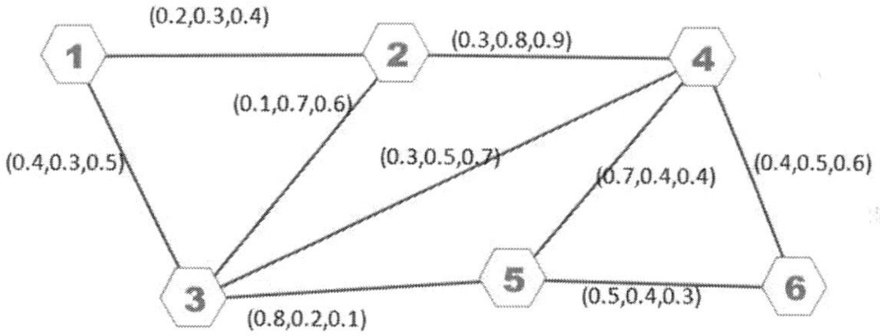

Step-1: Given neutrosophic graph G define in the matrix form refer by A,

$$A = \begin{bmatrix} \infty & (0.2,0.3,0.4) & (0.4,0.3,0.5) & - & - & - \\ (0.2,0.3,0.4) & \infty & (0.1,0.7,0.6) & (0.3,0.8,0.9) & - & - \\ (0.4,0.3,0.5) & (0.1,0.7,0.6) & \infty & (0.3,0.5,0.7) & (0.8,0.2,0.1) & - \\ - & (0.3,0.8,0.9) & (0.3,0.5,0.7) & \infty & (0.7,0.4,0.4) & (0.4,0.5,0.6) \\ - & - & (0.8,0.2,0.1) & (0.7,0.4,0.4) & \infty & (0.5,0.4,0.3) \\ - & - & - & (0.4,0.5,0.6) & (0.5,0.4,0.3) & \infty \end{bmatrix}$$

Step-2: Neutrosophic matrix A modified into a scoring matrix X_1, X_2 and X_3 by using different score functions Mullai et. al. (2017), Nancy et. al. (2016), and Zhang et al. (2010). for the single-valued neutrosophic set.

$$S_{ZHANG}(A)=\frac{(2+T-I-F)}{3}$$

$$S_{RIDVAN}(A)=\frac{(1+T-2I-F)}{2}$$

$$S_{NANCY}(A)=\frac{1+(1+T-2I-F)(2-T-F)}{2}$$

$$X_1 = \begin{bmatrix} 0 & 0.5 & 0.533 & 0 & 0 & 0 \\ 0.5 & 0 & 0.266 & 0.2 & 0 & 0 \\ 0.533 & 0.266 & 0 & 0.366 & 0.833 & 0 \\ 0 & 0.2 & 0.366 & 0 & 0.633 & 0.433 \\ 0 & 0 & 0.833 & 0.633 & 0 & 0.6 \\ 0 & 0 & 0 & 0.433 & 0.6 & 0 \end{bmatrix}$$

$$X_2 = \begin{bmatrix} 0 & 0.1 & 0.15 & 0 & 0 & 0 \\ 0.1 & 0 & -0.45 & -0.6 & 0 & 0 \\ 0.15 & -0.45 & 0 & -0.2 & 0.65 & 0 \\ 0 & -0.6 & -0.2 & 0 & 0.25 & -0.1 \\ 0 & 0 & 0.65 & 0.25 & 0 & 0.2 \\ 0 & 0 & 0 & -0.1 & 0.2 & 0 \end{bmatrix}$$

$$X_3 = \begin{bmatrix} 0 & 0.64 & 0.66 & 0 & 0 & 0 \\ 0.64 & 0 & -0.085 & 0.02 & 0 & 0 \\ 0.66 & -0.085 & 0 & 0.3 & 1.215 & 0 \\ 0 & 0.02 & 0.3 & 0 & 0.725 & 0.4 \\ 0 & 0 & 1.215 & 0.725 & 0 & 0.74 \\ 0 & 0 & 0 & 0.4 & 0.74 & 0 \end{bmatrix}$$

Step-3: Construct the PLM by Sort (i,:) in each matrix X_1, X_2 and X_3 then find the minimum value in each i.

Step-4: The preferred link matrix is constructed by referring to the above step.

$$PLM_1 = \begin{bmatrix} 0 & 0.5 & 0 & 0 & 0 & 0 \\ 0 & 0 & 0 & 0.2 & 0 & 0 \\ 0 & 0.266 & 0 & 0 & 0 & 0 \\ 0 & 0.2 & 0 & 0 & 0 & 0 \\ 0 & 0 & 0 & 0 & 0 & 0.6 \\ 0 & 0 & 0 & 0.433 & 0 & 0 \end{bmatrix}$$

Step-5: Constructed the node set matrix.

$$NSM_1 = \begin{bmatrix} 0 & (1,2) & 0 & 0 & 0 & 0 \\ 0 & 0 & 0 & (2,4) & 0 & 0 \\ 0 & (3,2) & 0 & 0 & 0 & 0 \\ 0 & (4,2) & 0 & 0 & 0 & 0 \\ 0 & 0 & 0 & 0 & 0 & (5,6) \\ 0 & 0 & 0 & (6,4) & 0 & 0 \end{bmatrix}$$

Step-6: Checking the condition $n \geq 3$ also tree does not form a cycle.
Step-7: Output of the MST shown in figure-3 and the crisp cost is minimized.

Figure 3. Final output of MST

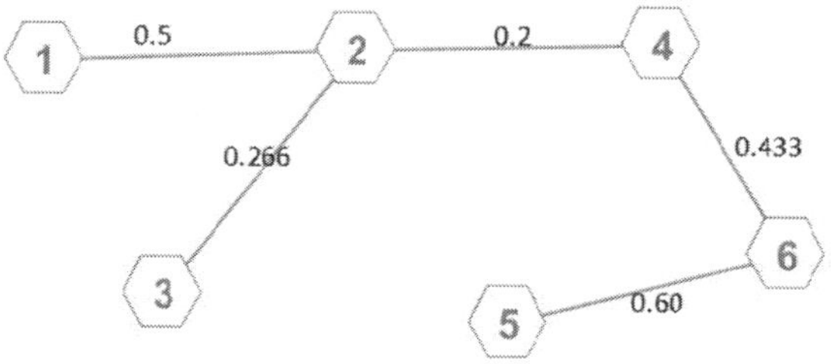

According to step -5 the crisp minimum cost of X_1 is 1.999 and the output of MST is {1,2}, {2,4}, {3,2}, {5,6}, {6,4} and the elapsed time calculated in MATLAB is 0.000511. Similarly, X_2 andX_3 get the output as shown below,

The crisp minimum cost X_2 is -0.55 and the output is {1,2}, {2,4}, {3,2}, {5,6}, {6,4} and the elapsed time is 0.000423.

The crisp minimum cost X_3 is 1.615 and the output MST is {1,2}, {2,3}, {3,2}, {4,2}, {5,4}, {6,4} and the elapsed time is 0.000602.

**Note: Negative cost does not consider any negative X_n cost.

Our proposed least edge weight algorithm are implemented and the results are displayed in the figure-4, where x-axis denotes the range of the optimal result and y-axis denotes the elapsed time for the obtained crisp cost

Figure 4. Time and crisp cost

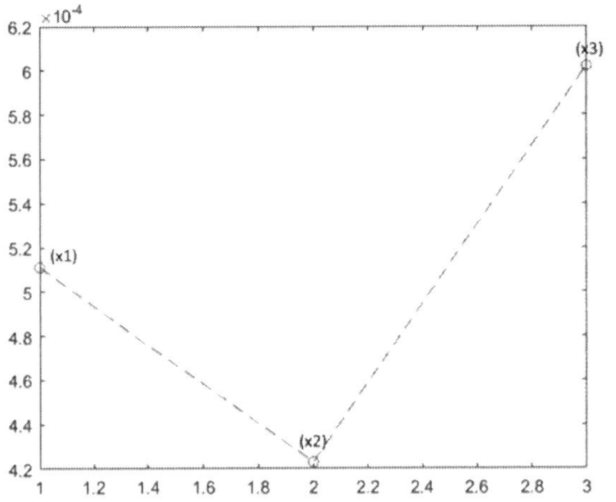

This figure-4 shows the running time for executing the given values of the crisp minimum cost of X_1, X_2 and X_3 .While using the different score functions.

COMPARATIVE STUDIES

This comparative section provides the results of MST problem with different score functions. Steps are followed to evaluate the proposed least edge weight algorithm with other algorithms. Some improved solution with time is shown in table-1. The difference between the proposed algorithm with broumi's and Mullai's algorithm. Broumi's algorithm also a matrix approach with Matlab and Mullai's algorithm is based on the comparison of edges in each iteration of the algorithm and this leads to high computation. Here our proposed approach which is based on the Matrix approach is simple to form a code in Matlab.

Our proposed approach is coded in MATLAB-2019b and executed with window11 version 22H2, operating system with an intel ® Core™ 8250 U processor running at 1.70 GHz and 8 GB of memory. Our proposed algorithm nicely evaluated when the number of nodes is increased.

Table 1. Comparative results with different score function with our proposed approach

Mullai's Result	Broumi's Result	Proposed Result
$S_{ZHANG}(A) = \frac{(2+T-I-F)}{3}$	$S_{ZHANG}(A) = \frac{(2+T-I-F)}{3}$	$S_{ZHANG}(A) = \frac{(2+T-I-F)}{3}$
{1, 2}, {2, 3}, {2, 4}, {4, 6},{6, 5} Cost=2	{1, 2}, {2, 3}, {2, 4}, {4, 6},{6, 5} Cost=2	{1,2}, {2,4},{3,2},{5,6},{6,4} Cost= 1.999. Elapsed time =0.00051.
		$S_{NANCY}(A) = \frac{1+(1+T-2I-F)(2-T-F)}{2}$ Cost=1.615. Elapsed time=0.000602.

The proposed results are shown in table-1 is compared with mullai's and broumi's results. The cost value is minimum when using the Nancy score function also noted the elapsed time between all the score functions with the results produced for MST.

CONCLUSION AND FUTURE WORK

This paper presents the neutrosophic MST with a Single valued neutrosophic number set having two or more score functions highlighting the fewer computation times and also the cost of MST. Computational works are also carried out with the given neutrosophic graph for knowing the strength of the crisp minimum cost.

In the future trying with multi-criteria neutrosophic edge weights with different algorithms to construct a minimum spanning tree. Ranking technics also trying to evaluate the score function for the new minimum spanning tree approaches in neutrosophic environment. In uncertainty model new technics and models are availed to create a new problem in the minimum spanning tree problems also in many other problems.

REFERENCES

Adhikary, K., Pal, P., & Poray, J. (2024). The Minimum Spanning Tree Problem on networks with Neutrosophic numbers. *Neutrosophic Sets and Systems*, *63*(1), 16.

Akbari Torkestani, J. (2013). Degree constrained minimum spanning tree problem: A learning automata approach. *The Journal of Supercomputing*, *64*(1), 226–249. DOI: 10.1007/s11227-012-0851-1

Archana, A., Deshpande, O. & Chaudhari. (2020). Fuzzy Approach to Compare a Minimal Spanning Tree Problem by Using Various Algorithm. *Advances in Fuzzy Mathematics*, *15*, 47–58.

Asady, B., & Zendehnam, A. (2007). Ranking fuzzy numbers by distance minimization. *Applied Mathematical Modelling*, *31*(11), 2589–2598. DOI: 10.1016/j.apm.2006.10.018

Atanassov, K. (1986). Intuitionistic fuzzy sets. *Fuzzy Sets and Systems*, *20*(1), 87–96. DOI: 10.1016/S0165-0114(86)80034-3

Broumi, S., Bakali, A., Talea, M., Smarandache, F., Dey, A., & Son, L. H. (2018). Spanning tree problem with neutrosophic edge weights. *Procedia Computer Science*, *127*, 190–199. DOI: 10.1016/j.procs.2018.01.114

Broumi, S., Bakali, A., Talea, M., Smarandache, F., & Uluçay, V. (2018). Minimum spanning tree in trapezoidal fuzzy neutrosophic environment. In *Innovations in Bio-Inspired Computing and Applications: Proceedings of the 8th International Conference on Innovations in Bio-Inspired Computing and Applications (IBICA 2017)* (pp. 25-35). Springer International Publishing. DOI: 10.1007/978-3-319-76354-5_3

Broumi, S., Smarandache, F., Talea, M., & Bakali, A. (2016). An introduction to bipolar single valued neutrosophic graph theory. *Applied Mechanics and Materials*, *841*, 184–191. DOI: 10.4028/www.scientific.net/AMM.841.184

Broumi, S., Talea, M., Bakali, A., & Smarandache, F. (2016). Single valued neutrosophic graphs. *Journal of New theory*, (10), 86-101.

Chakraborty, A., Mondal, S., & Broumi, S. (2019). *De-neutrosophication technique of pentagonal neutrosophic number and application in minimal spanning tree*. Infinite Study.

Chang, P. T., & Lee, E. S. (1999). Fuzzy decision networks and deconvolution. *Computers & Mathematics with Applications (Oxford, England)*, *37*(11-12), 53–63. DOI: 10.1016/S0898-1221(99)00143-1

de Almeida, T. A., Yamakami, A., & Takahashi, M. T. (2005, November). An evolutionary approach to solve minimum spanning tree problem with fuzzy parameters. In *International Conference on Computational Intelligence for Modelling, Control and Automation and International Conference on Intelligent Agents, Web Technologies and Internet Commerce (CIMCA-IAWTIC'06)* (Vol. 2, pp. 203-208). IEEE.

Deli, I., Ali, M., & Smarandache, F. (2015, August). Bipolar neutrosophic sets and their application based on multi-criteria decision making problems. In *2015 International conference on advanced mechatronic systems (ICAMechS)* (pp. 249-254). IEEE. DOI: 10.1109/ICAMechS.2015.7287068

Garg, H. (2016). An improved score function for ranking neutrosophic sets and its application to decision-making process. *International Journal for Uncertainty Quantification*, *6*(5).

Hassan, M. R. (2012). An efficient method to solve least-cost minimum spanning tree (LC-MST) problem. *Journal of King Saud University. Computer and Information Sciences*, *24*(2), 101–105. DOI: 10.1016/j.jksuci.2011.12.001

Itoh, T., & Ishii, H.ITO. (1996). An Approach Based on Necessity Measure to the Fuzzy Spanning Tree Problems. *Journal of the Operations Research Society of Japan*, *39*(2), 247–257. DOI: 10.15807/jorsj.39.247

Kandasamy, I. (2018). Double-valued neutrosophic sets, their minimum spanning trees, and clustering algorithm. *Journal of Intelligent Systems*, *27*(2), 163–182. DOI: 10.1515/jisys-2016-0088

Kumar, T., & Purusotham, S. (2018). The degree constrained k-cardinality minimum spanning tree problem: A lexi-search algorithm. *Decision Science Letters*, *7*(3), 301–310. DOI: 10.5267/j.dsl.2017.7.002

Mandal, K., & Basu, K. (2016). Improved similarity measure in neutrosophic environment and its application in finding minimum spanning tree. *Journal of Intelligent & Fuzzy Systems*, *31*(3), 1721–1730. DOI: 10.3233/JIFS-152082

Mandal, K., & Basu, K. (2016). Improved similarity measure in neutrosophic environment and its application in finding minimum spanning tree. *Journal of Intelligent & Fuzzy Systems*, *31*(3), 1721–1730. DOI: 10.3233/JIFS-152082

Mullai, M., Broumi, S., & Stephen, A. (2017). *Shortest path problem by minimal spanning tree algorithm using bipolar neutrosophic numbers*. Infinite Study.

. Smarandache, F. (1998). *Neutrosophy: neutrosophic probability, set, and logic: analytic synthesis & synthetic analysis*.

Smarandache, F., Broumi, S., Talea, M., & Bakali, A. (2016). Single Valued Neutrosophic Graphs: Degree, Order and Size. In *2016 IEEE International Conference on Fuzzy Systems (FUZZ)*. IEEE.

Sriram, R., Manimaran, G., & Murthy, C. S. R. (1998). Preferred link based delay-constrained least-cost routing in wide area networks. *Computer Communications*, *21*(18), 1655–1669. DOI: 10.1016/S0140-3664(98)00194-7

Turksen, I. B. (1986). Interval valued fuzzy sets based on normal forms. *Fuzzy Sets and Systems*, *20*(2), 191–210. DOI: 10.1016/0165-0114(86)90077-1

Wang, H., Smarandache, F., Zhang, Y., & Sunderraman, R. (2010). Single valued neutrosophic sets. *Infinite Study, 12*, 20110.

Ye, J. (2014). Single-valued neutrosophic minimum spanning tree and its clustering method. *Journal of Intelligent Systems*, *23*(3), 311–324. DOI: 10.1515/jisys-2013-0075

Zadeh, L. (1965). Fuzzy sets. *Information and Control*, *8*(3), 3338–3353. DOI: 10.1016/S0019-9958(65)90241-X

Zhang, H. Y., Wang, J. Q., & Chen, X. H. (2014). Interval neutrosophic sets and their application in multicriteria decision making problems. *TheScientificWorldJournal*, *2014*, 2014. DOI: 10.1155/2014/645953 PMID: 24695916

Chapter 22
Algorithms of Designing Decision Trees From Indeterm Soft Sets

Erick González Caballero
Asociación Latinoamericana de Ciencias Neutrosóficas, Cuba

Ketty Marilú Moscoso-Paucarchuco
Universidad Nacional Autónoma Altoandina de Tarma, Peru

Noel Batista Hernandez
Universidad Regional Autónoma de los Andes, Ecuador

Lorenzo Jovanny Cevallos Torres
Universidad Bolivariana del Ecuador, Ecuador

Maikel Leyva
https://orcid.org/0000-0001-7911-5879
Universidad de Guayaquil, Ecuador

Victor Gustavo Gómez Rodríguez
Universidad Bolivariana del Ecuador, Ecuador

ABSTRACT

Since their creation, soft sets have served as a technique to model different situations of uncertainty. Furthermore, they have been extended to or hybridized with other theories of uncertainty. One of them is the introduction of IndetermSoft sets, where indeterminacy is incorporated into at least one of the components of the soft sets. Decision trees are well-known tools that allow an object or entity to be classified by the data measured according to some attributes. The simplest algorithms for creating decision trees are ID3 and C4.5. The purpose of this chapter is to extend these two algorithms in the area of uncertainty and indeterminacy for data represented in the form of IndetermSoft sets. In this way it is possible to obtain decision rules when there is indeterminacy in the data.

DOI: 10.4018/979-8-3693-3204-7.ch022

INTRODUCTION

One of the methods to mathematically model problems that contain uncertainty is soft sets. This theory was introduced by the researcher D. Molodtsov in 1999 (Molodtsov, 1999). Soft sets consist of an ordered pair containing a set of parameters, and a mapping from this set of parameters into the power set of an initial universe set.

Molodtsov showed that L. Zadeh's fuzzy sets are a particular case of soft sets. Furthermore, soft sets have the advantage that they do not need a membership function to be defined, which is essential in the theory of fuzzy sets.

There are multiple applications of this theory in branches of both pure and applied mathematics. In mathematical analysis it is used to define soft numbers, soft limits, soft derivatives and soft integrals. In algebra the notions of soft groups, soft semirings, soft modules and certain types of soft algebras appear. Soft sets are associated with topological spaces of the General Topology as it was studied in some works of Molodtsov himself (Cagman, Enginoglu, & Citak, 2011; Maji, Biswas, & Roy, 2003; Molodtsov, 1999).

Soft sets have been hybridized with other theories of uncertainty such as rough sets, fuzzy sets, interval-valued fuzzy sets and intuitionistic fuzzy sets. The fields of application are decision-making, forecasting and game theory, among others. They have been applied in decision-making and medical diagnosis problems, in education, among others (Cagman, Enginoglu, & Citak, 2011; González-Caballero & Broumi, 2023; M. Voskoglou, Broumi, & Smarandache, 2022; M. G. Voskoglou, 2022).

One of the interesting results is the recognition of soft sets as a special class of information systems (Pei & Miao, 2005). The advantage of this acknowledgment is that we can use this theory to generate decision rules. In (Herawan & Deris, 2011), soft sets are discussed in the context of transactional datasets for association rules mining.

More recently F. Smarandache extended soft sets to the Neutrosophy framework. Neutrosophy is the branch of philosophy that studies indeterminacy, neutrality, the unknown, incoherence, inconsistency, contradiction and erroneousness, among others. So, in this framework, Smarandache defines IndetermSoft sets, HyperSoft sets, IndetermHyperSoft sets, MultiSoft sets and TreeSoft sets (Smarandache, 2022).

HyperSoft sets extend the function of parameters from a single attribute to multiple attributes (Abbas, Murtaza, & Smarandache, 2020; Ahsan, Sarwar, Lone, Almutlak, & Anwer, 2023; Musa & Asaad, 2021). IndetermSoft sets are soft sets where there is indeterminacy or ambiguity in some of their components (Smarandache, 2022). IndetermHyperSoft set combines indeterminacy in multi-parameter soft sets. The function of parameters in MultiSoft sets is based on the power set of the multiple parameters. Whereas TreeSoft sets allow the modeling of soft sets to be extended in the form of a tree structure, where there is a hierarchical level of the attributes (AL-baker, Elhenawy, & Mohamed, 2024; Gharib, Rajab, & Mohamed, 2023; Gharib, Smarandache, & Mohamed, 2024).

In this chapter we propose algorithms for designing decision trees in an area of uncertainty using IndetermSoft sets. A decision tree is a tree-shaped classification model where each leaf represents an alternative such that the leaves at the lowest level allow us to classify an object. Objects are classified according to their measured values concerning certain attributes. These trees can be represented in the form of decision rules (Budiman et al., 2018; Hssina, Merbouha, Ezzikouri, & Erritali, 2014).

Two of the basic algorithms for constructing decision trees are ID3 and C4.5, those developed by R. Quinlan (García-Pichardo, 2005; Hssina, Merbouha, Ezzikouri, & Erritali, 2014). Both of them start from n-tuples of values measured according to parameters, such that multiple training cases are used to create the decision tree, which will allow the classification of other examples. They are data mining techniques. Data mining or data exploration - it is the analysis stage of "Knowledge Discovery in Databases" or KDD - is a field of statistics and computer science referred to the process that attempts to discover patterns in large volumes of data sets (Maimon & Rokach, 2010).

Quinlan initially designed ID3 algorithm and later its improved variant C4.5, the last one allows processing discrete and continuous data, datasets with missing data and tree pruning as a heuristic.

In this chapter we adapt ID3 and C4.5 algorithms in situations where there is indeterminacy in the data when we are in the presence of IndetermSoft sets. We are based on the fact that soft sets can be converted into information systems as has been studied in (Pei & Miao, 2005). It is known that decision rules can be created from information systems according to the theory developed by Z. Pawlak (Thangavel & Pethalakshmi, 2009).

Specifically, we take advantage of the fact that these algorithms are based on the concept of entropy used in information theory as the measure of indeterminacy. Entropy is usually confused as a measure of disorder, but this is not the case. When the states of a system are more uncertain, which is manifested so that all of elements have more or less equal probabilities to appear, then there is more indeterminacy of the performance of the system (Gray, 1990).

For the development of this chapter we propose the following structure, we begin with a Background section where we recall the basic notions of the theories used in this chapter, such as soft sets, IndetermSoft sets, decision trees, and the description of ID3 and C4.5 algorithms. Then a section appears where we detail the algorithms proposed in the case of having IndetermSoft sets. Later, a section is dedicated to illustrating the proposed algorithms with an example. Next, we can read the section called Discussion. The chapter ends with the Conclusions.

BACKGROUND

This section contains the basic notions of the theories used in this chapter for the creation of the proposed algorithms. We begin with a reminder of soft sets definition.

Definition 1 (Molodtsov, 1999)**.** Given U is the initial universe set and E is the set of parameters, a pair *(F,E)* is called a *soft set* (over U) if and only if F is a mapping of E into the set of all subsets of U.

That is to say, having a set E of parameters and fixing a parameter $\varepsilon \in E$, then $F(\varepsilon) \in \mathcal{P}(U)$, where $\mathcal{P}(U)$ denotes the power set of U and $F(\varepsilon)$ is considered the set of ε-elements of the Soft Set *(F,E)* or the set of ε-approximate elements of the Soft Set.

It is not difficult to realize that fuzzy sets are soft sets. This is a consequence of the α-levels definition of a membership function μ_A where we have the following:

$$F(\alpha) = \{x \in U | \mu_A(x) \geq \alpha\},\ \alpha \in [0,1].$$

Thus, if we know the family F then we can reconstruct the function μ_A by using the following formula:

$$\mu_A(x) = \sup_{\substack{\alpha \in [0,1] \\ x \in F(\alpha)}} \alpha$$

Thus, a fuzzy set is a soft set *(F, [0, 1])*.

Given a binary operation $*$ for subsets of the set U, where *(F, A)* and *(G, B)* are soft sets over U. Then, the operation $*$ for soft sets is defined as follows:

$$(F, A)*(G, B)=(J, A\times B),$$

where $J(\alpha,\beta) = F(\alpha) * G(\beta)$; $\alpha \in A$, $\beta \in B$ and $A\times B$ is the Cartesian product of the sets A and B.

Definition 2 (Smarandache, 2022). Given U is the initial universe set and A is the set of parameters, also given $H \subseteq U$ and $H \neq \emptyset$. Then *(F, A)* is an *IndetermSoft set* if satisfies at least one of the following conditions:

1. A has some indeterminacy,
2. $\mathcal{P}(H)$ has some indeterminacy,
3. There exists at least an attribute value $v \in A$, such that $F(v)$ = indeterminate (unclear, incomplete, conflicting or not unique).

I.e., F is a NeutroFunction that is defined as a function that is simultaneously inner-defined, partially indeterminate and partially out-defined (Smarandache, 2021).

Apart from the uncertainty modeled by classic soft sets, IndetermSoft sets incorporate the indeterminacy that is part of the real world. A basic example that appears in (Smarandache, 2022) is the following:

Example 1. Let us assume that a town has many houses H. The attribute is a = *color* and A=*{white,red,green, blue, orange, yellow,...}* is the set of parameters.

The indeterminacy can be expressed in the following ways:

1. *Indeterminacy concerning the function*:
   a. The red-colored houses are F(red)= h_1 or h_2 we do not know which one of them.
   b. The green-colored houses are F(green)= not h_4, we do not know which one of them and we only know it is not h_4.
   c. The houses painted in white color are F(white)= either h_3 or h_4.
2. *Indeterminacy for the set H of houses:* The number of houses in the town is between 100 and 120.
3. *Indeterminacy for the set A of attributes*: We do not know the colors of the houses, we are only sure that they are painted in white, red or green.

Next, some notions of decision trees will be given, see Figure 1:

Figure 1. Decision tree whether to play or not to play tennis according to the weather situation

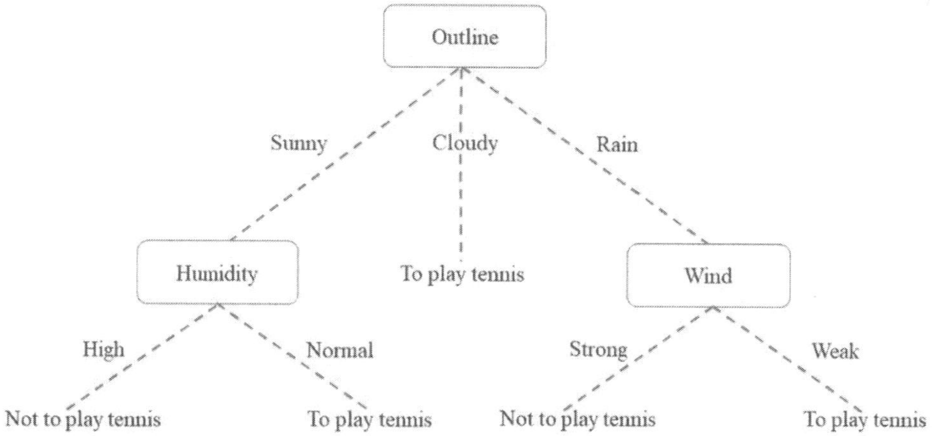

Figure 1 shows that if the weather is cloudy playing tennis is not recommended. If the weather is sunny, it is recommended to play tennis only if the humidity is normal. On the other hand, when the weather is rainy it is recommended to play tennis when the wind is weak.

To obtain the tree in Figure 1, a set of data was needed to be measured according to the parameters Outline = {sunny, cloudy, wind}, Humidity = {high, normal} and Wind = {strong, weak}. For example, data that appear in Table 1.

Table 1. Dataset with objects measured according to the attributes. The last column contains the measurements on the target class

Examples	Outline	Humidity	Wind	To play tennis
D_1	Sunny	High	Weak	No
D_2	Sunny	High	Strong	No
D_3	Cloudy	High	Weak	Yes
D_4	Rain	High	Weak	Yes
D_5	Rain	Normal	Weak	Yes
D_6	Rain	Normal	Strong	No
D_7	Cloudy	Normal	Strong	Yes
D_8	Sunny	High	Weak	No
D_9	Sunny	Normal	Weak	Yes
D_{10}	Rain	Normal	Weak	Yes
D_{11}	Sunny	Normal	Strong	Yes
D_{12}	Windy	High	Strong	Yes
D_{13}	Windy	Normal	Weak	Yes
D_{14}	Rain	High	Strong	No

To achieve this result, an algorithm such as those called ID3 or C4.5 is needed which we describe below in Table 2 (García-Pichardo, 2005; Hssina, Merbouha, Ezzikouri, & Erritali, 2014; Sher, Rehman, & Kim, 2022):

Table 2. Basic aspects of ID3 and C4.5 algorithms.

We have as inputs:
Algorithm(Examples, Target-Attribute, Attributes):
  If all examples are positive **Then** return a positive node
  If all examples are negative **Then** return a negative node
  If attributes are empty **Then** return the majority vote of the target-attribute in Examples.
  In another case,
    If *A* is an attribute THE BEST of attributes
    For each attribute value *v* do the following:
      Let *Examples(v)* be the subset of examples whose attribute value *A* is *v*,
      If *Examples(v)* is empty **Then** return a node with the majority vote of the Examples in Target-Attribute,
      If not Return **Algorithm(*Examples(v)*, Target-Attribute, Attributes\\{*A*})**

The algorithm in Table 2 means that it starts from a set of examples, which is a dataset that contains the evaluations of the data on a set of attributes (e.g., Outline, Humidity and Wind) and a target attribute (e.g., to play or not to play tennis).

The first cases are degenerate, which is when all the data have the same value for the target attribute or there are no values for the other attributes. Otherwise, the best attribute is selected and the majority vote for the examples concerning this attribute is returned. Then the algorithm is recursively repeated for the rest of the attributes if there is any, restricted to the data without the values of the attribute already studied. We must also emphasize that this is a dichotomous classification, where the values of the target class are Yes or No, Positive or Negative, Valid or Invalid and so on.

To select the best attribute in ID3 and C4.5 algorithms the value of entropy is used, which means the degree of indeterminacy of the system and depends on the probabilities associated with each element of the system in a specific state.

The entropy function is given according to Equation 1:

$$Entropy(S) = \sum_{i=1}^{n} -p(v_i)\log_2(p(v_i)) \qquad (1)$$

Here $p(v_i)$ is the probability of occurrence of the i-th value of the target attribute and $log_2(x)$ is the logarithm function in base 2.

ID3 algorithm selects the best attribute through the highest value of the information gain expressed in Equation 2.

$$IG(S,A) = Entropy(S) - \sum_{v \in A} \frac{|S_v|}{|S|} Entropy(S_v) \qquad (2)$$

So, given the attribute *A* then the entropies restricted to the possible values of this attribute are subtracted from the entropy of the entire system *S*; each of them is multiplied by the proportion of cases of each value concerning the total number of cases.

ID3 algorithm is only applied when we have nominal data and there are no missing values in the database. In addition, accuracy is lost when there is a large number of attributes or data.

To overcome these difficulties C4.5 algorithm was created, which has the advantage that it allows us to process both continuous and nominal data, allows lack of data and can also include a tree-pruning heuristic that limits the growth of the tree. This is achieved by replacing *IG(S, A)* with the gain ratio function, for this end the *SplitInfo function* is used, which is defined in Equation 3.

$$SplitInfo(A) = \sum_{i=1}^{n} -p(subset_i)\log_2(p(subset_i)) \tag{3}$$

This function is restricted to the ratios over the total number of examples, concerning the cardinality of *A* of possible values. For example, in Table 1 the Wind attribute has 8 values "Weak" and 6 values "Strong", therefore:

$$\begin{aligned} SplitInfo(Wind) &= -p(Weak)\log_2(p(Weak)) - p(Strong)\log_2(p(Strong)) \\ &= -\left(\frac{8}{14}\right)\log_2\left(\frac{8}{14}\right) - \left(\frac{6}{14}\right)\log_2\left(\frac{6}{14}\right) = 0.82447 \end{aligned}$$

Equation 4 is the one referring to the Gain Ratio:

$$GainRatio(S,A) = \frac{Gain(S,A)}{SplitInfo(A)} \tag{4}$$

This function allows us to calculate the weight of the attribute although it contains missing values in the database, as it can be restricted to the set of existing values. On the other hand, when there are continuous values, they are discretized by taking the number of values within specific defined classes, for example, the values ≥ 9, ≤ 5 and *(5, 9)* form three distinct disjoint classes.

ALGORITHMS ID3 AND C4.5 FOR INDETERMSOFT SETS

This section contains the details of the algorithms proposed in the chapter, as well as their rationale.

Let $\{(F_i, E_i)\}_{i=1}^{n}$ be a set of at least one IndetermSoft set. Each E_i is the set of possible values (parameters) of a given attribute. In particular, E_n corresponds to the set of values of the target attribute, which is dichotomous. F_i are the functions such that in at least a k and j it is satisfied $F_k(v_j)= h_1$ or h_2 or…or h_m with $m>1$ we do not know what of them.

Note that for now this type of indeterminacy is that will be accepted in the algorithm. Although in "not h$_1$ for sure" this type of indeterminacy can be expressed in the form of h_1 or h_2 or…or h_m for all h_p except h_l. This can also be extended when we have the condition "not h_l and h_q for sure" and so on.

Also, we will accept the indeterminacy on the set of parameters, when it is expressed by $h_l \in F_k(v_i)$ and $h_l \in F_k(v_j)$; in this case h_l is v_i or v_j, we do not know which one.

The first step of the algorithm is to convert IndetermSoft sets into a table similar to Table 1. When $F_k(v_j)= h_p$ we set the element v_j in the place corresponding to the k-th attribute and the p-th example.

On the other hand, when it is fulfilled $F_k(v_j)= h_1$ or h_2 or…or h_m it is placed v_j in each of the places of $h_1,h_2,\ldots,h_m$. Finally, $h_l \in F_k(v_i)$ and $h_l \in F_k(v_j)$ we set "v_i or v_j" in the place of the l-th example and the k-th attribute.

It is important to separate what is allowed in the extended ID3 algorithm from the conditions for performing calculations in C4.5 algorithm. In the first of them, there must be nominal values, never continuous and there must never be missing data. On the other hand, in the case of C4.5 the conditions of this classic algorithm continue to be met.

In the cases of indeterminacy, we will use Zadeh's idea of Σ-count in the entropy formulas for classical algorithms (Zadeh, 1984). In each of the formulas 1-4 of the algorithms a fractional number is used in the numerator instead of an integer. This comes from the probability formula used in the algorithms where:

$$p(B) = \frac{Numberofpositivecases}{Totalofcases} \tag{5}$$

To introduce indeterminacy keep in mind that in the classic formula each value is counted once for each time it appears within the column corresponding to the attribute being measured. In the case where there is indeterminacy and there is more than one parameter in the same cell (m>1 parameters), each of the elements is counted as $\frac{1}{m}$ instead of once and this is the value that is included in the formulas of the algorithm. Let us illustrate this with an example:

Example 2. Let us assume that we have some sets as indicated below corresponding to the attribute color for the set of houses $H = \{h_1,h_2,h_3,h_4\}$ such that $A=\{white, blue, green\}$: F(white)= $\{h_1,h_2\}$, F(blue)= h_3 or h_4 we do not know which of them and F(green)= h_3 or h_4 we do not know which of them.

The column in the Table corresponding to the attribute color is located as follows:

Table 3. Column corresponding to the color attribute of the houses in the given example

Examples	Color
h_1	White
h_2	White
h_3	Blue or Green
h_4	Blue or Green

Remark 1. Note that from the indeterminacy corresponding to objects we can derive an indeterminacy corresponding to parameters. For example, from F(blue)= h_3 or h_4 and F(green)= h_3 or h_4 an indeterminacy is achieved for each object in terms of its color. So "h_3 is blue or green"; in this case it is not known if h_3 is blue or green, or the color in which the house was painted is not easy to determine between blue or green, or it is believed that someone said one of the two colors, we are not sure of this, among other cases.

To form the probabilities of the appearance of the parameter white in the column, it is taken simply as it is done in classical algorithms, $p(white) = \frac{2}{4} = \frac{1}{2}$ this is calculated by counting the number of houses painted in white (2) divided by the total number of houses (4). However, for the houses h_3 and h_4 there is an indeterminacy and since there are two indeterminacy - two houses of which we are not sure about their color - then each of them is counted as $\frac{1}{2}$ times, that is:

$$p(blue) = \frac{\frac{1}{2}+\frac{1}{2}}{4} = \frac{1}{4} \text{ and } p(green) = \frac{\frac{1}{2}+\frac{1}{2}}{4} = \frac{1}{4}.$$

These formulas are based on the properties of entropy in the information theory and its intuitive meaning. First of all, let us remember that entropy measures the degree of indeterminacy of the system. As long as the probabilities of the elements of the system are equal or similar to each other, the greater the randomness of the system the greater the indeterminacy. In this case, when the probabilities are equal to each other, then the entropy is maximum, therefore the information gain is lower. For example, the condition $p(blue)=p(green)$ increases the entropy between these two values of color and the entropy in general, this is due to indeterminacy.

On the other hand, the more elements that are present in the indeterminacy the greater the entropy. This is calculated by the formula $Entropy(S)=log_2(N)$ when it is maximum, which occurs when $p_i = \frac{1}{N}$ for all $i=1,2,...,N$ (Gray, 1990). As it is known, $log_2(N)$ is an increasing function therefore the larger is N greater it is entropy.

E.g., let us assume that instead of Table 3 we have Table 4 where there is total ignorance.

Table 4. Column corresponding to the attribute color of the houses with total indeterminacy

Examples	Color
h_1	White or Blue or Green
h_2	White or Blue or Green
h_3	White or Blue or Green
h_4	White or Blue or Green

In Table 4 we have

$$p(white) = p(blue) = p(green) = \frac{\frac{1}{3}+\frac{1}{3}+\frac{1}{3}+\frac{1}{3}}{4} = \frac{4}{12} = \frac{1}{3}$$

and in this case the entropy is maximum equal to $\log_2 3 = 1.5850$, which means total lack of knowledge about the behavior of the attribute.

This algorithm is summarized in Table 5.

Table 5. General ID3 and C4.5 algorithms for IndetermSoft sets

Input: $\{(F_i,E_i)\}_{i=1}^{n}$ set of IndetermSoft sets or soft sets. At least one of them is an IndetermSoft set. (F_n,E_n) is the IndetermSoft set or soft set corresponding to the target attribute. 1. IndetermSoft sets or soft sets are converted into a table of Examples versus attribute values. The last column corresponds to the target attribute. **Note**: **If** we want to apply algorithm ID3 we must have nominal values within the Table. All cells must be covered. **If** we want to apply algorithm C4.5 we must have nominal or continuous values. Values may be missing within the table. 2. Apply the algorithm specified in Table 2. Keep in mind to use formulas 1-4 adapted to indeterminacy using the formula as we explained at the beginning of this section.

In the following section we will apply the algorithm to a case of IndetermSoft sets to illustrate its operation and show its practical usefulness.

AN ILLUSTRATIVE EXAMPLE

Example 3. Suppose we need to decide between buying or not buying a house, according to the data that appears in the following IndetermSoft sets or soft sets that correspond to the decision made by previous clients.

(F_1, E_1) where E_1=*{expensive, acceptable, cheap, very cheap}* corresponding to the attribute Cost, F_1(expensive)= $\{h_1,h_2,h_8,h_9,h_{11}\}$, F_1(acceptable)= $\{h_3,h_7\}$, F_1(cheap)= $\{h_4,h_5,h_6,h_7,h_{10},h_{14}\}$ and F_1(very cheap)= $\{h_{12},h_{13}\}$.

(F_2, E_2) where E_2=*{beautiful, acceptable, ugly}* corresponding to the attribute Beauty and F_2(beautiful)= $\{h_5,h_6,h_7,h_9\}$, F_2(acceptable)= $\{h_1,h_4,h_8,h_{10},h_{11},h_{12},h_{14}\}$, F_2(ugly)= $\{h_1,h_2,h_3,h_{13}\}$ and F_2(not beautiful)= $\{h_{13}\}$.

(F_3, E_3) where E_3=*{weak, strong}* corresponding to the attribute Materials, F_3(weak)= $\{h_1,h_2,h_3,h_4,h_8,h_{12},h_{14}\}$ and F_3(strong)= $\{h_2,h_5,h_6,h_7,h_9,h_{10},h_{11},h_{13}\}$.

(F_4, E_4) where E_4=*{good, not good}* corresponding to the attribute Neighborhood, F_4(good)= $\{h_1,h_3,h_4,h_5,h_8,h_9,h_{10},h_{13}\}$ and F_4(not good)= $\{h_2,h_6,h_7,h_{11},h_{12},h_{14}\}$.

(F_5, E_5) where E_5=*{no, yes}* corresponding to the target-attribute To buy, F_5(no)= $\{h_1,h_2,h_6,h_7,h_8,h_{14}\}$ and F_5(yes)= $\{h_3,h_4,h_5,h_7,h_9,h_{10},h_{11}\}$.

The aim is to help new clients for making the best decision by automating their possible choices according to certain attributes in the form of rules. In some cases, there is no necessary certainty and therefore they appear in an indeterminate form.

As we can see, there are functions where the same object is repeated for different parameters of the same attribute. For example, in terms of beauty h_1 is classified between ugly and acceptable, without being able to decide on a specific value. There is also indecision in purchasing h_7 that has yes and no values at the same time. For h_{13} the uncertainty is that it is not beautiful.

The previous data appears below in the form of a Table, which is the first step of the proposed algorithm.

Table 6. Dataset with the previous data of the IndetermSoft sets in the form of table

Examples	Cost	Beauty	Materials	Neighborhood	To buy
h_1	Expensive	Ugly or Acceptable	Weak	Good	No
h_2	Expensive	Ugly	Weak or Strong	Not good	No
h_3	Acceptable	Ugly	Weak	Good	Yes
h_4	Cheap	Acceptable	Weak	Good	Yes
h_5	Cheap	Beautiful	Strong	Good	Yes
h_6	Cheap	Beautiful	Strong	Not good	No
h_7	Acceptable or Cheap	Beautiful	Strong	Not good	Yes or No

continued on following page

Table 6. Continued

Examples	Cost	Beauty	Materials	Neighborhood	To buy
h_8	Expensive	Acceptable	Weak	Good	No
h_9	Expensive	Beautiful	Strong	Good	Yes
h_{10}	Cheap	Acceptable	Strong	Good	Yes
h_{11}	Expensive	Acceptable	Strong	Not good	Yes
h_{12}	Very Cheap	Acceptable	Weak	Not good	Yes
h_{13}	Very Cheap	Not Beautiful	Strong	Good	Yes
h_{14}	Cheap	Acceptable	Weak	Not good	No

The second step is to calculate *Entropy(S)* (Equation 1) and *IG(S, A)* (Equation 2) from the previous data and take into account the details of the probabilities when there is indeterminacy as we pointed out in the previous section.

$$Entropy(S) = -\left(\left(\frac{5+\frac{1}{2}}{14}\right)log_2\left(\frac{5+\frac{1}{2}}{14}\right) + \left(\frac{8+\frac{1}{2}}{14}\right)log_2\left(\frac{8+\frac{1}{2}}{14}\right)\right) = 0.96662.$$

Now the information gains are calculated according to each of the attributes:

$$IG(S,A_1) = Entropy(S) - \sum_{v=v_{11}}^{v_{14}} \frac{|S_v|}{|S|} Entropy(S_v).$$

Corresponding to "cost" where v_{11}*=expensive*, v_{12}*=acceptable*, v_{13}*=cheap* and v_{14}*=very cheap.*

$$IG(S,A_1) = 0.96662 - \left(-\left(\frac{5}{14}\right)\left[\left(\frac{2}{5}\right)log_2\left(\frac{2}{5}\right) + \left(\frac{3}{5}\right)log_2\left(\frac{3}{5}\right)\right] - \left(\frac{1+\frac{1}{2}}{14}\right)\right.$$

$$\left[\left(\frac{1+\frac{1}{4}}{1+\frac{1}{2}}\right)log_2\left(\frac{1+\frac{1}{4}}{1+\frac{1}{2}}\right) + \left(\frac{\frac{1}{4}}{1+\frac{1}{2}}\right)log_2\left(\frac{\frac{1}{4}}{1+\frac{1}{2}}\right)\right] - \left(\frac{5+\frac{1}{2}}{14}\right)\left[\left(\frac{3+\frac{1}{4}}{5+\frac{1}{2}}\right)\right.$$

$$\left.\left.log_2\left(\frac{3+\frac{1}{4}}{5+\frac{1}{2}}\right) + \left(\frac{2+\frac{1}{4}}{5+\frac{1}{2}}\right)log_2\left(\frac{2+\frac{1}{4}}{5+\frac{1}{2}}\right)\right] - \left(\frac{2}{14}\right)\left[\left(\frac{2}{2}\right)log_2\left(\frac{2}{2}\right) + \left(\frac{0}{2}\right)log_2\left(\frac{0}{2}\right)\right]\right)$$

$= 0.9666 - 0.34677 - 0.069645 - 0.38344 - 0 = 0.16677$

Thus, $IG(S,A_2)$ corresponding to beauty is calculated as:

$$IG(S,A_2) = Entropy(S) - \sum_{v=v_{21}}^{v_{23}} \frac{|S_v|}{|S|} Entropy(S_v),$$

Where v_{21}=*beautiful*,v_{22}=*acceptable* and v_{23}=*ugly*. So,

$$IG(S,A_2) = 0.9666 - \left(-\left(\frac{4}{14}\right)\left[\left(\frac{2+\frac{1}{2}}{4}\right)log_2\left(\frac{2+\frac{1}{2}}{4}\right) + \left(\frac{1+\frac{1}{2}}{4}\right)log_2\left(\frac{1+\frac{1}{2}}{4}\right)\right]\right.$$
$$-\left(\frac{6+\frac{1}{2}+\frac{1}{2}}{14}\right)\left[\left(\frac{4+\frac{1}{2}}{7}\right)log_2\left(\frac{4+\frac{1}{2}}{7}\right) + \left(\frac{2+\frac{1}{2}}{7}\right)log_2\left(\frac{2+\frac{1}{2}}{7}\right)\right] - \left(\frac{2+\frac{1}{2}+\frac{1}{2}}{14}\right)$$
$$\left.\left[\left(\frac{1+\frac{1}{2}}{3}\right)log_2\left(\frac{1+\frac{1}{2}}{3}\right) + \left(\frac{1+\frac{1}{2}}{3}\right)log_2\left(\frac{1+\frac{1}{2}}{3}\right)\right]\right)$$

$= 0.9666 - 0.27270 - 0.47014 - 0.21429 = 0.0094959.$

$IG(S,A_3)$ corresponding to Materials is calculated as:

$$IG(S,A_3) = Entropy(S) - \sum_{v=v_{31}}^{v_{32}} \frac{|S_v|}{|S|} Entropy(S_v),$$

where v_{31}=*weak* and v_{32}=*strong*. So,

$$IG(S,A_3) = 0.96662 - \left(-\left(\frac{6+\frac{1}{2}}{14}\right)\left[\left(\frac{3}{6+\frac{1}{2}}\right)log_2\left(\frac{3}{6+\frac{1}{2}}\right) + \left(\frac{3+\frac{1}{2}}{6+\frac{1}{2}}\right)log_2\left(\frac{3+\frac{1}{2}}{6+\frac{1}{2}}\right)\right]\right.$$

$$\left.-\left(\frac{7+\frac{1}{2}}{14}\right)\left[\left(\frac{5+\frac{1}{2}}{7+\frac{1}{2}}\right)log_2\left(\frac{5+\frac{1}{2}}{7+\frac{1}{2}}\right) + \left(\frac{2}{7+\frac{1}{2}}\right)log_2\left(\frac{2}{7+\frac{1}{2}}\right)\right]\right)$$

$= 0.9666 - 0.46230 - 0.44820 = 0.056118.$

$$IG(S,A_4) = Entropy(S) - \sum_{v=v_{41}}^{v_{42}} \frac{|S_v|}{|S|} Entropy(S_v)$$

Where v_{41}=*good* and v_{42}=*not good*. So,

$$IG(S,A_4) = 0.96662 - \left(-\left(\tfrac{8}{14}\right)\left[\left(\tfrac{6}{8}\right)log_2\left(\tfrac{6}{8}\right) + \left(\tfrac{2}{8}\right)log_2\left(\tfrac{2}{8}\right)\right] - \left(\tfrac{6}{14}\right)\left[\left(\frac{2+\frac{1}{2}}{6}\right)log_2\left(\frac{2+\frac{1}{2}}{6}\right) + \left(\frac{3+\frac{1}{2}}{6}\right)log_2\left(\frac{3+\frac{1}{2}}{6}\right)\right]\right)$$

$= 0.9666 - 0.46359 - 0.41994 = 0.083089$

Therefore, the attribute that contributes the most is "cost" according to the extended ID3 algorithm. Since we are dealing with nominal data and there is no missing data for using this algorithm. To gain simplicity we will not apply algorithm C4.5. To do this, the other indices would have to be calculated, which is not difficult to do, although it is cumbersome.

Let us now reduce Table 6 according to each of the cost values, see Tables 7-10.

Table 7. Dataset with the data for cost = expensive

Examples	Beauty	Materials	Neighborhood	To buy
h_1	Ugly or Acceptable	Weak	Good	No
h_2	Ugly	Weak or Strong	Not good	No
h_8	Acceptable	Weak	Good	No
h_9	Beautiful	Strong	Good	Yes
h_{11}	Acceptable	Strong	Not good	Yes

Table 8. Dataset with the data for cost = acceptable

Examples	Beauty	Materials	Neighborhood	To buy
h_3	Ugly	Weak	Good	Yes
h_7	Beautiful	Strong	Not good	Yes or No

Table 9. Dataset with the data for cost = cheap

Examples	Beauty	Materials	Neighborhood	To buy
h_4	Acceptable	Weak	Good	Yes
h_5	Beautiful	Strong	Good	Yes
h_6	Beautiful	Strong	Not good	No
h_7	Beautiful	Strong	Not good	Yes or No
h_{10}	Acceptable	Strong	Good	Yes
h_{14}	Acceptable	Weak	Not good	No

Table 10. Dataset with the data for cost = very cheap

Examples	Beauty	Materials	Neighborhood	To buy
h_{12}	Acceptable	Weak	Not good	Yes
h_{13}	Not Beautiful	Strong	Good	Yes

In Table 10 it is seen that when cost is very cheap it is recommended to buy. Now let us analyze each of the cases, in Table 7 we have

$$Entropy(S') = -\left(\left(\tfrac{2}{5}\right)log_2\left(\tfrac{2}{5}\right) + \left(\tfrac{3}{5}\right)log_2\left(\tfrac{3}{5}\right)\right) = 0.97095.$$

$$IG(S,A_2') = 0.97095 - \left(-\left(\tfrac{1}{5}\right)\left[\left(\tfrac{1}{1}\right)log_2\left(\tfrac{1}{1}\right) + \left(\tfrac{0}{1}\right)log_2\left(\tfrac{0}{1}\right)\right] - \left(\frac{2+\frac{1}{2}}{5}\right)\right.$$

$$\left[\left(\frac{1+\frac{1}{2}}{2+\frac{1}{2}}\right)log_2\left(\frac{1+\frac{1}{2}}{2+\frac{1}{2}}\right) + \left(\frac{1}{2+\frac{1}{2}}\right)log_2\left(\frac{1}{2+\frac{1}{2}}\right)\right]$$
$$\left.-\left(\frac{1+\frac{1}{2}}{5}\right)\left[\left(\frac{0}{1+\frac{1}{2}}\right)log_2\left(\frac{0}{1+\frac{1}{2}}\right) + \left(\frac{1+\frac{1}{2}}{1+\frac{1}{2}}\right)log_2\left(\frac{1+\frac{1}{2}}{1+\frac{1}{2}}\right)\right]\right)$$

$$= 0.97095 - 0 - 0.48548 - 0 = 0.48547.$$

$$IG(S,A_3') = 0.97095$$
$$-\left(-\left(\frac{2+\frac{1}{2}}{5}\right)\left[\left(\frac{0}{2+\frac{1}{2}}\right)log_2\left(\frac{0}{2+\frac{1}{2}}\right) + \left(\frac{2+\frac{1}{2}}{2+\frac{1}{2}}\right)log_2\left(\frac{2+\frac{1}{2}}{2+\frac{1}{2}}\right)\right]\right.$$

$$-\left(\frac{2+\frac{1}{2}}{5}\right)\left[\left(\frac{2}{2+\frac{1}{2}}\right)log_2\left(\frac{2}{2+\frac{1}{2}}\right)+\left(\frac{\frac{1}{2}}{2+\frac{1}{2}}\right)log_2\left(\frac{\frac{1}{2}}{2+\frac{1}{2}}\right)\right]-\left(\frac{1+\frac{1}{2}}{5}\right)$$
$$\left[\left(\frac{0}{1+\frac{1}{2}}\right)log_2\left(\frac{0}{1+\frac{1}{2}}\right)+\left(\frac{1+\frac{1}{2}}{1+\frac{1}{2}}\right)log_2\left(\frac{1+\frac{1}{2}}{1+\frac{1}{2}}\right)\right]\Bigg)$$

$= 0.97095 - 0\ 0.36096 - 0 = 0.60999.$

$$IG(S,A_4') = 0.97095 - \left(-\left(\tfrac{3}{5}\right)\left[\left(\tfrac{1}{3}\right)log_2\left(\tfrac{1}{3}\right)+\left(\tfrac{2}{3}\right)log_2\left(\tfrac{2}{3}\right)\right]\right.$$
$$\left.-\left(\tfrac{2}{5}\right)\left[\left(\tfrac{1}{2}\right)log_2\left(\tfrac{1}{2}\right)+\left(\tfrac{1}{2}\right)log_2\left(\tfrac{1}{2}\right)\right]\right)$$

$= 0.97095 - 0.55098 - 0.4 = 0.019970.$

Here the most influential is Materials and the cases are divided into two Tables, see Tables 11-12.

Table 11. Dataset with the data for cost = expensive and materials = weak

Examples	Beauty	Neighborhood	To buy
h_1	Ugly or Acceptable	Good	No
h_2	Ugly	Not good	No
h_8	Acceptable	Good	No

Table 12. Dataset with the data for cost = expensive and materials = strong

Examples	Beauty	Neighborhood	To buy
h_2	Ugly	Not good	No
h_9	Beautiful	Good	Yes
h_{11}	Acceptable	Not good	Yes

From Table 11 it is seen that it is decided Not to buy in this case.

From Table 12 we have to calculate again from which we obtain

$$Entropy(S'') = -\left(\left(\tfrac{2}{3}\right)log_2\left(\tfrac{2}{3}\right)+\left(\tfrac{1}{3}\right)log_2\left(\tfrac{1}{3}\right)\right) = 0.91830.$$

$$IG(S,A_2'') = 0.91830 - 0 - 0 - 0 = 0.91830.$$

$$IG(S,A_4'') = 0.91830$$
$$-\left(-\left(\tfrac{1}{3}\right)\left[\left(\tfrac{1}{1}\right)log_2\left(\tfrac{1}{1}\right) + \left(\tfrac{0}{1}\right)log_2\left(\tfrac{0}{1}\right)\right] - \left(\tfrac{2}{3}\right)\left[\left(\tfrac{1}{2}\right)log_2\left(\tfrac{1}{2}\right) + \left(\tfrac{1}{2}\right)log_2\left(\tfrac{1}{2}\right)\right]\right)$$

$$= 0.91830 - 0\ 0.66667 = 0.25163.$$

Here the dominant attribute is Beauty, which has a well-defined decision to make for each case.

Let us go back to Table 8 where we have

$$Entropy(S''') = -\left(\left(\frac{1+\frac{1}{2}}{2}\right)log_2\left(\frac{1+\frac{1}{2}}{2}\right) + \left(\frac{\frac{1}{2}}{2}\right)log_2\left(\frac{\frac{1}{2}}{2}\right)\right) = 0.81128.$$

$$IG(S,A_2''') = 0.81128 - \left(-\left(\tfrac{1}{2}\right)\left[\left(\tfrac{1}{1}\right)log_2\left(\tfrac{1}{1}\right) + \left(\tfrac{0}{1}\right)log_2\left(\tfrac{0}{1}\right)\right]\right.$$
$$\left. - \left(\tfrac{1}{2}\right)\left[\left(\frac{\frac{1}{2}}{1}\right)log_2\left(\frac{\frac{1}{2}}{1}\right) + \left(\frac{\frac{1}{2}}{1}\right)log_2\left(\frac{\frac{1}{2}}{1}\right)\right]\right) = 0.31128$$

In the same way it is fulfilled:

$$IG(S,A_3''') = IG(S,A_4''') = 0.31128,$$

therefore any of these attributes allows a final decision to be made. Let us take beauty as such an attribute, and see that indeterminacy occurs when the value Beautiful is reached. Therefore the user must choose what to do in this case. Furthermore, it is preferred to buy when cost is acceptable where probability is $\frac{1.5}{2} = 0.75$.

Finally, let us return to the data in Table 9.

$$Entropy(S^{iv}) = -\left(\left(\frac{3+\frac{1}{2}}{6}\right)log_2\left(\frac{3+\frac{1}{2}}{6}\right) + \left(\frac{2+\frac{1}{2}}{6}\right)log_2\left(\frac{2+\frac{1}{2}}{6}\right)\right) = 0.97987.$$

$$IG(S,A_2^{iv}) = 0.020722,$$

$$IG(S,A_3^{iv}) = 0.21319,$$

$$IG(S,A_4^{iv}) = 0.654863.$$

We choose neighborhood as the attribute that provides the most information, the data is divided into the results of the following two tables:

Table 13. Dataset with the data for neighborhood = good

Examples	Beauty	Materials	To buy
h_4	Acceptable	Weak	Yes
h_5	Beautiful	Strong	Yes
h_{10}	Acceptable	Strong	Yes

Table 14. Dataset with the data for neighborhood = not good

Examples	Beauty	Materials	To buy
h_6	Beautiful	Strong	No
h_7	Beautiful	Strong	Yes or No
h_{14}	Acceptable	Weak	No

From Table 13 there is a clear decision: always to buy when the neighborhood is Good.

Returning to Table 14 we have Entropy(S^v)=0.65002.

$$IG\left(S,A_3^v\right) = 0.54085,$$

$$IG\left(S,A_4^v\right) = 0.54085.$$

So, let us decide No with probability $\frac{2.5}{3} = 0.83333$.

Now it remains to represent the decision tree as seen in Figure 2.

Figure 2. Decision tree on whether to buy or not to buy a house according to its characteristics

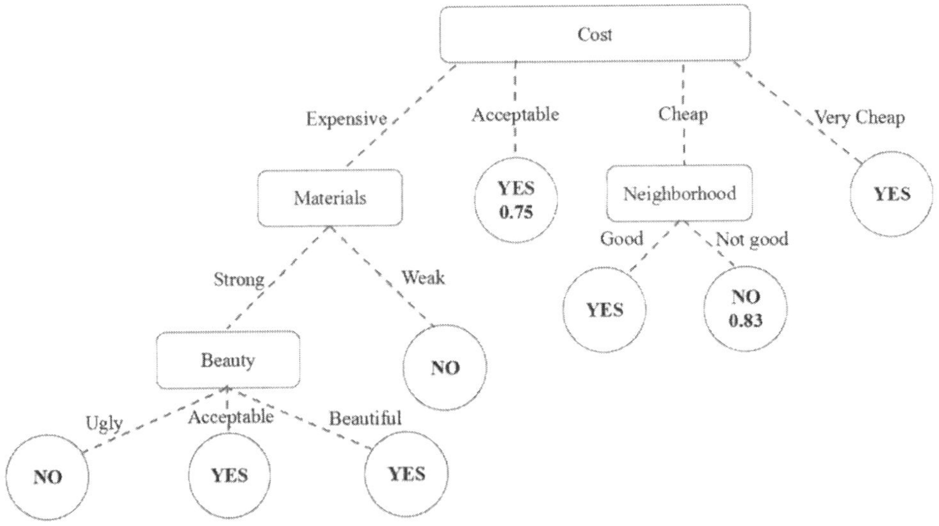

DISCUSSION

ID3 and C4.5 are some of the simplest algorithms to use for creating decision trees from examples. We have proven that they can be associated with IndetermSoft sets to generate decision trees trained for it. It is possible to incorporate other techniques such as pruning the tree or extending rules where the decision is not dichotomous.

Hitherto, the proposed algorithms present some conditions to be applied, e.g., the type of indeterminacy is of the form $F_k(v_j)= h_1$ or h_2 or…or h_m with m>1 we do not know what of them, or even $F_k(v_j)=$ not h_p among others. IndetermSoft sets are also allowed where the same object is contained in two different images of the function for different parameters. Moreover, the target class can contain indeterminacy.

In example 3, the feasibility of applying this algorithm to a house sale problem could be seen, similar to the classic example of soft sets created by Molodtsov. However, we can note that instead of more specific parameters such as wooden, cheap or expensive that constitute particular values of some attributes, we use a single variable called cost to include all these possible results. From this advantage, a greater number of possible variants of the same variable can be calculated and the algorithm processes more information, which is more reliable and permits greater uncertainty or indeterminacy.

The extended ID3 algorithm can be used when there is lack of some element, since it can be replaced by another as in the example in which the decision of a rule was made as No and Yes at the same time, and yet the algorithm can be applied. However, in this case there was indeterminacy. In any case, in these problems it is preferred to apply C4.5 algorithm.

In future works the authors intend to study new solutions to overcome the limitations of the defined algorithms. Hybridization between IndetermSoft sets can also be studied with other data mining algorithms.

CONCLUSION

IndetermSoft sets, as well as HyperSoft sets and Indeterm Hypersoft sets, among others defined by F. Smarandache, constitute new and recent variants of soft sets within the field of Neutrosophy. In this chapter, we define algorithms based on the classic ID3 and C4.5 to solve decision problems where the data are in the form of IndetermSoft sets and soft sets. To do this, we proposed some techniques to process indeterminacy and to obtain decision trees. In the end, we illustrated with an example of how these algorithms could be applied in a decision-making problem of selecting a house to buy. These algorithms can be utilized to designing recommendation systems or to solve problems under circumstances where we need to obtain decision rules containing not only uncertainty but also indeterminacy.

REFERENCES

Abbas, M., Murtaza, G., & Smarandache, F. (2020). Basic operations on hypersoft sets and hypersoft point. *Neutrosophic Sets and Systems*, *35*, 407–421.

Ahsan, M., Sarwar, M. A., Lone, S. A., Almutlak, S. A., & Anwer, S. (2023). Optimizing New Technology Implementation Through Fuzzy Hypersoft Set: A Framework Incorporating Entropy, Similarity Measure, and TOPSIS Techniques. *IEEE Access : Practical Innovations, Open Solutions*, *11*, 80680–80691. DOI: 10.1109/ACCESS.2023.3299861

AL-baker, S. F., El-henawy, I., & Mohamed, M. (2024). Pairing New Approach of Tree Soft with MCDM Techniques: Toward Advisory an Outstanding Web Service Provider Based on QoS Levels. *Neutrosophic Systems with Applications*, *14*, 17–29. DOI: 10.61356/j.nswa.2024.129

Budiman, E., Dengan, N., Kridalaksana, A. H., Wati, M., & Purnawansyah. (2018). Performance of Decision Tree C4.5 Algorithm in Student Academic Evaluation. In *Computational Science and Technology* (*Vol. 488*, pp. 380-389). Singapore: Springer Nature.

Cagman, N., Enginoglu, S., & Citak, F. (2011). Fuzzy Soft Set Theory and its Applications. *Iranian Journal of Fuzzy Systems*, *8*(3), 137–147.

García-Pichardo, V. H. (2005). *Algoritmo ID3 en la Detección de Ataques en Aplicaciones Web.* [Master Thesis, Instituto Tecnológico y de Estudios Superiores de Monterrey, Atizapán de Zaragoza, Mexico].

Gharib, M., Rajab, F., & Mohamed, M. (2023). Harnessing Tree Soft Set and Soft Computing Techniques' Capabilities in Bioinformatics: Analysis, Improvements, and Applications. *Neutrosophic Sets and Systems*, *61*, 579–597.

Gharib, M., Smarandache, F., & Mohamed, M. (2024). CSsEv: Modelling QoS Metrics in Tree Soft Toward Cloud Services Evaluator based on Uncertainty Environment. *International Journal of Neutrosophic Science*, *23*(2), 32–41. DOI: 10.54216/IJNS.230204

González-Caballero, E., & Broumi, S. (2023). New Software for Assessing Learning Skills in Education According to Models Based on Soft Sets, Grey Numbers, and Neutrosophic Numbers. In *Handbook of Research on the Applications of Neutrosophic Sets Theory and Their Extensions in Education* (pp. 260–278). IGI-Global. DOI: 10.4018/978-1-6684-7836-3.ch013

Gray, R. M. (1990). *Entropy and Information Theory*. Springer-Verlag. DOI: 10.1007/978-1-4757-3982-4

Herawan, T., & Deris, M. M. (2011). A soft set approach for association rules mining. *Knowledge-Based Systems*, *24*(1), 186–195. DOI: 10.1016/j.knosys.2010.08.005

Hssina, B., Merbouha, A., Ezzikouri, H., & Erritali, M. (2014). A comparative study of decision tree ID3 and C4.5. *International Journal of Advanced Computer Science and Applications*, *4*(2), 13–19. DOI: 10.14569/SpecialIssue.2014.040203

Maimon, O., & Rokach, L. (2010). *Data Mining and Knowledge Discovery Handbook*. Springer. DOI: 10.1007/978-0-387-09823-4

Maji, P. K., Biswas, R., & Roy, A. R. (2003). Soft Set Theory. *Computers & Mathematics with Applications (Oxford, England)*, *45*(4-5), 555–562. DOI: 10.1016/S0898-1221(03)00016-6

Molodtsov, D. (1999). Soft Set Theory-First Results. *Computers & Mathematics with Applications (Oxford, England)*, *37*(4-5), 19–31. DOI: 10.1016/S0898-1221(99)00056-5

Musa, S. Y., & Asaad, B. A. (2021). Bipolar Hypersoft Sets. *Mathematics*, *9*(15), 1826. DOI: 10.3390/math9151826

Pei, D., & Miao, D. (2005). From Soft Sets to Information Systems. *Paper presented at the 2005 IEEE International Conference on Granular Computing*. IEEE.

Sher, T., Rehman, A., & Kim, D. (2022). COVID-19 Outbreak Prediction by Using Machine Learning Algorithms. *Computers, Materials & Continua*, *74*(1), 1561–1574. DOI: 10.32604/cmc.2023.032020

Smarandache, F. (2021). Structure, NeutroStructure, and AntiStructure in Science. [IJNS]. *International Journal of Neutrosophic Science*, *13*(1), 28–33. DOI: 10.54216/IJNS.130104

Smarandache, F. (2022). Practical Applications of IndetermSoft Set and IndetermHyperSoft Set and Introduction to TreeSoft Set as an extension of the MultiSoft Set. *Neutrosophic Sets and Systems*, *51*, 939–947.

Thangavel, K., & Pethalakshmi, A. (2009). Dimensionality reduction based on rough set theory: A review. *Applied Soft Computing*, *9*(1), 1–12. DOI: 10.1016/j.asoc.2008.05.006

Voskoglou, M., Broumi, S., & Smarandache, F. (2022). A Combined Use of Soft and Neutrosophic Sets for Student Assessment with Qualitative Grades. *Journal of Neutrosophic and Fuzzy Systems*, *4*(1), 15–20. DOI: 10.54216/JNFS.040102

Voskoglou, M. G. (2022). Use of Soft Sets and the Bloom's Taxonomy for Assessing Learning Skills. *Transactions on Fuzzy Sets and Systems*, *1*(1), 106–113.

Zadeh, L. A. (1984). Fuzzy Probabilities. *Information Processing & Management*, *20*(3), 363–372. DOI: 10.1016/0306-4573(84)90067-0

ADDITIONAL READING

Bansal, M., Goyal, A., & Choudhary, A. (2022). A comparative analysis of K-nearest neighbor, genetic, support vector machine, decision tree, and long short term memory algorithms in machine learning. *Decision Analytics Journal*, *3*, 100071. DOI: 10.1016/j.dajour.2022.100071

Charbuty, B., & Abdulazeez, A. (2021). Classification based on decision tree algorithm for machine learning. *Journal of Applied Science and Technology Trends*, *2*(01), 20–28. DOI: 10.38094/jastt20165

Cintra, M. E., Monard, M. C., & Camargo, H. A. (2013). A fuzzy decision tree algorithm based on C4. 5. *Mathware & Soft Computing*, *20*(1), 56–62.

Saeed, M., Ahsan, M., Siddique, M. K., & Ahmad, M. R. (2020). *A study of the fundamentals of hypersoft set theory*. Infinite Study.

Smarandache, F. (2018). Extension of soft set to hypersoft set, and then to plithogenic hypersoft set. *Neutrosophic Sets and Systems, 22*(1), 168-170.

Smarandache, F. (2022). Introduction to the IndetermSoft Set and IndetermHyperSoft Set. *Neutrosophic Sets and Systems*, *50*, 629–650.

Smarandache, F. (2022). Soft set product extended to hypersoft set and indetermsoft set cartesian product extended to indetermhypersoft set. *Journal of Fuzzy Extension and Applications, 3*(4), 313-316.

Smarandache, F. (2023). New Types of Soft Sets: HyperSoft Set, IndetermSoft Set, IndetermHyperSoft Set, and TreeSoft Set-An Improved Version. *Neutrosophic Systems with Applications*, *8*, 35–41. DOI: 10.61356/j.nswa.2023.41

Tangirala, S. (2020). Evaluating the impact of GINI index and information gain on classification using decision tree classifier algorithm. *International Journal of Advanced Computer Science and Applications*, *11*(2), 612–619. DOI: 10.14569/IJACSA.2020.0110277

Zhou, H., Zhang, J., Zhou, Y., Guo, X., & Ma, Y. (2021). A feature selection algorithm of decision tree based on feature weight. *Expert Systems with Applications*, *164*, 113842. DOI: 10.1016/j.eswa.2020.113842

KEY TERMS AND DEFINITIONS

C4.5: This is a well-known algorithm developed by R. Quinlan to obtain decision trees from data collected on certain attributes. This is a Data Mining technique that emerged to overcome some limitations presented by ID3 algorithm. With this algorithm it is possible to obtain decision trees from data with a continuous variable, with missing data and in addition a tree pruning heuristic is incorporated that avoids the mismatches that occur in ID3 algorithm when the amount of data to process is too big.

Data Mining: It is a branch of artificial intelligence where patterns are discovered within large volumes of data. These patterns constitute understandable information or knowledge that is useful to the researcher within the branch of knowledge of the origin of such data.

Decision Tree: They are tree-shaped structures, where each leaf represents an attribute and for each edge, the different values of this attribute are proposed so that the last nodes are the values of a dichotomous target class. These values of the target class are of type Yes or No, To accept or To reject and so on. From them, it is possible to create decision rules.

Entropy: It is a concept that emerged from Thermodynamics to measure the indeterminacy in a system. It was extended to Information Theory through a well-known equation that allows the calculation of the indeterminacy regarding the information of the system, based on the probabilities of occurrence of each of the components. It is mistakenly understood popularly as a measure of disorder.

ID3: It is a well-known algorithm developed by R. Quinlan that allows the inductive creation of decision trees from examples of data collected on certain predefined attributes. It is an artificial intelligence and data mining technique that stands out for its simplicity. It is restricted to the conditions where the collected data are nominal and there cannot be missing data.

IndetermSoft Set: It is the extension of the soft sets defined by Professor F. Smarandache within the framework of Neutrosophy. Here, soft sets not only model uncertainty but also indeterminacy.

Soft Set: It is a theory developed in 1999 by the scientist D. Molodtsov to deal with problems of uncertainty. It is a generalization of fuzzy sets, where it is not necessary to define membership functions. It has a wide application in Algebra, Mathematical Analysis, Topology, Decision Making, among others.

Compilation of References

Abbas, M., Murtaza, G., & Smarandache, F. (2020). Basic operation on hypersoft sets and hypersoft point. *Neutrosophic Sets and System*, *35*, 407–421.

Abbas, M., Murtaza, G., & Smarandache, F. (2020). Basic operations on hypersoft sets and hypersoft point. *Neutrosophic Sets and Systems*, *35*, 407–421.

Abd El-Wahed, W. F., & Lee, S. M. (2006). Interactive fuzzy goal programming for multi-objective transportation problems. *Omega*, *34*(2), 158–166. DOI: 10.1016/j.omega.2004.08.006

Abdel-Basset, M., & Mohamed, M. (2021). Multicriteria group decision making based on neutrosophic analytic hierarchy process Suggested modifications. *Neutrosophic Sets and Systems, 43*.

Abdel-Basset, M., Gamal, A., Son, L.H., & Smarandache, F. (2020). A bipolar neutrosophic multi-criteria decision-making framework for professional selection. *Applied Sciences (Basel, Switzerland)*, *10*(4), 1–22.

Abdel-Basset, M., Mohamed, R., Zaied, Abd El-Nasser H., Gamal, A., & Smarandache, F. (2020). Solving the supply chain problem using the best-worst method based on a novel Plithogenic model. DOI: 10.1016/B978-0-12-819670-0.00001-9

Abdel-Basset, M., Ali, M., & Atef, A. (2020). Uncertainty assessments of linear time-cost tradeoffs using neutrosophic set. *Computers & Industrial Engineering*, *141*, 106286. DOI: 10.1016/j.cie.2020.106286

Abdel-Basset, M., Chang, V., & Gamal, A. (2019). Evaluation of the green supply chain management practices: A novel neutrosophic approach. *Computers in Industry*, *108*, 210–220. DOI: 10.1016/j.compind.2019.02.013

Abdel-Basset, M., Chang, V., Gamal, A., & Smarandache, F. (2019). An integrated neutrosophic ANP and VIKOR method for achieving sustainable supplier selection: A case study in importing field. *Computers in Industry*, *106*, 94–110. DOI: 10.1016/j.compind.2018.12.017

Abdel-Basset, M., Ding, W., Mohamed, R., & Metawa, N. (2020). An integrated plithogenic MCDM approach for financial performance evaluation of manufacturing industries. *Risk Management*, *22*, 192–218.

Abdel-Basset, M., Gamal, A., Manogaran, G., & Long, H. V. (2020). A novel group decision-making model based on neutrosophic sets for heart disease diagnosis. *Multimedia Tools and Applications*, *79*(15-16), 9977–10002. DOI: 10.1007/s11042-019-07742-7

Abdel-Basset, M., Manogaran, G., Gamal, A., & Chang, V. (2020). A Novel Intelligent Medical Decision Support Model Based on Soft Computing and IoT. *IEEE Internet of Things Journal*, *7*(5), 4160–4170. DOI: 10.1109/JIOT.2019.2931647

Abdel-Basset, M., Manogaran, G., Gamal, A., & Smarandache, F. (2019). A group decision making framework based on neutrosophic topsis approach for smart medical device selection. *Journal of Medical Systems*, *43*(2), 1–13. DOI: 10.1007/s10916-019-1156-1 PMID: 30627801

Abdel-Basset, M., & Mohamed, R. (2020). A novel plithogenic TOPSIS-CRITIC model for sustainable supply chain risk management. *Journal of Cleaner Production*, *247*, 119586.

Abdel-Basset, M., Mohamed, R., Sallam, K., & Elhoseny, M. (2020). A novel decision-making model for sustainable supply chain finance under uncertainty environment. *Journal of Cleaner Production*, *269*, 122324.

Abdel-Basset, M., Mohamed, R., Smarandache, F., & Elhoseny, M. (2021). A new decision-making model based on plithogenic set for supplier selection. *Computers, Materials & Continua*, *66*(3), 2751–2769.

Abdel-Basset, M., Saleh, M., Gamal, A., & Smarandache, F. (2019). An approach of TOPSIS technique for developing supplier selection with group decision making under type-2 neutrosophic number. *Applied Soft Computing*, *77*, 438–452. DOI: 10.1016/j.asoc.2019.01.035

Abdel-Basset, M., Smarandache, F., & Ye, J. (2019). Special issue on "Applications of neutrosophic theory in decision making-recent advances and future trends". *Complex & Intelligent Systems*, *5*(4), 363–364. DOI: 10.1007/s40747-019-00127-1

Abdel-Basst, M., Mohamed, R., & Elhoseny, M. (2020). A novel framework to evaluate innovation value proposition for smart product–service systems. *Environmental Technology & Innovation*, *20*, 101036.

AbdelMouty, A. M., Abdel-Monem, A., Aal, S. I. A., & Ismail, M. M. (2023). Analysis the Role of the Internet of Things and Industry 4.0 in Healthcare Supply Chain Using Neutrosophic Sets. *Neutrosophic Systems with Applications*, *4*, 33–42.

Abdulkareem, K. H., Arbaiy, N., Zaidan, A. A., Zaidan, B. B., Albahri, O. S., Alsalem, M. A., & Salih, M. M. (2020). A novel multiperspective benchmarking framework for selecting image dehazing intelligent algorithms based on BWM and group VIKOR techniques. *International Journal of Information Technology & Decision Making*, *19*(3), 909–957. DOI: 10.1142/S0219622020500169

Abdulkareem, K. H., Arbaiy, N., Zaidan, A. A., Zaidan, B. B., Albahri, O. S., Alsalem, M. A., & Salih, M. M. (2021). A new standardization and selection framework for real-time image dehazing algorithms from multi-foggy scenes based on fuzzy Delphi and hybrid multi-criteria decision analysis methods. *Neural Computing & Applications*, *33*(4), 1029–1054. DOI: 10.1007/s00521-020-05020-4

Abu Qamar, M., & Hassan, N. (2018). Generalized Q-neutrosophic soft expert set for decision under uncertainty. *Symmetry*, *10*(11), 621. DOI: 10.3390/sym10110621

Adam, F., & Hassan, N. (2016). Multi Q-fuzzy soft expert set and its application. *Journal of Intelligent & Fuzzy Systems*, *30*(2), 943–950. DOI: 10.3233/IFS-151816

Adhikary, K., Pal, P., & Poray, J. (2024). The Minimum Spanning Tree Problem on networks with Neutrosophic numbers. *Neutrosophic Sets and Systems*, *63*(1), 16.

Aggarwal, S., & Gupta, C. (2016). Solving intuitionistic fuzzy solid transportation problem via new ranking method based on signed distance. *International Journal of Uncertainty, Fuzziness and Knowledge-based Systems*, *24*(04), 483–501. DOI: 10.1142/S0218488516500240

Ahmad, F., Adhami, A. Y., & Smarandache, F. (2018). Single valued neutrosophic hesitant fuzzy computational algorithm for multiobjective nonlinear optimization problem. *Neutrosophic sets and systems, 22*, 76-86.

Ahmad, F., Smarandache, F., & Das, A. K. (2022). Neutrosophical fuzzy modeling and optimization approach for multiobjective four-index transportation problem.

Ahmad, B., & Kharal, A. (2009). On Fuzzy Soft Sets. *Advances in Fuzzy Systems*, *2009*, 586507.

Ahmad, F. (2022). Interactive neutrosophic optimization technique for multiobjective programming problems: An application to pharmaceutical supply chain management. *Annals of Operations Research, 311*(2), 551–585.

Ahmad, F., & Adhami, A. Y. (2019). Total cost measures with probabilistic cost function under varying supply and demand in transportation problem. *Opsearch, 56*(2), 583–602. DOI: 10.1007/s12597-019-00364-5

Ahmad, F., Adhami, A. Y., & Smarandache, F. (2020). Modified neutrosophic fuzzy optimization model for optimal closed-loop supply chain management under uncertainty. In *Optimization theory based on neutrosophic and plithogenic sets* (pp. 343–403). Elsevier. DOI: 10.1016/B978-0-12-819670-0.00015-9

Ahmad, F., & John, B. (2022). Modeling and optimization of multiobjective programming problems in neutrosophic hesitant fuzzy environment. *Soft Computing, 26*(12), 5719–5739. DOI: 10.1007/s00500-022-06953-9

Ahsan, M., Saeed, M., & Rahman, A. U. (2021). A theoretical and analytical approach for fundamental framework of composite mappings on fuzzy hypersoft classes. *Neutrosophic Sets and Systems, 45*(1), 18.

Ahsan, M., Sarwar, M. A., Lone, S. A., Almutlak, S. A., & Anwer, S. (2023). Optimizing New Technology Implementation Through Fuzzy Hypersoft Set: A Framework Incorporating Entropy, Similarity Measure, and TOPSIS Techniques. *IEEE Access : Practical Innovations, Open Solutions, 11*, 80680–80691. DOI: 10.1109/ACCESS.2023.3299861

Ahsan-ul-Haq, M. (2022). Neutrosophic Kumaraswamy Distribution with Engineering Application. *Neutrosophic Sets and Systems, 49*(1), 17.

Aiwua, Z., Jianguoa, D., & Hongjunb, G. (2015). Interval valued neutrosophic sets and multi-attribute decision-making based on generalized weighted aggregation operator. *Journal of Intelligent & Fuzzy Systems, 29*, 2697–2706.

Akbari Torkestani, J. (2013). Degree constrained minimum spanning tree problem: A learning automata approach. *The Journal of Supercomputing, 64*(1), 226–249. DOI: 10.1007/s11227-012-0851-1

Akram, M., & Adeel, A. (2019). TOPSIS approach for MAGDM based on interval-valued hesitant fuzzy N-soft environment. *International Journal of Fuzzy Systems, 21*(3), 993–1009. DOI: 10.1007/s40815-018-0585-1

Akram, M., Adeel, A., & Alcantud, J. C. R. (2019). Group decision-making methods based on hesitant N-soft sets. *Expert Systems with Applications, 115*, 95–105. DOI: 10.1016/j.eswa.2018.07.060

Akram, M., Luqman, A., & Alcantud, J. C. R. (2021). Risk evaluation in failure modes and effects analysis: Hybrid TOPSIS and ELECTRE I solutions with Pythagorean fuzzy information. *Neural Computing & Applications*, *33*(11), 5675–5703. DOI: 10.1007/s00521-020-05350-3

Aktaş, H., & Çağman, N. (2007). Soft sets and soft groups. *Information Sciences*, *177*(13), 2726–2735.

AL-baker, S. F., El-henawy, I., & Mohamed, M. (2024). Pairing New Approach of Tree Soft with MCDM Techniques: Toward Advisory an Outstanding Web Service Provider Based on QoS Levels. *Neutrosophic Systems with Applications*, *14*, 17–29. DOI: 10.61356/j.nswa.2024.129

Albassam, M., Khan, N., & Aslam, M. (2021). Neutrosophic D'Agostino Test of Normality: An Application to Water Data. *Journal of Mathematics*, *2021*, 1–5. DOI: 10.1155/2021/5582102

Alcantud, J. C. R., & Santos-García, G. (2017). A new criterion for soft set-based decision-making problems under incomplete information. *International Journal of Computational Intelligence Systems*, *10*(1), 394–404. DOI: 10.2991/ijcis.2017.10.1.27

Alcantud, J. C. R., & Torrecillas Muñoz, M. J. (2018). Intertemporal choice of fuzzy soft sets. *Symmetry*, *10*(9), 371. DOI: 10.3390/sym10090371

Alefeld, G., & Herzberger, J. (2012). *Introduction to interval computation*. Academic press.

Alhabib, R., Ranna, M. M., Farah, H., & Salama, A. (2018). Some neutrosophic probability distributions. *Neutrosophic Sets and Systems*, *22*, 30–38.

Alhasan, K. F. H., & Smarandache, F. *Neutrosophic Weibull distribution and neutrosophic family Weibull distribution*. Infinite Study.

Alhazaymeh, K., & Hassan, N. (2012). Interval-valued vague soft sets and its application. *Advances in Fuzzy Systems*, *2012*, 15–15.

Alhazaymeh, K., & Hassan, N. (2012). Possibility vague soft set and its application in decision making. *International Journal of Pure and Applied Mathematics*, *77*(4), 549–563.

Alhazaymeh, K., & Hassan, N. (2013). Possibility interval-valued vague soft set. *Applied Mathematical Sciences (Ruse)*, *7*(140), 6989–6994. DOI: 10.12988/ams.2013.310576

Alhazaymeh, K., & Hassan, N. (2014). Application of generalized vague soft expert set in decision making. *International Journal of Pure and Applied Mathematics, 93*(3), 361–367. DOI: 10.12732/ijpam.v93i3.6

Alhazaymeh, K., & Hassan, N. (2014). Generalized vague soft expert set. *International Journal of Pure and Applied Mathematics, 93*(3), 351–360.

Alhazaymeh, K., & Hassan, N. (2014). Mapping on generalized vague soft expert set. *International Journal of Pure and Applied Mathematics, 93*(3), 369–376. DOI: 10.12732/ijpam.v93i3.7

Alhazaymeh, K., Hassan, N., & Alhazaymeh, K. (2013). Generalized interval-valued vague soft set. *Applied Mathematical Sciences, 7*(140), 6983–6988.

Al-Hijjawi, S., & Alkhazaleh, S. (2023). Possibility Neutrosophic Hypersoft Set (PNHSS). *Neutrosophic Sets and Systems, 53*, 117–129.

Ali, M. I. (2011). A note on soft sets, rough soft sets and fuzzy soft sets. *Applied Soft Computing, 11*(4), 3329–3332.

Ali, M. I., Feng, F., Liu, X., Min, W. K., & Shabir, M. (2009). On some new operations in soft set theory. *Computers & Mathematics with Applications (Oxford, England), 57*(9), 1547–1553.

Alkhazaleh, S. (2015). Neutrosophic vague set theory. *Critical Review, 10*, 29–39.

Alkhazaleh, S., & Salleh, A. R. (2011). Soft Expert Sets. *Advances in Decision Sciences, 757868*, 1–13.

Alkhazaleh, S., & Salleh, A. R. (2014). Fuzzy soft expert set and its application. *Applied Mathematics, 5*(9), 1349–1368. DOI: 10.4236/am.2014.59127

Alkhazaleh, S., Salleh, A. R., & Hassan, N. (2011). Possibility fuzzy soft set. *Advances in Decision Sciences*, 2011.

Alkhazaleh, S., Salleh, A. R., Hassan, N., & Ahmad, A. G. (2010, November). Multisoft Sets. In *Proc. 2nd International Conference on Mathematical Sciences* (pp. 910-917).

Al-Qudah, Y., & Hassan, N. (2017). Bipolar fuzzy soft expert set and its application in decision making. *International Journal of Applied Decision Sciences, 10*(2), 175–191. DOI: 10.1504/IJADS.2017.084310

Al-Qudah, Y., & Hassan, N. (2017). Complex multi-fuzzy soft expert set and its application. *International Journal of Mathematics and Computer Science, 14*(1), 149–176.

Al-Quran, A., & Hassan, N. (2016). Fuzzy parameterised single valued neutrosophic soft expert set theory and its application in decision making. *International Journal of Applied Decision Science*, *9*(2), 212–227. DOI: 10.1504/IJADS.2016.080121

Al-Quran, A., & Hassan, N. (2016). Neutrosophic vague soft expert set theory. *Journal of Intelligent & Fuzzy Systems*, *30*(6), 3691–3702.

Al-Sharqi, , A. A.-Q. (2021). Interval-valued complex neutrosophic soft set and its applications in decision-making. *Neutrosophic Sets and Systems*, *40*, 149–168.

Aneja, Y. P., & Nair, K. P. (1979). Bicriteria transportation problem. *Management Science*, *25*(1), 73–78. DOI: 10.1287/mnsc.25.1.73

Ansari, Z. N., & Kant, R. (2021). A plithogenic based neutrosophic analytic hierarchy process framework to analyse the barriers hindering adoption of eco-innovation practices in supply chain. *International Journal of Sustainable Engineering*, *14*(6), 1509–1524.

Antony Crispin Sweety, A., & Janshi, R. (2021). Fermatean Neutrosophic Sets. *IJARCCE*, *10*(6), 24–27.

Arar, M., & Jafari, S. (2020). Neutrosophic μ-Topological spaces. *Neutrosophic Sets and Systems*, *38*, 51–66.

Arasteh, A. (2020). Supply chain management under uncertainty with the combination of fuzzy multi-objective planning and real options approaches. *Soft Computing*, *24*(7), 5177–5198. DOI: 10.1007/s00500-019-04271-1

Archana, A., Deshpande, O. & Chaudhari. (2020). Fuzzy Approach to Compare a Minimal Spanning Tree Problem by Using Various Algorithm. *Advances in Fuzzy Mathematics*, *15*, 47–58.

Asady, B., & Zendehnam, A. (2007). Ranking fuzzy numbers by distance minimization. *Applied Mathematical Modelling*, *31*(11), 2589–2598. DOI: 10.1016/j.apm.2006.10.018

Asif, M., Akram, M., & Ali, G. (2020). Pythagorean fuzzy matroids with application. *Symmetry*, *12*(3), 423. DOI: 10.3390/sym12030423

Aslam, M. (2019). Neutrosophic analysis of variance: Application to university students. *Complex & Intelligent Systems*, *5*(4), 403–407. DOI: 10.1007/s40747-019-0107-2

Aslam, M. (2021a). Analyzing wind power data using analysis of means under neutrosophic statistics. *Soft Computing*, *25*(10), 7087–7093. DOI: 10.1007/s00500-021-05661-0

Aslam, M. (2021b). Neutrosophic statistical test for counts in climatology. *Scientific Reports, 11*(1), 1–5. DOI: 10.1038/s41598-021-97344-x PMID: 34497328

Aslam, M. (2021c). A study on skewness and kurtosis estimators of wind speed distribution under indeterminacy. *Theoretical and Applied Climatology, 143*(3), 1227–1234. DOI: 10.1007/s00704-020-03509-5

Aslam, M. A. (2020). Neutrosophic Rayleigh distribution with some basic properties and application. In *Neutrosophic Sets in Decision Analysis and Operations Research* (pp. 119–128). IGI Global. DOI: 10.4018/978-1-7998-2555-5.ch006

Aslam, M., Khan, N., & Al-Marshadi, A. H. (2019). Design of variable sampling plan for pareto distribution using neutrosophic statistical interval method. *Symmetry, 11*(1), 80. DOI: 10.3390/sym11010080

Atanassov, K. T. (1986). Intuitionistic fuzzy sets. *Fuzzy Sets and Systems, 20*(1), 87–96. DOI: 10.1016/S0165-0114(86)80034-3

Atanassov, K. T. (2012). *On intuitionistic fuzzy sets theory* (Vol. 283). Springer.

Atanassov, K. T., & Gargov, G. (1989). Interval valued intuitionistic fuzzy sets. *Fuzzy Sets and Systems, 31*, 343–349.

Aytekin, A., Okoth, B. O., Korucuk, S., Karamaşa, Ç., & Tirkolaee, E. B. (2022). A neutrosophic approach to evaluate the factors affecting performance and theory of sustainable supply chain management: Application to textile industry. *Management Decision, 61*(2), 506–529.

Bageerathi, K., & Jeya Puvaneswari, P. (2019). Neutrosophic Feebly Connectedness and Compactness. *IOSR Journal of Polymer and Textile Engineering, 6*(3), 7–13.

Balachandran, V., & Perry, A. (1976). Transportation type problems with quantity discounts. *Naval Research Logistics Quarterly, 23*(2), 195–209. DOI: 10.1002/nav.3800230203

Balinski, M. L. (1961). Fixed-cost transportation problems. *Naval Research Logistics Quarterly, 8*(1), 41–54. DOI: 10.1002/nav.3800080104

Bao, C. P., Tsai, M. C., & Tsai, M. I. (2007). A new approach to study the multi-objective assignment problem. *WHAMPOA-An Interdisciplinary Journal, 53*, 123-132.

Basha, S. H., Tharwat, A., Ahmed, K., & Hassanien, A. E. (2018). A predictive model for seminal quality using neutrosophic rule-based classification system. International conference on advanced intelligent systems and informatics

Bashir, M., & Salleh, A.R. (2012). Fuzzy parameterized soft expert set. Abstract and Applied Analysis. DOI: 10.1155/2012/258361

Bashir, M., & Salleh, A. R. (2012). Possibility fuzzy soft expert set. *Ozean Journal of Applied Sciences*, *12*, 208–211.

Bashir, M., Salleh, A. R., & Alkhazaleh, S. (2012). Possibility intuitionistic fuzzy soft set. *Advances in Decision Sciences*, 2012.

Belacela, N., & Boulasselb, M. R. (2001). Multi-criteria fuzzy assignment problem: A useful tool to assist medical diagnosis. *Artificial Intelligence in Medicine*, *21*, 201–207.

Bellman, R. E., & Zadeh, L. A. (1970). Decision-making in a fuzzy environment. *Management Science*, *17*(4), B-141.

Bera, T., & Mahapatra, N. K. (2022). Neutrosophy-based transportation problem and its solution approach. *International Journal of Mathematics in Operational Research*, *22*(2), 252–281. DOI: 10.1504/IJMOR.2022.124041

Bharati, S. K. (2021). Transportation problem with interval-valued intuitionistic fuzzy sets: Impact of a new ranking. *Progress in Artificial Intelligence*, *10*(2), 129–145. DOI: 10.1007/s13748-020-00228-w

Bhardwaj, N., & Sharma, P. (2021). An advanced uncertainty measure using fuzzy soft sets: Application to decision-making problems. *Big Data Mining and Analytics*, *4*(2), 94–103. DOI: 10.26599/BDMA.2020.9020020

Bhat, A. S. (2021). A novel score and accuracy function for neutrosophic sets and their real- world applications to multi-criteria decision-making process. *Neutrosophic Sets and Systems*, *41*, 168–197.

Bhatia, H. L. (1981). Indefinite quadratic solid transportation problem. *Journal of Information and Optimization Sciences*, *2*(3), 297–303. DOI: 10.1080/02522667.1981.10698711

Bilgen, B. (2010). Application of fuzzy mathematical programming approach to the production allocation and distribution supply chain network problem. *Expert Systems with Applications*, *37*(6), 4488–4495. DOI: 10.1016/j.eswa.2009.12.062

Biswas, P., Pramanik, S., & Giri, B. C. (2016). Topsismethod for multi-attribute-groupdecision-makingundersingle-valuedneutrosophicenvironment. *Neural Computing & Applications*, *27*(3), 727–737. DOI: 10.1007/s00521-015-1891-2

Biswas, R. (1990). Fuzzy subgroups and anti fuzzy subgroups. *Fuzzy Sets and Systems*, *35*(1), 121–124. DOI: 10.1016/0165-0114(90)90025-2

Bit, A. K., Biswal, M. P., & Alam, S. (1992). Fuzzy programming approach to multicriteria decision making transportation problem. *Fuzzy Sets and Systems*, *50*(2), 135–141.

Brauers, W. K., & Zavadskas, E. K. (2006). The MOORA method and its application to privatization in a transition economy. *Control and Cybernetics*, *35*(2), 445–469.

Broumi, S., & Smarandache, F. (2015). New operations on interval neutrosophic sets. *Journal of New Theory*, (1), 24-37.

Broumi, S., Bakali, A., Talea, M., Smarandache, F., & Uluçay, V. (2018). Minimum spanning tree in trapezoidal fuzzy neutrosophic environment. In *Innovations in Bio-Inspired Computing and Applications: Proceedings of the 8th International Conference on Innovations in Bio-Inspired Computing and Applications (IBICA 2017)* (pp. 25-35). Springer International Publishing. DOI: 10.1007/978-3-319-76354-5_3

Broumi, S., Talea, M., Bakali, A., & Smarandache, F. (2016). Single valued neutrosophic graphs. *Journal of New theory*, (10), 86-101.

Broumi, S. (2023). Fermatean Neutrosophic Matrices and Their Basic Operations. *Neutrosophic Sets and Systems*, *58*(1), 35.

Broumi, S., Bakali, A., Talea, M., & Smarandache, F. (2019). Computation of shortest path problem in a network with sv-triangular neutrosophic numbers. *Uluslararası Yönetim Bilişim Sistemleri Ve Bilgisayar Bilimleri Dergisi*, *3*(2), 41–51. DOI: 10.33461/uybisbbd.588290

Broumi, S., Bakali, A., Talea, M., Smarandache, F., Dey, A., & Son, L. H. (2018). Spanning tree problem with neutrosophic edge weights. *Procedia Computer Science*, *127*, 190–199. DOI: 10.1016/j.procs.2018.01.114

Broumi, S., Raut, P. K., & Behera, S. P. (2023). Solving shortest path problems using an ant colony algorithm with triangular neutrosophic arc weights. *International Journal of Neutrosophic Science*, *20*(4), 128–137.

Broumi, S., Smarandache, F., Talea, M., & Bakali, A. (2016). An introduction to bipolar single valued neutrosophic graph theory. *Applied Mechanics and Materials*, *841*, 184–191. DOI: 10.4028/www.scientific.net/AMM.841.184

Broumi, S., Talea, M., Bakali, A., Smarandache, F., & Kishore Kumar, P. K. (2017). Shortest path problem on single valued neutrosophic graphs. *2017 International Symposium on Networks, Computers and Communications (ISNCC)*. IEEE. DOI: 10.1109/ISNCC.2017.8071993

Budiman, E., Dengan, N., Kridalaksana, A. H., Wati, M., & Purnawansyah. (2018). Performance of Decision Tree C4.5 Algorithm in Student Academic Evaluation. In *Computational Science and Technology* (*Vol. 488*, pp. 380-389). Singapore: Springer Nature.

Bukajh J.S. (1998). *Operations Research Book translated into Arabic*. The Arab Center for Arabization, Translation, Authoring and Publishing -Damascus.

Cagman, N., Enginoglu, S., & Citak, F. (2011). Fuzzy soft set theory and its applications. *Iranian Journal of Fuzzy Systems, 8*(3), 137-147.

Cagman, N., Enginoglu, S., & Citak, F. (2011). Fuzzy Soft Set Theory and its Applications. *Iranian Journal of Fuzzy Systems*, *8*(3), 137–147.

Celik, Y., Ekiz, C., & Yamak, S. (2013). Applications of fuzzy soft sets in ring theory. *Annals Fuzzy Mathematics and Informatics*, *5*(3), 451–462.

Chakraborty, A., Mondal, S., & Broumi, S. (2019). *De-neutrosophication technique of pentagonal neutrosophic number and application in minimal spanning tree*. Infinite Study.

Chanas, S., & Kuchta, D. (1996). A concept of the optimal solution of the transportation problem with fuzzy cost coefficients. *Fuzzy Sets and Systems*, *82*(3), 299–305. DOI: 10.1016/0165-0114(95)00278-2

Chanas, S., & Zieliński, P. (2001). Critical path analysis in the network with fuzzy activity times. *Fuzzy Sets and Systems*, *122*(2), 195–204. DOI: 10.1016/S0165-0114(00)00076-2

Chang, P. T., & Lee, E. S. (1999). Fuzzy decision networks and deconvolution. *Computers & Mathematics with Applications (Oxford, England)*, *37*(11-12), 53–63. DOI: 10.1016/S0898-1221(99)00143-1

Charnes, A., & Cooper, W. W. (1957). Management models and industrial applications of linear programming. *Management Science*, *4*(1), 38–91. DOI: 10.1287/mnsc.4.1.38

Chen, C. T. (2000). Extensions of the topsis for group decision-making under fuzzy environment. *Fuzzy Sets and Systems*, *114*(1), 1–9. DOI: 10.1016/S0165-0114(97)00377-1

Chen, D., Tsang, E. C. C., Yeung, D. S., & Wang, X. (2005). The parameterization reduction of soft sets and its applications. *Computers & Mathematics with Applications (Oxford, England)*, *49*(5-6), 757–763. DOI: 10.1016/j.camwa.2004.10.036

Chen, S. J., Hwang, C. L., Chen, S. J., & Hwang, C. L. (1992). *Fuzzy multiple attribute decision making methods*. Springer Berlin Heidelberg. DOI: 10.1007/978-3-642-46768-4

Chen, S. P. (2007). Analysis of critical paths in a project network with fuzzy activity times. *European Journal of Operational Research, 183*(1), 442–459. DOI: 10.1016/j.ejor.2006.06.053

Chen, S.-M., & Chang, T.-H. (2001). Finding multiple possible critical paths using fuzzy PERT. *IEEE Transactions on Systems, Man, and Cybernetics. Part B, Cybernetics, 31*(6), 930–937. DOI: 10.1109/3477.969496 PMID: 18244858

Chen, S.-P., & Chang, P.-C. (2006). A mathematical programming approach to supply chain models with fuzzy parameters. *Engineering Optimization, 38*(6), 647–669. DOI: 10.1080/03052150600716116

Chen, W., Shen, Y., & Wang, Y. (2018). Evaluation of economic transformation and upgrading of resource-based cities in Shaanxi province based on an improved TOPSIS method. *Sustainable Cities and Society, 37*, 232–240. DOI: 10.1016/j.scs.2017.11.019

Chen, Y., Liu, J., Chen, Z., & Zhang, Y. (2020). Group decision-making method based on generalized vague N-soft sets. In *Chinese Control And Decision Conference (CCDC),* (pp. 4010–4015). IEEE. DOI: 10.1109/CCDC49329.2020.9164602

Chhibber, D., Bisht, D. C., & Srivastava, P. K. (2021). Pareto-optimal solution for fixed-charge solid transportation problem under intuitionistic fuzzy environment. *Applied Soft Computing, 107*, 107368. DOI: 10.1016/j.asoc.2021.107368

Chinnadurai, V., & Arulselvam, A. (2021). Pythagorean Neutrosophic Ideals in Semigroups. *Neutrosophic Sets and Systems, 41*, 258–269.

Chinneck, J. W., & Ramadan, K. (2000). Linear programming with interval coefficients. *The Journal of the Operational Research Society, 51*(2), 209–220. DOI: 10.1057/palgrave.jors.2600891

Chopra, S., & Meindl, P. (2007). *Supply chain management. Strategy, planning & operation*. Springer.

Choudhary, A., & Yadav, S. P. (2022). An approach to solve interval valued intuitionistic fuzzy transportation problem of Type-2. *International Journal of System Assurance Engineering and Management, 13*(6), 2992–3001. DOI: 10.1007/s13198-022-01771-6

Chu, P. C., & Beasley, J. E. (1997). A general algorithm for the generalized assignment problem. *Computers & Operations Research*, *24*, 17–23.

Colhon, M., Tilea, M., González-Marcos, A., Reşceanu, A., Smarandache, F., & Navaridas-Nalda, F. (2023). A Neutrosophic Decision-Making Model for Determining Young People's Active Engagement. *International Journal of Information Technology & Decision Making*.

Csaszar, A. (2002). Generalized topology, generalized continuity. *Acta Mathematica Hungarica*, *96*, 351–357.

Csaszar, A. (2004). Extremally disconnected generalized topologies. *Annales Univ. Sci. Budapest.*, *47*, 91–96.

Dahiya, K., & Verma, V. (2007). Capacitated transportation problem with bounds on RIM conditions. *European Journal of Operational Research*, *178*(3), 718–737.

Dalkılıç, O. (2021). A novel approach to soft set theory in decision-making under uncertainty. *International Journal of Computer Mathematics*, *98*(10), 1935–1945. DOI: 10.1080/00207160.2020.1868445

Dalkiliç, O. (2022). On topological structures of virtual fuzzy parametrized fuzzy soft sets. *Complex & Intelligent Systems*, *8*(1), 337–348. DOI: 10.1007/s40747-021-00378-x

Dalkılıç, O., & Demirtaş, N. (2021). Bipolar fuzzy soft D-metric spaces. *Communications Faculty of Sciences University of Ankara Series A1 Mathematics and Statistics*, *70*(1), 64–73. DOI: 10.31801/cfsuasmas.774658

Dalkılıç, O., & Demirtaş, N. (2023). Algorithms for Covid-19 outbreak using soft set theory: Estimation and application. *Soft Computing*, *27*(6), 3203–3211. DOI: 10.1007/s00500-022-07519-5 PMID: 36268457

Darvish Falehi, A., & Torkaman, H. (2023). Optimal fractional order interval type-2 fuzzy controller for upside-down asymmetric multilevel inverter based dynamic voltage restorer to accurately compensate faulty network voltage. *Journal of Ambient Intelligence and Humanized Computing*, *14*(12), 16683–16701. DOI: 10.1007/s12652-023-04673-y

Das, A. K., & Granados, C. (2021). An advanced approach to fuzzy soft group decision-making using weighted average ratings. *SN Computer Science*, *2*(6), 471. DOI: 10.1007/s42979-021-00873-5

Das, A. K., & Granados, C. (2022). A new fuzzy parameterized intuitionistic fuzzy soft multiset theory and group decision-making. *Journal of Current Science and Technology*, *12*, 547–567.

Das, A. K., & Granados, C. (2023). IFP-intuitionistic multi fuzzy N-soft set and its induced IFP-hesitant N-soft set in decision-making. *Journal of Ambient Intelligence and Humanized Computing*, *14*(8), 10143–10152. DOI: 10.1007/s12652-021-03677-w

Das, A. K., Granados, C., & Bhattacharya, J. (2022). Some new operations on fuzzy soft sets and their applications in decision-making. *Songklanakarin Journal of Science and Technology*, *44*(2), 440–449.

Das, A. K., Gupta, N., & Granados, C. (2023). Weighted hesitant bipolar-valued fuzzy soft set in decision-making. *Songklanakarin Journal of Science and Technology*, *45*(6), 681–690.

Das, S. K., Goswami, A., & Alam, S. S. (1999). Multiobjective transportation problem with interval cost, source and destination parameters. *European Journal of Operational Research*, *117*(1), 100–112. DOI: 10.1016/S0377-2217(98)00044-7

de Almeida, T. A., Yamakami, A., & Takahashi, M. T. (2005, November). An evolutionary approach to solve minimum spanning tree problem with fuzzy parameters. In *International Conference on Computational Intelligence for Modelling, Control and Automation and International Conference on Intelligent Agents, Web Technologies and Internet Commerce (CIMCA-IAWTIC'06)* (Vol. 2, pp. 203-208). IEEE.

Debnath, S. (2021). Fuzzy hypersoft sets and its weightage operator for decision making. *Journal of Fuzzy Extension and Applications*, *2*(2), 163–170.

Deb, S. C., & Islam, S. (2023). Application of Neutrosophic Interval valued Goal Programming to a Supply Chain Inventory Model for Deteriorating Items with Time Dependent Demand. *Neutrosophic Sets and Systems*, *53*(1), 35.

Deli, I., Ali, M., & Smarandache, F. (2015, August). Bipolar neutrosophic sets and their application based on multi-criteria decision making problems. In *2015 International conference on advanced mechatronic systems (ICAMechS)* (pp. 249-254). IEEE. DOI: 10.1109/ICAMechS.2015.7287068

Deng, H., Yeh, C. H., & Willis, R. J. (2000). Inter-company comparison using modified TOPSIS with objective weights. *Computers & Operations Research*, *27*(10), 963–973. DOI: 10.1016/S0305-0548(99)00069-6

De, P. K., & Yadav, B. (2011). An algorithm to solve multi-objective assignment problem using interactive fuzzy goal programming approach. *International Journal of Contemporary Mathematical Sciences*, *6*(34), 1651–1662.

Deveci, K., Cin, R., & Kağızman, A. (2020). A modified interval valued intuitionistic fuzzy CODAS method and its application to multi-criteria selection among renewable energy alternatives in Turkey. *Applied Soft Computing*, *96*, 106660. DOI: 10.1016/j.asoc.2020.106660

Dıaz, J. A., & Fernández, E. (2001). A tabu search heuristic for the generalized assignment problem. *European Journal of Operational Research*, *132*(1), 22–38.

Dohale, V., Ambilkar, P., Kumar, A., Mangla, S. K., & Bilolikar, V. (2023). Analyzing the enablers of circular supply chain using Neutrosophic-ISM method: Lessons from the Indian apparel industry. *International Journal of Logistics Management*, *34*(3), 611–643.

Dong, B., Duan, M., & Li, Y. (2022). Exploration of Joint Optimization and Visualization of Inventory Transportation in Agricultural Logistics Based on Ant Colony Algorithm. *Computational Intelligence and Neuroscience*. PMID: 35755759

Dorfeshan, Y., & Mousavi, S. M. (2019). A group TOPSIS-COPRAS methodology with Pythagorean fuzzy sets considering weights of experts for project critical path problem. *Journal of Intelligent & Fuzzy Systems*, *36*(2), 1375–1387. DOI: 10.3233/JIFS-172252

Dubois, D. J. (1980). *Fuzzy sets and systems: theory and applications* (Vol. 144). Academic press.

Dubois, D., Fargier, H., & Galvagnon, V. (2003). On latest starting times and floats in activity networks with ill-known durations. *European Journal of Operational Research*, *147*(2), 266–280. DOI: 10.1016/S0377-2217(02)00560-X

Dubois, D., & Prade, H. (1988c). Default reasoning and possibility theory. *Artificial Intelligence*, *35*(2), 243–257. DOI: 10.1016/0004-3702(88)90014-8

Ebrahimnejad, A., & Verdegay, J. L. (2018). A new approach for solving fully intuitionistic fuzzy transportation problems. *Fuzzy Optimization and Decision Making*, *17*(4), 447–474. DOI: 10.1007/s10700-017-9280-1

Edalatpanah, S. (2020). Data envelopment analysis based on triangular neutrosophic numbers. *CAAI Transactions on Intelligence Technology*, *5*(2), 94–98. DOI: 10.1049/trit.2020.0016

Ehrgott, M., Gandibleux, X., & Przybylski, A. (2016). Exact methods for multi-objective combinatorial optimisation. *Multiple criteria decision analysis: State of the art surveys*, 817-850.

Ehrgott, M. (2005). *Multicriteria optimization* (Vol. 491). Springer Science & Business Media.

El Sayed, M. A., & Baky, I. A. (2023). Multi-choice fractional stochastic multi-objective transportation problem. *Soft Computing*, *27*(16), 11551–11567. DOI: 10.1007/s00500-023-08101-3

Emrouznejad, A., & Zerafat Angiz,, L. M., & Ho, W. (2012). An alternative formulation for the fuzzy assignment problem. *The Journal of the Operational Research Society*, *63*(1), 59–63.

Fallah, M., & Nozari, H. (2021). Neutrosophic mathematical programming for optimization of multi-objective sustainable biomass supply chain network design. *Computer Modeling in Engineering & Sciences*, *129*(2), 927–951. DOI: 10.32604/cmes.2021.017511

Farid, H. M. A., & Riaz, M. (2023). Single-valued neutrosophic dynamic aggregation information with time sequence preference for IoT technology in supply chain management. *Engineering Applications of Artificial Intelligence*, *126*, 106940.

Faris, & Smarandache. (2016). New Neutrosophic Sets via Neutrosophic Topological Spaces. *New Trends in Neutrosophic Theory and Applications*, (2), 1–10.

Farnam, M., Darehmiraki, M., & Behdani, Z. (2024). Neutrosophic data envelopment analysis based on parametric ranking method. *Applied Soft Computing*, *153*, 111297. DOI: 10.1016/j.asoc.2024.111297

Fatimah, F., & Alcantud, J. C. R. (2021). The multi-fuzzy N-soft set and its applications to decision-making. *Neural Computing & Applications*, *33*(17), 11437–11446. DOI: 10.1007/s00521-020-05647-3

Fletcher, C. R. (1986). Applied abstract algebra, by R. Lidl and G. Pilz. Pp 545. DM 136. 1984. ISBN 3-540-96035-X (Springer). *Mathematical Gazette*, *70*(453), 246–247. DOI: 10.2307/3615715

Gaber, M., Alharbi, M. G., Dagestani, A. A., & Ammar, E. S. (2021). Optimal Solutions for Constrained Bimatrix Games with Payoffs Represented by Single-Valued Trapezoidal Neutrosophic Numbers. *Journal of Mathematics*, *2021*, 1–13. DOI: 10.1155/2021/5594623

Gai, L., Liu, H. C., Wang, Y., & Xing, Y. (2023). Green supplier selection and order allocation using linguistic Z-numbers MULTIMOORA method and bi-objective non-linear programming. *Fuzzy Optimization and Decision Making*, *22*(2), 267–288. DOI: 10.1007/s10700-022-09392-1

García-Pichardo, V. H. (2005). *Algoritmo ID3 en la Detección de Ataques en Aplicaciones Web.* [Master Thesis, Instituto Tecnológico y de Estudios Superiores de Monterrey, Atizapán de Zaragoza, Mexico].

Garfinkel, R. S., & Rao, M. R. (1971). The bottleneck transportation problem. *Naval Research Logistics Quarterly*, *18*(4), 465–472. DOI: 10.1002/nav.3800180404

Garg, H. (2016). An improved score function for ranking neutrosophic sets and its application to decision-making process. *International Journal for Uncertainty Quantification*, *6*(5).

Gau, W. L., & Buehrer, D. J. (1993). Vague sets. *IEEE Transactions on Systems, Man, and Cybernetics*, *23*(2), 610–614.

Gayen, S., Smarandache, F., Jha, S., Singh, M. K., Broumi, S., & Kumar, R. (2020). Introduction to Plithogenic Subgroup. In Smarandache, F., & Broumi, S. (Eds.), *Neutrosophic Graph Theory and Algorithms* (pp. 213–259). IGI Global. DOI: 10.4018/978-1-7998-1313-2.ch008

Geetha, S., VLG, N., & Ponalagusamy, R. (2013). Multi-criteria interval valued intuitionistic fuzzy decision making using a new score function. *KIM 2013 knowledge and information management conference.*

Geetha, S., & Nair, K. P. K. (1993). A variation of the assignment problem. *European Journal of Operational Research*, *68*(3), 422–426.

Gharib, M., Rajab, F., & Mohamed, M. (2023). Harnessing Tree Soft Set and Soft Computing Techniques' Capabilities in Bioinformatics: Analysis, Improvements, and Applications. *Neutrosophic Sets and Systems*, *61*, 579–597.

Gharib, M., Smarandache, F., & Mohamed, M. (2024). CSsEv: Modelling QoS Metrics in Tree Soft Toward Cloud Services Evaluator based on Uncertainty Environment. *International Journal of Neutrosophic Science*, *23*(2), 32–41. DOI: 10.54216/IJNS.230204

Gholamian, N., Mahdavi, I., Tavakkoli-Moghaddam, R., & Mahdavi-Amiri, N. (2015). Comprehensive fuzzy multi-objective multi-product multi-site aggregate production planning decisions in a supply chain under uncertainty. *Applied Soft Computing*, *37*, 585–607. DOI: 10.1016/j.asoc.2015.08.041

Ghosh, S., & Roy, S. K. (2023). Closed-loop multi-objective waste management through vehicle routing problem in neutrosophic hesitant fuzzy environment. *Applied Soft Computing*, *148*, 110854. DOI: 10.1016/j.asoc.2023.110854

Ghosh, S., Roy, S. K., & Verdegay, J. L. (2022). Fixed-charge solid transportation problem with budget constraints based on carbon emission in neutrosophic environment. *Soft Computing*, *26*(21), 11611–11625. DOI: 10.1007/s00500-022-07442-9

Gilchrist, W. (2000). *Statistical modelling with quantile functions*. CRC Press. DOI: 10.1201/9781420035919

Giri, B. K., & Roy, S. K. (2022). Neutrosophic multi-objective green four-dimensional fixed-charge transportation problem. *International Journal of Machine Learning and Cybernetics*, *13*(10), 3089–3112. DOI: 10.1007/s13042-022-01582-y

Gomathy, S., Deivanayagampillai, N., & Said, B. (2002). Plithogenic sets and their application in decision making. *Neutrosophic Sets and Systems.*, *38*. Advance online publication. DOI: 10.5281/zenodo.4300565

Gonul Bilgin, N., Pamučar, D., & Riaz, M. (2022). Fermatean neutrosophic topological spaces and an application of neutrosophic kano method. *Symmetry*, *14*(11), 2442. DOI: 10.3390/sym14112442

González-Caballero, E., & Broumi, S. (2023). New Software for Assessing Learning Skills in Education According to Models Based on Soft Sets, Grey Numbers, and Neutrosophic Numbers. In *Handbook of Research on the Applications of Neutrosophic Sets Theory and Their Extensions in Education* (pp. 260–278). IGI-Global. DOI: 10.4018/978-1-6684-7836-3.ch013

Gope, S., & Das, S. (2014). Fuzzy Dominance Matrix and its Application in Decision Making Problems. *International Journal of Soft Computing and Engineering, 4*.

Granados, C., Das, A.K. & Das, B. (2022). Some continuous neutrosophic distributions with neutrosophic parameters based on neutrosophic random variables. *Advances in the Theory of Nonlinear Analysis and its Applications, 33*(2) 293-308.

Granados, C., Das, A.K. & Osu, B. (2023). Weighted neutrosophic soft multiset and its application to decision making. *Yugoslav Journal of Operations Research, 33*(2), 293-308.

Gray, P. (1971). Exact solution of the fixed-charge transportation problem. *Operations Research*, *19*(6), 1529–1538. DOI: 10.1287/opre.19.6.1529

Gray, R. M. (1990). *Entropy and Information Theory*. Springer-Verlag. DOI: 10.1007/978-1-4757-3982-4

Grida, M., Mohamed, R., & Zaied, , A. N. (2020). A Novel Plithogenic MCDM Framework for Evaluating the Performance of IoT Based Supply Chain. *Neutrosophic Sets and Systems, 33*, 321–338.

Gul, S. (2021). Fermatean fuzzy set extensions of SAW, ARAS, and VIKOR with applications in COVID-19 testing laboratory selection problem. *Expert Systems: International Journal of Knowledge Engineering and Neural Networks, 38*(8), e12769. DOI: 10.1111/exsy.12769 PMID: 34511690

Gümüs, A. T., & Güneri, A. F. (2007). Multi-echelon inventory management in supply chains with uncertain demand and lead times: Literature review from an operational research perspective. *Proceedings of the Institution of Mechanical Engineers. Part B, Journal of Engineering Manufacture, 221*(10), 1553–1570. DOI: 10.1243/09544054JEM889

Gupta, G., Shivani, & Rani, D. (2024). Neutrosophic goal programming approach for multi-objective fixed-charge transportation problem with neutrosophic parameters. *OPSEARCH*, 1-27. DOI: 10.1007/s12597-024-00747-3

Habiba, U., & Quddoos, A. (2022). Pentagonal Neutrosophic Transportation Problems with Interval Cost. *Neutrosophic Sets and Systems, 51*(i), 896–90. DOI: 10.5281/zenodo.7135436

Haddad, H., Mohammadi, H., & Pooladkhan, H. (2002). Two models for the generalized assignment problem in uncertain environment. *Management Science Letters, 2*(2), 623-630.

Haiyan, Z., & Jingjing, J. (2015, December). Fuzzy soft relation and its application in decision making. In 2015 7th International Conference on Modelling, Identification and Control (ICMIC) (pp. 1-4). IEEE.

Hamadameen, O. A. (2018). A noval technique for solving multi- objective linear programming problems. *Aro - The Scientific Journal of Koya University, 5*(2), 1-8

Hammer, P. L. (1969). Time-minimizing transportation problems. *Naval Research Logistics Quarterly, 16*(3), 345–357. DOI: 10.1002/nav.3800160307

Hammer, P. L. (1971). Communication on "the bottleneck transportation problem" and "some remarks on the time transportation problem". *Naval Research Logistics Quarterly, 18*(4), 487–490. DOI: 10.1002/nav.3800180406

Hamou, A., & El-Amine, C. (2018) An exact method for the multi- objective assignment problem. *Les Annales RECITS, 5*, 31-36.

Han, J., Yang, C., Lim, C.-C., Zhou, X., Shi, P., & Gui, W. (2020). Power scheduling optimization under single-valued neutrosophic uncertainty. *Neurocomputing*, *382*, 12–20. DOI: 10.1016/j.neucom.2019.11.089

Han, L., & Wei, C. (2020). An extended EDAS method for multicriteria decision-making based on multivalued neutrosophic sets. *Complexity*, *2020*, 1–9. DOI: 10.1155/2020/7578507

Hapke, M., & Slowinski, R. (1996). Fuzzy priority heuristics for project scheduling. *Fuzzy Sets and Systems*, *83*(3), 291–299. DOI: 10.1016/0165-0114(95)00338-X

Hassan, M. R. (2012). An efficient method to solve least-cost minimum spanning tree (LC-MST) problem. *Journal of King Saud University. Computer and Information Sciences*, *24*(2), 101–105. DOI: 10.1016/j.jksuci.2011.12.001

Hassan, N., Ulucay, V., & Sahin, M. (2018). Q-neutrosophic soft expert set and its application in decision making. *International Journal of Fuzzy System Applications*, *7*(4), 37–61. DOI: 10.4018/IJFSA.2018100103

Hazaymeh, A.A., Abdullah, I.B., Balkhi, Z.T., & Ibrahim, R.I. (2012). Generalized fuzzy soft expert set. *Journal of Applied Mathematics*.

He, T., Wei, G., Wei, C., & Wang, J. (2019). CODAS method for Pythagorean 2-tuple linguistic multiple attribute group decision making. *IEEE Access*. IEEE.

Hema, R., Sudharani, R., & Kavitha, M. (2023). A Novel Approach on Plithogenic Interval Valued Neutrosophic Hyper-soft Sets and its Application in Decision Making. *Indian Journal of Science and Technology*, *16*(32), 2494–2502.

Herawan, T., & Deris, M. M. (2011). A soft set approach for association rules mining. *Knowledge-Based Systems*, *24*(1), 186–195. DOI: 10.1016/j.knosys.2010.08.005

Hezam, I. M., Nayeem, M. K., Foul, A., & Alrasheedi, A. F. (2021). Covid-19 vaccine: A neutrosophic mcdm approach for determining the priority groups. *Results in Physics*, *20*, 103654. DOI: 10.1016/j.rinp.2020.103654 PMID: 33520620

Hitchcock, F. L. (1941). The distribution of a product from several sources to numerous localities. *Journal of Mathematics and Physics*, *20*(1-4), 224–230. DOI: 10.1002/sapm1941201224

Hong, T., & Chuang, T. (1999). New triangular fuzzy Johnson algorithm. *Computers & Industrial Engineering*, *36*(1), 179–200. DOI: 10.1016/S0360-8352(99)00008-X

Hssina, B., Merbouha, A., Ezzikouri, H., & Erritali, M. (2014). A comparative study of decision tree ID3 and C4.5. *International Journal of Advanced Computer Science and Applications*, *4*(2), 13–19. DOI: 10.14569/SpecialIssue.2014.040203

Hwang, C. L., & Yoon, K. (1981). Methods for multiple attribute decision making. *Multiple attribute decision making: methods and applications a state-of-the-art survey*, 58-191.

Hwang, F. P., Chen, S. J., & Hwang, C. L. (1992). *Fuzzy multiple attribute decision making: Methods and applications*. Springer Berlin/Heidelberg.

Hwang, C. L., Lai, Y. J., & Liu, T. Y. (1993). A new approach for multiple objective decision making. *Computers & Operations Research, 20*(8), 889–899. DOI: 10.1016/0305-0548(93)90109-V

Ihsan, M., Rahman, A. U., & Saeed, M. (2021). Fuzzy Hypersoft Expert Set with Application in Decision Making for the Best Selection of Product. *Neutrosophic Sets and Systems, 46*, 318–336.

Ihsan, M., Rahman, A. U., & Saeed, M. (2021). Hypersoft Expert Set With Application in Decision Making for Recruitment Process. *Neutrosophic Sets and Systems, 42*, 191–207.

Ihsan, M., Saeed, M., Rahman, A. U., Khalifa, H. A. E.-W., & El-Morsy, S. (2022). An intelligent fuzzy parameterized multi-criteria decision-support system based on intuitionistic fuzzy hypersoft expert set for automobile evaluation. *Advances in Mechanical Engineering, 14*(7). Advance online publication. DOI: 10.1177/16878132221110005

Ihsan, M., Saeed, M., Rahman, A. U., & Smarandache, F. (2022). Multi-Attribute Decision Support Model Based on Bijective Hypersoft Expert Set. *Punjab University Journal of Mathematics, 54*(1), 55–73. DOI: 10.52280/pujm.2022.540105

Ishibuchi, H., & Tanaka, H. (1990). Multiobjective programming in optimization of the interval objective function. *European Journal of Operational Research, 48*(2), 219–225.

Ismail, M. M., Ibrahim, M. M., & Zaki, S. (2023). A Neutrosophic Approach for Multi-Factor Analysis of Uncertainty and Sustainability of Supply Chain Performance. *Neutrosophic Sets and Systems, 58*(1), 16.

Itoh, T., & Ishii, H.ITO. (1996). An Approach Based on Necessity Measure to the Fuzzy Spanning Tree Problems. *Journal of the Operations Research Society of Japan, 39*(2), 247–257. DOI: 10.15807/jorsj.39.247

Ivanov, D., Tsipoulanidis, A., & Schönberger, J. (2021). *Global supply chain and operations management*. Springer.

Jain, A., Pal Nandi, B., Gupta, C., & Tayal, D. K. (2020). Senti-NSetPSO: Large-sized document-level sentiment analysis using Neutrosophic Set and particle swarm optimization. *Soft Computing*, *24*(1), 3–15. DOI: 10.1007/s00500-019-04209-7

Jansi Rani, J., Dhanasekar, S., Micheal, D. R., & Manivannan, A. (2023). On solving fully intuitionistic fuzzy transportation problem via branch and bound technique. *Journal of Intelligent & Fuzzy Systems*, *44*(4), 6219–6229. DOI: 10.3233/JIFS-221345

Jansi, R., Mohana, K., & Smarandache, F. (2019). Correlation Measure for Pythagorean Neutrosophic Sets with T and F as Dependent Neutrosophic Components. *Neutrosophic Sets and Systems*, *30*, 202–212.

Jayalakshmi, M., & Sujatha, V. (2018). A new algorithm to solve multi-objective assignment problem. *International Journal of Pure and Applied Mathematics*, *119*(16), 719–724.

Jdid, M. (2022). Neutrosophic Treatment of the Static Model of Inventory Management with Deficit. *International Journal of Neutrosophic Science, 18*(1), 20-29. DOI: 10.54216/IJNS.180103

Jdid, M. (2023). Important Neutrosophic Economic Indicators of the Static Model of Inventory Management without Deficit. *Journal of Neutrosophic and Fuzzy Systems, 5*(1), 08-14. DOI: 10.54216/JNFS.050101

Jdid, M., & Models, N. N. (2023). *Prospects for Applied Mathematics and Data Analysis, 2*(1), 42-46. DOI: 10.54216/PAMDA.020104

Jdid, M., & Smarandache, F. (2023). Neutrosophic Static Model without Deficit and Variable Prices. *Neutrosophic Sets and Systems, 60,* 124-132. https://fs.unm.edu/nss8/index.php/111/article/view/3744

Jdid, M., Alhabib, R., & Bahbouh, O. (2022). The Neutrosophic Treatment for Multiple Storage Problem of Finite Materials and Volumes. *International Journal of Neutrosophic Science, 18*(1), 42-56. DOI: 10.54216/IJNS.180105

Jdid, M., Alhabib, R., & Salama, A. (2021). The static model of inventory management without a deficit with Neutrosophic logic. *International Journal of Neutrosophic Science, 16*(1). DOI: 10.54216/IJNS.160104

Jdid, M., Alhabib, R., Khalid, H. & Salama, A. (2022). The Neutrosophic Treatment of the Static Model for the Inventory Management with Safety Reserve. *InternationalJournal of Neutrosophic Science, 18*(2), 262-271. DOI: 10.54216/IJNS.180209

Jdid, M., & Smarandache, F. (2023). Graphical Method for Solving Neutrosophical Nonlinear Programming Models, / Int.J.Data. *Sci. & Big Data Anal.*, *3*(2), 66–72. DOI: 10.51483/IJDSBDA.3.2.2023.66-72

Jdid, M., & Smarandache, F. (2023). Lagrange Multipliers and Neutrosophic Nonlinear Programming Problems Constrained by Equality Constraints. *Neutrosophic Systems with Applications*, *6*, 25–31. DOI: 10.61356/j.nswa.2023.35

Jdid, M., & Smarandache, F. (2023). *Optimal Agricultural Land Use: An Efficient Neutrosophic Linear Programming Method.* Neutrosophic Systems With Applications.

Jeya Puvaneswari, P., & Bageerathi, K. (2017). On Neutrosophic Feebly open sets in Neutrosophic topological spaces. *International Journal of Mathematics Trends and Technology*, *41*(3), 230–237.

Jeya Puvaneswari, P., & Bageerathi, K. (2019). Some Functions Concerning Neutrosophic Feebly Open & Closed Sets. *International Journal of Scientific Research and Reviews*, *8*(2), 1546–1559.

Jia-hua, D., Zhang, H., & He, Y. (2019). Possibility Pythagorean fuzzy soft set and its application. *Journal of Intelligent & Fuzzy Systems*, *36*(1), 413–421. DOI: 10.3233/JIFS-181649

Johnson, S. M. (1954). Optimal Two and Three Stage Production schedules with setup Times Included. *Naval Research Logistics Quarterly*, *1*(1), 61–68. DOI: 10.1002/nav.3800010110

Kagade, K. L., & Bajaj, V. H. (2010). Fuzzy method for solving multi-objective assignment problem with interval cost. *Journal of Statistics and Mathematics, 1*(1), 1-9.

Kalyan Mondal, S. P. (2021). NN-TOPSIS strategy for MADM in neutrosophic number setting. *Neutrosophic sets and system, 47*, 66-92.

Kamacı, H., & Saqlain, M. (2021). n-ary Fuzzy Hypersoft Expert Sets. *Neutrosophic Sets and Systems*, *43*, 180–211.

Kamal, M., Modibbo, U. M., AlArjani, A., & Ali, I. (2021). Neutrosophic fuzzy goal programming approach in selective maintenance allocation of system reliability. *Complex & Intelligent Systems*, *7*(2), 1045–1059. DOI: 10.1007/s40747-021-00269-1

Kanchana, A., Nagarajan, D., Broumi, S., & Prastyo, D. D. (2024). A Comprehensive Analysis of Neutrosophic Bonferroni Mean Operator. *Neutrosophic Sets and Systems*, *64*, 57–76.

Kandasamy, I. (2018). Double-valued neutrosophic sets, their minimum spanning trees, and clustering algorithm. *Journal of Intelligent Systems*, *27*(2), 163–182. DOI: 10.1515/jisys-2016-0088

Kar, C., Samim Aktar, M., Maiti, M., & Das, P. (2023). Solving Fully Neutrosophic Incompatible Multi-Item Fixed Charge Four-Dimensional Transportation Problem with Volume Constraints. *New Mathematics and Natural Computation*, 1-29. DOI: 10.1142/S1793005724500054

Karaaslan, F. (2016a). Correlation coefficient between possibility neutrosophic soft sets. *Math. Sci. Lett*, *5*(1), 71–74. DOI: 10.18576/msl/050109

Karaaslan, F. (2016b). Similarity measure between possibility neutrosophic soft sets and its applications. *University Politehnica of Bucharest Scientific Bulletin-Series A-Applied Mathematics and Physics*, *78*(3), 155–162.

Karaaslan, F. (2017). Possibility neutrosophic soft sets and PNS-decision making method. *Applied Soft Computing*, *54*, 403–414. DOI: 10.1016/j.asoc.2016.07.013

Karasan, A., & Kaya, I. (2020b). Neutrosophic TOPSIS method for technology evaluation of unmanned aerial vehicles (UAVs). In *Intelligent and Fuzzy Techniques in Big Data Analytics and Decision Making:Proceedings of the INFUS 2019 Conference* (pp. 665-673). Springer International Publishing.

Karaşan, A., & Kahraman, C. (2020a). Selection of the most appropriate renewable energy alternatives by using a novel interval-valued neutrosophic ELECTRE I method. *Informatica (Vilnius)*, *31*(2), 225–248. DOI: 10.15388/20-INFOR388

Kashihara, K. (1996). *Comments and topics on Smarandache notions and problems*. Infinite Study.

Kaur, A., & Kumar, A. (2011). A new method for solving fuzzy transportation problems using ranking function. *Applied Mathematical Modelling*, *35*(12), 5652–5661. DOI: 10.1016/j.apm.2011.05.012

Keshavarz Ghorabaee, M., Zavadskas, E. K., Olfat, L., & Turskis, Z. (2015). Multi-criteria inventory classification using a new method of evaluation based on distance from average solution (EDAS). *Informatica (Vilnius)*, *26*(3), 435–451. DOI: 10.15388/Informatica.2015.57

Khalifa, H. A., Al-Shabi, M., & Mukherjee, S. (2018). An interactive approach for solving fuzzy multi-objective assignment problems. *Journal of advances in mathematics and computer science, 28*(6), 1-12.

Khalifa, H. A. (2020). An approach to the optimization of multi- objective assignment problem with neutrosophic numbers. *International Journal of Industrial Engineering& Production Research, 31*(2), 287–294.

Khalifa, H. A. E. W., Broumi, S., Edalatpanah, S. A., & Alburaikan, A. (2023). A novel approach for solving neutrosophic fractional transportation problem with non-linear discounting cost. *Neutrosophic Sets and Systems, 61*(1), 10.

Khalifa, H. A. E. W., Kumar, P., & Smarandache, F. (2020). *On optimizing neutrosophic complex programming using lexicographic order*. DOI: 10.5281/zenodo.3723173

Khalifa, N. E., Loey, M., Chakrabortty, R. K., & Taha, M. H. N. (2022). Within the Protection of COVID-19 Spreading: A Face Mask Detection Model Based on the Neutrosophic RGB with Deep Transfer Learning. *Neutrosophic Sets and Systems, 50*, 320–335.

Khalil, A. M., Li, S. G., Li, H. X., & Ma, S. Q. (2019). Possibility m-polar fuzzy soft sets and its application in decision-making problems. *Journal of Intelligent & Fuzzy Systems, 37*(1), 929–940. DOI: 10.3233/JIFS-181769

Khameneh, A. Z., Kiliçman, A., & Salleh, A. R. (2014). Fuzzy soft product topology. *Ann. Fuzzy Math.Inform (Champaign, Ill.), 7*(6), 935–947.

Khandelwal, A., & Kumar, A. (2024). A modified method for solving the unbalanced TP. *International Journal of Operations Research, 49*(1), 1–18. DOI: 10.1504/IJOR.2024.136005

Khan, M., Son, L. H., Ali, M., Chau, H. T. M., Na, N. T. N., & Smarandache, F. (2018). Systematic review of decision making algorithms in extended neutrosophic sets. *Symmetry, 10*(8), 314. DOI: 10.3390/sym10080314

Khan, S., Gulistan, M., & Wahab, H. A. (2021). Development of the structure of q-Rung Orthopair Fuzzy Hypersoft Set with basic Operations. *Punjab University Journal of Mathematics, 53*(12), 881–892. DOI: 10.52280/pujm.2021.531204

Khan, S., Haleem, A., & Khan, M. I. (2023). A risk assessment framework using neutrosophic theory for the halal supply chain under an uncertain environment. *Arab Gulf Journal of Scientific Research.*

Khan, Z., Amin, A., Khan, S. A., & Gulistan, M. (2021). Statistical Development of the Neutrosophic Lognormal Model with Application to Environmental Data. *Neutrosophic Sets and Systems, 47*(1), 1.

Kharal, A., & Ahmad, B. (2009). Mappings on fuzzy soft classes. *Advances in Fuzzy Systems, 2009*, 1–6.

Khorsand, R., & Ramezanpour, M. (2020). An energy-efficient task-scheduling algorithm based on a multi-criteria decision-making method in cloud computing. *International Journal of Communication Systems*, *33*(9), e4379. DOI: 10.1002/dac.4379

Kirişci, M., Demir, I., & Şimşek, N. (2022). Fermatean fuzzy ELECTRE multi-criteria group decision-making and most suitable biomedical material selection. *Artificial Intelligence in Medicine*, *127*, 102278. DOI: 10.1016/j.artmed.2022.102278 PMID: 35430046

Kiruthiga, M., & Loganathan, C. (2015). Fuzzy multi-objective linear programming problem using membership function. *International Journal of Science, Engineering, and Technology, Applied Sciences*, *5*(8), 1171-1178.

Kishorekumar, M., Karpagadevi, M., Mariappan, R., Krishnaprakash, S., & Revathy, A. (2023, February). Interval-valued picture fuzzy geometric Bonferroni mean aggregation operators in multiple attributes. In *2023 Fifth International Conference on Electrical, Computer and Communication Technologies (ICECCT)* (pp. 1-8). IEEE.

Klingman, D., & Russell, R. (1975). Solving constrained transportation problems. *Operations Research*, *23*(1), 91–106. DOI: 10.1287/opre.23.1.91

Koopmans, T. C. (1949). Optimum utilization of the transportation system. *Econometrica*, *17*, 136–146. DOI: 10.2307/1907301

Korucuk, S., Demir, E., Karamasa, C., & Stević, Ž. (2020). Determining the dimensions of the innovation ability in logistics sector by using plithogenic-critic method: An application in Sakarya Province. *International Review (Steubenville, Ohio)*, (1-2), 119–127.

Kour, D., & Basu, K. (2015). Application of extended fuzzy programming technique to a real-life transportation problem in neutrosophic environment. *Neutrosophic Sets and Systems*, *10*, 74–86.

Kousar, S., Sangi, M. N., Kausar, N., Pamučar, D., Ozbilge, E., & Cagin, T. (2023). Multi-objective optimization model for uncertain crop production under neutrosophic fuzzy environment: A case study. *AIMS Mathematics*, *8*(3), 7584–7605. DOI: 10.3934/math.2023380

Krishankumar, R., Premaladha, J., Ravichandran, K. S., Sekar, K. R., Manikandan, R., & Gao, X. Z. (2020). A novel extension to VIKOR method under intuitionistic fuzzy context for solving personnel selection problem. *Soft Computing*, *24*(2), 1063–1081. DOI: 10.1007/s00500-019-03943-2

Krishna Prabha, S., Hema, P., Balaji, P., & Kalaiselvi, B. (2022). Interval Value Intuitionistic Fuzzy Transportation Problem Via Modified ATM. *Journal of Emerging Technologies and Innovative Research*, *4*, 633–643.

Kumar, A., Chopra, R., & Saxena, R. R. (2021). An Efficient Enumeration Technique for a Transportation Problem in Neutrosophic Environment. *Neutrosophic Sets and Systems*, *47*, 354–365.

Kumar, A., & Gupta, A. (2011). Methods for solving fuzzy assignment problems and fuzzy travelling salesman problems with different membership functions. *Fuzzy Information and Engineering*, *3*, 3–21.

Kumar, M., Vrat, P., & Shankar, R. (2006). A fuzzy programming approach for vendor selection problem in a supply chain. *International Journal of Production Economics*, *101*(2), 273–285. DOI: 10.1016/j.ijpe.2005.01.005

Kumar, P. S. (2018). PSK method for solving intuitionistic fuzzy solid transportation problems. *International Journal of Fuzzy System Applications*, *7*(4), 62–99. DOI: 10.4018/IJFSA.2018100104

Kumar, R., Edalatpanah, S. A., Jha, S., & Singh, R. (2019). A Pythagorean fuzzy approach to the transportation problem. *Complex & Intelligent Systems*, *5*(2), 255–263.

Kumar, T., & Purusotham, S. (2018). The degree constrained k-cardinality minimum spanning tree problem: A lexi-search algorithm. *Decision Science Letters*, *7*(3), 301–310. DOI: 10.5267/j.dsl.2017.7.002

Kuroki, N. (1981). On fuzzy ideals and fuzzy bi-ideals in semigroups. *Fuzzy Sets and Systems*, *5*(2), 203–215. DOI: 10.1016/0165-0114(81)90018-X

Kuroki, N. (1982). Fuzzy semiprime ideals in semigroups. *Fuzzy Sets and Systems*, *8*(1), 71–79. DOI: 10.1016/0165-0114(82)90031-8

Kuroki, N. (1991). On fuzzy semigroups. *Information Sciences*, *53*(3), 203–236. DOI: 10.1016/0020-0255(91)90037-U

Lakshmana Gomathi Nayagam, , B. R. (2023). A Total Order on Single Valued and Interval Valued Neutrosophic Triplets. *Neutrosophic Sets and Systems*, *55*, 20.

Leberling, H. (1981). On finding compromise solutions in multicriteria problems using the fuzzy min-operator. *Fuzzy Sets and Systems*, *6*(2), 105–118.

Lee, S. M., & Moore, L. J. (1973). Optimizing transportation problems with multiple objectives. *AIIE Transactions*, *5*(4), 333–338. DOI: 10.1080/05695557308974920

Li, X. & Lu M, (2019). Some Novel Similarity and distance and Measures of Pythagorean Fuzzy Sets and their Applications. *J. Intel, Fuzzy Syst. 37*, 1781-1799.

Liang, T.-F. (2008). Integrating production-transportation planning decision with fuzzy multiple goals in supply chains. *International Journal of Production Research, 46*(6), 1477–1494. DOI: 10.1080/00207540600597211

Li, J. W.-h. (2019). Multi-criteria decision-making method based on dominance degree and BWM with probabilistic hesitant fuzzy information. *International Journal of Machine Learning and Cybernetics.*

Lin, C. J., & Wen, U. P. (2004). A labeling algorithm for the fuzzy assignment problem. *Fuzzy Sets and Systems, 142*(3), 373–391.

Lin, Y. L., Ho, L. H., Yeh, S. L., & Chen, T. Y. (2019). A pythagoreanfuzzytopsis method based on novel correlation measures and its application to multiple criteria decision analysis of inpatient stroke rehabilitation. *International Journal of Computational Intelligence Systems, 12*(1), 410–425. DOI: 10.2991/ijcis.2018.125905657

Liu, D., & Hu, C. (2020). A dynamic critical path method for project scheduling based on a generalised fuzzy similarity. *The Journal of the Operational Research Society, 72*(2), 458–470. DOI: 10.1080/01605682.2019.1671150

Liu, P., & Wang, Y. (2014). Multiple attribute decision-making method based on single-valued neutrosophic normalized weighted bonferroni mean. *Neural Computing & Applications, 25*(7), 2001–2010. DOI: 10.1007/s00521-014-1688-8

Liu, P., & Zhang, L. (2017). An extended multiple criteria decision making method based on neutrosophic hesitant fuzzy information. *Journal of Intelligent & Fuzzy Systems, 32*(6), 4403–4413. DOI: 10.3233/JIFS-16136

Liu, P., & Zhang, L. (2017). Multiple criteria decision-making method based on neutrosophic hesitant fuzzy Heronian mean aggregation operators. *Journal of Intelligent & Fuzzy Systems, 32*(1), 303–319. DOI: 10.3233/JIFS-151760

Liu, S. T. (2016). Fractional transportation problem with fuzzy parameters. *Soft Computing, 20*(9), 3629–3636. DOI: 10.1007/s00500-015-1722-5

Lohgaonkar, M. H., & Bajaj, V. H. (2010). Fuzzy approach to solve multi-objective capacitated transportation problem. *International Journal of Bioinformatics Research, 2*(1), 10–14.

Luo, S., & Liu, J. (2022). An innovative index system and HFFS-MULTIMOORA method based group decision-making framework for regional green development level evaluation. *Expert Systems with Applications*, *189*, 116090. DOI: 10.1016/j.eswa.2021.116090

Lupiáñez, F. G. (2009). Interval neutrosophic sets and topology. *Kybernetes*, *38*(3/4), 621–624.

Ma, X., Sun, H., Qin, H., Wang, Y., & Zheng, Y. (2024). A novel multi-attribute decision making method based on interval-valued fermatean fuzzy bonferroni mean operators. *Journal of Intelligent & Fuzzy Systems*, 1-21.

Mahdavi, A., & Kundu, D. (2017). A new method for generating distributions with an application to exponential distribution. *Communications in Statistics. Theory and Methods*, *46*(13), 6543–6557. DOI: 10.1080/03610926.2015.1130839

Maimon, O., & Rokach, L. (2010). *Data Mining and Knowledge Discovery Handbook*. Springer. DOI: 10.1007/978-0-387-09823-4

Maiti, I., Mandal, T., & Pramanik, S. (2020). Neutrosophic goal programming strategy for multi-level multi-objective linear programming problem. *Journal of Ambient Intelligence and Humanized Computing*, *11*(8), 3175–3186. DOI: 10.1007/s12652-019-01482-0

Maji, P. K., Biswas, R., & Roy, A. R. (2001). Fuzzy soft sets. *Journal of Fuzzy Mathematics*, *9*(3), 589–602.

Maji, P. K., Biswas, R., & Roy, A. R. (2001). Fuzzy soft sets. *Journal of Fuzzy Mathematics.*, *9*(3), 589–602.

Maji, P. K., Biswas, R., & Roy, A. R. (2002). An application of soft sets in decision-making problem. *Computers & Mathematics with Applications (Oxford, England)*, *44*(8–9), 1077–1083. DOI: 10.1016/S0898-1221(02)00216-X

Maji, P. K., Biswas, R., & Roy, A. R. (2003). Soft set theory. *Computers & Mathematics with Applications (Oxford, England)*, *45*(4-5), 555–562.

Maji, P. K., Biswas, R., & Roy, A. R. (2003). Soft Set Theory. *Computers & Mathematics with Applications (Oxford, England)*, *45*(4-5), 555–562. DOI: 10.1016/S0898-1221(03)00016-6

Maji, P. K., Roy, A. R., & Biswas, R. (2002). An application of soft sets in a decision making problem. *Computers & Mathematics with Applications (Oxford, England)*, *44*(8-9), 1077–1083.

Mandal, K., & Basu, K. (2016). Improved similarity measure in neutrosophic environment and its application in finding minimum spanning tree. *Journal of Intelligent & Fuzzy Systems, 31*(3), 1721–1730. DOI: 10.3233/JIFS-152082

Manikantan, T., & Ramkumar, S. (2021). Cartesian product of the extensions of fuzzy soft ideals over near-rings. *International Journal of Dynamical Systems and Differential Equations, 11*(5-6), 426–447.

Mao, X., Guoxi, Z., Fallah, M., & Edalatpanah, S. (2020). A neutrosophic-based approach in data envelopment analysis with undesirable outputs. *Mathematical Problems in Engineering, 2020*, 1–8. DOI: 10.1155/2020/7626102

Margaret, M. A., Pricilla, T. M., & Alkhazaleh, S. (2019). *Neutrosophic Vague Topological Spaces, Neutrosophic Sets and Systems, 28.* University of New Mexico.

Martin, N. (2022). Plithogenic SWARA-TOPSIS Decision Making on Food Processing Methods with Different Normalization Techniques. *Advances in Decision Making, 69.*

Martin, N., Smarandache, F., & Priya, R. (2022). Introduction to Plithogenic Sociogram with preference representations by Plithogenic Number. *Journal of Fuzzy Extension and Applications, 3*(1), 96-108.

Martin, N., Smarandache, F., & Broumi, S. (2021). PROMTHEE plithogenic pythagorean hypergraphic approach in smart materials selection. *Int J Neutrosophic Sci, 13*, 52–60.

Martin, N., Smarandache, F., & Priya, R. (2022). Introduction to plithogenic sociogram with preference representations by plithogenic number. *Journal of Fuzzy Extension and Applications, 3*(1), 96–108. DOI: 10.22105/jfea.2021.288057.1151

Martin, N., Smarandache, F., & Sudha, S. (2023). A novel method of decision making based on plithogenic contradictions. *Neutrosophic Systems with Applications, 10*, 12–24.

Mazlum, M., & Güneri, A. F. (2015). CPM, PERT and Project Management with Fuzzy Logic Technique and Implementation on a Business. *Procedia: Social and Behavioral Sciences, 210*, 348–357. DOI: 10.1016/j.sbspro.2015.11.378

McCahon, C. S., & Lee, E. S. (1990). Job Sequencing with fuzzy processing times, computers &. *Mathematics for Applications, 19*(7), 31–41.

Medvedeva, O. A., & Medvedev, S. N. (2018, March). A dual approach to solving a multi-objective assignment problem. *Journal of Physics: Conference Series, 973*(1), 012039.

Miari, M., Anan, M. T., & Zeina, M. B. (2022). Neutrosophic two way ANOVA. *International Journal of Neutrosophic Science*, *18*(3), 73–83. DOI: 10.54216/IJNS.180306

Mishra, A. R., Liu, P., & Rani, P. (2022). COPRAS method based on interval-valued hesitant Fermatean fuzzy sets and its application in selecting desalination technology. *Applied Soft Computing*, *119*, 108570. DOI: 10.1016/j.asoc.2022.108570

Mishra, A. R., Mardani, A., Rani, P., & Zavadskas, E. K. (2020a). A novel EDAS approach on intuitionistic fuzzy set for assessment of health-care waste disposal technology using new parametric divergence measures. *Journal of Cleaner Production*, *272*, 122807. DOI: 10.1016/j.jclepro.2020.122807

Mishra, A. R., & Rani, P. (2018). Interval-valued intuitionistic fuzzy WASPAS method: Application in reservoir flood control management policy. *Group Decision and Negotiation*, *27*(6), 1047–1078. DOI: 10.1007/s10726-018-9593-7

Mishra, A. R., & Rani, P. (2021). Multi-criteria healthcare waste disposal location selection based on Fermatean fuzzy WASPAS method. *Complex & Intelligent Systems*, *7*(5), 2469–2484. DOI: 10.1007/s40747-021-00407-9 PMID: 34777968

Mishra, A. R., Rani, P., Mardani, A., Pardasani, K. R., Govindan, K., & Alrasheedi, M. (2020b). Healthcare evaluation in hazardous waste recycling using novel interval-valued intuitionistic fuzzy information based on complex proportional assessment method. *Computers & Industrial Engineering*, *139*, 106140. DOI: 10.1016/j.cie.2019.106140

Mishra, A. R., Rani, P., & Pandey, K. (2022). Fermatean fuzzy CRITIC-EDAS approach for the selection of sustainable third-party reverse logistics providers using improved generalized score function. *Journal of Ambient Intelligence and Humanized Computing*, *13*(1), 1–17. DOI: 10.1007/s12652-021-02902-w PMID: 33584868

Mishra, A. R., Singh, R. K., & Motwani, D. (2020c). Intuitionistic fuzzy divergence measure-based ELECTRE method for performance of cellular mobile telephone service providers. *Neural Computing & Applications*, *32*(8), 3901–3921. DOI: 10.1007/s00521-018-3716-6

Mohamed, Z., Ismail, M. M., & Abd El-Gawad, A. F. (2023). Neutrosophic Model to Examine the Challenges Faced by Manufacturing Businesses in Adopting Green Supply Chain Practices and to Provide Potential Solutions. *Neutrosophic Systems With Applications*, *3*, 45–52.

Mohammed, M. A., Abdulkareem, K. H., Al-Waisy, A. S., Mostafa, S. A., Al-Fahdawi, S., Dinar, A. M., Alhakami, W., Baz, A., Al-Mhiqani, M. N., Alhakami, H., Arbaiy, N., Maashi, M. S., Mutlag, A. A., Garcia-Zapirain, B., & De La Torre Diez, I. (2020). Benchmarking methodology for selection of optimal COVID-19 diagnostic model based on entropy and TOPSIS methods. *IEEE Access : Practical Innovations, Open Solutions*, *8*, 99115–99131. DOI: 10.1109/ACCESS.2020.2995597

Mollaoglu, M., Bucak, U., Demirel, H., & Balin, A. (2022). Evaluationofvarious fuelalternativesintermsofsustainabilityfortheshipinvestmentdecisionusingsinglevalued neutrosophic numbers with topsis methods. *Proceedings of the Institution of Mechanical Engineers, PartM:Journal of Engineering for the Maritime Environment*, *237*(1), 215–226.

Molodtsov, D. (1999). Soft set theory-first results. *Computers & Mathematics with Applications (Oxford, England)*, *37*(4-5), 19–31.

Molodtsov, D. (1999). Soft set theory—First results. *Computers & Mathematics with Applications (Oxford, England)*, *37*(4-5), 19–31. DOI: 10.1016/S0898-1221(99)00056-5

Moore, R. E. (1979). *Method and Applications of Interval Analysis*. SLAM. DOI: 10.1137/1.9781611970906

Muddineni, V. P., Sandepudi, S. R., & Bonala, A. K. (2017). Improved weighting factor selection for predictive torque control of induction motor drive based on a simple additive weighting method. *Electric Power Components and Systems*, *45*(13), 1450–1462. DOI: 10.1080/15325008.2017.1347215

Mukherjee, S., & Basu, K. (2010). Application of fuzzy ranking method for solving assignment problems with fuzzy costs. *International Journal of Computational and Applied Mathematics*, *5*(3), 359-369.

Mukherjee, A., & Das, A. K. (2018). Einstein operations on fuzzy soft multisets and decision making. *Boletim da Sociedade Paranaense de Matematica*, *40*, 1–10.

Mullai, M., Broumi, S., & Stephen, A. (2017). *Shortest path problem by minimal spanning tree algorithm using bipolar neutrosophic numbers*. Infinite Study.

Munn, W. D. (1964). The Algebraic Theory of Semigroups, Vol. I. By A. H. Clifford and G. B. Preston. Pp. xv + 224. $10.60. 1961. (American Mathematical Society, Providence.). *Mathematical Gazette*, *48*(363), 122–122. DOI: 10.2307/3614367

Musa, S. Y., & Asaad, B. A. (2021). Bipolar Hypersoft Sets. *Mathematics*, *9*(15), 1826. DOI: 10.3390/math9151826

Musa, S. Y., Mohammed, R. A., & Asaad, B. A. (2023). N-Hypersoft Sets: An Innovative Extension of Hypersoft Sets and Their Applications. *Symmetry, 15*(9), 1795. DOI: 10.3390/sym15091795

Nabeeh, N. A., Mohamed, M., Abdel-Monem, A., Abdel-Basset, M., Sallam, K. M., El-Abd, M., & Wagdy, A. (2022, July). A Neutrosophic Evaluation Model for Blockchain Technology in Supply Chain Management. In *2022 IEEE International Conference on Fuzzy Systems (FUZZ-IEEE)* (pp. 1-8). IEEE.

Nafei, A. H., Javadpour, A., Nasseri, H., & Yuan, W. (2021). Optimizedscore function and its application in group multiattribute decision making based on fuzzy neutrosophic sets. *International Journal of Intelligent Systems, 36*(12), 7522–7543. DOI: 10.1002/int.22597

Nair, N. U., & Sankaran, P. (2009). Quantile-based reliability analysis. *Communications in Statistics. Theory and Methods, 38*(2), 222–232. DOI: 10.1080/03610920802187430

Narayanan, A. L., & Meenakshi, A. R. (2003). Fuzzy M-semigroup. *The Journal of Fuzzy Mathematics, 11*(1), 41–52.

Nasseri, H., Morteznia, M., & Mirmohseni, M. (2017). A new method for solving fully fuzzy multi objective supplier selection problem. *International journal of research in industrial engineering, 6*(3), 214-227.

Nasution, S. (1994). Fuzzy critical path method. *IEEE Transactions on Systems, Man, and Cybernetics, 24*(1), 48–57. DOI: 10.1109/21.259685

Nayagam, V. L., & Bharanidharan, R. (2023). A Total Ordering on n - Valued Refined Neutrosophic Sets using Dictionary Ranking based on Total ordering on n - Valued Neutrosophic Tuplets. *Neutrosophic Sets and Systems, 58.*

Nayagam, V. L., Jeevaraj, & Sivaraman, G. (2016). Total ordering for intuitionistic fuzzy numbers. *Complexity, 21*, 54–66. DOI: 10.1002/cplx.21783

Nayagam, V. L., Jeevaraj, S., & Dhanasekaran, P. (2016). A linear ordering on the class of Trapezoidal intuitionistic fuzzy numbers. *Expert Systems with Applications, 60*, 269–279. DOI: 10.1016/j.eswa.2

Nayagam, V. L., & Sivaraman, G. (2011). Ranking of interval-valued intuitionistic fuzzy sets. *Applied Soft Computing, 11*, 3368–3372. DOI: 10.1016/j.asoc.2011.01.008

Nayagam, V., Ponnialagan, D., & Jeevaraj, S. (2020). Similarity measure on incomplete imprecise interval information and its applications. *Neural Computing & Applications, 32*, 3749–3761. DOI: 10.1007/s00521-019-04277-8

Nowpada & Saradhi. (2011). Fuzzy critical path method in interval valued activity networks. *International Journal of Pure and Applied Sciences and Technology, 3,* 72-79.

Nozari, H., Tavakkoli-Moghaddam, R., & Gharemani-Nahr, J. (2022). A neutrosophic fuzzy programming method to solve a multi-depot vehicle routing model under uncertainty during the covid-19 pandemic. *International Journal of Engineering, 35*(2), 360–371.

Nzei, L. C., Eghwerido, J. T., & Ekhosuehi, N. (2020). Topp-Leone Gompertz Distribution: Properties and Applications. *Journal of Data Science : JDS, 18*(4), 782–794. DOI: 10.6339/JDS.202010_18(4).0012

ÓhÉigeartaigh, M. (1982). A fuzzy transportation algorithm. *Fuzzy Sets and Systems, 8*(3), 235–243. DOI: 10.1016/S0165-0114(82)80002-X

Olgun, N., & Hatip, A. *The effect of the neutrosophic logic on the decision tree* (Vol. 7).

Opricovic, S., & Tzeng, G. H. (2004). Compromise solution by MCDM methods: A comparative analysis of VIKOR and TOPSIS. *European Journal of Operational Research, 156*(2), 445–455. DOI: 10.1016/S0377-2217(03)00020-1

Oudouar, F., & Zaoui, E. M. (2021, November). A novel hybrid heuristic based on ant colony algorithm for solving multi-product inventory routing problem. In *International Conference on Advanced Technologies for Humanity* (pp. 519-529). Cham: Springer International Publishing.

Ozlu, S., & Karaaslan, F. (2022). Hybrid similarity measures of single-valued neutrosophic type-2 fuzzy sets and their application to MCDM based on TOPSIS. *Soft Computing, 26*(9), 4059–4080. DOI: 10.1007/s00500-022-06824-3

Paik, B., & Mondal, S. K. (2021). A distance-similarity method to solve fuzzy sets and fuzzy soft sets based decision-making problems. *Soft Computing, 24*(7), 5217–5229. DOI: 10.1007/s00500-019-04273-z

Paik, B., & Mondal, S. K. (2021). Representation and application of Fuzzy soft sets in type-2 environment. *Complex & Intelligent Systems, 7*(3), 1597–1617. DOI: 10.1007/s40747-021-00286-0

Palanikumar, M., Iampan, A., & Broumi, S. (2022). MCGDM based on VIKOR and TOPSIS proposes neutrsophic Fermatean fuzzy soft with aggregation operators. *International Journal of Neutrosophic Science, 19*(3), 85–94. DOI: 10.54216/IJNS.190308

Pandian, P., & Natarajan, G. (2010). A new method for finding an optimal solution for transportation problems. *International Journal of Mathematical Sciences and engineering applications, 4*(2), 59-65.

Pandian, P., & Anuradha, D. (2010). A new approach for solving solid transportation problems. *Applied Mathematical Sciences, 4*(72), 3603–3610.

Paneerselvam, R. (2008). *Operations research* (3rd ed.). PHI Learning Private Ltd.

Parzen, E. (1979). Density quantile estimation approach to statistical data modelling. In *Smoothing Techniques for Curve Estimation* (pp. 155–180). Springer. DOI: 10.1007/BFb0098495

Patro, S., & Smarandache, F. *The neutrosophic statistical distribution, more problems, more solutions*. Infinite Study.

Pawlak, Z. (1982). Rough sets. *International Journal of Computer & Information Sciences, 11*(5), 341-356.

Pei, D., & Miao, D. (2005). From Soft Sets to Information Systems. *Paper presented at the 2005 IEEE International Conference on Granular Computing*. IEEE.

Peidro, D., Mula, J., Poler, R., & Verdegay, J.-L. (2009). Fuzzy optimization for supply chain planning under supply, demand and process uncertainties. *Fuzzy Sets and Systems, 160*(18), 2640–2657. DOI: 10.1016/j.fss.2009.02.021

Peng, X., & Yang, Y. (2016). Fundamental Properties of Interval-Valued Pythagorean Fuzzy Aggregation Operators. *International Journal of Intelligent Systems, 31*(5), 444–487. DOI: 10.1002/int.21790

Peng, X., & Yang, Y. (2016). Pythagorean Fuzzy Choquet Integral Based Mabac Method for Multiple attribute Group Decision Making. *International Journal of Intelligent Systems, 31*(10), 989–1020. DOI: 10.1002/int.21814

Petchimuthu, S., Garg, H., Kamacı, H., & Atagün, A. O. (2020). The mean operators and generalized products of fuzzy soft matrices and their applications in MCGDM. *Computational & Applied Mathematics, 39*(2), 1–32. DOI: 10.1007/s40314-020-1083-2

Petrovic, D., Roy, R., & Petrovic, R. (1999). Supply chain modelling using fuzzy sets. *International Journal of Production Economics, 59*(1-3), 443–453. DOI: 10.1016/S0925-5273(98)00109-1

Pouresmaeil, H., Khorram, E., & Shivanian, E. (2022). A parametric scoring function and the associated method for interval neutrosophic multi-criteria decision-making. *Evolving Systems, 13*(2), 347–359. DOI: 10.1007/s12530-021-09394-1

Pramanik, S., & Banerjee, D. (2018). *Neutrosophic number goal programming for multi-objective linear programming problem in neutrosophic number environment.*

Pramanik, S. (2023). Interval pentapartitioned neutrosophic sets. *Neutrosophic Sets and Systems*, *55*, 232–246.

Pramanik, S., & Biswas, P. (2012). Multi-objective assignment problem with generalized trapezoidal fuzzy numbers. *International Journal of Applied Information Systems*, *2*(6), 13–20.

Pramanik, S., Dey, P. P., & Giri, B. C. (2015). TOPSIS for single valued neutrosophic soft expert set based multi-attribute decision making problems. *Neutrosophic Sets and Systems*, *10*, 88–95.

Pramanik, S., & Mondal, K. (2015). Interval neutrosophic multi-attribute decision-making based on grey relational analysis. *Neutrosophic Sets and Systems*, *9*(1), 13–22.

Pratihar, J., Kumar, R., Dey, A., & Broumi, S. (2020). Transportation problem in neutrosophic environment. In *Neutrosophic graph theory and algorithms* (pp. 180–212). IGI Global. DOI: 10.4018/978-1-7998-1313-2.ch007

Pratihar, J., Kumar, R., Edalatpanah, S. A., & Dey, A. (2021). Modified Vogel's approximation method for transportation problem under uncertain environment. *Complex & Intelligent Systems*, *7*(1), 29–40. DOI: 10.1007/s40747-020-00153-4

Prema, R., & Radha, R. (2022). Generalized Neutrosophic Pythagorean Set. *International Research Journal of Modernization in Engineering Technology and Science*, *4*(11), 1571–1575. https://www.doi.org/10.56726/IRJMETS31596

Priyadharshini, S. P., & Irudayam, F. N. (2023). An analysis of obesity in school children during the pandemic COVID-19 using plithogenic single valued fuzzy sets. *Neutrosophic Systems with Applications*, *9*, 24–28.

Qiuping, N., Yuanxiang, T., Broumi, S., & Uluçay, V. (2023). A parametric neutrosophic model for the solid transportation problem. *Management Decision*, *61*(2), 421–442. DOI: 10.1108/MD-05-2022-0660

Quek, S. G., Selvachandran, G., Smarandache, F., Vimala, J., Le, S. H., Bui, Q.-T., & Gerogiannis, V. C. (2020). Entropy Measures for Plithogenic Sets and Applications in Multi-Attribute Decision Making. *Mathematics*, *8*(6), 965. DOI: 10.3390/math8060965

Radha, R., & Stanis Arul Mary, A. (2021). Pentapartitioned Neutrosophic Pythagorean set. *IRJASH*, *3*(2S), 62–82. DOI: 10.47392/irjash.2021.041

Radhika, K., & Arun Prakash, K. (2022). Multi-objective optimization for multi-type transportation problem in intuitionistic fuzzy environment. *Journal of Intelligent & Fuzzy Systems*, *43*(1), 1439–1452. DOI: 10.3233/JIFS-213517

Rahman, A. U., Saeed, M., & Abd El-Wahed Khalifa, H. (2022). Multi-attribute decision-making based on aggregations and similarity measures of neutrosophic hypersoft sets with possibility setting. *Journal of Experimental & Theoretical Artificial Intelligence*, 1–26.

Rahman, A. U., Saeed, M., & Garg, H. (2022). An innovative decisive framework for optimized agri-automobile evaluation and HRM pattern recognition via possibility fuzzy hypersoft setting. *Advances in Mechanical Engineering*, *14*(10). DOI: 10.1177/16878132221132146

Rahman, A. U., Saeed, M., Khalifa, H. A. E. W., & Afifi, W. A. (2022). Decision making algorithmic techniques based on aggregation operations and similarity measures of possibility intuitionistic fuzzy hypersoft sets. *AIMS Mathematics*, *7*(3), 3866–3895. DOI: 10.3934/math.2022214

Rahman, A. U., Saeed, M., Mohammed, M. A., Abdulkareem, K. H., Nedoma, J., & Martinek, R. (2023). Fppsv-NHSS: Fuzzy parameterized possibility single valued neutrosophic hypersoft set to site selection for solid waste management. *Applied Soft Computing*, *140*, 110273. DOI: 10.1016/j.asoc.2023.110273

Rahman, A. U., Saeed, M., Mohammed, M. A., Krishnamoorthy, S., Kadry, S., & Eid, F. (2022). An integrated algorithmic MADM approach for heart diseases' diagnosis based on neutrosophic hypersoft set with possibility degree-based setting. *Life (Chicago, Ill.)*, *12*(5), 729. DOI: 10.3390/life12050729 PMID: 35629396

Rahman, A. U., Saeed, M., & Smarandache, F. (2020). Convex and Concave Hypersoft Sets with Some Properties. *Neutrosophic Sets and Systems*, *38*, 497–508.

Rahman, A. U., Saeed, M., Smarandache, F., & Ahmad, M. R. (2020). Development of Hybrids of Hypersoft Set with Complex Fuzzy Set, Complex Intuitionistic Fuzzy set and Complex Neutrosophic Set. *Neutrosophic Sets and Systems*, *1*, 334.

Rajarajeswari, P., Sudha, A. S., & Karthika, R. (2013). A new operation on hexagonal fuzzy number. *International Journal of Fuzzy Logic Systems, 3*(3), 15-26.

Raksha Ben, N., & Hari Siva Annam, G. (2021). Generalized Topological Spaces via Neutrosophic Sets. *J. Math. Comput. Sci.*, *11*, 716–734.

Raksha Ben, N., & Hari Siva Annam, G. (2021). Some new open sets in μ_N topological space. *Malaya Journal of Matematik*, *9*(1), 89–94.

Rani, J. J., & Manivannan, A. (2024). An application of generalized symmetric type-2 intuitionistic fuzzy variables to a transportation problem with the effect of a new ranking function. *Expert Systems with Applications*, *237*, 121384. DOI: 10.1016/j.eswa.2023.121384

Rani, P., & Mishra, A. R. (2021). Fermatean fuzzy Einstein aggregation operators-based MULTIMOORA method for electric vehicle charging station selection. *Expert Systems with Applications*, *182*, 115267. DOI: 10.1016/j.eswa.2021.115267

Rani, P., Mishra, A. R., & Mardani, A. (2020a). An extended Pythagorean fuzzy complex proportional assessment approach with new entropy and score function: Application in pharmacological therapy selection for type 2 diabetes. *Applied Soft Computing*, *94*, 106441. DOI: 10.1016/j.asoc.2020.106441

Rani, P., Mishra, A. R., & Pardasani, K. R. (2020b). A novel WASPAS approach for multi-criteria physician selection problem with intuitionistic fuzzy type-2 sets. *Soft Computing*, *24*(3), 2355–2367. DOI: 10.1007/s00500-019-04065-5

Rani, P., Mishra, A. R., Pardasani, K. R., Mardani, A., Liao, H., & Streimikiene, D. (2019). A novel VIKOR approach based on entropy and divergence measures of Pythagorean fuzzy sets to evaluate renewable energy technologies in India. *Journal of Cleaner Production*, *238*, 117936. DOI: 10.1016/j.jclepro.2019.117936

Rathnasabapathy, P., & Palanisami, D. (2023). A theoretical development of improved cosine similarity measure for interval valued intuitionistic fuzzy sets and its applications. *Journal of Ambient Intelligence and Humanized Computing*, *14*(12), 16575–16587. DOI: 10.1007/s12652-022-04019-0 PMID: 35789601

Raut, P. K., Behera, S. P., Broumi, S., & Mishra, D. (2023). Calculation of Fuzzy shortest path problem using Multi-valued Neutrosophic number under fuzzy environment. *Neutrosophic Sets and Systems*, *57*. https://digitalrepository.unm.edu/nss_journal/vol57/iss1/24

Renee Miriam, M., Martin, N., Aleeswari, A., & Broumi, S. (2023). Rework Warehouse Inventory Model for Product Distribution with Quality Conservation in Neutrosophic Environment. *International Journal of Neutrosophic Science*, *21*(2), 177–195.

Revathy, A., Inthumathi, V., Krishnaprakash, S., & Kishorekumar, M. (2023). Fermatean fuzzy normalised Bonferroni mean operator in multi criteria decision making on selection of electric bike. In *2023 Fifth International Conference on Electrical, Computer and Communication Technologies (ICECCT)* (pp. 1-7). IEEE.

Ridvan, S., Aslan, F., & Gokec, D. K. (2021). A single-valued neutrosophic multi-criteria group decision approach with DPL-TOPSIS method based on optimization. *International Journal of Intelligent Systems*, *36*(7), 3339–3366. DOI: 10.1002/int.22418

Rivieccio, U. (2008). Neutrosophic logics: Prospects and problems. *Fuzzy Sets and Systems*, *159*(14), 1860–1868. DOI: 10.1016/j.fss.2007.11.011

Rizk-Allah, R. M., Hassanien, A. E., & Elhoseny, M. (2018). A multi-objective transportation model under neutrosophic environment. *Computers & Electrical Engineering*, *69*, 705–719.

Rodríguez, A., Ortega, F., & Concepción, R. (2017). An intuitionistic method for the selection of a risk management approach to information technology projects. *Information Sciences*, *375*, 202–218.

Rodríguez, R. M., Labella, Á., & Martínez, L. (2016). An overview on fuzzy modelling of complex linguistic preferences in decision making. *International Journal of Computational Intelligence Systems*, *9*(Suppl 1), 81–94.

Rosenfeld, A. (1971). Fuzzy groups. *Journal of Mathematical Analysis and Applications*, *35*(3), 512–517. DOI: 10.1016/0022-247X(71)90199-5

Roy, A. R., & Maji, P. K. (2007). A fuzzy soft set theoretic approach to decision making problems. *Journal of Computational and Applied Mathematics*, *203*(2), 412–418.

Roy, S., & Samanta, T. K. (2013). An introduction to open and closed sets on fuzzy soft topological spaces. *Annals of Fuzzy Mathematics and Informatics*, *6*(2), 425–431.

Saeed, M., Ahsan, M., Saeed, M. H., Rahman, A. U., Mehmood, A., Mohammed, M. A., & Damaševičius, R. (2022). An optimized decision support model for COVID-19 diagnostics based on complex fuzzy hypersoft mapping. *Mathematics*, *10*(14), 2472.

Saeed, M., Ahsan, M., Saeed, M. H., Rahman, A. U., Mohammed, M. A., Nedoma, J., & Martinek, R. (2023). An algebraic modeling for tuberculosis disease prognosis and proposed potential treatment methods using fuzzy hypersoft mappings. *Biomedical Signal Processing and Control*, *80*, 104267.

Saeed, M., Ahsan, M., Siddique, M. K., & Ahmad, M. R. (2020). A Study of The Fundamentals of Hypersoft Set Theory. *International Journal of Scientific and Engineering Research*, *11*, 320–329.

Saeed, M., Ahsan, M., Ur Rahman, A., Saeed, M. H., & Mehmood, A. (2021b). An application of neutrosophic hypersoft mapping to diagnose brain tumor and propose appropriate treatment. *Journal of Intelligent & Fuzzy Systems, 41*(1), 1677–1699.

Saeed, M., Rahman, A. U., Ahsan, M., & Smarandache, F. (2021a). An inclusive study on fundamentals of hypersoft set. *Theory and Application of Hypersoft Set, 1*, 1–23.

Saeed, M., Rahman, A. U., Ahsan, M., & Smarandache, F. (2022). Theory of hypersoft sets: Axiomatic properties, aggregation operations, relations, functions and matrices. *Neutrosophic Sets and Systems, 51*(1), 46.

Saeed, M., Rahman, A. U., Ahsan, M., & Smarandache, F. (2022). Theory of Hypersoft Sets: Axiomatic Properties, Aggregation Operations, Relations, Functions and Matrices. *Neutrosophic Sets and Systems, 51*, 744–765.

Şahin, M., Kargın, A., & Smarandache, F. (2018). *Generalized single valued triangular neutrosophic numbers and aggregation operators for application to multi-attribute group decision making*. Infinite Study.

Sahin, M., Alkhazaleh, S., & Ulucay, V. (2015). Neutrosophic soft expert sets. *Applied Mathematics, 6*(1), 116–127. DOI: 10.4236/am.2015.61012

Sahin, R., & Yigider, M. (2016). A multi-criteria neutrosophic group decision making method based topsis for supplier selection. *Applied Mathematics & Information Sciences, 10*(5), 1–10. DOI: 10.18576/amis/100525

Sahoo, L. (2017). Solving Job Sequencing problem with fuzzy processing times. *IJARIIE, 3*(4), 3326–3329.

Sakawa, M., & Yano, H. (1989). Interactive decision making for multiobjective nonlinear programming problems with fuzzy parameters. *Fuzzy Sets and Systems, 29*(3), 315–326.

Salama, A. (2018). Neutrosophic approach to grayscale images domain.

Salama, A., & Alblowi, S. (2012). Neutrosophic set and neutrosophic topological spaces. *IOSR Journal of Mathematics (IOSR-JM), 3*(4).

Salama, A., El-Ghareeb, H. A., Manie, A. M., & Smarandache, F. Introduction to develop some software programs for dealing with neutrosophic sets.

Salama, A., Smarandache, F., & Alblowi, S. *New neutrosophic crisp topological concepts*. Infinite Study.

Salama, A., Smarandache, F., & Eisa, M. (2014). *Introduction to image processing via neutrosophic techniques*. Infinite Study.

Salama, A., Smarandache, F., & Kromov, V. (2022). Neutrosophic closed set and neutrosophic continuous functions. *Collected Papers. Volume IX: On Neutrosophic Theory and Its Applications in Algebra*, 25.

Salama, A., Smarandache, F., & Kroumov, V. (2014). *Neutrosophic crisp sets & neutrosophic crisp topological spaces*. Infinite Study.

Salama, A. A., & Albowi, S. A. (2012). Neutrosophic set and Neutrosophic Topological Space. *ISOR J. Mathematics*, *3*(4), 31–35.

Samanta, S., Chakraborty, D., & Jana, D. K. (2024). Neutrosophic multi-period two stage four-dimensional transportation problem for breakable items. *Expert Systems with Applications*, *246*, 123266. DOI: 10.1016/j.eswa.2024.123266

Selvachandran, G., Maji, P. K., Abed, I. E., & Salleh, A. R. (2016). Complex vague soft sets and its distance measures. *Journal of Intelligent & Fuzzy Systems*, *31*(1), 55–68.

Selvakumari, K., & Santhi, S. (2018). An Approach for solving Fuzzy Sequencing Problems with octagonal Fuzzy numbers using Robust Ranking Techniques. *Int. Jour. Of Mathematics Trends and Technology*, *56*(3), 148–152. DOI: 10.14445/22315373/IJMTT-V56P521

Senapati, T., & Yager, R. R. (2020). Fermatean fuzzy sets. *Journal of Ambient Intelligence and Humanized Computing*, *11*(2), 663–674. DOI: 10.1007/s12652-019-01377-0

Sennaroglu, B., & Celebi, G. V. (2018). A military airport location selection by AHP integrated PROMETHEE and VIKOR methods. *Transportation Research Part D, Transport and Environment*, *59*, 160–173. DOI: 10.1016/j.trd.2017.12.022

Sezgin, A., & Atagün, A. O. (2011). On operations of soft sets. *Computers & Mathematics with Applications (Oxford, England)*, *61*(5), 1457–1467.

Shafi Salimi, P., & Edalatpanah, S. A. (2020). Supplier selection using fuzzy AHP method and D-Numbers. *Journal of fuzzy extension and applications, 1*(1), 1-14.

Shannon, C. E. (2001). A mathematical theory of communication. *Mobile Computing and Communications Review*, *5*(1), 3–55. DOI: 10.1145/584091.584093

Shaocheng, T. (1994). Interval number and fuzzy number linear programming. *Fuzzy Sets and Systems*, *66*(3), 301–306. DOI: 10.1016/0165-0114(94)90097-3

Sharma, J. K. (2016). Operations Research Theory and Applications. Am imprint of Laxmi Publications.

Sharma, J., & Swarup, K. (1977). Indefinite quadratic programming and transportation technique.

Sher, T., Rehman, A., & Kim, D. (2022). COVID-19 Outbreak Prediction by Using Machine Learning Algorithms. *Computers, Materials & Continua*, *74*(1), 1561–1574. DOI: 10.32604/cmc.2023.032020

Sherwani, R. A. K., Aslam, M., Raza, M. A., Farooq, M., Abid, M., & Tahir, M. (2021). Neutrosophic Normal Probability Distribution—A Spine of Parametric Neutrosophic Statistical Tests: Properties and Applications. In *Neutrosophic Operational Research* (pp. 153-169). Springer.

Sherwani, R. A. K., Naeem, M., Aslam, M., Raza, M., Abid, M., & Abbas, S. *Neutrosophic beta distribution with properties and applications*. Infinite Study.

Shio Gai Quek, H. G. (2023). VIKOR and TOPSIS framework with a truthful-distance measure for the (t, s)-regulated interval-valued neutrosophic soft set. *Soft Computing*.

Simic, V., Gokasar, I., Deveci, M., & Isik, M. (2021a). Fermatean fuzzy group decision-making based CODAS approach for taxation of public transit investments. *IEEE Transactions on Engineering Management*.

Simic, V., Karagoz, S., Deveci, M., & Aydin, N. (2021b). Picture fuzzy extension of the CODAS method for multi-criteria vehicle shredding facility location. *Expert Systems with Applications*, *175*, 114644. DOI: 10.1016/j.eswa.2021.114644

Simić, V., Milovanović, B., Pantelić, S., Pamučar, D., & Tirkolaee, E. B. (2023). Sustainable route selection of petroleum transportation using a type-2 neutrosophic number based ITARA-EDAS model. *Information Sciences*, *622*, 732–754. DOI: 10.1016/j.ins.2022.11.105

Singh, A., Arora, R., & Arora, S. (2022). Bilevel transportation problem in neutrosophic environment. *Computational & Applied Mathematics*, *41*(1), 1–25. DOI: 10.1007/s40314-021-01711-3

Singh, A., Arora, R., & Arora, S. (2023). A Novel Fully Interval-Valued Intuitionistic Fuzzy Multi-objective Indefinite Quadratic Transportation Problem with an Application to Cost and Wastage Management in the Food Industry. In *Fuzzy Optimization, Decision-making and Operations Research: Theory and Applications* (pp. 87–110). Springer International Publishing. DOI: 10.1007/978-3-031-35668-1_5

Singh, A., Kumar, A., & Appadoo, S. S. (2017). Modified approach for optimization of real life transportation problem in neutrosophic environment. *Mathematical Problems in Engineering, 2017*, 1–9. Advance online publication. DOI: 10.1155/2017/2139791

Sivaramakrishnan, S., & Suresh, K. (2019). Interval-valued anti fuzzy subring. *Journal of Applied Science and Computations*, *6*(1), 521–526.

Sivaramakrishnan, S., & Vijayabalaji, S. (2020). Interval-valued anti fuzzy linear space. *IEEE Xplore*, *839-841*, 1–3. DOI: 10.1109/ICSCAN49426.2020.9262371

Sivaramakrishnan, S., Vijayabalaji, S., & Balaji, P. (2024). Neutrosophic interval-valued anti-fuzzy linear space. *Neutrosophic Sets and Systems*, *63*, 271–284. DOI: 10.5281/zenodo.10531827

Sivaraman, G., Nayagam, V. L., & Ponalagusamy, R. (2012)., Intuitionistic Fuzzy Interval Information System. *International Journal of Computer Theory and Engineering, 4*.

Sivaraman, G., Nayagam, V. L., & Ponalagusamy, R. (2014). A complete ranking of incomplete interval information. *Expert Systems with Applications*, *41*, 1947–1954. DOI: 10.1016/j.eswa

Smarandache, F. (1998). *Neutrosophy: neutrosophic probability, set, and logic: analytic synthesis & synthetic analysis.*

Smarandache, F. (1999). A Unifying Field in Logics. Neutrosophy: Neutrosophic Probability, Set and Logic. American research Press, Rehoboth.

Smarandache, F. (1999). A unifying field in Logics: Neutrosophic Logic. In *Philosophy* (pp. 1-141). American Research Press.

Smarandache, F. (2000). Neutrosophy. *arXiv preprint math/0010099.*

Smarandache, F. (2004). A geometric interpretation of the neutrosophic set-A generalization of the intuitionistic fuzzy set. *arXiv preprint math/0404520.*

Smarandache, F. (2010, July). α-discounting method for multi-criteria decision making (α-d MCDM). In *2010 13th International Conference on Information Fusion* (pp. 1-7). IEEE.

Smarandache, F. (2014). *Introduction to Neutrosophic statistics*. Sitech & Education Publishing.

Smarandache, F. (2014). *Neutrosophic Theory and Its Applications, Vol. I: Collected Papers*. Infinite Study.

Smarandache, F. (2014). Refined neutrosophic logic and its application to physics.

Smarandache, F. (2018). Extension of soft set to hypersoft set, and then to plithogenic hypersoft set. *Neutrosophic Sets and Systems, 22*(1), 168-170.

Smarandache, F. (2018). Extension of soft set to hypersoft set, and then to plitho-genichypersoft set. *Neutrosophic Sets and Systems, 22*(1), 168-170.

Smarandache, F. (2018). Plithogeny, plithogenic set, logic, probability, and statistics. *arXiv preprint arXiv:1808.03948.*

Smarandache, F. (2020). The Score, Accuracy, and Certainty Functions determine a Total Order on the Set of Neutrosophic Triplets (T, I, F). *Neutrosophic Sets and Systems,38*, 1. https://digitalrepository.unm.edu/nss_journal/vol38/iss1/1

Smarandache, F. (2022b). Soft set product extended to hypersoft set and indetermsoft set cartesian product extended to indetermhypersoft set. *Journal of Fuzzy Extension and Applications, 3*(4), 313-316.

Smarandache, F., & Jdid, M. (2023). On Overview of Neutrosophic and Plithogenic Theories and Applications. *Prospects for Applied Mathematics and Data Analysis, 2*(1), 19-26. DOI: 10.54216/PAMDA.020102

Smarandache, F., & Pramanik, S. (2016). *New trends in neutrosophic theory and applications* (Vol. 1). Infinite Study.

Smarandache, F. (1998). *A Unifying Field in Logics. Neutrosophy: Neutrosophic Probability. Set and Logic*. American Research Press.

Smarandache, F. (1998). *A unifying field of logics. Neutrosophy: neutrosophic probability, set and logic*. American Research Press.

Smarandache, F. (1999).*A Unifying Field in Logic: Neutrosophic Logic, Neutrosophy, Neutrosophic set, Neutrosophic Probability*. American Research Press.

Smarandache, F. (1999). A unifying field in logics: Neutrosophic logic. In *Philosophy* (pp. 1–141). American Research Press.

Smarandache, F. (2002). Neutrosophy and Neutrosophic Logic. *First International Conference on Neutrosophy*, Neutrosophic Logic, Set, Probability, and Statistics University of New Mexico.

Smarandache, F. (2003). Definiton of neutrosophic logic-a generalization of the intuitionistic fuzzy logic. *EUSFLAT Conf.*

Smarandache, F. (2004). Neutrosophic set a generalization of the intuitionistic fuzzy set. *International Journal of Pure and Applied Mathematics*.

Smarandache, F. (2005). Generalization of the intuitionistic fuzzy logic to the neutrosophic fuzzy set. *International Journal of Pure and Applied Mathematics*, *24*(3), 287–297.

Smarandache, F. (2006). Neutrosophic set - A generalization of the intuitionistic fuzzy set. In *2006 IEEE International Conference on Granular Computing* (pp. 38–42). IEEE. DOI: 10.1109/GRC.2006.1635754

Smarandache, F. (2006, May). *Neutrosophic set-a generalization of the intuitionistic fuzzy set. In 2006 IEEE international conference on granular computing*. IEEE.

Smarandache, F. (2010). Neutrosophic Set: A Generalization of Intuitionistic Fuzzy set. *Journal of Defense Resources Management*, *1*, 107–116.

Smarandache, F. (2013). *Introduction to neutrosophic measure, neutrosophic integral, and neutrosophic probability*. Sitech and Education Publisher.

Smarandache, F. (2014). *Introduction to Neutrosophic statistics*. Sitech & Education Publishing.

Smarandache, F. (2014). *Introduction to neutrosophic statistics*. Sitech and Education Publishing.

Smarandache, F. (2018). Extension of Soft Set to Hypersoft Set, and then to Plithogenic Hypersoft Set. *Neutrosophic Sets and Systems*, *22*, 168–170.

Smarandache, F. (2018). Plithogenic Set, an Extension of Crisp, Fuzzy, Intuitionistic Fuzzy, and Neutrosophic Sets - Revisited. *Neutrosophic Sets and Systems.*, *21*, 153–166.

Smarandache, F. (2020). The score, accuracy, and certainty functions determine a total order on the set of neutrosophic triplets (t, i, f). *Neutrosophic Sets and Systems*, *38*, 1–14.

Smarandache, F. (2021). Structure, NeutroStructure, and AntiStructure in Science. [IJNS]. *International Journal of Neutrosophic Science*, *13*(1), 28–33. DOI: 10.54216/IJNS.130104

Smarandache, F. (2022a). Introduction to the IndetermSoft Set and IndetermHyperSoft Set. *Neutrosophic Sets and Systems*, *50*, 629–650.

Smarandache, F. (2022c). Practical Applications of IndetermSoft Set and IndetermHyperSoft Set and Introduction to TreeSoft Set as an extension of the MultiSoft Set. *Neutrosophic Sets and Systems*, *51*, 939–947.

Smarandache, F. (2023). New Types of Soft Sets: HyperSoft Set, IndetermSoft Set, IndetermHyperSoft Set, and TreeSoft Set. *International Journal of Neutrosophic Science, 20*(4), 58–64. DOI: 10.54216/IJNS.200404

Smarandache, F. (2023). New Types of Soft Sets" HyperSoft Set, IndetermSoft Set, IndetermHyperSoft Set, and TreeSoft Set": An Improved Version. *Neutrosophic Systems with Applications, 8*, 35–41. DOI: 10.61356/j.nswa.2023.41

Smarandache, F., Abdel-Basset, M., & Broumi, S. (2022). Neutrosophic Sets and Systems, Vol. 50, 2022. *Neutrosophic Sets and Systems, 50*(1), 40.

Smarandache, F., Broumi, S., Talea, M., & Bakali, A. (2016). Single Valued Neutrosophic Graphs: Degree, Order and Size. In *2016 IEEE International Conference on Fuzzy Systems (FUZZ)*. IEEE.

Smarandache, F., Deli, I., & Said, B. (2014). N - valued interval neutrosophic sets and their application in medical diagnosis. *Critical Review*.

Smith, R. D., & Dudek, R. D. (1967). A General Algorithm for solution of the N-jobs, M-Machine sequencing problem of the Flow Shop. *Operations Research, 21*(1).

Songsaeng, M., & Iampan, A. (2020). Neutrosophic cubic set theory applied to up- algebras. *Thai Journal of Mathematics, 18*(3), 1447–1474.

Sriram, R., Manimaran, G., & Murthy, C. S. R. (1998). Preferred link based delay-constrained least-cost routing in wide area networks. *Computer Communications, 21*(18), 1655–1669. DOI: 10.1016/S0140-3664(98)00194-7

Stancu-Minasian, I. M. (1997). Fractional transportation problem. In *Fractional Programming: Theory, Methods and Applications* (pp. 336–364). Springer Netherlands. DOI: 10.1007/978-94-009-0035-6_11

Subha, R., & Mohana, K. (2023). Fermatean pentapartitioned neutrosophic sets. *Indian Journal of Natural Sciences, 14*.

Sudha, S., & Martin, N. (2023, June). Comparative analysis of Plithogenic neutrosophic PIPRECIA over neutrosophic AHP in criteria ordering of logistics selection. In *AIP Conference Proceedings* (*Vol. 2649*, No. 1). AIP Publishing.

Sudha, S., Martin, N., & Smarandache, F. (2023). Applications of Extended Plithogenic Sets in Plithogenic Sociogram. *International Journal of Neutrosophic Science, 20*(4), 8-35.

Sudha, S., Martin, N., Anand, M. C. J., Palanimani, P. G., Thirunamakkani, T., & Ranjitha, B. (2023). MACBETH-MAIRCA Plithogenic Decision-Making on Feasible Strategies of Extended Producer's Responsibility towards Environmental Sustainability. [IJNS]. *International Journal of Neutrosophic Science, 22*(2), 114–130.

Sudha, S., Martin, N., & Broumi, S. (2022). Plithogenic CRITIC-MAIRCA Ranking of Feasible Livestock Feeding Stuffs. *International Journal of Neutrosophic Science, 18*(4), 160–173.

Sudha, S., Martin, N., & Smarandache, F. (2023). State of Art of Plithogeny Multi Criteria Decision Making Methods. *Neutrosophic Sets and Systems, 56*(1), 27.

Sultana, F., Gulistan, M., Ali, M., Yaqoob, N., Khan, M., Rashid, T., & Ahmed, T. (2022, April 4). A study of plithogenic graphs: Applications in spreading coronavirus disease (COVID-19) globally. *Journal of Ambient Intelligence and Humanized Computing*, 1–21. Advance online publication. DOI: 10.1007/s12652-022-03772-6 PMID: 35401852

Sushil. (2018). Interpretive Multi-Criteria Valuation of Flexibility Initiatives on Direct Value Chain. *Benchmarking An International Journal, 25*, 3720-3742.

Szmelter-Jarosz, A., Ghahremani-Nahr, J., & Nozari, H. (2021). A neutrosophic fuzzy optimisation model for optimal sustainable closed-loop supply chain network during COVID-19. *Journal of Risk and Financial Management, 14*(11), 519.

Tanay, B., & Kandemir, M. B. (2011). Topological structure of fuzzy soft sets. *Computers & Mathematics with Applications (Oxford, England), 61*(10), 2952–2957.

Tang, J., Fung, R. Y., & Yung, K.-L. (2003). Fuzzy modelling and simulation for aggregate production planning. *International Journal of Systems Science, 34*(12-13), 661–673. DOI: 10.1080/00207720310001624113

Tapia, J. F. D. (2021). Evaluating negative emissions technologies using neutrosophic data envelopment analysis. *Journal of Cleaner Production, 286*, 125494. DOI: 10.1016/j.jclepro.2020.125494

Tayal, D. K., Yadav, S. K., & Arora, D. (2023). Personalized ranking of products using aspect-based sentiment analysis and Plithogenic sets. *Multimedia Tools and Applications, 82*(1), 1261–1287.

Tayur, S., Ganeshan, R., & Magazine, M. (2012). *Quantitative models for supply chain management, 17*. Springer Science & Business Media.

Thamaraiselvi, A., & Santhi, R. (2015). Optimal solution of fuzzy transportation problem using hexagonal fuzzy numbers. *International Journal of Scientific and Engineering Research*, *6*(3), 40–45.

Thangavel, K., & Pethalakshmi, A. (2009). Dimensionality reduction based on rough set theory: A review. *Applied Soft Computing*, *9*(1), 1–12. DOI: 10.1016/j.asoc.2008.05.006

Tu, A., Ye, J., & Wang, B. (2021). Neutrosophic Number Optimization Models and Their Application in the Practical Production Process. *Journal of Mathematics*, *2021*, 1–8. DOI: 10.1155/2021/6668711

Turksen, I. B. (1986). Interval valued fuzzy sets based on normal forms. *Fuzzy Sets and Systems*, *20*(2), 191–210. DOI: 10.1016/0165-0114(86)90077-1

Ulucay, V., Sahin, M., & Hassan, N. (2018). Generalized neutrosophic soft expert set for multiple-criteria decision-making. *Symmetry*, *10*(10), 437. DOI: 10.3390/sym10100437

Vasantha Kandasamy, W. (2006). Smarandache Neutrosophic algebraic structures. *arXiv Mathematics e-prints*, math/0603708.

Veeramani, C., Edalatpanah, S. A., & Sharanya, S. (2021). Solving the multi-objective fractional transportation problem through the neutrosophic goal programming approach. *Discrete Dynamics in Nature and Society*, *2021*, 1–17. DOI: 10.1155/2021/7308042

Velu, L., & Ramalingam, B. (2023). Total Ordering on Generalized 'n' Gonal Linear Fuzzy Numbers. *Int J Comput Intell Syst*, *16*, 23. DOI: 10.1007/s44196-022-00180-8

Verma, R., Biswal, M. P., & Biswas, A. (1997). Fuzzy programming technique to solve multi-objective transportation problems with some non-linear membership functions. *Fuzzy Sets and Systems*, *91*(1), 37–43. DOI: 10.1016/S0165-0114(96)00148-0

Vidhya, R., Hepzibah, I., & Gani, N. (2017). Neutrosophic multi-objective linear programming problems. *Global Journal of Pure and Applied Mathematics*, *13*(2), 265–280.

Vijayabalaji, S. (2012). Cartesian Product and Homomorphism of Interval-Valued Fuzzy Linear Space. *International Journal of Open Problems in Computer Science and Mathematics*, *5*(4), 93–103. DOI: 10.12816/0006141

Vijayabalaji, S., & Balaji, P. (2015). MCDM method in cricket by rough matrix theory. *International Journal of Mathematical Analysis*, *9*, 869–875. DOI: 10.12988/ijma.2015.5380

Vijayabalaji, S., & Balaji, P. (2020). Best'11 strategy in cricket using MCDM, rough matrix and assignment model. *Journal of Intelligent & Fuzzy Systems, 39*(5), 7431–7447. DOI: 10.3233/JIFS-200784

Vijayabalaji, S., & Sivaramakrishnan, S. (2012, September 26). Sivaramakrishnan; Anti fuzzy M-semigroup. *AIP Conference Proceedings, 1482*(1), 446–448. DOI: 10.1063/1.4757511

Voskoglou, M. G. (2022). Use of Soft Sets and the Bloom's Taxonomy for Assessing Learning Skills. *Transactions on Fuzzy Sets and Systems, 1*(1), 106–113.

Voskoglou, M., Broumi, S., & Smarandache, F. (2022). A Combined Use of Soft and Neutrosophic Sets for Student Assessment with Qualitative Grades. *Journal of Neutrosophic and Fuzzy Systems, 4*(1), 15–20. DOI: 10.54216/JNFS.040102

Wang, H., Smarandache, F., Sunderraman, R., & Zhang, Y.-Q. (2005). *interval neutrosophic sets and logic: theory and applications in computing: Theory and applications in computing* (Vol. 5). Infinite Study.

Wang, H., Smarandache, F., Zhang, Y., & Sunderraman, R. (2010). Single valued neutrosophic sets. *Infinite study, 12*, 20110.

Wang, H., Smarandache, F., Zhang, Y., & Sunderraman, R. (2010). Single valued neutrosophic sets. *Infinite Study, 12*, 20110.

Wang, J. J., Jing, Y. Y., Zhang, C. F., & Zhao, J. H. (2009). Review on multi-criteria decision analysis aid in sustainable energy decision-making. *Renewable & Sustainable Energy Reviews, 13*(9), 2263–2278.

Wang, L., Zhang, H. Y., Wang, J. Q., & Li, L. (2018). Picture fuzzy normalized projection-based VIKOR method for the risk evaluation of construction project. *Applied Soft Computing, 64*, 216–226. DOI: 10.1016/j.asoc.2017.12.014

Wang, P., Lin, Y., Fu, M., & Wang, Z. (2023). VIKOR method for plithogenic probabilistic linguistic MAGDM and application to sustainable supply chain financial risk evaluation. *International Journal of Fuzzy Systems, 25*(2), 780–793.

Wang, Q., Huang, Y., Kong, S., Ma, X., Liu, Y., Das, S. K., & Edalatpanah, S. A. (2021). A Novel Method for Solving Multiobjective Linear Programming Problems with Triangular Neutrosophic Numbers. *Journal of Mathematics, 2021*, 1–8. DOI: 10.1155/2021/6631762

Wang, Y. (2022). Inventory Path Optimization of VMI Large Logistics Enterprises Based on Ant Colony Algorithm. *Mobile Information Systems*, 2022.

Wei, G. (2019). The Novel Generalized Exponential *Entropy for intuitionistic Fuzzy Sets and Interval Valued Intuitionistic,* 2339.

Wei, G., & Lu, M. (2018). Pythagorean Fuzzy Maclaurin Symmetric Mean Operators in Multiple Attribute decision Making. *IEEE Access : Practical Innovations, Open Solutions*, *6*, 7866–7884. DOI: 10.1109/ACCESS.2018.2877725

Weihrich, H., Koontz, H., Cannice, M., & SDR Printers. (2013). *Management : a global, innovative, and entrepreneurial perspective.* McGraw-Hill Education Publishing Company.

Williams, A. C. (1963). A stochastic transportation problem. *Operations Research*, *11*(5), 759–770. DOI: 10.1287/opre.11.5.759

Xie, F., Butt, M. M., Li, Z., & Zhu, L. (2017). An upper bound on the minimal total cost of the transportation problem with varying demands and supplies. *Omega*, *68*, 105–118. DOI: 10.1016/j.omega.2016.06.007

Xu, W., Ma, J., Wang, S., & Hao, G. (2010). Vague soft sets and their properties. *Computers & Mathematics with Applications (Oxford, England)*, *59*(2), 787–794.

Yager, R. R. (1981). A procedure for ordering fuzzy subsets of the unit interval. *Information Sciences*, *24*(2), 143–161.

Yager, R. R. (1981). A procedure for ordering fuzzy subsets of the unit interval. *Information Sciences*, *24*(2), 143–161. DOI: 10.1016/0020-0255(81)90017-7

Yager, R. R. (1988). On ordered weighted averaging aggregation operators in multicriteria decisionmaking. *IEEE Transactions on Systems, Man, and Cybernetics*, *18*(1), 183–190. DOI: 10.1109/21.87068

Yager, R. R. (2013). Pythagorean fuzzy subsets. In *Proceedings of the 2013 Joint IFSA World Congress and NAFIPS Annual Meeting, IFSA/NAFIPS 2013* (pp. 57–61). IEEE. DOI: 10.1109/IFSA-NAFIPS.2013.6608375

Yager, R. R. (2014). Pythagorean membership grades in multicriteria decision making. *IEEE Transactions on Fuzzy Systems*, *22*(4), 958–965. DOI: 10.1109/TFUZZ.2013.2278989

Yager, R. R., & Abbasov, A. M. (2013). Pythagorean membership grades, complex numbers, and decision making. *International Journal of Intelligent Systems*, *28*(5), 436–452. DOI: 10.1002/int.21584

Yang, L., & Liu, B. (2005, May). A multi-objective fuzzy assignment problem: New model and algorithm. In *The 14th IEEE International Conference on Fuzzy Systems, 2005. FUZZ'05.* (pp. 551-556). IEEE.

Yang, C., Wang, Q., Pan, M., Hu, J., Peng, W., Zhang, J., & Zhang, L. (2022). A linguistic Pythagorean hesitant fuzzy MULTIMOORA method for third-party reverse logistics provider selection of electric vehicle power battery recycling. *Expert Systems with Applications*, *198*, 116808. DOI: 10.1016/j.eswa.2022.116808

Yang, L., Li, D., & Tan, R. (2020). Research on the shortest path solution method of interval valued neutrosophic graphs based on the ant colony algorithm. *IEEE Access : Practical Innovations, Open Solutions*, *8*, 88717–88728.

Yang, L., & Liu, L. (2007). Fuzzy fixed charge solid transportation problem and algorithm. *Applied Soft Computing*, *7*(3), 879–889. DOI: 10.1016/j.asoc.2005.11.011

Yao, J.-S., & Lin, F.-T. (2000). Fuzzy critical path method based on signed distance ranking of fuzzy numbers. *IEEE Transactions on Systems, Man, and Cybernetics. Part A, Systems and Humans*, *30*(1), 76–82. DOI: 10.1109/3468.823483

Yassein, H. R., & Mohammed, A. H. (2011). *Antifuzzy bi-Γ-ideals of Γ- semigroups*. Research Gate. https://www.researchgate.net/publication/321706124

Ye, J. (2014). Similarity measures between interval neutrosophic sets and their applications in multicriteria decision-making. *Journal of Intelligent & Fuzzy Systems*, *26*(1), 165–172.

Ye, J. (2014). Single-valued neutrosophic minimum spanning tree and its clustering method. *Journal of Intelligent Systems*, *23*(3), 311–324. DOI: 10.1515/jisys-2013-0075

Ye, J. (2016). Multiple-attribute group decision-making method under a neutrosophic number environment. *Journal of Intelligent Systems*, *25*(3), 377–386. DOI: 10.1515/jisys-2014-0149

Yolcu, A., & Ozturk, T. Y. (2021). Fuzzy Hypersoft Sets and It's Application to Decision-Making. *Theory and Application of Hypersoft Set* (pp. 50-64), Pons Publishing House, Brussel.

Yolcu, A., Smarandache, F., & Öztürk, T. Y. (2021). Intuitionistic fuzzy hypersoft sets. *Communications Faculty of Sciences University of Ankara Series A1 Mathematics and Statistics, 70*(1), 443-455.

Yolcu, A. (2023). Intuitionistic fuzzy hypersoft topology and ıts applications to multi-criteria decision-making. *Sigma*, *41*(1), 106–118.

Yolcu, A., & Öztürk, T. Y. (2021). Fuzzy hypersoft sets and it's application to decision-making. In Smarandache, F., Saeed, M., Abdel-Baset, M., & Saqlain, M. (Eds.), *Theory and Application of Hypersoft Set* (pp. 50–64). Pons Publishing House.

Yolcu, A., & Ozturk, T. Y. (2022). On fuzzy hypersoft topological spaces. *Caucasian Journal of Science*, *9*(1), 1–19.

Yolcu, A., Smarandache, F., & Öztürk, T. Y. (2021). Intuitionistic fuzzy hypersoft sets. *Communications Faculty of Sciences University of Ankara Series A1 Mathematics and Statistics*, *70*(1), 443–455. DOI: 10.31801/cfsuasmas.788329

Yon-Delgado, J. C., Yon-Delgado, M. R., Aguirre-Baique, N., Gamarra-Salinas, R., & Ponce-Bardales, Z. E. (2023). Neutrosophic Plithogenic AHP Model for Inclusive Higher Education Program Selection. *International Journal of Neutrosophic Science*, *21*(1), 50–0.

Yoon, K. P., & Hwang, C. L. (1995). *Multiple attribute decision making: an introduction*. Sage publications. DOI: 10.4135/9781412985161

Zadeh, L. A. (1965). Fuzzy set. *Information and Control*, *8*, 338–353.

Zadeh, L. A. (1965). Fuzzy sets. *Information and Control*, *8*(3), 338–353. DOI: 10.1016/S0019-9958(65)90241-X

Zadeh, L. A. (1975). The concept of a linguistic variable and its application to approximate reasoning-I. *Information Sciences*, *8*(3), 199–249. DOI: 10.1016/0020-0255(75)90036-5

Zadeh, L. A. (1978). Fuzzy sets as a basis for a theory of possibility. *Fuzzy Sets and Systems*, *1*(1), 3–28. DOI: 10.1016/0165-0114(78)90029-5

Zadeh, L. A. (1984). Fuzzy Probabilities. *Information Processing & Management*, *20*(3), 363–372. DOI: 10.1016/0306-4573(84)90067-0

Zavadskas, E. K., Turskis, Z., Antucheviciene, J., & Zakarevicius, A. (2012). Optimization of weighted aggregated sum product assessment. *Elektronika ir Elektrotechnika*, *122*(6), 3–6. DOI: 10.5755/j01.eee.122.6.1810

Zeema, J. L., & Christopher, D. F. X. (2019). Evolving optimized neutrosophic C means clustering using behavioral inspiration of artificial bacterial foraging (ONCMC-ABF) in the prediction of Dyslexia. *Journal of King Saud University. Computer and Information Sciences*.

Zeina, M. B. (2020). Erlang Service Queueing Model with Neutrosophic Parameters. *International Journal of Neutrosophic Science*, *6*(2), 106–112. DOI: 10.54216/IJNS.060202

Zhang, D., Zhang, J., Lai, K. K., & Lu, Y. (2009). An novel approach to supplier selection based on vague sets group decision. *Expert Systems with Applications*, *36*(5), 9557–9563.

Zhang, H. D., & Shu, L. (2014). Possibility multi-fuzzy soft set and its application in decision making. *Journal of Intelligent & Fuzzy Systems*, *27*(4), 2115–2125. DOI: 10.3233/IFS-141176

Zhang, H. Y., Wang, J. Q., & Chen, X. H. (2014). Interval neutrosophic sets and their application in multicriteria decision making problems. *TheScientificWorldJournal*, *2014*, 2014. DOI: 10.1155/2014/645953 PMID: 24695916

Zhang, M., Cheng, M. X., & Tarn, T. J. (2006). A mathematical formulation of DNA computation. *IEEE Transactions on Nanobioscience*, *5*(1), 32–40. DOI: 10.1109/TNB.2005.864017 PMID: 16570871

Zhang, S., Gao, H., Wei, G., Wei, Y., & Wei, C. (2019). Evaluation based on distance from average solution method for multiple criteria group decision making under picture 2-tuple linguistic environment. *Mathematics*, *7*(3), 243. DOI: 10.3390/math7030243

Zhang, X., & Xu, Z. (2014). Extensioon of TOPSIS to Multicriteria decision Making with Pythagorean Fuzzy Sets. *International Journal of Intelligent Systems*, *29*(12), 1061–1078. DOI: 10.1002/int.21676

Zhang, Z., Yang, J. Ye. Y., Hu, Y., & Zhang, Q. (2012). A type of score function on intuitionistic fuzzy set with double parameters and its application to pattern recognition and medical diagnosis. *Procedia Engineering*, *29*, 4336–4342. DOI: 10.1016/j.proeng.2012.01.667

Zhan, J., Wang, W., Carlos, A. R., & Zhan, J. (2022). A three-way decision approach with prospect-regret theory via fuzzy set pair dominance degrees for incomplete information systems. *Information Sciences*, *617*, 310–330.

Zhao, J., Li, B., Rahman, A. U., & Saeed, M. (2023). An intelligent multiple-criteria decision-making approach based on sv-neutrosophic hypersoft set with possibility degree setting for investment selection. *Management Decision*, *61*(2), 472–485. DOI: 10.1108/MD-04-2022-0462

Zimmermann, H. J. (1978). Fuzzy programming and linear programming with several objective functions. *Fuzzy Sets and Systems*, *1*(1), 45–55.

Zulqarnain, M., Siddique, I., Ali, R., Awrejcewicz, J., Karamti, H., Grzelczyk, D., Iampan, A., & Asif, M. (2022). Einstein Ordered Weighted Aggregation Operators for Pythagorean Fuzzy Hypersoft Set With Its Application to Solve MCDM Problem. *IEEE Access : Practical Innovations, Open Solutions*, *10*, 95294–95320. DOI: 10.1109/ACCESS.2022.3203717

About the Contributors

Florentin Smarandache, PhD, PostDocs, is an Emeritus Professor of Mathematics at the University of New Mexico, United States. He got his MSc in Mathematics and Computer Science from the University of Craiova, Romania, PhD in Mathematics from the State University of Kishinev, and Postdoctoral in Applied Mathematics from Okayama University of Sciences, Japan, and The Guangdong University of Technology, Guangzhou, China. He is the founder of neutrosophy (generalization of dialectics), neutrosophic set, logic, probability and statistics since 1995 and has published hundreds of papers and books on neutrosophic physics, superluminal and instantaneous physics, unmatter, quantum paradoxes, absolute theory of relativity, redshift and blueshift due to the medium gradient and refraction index besides the Doppler effect, paradoxism, outerart, neutrosophy as a new branch of philosophy, Law of Included Multiple-Middle, multispace and multistructure, HyperSoft set, TreeSoft Set, IndetermSoft Set and IndetermHyperSoft Set, SuperHyperGraph, SuperHyperTopology, SuperHyperAlgebra, SuperHyperFunction, Neutrosophic SuperHyperAlgebra, degree of dependence and independence between neutrosophic components, refined neutrosophic set, neutrosophic over-under-off-set, plithogenic set / logic / probability / statistics, symbolic plithogenic algebraic structures, neutrosophic triplet and duplet structures, quadruple neutrosophic structures, extension of algebraic structures to NeutroAlgebra and AntiAlgebra, NeutroGeometry and AntiGeometry, NeutroTopology and AntiTopology, Refined Neutrosophic Topology, Refined Neutrosophic Crisp Topology, Dezert-Smarandache Theory and so on to many peer-reviewed international journals and many books and he presented papers and plenary lectures to many international conferences around the world. In addition, he published many books of poetry, dramas, children stories, translations, essays, a novel, folklore collections, traveling memories, and art albums [http://fs.unm.edu/FlorentinSmarandache.htm].

Maikel Y. Leyva Vázquez, PhD, has a notable career spanning several universities in Cuba and Ecuador. He has held positions such as the General Director of the Babahoyo Campus at Universidad Regional Autónoma de los Andes (UNIANDES) and Vice Dean of Research at Universidad de las Ciencias Informáticas. His academic qualifications include a Bachelor's degree in Informatics Engineering from Universidad Tecnológica de La Habana "José Antonio Echeverría," Cujae, an MSc in Bioinformatics from El Instituto Superior de Tecnologías y Ciencias Aplicadas (InSTeC), and a PhD in Technical Sciences from Universidad de las Ciencias Informáticas. Additionally, as a member of the Grupo de Investigación en Inteligencia Artificial y Reconocimiento Facial at Universidad Politécnica Salesiana, Dr. Leyva Vázquez contributed to artificial intelligence. He is deeply involved in the scientific community, presenting at international forums, publishing research, and mentoring PhD students.He/She serves as the president of the Latin American Association of Neutrosophic Sciences and as editor of several academic journals, significantly contributing to intellectual exchange in the academic realm.

* * *

Ritu Arora is presently working as Professor in the Department of Mathematics, Keshav Mahavidyalaya, University of Delhi. She has more than 22 years of teaching experience. She did her Masters and Ph.D. in Mathematics from University of Delhi. Her areas of research interest include Mathematical Programming/Optimization and its application to various allocation problems. She has published more than 25 research papers in National and International Journals of repute. She has authored the book "Linear Programming and Game Theory" by Narosa Publishing House. She is also one of the editors in the book series, Advances in Metaheuristics: "Combinatorial Optimization under Uncertainty – Real Life Scenarios in Allocation Problems", Edited by, CRC Press, Taylor and Francis Group, 2023.

Shalini Arora is presently working as Professor in Mathematics and Head at Applied Sciences and Humanities Department, Indira Gandhi Delhi Technical University for Women (IGDTUW). She has more than 22 years of Teaching experience. She did her Masters and Ph.D. in Mathematics from IIT Delhi. Prior to joining IGDTUW She worked as Assistant Professor at the Operations Management Group, IIM Calcutta and Department of Mathematics at Lady Shri Ram College. She is a recipient of the 'Young Scientist Award' by the SERC division of DST. Her areas of research interest include Mathematical Programming, Allocation Problems viz., Transportation and Assignment Problems, Combinatorial optimization etc. She has published more than 30 research papers in Journals of International and national repute. She is also one of the editors in the Book Series: Advances in Metaheuristics-

1st Edition "Combinatorial Optimization under Uncertainty – Real Life Scenarios in Allocation Problems", Edited by, CRC Press, Taylor and Francis Group, May 2023.

Sima Das is an Assistant Professor in the Department of Computer Science and Engineering at Bengal College of Engineering and Technology, Durgapur, India. She previously served as an Assistant Professor in the Department of Computer Science and Engineering at Camellia Institute of Technology and Management in Hooghly, West Bengal, India. Sima Das completed her M. Tech in Computer Science and Engineering in August 2020 from Maulana Abul Kalam Azad University of Technology (Main Campus), located in West Bengal, India. She is Doctoral Fellow in the Department of Computer Science and Engineering at NIT Rourkela, Odisha, India. Her areas of expertise and interest include Artificial Intelligence, Machine Learning, Deep Learning, Internet of Things, Cybersecurity, Smart Healthcare, and Agriculture. Sima Das has been recognized for her outstanding research contributions with the Research Excellence Award from the Global Innovation & Excellence Award in 2021. She is an Associate Member of the Institute of Engineers and a Professional Member of IEEE. Sima Das has authored numerous books, book chapters, conference papers, patents, and journal articles. Additionally, she actively participates.

Seyyed Ahmad Edalatpanah is Associate Professor at the Ayandegan Institute of Higher Education, Tonekabon, Iran. S. A. Edalatpanah received his Ph.D. degree in Applied Mathematics from the University of Guilan, Rasht, Iran. He is currently working as the Chief of R&D at the Ayandegan Institute of Higher Education, Iran. He is also an academic member of Guilan University and the Islamic Azad University of Iran. Dr. Edalatpanah's fields of interest include numerical computations, operational research, uncertainty, fuzzy set and its extensions, numerical linear algebra, soft computing, and optimization. He has published over 150 journal and conference proceedings papers in the above research areas. He serves on the editorial boards of several international journals. He is also the Director-in-Charge of the Journal of Fuzzy Extension & Applications at: http://www.journal-fea.com/ . Currently, he is president of "International Society of Fuzzy Set Extensions and Applications" (ISFSEA) at: https://isfsea.org.Edalatpanah's research is widely recognized internationally, he has been featured in the list of the Top 2% scientists in the world published by Stanford University from 2021.

Lakshmana Gomathi Nayagam received his MSc and PhD in Mathematics from Madurai Kamaraj University in India in 1994 and 2002, respectively. He has been a visiting researcher at universities in the United Kingdom, New Zealand, Switzerland, Dubai, Hong Kong, and Taiwan. He is currently an professor in the

Department of Mathematics at the National Institute of Technology, Tiruchirappalli, India. His current research interests include Fuzzy Topological Structures, Fuzzy Alge- braic Structures, Fuzzy Decision Making Problems, Neutrosophic sets, Fuzzy Clustering Techniques, and Fuzzy Control Systems. He is the author or co-author of over 50 papers published in refereed indexed international journals and conferences. He is also a regular reviewer for many prestigious international journals and has organized and executed many national confer- ences/workshops/ seminars and delivered nearly 200 special lectures in various institutions in various capacities. V. Lakshmana Gomathi Nayagam is a senior member of IEEE, IEEE Computational Intelligence Society, Control Systems Society and Information Theory Society, annual member of the European Society for Fuzzy Logic and Technology (EUSFLAT), Life member of Taiwan Fuzzy Systems Association, Indian Mathematics Society, and Indian Society for Technical Education.

Mohana K. graduated from the University of Bharathiyar with B. Sc., M. Sc., M. Phil., Ph. D., PGDCA., PGDOR., degrees in Mathematics. Assistant Professor in Mathematics department since 2014. Published more than 60 research papers in quality Journals. Produced 4 M. Phil's and currently working in the field of topology with 4 Ph. D scholars.

Khuram Ali Khan is an Associate Professor at Department of Mathematics, University of Sargodha, Sargodha, Pakistan. He has published more 100 papers in peer reviewed journals with citations more than 700. He is the reviewer of many peer reviewed journals. His areas of research are: Mathematical Analysis, Mathematical Inequalities, Time scales Calculus, Hypersoft Set Theory.

Sivasakthi M. is currently an Associate Professor of Mathematics at Krishnasamy College of Science, Arts, and Management for Women in Cuddalore. She received her M.Sc. from Thiruvallur University, her M.Phil. from Periyar University, and her Ph.D. from Thiruvallur University, where her dissertation was on Graph Theory - Harmonic Mean Labelling of Graphs. She has 15 years of undergraduate teaching experience, 14 years of graduate teaching experience, and 7 years of guidance for the M.Phil course. She has published 11 international journals, including UGC-CARE and Scopus. Her current research focuses on graph labelling.

Manoj Kumar Mondal completed his Ph.D. from IIEST, Shibpur, India in 2015. He works on Solid Mechanics, Visco-elasticity, Fuzzy sets, Fuzzy-MCDM etc. He has published several research articles in different reputed national and international journals. Currently, he is working at the Haldia Institute of Technology, Haldia, India.

Taha Yasin Öztürk is BSc (2008) and MSc (2010) from Kafkas University Turkey. PhD (2013) from Ataturk University Turkey. Currently, professor in Department of Mathematics, Faculty of Sciences and Letter, Kafkas University, Turkey. Now he is working as Professor at Kafkas University. His research interest focuses soft set, fuzzy set, bipolar soft set, neutrosophic set and topological structures of this notions.

Daniel P. received his bachelor's degree in Mathematics from Bharathidasan University in India in 2014, and his master's degree from the Department of Mathematics at Pondicherry University in Pondicherry, India, in 2019. He has worked as a Research Fellow at the Central University of Tamil Nadu, Thiruvarur, India. He is currently working as an assistant professor in the department of mathematics at St. Xavier's college, palayamkottai and pursuing doctorate in part time in the Department of Mathematics at the National Institute of Technology, Tiruchirappalli, India. His research interests include fuzzy control systems, Fuzzy Decision Making Problems, Fuzzy Information System and Neutrosophic Information Systems.

Balaji Parthasarathy is an Assistant Professor of Mathematics at MEASI Academy of Architecture, Chennai. He received his Ph.D. from Anna University in Chennai. His research interests include rough sets, rough matrices, sigmoid valued fuzzy soft sets, neutrosophic sets and soft sets. Furthermore, he has 17 years of teaching and 10 years of research experience. His research papers have been published in several reputed international journals. He authored 15 academic articles and two book chapters. He delivered 5 invited talks.

Bharanidharan R. received his M.Sc Mathematics degree from Bharathidasan University and M.Phil Mathematics degree from Periyar university. He cleared CSIR Net Exam in 2018. Now currently pursuing his research in NIT Trichy.

Atiqe Ur Rahman is working as Head of Department (Mathematics) at the Institute of Leadership and Management, Jauharabad Campus, Khushab, Punjab, Pakistan, and has affiliation with the Department of Mathematics, University of Management and Technology, Lahore, Pakistan, where he is attached to Dr. Muhammad Saeed's Lab for research in the following areas: 1. Fuzzy Set, Intuitionistic Fuzzy Set, Neutrosophic Set, Soft Set, Hypersoft Set with their Hybrids and Algebraic Structures 2. Complexity and Convexity in Fuzzy and Soft like Environments 3. Mathematical Inequalities and Time Scales Calculus Web of Science Researcher ID: ABG-8247-2020 Scopus Author ID: 57221292719 https://sciprofiles.com/profile/aurkhb.

Muhammad Saeed is the full professor at the Department of Mathematics, University of Management and Technology, Lahore, Pakistan. He has published more than 200 research papers with more than 2000 citations. His areas of research are: Fuzzy Sets, Rough Sets, Soft Set, Hypersoft Set, Neutrosophic Sets, Multicriteria Decision Making, Optimizations, Artificial Intelligence, Pattern recognition and optimization under convex environments, Graph theory in fuzzy-like, soft-like and hypersoft-like environments, similarity, distance measures and their relevant operators in multipolar hybrid structures.

Aakanksha Singh is currently a Research Scholar in Department of Applied Science and Humanities, Indira Gandhi Delhi Technical University for Women, Delhi. Her area of research is transportation problem. She is also an Assistant Professor in Department of Mathematics, Aryabhatta College, University of Delhi, India. She has a teaching experience of more than 12 years. She holds a Master's degree in Pure Mathematics from Miranda House, University of Delhi, Delhi, India.

Biplab Sinha Mahapatra completed his PhD from Jadavpur University, Kolkata in the year 2016. His research area includes: Reliability optimization, multi-objective mathematical programming, Fuzzy logic, Multi-criteria decision analysis, Geometric programming, and Genetic algorithm. He has published several research papers in international journals such as: Annals of operations research, Expert system with application, Journal of intelligent and fuzzy systems, Fuzzy information and engineering. He is currently working at Haldia Institute of Technology.

S. Sivaramakrishnan is an Associate Professor and Head of Mathematics at Manakula Vinayagar Institute of Technology in Kalitheerthal Kuppam, Puducherry. He received his Ph.D. from Anna University in Chennai. His research interests include cubic sets, neutrosophic sets, fuzzy algebra, and soft sets. Furthermore, He has 20 years of teaching and 13 years of research experience. His research papers have been published in several reputable international journals. He authored more than 20 academic articles and two book chapters. He delivered 5 invited talks and acting as doctoral committee member for Anna University, Chennai.

G. Venkadesh is an Assistant Professor of Mathematics at Krishnasamy College of Engineering and Technology, S. Kumarapuram, Cuddalore. His research interests include Fuzzy set, soft set, soft matrices, Pythagorean fuzzy set, neutrosophic sets. Furthermore, he has 14years of teaching and 1 year of research experience. He presented 1 paper in an International conference and abstract published in conference proceedings.

Srinivasan Vijayabalaji is presently working as an Assistant Professor (Senior Grade) in the Department of Mathematics (S & H), University College of engineering Panruti (A Constituent College of Anna University), Panruti-607106, Tamil Nadu, India. He is having more than 23 years of teaching and research experience. His research area includes functional analysis, fuzzy n- normed linear space, fuzzy algebra, soft sets, rough sets and neutrosophic sets. He has guided 7 Ph.D students and published around 58 papers in reputed international and national journals.

Adem Yolcu graduated from the Department of Mathematics, Faculty of Science and Letter, Kafkas University, Kars, Turkey in 2014. He received MSc from Kafkas University in 2016. Then, he received PhD from Kafkas University in 2020. Now he is working as Associate Professor at Kafkas University. His research interest focuses mainly soft set, fuzzy set, intuitionistic fuzzy set, neutrosophic sets and topological structures of this notions.

Index

A

C

D

E

F

G

O

P

Q

R

S

T

U

V

W

Printed in the United States
by Baker & Taylor Publisher Services